Comprehensive
Natural Products
Chemistry

Comprehensive Natural Products Chemistry

Editors-in-Chief
Sir Derek Barton†
Texas A&M University, USA

Koji Nakanishi
Columbia University, USA

Executive Editor
Otto Meth-Cohn
University of Sunderland, UK

Volume 5
ENZYMES, ENZYME MECHANISMS, PROTEINS,
AND ASPECTS OF NO CHEMISTRY

Volume Editor
C. Dale Poulter
University of Utah, USA

WITHDRAWAL   1999

ELSEVIER

AMSTERDAM – LAUSANNE – NEW YORK – OXFORD – SHANNON – SINGAPORE – TOKYO

Elsevier Science Ltd., The Boulevard, Langford Lane, Kidlington, Oxford, OX5 1GB, UK

First edition 1999

Library of Congress Cataloging-in-Publication Data
Comprehensive natural products chemistry / editors-in-chief, Sir Derek Barton, Koji Nakanishi ; executive editor, Otto Meth-Cohn. -- 1st ed.
   p. cm.
  Includes index.
  Contents: v. 5. Enzymes, enzyme mechanisms, proteins, and aspects of NO chemistry / volume editor C. Dale Poulter
  1. Natural products. I. Barton, Derek, Sir, 1918-1998. II. Nakanishi, Koji, 1925- . III. Meth-Cohn, Otto.
QD415.C63 1999
547.7--dc21   98-15249

British Library Cataloguing in Publication Data
Comprehensive natural products chemistry
  1. Organic compounds
  I. Barton, Sir Derek, 1918-1998 II. Nakanishi Koji III. Meth-Cohn Otto
572.5

ISBN 0-08-042709-X (set : alk. paper)
ISBN 0-08-043157-7 (Volume 5 : alk. paper)

∞™ The paper used in this publication meets the minimum requirements of the American National Standard for Information Sciences—Permanence of Paper for Printed Library Materials, ANSI Z39.48–1984.

Typeset by BPC Digital Data Ltd., Glasgow, UK.
Printed and bound in Great Britain by BPC Wheatons Ltd., Exeter, UK.

Contents

Introduction

For many decades, Natural Products Chemistry has been the principal driving force for progress in Organic Chemistry.

In the past, the determination of structure was arduous and difficult. As soon as computing became easy, the application of X-ray crystallography to structural determination quickly surpassed all other methods. Supplemented by the equally remarkable progress made more recently by Nuclear Magnetic Resonance techniques, determination of structure has become a routine exercise. This is even true for enzymes and other molecules of a similar size. Not to be forgotten remains the progress in mass spectrometry which permits another approach to structure and, in particular, to the precise determination of molecular weight.

There have not been such revolutionary changes in the partial or total synthesis of Natural Products. This still requires effort, imagination and time. But remarkable syntheses have been accomplished and great progress has been made in stereoselective synthesis. However, the one hundred percent yield problem is only solved in certain steps in certain industrial processes. Thus there remains a great divide between the reactions carried out in living organisms and those that synthetic chemists attain in the laboratory. Of course Nature edits the accuracy of DNA, RNA, and protein synthesis in a way that does not apply to a multi-step Organic Synthesis.

Organic Synthesis has already a significant component that uses enzymes to carry out specific reactions. This applies particularly to lipases and to oxidation enzymes. We have therefore, given serious attention to enzymatic reactions.

No longer standing in the wings, but already on-stage, are the wonderful tools of Molecular Biology. It is now clear that multi-step syntheses can be carried out in one vessel using multiple cloned enzymes. Thus, Molecular Biology and Organic Synthesis will come together to make economically important Natural Products.

From these preliminary comments it is clear that Natural Products Chemistry continues to evolve in different directions interacting with physical methods, Biochemistry, and Molecular Biology all at the same time.

This new Comprehensive Series has been conceived with the common theme of "How does Nature make all these molecules of life?" The principal idea was to organize the multitude of facts in terms of Biosynthesis rather than structure. The work is not intended to be a comprehensive listing of natural products, nor is it intended that there should be any detail about biological activity. These kinds of information can be found elsewhere.

The work has been planned for eight volumes with one more volume for Indexes. As far as we are aware, a broad treatment of the whole of Natural Products Chemistry has never been attempted before. We trust that our efforts will be useful and informative to all scientific disciplines where Natural Products play a role.

D. H. R. Barton†   K. Nakanishi   O. Meth-Cohn

Preface

It is surprising indeed that this work is the first attempt to produce a "comprehensive" overview of Natural Products beyond the student text level. However, the awe-inspiring breadth of the topic, which in many respects is still only developing, is such as to make the job daunting to anyone in the field. Fools rush in where angels fear to tread and the particular fool in this case was myself, a lifelong enthusiast and reader of the subject but with no research base whatever in the field!

Having been involved in several of the *Comprehensive* works produced by Pergamon Press, this omission intrigued me and over a period of gestation I put together a rough outline of how such a work could be written and presented it to Pergamon. To my delight they agreed that the project was worthwhile and in short measure Derek Barton was approached and took on the challenge of fleshing out this framework with alacrity. He also brought his long-standing friend and outstanding contributor to the field, Koji Nakanishi, into the team. With Derek's knowledge of the whole field, the subject was broken down into eight volumes and an outstanding team of internationally recognised Volume Editors was appointed.

We used Derek's 80th birthday as a target for finalising the work. Sadly he died just a few months before reaching this milestone. This work therefore is dedicated to the memory of Sir Derek Barton, Natural Products being the area which he loved best of all.

OTTO METH-COHN
Executive Editor

SIR DEREK BARTON

Sir Derek Barton, who was Distinguished Professor of Chemistry at Texas A&M University and holder of the Dow Chair of Chemical Invention died on March 16, 1998 in College Station, Texas of heart failure. He was 79 years old and had been Chairman of the Executive Board of Editors for Tetrahedron Publications since 1979.

Barton was considered to be one of the greatest organic chemists of the twentieth century whose work continues to have a major influence on contemporary science and will continue to do so for future generations of chemists.

Derek Harold Richard Barton was born on September 8, 1918 in Gravesend, Kent, UK and graduated from Imperial College, London with the degrees of B.Sc. (1940) and Ph.D. (1942). He carried out work on military intelligence during World War II and after a brief period in industry, joined the faculty at Imperial College. It was an early indication of the breadth and depth of his chemical knowledge that his lectureship was in physical chemistry. This research led him into the mechanism of elimination reactions and to the concept of molecular rotation difference to correlate the configurations of steroid isomers. During a sabbatical leave at Harvard in 1949–1950 he published a paper on the "Conformation of the Steroid Nucleus" (Experientia, 1950, **6**, 316) which was to bring him the Nobel Prize in Chemistry in 1969, shared with the Norwegian chemist, Odd Hassel. This key paper (only four pages long) altered the way in which chemists thought about the shape and reactivity of molecules, since it showed how the reactivity of functional groups in steroids depends on their axial or equatorial positions in a given conformation. Returning to the UK he held Chairs of Chemistry at Birkbeck College and Glasgow University before returning to Imperial College in 1957, where he developed a remarkable synthesis of the steroid hormone, aldosterone, by a photochemical reaction known as the Barton Reaction (nitrite photolysis). In 1978 he retired from Imperial College and became Director of the Natural Products Institute (CNRS) at Gif-sur-Yvette in France where he studied the invention of new chemical reactions, especially the chemistry of radicals, which opened up a whole new area of organic synthesis involving Gif chemistry. In 1986 he moved to a third career at Texas A&M University as Distinguished Professor of Chemistry and continued to work on novel reactions involving radical chemistry and the oxidation of hydrocarbons, which has become of great industrial importance. In a research career spanning more than five decades, Barton's contributions to organic chemistry included major discoveries which have profoundly altered our way of thinking about chemical structure and reactivity. His chemistry has provided models for the biochemical synthesis of natural products including alkaloids, antibiotics, carbohydrates, and DNA. Most recently his discoveries led to models for enzymes which oxidize hydrocarbons, including methane monooxygenase.

The following are selected highlights from his published work:

The 1950 paper which launched Conformational Analysis was recognized by the Nobel Prize Committee as the key contribution whereby the third dimension was added to chemistry. This work alone transformed our thinking about the connection between stereochemistry and reactivity, and was later adapted from small molecules to macromolecules e.g., DNA, and to inorganic complexes.

Barton's breadth and influence is illustrated in "Biogenetic Aspects of Phenol Oxidation" (*Festschr. Arthur Stoll*, 1957, 117). This theoretical work led to many later experiments on alkaloid biosynthesis and to a set of rules for *ortho-para*-phenolic oxidative coupling which allowed the predication of new natural product systems before they were actually discovered and to the correction of several erroneous structures.

In 1960, his paper on the remarkably short synthesis of the steroid hormone aldosterone (*J. Am. Chem. Soc.*, 1960, **82**, 2641) disclosed the first of many inventions of new reactions—in this case nitrite photolysis—to achieve short, high yielding processes, many of which have been patented and are used worldwide in the pharmaceutical industry.

Moving to 1975, by which time some 500 papers had been published, yet another "Barton reaction" was born—"The Deoxygenation of Secondary Alcohols" (*J. Chem. Soc. Perkin Trans. 1*, 1975, 1574), which has been very widely applied due to its tolerance of quite hostile and complex local environments in carbohydrate and nucleoside chemistry. This reaction is the chemical counterpart to ribonucleotide→ deoxyribonucleotide reductase in biochemistry and, until the arrival of the Barton reaction, was virtually impossible to achieve.

In 1985, "Invention of a New Radical Chain Reaction" involved the generation of carbon radicals from carboxylic acids (*Tetrahedron*, 1985, **41**, 3901). The method is of great synthetic utility and has been used many times by others in the burgeoning area of radicals in organic synthesis.

These recent advances in synthetic methodology were remarkable since his chemistry had virtually no precedent in the work of others. The radical methodology was especially timely in light of the significant recent increase in applications for fine chemical syntheses, and Barton gave the organic community an entrée into what will prove to be one of the most important methods of the twenty-first century. He often said how proud he was, at age 71, to receive the ACS Award for Creativity in Organic Synthesis for work published in the preceding five years.

Much of Barton's more recent work is summarized in the articles "The Invention of Chemical Reactions—The Last 5 Years" (*Tetrahedron*, 1992, **48**, 2529) and "Recent Developments in Gif Chemistry" (*Pure Appl. Chem.*, 1997, **69**, 1941).

Working 12 hours a day, Barton's stamina and creativity remained undiminished to the day of his death. The author of more than 1000 papers in chemical journals, Barton also held many successful patents. In addition to the Nobel Prize he received many honors and awards including the Davy, Copley, and Royal medals of the Royal Society of London, and the Roger Adams and Priestley Medals of the American Chemical Society. He held honorary degrees from 34 universities. He was a Fellow of the Royal Societies of London and Edinburgh, Foreign Associate of the National Academy of Sciences (USA), and Foreign Member of the Russian and Chinese Academies of Sciences. He was knighted by Queen Elizabeth in 1972, received the Légion d'Honneur (Chevalier 1972; Officier 1985) from France, and the Order of the Rising Sun from the Emperor of Japan. In his long career, Sir Derek trained over 300 students and postdoctoral fellows, many of whom now hold major positions throughout the world and include some of today's most distinguished organic chemists.

For those of us who were fortunate to know Sir Derek personally there is no doubt that his genius and work ethic were unique. He gave generously of his time to students and colleagues wherever he traveled and engendered such great respect and loyalty in his students and co-workers, that major symposia accompanied his birthdays every five years beginning with the 60th, ending this year with two celebrations just before his 80th birthday.

With the death of Sir Derek Barton, the world of science has lost a major figure, who together with Sir Robert Robinson and Robert B. Woodward, the cofounders of *Tetrahedron*, changed the face of organic chemistry in the twentieth century.

Professor Barton is survived by his wife, Judy, and by a son, William from his first marriage, and three grandchildren.

A. I. SCOTT
Texas A&M University

Reprinted from *Tetrahedron*, 1998, **54**, 8847
Photograph courtesy of Library and Information Centre, Royal Society of Chemistry. © The Nobel Foundation

Contributors to Volume 5

Professor R. N. Armstrong
Department of Biochemistry, Vanderbilt University School of Medicine, 23rd Avenue South & Pierce, Nashville, TN 37232-0146, USA

Ms. C. K. Bagdassarian
Department of Chemistry, The College of William and Mary, PO Box 8795, Williamsburg, VA 23187, USA

Dr. N. R. Baker
Department of Chemistry, University of Iowa, Iowa City, IA 52242, USA

Dr. T. P. Begley
Department of Chemistry, 120 Baker Laboratory, Cornell University, Ithaca, NY 14853, USA

Dr. S. D. Copley
Department of Chemistry and Biochemistry, University of Colorado, Campus Box 215, Boulder, CO 80309, USA

Dr. J. M. Dolence
Department of Chemistry, University of Utah, Salt Lake City, UT 84112, USA

Dr. P. A. Frey
Institute for Enzyme Research, University of Wisconsin, 1710 University Avenue, Madison, WI 53705-4098, USA

Dr. B. Ganem
Department of Chemistry, Baker Laboratory, Cornell University, Ithaca, NY 14853-1301, USA

Dr. J. A. Gerlt
Department of Biochemistry, University of Illinois, 600 South Mathews Avenue, Urbana, IL 61801, USA

Dr. D. C. Goodwin
Department of Biochemistry, Vanderbilt University School of Medicine, Nashville, TN 37232-0146, USA

Dr. C. B. Grissom
Department of Chemistry, University of Utah, Salt Lake City, UT 84112, USA

Professor D. Hilvert
Laboratorium für Organische Chemie, Eidgenössische Technische Hochschule, Universitätstrasse 16, CH-8092 Zurich, Switzerland

Dr. A. S. Kalgutkar
Department of Biochemistry, Vanderbilt University School of Medicine, Nashville, TN 37232-0146, USA

Dr. S. Licht
Department of Chemistry, Massachusetts Institute of Technology, Building 18, Room 480 B, 77 Massachusetts Avenue, Cambridge, MA 02139-4307, USA

Professor L. J. Marnett
Department of Biochemistry, Vanderbilt University School of Medicine, 23rd Avenue South & Pierce, Nashville, TN 37232-0146, USA

Dr. R. Medhekar
Department of Chemistry, University of Iowa, Iowa City, IA 52242, USA

Professor C. D. Poulter
Department of Chemistry, University of Utah, Salt Lake City, UT 84112, USA

Professor D. M. Quinn
Department of Chemistry, University of Iowa, Iowa City, IA 52242-0001, USA

Dr. S. W. Rowlinson
Department of Biochemistry, Vanderbilt University School of Medicine, Nashville, TN 37232-0146, USA

Dr. V. L. Schramm
Department of Biochemistry, Albert Einstein College of Medicine of Yeshiva University, 1300 Morris Park Avenue, Bronx, NY 10461, USA

Dr. J. Stubbe
Departments of Chemistry and Biology, Massachusetts Institute of Technology, Building 18, Room 480 B, 77 Massachusetts Avenue, Cambridge, MA 02139-4307, USA

Dr. W. Taylor
Department of Chemistry, Indiana University, Bloomington, IN 47405, USA

Dr. C. P. Whitman
Medicinal Chemistry Division, College of Pharmacy, University of Texas, PHR 4.220B, Austin, TX 78712, USA

Dr. T. S. Widlanski
Department of Chemistry, Indiana University, Bloomington, IN 47405, USA

Dr. S. G. Withers
Department of Chemistry, University of British Columbia, 2036 Main Mall, Vancouver, BC V6T 1Z1, Canada

Dr. Y. Xu
Department of Chemistry, University of Utah, Salt Lake City, UT 84112, USA

Dr. D. L. Zechel
Department of Chemistry, University of British Columbia, 2036 Main Mall, Vancouver, BC V6T 1Z1, Canada

Abbreviations

The most commonly used abbreviations in *Comprehensive Natural Products Chemistry* are listed below. Please note that in some instances these may differ from those used in other branches of chemistry

A	adenine
ABA	abscisic acid
Ac	acetyl
ACAC	acetylacetonate
ACTH	adrenocorticotropic hormone
ADP	adenosine 5'-diphosphate
AIBN	2,2'-azobisisobutyronitrile
Ala	alanine
AMP	adenosine 5'-monophosphate
APS	adenosine 5'-phosphosulfate
Ar	aryl
Arg	arginine
ATP	adenosine 5'-triphosphate
B	nucleoside base (adenine, cylosine, guanine, thymine or uracil)
9-BBN	9-borabicyclo[3.3.1]nonane
BOC	*t*-butoxycarbonyl (or carbo-*t*-butoxy)
BSA	*N,O*-bis(trimethylsilyl)acetamide
BSTFA	*N,O*-bis(trimethylsilyl)trifluoroacetamide
Bu	butyl
Bu^n	*n*-butyl
Bu^i	isobutyl
Bu^s	*s*-butyl
Bu^t	*t*-butyl
Bz	benzoyl
CAN	ceric ammonium nitrate
CD	cyclodextrin
CDP	cytidine 5'-diphosphate
CMP	cytidine 5'-monophosphate
CoA	coenzyme A
COD	cyclooctadiene
COT	cyclooctatetraene
Cp	η^5-cyclopentadiene
Cp*	pentamethylcyclopentadiene
12-Crown-4	1,4,7,10-tetraoxacyclododecane
15-Crown-5	1,4,7,10,13-pentaoxacyclopentadecane
18-Crown-6	1,4,7,10,13,16-hexaoxacyclooctadecane
CSA	camphorsulfonic acid
CSI	chlorosulfonyl isocyanate
CTP	cytidine 5'-triphosphate
cyclic AMP	adenosine 3',5'-cyclic monophosphoric acid
CySH	cysteine
DABCO	1,4-diazabicyclo[2.2.2]octane
DBA	dibenz[*a,h*]anthracene
DBN	1,5-diazabicyclo[4.3.0]non-5-ene

xvi   *Abbreviations*

DBU	1,8-diazabicyclo[5.4.0]undec-7-ene
DCC	dicyclohexylcarbodiimide
DEAC	diethylaluminum chloride
DEAD	diethyl azodicarboxylate
DET	diethyl tartrate (+ or -)
DHET	dihydroergotoxine
DIBAH	diisobutylaluminum hydride
Diglyme	diethylene glycol dimethyl ether (or bis(2-methoxyethyl)ether)
DiHPhe	2,5-dihydroxyphenylalanine
Dimsyl Na	sodium methylsulfinylmethide
DIOP	2,3-*O*-isopropylidene-2,3-dihydroxy-1,4-bis(diphenylphosphino)butane
dipt	diisopropyl tartrate (+ or -)
DMA	dimethylacetamide
DMAD	dimethyl acetylenedicarboxylate
DMAP	4-dimethylaminopyridine
DME	1,2-dimethoxyethane (glyme)
DMF	dimethylformamide
DMF-DMA	dimethylformamide dimethyl acetal
DMI	1,3-dimethyl-2-imidazalidinone
DMSO	dimethyl sulfoxide
DMTSF	dimethyl(methylthio)sulfonium fluoroborate
DNA	deoxyribonucleic acid
DOCA	deoxycorticosterone acetate
EADC	ethylaluminum dichloride
EDTA	ethylenediaminetetraacetic acid
EEDQ	*N*-ethoxycarbonyl-2-ethoxy-1,2-dihydroquinoline
Et	ethyl
EVK	ethyl vinyl ketone
FAD	flavin adenine dinucleotide
Fl	flavin
FMN	flavin mononucleotide
G	guanine
GABA	4-aminobutyric acid
GDP	guanosine 5'-diphosphate
GLDH	glutamate dehydrogenase
gln	glutamine
Glu	glutamic acid
Gly	glycine
GMP	guanosine 5'-monophosphate
GOD	glucose oxidase
G-6-P	glucose-6-phosphate
GTP	guanosine 5'-triphosphate
Hb	hemoglobin
His	histidine
HMPA	hexamethylphosphoramide (or hexamethylphosphorous triamide)
Ile	isoleucine
INAH	isonicotinic acid hydrazide
IpcBH	isopinocampheylborane
Ipc$_2$BH	diisopinocampheylborane
KAPA	potassium 3-aminopropylamide
K-Slectride	potassium tri-*s*-butylborohydride

LAH	lithium aluminum hydride
LAP	leucine aminopeptidase
LDA	lithium diisopropylamide
LDH	lactic dehydrogenase
Leu	leucine
LICA	lithium isopropylcyclohexylamide
L-Selectride	lithium tri-*s*-butylborohydride
LTA	lead tetraacetate
Lys	lysine
MCPBA	*m*-chloroperoxybenzoic acid
Me	methyl
MEM	methoxyethoxymethyl
MEM-Cl	ß-methoxyethoxymethyl chloride
Met	methionine
MMA	methyl methacrylate
MMC	methyl magnesium carbonate
MOM	methoxymethyl
Ms	mesyl (or methanesulfonyl)
MSA	methanesulfonic acid
MsCl	methanesulfonyl chloride
MVK	methyl vinyl ketone
NAAD	nicotinic acid adenine dinucleotide
NAD	nicotinamide adenine dinucleotide
NADH	nicotinamide adenine dinucleotide phosphate, reduced
NBS	*N*-bromosuccinimider
NMO	*N*-methylmorpholine *N*-oxide monohydrate
NMP	*N*-methylpyrrolidone
PCBA	*p*-chlorobenzoic acid
PCBC	*p*-chlorobenzyl chloride
PCBN	*p*-chlorobenzonitrile
PCBTF	*p*-chlorobenzotrifluoride
PCC	pyridinium chlorochromate
PDC	pyridinium dichromate
PG	prostaglandin
Ph	phenyl
Phe	phenylalanine
Phth	phthaloyl
PPA	polyphosphoric acid
PPE	polyphosphate ester (or ethyl *m*-phosphate)
Pr	propyl
Pr^i	isopropyl
Pro	proline
Py	pyridine
RNA	ribonucleic acid
Rnase	ribonuclease
Ser	serine
Sia$_2$BH	disiamylborane
TAS	tris(diethylamino)sulfonium
TBAF	tetra-*n*-butylammonium fluoroborate
TBDMS	*t*-butyldimethylsilyl
TBDMS-Cl	*t*-butyldimethylsilyl chloride
TBDPS	*t*-butyldiphenylsilyl
TCNE	tetracyanoethene

TES   triethylsilyl
TFA   trifluoracetic acid
TFAA   trifluoroacetic anhydride
THF   tetrahydrofuran
THF   tetrahydrofolic acid
THP   tetrahydropyran (or tetrahydropyranyl)
Thr   threonine
TMEDA   *N,N,N',N'*,tetramethylethylenediamine[1,2-bis(dimethylamino)ethane]
TMS   trimethylsilyl
TMS-Cl   trimethylsilyl chloride
TMS-CN   trimethylsilyl cyanide
Tol   toluene
TosMIC   tosylmethyl isocyanide
TPP   tetraphenylporphyrin
Tr   trityl (or triphenylmethyl)
Trp   tryptophan
Ts   tosyl (or *p*-toluenesulfonyl)
TTFA   thallium trifluoroacetate
TTN   thallium(III) nitrate
Tyr   tyrosine
Tyr-OMe   tyrosine methyl ester

U   uridine
UDP   uridine 5'-diphosphate
UMP   uridine 5'-monophosphate

Contents of All Volumes

An Historical Perspective of Natural Products Chemistry

KOJI NAKANISHI

Columbia University, New York, USA

To give an account of the rich history of natural products chemistry in a short essay is a daunting task. This brief outline begins with a description of ancient folk medicine and continues with an outline of some of the major conceptual and experimental advances that have been made from the early nineteenth century through to about 1960, the start of the modern era of natural products chemistry. Achievements of living chemists are noted only minimally, usually in the context of related topics within the text. More recent developments are reviewed within the individual chapters of the present volumes, written by experts in each field. The subheadings follow, in part, the sequence of topics presented in Volumes 1–8.

1. ETHNOBOTANY AND "NATURAL PRODUCTS CHEMISTRY"

Except for minerals and synthetic materials our surroundings consist entirely of organic natural products, either of prebiotic organic origins or from microbial, plant, or animal sources. These materials include polyketides, terpenoids, amino acids, proteins, carbohydrates, lipids, nucleic acid bases, RNA and DNA, etc. Natural products chemistry can be thought of as originating from mankind's curiosity about odor, taste, color, and cures for diseases. Folk interest in treatments for pain, for food-poisoning and other maladies, and in hallucinogens appears to go back to the dawn of humanity

For centuries China has led the world in the use of natural products for healing. One of the earliest health science anthologies in China is the Nei Ching, whose authorship is attributed to the legendary Yellow Emperor (thirtieth century BC), although it is said that the dates were backdated from the third century by compilers. Excavation of a Han Dynasty (206 BC–AD 220) tomb in Hunan Province in 1974 unearthed decayed books, written on silk, bamboo, and wood, which filled a critical gap between the dawn of medicine up to the classic Nei Ching; Book 5 of these excavated documents lists 151 medical materials of plant origin. Generally regarded as the oldest compilation of Chinese herbs is Shen Nung Pen Ts'ao Ching (Catalog of Herbs by Shen Nung), which is believed to have been revised during the Han Dynasty; it lists 365 materials. Numerous revisions and enlargements of Pen Ts'ao were undertaken by physicians in subsequent dynasties, the ultimate being the Pen Ts'ao Kang Mu (General Catalog of Herbs) written by Li Shih-Chen over a period of 27 years during the Ming Dynasty (1573–1620), which records 1898 herbal drugs and 8160 prescriptions. This was circulated in Japan around 1620 and translated, and has made a huge impact on subsequent herbal studies in Japan; however, it has not been translated into English. The number of medicinal herbs used in 1979 in China numbered 5267. One of the most famous of the Chinese folk herbs is the ginseng root *Panax ginseng*, used for health maintenance and treatment of various diseases. The active principles were thought to be the saponins called ginsenosides but this is now doubtful; the effects could well be synergistic between saponins, flavonoids, etc. Another popular folk drug, the extract of the Ginkgo tree, *Ginkgo biloba* L., the only surviving species of the Paleozoic era (250 million years ago) family which became extinct during the last few million years, is mentioned in the Chinese Materia Medica to have an effect in improving memory and sharpening mental alertness. The main constituents responsible for this are now understood to be ginkgolides and flavonoids, but again not much else is known. Clarifying the active constituents and mode of (synergistic) bioactivity of Chinese herbs is a challenging task that has yet to be fully addressed.

The Assyrians left 660 clay tablets describing 1000 medicinal plants used around 1900–400 BC, but the best insight into ancient pharmacy is provided by the two scripts left by the ancient Egyptians, who were masters of human anatomy and surgery because of their extensive mummification practices. The Edwin Smith Surgical Papyrus purchased by Smith in 1862 in Luxor (now in the New York Academy of Sciences collection), is one of the most important medicinal documents of the ancient Nile Valley, and describes the healer's involvement in surgery, prescription, and healing practices using plants, animals, and minerals. The Ebers Papyrus, also purchased by Edwin Smith in 1862, and then acquired by Egyptologist George Ebers in 1872, describes 800 remedies using plants, animals, minerals, and magic. Indian medicine also has a long history, possibly dating back to the second millennium BC. The Indian materia medica consisted mainly of vegetable drugs prepared from plants but also used animals, bones, and minerals such as sulfur, arsenic, lead, copper sulfate, and gold. Ancient Greece inherited much from Egypt, India, and China, and underwent a gradual transition from magic to science. Pythagoras (580–500 BC) influenced the medical thinkers of his time, including Aristotle (384–322 BC), who in turn affected the medical practices of another influential Greek physician Galen (129–216). The Iranian physician Avicenna (980–1037) is noted for his contributions to Aristotelian philosophy and medicine, while the German-Swiss physician and alchemist Paracelsus (1493–1541) was an early champion who established the role of chemistry in medicine.

The rainforests in Central and South America and Africa are known to be particularly abundant in various organisms of interest to our lives because of their rich biodiversity, intense competition, and the necessity for self-defense. However, since folk-treatments are transmitted verbally to the next generation via shamans who naturally have a tendency to keep their plant and animal sources confidential, the recipes tend to get lost, particularly with destruction of rainforests and the encroachment of "civilization." Studies on folk medicine, hallucinogens, and shamanism of the Central and South American Indians conducted by Richard Schultes (Harvard Botanical Museum, emeritus) have led to renewed activity by ethnobotanists, recording the knowledge of shamans, assembling herbaria, and transmitting the record of learning to the village.

Extracts of toxic plants and animals have been used throughout the world for thousands of years for hunting and murder. These include the various arrow poisons used all over the world. *Strychnos* and *Chondrodendron* (containing strychnine, etc.) were used in South America and called "curare," *Strophanthus* (strophantidine, etc.) was used in Africa, the latex of the upas tree *Antiaris toxicaria* (cardiac glycosides) was used in Java, while *Aconitum napellus*, which appears in Greek mythology (aconitine) was used in medieval Europe and Hokkaido (by the Ainus). The Colombian arrow poison is from frogs (batrachotoxins; 200 toxins have been isolated from frogs by B. Witkop and J. Daly at NIH). Extracts of *Hyoscyamus niger* and *Atropa belladonna* contain the toxic tropane alkaloids, for example hyoscyamine, belladonnine, and atropine. The belladonna berry juice (atropine) which dilates the eye pupils was used during the Renaissance by ladies to produce doe-like eyes (belladona means beautiful woman). The Efik people in Calabar, southeastern Nigeria, used extracts of the calabar bean known as esere (physostigmine) for unmasking witches. The ancient Egyptians and Chinese knew of the toxic effect of the puffer fish, fugu, which contains the neurotoxin tetrodotoxin (Y. Hirata, K. Tsuda, R. B. Woodward).

When rye is infected by the fungus *Claviceps purpurea*, the toxin ergotamine and a number of ergot alkaloids are produced. These cause ergotism or the "devil's curse," "St. Anthony's fire," which leads to convulsions, miscarriages, loss of arms and legs, dry gangrene, and death. Epidemics of ergotism occurred in medieval times in villages throughout Europe, killing tens of thousands of people and livestock; Julius Caesar's legions were destroyed by ergotism during a campaign in Gaul, while in AD 994 an estimated 50,000 people died in an epidemic in France. As recently as 1926, a total of 11,000 cases of ergotism were reported in a region close to the Urals. It has been suggested that the witch hysteria that occurred in Salem, Massachusetts, might have been due to a mild outbreak of ergotism. Lysergic acid diethylamide (LSD) was first prepared by A. Hofmann, Sandoz Laboratories, Basel, in 1943 during efforts to improve the physiological effects of the ergot alkaloids when he accidentally inhaled it. "On Friday afternoon, April 16, 1943," he wrote, "I was seized by a sensation of restlessness... ." He went home from the laboratory and "perceived an uninterrupted stream of fantastic dreams ... ." (*Helvetica Chimica Acta*).

Numerous psychedelic plants have been used since ancient times, producing visions, mystical fantasies (cats and tigers also seem to have fantasies?, see nepetalactone below), sensations of flying, glorious feelings in warriors before battle, etc. The ethnobotanists Wasson and Schultes identified "ololiqui," an important Aztec concoction, as the seeds of the morning glory *Rivea corymbosa* and gave the seeds to Hofmann who found that they contained lysergic acid amides similar to but less potent than LSD. Iboga, a powerful hallucinogen from the root of the African shrub *Tabernanthe iboga*, is used by the Bwiti cult in Central Africa who chew the roots to obtain relief from fatigue and hunger; it contains the alkaloid ibogamine. The powerful hallucinogen used for thousands of years by the American Indians, the peyote cactus, contains mescaline and other alkaloids. The Indian hemp plant, *Cannabis sativa*, has been used for making rope since 3000 BC, but when it is used for its pleasure-giving effects it is called cannabis and has been known in central Asia, China, India, and the Near East since ancient times. Marijuana, hashish (named after the Persian founder of the Assassins of the eleventh century, Hasan-e Sabbah), charas, ghanja, bhang, kef, and dagga are names given to various preparations of the hemp plant. The constituent responsible for the mind-altering effect is 1-tetrahydrocannabinol (also referred to as 9-THC) contained in 1%. R. Mechoulam (1930–, Hebrew University) has been the principal worker in the cannabinoids, including structure determination and synthesis of 9-THC (1964 to present); the Israeli police have also made a contribution by providing Mechoulam with a constant supply of marijuana. Opium (morphine) is another ancient drug used for a variety of pain-relievers and it is documented that the Sumerians used poppy as early as 4000 BC; the narcotic effect is present only in seeds before they are fully formed. The irritating secretion of the blister beetles, for example *Mylabris* and the European species *Lytta vesicatoria*, commonly called Spanish fly, was used medically as a topical skin irritant to remove warts but was also a major ingredient in so-called love potions (constituent is cantharidin, stereospecific synthesis in 1951, G. Stork, 1921–; prep. scale high-pressure Diels–Alder synthesis in 1985, W. G. Dauben, 1919–1996).

Plants have been used for centuries for the treatment of heart problems, the most important being the foxgloves *Digitalis purpurea* and *D. lanata* (digitalin, diginin) and *Strophanthus gratus* (ouabain). The bark of cinchona *Cinchona officinalis* (called quina-quina by the Indians) has been used widely among the Indians in the Andes against malaria, which is still one of the major infectious diseases; its most important alkaloid is quinine. The British protected themselves against malaria during the occupation of India through gin and tonic (quinine!). The stimulant coca, used by the Incas around the tenth century, was introduced into Europe by the conquistadors; coca beans are also commonly chewed in West Africa. Wine making was already practiced in the Middle East 6000–8000 years ago; Moors made date wines, the Japanese rice wine, the Vikings honey mead, the Incas maize chicha. It is said that the Babylonians made beer using yeast 5000–6000 years ago. As shown above in parentheses, alkaloids are the major constituents of the herbal plants and extracts used for centuries, but it was not until the early nineteenth century that the active principles were isolated in pure form, for example morphine (1816), strychnine (1817), atropine (1819), quinine (1820), and colchicine (1820). It was a century later that the structures of these compounds were finally elucidated.

2. DAWN OF ORGANIC CHEMISTRY, EARLY STRUCTURAL STUDIES, MODERN METHODOLOGY

The term "organic compound" to define compounds made by and isolated from living organisms was coined in 1807 by the Swedish chemist Jons Jacob Berzelius (1779–1848), a founder of today's chemistry, who developed the modern system of symbols and formulas in chemistry, made a remarkably accurate table of atomic weights and analyzed many chemicals. At that time it was considered that organic compounds could not be synthesized from inorganic materials *in vitro*. However, Friedrich Wöhler (1800–1882), a medical doctor from Heidelberg who was starting his chemical career at a technical school in Berlin, attempted in 1828 to make "ammonium cyanate," which had been assigned a wrong structure, by heating the two inorganic salts potassium cyanate and ammonium sulfate; this led to the unexpected isolation of white crystals which were identical to the urea from urine, a typical organic compound. This well-known incident marked the beginning of organic chemistry. With the preparation of acetic acid from inorganic material in 1845 by Hermann Kolbe (1818–1884) at Leipzig, the myth surrounding organic compounds, in which they were associated with some vitalism was brought to an end and organic chemistry became the chemistry of carbon compounds. The same Kolbe was involved in the development of aspirin, one of the earliest and most important success stories in natural products chemistry. Salicylic acid from the leaf of the wintergreen plant had long been used as a pain reliever, especially in treating arthritis and gout. The inexpensive synthesis of salicylic acid from sodium phenolate and carbon dioxide by Kolbe in 1859 led to the industrial production in 1893 by the Bayer Company of acetylsalicylic acid "aspirin," still one of the most popular drugs. Aspirin is less acidic than salicylic acid and therefore causes less irritation in the mouth, throat, and stomach. The remarkable mechanism of the anti-inflammatory effect of aspirin was clarified in 1974 by John Vane (1927–) who showed that it inhibits the biosynthesis of prostaglandins by irreversibly acetylating a serine residue in prostaglandin synthase. Vane shared the 1982 Nobel Prize with Bergström and Samuelsson who determined the structure of prostaglandins (see below).

In the early days, natural products chemistry was focused on isolating the more readily available plant and animal constituents and determining their structures. The course of structure determination in the 1940s was a complex, indirect process, combining evidence from many types of experiments. The first effort was to crystallize the unknown compound or make derivatives such as esters or 2,4-dinitrophenylhydrazones, and to repeat recrystallization until the highest and sharp melting point was reached, since prior to the advent of isolation and purification methods now taken for granted, there was no simple criterion for purity. The only chromatography was through special grade alumina (first used by M. Tswett in 1906, then reintroduced by R. Willstätter). Molecular weight estimation by the Rast method which depended on melting point depression of a sample/camphor mixture, coupled with Pregl elemental microanalysis (see below) gave the molecular formula. Functionalities such as hydroxyl, amino, and carbonyl groups were recognized on the basis of specific derivatization and crystallization, followed by redetermination of molecular formula; the change in molecular composition led to identification of the functionality. Thus, sterically hindered carbonyls, for example the 11-keto group of cortisone, or tertiary hydroxyls, were very difficult to pinpoint, and often had to depend on more searching experiments. Therefore, an entire paper describing the recognition of a single hydroxyl group in a complex natural product would occasionally appear in the literature. An oxygen function suggested from the molecular formula but left unaccounted for would usually be assigned to an ether.

Determination of C-methyl groups depended on Kuhn–Roth oxidation which is performed by drastic oxidation with chromic acid/sulfuric acid, reduction of excess oxidant with hydrazine, neutralization with alkali, addition of phosphoric acid, distillation of the acetic acid originating from the C-methyls, and finally its titration with alkali. However, the results were only approximate, since *gem*-dimethyl groups only yield one equivalent of acetic acid, while primary, secondary, and tertiary methyl groups all give different yields of acetic acid. The skeletal structure of polycyclic compounds were frequently deduced on the basis of dehydrogenation reactions. It is therefore not surprising that the original steroid skeleton put forth by Wieland and Windaus in 1928, which depended a great deal on the production of chrysene upon Pd/C dehydrogenation, had to be revised in 1932 after several discrepancies were found (they received the Nobel prizes in 1927 and 1928 for this "extraordinarily difficult structure determination," see below).

In the following are listed some of the Nobel prizes awarded for the development of methodologies which have contributed critically to the progress in isolation protocols and structure determination. The year in which each prize was awarded is preceded by "Np."

Fritz Pregl, 1869–1930, Graz University, Np 1923. Invention of carbon and hydrogen microanalysis. Improvement of Kuhlmann's microbalance enabled weighing at an accuracy of 1 µg over a 20 g range, and refinement of carbon and hydrogen analytical methods made it possible to perform analysis with 3–4 mg of sample. His microbalance and the monograph *Quantitative Organic Microanalysis* (1916) profoundly influenced subsequent developments in practically all fields of chemistry and medicine.

The Svedberg, 1884–1971, Uppsala, Np 1926. Uppsala was a center for quantitative work on colloids for which the prize was awarded. His extensive study on ultracentrifugation, the first paper of which was published in the year of the award, evolved from a spring visit in 1922 to the University of Wisconsin. The ultracentrifuge together with the electrophoresis technique developed by his student Tiselius, have profoundly influenced subsequent progress in molecular biology and biochemistry.

Arne Tiselius, 1902–1971, Ph.D. Uppsala (T. Svedberg), Uppsala, Np 1948. Assisted by a grant from the Rockefeller Foundation, Tiselius was able to use his early electrophoresis instrument to show four bands in horse blood serum, alpha, beta and gamma globulins in addition to albumin; the first paper published in 1937 brought immediate positive responses.

Archer Martin, 1910–, Ph.D. Cambridge; Medical Research Council, Mill Hill, and Richard Synge, 1914–1994, Ph.D. Cambridge; Rowett Research Institute, Food Research Institute, Np 1952. They developed chromatography using two immiscible phases, gas–liquid, liquid–liquid, and paper chromatography, all of which have profoundly influenced all phases of chemistry.

Frederick Sanger, 1918–, Ph.D. Cambridge (A. Neuberger), Medical Research Council, Cambridge, Np 1958 and 1980. His confrontation with challenging structural problems in proteins and nucleic acids led to the development of two general analytical methods, 1,2,4-fluorodinitrobenzene (DNP) for tagging free amino groups (1945) in connection with insulin sequencing studies, and the dideoxynucleotide method for sequencing DNA (1977) in connection with recombinant DNA. For the latter he received his second Np in chemistry in 1980, which was shared with Paul Berg (1926–, Stanford University) and Walter Gilbert (1932–, Harvard University) for their contributions, respectively, in recombinant DNA and chemical sequencing of DNA. The studies of insulin involved usage of DNP for tagging disulfide bonds as cysteic acid residues (1949), and paper chromatography introduced by Martin and Synge 1944. That it was the first elucidation of any protein structure lowered the barrier for future structure studies of proteins.

Stanford Moore, 1913–1982, Ph.D. Wisconsin (K. P. Link), Rockefeller, Np 1972; and William Stein, 1911–1980, Ph.D. Columbia (E. G. Miller); Rockefeller, Np 1972. Moore and Stein cooperatively developed methods for the rapid quantification of protein hydrolysates by combining partition chromatography, ninhydrin coloration, and drop-counting fraction collector, i.e., the basis for commercial amino acid analyzers, and applied them to analysis of the ribonuclease structure.

Bruce Merrifield, 1921–, Ph.D. UCLA (M. Dunn), Rockefeller, Np 1984. The concept of solid-phase peptide synthesis using porous beads, chromatographic columns, and sequential elongation of peptides and other chains revolutionized the synthesis of biopolymers.

High-performance liquid chromatography (HPLC), introduced around the mid-1960s and now coupled on-line to many analytical instruments, for example UV, FTIR, and MS, is an indispensable daily tool found in all natural products chemistry laboratories.

3. STRUCTURES OF ORGANIC COMPOUNDS, NINETEENTH CENTURY

The discoveries made from 1848 to 1874 by Pasteur, Kekulé, van't Hoff, Le Bel, and others led to a revolution in structural organic chemistry. Louis Pasteur (1822–1895) was puzzled about why the potassium salt of tartaric acid (deposited on wine casks during fermentation) was dextrorotatory while the sodium ammonium salt of racemic acid (also deposited on wine casks) was optically inactive although both tartaric acid and "racemic" acid had identical chemical compositions. In 1848, the 25 year old Pasteur examined the racemic acid salt under the microscope and found two kinds of crystals exhibiting a left- and right-hand relation. Upon separation of the left-handed and right-handed crystals, he found that they rotated the plane of polarized light in opposite directions. He had thus performed his famous resolution of a racemic mixture, and had demonstrated the phenomenon of chirality. Pasteur went on to show that the racemic acid formed two kinds of salts with optically active bases such as quinine; this was the first demonstration of diastereomeric resolution. From this work Pasteur concluded that tartaric acid must have an element of asymmetry within the molecule itself. However, a three-dimensional understanding of the enantiomeric pair was only solved 25 years later (see below). Pasteur's own interest shifted to microbiology where he made the crucial discovery of the involvement of "germs" or microorganisms in various processes and proved that yeast induces alcoholic fermentation, while other microorganisms lead to diseases; he thus saved the wine industries of France, originated the process known as "pasteurization," and later developed vaccines for rabies. He was a genius who made many fundamental discoveries in chemistry and in microbiology.

The structures of organic compounds were still totally mysterious. Although Wöhler had synthesized urea, an isomer of ammonium cyanate, in 1828, the structural difference between these isomers was not known. In 1858 August Kekulé (1829–1896; studied with André Dumas and C. A. Wurtz in Paris, taught at Ghent, Heidelberg, and Bonn) published his famous paper in Liebig's *Annalen der Chemie* on the structure of carbon, in which he proposed that carbon atoms could form C–C bonds with hydrogen and other atoms linked to them; his dream on the top deck of a London bus led him to this concept. It was Butlerov who introduced the term "structure theory" in 1861. Further, in 1865 Kekulé conceived the cyclo-hexa-1:3:5-triene structure for benzene (C_6H_6) from a dream of a snake biting its own tail. In 1874, two young chemists, van't Hoff (1852–1911, Np 1901) in Utrecht, and Le Bel (1847–1930) in Paris, who had met in 1874 as students of C. A. Wurtz, published the revolutionary three-dimensional (3D) structure of the tetrahedral carbon Cabcd to explain the enantiomeric behavior of Pasteur's salts. The model was welcomed by J. Wislicenus (1835–1902, Zürich, Würzburg, Leipzig) who in 1863 had demonstrated the enantiomeric nature of the two lactic acids found by Scheele in sour milk (1780) and by Berzelius in muscle tissue (1807). This model, however, was criticized by Hermann Kolbe (1818–1884, Leipzig) as an "ingenious but in reality trivial and senseless natural philosophy." After 10 years of heated controversy, the idea of tetrahedral carbon was fully accepted, Kolbe had died and Wislicenus succeeded him in Leipzig.

Emil Fischer (1852–1919, Np 1902) was the next to make a critical contribution to stereochemistry. From the work of van't Hoff and Le Bel he reasoned that glucose should have 16 stereoisomers. Fischer's doctorate work on hydrazines under Baeyer (1835–1917, Np 1905) at Strasbourg had led to studies of osazones which culminated in the brilliant establishment, including configurations, of the Fischer sugar tree starting from D-(+)-glyceraldehyde all the way up to the aldohexoses, allose, altrose, glucose, mannose, gulose, idose, galactose, and talose (from 1884 to 1890). Unfortunately Fischer suffered from the toxic effects of phenylhydrazine for 12 years. The arbitrarily but luckily chosen absolute configuration of D-(+)-glyceraldehyde was shown to be correct sixty years later in 1951 (Johannes-Martin Bijvoet, 1892–1980). Fischer's brilliant correlation of the sugars comprising the Fischer sugar tree was performed using the Kiliani (1855–1945)–Fischer method via cyanohydrin intermediates for elongating sugars. Fischer also made remarkable contributions to the chemistry of amino acids and to nucleic acid bases (see below).

4. STRUCTURES OF ORGANIC COMPOUNDS, TWENTIETH CENTURY

The early concept of covalent bonds was provided with a sound theoretical basis by Linus Pauling (1901–1994, Np 1954), one of the greatest intellects of the twentieth century. Pauling's totally interdisciplinary research interests, including proteins and DNA is responsible for our present understanding of molecular structures. His books *Introduction to Quantum Mechanics* (with graduate student E. B. Wilson, 1935) and *The Nature of the Chemical Bond* (1939) have had a profound effect on our understanding of all of chemistry.

The actual 3D shapes of organic molecules which were still unclear in the late 1940s were then brilliantly clarified by Odd Hassel (1897–1981, Oslo University, Np 1969) and Derek Barton (1918–1998, Np 1969). Hassel, an X-ray crystallographer and physical chemist, demonstrated by electron diffraction that cyclohexane adopted the chair form in the gas phase and that it had two kinds of bonds, "standing (axial)" and "reclining (equatorial)" (1943). Because of the German occupation of Norway in 1940, instead of publishing the result in German journals, he published it in a Norwegian journal which was not abstracted in English until 1945. During his 1949 stay at Harvard, Barton attended a seminar by Louis Fieser on steric effects in steroids and showed Fieser that interpretations could be simplified if the shapes ("conformations") of cyclohexane rings were taken into consideration; Barton made these comments because he was familiar with Hassel's study on *cis*- and *trans*-decalins. Following Fieser's suggestion Barton published these ideas in a four-page *Experientia* paper (1950). This led to the joint Nobel prize with Hassel (1969), and established the concept of conformational analysis, which has exerted a profound effect in every field involving organic molecules.

Using conformational analysis, Barton determined the structures of many key terpenoids such as ß-amyrin, cycloartenone, and cycloartenol (Birkbeck College). At Glasgow University (from 1955) he collaborated in a number of cases with Monteath Robertson (1900–1989) and established many challenging structures: limonin, glauconic acid, byssochlamic acid, and nonadrides. Barton was also associated with the Research Institute for Medicine and Chemistry (RIMAC), Cambridge, USA founded by the Schering company, where with J. M. Beaton, he produced 60 g of aldosterone at a time when the world supply of this important hormone was in mg quantities. Aldosterone synthesis ("a good problem") was achieved in 1961 by Beaton ("a good experimentalist") through a nitrite photolysis, which came to be known as the Barton reaction ("a good idea") (quotes from his 1991 autobiography published by the American Chemical Society). From Glasgow, Barton went on to Imperial College, and a year before retirement, in 1977 he moved to France to direct the research at ICSN at Gif-sur-Yvette where he explored the oxidation reaction selectivity for unactivated C–H. After retiring from ICSN he made a further move to Texas A&M University in 1986, and continued his energetic activities, including chairman of the *Tetrahedron* publications. He felt weak during work one evening and died soon after, on March 16, 1998. He was fond of the phrase "gap jumping" by which he meant seeking generalizations between facts that do not seem to be related: "In the conformational analysis story, one had to jump the gap between steroids and chemical physics" (from his autobiography). According to Barton, the three most important qualities for a scientist are "intelligence, motivation, and honesty." His routine at Texas A&M was to wake around 4 a.m., read the literature, go to the office at 7 a.m. and stay there until 7 p.m.; when asked in 1997 whether this was still the routine, his response was that he wanted to wake up earlier because sleep was a waste of time—a remark which characterized this active scientist approaching 80!

Robert B. Woodward (1917–1979, Np 1965), who died prematurely, is regarded by many as the preeminent organic chemist of the twentieth century. He made landmark achievements in spectroscopy, synthesis, structure determination, biogenesis, as well as in theory. His solo papers published in 1941–1942 on empirical rules for estimating the absorption maxima of enones and dienes made the general organic chemical community realize that UV could be used for structural studies, thus launching the beginning of the spectroscopic revolution which soon brought on the applications of IR, NMR, MS, etc. He determined the structures of the following compounds: penicillin in 1945 (through joint UK–USA collaboration, see Hodgkin), strychnine in 1948, patulin in 1949, terramycin, aureomycin, and ferrocene (with G. Wilkinson, Np 1973—shared with E. O. Fischer for sandwich compounds) in 1952, cevine in 1954 (with Barton Np 1966, Jeger and Prelog, Np 1975), magnamycin in 1956, gliotoxin in 1958, oleandomycin in 1960, streptonigrin in 1963, and tetrodotoxin in 1964. He synthesized patulin in 1950, cortisone and cholesterol in 1951, lanosterol, lysergic acid (with Eli Lilly), and strychnine in 1954, reserpine in 1956, chlorophyll in 1960, a tetracycline (with Pfizer) in 1962, cephalosporin in 1965, and vitamin B_{12} in 1972 (with A. Eschenmoser, 1925–, ETH Zürich). He derived biogenetic schemes for steroids in 1953 (with K. Bloch, see below), and for macrolides in 1956, while the Woodward–Hoffmann orbital symmetry rules in 1965 brought order to a large class of seemingly random cyclization reactions.

Another central figure in stereochemistry is Vladimir Prelog (1906–1998, Np 1975), who succeeded Leopold Ruzicka at the ETH Zürich, and continued to build this institution into one of the most active and lively research and discussion centers in the world. The core group of intellectual leaders consisted of P. Plattner (1904–1975), O. Jeger, A. Eschenmoser, J. Dunitz, D. Arigoni, and A. Dreiding (from Zürich University). After completing extensive research on alkaloids, Prelog determined the structures of nonactin, boromycin, ferrioxamins, and rifamycins. His seminal studies in the synthesis and properties of 8–12 membered rings led him into unexplored areas of stereochemisty and chirality. Together with Robert Cahn (1899–1981, London Chemical Society) and Christopher Ingold (1893–1970, University College, London; pioneering mechanistic interpretation of organic reactions), he developed the Cahn–Ingold–Prelog (CIP) sequence rules for the unambiguous specification of stereoisomers. Prelog was an excellent story teller, always had jokes to tell, and was respected and loved by all who knew him.

4.1 Polyketides and Fatty Acids

Arthur Birch (1915–1995) from Sydney University, Ph.D. with Robert Robinson (Oxford University), then professor at Manchester University and Australian National University, was one of the earliest chemists to perform biosynthetic studies using radiolabels; starting with polyketides he studied the biosynthesis of a variety of natural products such as the C_6–C_3–C_6 backbone of plant phenolics, polyene macrolides, terpenoids, and alkaloids. He is especially known for the Birch reduction of aromatic rings, metal–ammonia reductions leading to 19-norsteroid hormones and other important products (1942–) which were of industrial importance. Feodor Lynen (1911–1979, Np 1964) performed studies on the intermediary metabolism of the living cell that led him to the demonstration of the first step in a chain of reactions resulting in the biosynthesis of sterols and fatty acids.

Prostaglandins, a family of 20-carbon, lipid-derived acids discovered in seminal fluids and accessory genital glands of man and sheep by von Euler (1934), have attracted great interest because of their extremely diverse biological activities. They were isolated and their structures elucidated from 1963 by S. Bergström (1916–, Np 1982) and B. Samuelsson (1934–, Np 1982) at the Karolinska Institute, Stockholm. Many syntheses of the natural prostaglandins and their nonnatural analogues have been published.

Tetsuo Nozoe (1902–1996) who studied at Tohoku University, Sendai, with Riko Majima (1874–1962, see below) went to Taiwan where he stayed until 1948 before returning to Tohoku University. At National Taiwan University he isolated hinokitiol from the essential oil of *taiwanhinoki*. Remembering the resonance concept put forward by Pauling just before World War II, he arrived at the seven-membered nonbenzenoid aromatic structure for hinokitiol in 1941, the first of the troponoids. This highly original work remained unknown to the rest of the world until 1951. In the meantime, during 1945–1948, nonbenzenoid aromatic structures had been assigned to stipitatic acid (isolated by H. Raistrick) by Michael J. S. Dewar (1918–) and to the thujaplicins by Holger Erdtman (1902–1989); the term tropolones was coined by Dewar in 1945. Nozoe continued to work on and discuss troponoids, up to the night before his death, without knowing that he had cancer. He was a remarkably focused and warm scientist, working unremittingly. Erdtman (Royal Institute of Technology, Stockholm) was the central figure in Swedish natural products chemistry who, with his wife Gunhild Aulin Erdtman (dynamic General Secretary of the Swedish Chemistry Society), worked in the area of plant phenolics.

As mentioned in the following and in the concluding sections, classical biosynthetic studies using radioactive isotopes for determining the distribution of isotopes has now largely been replaced by the use of various stable isotopes coupled with NMR and MS. The main effort has now shifted to the identification and cloning of genes, or where possible the gene clusters, involved in the biosynthesis of the natural product. In the case of polyketides (acyclic, cyclic, and aromatic), the focus is on the polyketide synthases.

4.2 Isoprenoids, Steroids, and Carotenoids

During his time as an assistant to Kekulé at Bonn, Otto Wallach (1847–1931, Np 1910) had to familiarize himself with the essential oils from plants; many of the components of these oils were compounds for which no structure was known. In 1891 he clarified the relations between 12 different monoterpenes related to pinene. This was summarized together with other terpene chemistry in book form in 1909, and led him to propose the "isoprene rule." These achievements laid the foundation for the future development of terpenoid chemistry and brought order from chaos.

The next period up to around 1950 saw phenomenal advances in natural products chemistry centered on isoprenoids. Many of the best natural products chemists in Europe, including Wieland, Windaus, Karrer, Kuhn, Butenandt, and Ruzicka contributed to this breathtaking pace. Heinrich Wieland (1877–1957) worked on the bile acid structure, which had been studied over a period of 100 years and considered to be one of the most difficult to attack; he received the Nobel Prize in 1927 for these studies. His friend Adolph Windaus (1876–1959) worked on the structure of cholesterol for which he also received the Nobel Prize in 1928. Unfortunately, there were chemical discrepancies in the proposed steroidal skeletal structure, which had a five-membered ring B attached to C-7 and C-9. J. D. Bernal, Mineralogical Museums, Cambridge University, who was examining the X-ray patterns of ergosterol (1932) noted that the dimensions were inconsistent with the Wieland–Windaus formula. A reinterpretation of the production of chrysene from sterols by Pd/C dehydrogenation reported by Diels (see below) in 1927 eventually led Rosenheim and King and Wieland and Dane to deduce the correct structure in 1932. Wieland also worked on the structures of morphine/strychnine alkaloids, phalloidin/amanitin cyclopeptides of toxic mushroom *Amanita phalloides*, and pteridines, the important fluorescent pigments of butterfly wings. Windaus determined the structure of ergosterol and continued structural studies of its irradiation product which exhibited antirachitic activity "vitamin D." The mechanistically complex photochemistry of ergosterol leading to the vitamin D group has been investigated in detail by Egbert Havinga (1927–1988, Leiden University), a leading photochemist and excellent tennis player.

Paul Karrer (1889–1971, Np 1937), established the foundations of carotenoid chemistry through structural determinations of lycopene, carotene, vitamin A, etc. and the synthesis of squalene, carotenoids, and others. George Wald (1906–1997, Np 1967) showed that vitamin A was the key compound in vision during his stay in Karrer's laboratory. Vitamin K (K from "Koagulation"), discovered by Henrik Dam (1895–1976, Polytechnic Institute, Copenhagen, Np 1943) and structurally studied by Edward Doisy (1893–1986, St. Louis University, Np 1943), was also synthesized by Karrer. In addition, Karrer synthesized riboflavin (vitamin B_2) and determined the structure and role of nicotinamide adenine dinucleotide phosphate (NADP$^+$) with Otto Warburg. The research on carotenoids and vitamins of Karrer who was at Zürich University overlapped with that of Richard Kuhn (1900–1967, Np 1938) at the ETH Zürich, and the two were frequently rivals. Richard Kuhn, one of the pioneers in using UV-vis spectroscopy for structural studies, introduced the concept of "atropisomerism" in diphenyls, and studied the spectra of a series of diphenyl polyenes. He determined the structures of many natural carotenoids, proved the structure of riboflavin-5-phosphate (flavin-adenine-dinucleotide-5-phosphate) and showed that the combination of NAD-5-phosphate with the carrier protein yielded the yellow oxidation enzyme, thus providing an understanding of the role of a prosthetic group. He also determined the structures of vitamin B complexes, i.e., pyridoxine, *p*-aminobenzoic acid, pantothenic acid. After World War II he went on to structural studies of nitrogen-containing oligosaccharides in human milk that provide immunity for infants, and brain gangliosides. Carotenoid studies in Switzerland were later taken up by Otto Isler (1910–1993), a Ruzicka student at Hoffmann-La Roche, and Conrad Hans Eugster (1921–), a Karrer student at Zürich University.

Adolf Butenandt (1903–1998, Np 1939) initiated and essentially completed isolation and structural studies of the human sex hormones, the insect molting hormone (ecdysone), and the first pheromone, bombykol. With help from industry he was able to obtain large supplies of urine from pregnant women for estrone, sow ovaries for progesterone, and 4,000 gallons of male urine for androsterone (50 mg, crystals). He isolated and determined the structures of two female sex hormones, estrone and progesterone, and the male hormone androsterone all during the period 1934–1939 (!) and was awarded the Nobel prize in 1939. Keen intuition and use of UV data and Pregl's microanalysis all played important roles. He was appointed to a professorship in Danzig at the age of 30. With Peter Karlson he isolated from 500 kg of silkworm larvae 25 mg of α-ecdysone, the prohormone of insect and crustacean molting hormone, and determined its structure as a polyhydroxysteroid (1965); 20-hydroxylation gives the insect and crustacean molting hormone or ß-ecdysone (20-hydroxyecdysteroid). He was also the first to isolate an insect pheromone, bombykol, from female silkworm moths (with E. Hecker). As president of the Max Planck Foundation, he strongly influenced the postwar rebuilding of German science.

The successor to Kuhn, who left ETH Zürich for Heidelberg, was Leopold Ruzicka (1887–1967, Np 1939) who established a close relationship with the Swiss pharmaceutical industry. His synthesis of the 17- and 15-membered macrocyclic ketones, civetone and muscone (the constituents of musk) showed that contrary to Baeyer's prediction, large alicyclic rings could be strainless. He reintroduced and refined the isoprene rule proposed by Wallach (1887) and determined the basic structures of many sesqui-, di-, and triterpenes, as well as the structure of lanosterol, the key intermediate in cholesterol biosynthesis. The "biogenetic isoprene rule" of the ETH group, Albert Eschenmoser, Leopold Ruzicka, Oskar Jeger, and Duilio Arigoni, contributed to a concept of terpenoid cyclization (1955), which was consistent with the mechanistic considerations put forward by Stork as early as 1950. Besides making the ETH group into a center of natural products chemistry, Ruzicka bought many seventeenth century Dutch paintings with royalties accumulated during the war from his Swiss and American patents, and donated them to the Zürich Kunsthaus.

Studies in the isolation, structures, and activities of the antiarthritic hormone, cortisone and related compounds from the adrenal cortex were performed in the mid- to late 1940s during World War II by Edward Kendall (1886–1972, Mayo Clinic, Rochester, Np 1950), Tadeus Reichstein (1897–1996, Basel University, Np 1950), Philip Hench (1896–1965, Mayo Clinic, Rochester, Np 1950), Oskar Wintersteiner (1898–1971, Columbia University, Squibb) and others initiated interest as an adjunct to military medicine as well as to supplement the meager supply from beef adrenal glands by synthesis. Lewis Sarett (1917–, Merck & Co., later president) and co-workers completed the cortisone synthesis in 28 steps, one of the first two totally stereocontrolled syntheses of a natural product; the other was cantharidin (Stork 1951) (see above). The multistep cortisone synthesis was put on the production line by Max Tishler (1906–1989, Merck & Co., later president) who made contributions to the synthesis of a number of drugs, including riboflavin. Besides working on steroid reactions/synthesis and antimalarial agents, Louis F. Fieser (1899–1977) and Mary Fieser (1909–1997) of Harvard University made huge contributions to the chemical community through their outstanding books *Natural Products related to Phenanthrene* (1949), *Steroids* (1959), *Advanced Organic Chemistry* (1961), and *Topics in Organic Chemistry* (1963), as well as their textbooks and an important series of books on Organic Reagents. Carl Djerassi (1923–, Stanford University), a prolific chemist, industrialist, and more recently a novelist, started to work at the Syntex laboratories in Mexico City where he directed the work leading to the first oral contraceptive ("the pill") for women.

Takashi Kubota (1909–, Osaka City University), with Teruo Matsuura (1924–, Kyoto University), determined the structure of the furanoid sesquiterpene, ipomeamarone, from the black rotted portion of spoiled sweet potatoes; this research constitutes the first characterization of a phytoallexin, defense substances produced by plants in response to attack by fungi or physical damage. Damaging a plant and characterizing the defense substances produced may lead to new bioactive compounds. The mechanism of induced biosynthesis of phytoallexins, which is not fully understood, is an interesting biological mechanistic topic that deserves further investigation. Another center of high activity in terpenoids and nucleic acids was headed by Frantisek Sorm (1913–1980, Institute of Organic and Biochemistry, Prague), who determined the structures of many sesquiterpenoids and other natural products; he was not only active scientifically but also was a central figure who helped to guide the careers of many Czech chemists.

The key compound in terpenoid biosynthesis is mevalonic acid (MVA) derived from acetyl-CoA, which was discovered fortuitously in 1957 by the Merck team in Rahway, NJ headed by Karl Folkers (1906–1998). They soon realized and proved that this C_6 acid was the precursor of the C_5 isoprenoid unit isopentenyl diphosphate (IPP) that ultimately leads to the biosynthesis of cholesterol. In 1952 Konrad Bloch (1912–, Harvard, Np 1964) with R. B. Woodward published a paper suggesting a mechanism of the cyclization of squalene to lanosterol and the subsequent steps to cholesterol, which turned out to be essentially correct. This biosynthetic path from MVA to cholesterol was experimentally clarified in stereochemical detail by John Cornforth (1917–, Np 1975) and George Popják. In 1932, Harold Urey (1893–1981, Np 1934) of Columbia University discovered heavy hydrogen. Urey showed, contrary to common expectation, that isotope separation could be achieved with deuterium in the form of deuterium oxide by fractional electrolysis of water. Urey's separation of the stable isotope deuterium led to the isotopic tracer methodology that revolutionized the protocols for elucidating biosynthetic processes and reaction mechanisms, as exemplified beautifully by the cholesterol studies. Using MVA labeled chirally with isotopes, including chiral methyl, i.e., -CHDT, Cornforth and Popják clarified the key steps in the intricate biosynthetic conversion of mevalonate to cholesterol in stereochemical detail. The chiral methyl group was also prepared independently by Duilio Arigoni (1928–, ETH, Zürich). Cornforth has had great difficulty in hearing and speech since childhood but has been helped expertly by his chemist wife Rita; he is an excellent tennis and chess player, and is renowned for his speed in composing occasional witty limericks.

Although MVA has long been assumed to be the only natural precursor for IPP, a non-MVA pathway in which IPP is formed via the glyceraldehyde phosphate-pyruvate pathway has been discovered (1995–1996) in the ancient bacteriohopanoids by Michel Rohmer, who started working on them with Guy Ourisson (1926–, University of Strasbourg, terpenoid studies, including prebiotic), and by Duilio Arigoni in the ginkgolides, which are present in the ancient *Ginkgo biloba* tree. It is possible that many other terpenoids are biosynthesized via the non-MVA route. In classical biosynthetic experiments, ^{14}C-labeled acetic acid was incorporated into the microbial or plant product, and location or distribution of the ^{14}C label was deduced by oxidation or degradation to specific fragments including acetic acid; therefore, it was not possible or extremely difficult to map the distribution of all radioactive carbons. The progress in ^{13}C NMR made it possible to incorporate ^{13}C-labeled acetic acid and locate all labeled carbons. This led to the discovery of the nonmevalonate pathway leading to the IPP units. Similarly, NMR and MS have made it possible to use the stable isotopes, e.g., ^{18}O, ^{2}H, ^{15}N, etc., in biosynthetic studies. The current trend of biosynthesis has now shifted to genomic approaches for cloning the genes of various enzyme synthases involved in the biosynthesis.

4.3 Carbohydrates and Cellulose

The most important advance in carbohydrate structures following those made by Emil Fischer was the change from acyclic to the current cyclic structure introduced by Walter Haworth (1883–1937). He noticed the presence of α- and ß-anomers, and determined the structures of important disaccharides including cellobiose, maltose, and lactose. He also determined the basic structural aspects of starch, cellulose, inulin, and other polysaccharides, and accomplished the structure determination and synthesis of vitamin C, a sample of which he had received from Albert von Szent-Györgyi (1893–1986, Np 1937). This first synthesis of a vitamin was significant since it showed that a vitamin could be synthesized in the same way as any other organic compound. There was strong belief among leading scientists in the 1910s that cellulose, starch, protein, and rubber were colloidal aggregates of small molecules. However, Hermann Staudinger (1881–1965, Np 1953) who succeeded R. Willstätter and H. Wieland at the ETH Zürich and Freiburg, respectively, showed through viscosity measurements and various molecular weight measurements that macromolecules do exist, and developed the principles of macromolecular chemistry.

In more modern times, Raymond Lemieux (1920–, Universities of Ottawa and Alberta) has been a leader in carbohydrate research. He introduced the concept of *endo-* and *exo-*anomeric effects, accomplished the challenging synthesis of sucrose (1953), pioneered in the use of NMR coupling constants in configuration studies, and most importantly, starting with syntheses of oligosaccharides responsible for human blood group determinants, he prepared antibodies and clarified fundamental aspects of the binding of oligosaccharides by lectins and antibodies. The periodate–potassium permanganate cleavage of double bonds at room temperature (1955) is called the Lemieux reaction.

4.4 Amino Acids, Peptides, Porphyrins, and Alkaloids

It is fortunate that we have China's record and practice of herbal medicine over the centuries, which is providing us with an indispensable source of knowledge. China is rapidly catching up in terms of infrastructure and equipment in organic and bioorganic chemistry, and work on isolation, structure determination, and synthesis stemming from these valuable sources has picked up momentum. However, as mentioned above, clarification of the active principles and mode of action of these plant extracts will be quite a challenge since in many cases synergistic action is expected. Wang Yu (1910–1997) who headed the well-equipped Shanghai Institute of Organic Chemistry surprised the world with the total synthesis of bovine insulin performed by his group in 1965; the human insulin was synthesized around the same time by P. G. Katsoyannis, A. Tometsko, and C. Zaut of the Brookhaven National Laboratory (1966).

One of the giants in natural products chemistry during the first half of this century was Robert Robinson (1886–1975, Np 1947) at Oxford University. His synthesis of tropinone, a bicyclic amino ketone related to cocaine, from succindialdehyde, methylamine, and acetone dicarboxylic acid under Mannich reaction conditions was the first biomimetic synthesis (1917). It reduced Willstätter's 1903 13-step synthesis starting with suberone into a single step. This achievement demonstrated Robinson's analytical prowess. He was able to dissect complex molecular structures into simple biosynthetic building blocks, which allowed him to propose the biogenesis of all types of alkaloids and other natural products. His laboratory at Oxford, where he developed the well-known Robinson annulation reaction (1937) in connection with his work on the synthesis of steroids became a world center for natural products study. Robinson was a pioneer in the so-called electronic theory of organic reactions, and introduced the use of curly arrows to show the movements of electrons. His analytical power is exemplified in the structural studies of strychnine and brucine around 1946–1952. Barton clarified the biosynthetic route to the morphine alkaloids, which he saw as an extension of his biomimetic synthesis of usnic acid through a one-electron oxidation; this was later extended to a general phenolate coupling scheme. Morphine total synthesis was brilliantly achieved by Marshall Gates (1915–, University of Rochester) in 1952.

The yield of the Robinson tropinone synthesis was low but Clemens Schöpf (1899–1970) , Ph.D. Munich (Wieland), Universität Darmstadt, improved it to 90% by carrying out the reaction in buffer; he also worked on the stereochemistry of morphine and determined the structure of the steroidal alkaloid salamandarine (1961), the toxin secreted from glands behind the eyes of the salamander.

Roger Adams (1889–1971, University of Illinois), was the central figure in organic chemistry in the USA and is credited with contributing to the rapid development of its chemistry in the late 1930s and 1940s, including training of graduate students for both academe and industry. After earning a Ph.D. in 1912 at Harvard University he did postdoctoral studies with Otto Diels (see below) and Richard Willstätter (see below) in 1913; he once said that around those years in Germany he could cover all *Journal of the American Chemical Society* papers published in a year in a single night. His important work include determination of the structures of tetrahydrocannabinol in marijuana, the toxic gossypol in cottonseed oil, chaulmoogric acid used in treatment of leprosy, and the Senecio alkaloids with Nelson Leonard (1916–, University of Illinois, now at Caltech). He also contributed to many fundamental organic reactions and syntheses. The famous Adams platinum catalyst is not only important for reducing double bonds in industry and in the laboratory, but was central for determining the number of double bonds in a structure. He was also one of the founders of the *Organic Synthesis* (started in 1921) and the *Organic Reactions* series. Nelson Leonard switched interests to bioorganic chemistry and biochemistry, where he has worked with nucleic acid bases and nucleotides, coenzymes, dimensional probes, and fluorescent modifications such as ethenoguanine.

The complicated structures of the medieval plant poisons aconitine (from *Aconitum*) and delphinine (from *Delphinium*) were finally characterized in 1959–1960 by Karel Wiesner (1919–1986, University of New Brunswick), Leo Marion (1899–1979, National Research Council, Ottawa), George Büchi (1921–, mycotoxins, aflatoxin/DNA adduct, synthesis of terpenoids and nitrogen-containing bioactive compounds, photochemistry), and Maria Przybylska (1923–, X-ray).

The complex chlorophyll structure was elucidated by Richard Willstätter (1872–1942, Np 1915). Although he could not join Baeyer's group at Munich because the latter had ceased taking students, a close relation developed between the two. During his chlorophyll studies, Willstätter reintroduced the important technique of column chromatography published in Russian by Michael Tswett (1906). Willstätter further demonstrated that magnesium was an integral part of chlorophyll, clarified the relation between chlorophyll and the blood pigment hemin, and found the wide distribution of carotenoids in tomato, egg yolk, and bovine corpus luteum. Willstätter also synthesized cyclooctatetraene and showed its properties to be wholly unlike benzene but close to those of acyclic polyenes (around 1913). He succeeded Baeyer at Munich in 1915, synthesized the anesthetic cocaine, retired early in protest of anti-Semitism, but remained active until the Hitler era, and in 1938 emigrated to Switzerland.

The hemin structure was determined by another German chemist of the same era, Hans Fischer (1881–1945, Np 1930), who succeeded Windaus at Innsbruck and at Munich. He worked on the structure of hemin from the blood pigment hemoglobin, and completed its synthesis in 1929. He continued Willstätter's structural studies of chlorophyll, and further synthesized bilirubin in 1944. Destruction of his institute at Technische Hochschule München, during World War II led him to take his life in March 1945. The biosynthesis of hemin was elucidated largely by David Shemin (1911–1991).

In the mid 1930s the Department of Biochemistry at Columbia Medical School, which had accepted many refugees from the Third Reich, including Erwin Chargaff, Rudolf Schoenheimer, and others on the faculty, and Konrad Bloch (see above) and David Shemin as graduate students, was a great center of research activity. In 1940, Shemin ingested 66 g of ^{15}N-labeled glycine over a period of 66 hours in order to determine the half-life of erythrocytes. David Rittenberg's analysis of the heme moiety with his home-made mass spectrometer showed all four pyrrole nitrogens came from glycine. Using ^{14}C (that had just become available) as a second isotope (see next paragraph), doubly labeled glycine ^{15}NH$_2$^{14}CH$_2$COOH and other precursors, Shemin showed that glycine and succinic acid condensed to yield δ-aminolevulinate, thus elegantly demonstrating the novel biosynthesis of the porphyrin ring (around 1950). At this time, Bloch was working on the other side of the bench.

Melvin Calvin (1911–1997, Np 1961) at University of California, Berkeley, elucidated the complex photosynthetic pathway in which plants reduce carbon dioxide to carbohydrates. The critical ^{14}CO$_2$ had just been made available at Berkeley Lawrence Radiation Laboratory as a result of the pioneering research of Martin Kamen (1913–), while paper chromatography also played crucial roles. Kamen produced ^{14}C with Sam Ruben (1940), used ^{18}O to show that oxygen in photosynthesis comes from water and not from carbon dioxide, participated in the *Manhattan* project, testified before the House UnAmerican Activities Committee (1947), won compensatory damages from the US Department of State, and helped build the University of California, La Jolla (1957). The entire structure of the photosynthetic reaction center (>10 000 atoms) from the purple bacterium *Rhodopseudomonas viridis* has been established by X-ray crystallography in the landmark studies performed by Johann Deisenhofer (1943–), Robert Huber (1937–), and Hartmut Michel (1948–) in 1989; this was the first membrane protein structure determined by X-ray, for which they shared the 1988 Nobel prize. The information gained from the full structure of this first membrane protein has been especially rewarding.

The studies on vitamin B_{12}, the structure of which was established by crystallographic studies performed by Dorothy Hodgkin (1910–1994, Np 1964), are fascinating. Hodgkin also determined the structure of penicillin (in a joint effort between UK and US scientists during World War II) and insulin. The formidable total synthesis of vitamin B_{12} was completed in 1972 through collaborative efforts between Woodward and Eschenmoser, involving 100 postdoctoral fellows and extending over 10 years. The biosynthesis of fascinating complexity is almost completely solved through studies performed by Alan Battersby (1925–, Cambridge University), Duilio Arigoni, and Ian Scott (1928–, Texas A&M University) and collaborators where advanced NMR techniques and synthesis of labeled precursors is elegantly combined with cloning of enzymes controlling each biosynthetic step. This work provides a beautiful demonstration of the power of the combination of bioorganic chemistry, spectroscopy and molecular biology, a future direction which will become increasingly important for the creation of new "unnatural" natural products.

4.5 Enzymes and Proteins

In the early days of natural products chemistry, enzymes and viruses were very poorly understood. Thus, the 1926 paper by James Sumner (1887–1955) at Cornell University on crystalline urease was received with ignorance or skepticism, especially by Willstätter who believed that enzymes were small molecules and not proteins. John Northrop (1891–1987) and co-workers at the Rockefeller Institute went on to crystallize pepsin, trypsin, chymotrypsin, ribonuclease, deoyribonuclease, carboxypeptidase, and other enzymes between 1930 and 1935. Despite this, for many years biochemists did not recognize the significance of these findings, and considered enzymes as being low molecular weight compounds adsorbed onto proteins or colloids. Using Northrop's method for crystalline enzyme preparations, Wendell Stanley (1904–1971) at Princeton obtained tobacco mosaic virus as needles from one ton of tobacco leaves (1935). Sumner, Northrop, and Stanley shared the 1946 Nobel prize in chemistry. All these studies opened a new era for biochemistry.

Meanwhile, Linus Pauling, who in mid-1930 became interested in the magnetic properties of hemoglobin, investigated the configurations of proteins and the effects of hydrogen bonds. In 1949 he showed that sickle cell anemia was due to a mutation of a single amino acid in the hemoglobin molecule, the first correlation of a change in molecular structure with a genetic disease. Starting in 1951 he and colleagues published a series of papers describing the alpha helix structure of proteins; a paper published in the early 1950s with R. B. Corey on the structure of DNA played an important role in leading Francis Crick and James Watson to the double helix structure (Np 1962).

A further important achievement in the peptide field was that of Vincent Du Vigneaud (1901–1978, Np 1955), Cornell Medical School, who isolated and determined the structure of oxytocin, a posterior pituitary gland hormone, for which a structure involving a disulfide bond was proposed. He synthesized oxytocin in 1953, thereby completing the first synthesis of a natural peptide hormone.

Progress in isolation, purification, crystallization methods, computers, and instrumentation, including cyclotrons, have made X-ray crystallography the major tool in structural. Numerous structures including those of ligand/receptor complexes are being published at an extremely rapid rate. Some of the past major achievements in protein structures are the following. Max Perutz (1914, Np 1962) and John Kendrew (1914–1997, Np 1962), both at the Laboratory of Molecular Biology, Cambridge University, determined the structures of hemoglobin and myoglobin, respectively. William Lipscomb (1919–, Np 1976), Harvard University, who has trained many of the world's leaders in protein X-ray crystallography has been involved in the structure determination of many enzymes including carboxypeptidase A (1967); in 1965 he determined the structure of the anticancer bisindole alkaloid, vinblastine. Folding of proteins, an important but still enigmatic phenomenon, is attracting increasing attention. Christian Anfinsen (1916–1995, Np 1972), NIH, one of the pioneers in this area, showed that the amino acid residues in ribonuclease interact in an energetically most favorable manner to produce the unique 3D structure of the protein.

4.6 Nucleic Acid Bases, RNA, and DNA

The "Fischer indole synthesis" was first performed in 1886 by Emil Fischer. During the period 1881–1914, he determined the structures of and synthesized uric acid, caffeine, theobromine, xanthine, guanine, hypoxanthine, adenine, guanine, and made theophylline-D-glucoside phosphoric acid, the first synthetic nucleotide. In 1903, he made 5,5-diethylbarbituric acid or Barbital, Dorminal, Veronal, etc. (sedative), and in 1912, phenobarbital or Barbipil, Luminal, Phenobal, etc. (sedative). Many of his syntheses formed the basis of German industrial production of purine bases. In 1912 he showed that tannins are gallates of sugars such as maltose and glucose. Starting in 1899, he synthesized many of the 13 α-amino acids known at that time, including the L- and D-forms, which were separated through fractional crystallization of their salts with optically active bases. He also developed a method for synthesizing fragments of proteins, namely peptides, and made an 18-amino acid peptide. He lost his two sons in World War I, lost his wealth due to postwar inflation, believed he had terminal cancer (a misdiagnosis), and killed himself in July 1919. Fischer was a skilled experimentalist, so that even today, many of the reactions performed by him and his students are so delicately controlled that they are not easy to reproduce. As a result of his suffering by inhaling diethylmercury, and of the poisonous effect of phenylhydrazine, he was one of the first to design fume hoods. He was a superb teacher and was also influential in establishing the Kaiser Wilhelm Institute, which later became the Max Planck Institute. The number and quality of his accomplishments and contributions are hard to believe; he was truly a genius.

Alexander Todd (1907–1997, Np 1957) made critical contributions to the basic chemistry and synthesis of nucleotides. His early experience consisted of an extremely fruitful stay at Oxford in the Robinson group, where he completed the syntheses of many representative anthocyanins, and then at Edinburgh where he worked on the synthesis of vitamin B_1. He also prepared the hexacarboxylate of vitamin B_{12} (1954), which was used by D. Hodgkin's group for their X-ray elucidation of this vitamin (1956). M. Wiewiorowski (1918–), Institute for Bioorganic Chemistry, in Poznan, has headed a famous group in nucleic acid chemistry, and his colleagues are now distributed worldwide.

4.7 Antibiotics, Pigments, and Marine Natural Products

The concept of one microorganism killing another was introduced by Pasteur who coined the term antibiosis in 1877, but it was much later that this concept was realized in the form of an actual antibiotic. The bacteriologist Alexander Fleming (1881–1955, University of London, Np 1945) noticed that an airborne mold, a *Penicillium* strain, contaminated cultures of *Staphylococci* left on the open bench and formed a transparent circle around its colony due to lysis of *Staphylococci*. He published these results in 1929. The discovery did not attract much interest but the work was continued by Fleming until it was taken up further at Oxford University by pathologist Howard Florey (1898–1968, Np 1945) and biochemist Ernst Chain (1906–1979, Np 1945). The bioactivities of purified "penicillin," the first antibiotic, attracted serious interest in the early 1940s in the midst of World War II. A UK/USA team was formed during the war between academe and industry with Oxford University, Harvard University, ICI, Glaxo, Burroughs Wellcome, Merck, Shell, Squibb, and Pfizer as members. This project resulted in the large scale production of penicillin and determination of its structure (finally by X-ray, D. Hodgkin). John Sheehan (1915–1992) at MIT synthesized 6-aminopenicillanic acid in 1959, which opened the route for the synthesis of a number of analogues. Besides being the first antibiotic to be discovered, penicillin is also the first member of a large number of important antibiotics containing the ß-lactam ring, for example cephalosporins, carbapenems, monobactams, and nocardicins. The strained ß-lactam ring of these antibiotics inactivates the transpeptidase by acylating its serine residue at the active site, thus preventing the enzyme from forming the link between the pentaglycine chain and the D-Ala-D-Ala peptide, the essential link in bacterial cell walls. The overuse of ß-lactam antibiotics, which has given rise to the disturbing appearance of microbial resistant strains, is leading to active research in the design of synthetic ß-lactam analogues to counteract these strains. The complex nature of the important penicillin biosynthesis is being elucidated through efforts combining genetic engineering, expression of biosynthetic genes as well as feeding of synthetic precursors, etc. by Jack Baldwin (1938–, Oxford University), José Luengo (Universidad de León, Spain) and many other groups from industry and academe.

Shortly after the penicillin discovery, Selman Waksman (1888–1973, Rutgers University, Np 1952) discovered streptomycin, the second antibiotic and the first active against the dreaded disease tuberculosis. The discovery and development of new antibiotics continued throughout the world at pharmaceutical companies in Europe, Japan, and the USA from soil and various odd sources: cephalosporin from sewage in Sardinia, cyclosporin from Wisconsin and Norway soil which was carried back to Switzerland, avermectin from the soil near a golf course in Shizuoka Prefecture. People involved in antibiotic discovery used to collect soil samples from various sources during their trips but this has now become severely restricted to protect a country's right to its soil. M. M. Shemyakin (1908–1970, Institute of Chemistry of Natural Products, Moscow) was a grand master of Russian natural products who worked on antibiotics, especially of the tetracycline class; he also worked on cyclic antibiotics composed of alternating sequences of amides and esters and coined the term depsipeptide for these in 1953. He died in 1970 of a sudden heart attack in the midst of the 7th IUPAC Natural Products

Symposium held in Riga, Latvia, which he had organized. The Institute he headed was renamed the Shemyakin Institute.

Indigo, an important vat dye known in ancient Asia, Egypt, Greece, Rome, Britain, and Peru, is probably the oldest known coloring material of plant origin, Indigofera and Isatis. The structure was determined in 1883 and a commercially feasible synthesis was performed in 1883 by Adolf von Baeyer (see above, 1835–1917, Np 1905), who founded the German Chemical Society in 1867 following the precedent of the Chemistry Society of London. In 1872 Baeyer was appointed a professor at Strasbourg where E. Fischer was his student, and in 1875 he succeeded J. Liebig in Munich. Tyrian (or Phoenician) purple, the dibromo derivative of indigo which is obtained from the purple snail Murex bundaris, was used as a royal emblem in connection with religious ceremonies because of its rarity; because of the availability of other cheaper dyes with similar color, it has no commercial value today. K. Venkataraman (1901–1981, University of Bombay then National Chemical Laboratory) who worked with R. Robinson on the synthesis of chromones in his early career, continued to study natural and synthetic coloring matters, including synthetic anthraquinone vat dyes, natural quinonoid pigments, etc. T. R. Seshadri (1900–1975) is another Indian natural products chemist who worked mainly in natural pigments, dyes, drugs, insecticides, and especially in polyphenols. He also studied with Robinson, and with Pregl at Graz, and taught at Delhi University. Seshadri and Venkataraman had a huge impact on Indian chemistry. After a 40 year involvement, Toshio Goto (1929–1990) finally succeeded in solving the mysterious identity of commelinin, the deep-blue flower petal pigment of the Commelina communis isolated by Kozo Hayashi (1958) and protocyanin, isolated from the blue cornflower Centaurea cyanus by E. Bayer (1957). His group elucidated the remarkable structure in its entirety which consisted of six unstable anthocyanins, six flavones and two metals, the molecular weight approaching 10 000; complex stacking and hydrogen bonds were also involved. Thus the pigmentation of petals turned out to be far more complex than the theories put forth by Willstätter (1913) and Robinson (1931). Goto suffered a fatal heart attack while inspecting the first X-ray structure of commelinin; commelinin represents a pinnacle of current natural products isolation and structure determination in terms of subtlety in isolation and complexity of structure.

The study of marine natural products is understandably far behind that of compounds of terrestrial origin due to the difficulty in collection and identification of marine organisms. However, it is an area which has great potentialities for new discoveries from every conceivable source. One pioneer in modern marine chemistry is Paul Scheuer (1915–, University of Hawaii) who started his work with quinones of marine origin and has since characterized a very large number of bioactive compounds from mollusks and other sources. Luigi Minale (1936–1997, Napoli) started a strong group working on marine natural products, concentrating mainly on complex saponins. He was a leading natural products chemist who died prematurely. A. Gonzalez Gonzalez (1917–) who headed the Organic Natural Products Institute at the University of La Laguna, Tenerife, was the first to isolate and study polyhalogenated sesquiterpenoids from marine sources. His group has also carried out extensive studies on terrestrial terpenoids from the Canary Islands and South America. Carotenoids are widely distributed in nature and are of importance as food coloring material and as antioxidants (the detailed mechanisms of which still have to be worked out); new carotenoids continue to be discovered from marine sources, for example by the group of Synnove Liaaen-Jensen, Norwegian Institute of Technology). Yoshimasa Hirata (1915–), who started research at Nagoya University, is a champion in the isolation of nontrivial natural products. He characterized the bioluminescent luciferin from the marine ostracod *Cypridina hilgendorfii* in 1966 (with his students, Toshio Goto, Yoshito Kishi, and Osamu Shimomura); tetrodotoxin from the fugu fish in 1964 (with Goto and Kishi and co-workers), the structure of which was announced simultaneously by the group of Kyosuke Tsuda (1907–, tetrodotoxin, matrine) and Woodward; and the very complex palytoxin, $C_{129}H_{223}N_3O_{54}$ in 1981–1987 (with Daisuke Uemura and Kishi). Richard E. Moore, University of Hawaii, also announced the structure of palytoxin independently. Jon Clardy (1943–, Cornell University) has determined the X-ray structures of many unique marine natural products, including brevetoxin B (1981), the first of the group of toxins with contiguous *trans*-fused ether rings constituting a stiff ladder-like skeleton. Maitotoxin, $C_{164}H_{256}O_{68}S_2Na_2$, MW 3422, produced by the dinoflagellate *Gambierdiscus toxicus* is the largest and most toxic of the nonbiopolymeric toxins known; it has 32 alicyclic 6- to 8-membered ethereal rings and acyclic chains. Its isolation (1994) and complete structure determination was accomplished jointly by the groups of Takeshi Yasumoto (Tohoku University), Kazuo Tachibana and Michio Murata (Tokyo University) in 1996. Kishi, Harvard University, also deduced the full structure in 1996.

The well-known excitatory agent for the cat family contained in the volatile oil of catnip, *Nepeta cataria*, is the monoterpene nepetalactone, isolated by S. M. McElvain (1943) and structure determined by Jerrold Meinwald (1954); cats, tigers, and lions start purring and roll on their backs in response to this lactone. Takeo Sakan (1912–1993) investigated the series of monoterpenes neomatatabiols, etc.

from Actinidia, some of which are male lacewing attractants. As little as 1 fg of neomatatabiol attracts lacewings.

The first insect pheromone to be isolated and characterized was bombykol, the sex attractant for the male silkworm, *Bombyx mori* (by Butenandt and co-workers, see above). Numerous pheromones have been isolated, characterized, synthesized, and are playing central roles in insect control and in chemical ecology. The group at Cornell University have long been active in this field: Tom Eisner (1929–, behavior), Jerrold Meinwald (1927–, chemistry), Wendell Roeloff (1939–, electrophysiology, chemistry). Since the available sample is usually minuscule, full structure determination of a pheromone often requires total synthesis; Kenji Mori (1935–, Tokyo University) has been particularly active in this field. Progress in the techniques for handling volatile compounds, including collection, isolation, GC/MS, etc., has started to disclose the extreme complexity of chemical ecology which plays an important role in the lives of all living organisms. In this context, natural products chemistry will be play an increasingly important role in our grasp of the significance of biodiversity.

5. SYNTHESIS

Synthesis has been mentioned often in the preceding sections of this essay. In the following, synthetic methods of more general nature are described. The Grignard reaction of Victor Grignard (1871–1935, Np 1912) and then the Diels–Alder reaction by Otto Diels (1876–1954, Np 1950) and Kurt Alder (1902–1956, Np 1950) are extremely versatile reactions. The Diels–Alder reaction can account for the biosynthesis of several natural products with complex structures, and now an enzyme, a Diels–Alderase involved in biosynthesis has been isolated by Akitami Ichihara, Hokkaido University (1997).

The hydroboration reactions of Herbert Brown (1912–, Purdue University, Np 1979) and the Wittig reactions of Georg Wittig (1897–1987, Np 1979) are extremely versatile synthetic reactions. William S. Johnson (1913–1995, University of Wisconsin, Stanford University) developed efficient methods for the cyclization of acyclic polyolefinic compounds for the synthesis of corticoid and other steroids, while Gilbert Stork (1921–, Columbia University) introduced enamine alkylation, regiospecific enolate formation from enones and their kinetic trapping (called "three component coupling" in some cases), and radical cyclization in regio- and stereospecific constructions. Elias J. Corey (1928–, Harvard University, Np 1990) introduced the concept of retrosynthetic analysis and developed many key synthetic reactions and reagents during his synthesis of bioactive compounds, including prostaglandins and gingkolides. A recent development is the ever-expanding supramolecular chemistry stemming from 1967 studies on crown ethers by Charles Pedersen (1904–1989), 1968 studies on cryptates by Jean-Marie Lehn (1939–), and 1973 studies on host–guest chemistry by Donald Cram (1919–); they shared the chemistry Nobel prize in 1987.

6. NATURAL PRODUCTS STUDIES IN JAPAN

Since the background of natural products study in Japan is quite different from that in other countries, a brief history is given here. Natural products is one of the strongest areas of chemical research in Japan with probably the world's largest number of chemists pursuing structural studies; these are joined by a healthy number of synthetic and bioorganic chemists. An important Symposium on Natural Products was held in 1957 in Nagoya as a joint event between the faculties of science, pharmacy, and agriculture. This was the beginning of a series of annual symposia held in various cities, which has grown into a three-day event with about 50 talks and numerous papers; practically all achievements in this area are presented at this symposium. Japan adopted the early twentieth century German or European academic system where continuity of research can be assured through a permanent staff in addition to the professor, a system which is suited for natural products research which involves isolation and assay, as well as structure determination, all steps requiring delicate skills and much expertise.

The history of Japanese chemistry is short because the country was closed to the outside world up to 1868. This is when the Tokugawa shogunate which had ruled Japan for 264 years was overthrown and the Meiji era (1868–1912) began. Two of the first Japanese organic chemists sent abroad were Shokei Shibata and Nagayoshi Nagai, who joined the laboratory of A. W. von Hoffmann in Berlin. Upon return to Japan, Shibata (Chinese herbs) started a line of distinguished chemists, Keita and Yuji Shibata (flavones) and Shoji Shibata (1915–, lichens, fungal bisanthraquinonoid pigments, ginsenosides); Nagai returned to Tokyo Science University in 1884, studied ephedrine, and left a big mark in the embryonic era of organic chemistry. Modern natural products chemistry really began when three extraordinary organic chemists returned from Europe in the 1910s and started teaching and research at their respective faculties:

Riko Majima, 1874–1962, C. D. Harries (Kiel University); R. Willstätter (Zürich): Faculty of Science, Tohoku University; studied urushiol, the catecholic mixture of poison ivy irritant.

Yasuhiko Asahina, 1881–1975, R. Willstätter: Faculty of pharmacy, Tokyo University; lichens and Chinese herb.

Umetaro Suzuki, 1874–1943, E. Fischer: Faculty of agriculture, Tokyo University; vitamin B_1(thiamine).

Because these three pioneers started research in three different faculties (i.e., science, pharmacy, and agriculture), and because little interfaculty personnel exchange occurred in subsequent years, natural products chemistry in Japan was pursued independently within these three academic domains; the situation has changed now. The three pioneers started lines of first-class successors, but the establishment of a strong infrastructure takes many years, and it was only after the mid-1960s that the general level of science became comparable to that in the rest of the world; the 3rd IUPAC Symposium on the Chemistry of Natural Products, presided over by Munio Kotake (1894–1976, bufotoxins, see below), held in 1964 in Kyoto, was a clear turning point in Japan's role in this area.

Some of the outstanding Japanese chemists not already quoted are the following. Shibasaburo Kitazato (1852–1931), worked with Robert Koch (Np 1905, tuberculosis) and von Behring, antitoxins of diphtheria and tetanus which opened the new field of serology, isolation of microorganism causing dysentery, founder of Kitazato Institute; Chika Kuroda (1884–1968), first female Ph.D., structure of the complex carthamin, important dye in safflower (1930) which was revised in 1979 by Obara *et al.*, although the absolute configuration is still unknown (1998); Munio Kotake (1894–1976), bufotoxins, tryptophan metabolites, nupharidine; Harusada Suginome (1892–1972), aconite alkaloids; Teijiro Yabuta (1888–1977), kojic acid, gibberrelins; Eiji Ochiai (1898–1974), aconite alkaloids; Toshio Hoshino (1899–1979), abrine and other alkaloids; Yusuke Sumiki (1901–1974), gibberrelins; Sankichi Takei (1896–1982), rotenone; Shiro Akabori (1900–1992), peptides, C-terminal hydrazinolysis of amino acid ; Hamao Umezawa (1914–1986), kanamycin, bleomycin, numerous antibiotics; Shojiro Uyeo (1909–1988), lycorine; Tsunematsu Takemoto (1913–1989), inokosterone, kainic acid, domoic acid, quisqualic acid; Tomihide Shimizu (1889–1958), bile acids; Kenichi Takeda (1907–1991), Chinese herbs, sesquiterpenes; Yoshio Ban (1921–1994), alkaloid synthesis; Wataru Nagata (1922–1993), stereocontrolled hydrocyanation.

7. CURRENT AND FUTURE TRENDS IN NATURAL PRODUCTS CHEMISTRY

Spectroscopy and X-ray crystallography has totally changed the process of structure determination, which used to generate the excitement of solving a mystery. The first introduction of spectroscopy to the general organic community was Woodward's 1942–1943 empirical rules for estimating the UV maxima of dienes, trienes, and enones, which were extended by Fieser (1959). However, Butenandt had used UV for correctly determining the structures of the sex hormones as early as the early 1930s, while Karrer and Kuhn also used UV very early in their structural studies of the carotenoids. The Beckman DU instruments were an important factor which made UV spectroscopy a common tool for organic chemists and biochemists. With the availability of commercial instruments in 1950, IR spectroscopy became the next physical tool, making the 1950 Colthup IR correlation chart and the 1954 Bellamy monograph indispensable. The IR fingerprint region was analyzed in detail in attempts to gain as much structural information as possible from the molecular stretching and bending vibrations. Introduction of NMR spectroscopy into organic chemistry, first for protons and then for carbons, has totally changed the picture of structure determination, so that now IR is used much less frequently; however, in biopolymer studies, the techniques of difference FTIR and resonance Raman spectroscopy are indispensable.

The dramatic and rapid advancements in mass spectrometry are now drastically changing the protocol of biomacromolecular structural studies performed in biochemistry and molecular biology. Herbert Hauptman (mathematician, 1917–, Medical Foundation, Buffalo, Np 1985) and Jerome Karle (1918–, US Naval Research Laboratory, Washington, DC, Np 1985) developed direct methods for the determination of crystal structures devoid of disproportionately heavy atoms. The direct method together with modern computers revolutionized the X-ray analysis of molecular structures, which has become routine for crystalline compounds, large as well as small. Fred McLafferty (1923–, Cornell University) and Klaus Biemann (1926–, MIT) have made important contributions in the development of organic and bioorganic mass spectrometry. The development of cyclotron-based facilities for crystallographic biology studies has led to further dramatic advances enabling some protein structures to be determined in a single day, while cryoscopic electron micrography developed in 1975 by Richard Henderson and Nigel Unwin has also become a powerful tool for 3D structural determinations of membrane proteins such as bacteriorhodopsin (25 kd) and the nicotinic acetylcholine receptor (270 kd).

Circular dichroism (c.d.), which was used by French scientists Jean B. Biot (1774–1862) and Aimé Cotton during the nineteenth century "deteriorated" into monochromatic measurements at 589 nm after R.W. Bunsen (1811–1899, Heidelberg) introduced the Bunsen burner into the laboratory which readily emitted a 589 nm light characteristic of sodium. The 589 nm $[\alpha]_D$ values, remote from most chromophoric maxima, simply represent the summation of the low-intensity readings of the decreasing end of multiple Cotton effects. It is therefore very difficult or impossible to deduce structural information from $[\alpha]_D$ readings. Chiroptical spectroscopy was reintroduced to organic chemistry in the 1950s by C. Djerassi at Wayne State University (and later at Stanford University) as optical rotatory dispersion (ORD) and by L. Velluz and M. Legrand at Roussel-Uclaf as c.d. Günther Snatzke (1928–1992, Bonn then Ruhr University Bochum) was a major force in developing the theory and application of organic chiroptical spectroscopy. He investigated the chiroptical properties of a wide variety of natural products, including constituents of indigenous plants collected throughout the world, and established semiempirical sector rules for absolute configurational studies. He also established close collaborations with scientists of the former Eastern bloc countries and had a major impact in increasing the interest in c.d. there.

Chiroptical spectroscopy, nevertheless, remains one of the most underutilized physical measurements. Most organic chemists regard c.d. (more popular than ORD because interpretation is usually less ambiguous) simply as a tool for assigning absolute configurations, and since there are only two possibilities in absolute configurations, c.d. is apparently regarded as not as crucial compared to other spectroscopic methods. Moreover, many of the c.d. correlations with absolute configuration are empirical. For such reasons, chiroptical spectroscopy, with its immense potentialities, is grossly underused. However, c.d. curves can now be calculated nonempirically. Moreover, through-space coupling between the electric transition moments of two or more chromophores gives rise to intense Cotton effects split into opposite signs, exciton-coupled c.d.; fluorescence-detected c.d. further enhances the sensitivity by 50- to 100-fold. This leads to a highly versatile nonempirical microscale solution method for determining absolute configurations, etc.

With the rapid advances in spectroscopy and isolation techniques, most structure determinations in natural products chemistry have become quite routine, shifting the trend gradually towards activity-monitored isolation and structural studies of biologically active principles available only in microgram or submicrogram quantities. This in turn has made it possible for organic chemists to direct their attention towards clarifying the mechanistic and structural aspects of the ligand/biopolymeric receptor interactions on a more well-defined molecular structural basis. Until the 1990s, it was inconceivable and impossible to perform such studies.

Why does sugar taste sweet? This is an extremely challenging problem which at present cannot be answered even with major multidisciplinary efforts. Structural characterization of sweet compounds and elucidation of the amino acid sequences in the receptors are only the starting point. We are confronted with a long list of problems such as cloning of the receptors to produce them in sufficient quantities to investigate the physical fit between the active factor (sugar) and receptor by biophysical methods, and the time-resolved change in this physical contact and subsequent activation of G-protein and enzymes. This would then be followed by neurophysiological and ultimately physiological and psychological studies of sensation. How do the hundreds of taste receptors differ in their structures and their physical contact with molecules, and how do we differentiate the various taste sensations? The same applies to vision and to olfactory processes. What are the functions of the numerous glutamate receptor subtypes in our brain? We are at the starting point of a new field which is filled with exciting possibilities.

Familiarity with molecular biology is becoming essential for natural products chemists to plan research directed towards an understanding of natural products biosynthesis, mechanisms of bioactivity triggered by ligand–receptor interactions, etc. Numerous genes encoding enzymes have been cloned and expressed by the cDNA and/or genomic DNA-polymerase chain reaction protocols. This then leads to the possible production of new molecules by gene shuffling and recombinant biosynthetic techniques. Monoclonal catalytic antibodies using haptens possessing a structure similar to a high-energy intermediate of a proposed reaction are also contributing to the elucidation of biochemical mechanisms and the design of efficient syntheses. The technique of photoaffinity labeling, brilliantly invented by Frank Westheimer (1912–, Harvard University), assisted especially by advances in mass spectrometry, will clearly be playing an increasingly important role in studies of ligand–receptor interactions including enzyme–substrate reactions. The combined and sophisticated use of various spectroscopic means, including difference spectroscopy and fast time-resolved spectroscopy, will also become increasingly central in future studies of ligand–receptor studies.

Organic chemists, especially those involved in structural studies have the techniques, imagination, and knowledge to use these approaches. But it is difficult for organic chemists to identify an exciting and worthwhile topic. In contrast, the biochemists, biologists, and medical doctors are daily facing exciting life-related phenomena, frequently without realizing that the phenomena could be understood or at least clarified on a chemical basis. Broad individual expertise and knowledge coupled with multidisciplinary research collaboration thus becomes essential to investigate many of the more important future targets successfully. This approach may be termed "dynamic," as opposed to a "static" approach, exemplified by isolation and structure determination of a single natural product. Fortunately for scientists, nature is extremely complex and hence all the more challenging. Natural products chemistry will be playing an absolutely indispensable role for the future. Conservation of the alarming number of disappearing species, utilization of biodiversity, and understanding of the intricacies of biodiversity are further difficult, but urgent, problems confronting us.

That natural medicines are attracting renewed attention is encouraging from both practical and scientific viewpoints; their efficacy has often been proven over the centuries. However, to understand the mode of action of folk herbs and related products from nature is even more complex than mechanistic clarification of a single bioactive factor. This is because unfractionated or partly fractionated extracts are used, often containing mixtures of materials, and in many cases synergism is most likely playing an important role. Clarification of the active constituents and their modes of action will be difficult. This is nevertheless a worthwhile subject for serious investigations.

Dedicated to Sir Derek Barton whose amazing insight helped tremendously in the planning of this series, but who passed away just before its completion. It is a pity that he was unable to write this introduction as originally envisaged, since he would have had a masterful overview of the content he wanted, based on his vast experience. I have tried to fulfill his task, but this introduction cannot do justice to his original intention.

ACKNOWLEDGMENT

I am grateful to current research group members for letting me take quite a time off in order to undertake this difficult writing assignment with hardly any preparation. I am grateful to Drs. Nina Berova, Reimar Bruening, Jerrold Meinwald, Yoko Naya, and Tetsuo Shiba for their many suggestions.

8. BIBLIOGRAPHY

"A 100 Year History of Japanese Chemistry," Chemical Society of Japan, Tokyo Kagaku Dojin, 1978.
K. Bloch, *FASEB J.*, 1996, **10**, 802.
"Britannica Online," 1994–1998.
Bull. Oriental Healing Arts Inst. USA, 1980, **5**(7).
L. F. Fieser and M. Fieser, "Advanced Organic Chemistry," Reinhold, New York, 1961.
L. F. Fieser and M. Fieser, "Natural Products Related to Phenanthrene," Reinhold, New York, 1949.
M. Goodman and F. Morehouse, "Organic Molecules in Action," Gordon & Breach, New York, 1973.
L. K. James (ed.), "Nobel Laureates in Chemistry," American Chemical Society and Chemistry Heritage Foundation, 1994.
J. Mann, "Murder, Magic and Medicine," Oxford University Press, New York, 1992.
R. M. Roberts, "Serendipity, Accidental Discoveries in Science," Wiley, New York, 1989.
D. S. Tarbell and T. Tarbell, "The History of Organic Chemistry in the United States, 1875–1955," Folio, Nashville, TN, 1986.

5.01
Biological Reactions—Mechanisms and Catalysts: An Overview

C. DALE POULTER
University of Utah, Salt Lake City, UT, USA

5.01.1 INTRODUCTION

Each living cell relies on a complex set of chemical reactions to provide thousands of different molecules necessary to sustain life. For the most part, biological compounds are composed of carbon, hydrogen, nitrogen, oxygen, sulfur, and phosphorus atoms, those elements typically encountered in "organic" compounds. Thus, it is not surprising that the chemical reactions involved in the synthesis and degradation of naturally occurring molecules are familiar to organic chemists. Also not surprising, but truly impressive nonetheless, is the incredible diversity of chemical structures found in nature. The "small" organic molecules typically classified as natural products fulfill roles essential for survival of the host organism too numerous to list. A few examples, admittedly selected at random, include hormones acting as messengers for intercellular communication; structural components of large noncovalent assemblies such as membranes; integral components of proteins involved in vision, regulation of vesicle fusion, and signal transduction networks for cell division; defensive agents against predation; attractants for reproduction by plants and animals; and cofactors required by proteins for binding and catalysis.

As organisms have evolved from simpler ancestors, so have metabolic pathways evolved from simpler predecessors. A few simple building blocks, such as acetate, pyruvate, glyceraldehyde, and glycolate, are the sources for the carbon skeletons of natural products. The synthesis of individual natural products from simple precursors typically requires long intricately convergent sequences that are as elegant as any synthetic strategy conceived by man. Given the requirements and limitations imposed by a living cell, the synthesis of individual compounds must be tightly regulated in order to deliver a precise dose at the appropriate moment, fast, energy efficient, and accomplished under unfavorable conditions of low reactant concentrations and restricted temperatures. These stringent requirements have been overcome by the co-evolution of protein catalysts. Typically, each reaction is catalyzed by an enzyme dedicated specifically to that purpose. The chapters in this volume present a cross-section of biological reactions and protein catalysts that illustrates the diversity of the chemistry used to synthesize natural molecules.

5.01.2 BIOLOGICAL REACTIONS

One might think that biological reactions are confined to a rather narrow subset of transformations that have inherently low barriers and do not involve formation of highly reactive intermediates which could react with the protein catalysts. Developments in mechanistic enzymology, stimulated by the increased availability of enzymes through recombinant DNA technology, have revealed that biological reactions are as rich and varied at a fundamental mechanistic level as those used by organic chemists to synthesize molecules in the laboratory.

The construction of carbon-carbon bonds in nature illustrates the wide range of different mechanisms that are found. Aldol and Claisen condensations normally require powerful bases to generate enolate intermediates. Similar reactions are found in the polyketide, isoprenoid, fatty acid, and carbohydrate biosynthetic pathways where the reactive enolates are stabilized by hydrogen bonds (Chapters 5.02 and 5.03). Electrophilic alkylations are used to connect sugar units in polysaccharide biosynthesis (Chapter 5.12) and to join isoprenoid units to one another (Chapter 5.13). In the chain elongation and cyclization steps in isoprenoid biosynthesis, the reactions involve formation of highly reactive carbocationic intermediates. The enzymes that catalyze these transformations must generate and process the carbocations without suffering alkylation of side chain nucleophiles or the amide linkages in the protein backbone.

Many of the functional group manipulations in natural products biosynthesis also involve highly reactive species. Purines and pyrimidines are deaminated by nucleophilic aromatic substitutions that proceed through unstable Meisenheimer intermediates (Chapter 5.05) and nucleophilic substitution is commonly seen at saturated carbons (Chapters 5.04 and 5.16). Free radicals are involved in a large number of enzyme catalyzed reactions. Some of the prominent examples discussed in this volume are the reduction of nucleotides to deoxynucleotides by ribonucleotide reductase (Chapter 5.08), the rearrangement catalyzed by lysine 2,3-aminomutase (Chapter 5.09), oxygenation reactions in the biosynthesis of prostaglandins (Chapter 5.10), and some cobalamin-dependent reactions (Chapter 5.11). During the cobalamin-dependent processes, the free radical intermediates are generated by rupture of a Co-carbon bond in the organometallic cofactor. Exposure of DNA to UV light results in the formation of cyclobutane dimers by a 2 + 2 cycloaddition of the double bonds in thymidine residues. One of the repair mechanisms for these lesions involves the light-induced formation of a dimer radical anion that rapidly undergoes cycloreversion to restore the original DNA structure (Chapter 5.15).

5.01.3 CATALYSTS

For most of the past 100 years, natural products chemists have focused their efforts on determining the structures of natural products, elucidating the steps in biosynthetic pathways, and studying the mechanisms of biosynthetic reactions. These studies have provided an elegant and extensive description of the various biosynthetic pathways found in living cells. This information is often summarized in complex charts showing the intricate branches and interconnections that have evolved in nature. Issues related to regulation of the pathways, the properties of the enzymes that catalyze biological reactions, and the structures of the catalysts themselves have received much less attention. Only since the mid-1980s following the introduction of recombinant DNA technology has it become practical to deal with many of these issues. The virtual explosion of work on biosynthetic enzymes can be directly attributed to the availability of recombinant forms of the proteins. In most instances one can now overproduce recombinant enzymes that have identical catalytic properties to their counterparts isolated directly, often in minuscule quantities, from wild-type host organisms. Furthermore, any specific amino acid or specific combination of amino acids can be substituted at will with any of the other 19 naturally occurring amino acids. Site-directed mutagenesis has become an indispensable tool for evaluating the role of a particular amino acid in a complex protein by allowing side chain substitutions to be made at will. The construction of extensive DNA databases has followed close on the heels of advancements in DNA sequencing technology made during the 1990s. These databases are readily accessible on the World Wide Web and can be searched to locate conserved regions in a collection of related enzymes that often denote amino acids important for substrate binding or catalysis. The searches can also reveal phylogenetic relationships that can provide insights about the evolution of enzymes and metabolic pathways.

The availability of milligram to gram quantities of recombinant biosynthetic enzymes, in combination with new X-ray and NMR techniques, has fueled a major effort to determine the structures of individual proteins and multiprotein complexes at atomic resolution. The insights about structure, mechanism, and function obtained from these studies are common themes in all of the chapters in this volume. Some enzymes, such as the glycosyl hydrolases and transferases described in Chapter 5.12 catalyze reactions that are found in many different metabolic pathways. Over 60 different families of glycosidases encompassing over 1000 individual enzymes have been identified from amino acid sequence comparisons, and new enzymes are being added to the list almost daily. Not surprisingly, enzymes clustered within a family show substantial conservation in their amino acid sequences, catalyze the same reaction although with different substrate selectivities, and have similar overall structures, especially in the regions near their catalytic sites. In contrast, enzymes from different families may catalyze reactions by similar mechanisms but usually have unrelated structures. This situation is dramatically illustrated by comparing the all α-helix structures of glycosidases belonging to Family 8 with the β-sheet structures of the members of Family 11.

During the 1990s the simple but brilliant proposal made by Jencks in the 1960s for generating antibodies with catalytic capability was reduced to practice. Chemically stable compounds that mimic the stereoelectronic properties of transition states for reactions were used as haptens to elicit the immune response (Chapter 5.17). One can screen for and identify specific antibodies that accelerate the chemical reactions among the polyclonal collection produced by the immune system. Antibody catalysts have been generated for reactions that are commonly encountered in living cells such as the rearrangement of chorismate to prephenate in the aromatic amino acid biosynthetic pathway (Chapter 5.14), ester hydrolysis (Chapter 5.06), and aldol condensations (Chapters 5.02 and 5.03). However, the technique for generating catalytic antibodies can, in principle, be extended to a wide range of chemical transformations, including reactions such as the Diels–Alder cycloaddition commonly used by synthetic organic chemists but rarely encountered in nature.

The chapters that follow in this volume present authoritative accounts of the rapidly expanding developments in the area of mechanistic enzymology. The 15-year period from the mid-1980s has seen the development and use of powerful new tools for studying biosynthetic reactions. It is also evident that much remains to be done. Connections between structure and function have only been made for a few of numerous biosynthetic enzymes found in cells. Little is known about how these enzymes have evolved and, as a corollary, how biosynthetic pathways themselves have evolved to provide new families of metabolites for cells. Mechanisms for regulation of biosynthetic pathways are also poorly understood. During the late 1990s and into the early twenty-first century we should see major inroads in this important area as the new techniques now being used to study signal transduction networks in cells are adapted to investigating the regulation of natural products biosynthesis. The chapters in this volume provide an account of recent developments and a glimpse of important unresolved questions for future research.

5.02
Stabilization of Reactive Intermediates and Transition States in Enzyme Active Sites by Hydrogen Bonding

JOHN A. GERLT
University of Illinois, Urbana, IL, USA

5.02.1 INTRODUCTION

The mechanisms of many enzyme-catalyzed reactions involve intermediates that, in solution, are too unstable to exist, e.g., carbanions and carbenium ions.[1,2] In order for an intermediate to be kinetically competent and consistent with the measured rapid rates (k_{cat}) of the catalyzed reaction, functional groups in an active site must be able to stabilize both the intermediate as well as the "flanking" transition states that lead to both its formation and subsequent reaction to products, otherwise, enzymes would not be proficient catalysts.[3]

The number of high-resolution structures available for enzymes has increased dramatically, and the pace at which new structures are being deposited in the Protein Data Base has not abated. Thus, the identities and spatial disposition of the functional groups that interact with bound substrates, intermediates, and flanking transition states can now be specified for many enzyme active sites. With this information, mechanistic enzymologists and physical organic chemists ought to be able to provide *quantitative* explanations for the large rate accelerations that characterize enzyme-catalyzed reactions. Accordingly, the nature of the interactions by which reactive intermediates and transition states interact with active site functional groups remains an area of intense interest in mechanistic enzymology.

This chapter will focus on an emerging general strategy by which enzymes can stabilize reactive anionic intermediates and the flanking transition states that lead to their formation and further reaction. While hydrogen bonding has long been acknowledged as important in determining substrate specificity,[4-6] the magnitudes of the potential changes in hydrogen bond strength between active site functional groups and bound substrates as they are converted to transient intermediates have not been considered to be sufficient to contribute significantly to the observed rate accelerations. However, proposals,[7-9] now supported by model studies in nonaqueous solvents,[10-13] espouse the view that, when secluded from bulk aqueous solvent, hydrogen bond strengths can increase much more significantly than previously recognized. Consequently, hydrogen bonding can, in fact, be responsible for a major fraction of the observed rate accelerations. The sizable changes in hydrogen bond strength are the consequence of an increase in the proton affinity of a hydrogen bond acceptor of the substrate as the intermediate is formed, so that it approaches the proton affinity of one or more hydrogen bond donors provided by active site functional groups. (In aqueous solution, the proton affinity of a hydrogen bond donor or acceptor is described by its pK_a value (donor) or that of its conjugate acid (acceptor). Because the active sites of enzymes are removed from bulk solvent, the pK_a values of acids determined in aqueous solution are not necessarily measures of the proton affinities of hydrogen bond donors and acceptors in active sites. However, other quantitative measures of proton affinities in nonaqueous environments are not readily available, so the values of the aqueous pK_a of substrates, intermediates, and active site functional groups are used in this chapter to quantitate the proton affinities of hydrogen bond donors and acceptors.)

This strategy is conceptually simple and, as a result, has the potential of being widely applicable to understanding how enzymes have acquired the ability to be proficient catalysts that selectively bind/stabilize reactive intermediates and their flanking rate-limiting transition states. This strategy is the subject of this chapter.

5.02.2 INTERMEDIATES IN ENZYME-CATALYZED REACTIONS

The common physicochemical characteristic shared by the mechanisms of many enzyme-catalyzed reactions is that, as the bound substrate is converted to bound product via a reactive intermediate, the electron density/anionic charge on an oxygen atom of the substrate increases as the intermediate is formed and then decreases as the product is formed. This increase in electron density increases the proton affinity of the oxygen atom.

These reactions include, but are not limited to: (i) enolization reactions in which an active site base abstracts the proton from a carbon adjacent to a carbonyl or carboxylic acid group

(α-proton of a carbon acid) to generate a carbanionic/enolate anion intermediate; and (ii) substitution reactions in which a nucleophile adds to the carbonyl group to generate an anionic tetrahedral intermediate.

In each of these reaction types, the intermediate is unstable as a consequence of the increase in charge on the oxygen. Yet, the available mechanistic data suggest that these intermediates exist on the reaction coordinates of many enzyme-catalyzed reactions as substrate is converted to product.

In the remainder of this section the proton affinities of the substrates and intermediates in these two types of common enzyme-catalyzed reactions are considered. In following sections, the mechanisms by which the increases in proton affinity can be used to stabilize the intermediates will be developed and discussed.

5.02.2.1   Enolate Anions: Enolization Reactions of Carbon Acids

Many familiar enzyme-catalyzed reactions, including several in glycolysis, the citric acid cycle, and β-oxidation of fatty acids, involve the initial enolization of a ketone, aldehyde, thioester, or carboxylate anion substrate (carbon acid) via general base-catalyzed abstraction of the α-proton, as illustrated in Equation (1). The resulting intermediates can form products in a variety of overall reactions, including racemization (1,1-proton transfer), isomerization (1,2- and 1,3-proton transfer), Claisen condensation, and vinylogous β-elimination. Examples of these include mandelate racemase, triose phosphate and ketosteroid isomerases, citrate synthase, and enolase, respectively.

$$\text{B:} \quad \text{H}\!\!\diagdown\!\!\diagup\!\!\diagdown\!\!\overset{\text{O}}{\diagup} \quad \rightleftharpoons \quad \text{BH}^+ \quad \diagdown\!\!\diagup\!\!\diagdown\!\!\overset{\text{O}^-}{\diagup} \tag{1}$$

The pK_a values of the cationic conjugate acids of the substrates for enolization reactions range from ~ -2 for substrates that are aldehyde and ketone carbon acids[14-16] to ~ -8 for carboxylate carbon acids.[17] In the latter case, the pK_a value for the cationic conjugate acid of the neutral carboxylic acid functional groups is used for understanding the effects of hydrogen bond strengths on catalysis, since charge neutralization of the bound carboxylate anion by cationic groups in the active site invariably occurs (see Section 5.02.5) and makes this the appropriate measure of proton affinity for the carboxylate group of the bound substrates. These pK_a values illustrate that these functional groups are weak hydrogen bond acceptors and that the proton affinities of the carbonyl/ carboxyl groups of substrates are very low.

The pK_a values for the neutral enol tautomers of the carbon acid substrates range from ~ 7 for those of carboxylic acids[18] to ~ 11 for those of aldehydes and ketones.[19] Thus, as the reaction coordinate for formation of an enolate anion is traversed, the proton affinity of the carbonyl/enolate oxygen atom also increases by ~ 13–15 pK_a units. As developed in Section 5.02.6, the increase in proton affinity as the keto tautomer of the substrate is converted to its enol tautomer (the intermediate) is considerable and can serve as a convenient handle for stabilization of the otherwise too unstable enolate intermediate by one or more functional groups in an enzyme active site.

5.02.2.2   Tetrahedral Intermediates: Nucleophilic Substitution of Carbonyl Compounds

Addition of a nucleophile to a carbonyl group results in the formation of a tetrahedral intermediate, as illustrated in Equation (2). Decomposition of these intermediates leads to the formation of products in several types of overall reactions, although most of these occur in reaction in which an amide or peptide bond is hydrolyzed, e.g., the reactions catalyzed by proteases and esterases.

$$\text{Nuc:} \quad \text{R}\!\!-\!\!\overset{\text{O}}{\underset{\text{X}}{\diagup\!\!\diagdown}} \quad \rightleftharpoons \quad \text{R}\!\!-\!\!\overset{\text{O}^-}{\underset{\text{Nuc}}{\diagdown\!\!\!\!\diagup\text{X}}} \tag{2}$$

The pK_a values of the conjugate acids of the substrates for these reactions are $\leqslant -2$.[20] Like the keto tautomers of carbon acids, the oxygen atoms of the substrates are weak hydrogen bond acceptors.

The pK_a values for neutral tetrahedral intermediates are estimated to be ~ 13.[21] Thus, as the reaction coordinate is traversed for formation of the tetrahedral intermediate, the proton affinity of the oxygen increases by ~ 15 pK_a units. Again, the increase is considerable and provides a mechanism by which the unstable tetrahedral intermediate can be stabilized in enzyme active sites.

5.02.3 QUANTITATIVE UNDERSTANDING OF REACTION PROFILES

5.02.3.1 The Albery–Knowles Analysis of Triose Phosphate Isomerase

The k_{cat} for an enzyme-catalyzed reaction is determined by the relative free energies of substrates, intermediates, and transition states that are bound in the active site and can be no larger than the rate of formation of the stabilized intermediate from bound substrate.

The pioneering work of Albery and Knowles and their co-workers that provided a quantitative description[22] of the energetics of the reaction catalyzed by triose phosphate isomerase (TIM) (Equation (3)) has had tremendous impact on mechanistic enzymology. This analysis (Figure 1)[23–29] revealed that the transition state for a physical step, the dissociation of glyceraldehyde 3-phosphate from the enzyme, is rate determining in the overall reaction, demonstrating that in the course of evolution of catalytic efficiency nature was able to stabilize both the enediolate intermediate and the flanking transition states for formation of the intermediate via general base-catalyzed enolization and formation of the products via general acid-catalyzed ketonization reactions.

$$
\begin{array}{ccc}
\text{HO}\quad\text{O} & \xrightarrow{\text{TIM}} & \text{O}\quad\text{OH} \\
\text{H}\cdots\!\!\diagdown\!\!\diagup & & \diagdown\!\!\diagup\!\!\cdots\text{H} \\
{}^{2-}\text{O}_3\text{POH}_2\text{C}\quad\text{H} & & {}^{2-}\text{O}_3\text{POH}_2\text{C}\quad\text{H}
\end{array}
\qquad (3)
$$

The Albery–Knowles analysis suggested several sequential steps that have been widely accepted in understanding the evolution of catalytic efficiency and perfection in enzymes:[30,31]

   (i) initial uniform binding of the intermediates and transition states that occur in the non-enzymatic reactions;

   (ii) subsequent differential binding of bound substrates, intermediates, and products to achieve internal equilibrium constants of near unity; and

   (iii) catalysis of elementary steps (i.e., differential binding of transition states) to allow the transition states for formation and breakdown of the intermediate to be nearly isoenergetic with the transition states for substrate binding/product dissociation.

Each of these steps in enhancing catalytic efficiency must be accomplished by interactions of the bound substrate, intermediates, and transition states with active site functional groups. These fundamental concepts have weathered the test of time, although the general strategy described in this chapter leads to the proposal that intermediates should not be stabilized to achieve an equilibrium constant of unity with bound substrate. If such stabilization occurred, the oxygen atom of the transition state that originates as the carbonyl oxygen in the substrate and becomes the enolate oxygen of the intermediate will have a significantly lower proton affinity than the enolate oxygen. Thus, the interactions with electrophilic groups (hydrogen bond donors and cations) that stabilize the intermediate will necessarily provide less stabilization of the transition states that flank the intermediate.

5.02.3.2 Marcus Formalism

Despite the success of the principles espoused by Albery and Knowles in understanding the catalytic efficiency of enzymes, the author believes that additional insights into the evolution of catalytic efficiency can be obtained by an explicit deconvolution of the energetic factors that are contained in the free energy profiles for enzyme-catalyzed reactions. In particular, a dissection of

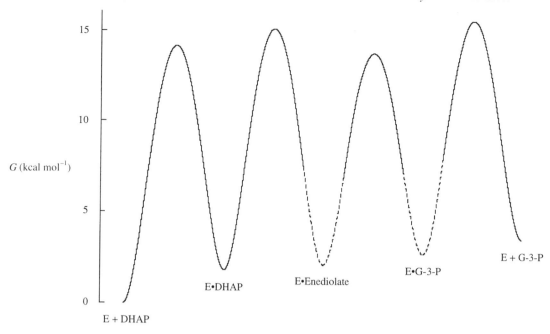

Figure 1 The free energy diagram for the TIM-catalyzed reaction. The dashed lines indicate uncertainties in the concentrations of bound enediolate intermediate (EZ) and bound G3P product (EP) (after Albery and Knowles[22]).

the activation energy for formation of a bound, stabilized intermediate from a bound substrate into thermodynamic and intrinsic kinetic free energy components is useful and can be performed using a Marcus formalism. With the insights that are provided by this analysis, the problems encountered in relating active site structure to reaction mechanisms and rates can be defined more precisely.[7]

In Marcus formalism, the free-energy profile for a conversion of bound substrate to the bound intermediate via a rate-determining transition state is described by Equation (4) for an inverted parabola that specifies the free energy, G, as a function of the reaction coordinate, x, where $x = 0$ for the bound substrate and $x = 1$ for the bound intermediate:[32-34]

$$G = -4\Delta G^{\ddagger}_{int}(x-0.5)^2 + \Delta G^{\circ}(x-0.5) \tag{4}$$

where ΔG° and $\Delta G^{\ddagger}_{int}$ are independent parameters that describe the thermodynamic and intrinsic kinetic barriers for formation of the intermediate, respectively. The equation for an inverted parabola requires that $\Delta G^{\circ}/4 \leqslant \Delta G^{\ddagger}_{int} \leqslant 4\Delta G^{\circ}$.

G is a maximum at the transition state, so the position of the transition state on the reaction coordinate, $x^{\ddagger}$, is given by Equation (5):

$$x^{\ddagger} = 0.5 + \Delta G^{\circ}/8\Delta G^{\ddagger}_{int} \tag{5}$$

From Equations (4) and (5), $\Delta G^{\ddagger}$, the activation energy for the conversion of bound substrate to the bound intermediate via the transition state, is given by Equation (6):

$$\Delta G^{\ddagger} = \Delta G^{\ddagger}_{int}(1 + \Delta G^{\circ}/4\Delta G^{\ddagger}_{int})^2 \tag{6}$$

From Equation (6), $\Delta G^{\ddagger}$ is partitioned between the value of ΔG° and a second energy that is associated with but not necessarily equal to $\Delta G^{\ddagger}_{int}$. When the bound substrate and intermediate are isoenergetic, i.e., $\Delta G^\circ = 0$, an activation barrier remains that is numerically equal to $\Delta G^{\ddagger}_{int}$ (Figure 2, curve a). When $\Delta G^\circ > 0$, the contribution of the intrinsic barrier to $\Delta G^{\ddagger}$ is less than $\Delta G^{\ddagger}_{int}$ (Figure 2, curve b), with the amount of the contribution depending upon the value of ΔG°.

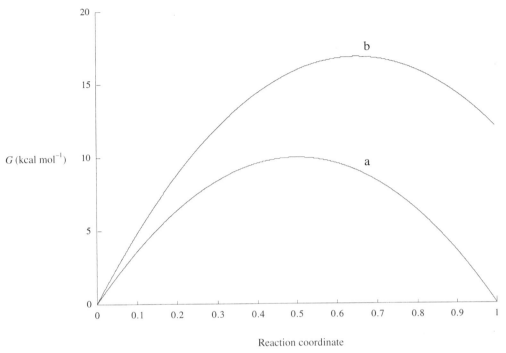

Reaction coordinate

Figure 2   Reaction coordinates described by Equation (4) that describe the dependence of G on the position of the reaction coordinate, x, where $x = 0$ for bound substrate and $x = 1$ for bound intermediate. When $\Delta G^\circ = 0$, $\Delta G^{\ddagger} = \Delta G^{\ddagger}_{int}$ (curve a); when $\Delta G^\circ > 0$, $(\Delta G^{\ddagger} - \Delta G^\circ) < \Delta G^{\ddagger}_{int}$ (curve b). The reaction coordinates were calculated assuming (i) $\Delta G^\circ = 0$ kcal mol^{-1} and $\Delta G^{\ddagger}_{int} = 10$ kcal mol^{-1}; and (ii) $\Delta G^\circ = 12$ kcal mol^{-1} and $\Delta G^{\ddagger}_{int} = 10$ kcal mol^{-1}. For curve a, $(\Delta G^{\ddagger} - \Delta G^\circ) = \Delta G^{\ddagger}_{int} = 10$ kcal mol^{-1}; for curve b, $(\Delta G^{\ddagger} - \Delta G^\circ) = 4.91$ kcal mol^{-1}.

The dissection of $\Delta G^{\ddagger}$ into ΔG° and a contribution associated with $\Delta G^{\ddagger}_{int}$ is a mathematical description of the differential binding of intermediates (ΔG°) and of transition states ($\Delta G^{\ddagger}_{int}$) proposed by Albery and Knowles.[22]

This application of Marcus formalism reveals that $\Delta G^{\ddagger}$ can be reduced by active site-induced reductions in ΔG° and/or $\Delta G^{\ddagger}_{int}$ from those values that characterize the nonenzymatic reactions. However, this mathematical analysis does not reveal the structural strategies that are used to reduce ΔG° and/or $\Delta G^{\ddagger}_{int}$ in enzyme active sites. Those are suggested by an examination of the structures of active sites.

5.02.4   RELATIONSHIP OF INTERMEDIATES AND TRANSITION STATES

The "principle" of differential binding of bound substrate, intermediate, and product to achieve internal equilibrium constants of unity is the second of three steps that Albery and Knowles proposed in their description of the evolution of catalytic efficiency. Adjusting the internal thermodynamics of the reaction to allow bound substrate and product to be isoenergetic is certainly important since this will tend to equalize the energies of the intervening transition states.[35]

However, adjustment of the thermodynamics of the bound species to decrease the energy of the intermediate such that it is isoenergetic with the substrate and product may not be an optimal solution for effective catalysis. By equalizing the energies of the substrate, intermediate, and product, the physicochemical properties of the transition states flanking the intermediate will differ from those of the intermediate, thereby enforcing the enzyme to use completely different mechanisms to stabilize both the intermediate and the transition states. For example, if the change in the proton affinity of an oxygen atom of the substrate as it is converted to the intermediate is used to stabilize the intermediate by enhanced hydrogen bonding, considerably less stabilization of the transition state by hydrogen bonding will be possible if the substrate and intermediate are isoenergetic.

This point is illustrated in Figure 3 for a hypothetical enolization reaction of a carbon acid. Curve a represents an uncatalyzed reaction in which $\Delta G^{\circ} = 30$ kcal mol^{-1} and $\Delta G^{\ddagger}_{\mathrm{int}} = 12$ kcal mol^{-1}; using Equation (3) $\Delta G^{\ddagger} = 31.7$ kcal mol^{-1} and the value of $x^{\ddagger}$, the reaction coordinate of the transition state, is 0.81. Differential binding of the intermediate so that ΔG° is reduced to 0 (curve b) decreases $\Delta G^{\ddagger}$ to 12 kcal mol^{-1}, the value of $\Delta G^{\ddagger}_{\mathrm{int}}$; the value of $x^{\ddagger}$ is 0.50, so the physicochemical properties of the transition state are expected to be midway between those of the substrate and intermediate, including the proton affinity of the oxygen that is changing from a carbonyl oxygen to an enolate oxygen.

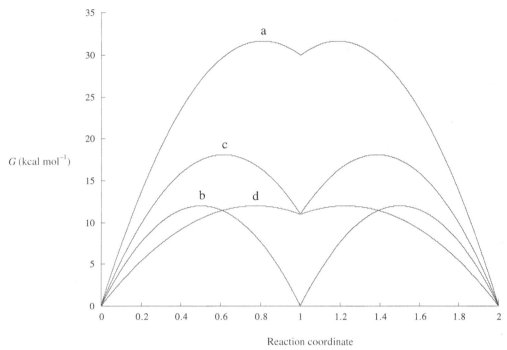

Figure 3   Reaction coordinates that illustrate the interdependence of ΔG° and $\Delta G^{\ddagger}_{\mathrm{int}}$ in determining both $\Delta G^{\ddagger}$ (Equation (4)) and $x^{\ddagger}$ (Equation (6)). Curve a, $\Delta G^{\circ} = 30$ kcal mol^{-1}, $\Delta G^{\ddagger}_{\mathrm{int}} = 12$ kcal mol^{-1}, $\Delta G^{\ddagger} = 31.7$ kcal mol^{-1}, and $x^{\ddagger} = 0.81$; curve b, $\Delta G^{\circ} = 0$ kcal mol^{-1}, $\Delta G^{\ddagger}_{\mathrm{int}} = 12$ kcal mol^{-1}, $\Delta G^{\ddagger} = 12$ kcal mol^{-1}, and $x^{\ddagger} = 0.50$; curve c, $\Delta G^{\circ} = 11$ kcal mol^{-1}, $\Delta G^{\ddagger}_{\mathrm{int}} = 12$ kcal mol^{-1}, $\Delta G^{\ddagger} = 18.1$ kcal mol^{-1}, and $x^{\ddagger} = 0.83$; curve d, $\Delta G^{\circ} = 11$ kcal mol^{-1}, $\Delta G^{\ddagger}_{\mathrm{int}} = 5$ kcal mol^{-1}, $\Delta G^{\ddagger} = 30$ kcal mol^{-1}, and $x^{\ddagger} = 0.81$.

According to the Hammond postulate, one mechanism for equalizing the physicochemical properties of the intermediate and flanking transition states is to increase ΔG° for formation of the intermediate without decreasing $\Delta G^{\ddagger}_{\mathrm{int}}$. However, this approach is not an effective mechanism for also decreasing $\Delta G^{\ddagger}$. As illustrated in curve c of Figure 3, an increase in ΔG° to 11 kcal mol^{-1} increases $x^{\ddagger}$ to 0.83, i.e., the transition state is now similar to the intermediate, but $G^{\ddagger}$ is increased to 18.1 kcal mol^{-1}. Thus, while the transition state now resembles the intermediate, the increase in $G^{\ddagger}$ is catalytically counterproductive.

Equations (4) and (6) suggest an alternative approach for equalizing the physicochemical properties of the intermediate and flanking transition states without large increases in $G^{\ddagger}$:[7] by decreasing $\Delta G^{\ddagger}_{\mathrm{int}}$ the intermediate and flanking transition states can be nearly isoenergetic, *without* requiring that the substrate and intermediate be isoenergetic. Since $\Delta G^{\ddagger}_{\mathrm{int}}$ describes the (excess) kinetic barrier to the reaction, it is reasonable to propose that its numerical value can be reduced by preorganization of the general basic and general acidic catalysts relative to the substrate, thereby reducing the entropic requirements for formation of the intermediate via the flanking, rate-limiting transition states. As illustrated by curve d of Figure 3, if the value of $\Delta G^{\ddagger}_{\mathrm{int}}$ is decreased to 5.0 kcal mol^{-1} while the value of ΔG° remains 11 kcal mol^{-1}, the value of $x^{\ddagger}$ is 0.77, and that of $G^{\ddagger}$ is 12 kcal mol^{-1}, the same as that achieved by making the substrate and intermediate isoenergetic.

In those reactions in which the intermediates are *very* unstable in aqueous solution, the "requirement" that the equilibrium constant for formation of the intermediate from bound substrate approaches unity would require an extraordinary amount of differential stabilization of the bound intermediate relative to the bound substrate. In contrast, if the intermediate were nearly isoenergetic with the transition state, considerably less stabilization of the intermediate would be required to account for the observed k_{cat}.

Adjustment of both ΔG° and $\Delta G^{\ddagger}_{int}$ to achieve intermediates and flanking intermediates that have similar proton affinities leads to the prediction that the equilibrium constant for formation of the intermediate will be significantly less than unity ($\Delta G^{\circ} \gg 0$; e.g., curve d in Figure 3). While, it should be possible to measure ΔG° for the formation of the bound intermediate from bound substrate, such measurements rarely have been performed, so distinction between the proposals made by Albery and Knowles[22] and those described in this section[7] are not yet generally possible.

5.02.5 STRUCTURES OF ENZYME ACTIVE SITES

The availability of a large number of high-resolution structures of active sites was crucial in recognizing the importance of hydrogen bonding as a general strategy for stabilizing intermediates and transition states in enzyme active sites. In the following sections, the structures of representative active sites will be analyzed to reveal the presence of electrophilic functional groups, both hydrogen bond donors and cationic Lewis acid catalysts, that are appropriately positioned to stabilize reactive intermediates. A large number of structures are available for other active sites that catalyze the formation of unstable intermediates; these contain the same types of electrophilic functional groups and will not be discussed in this chapter.

5.02.5.1 Stabilization of Enolate Anions: Triose Phosphate Isomerase

The work of Albery and Knowles on understanding the energetics of the reaction catalyzed by TIM has made this enzyme one of the major paradigms in understanding the interrelationships among active site structure, mechanism, and catalytic efficiency.[36]

A number of structures are available for TIM, including that for a complex with phosphoglycolohydroxamate (PGH) (1), a competitive inhibitor that is a structural analogue of the enediolate intermediate (2) on the reaction pathway.[37]

$$\text{(1)} \qquad\qquad \text{(2)}$$

The active site structure, shown in Figure 4, reveals the presence not only of Glu165 that functions as the general basic catalyst that mediates the 1,2-proton transfer reaction (in competition with exchange of the substrate-derived proton with solvent[38,39]) but also the presence of His95 that is hydrogen-bonded to both oxygens of PGH that mimic the enolate oxygens of the intermediate.

Considerable effort has been expended in understanding the role of His95 in the TIM-catalyzed reaction. Kinetic studies of wild type TIM as well as the H95Q and H95N site-directed substitutions suggest that His95 is important for efficient catalysis (k_{cat} is decreased $\sim$ 200-fold by either substitution).[40] Interestingly, the mechanism of the H95Q-catalyzed reaction differs in significant detail from that catalyzed by wild type TIM. Whereas 94–97% of the substrate-derived proton abstracted by Glu165 exchanges with solvent hydrogen in the wild type-catalyzed reaction (or 3–6% of the substrate-derived proton is transferred to the product[24]), no exchange with solvent hydrogen is observed in the H95Q-catalyzed reaction.[40] The absence of exchange suggests (i) that the enediolate intermediate is so unstable/reactive that the substrate-derived proton is quantitatively transferred to the enediolate oxygens, and (ii) that His95 mediates the proton transfer between the enediolate oxygens that is required for the isomerization reaction.

In addition to this direct catalytic role, His95 also serves as the electrophile that polarizes the carbonyl group of dihydroxyacetone phosphate (DHAP). The evidence for this function is that when bound in the active site of wild type TIM the carbonyl stretch of DHAP is shifted by 19 cm^{-1} relative to that in solution;[41] no such polarization is observed when DHAP is bound in the active site of either H95Q or H95N.[42]

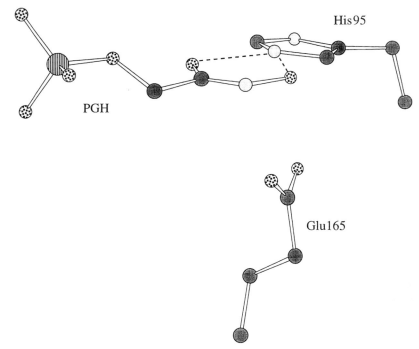

Figure 4 The active site of triose phosphate isomerase showing the relative orientations of Glu165, the general basic catalyst, His95, the neutral electrophilic catalyst, Lys13, a cationic group that facilitates binding of the anionic substrates to the active site, and phosphoglycolohydroxamate (PGH), a competitive inhibitor that mimics the enediolate intermediate. The coordinates used in construction of this figure were obtained from the Protein Data Base, file 1TPH.

Interestingly and importantly, the neutral imidazole functional group of His95 is the electrophilic catalyst. The evidence for this is twofold: (i) the high-resolution structure reveals that the functional group is hydrogen bonded to a backbone NH;[37] and (ii) ^{15}N NMR of isotopically labeled enzyme demonstrates that the nitrogen atoms are neutral rather than cationic over the pH range from 5.5 to 10, requiring that the pK_a of the conjugate acid of the neutral functional group is < 5.[43] This pK_a value, in turn, suggests that the value of the pK_a of the neutral imidazole functional group to form imidazolate anion is depressed to < 12 in the active site, since the values of these two pK_a of substituted imidazoles are influenced by similar amounts by substitutions.[44] Presumably, the values of these pK_a are perturbed from those normally associated with histidine (6.8 and 14, respectively) by the proximity of the functional group of His95 to the positive N-terminal end of an α-helix.[45]

Based upon these structural considerations, a reasonable mechanism for the TIM-catalyzed reaction is shown in Figure 5,[36] including hydrogen bonding interactions between the substrate, enediolate intermediate, and neutral His95 that are likely to be responsible for differential stabilization of the enediolate intermediate. In Figure 5, the hydrogen bond between the substrate carbonyl group and the neutral His95 is a "normal" hydrogen bond whereas those involving the enediolate intermediate(s) are stronger as the proton affinities of the donor and acceptor atoms become similar (see Section 5.02.7).

The conclusion that neutral His95 functions as the electrophilic catalyst provided a very important clue for understanding the role of hydrogen bonding in stabilizing intermediates and transition states in enzyme active sites. His95 must be able to provide significant stabilization of these species *without* the complete transfer of a proton to the enediolate intermediate.[7,9]

5.02.5.2 Stabilization of Enolate Anions: 4-Chlorobenzoyl CoA Dehalogenase and the Crotonase Superfamily

While the overall reactions catalyzed by crotonase (Equation (7)) and 4-chlorobenzoyl CoA dehalogenase (Equation (8)) appear distinct, the mechanisms of each could involve the formation of a transiently stable enolate of the thioester substrate. However, alignment of their primary sequences, as well as those of other enzymes in the "crotonase superfamily" that bind CoA esters

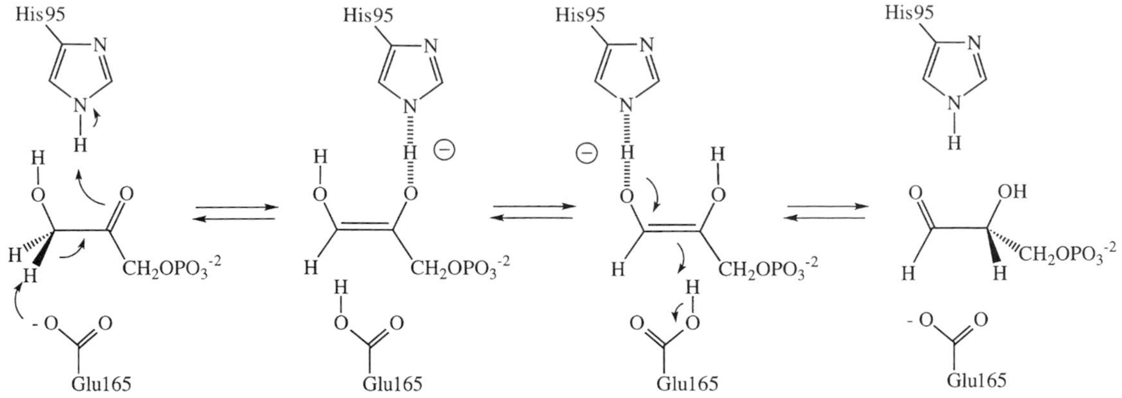

Figure 5 Proposed mechanism for the TIM-catalyzed reaction.

and catalyze reactions that are likely to involve the formation of either an enolate anion intermediate (in 1,3-proton transfer, racemization/epimerization, or Claisen condensation reactions) or a tetrahedral intermediate (in CoA ester hydrolysis reactions), suggests that crotonase and the dehalogenase are related by divergent evolution.[46–48] High-resolution structures have been reported for both the dehalogenase[49] and crotonase,[50] and these reveal that the structures of the active sites are, in fact, remarkably similar. In particular, a common hydrogen bonding mechanism is used by members of this superfamily to stabilize anionic enolate anion or tetrahedral intermediates.

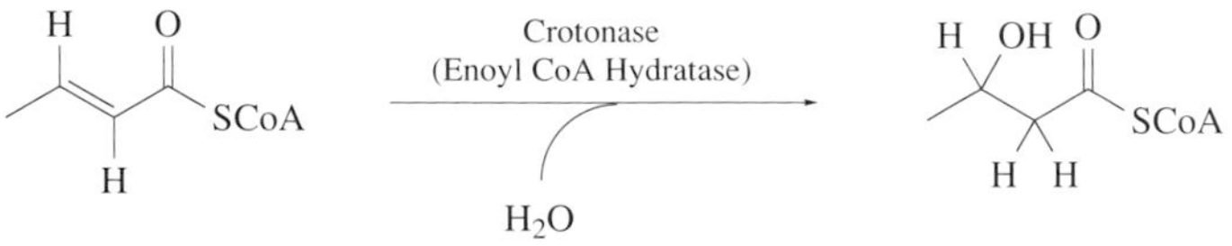

(7)

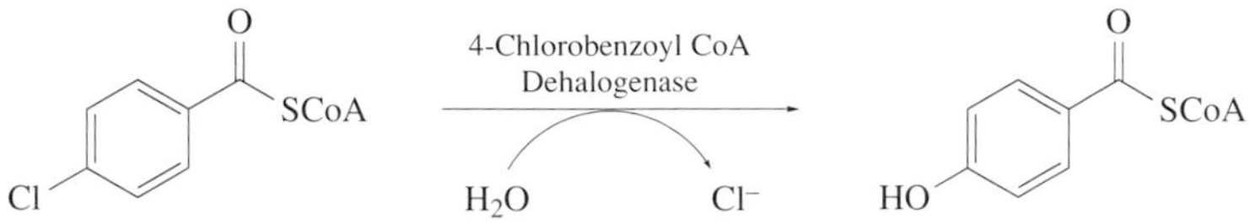

(8)

The high-resolution structure of the active site of the dehalogenase complexed with the product 4-hydroxybenzoyl CoA is shown in Figure 6.[49] The carbonyl group of the thioester is hydrogen bonded to the backbone NH groups of Phe64 and Gly114. The carboxylate group of Asp145 is positioned for nucleophilic addition to the aromatic ring of the substrate to generate an anionic Meisenheimer intermediate that is a structural analogue of the enolate anion intermediate that would be formed in the reaction catalyzed by crotonase. The Meisenheimer complex must be stabilized, and the hydrogen bonds in the "oxyanion hole" provided by peptide NH groups of Phe64 and Gly114 are likely to accomplish this function (see Section 5.02.6).

A reasonable mechanism for the dehalogenase-catalyzed reaction is shown in Figure 7, including hydrogen bonding interactions between the substrate, Meisenheimer intermediate, and the backbone NH groups that are likely to be responsible for differential stabilization of the intermediate.[51–55] In Figure 7, the hydrogen bonds involving the substrate carbonyl group are "normal" hydrogen bonds whereas those involving the intermediate are stronger (perhaps of the "short strong/low barrier" type) as the proton affinities of the donor and acceptor atoms become similar (see Section 5.02.7).

In the active site of the homologous crotonase,[50] the thioester carbonyl oxygen of the inhibitor acetoacetyl CoA is hydrogen bonded to the peptide NH group of Gly141 (the sequence homologue of Gly114 in the dehalogenase), and Glu144 and Glu164 are appropriately positioned to be general basic and general acidic catalysts, respectively, in the addition of water to the double bond in crotonyl CoA. While the question remains as to whether an enolate anion intermediate is on the reaction pathway,[56–58] the results of kinetic isotope effect studies suggest that the hydration/dehydration reaction is "concerted" (the analogy to the mechanism of the dehalogenase-catalyzed reaction does support the notion that a transiently stable enolate anion is stabilized by the analogous hydrogen bond donors in the active site of crotonase).

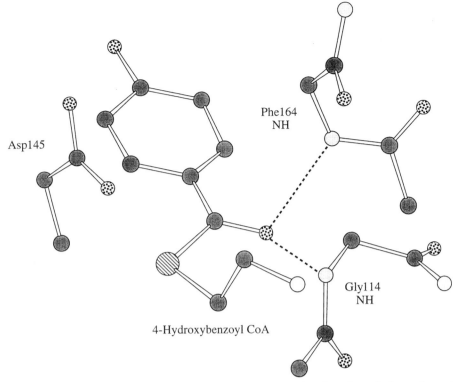

Figure 6 The active site of 4-chlorobenzoyl CoA dehalogenase showing the relative orientations of the electrophilic NH groups of Phe64 and Gly114 and 4-hydroxybenzoyl CoA, the product of the dehalogenase-catalyzed reactions. The NH groups stabilize the Meisenheimer complex formed by addition of Asp145 to the substrate. The coordinates used in the construction of this figure were generously provided by Professors Hazel M. Holden and Debra Dunaway-Mariano; these have been deposited in the Protein Data Base, file 1NZY.

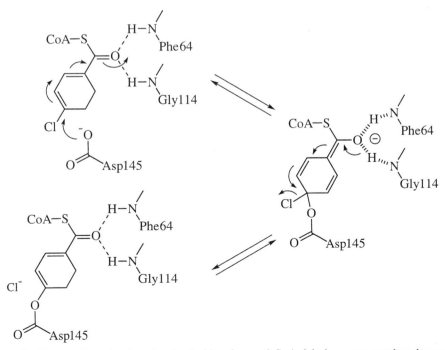

Figure 7 Proposed mechanism for the 4-chlorobenzoyl CoA dehalogenase-catalyzed reaction.

A homologue of the "oxyanion hole" is present in the active site of β-hydroxyisobutyryl-CoA hydrolase,[59] a member of the crotonase superfamily even though the reaction involves hydrolysis of the thioester bond rather than formation of an enolate anion. While the mechanism of the reaction catalyzed by this hydrolase has not been investigated, addition of a nucleophile to the thioester is expected to involve the formation of a tetrahedral intermediate that must be stabilized to be kinetically competent. Thus, the evolution of this superfamily apparently was dominated by the requirement that an unstable oxyanion intermediate, either enolate anion or tetrahedral intermediate, be stabilized by a common hydrogen bonding mechanism.

5.02.5.3   Stabilization of Enolate Anions: Mandelate Racemase

Whereas the active sites of the previously discussed enzymes stabilize enolate anions with neutral, weakly acidic hydrogen bond donors, a number of enzymes additionally use a single divalent metal ion to assist in the stabilization of highly reactive intermediates. An example is the Mg^{2+}-dependent mandelate racemase (MR) that catalyzes the 1,1-proton transfer reaction that equilibrates the enantiomers of mandelate, as illustrated in Equation (9).[60] This reaction is known to involve the formation of a transiently stable enolate anion derived by abstraction of the α-proton from the substrate carboxylate anion.[61] The α-protons of such carboxylate anions ($pK_a \geqslant 29$) are less acidic than those of aldehydes, ketones, and thioesters ($pK_a \leqslant 25$), suggesting that an additional amount of stabilization may be required so that the enolate anions can be kinetically competent.[62] Thus, the electrostatic interaction provided by a proximal divalent cation may provide this stabilization, although these are coordinated in the active site via three carboxylate ligands so the effective positive charge is reduced.

$$\text{(9)}$$

The high-resolution structure of the active site of MR complexed with the competitive inhibitor (S)-atrolactate ((S)-α-methylmandelate) is shown in Figure 8.[63] This structure allowed identification of Lys166[63,64] and His297[65,66] as the general basic catalysts that catalyze proton abstraction from (S)- and (R)-mandelates, respectively.

In addition, this structure reveals the presence of three groups that potentially participate as electrophilic catalysts. One carboxylate oxygen of the substrate analogue is directly coordinated to the essential Mg^{2+} as well as hydrogen bonded to the ε-ammonium group of Lys164; by neutralizing the charge on the carboxylate group these cation–anion interactions presumably reduce the pK_a value of the α-proton from 29 (that of mandelate *anion*) toward 22 (that of mandelic *acid*) (actually accomplished by differential stabilization of the enolate anion intermediate). The other carboxylate oxygen of the inhibitor accepts a hydrogen bond from the carboxylic acid group of Glu317. This neutral hydrogen bond donor probably provides stabilization of the enolate anion intermediate and the flanking transition states by hydrogen bonding interactions.[67]

A reasonable mechanism for the MR-catalyzed reaction is shown in Figure 9, including the interactions between the substrate, enolate anion intermediate, the hydrogen bond donor Glu317, and the cationic groups Mg^{2+} and Lys164.[60] These interactions probably are responsible for differential stabilization of the intermediate. In Figure 9, the hydrogen bond involving the substrate carbonyl group and Glu317 is a "normal" hydrogen bond whereas that involving the intermediate is stronger (perhaps of the short strong/low barrier type) as the proton affinities of the donor and acceptor atoms become similar (see 5.02.7).

5.02.5.4   Stabilization of Enolate Anions: Enolase

The glycolytic enzyme enolase initiates the Mg^{2+}-dependent dehydration of 2-phosphoglycerate (2-PGA, (3)) to form phospho*enol* pyruvate (PEP) (4) by abstraction of the α-proton, as illustrated in Equation (10). The pK_a value of the α-proton of (3) is $\geqslant 32$,[58] and an enolate anion intermediate is known to be on the reaction pathway.[68]

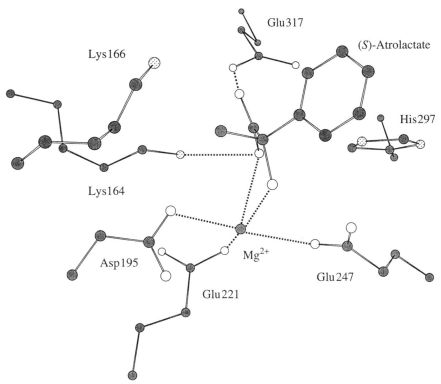

Figure 8 The active site of mandelate racemase showing the relative orientations of (*S*)-atrolactate, a competitive inhibitor, Lys166 and His297, the general basic catalysts that catalyze proton abstraction from (*S*)- and (*R*)-mandelates, respectively, as well as three electrophilic catalysts, Lys164, Mg^{2+}, and Glu317. The coordinates used in the construction of this figure were obtained from the Protein Data Base, file 1MDR.

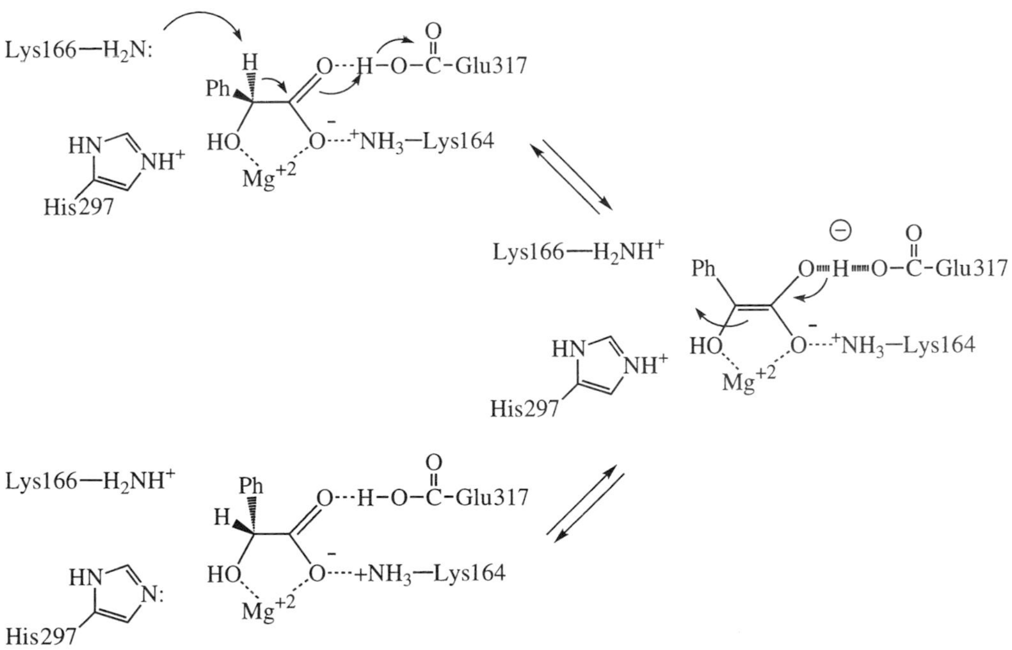

Figure 9 Proposed mechanism for the MR-catalyzed reaction.

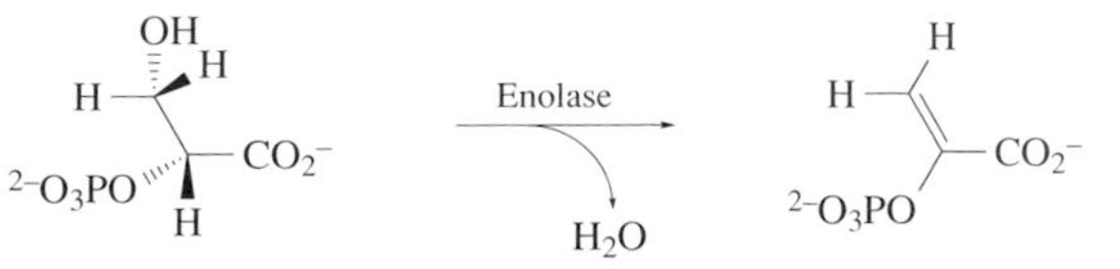

$$(10)$$

The high-resolution structure of enolase with an equilibrium mixture ($\sim 1:1$) of 2-PGA and PEP has been determined, and the structures of the complexes with either 2-PGA or PEP bound have been deconvoluted.[69] The structure of the active site of enolase complexes with 2-PGA is shown in Figure 10. This structure allowed identification of Lys345 as the general basic catalyst that mediates α-proton abstraction from 2-PGA and of Glu211 as the general acidic catalyst that facilitates the departure of the OH group.[70]

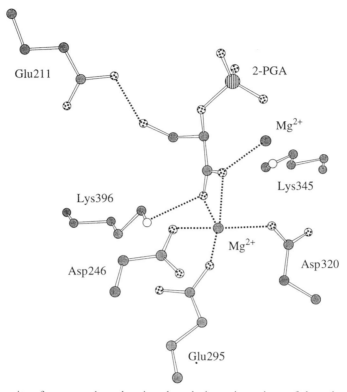

Figure 10   The active site of yeast enolase showing the relative orientations of the substrate 2-PGA, Lys345, the general basic catalyst, Glu211, the general acidic catalyst, and three electrophilic catalysts, Lys396 and two Mg^{2+} ions. The coordinates used in the construction of this figure were obtained from the Protein Data Base, file 1ONE.

This structure also reveals the positions of two Mg^{2+} ions that are required for catalytic activity. Both carboxylate oxygens of 2-PGA are coordinated to one Mg^{2+} ion. One of these is also hydrogen bonded to the ε-ammonium group of Lys396; the other carboxylate oxygen is coordinated to the second Mg^{2+} ion. In aggregate, these multiple cation–anion interactions allow substantial differential stabilization of the highly unstable enolate anion intermediate. In contrast to the active site of MR, the active site of enolase contains *no* neutral hydrogen bond donor that contributes to stabilization of the intermediate.

A reasonable mechanism for the enolase-catalyzed reaction is shown in Figure 11, including the interactions between the substrate, enolate anion intermediate, both Mg^{2+} ions, and Lys396.[70] These electrostatic interactions are likely to be responsible for differential stabilization of the intermediate (see Section 5.02.10).

5.02.5.5   Stabilization of Tetrahedral Intermediates: Chymotrypsin

The members of the trypsin/chymotrypsin/elastase superfamily catalyze the hydrolysis of peptide bonds with the assistance of a catalytic triad consisting of hydrogen bonded Asp-His-Ser residues. The mechanism involves the formation of transiently stable tetrahedral intermediates. High-resolution structures are available for members of the family complexed with inhibitors that reveal the interactions of substrates and tetrahedral intermediates with hydrogen bond donors in the active site.

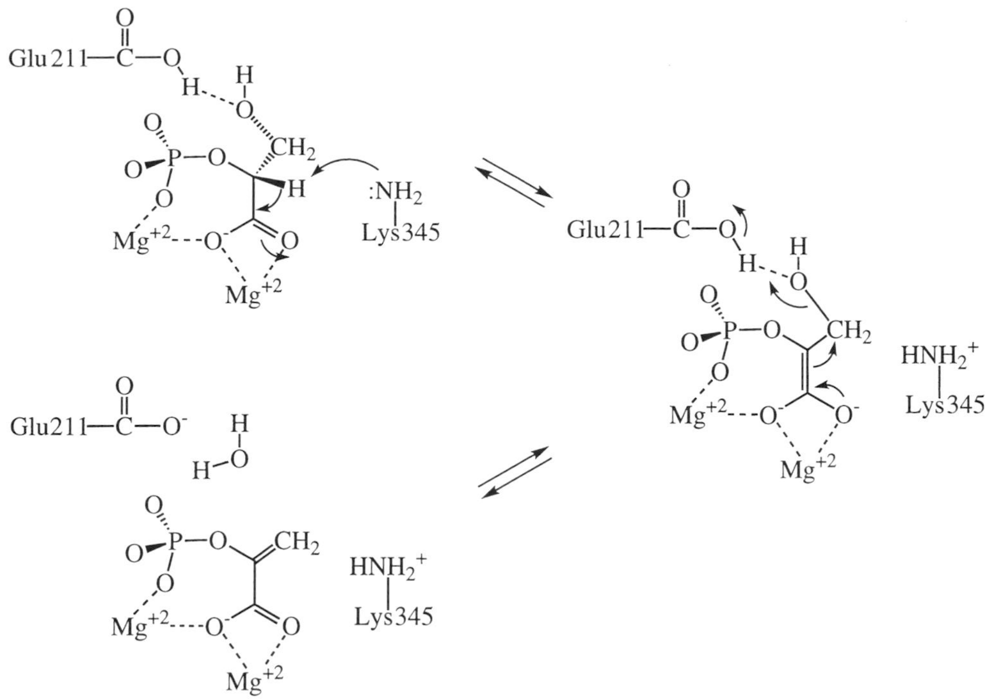

Figure 11 Proposed mechanism for the enolase-catalyzed reaction.

The active site of chymotrypsin complexed with the covalent inhibitor *N*-acetyl-Leu-Phe-trifluoromethyl ketone is illustrated in Figure 12.[71] This structure reveals not only the hydrogen bond geometry between the Asp102 and His57 (protonated in this complex), but also between the anionic oxygen of the tetrahedral intermediate and the neutral, very weakly acidic backbone NH groups of Gly193 and Ser195 that constitute the "oxyanion" hole. Each of these hydrogen bonding arrays probably contributes significantly to the overall stabilization that is required to allow the tetrahedral intermediate to be kinetically competent.

A reasonable mechanism for the chymotrypsin-catalyzed reaction is shown in Figure 13.[8,72,73] The hydrogen bonding interactions between Asp102 and His57 and also between the carbonyl group of the substrate and the NH groups of the oxyanion hole are likely to be responsible for differential stabilization of the intermediate. In Figure 13, the hydrogen bonds involving the bound substrate are "normal" hydrogen bonds whereas those involving the bound intermediate are stronger (perhaps of the short strong/low barrier type) as the proton affinities of the donor and acceptor atoms become similar (see Section 5.02.7).

5.02.6 IMPORTANCE OF HYDROGEN BONDS IN STABILIZING REACTIVE INTERMEDIATES

The hydrogen bond donors that interact with the carbonyl oxygens of substrates are invariably weakly acidic and neutral.[7-9] As a consequence, the proton affinities of the hydrogen bond donor and substrate acceptor are markedly different. However, as the reactive intermediate is formed, the proton affinity of the acceptor increases markedly to become similar to that of the donor, although the magnitude of the change is not so large that proton transfer from the donor to the acceptor to generate a neutral intermediate and an anionic conjugate base is expected to occur. Unlike the hydrogen bonded substrate, the hydrogen bonded intermediate has a net negative charge.

In the case of the catalytic triad in the proteases, the hydrogen bond donor, the neutral His, is weakly acidic and the hydrogen bond acceptor, the anionic Asp, is weakly basic, so their proton affinities are also markedly different. However, when the tetrahedral intermediate is formed, the His becomes protonated, and the proton affinities of the donor (now the protonated His) and the acceptor (the conjugate acid of the Asp) are more similar.[72] In contrast to the situation discussed in the previous paragraph, the proton affinity of the donor markedly *decreases* as the intermediate is formed, and the net charge on the hydrogen bonded members of the catalytic triad is zero.

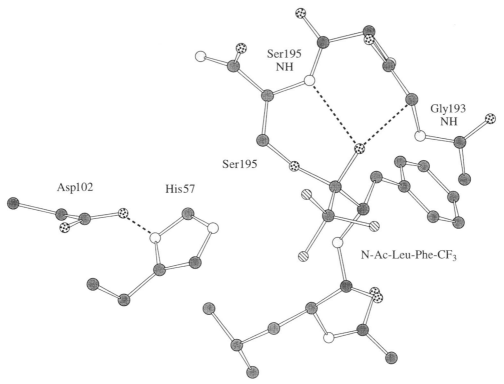

Figure 12 The active site of chymotrypsin showing the interaction of the active site-directed inhibitor *N*-acetyl-Leu-Phe-trifluoromethyl ketone with Ser195, His57, Asp102, and the electrophilic NH groups of Gly193 and Ser195. The coordinates used in the construction of this figure were obtained from the Protein Data Base, file 7GCH.

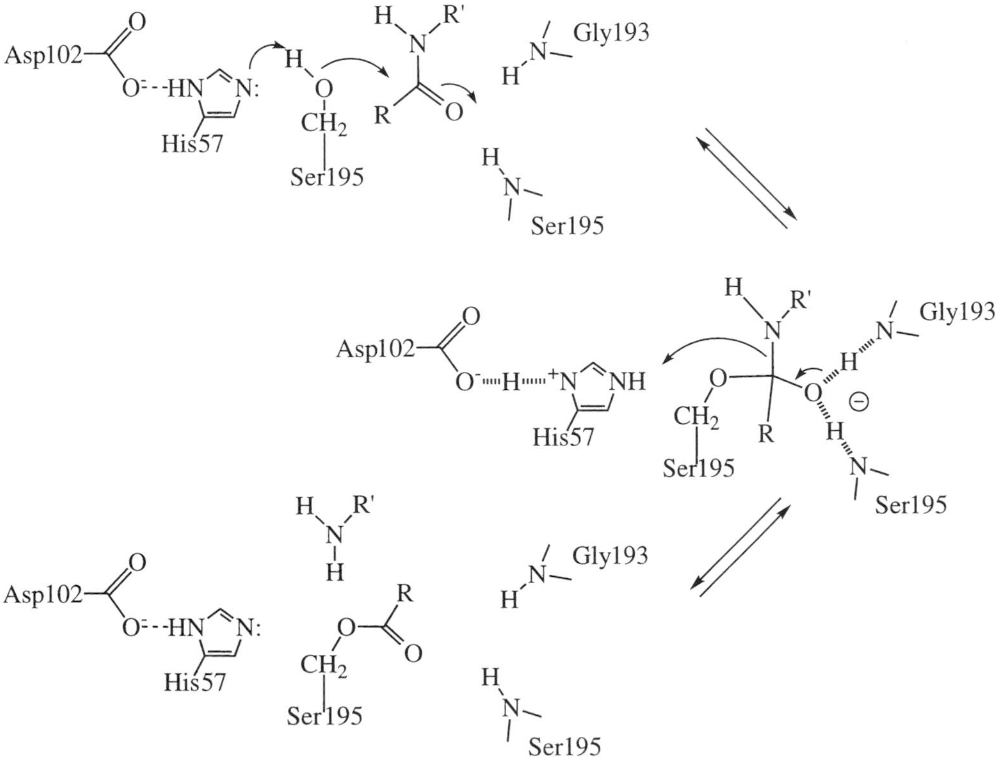

Figure 13 Proposed mechanism for the chymotrypsin-catalyzed reaction.

The large changes in proton affinities of the hydrogen acceptor or donor form the basis of the proposal that significant stabilization of reactive intermediates can be achieved by increases in the strengths of hydrogen bonds as bound substrate is converted to bound intermediate.

5.02.7   SHORT, STRONG HYDROGEN BONDS

Gerlt and Gassman[7,8] and, later, Cleland and Kreevoy[9] proposed that the changes in proton affinity that occur as bound substrate is converted to bound intermediate (often via the rate-limiting transition state) are the major source of differential stabilization of both the bound intermediate (decrease in ΔG°) and flanking transition states (decrease in $\Delta G^{\ddagger}_{\mathrm{int}}$). The similar proton affinities of the hydrogen bond acceptor and donor suggested the involvement of short strong (the terminology of Gerlt and Gassman) or low barrier (the terminology of Cleland[74]) hydrogen bonds in stabilization of the intermediates and transition states. Short strong/low barrier hydrogen bonds[75] are described by potential functions in which the barrier for proton transfer between the donor and acceptor is less than that of the zero point energy of the proton, so the proton can reside midway between donor and acceptor, hence the term "low barrier" (Figure 14). As a result of this potential function, these hydrogen bonds have covalent character[76] and their strengths as assessed by ΔH exceed those of "normal" hydrogen bonds in which the hydrogen bond proton is bonded only to the acceptor,[77] hence the term "short strong."

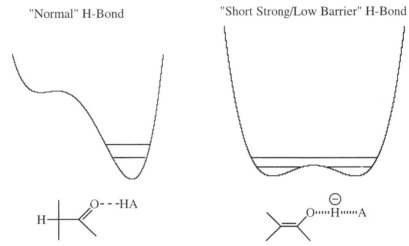

Figure 14   A comparison of the potential functions for "normal" and "short strong/low barrier" hydrogen bonds.

Short strong/low barrier hydrogen bonds are well known to occur both in the crystalline state and in nonhydrogen bonding solvents as judged from IR and NMR spectroscopies.[77] In virtually all cases, the structurally (i.e., crystallographically) characterized short strong/low barrier hydrogen bonds are symmetrical, so the difference in the proton affinities of the donor and acceptor is necessarily zero (most conveniently described as $\Delta pK_a = 0$). However, hydrogen bonds having covalent character certainly do not form uniquely when $\Delta pK_a = 0$: as pK_a increases from 0, the potential function will gradually include an increased contribution from the ionic interactions that characterize "normal" hydrogen bonds concomitant with a decreased contribution from the covalent interactions that characterize the symmetrical hydrogen bonds.

The "problems" associated with this proposal are twofold: (i) the strengths of individual hydrogen bonds are difficult, if not impossible, to measure in the active sites of enzymes, so the results of physical organic models must be used to assess the magnitude of the increase in hydrogen bond strength that can occur as bound substrate is converted to bound intermediate as a result of a large change in ΔpK_a; and (ii) neither an experimental nor a theoretical quantitative partitioning of the distribution of covalent and electrostatic/ionic contributions to hydrogen bonds as ΔpK_a is varied is available.

Not surprisingly, given these gaps in our knowledge, the proposal that changes in hydrogen bond strength are responsible for much of the required stabilization of reactive intermediates was, from the outset[78,79] and continued to be,[10,11,80,81] met with considerable controversy. For example, claims were made in the literature that this proposal implied that a special stabilization was available only when $\Delta pK_a = 0$, i.e., a plot of the dependence of hydrogen bond strength on ΔpK_a would be discontinuous: if the proton affinities of the donor and acceptor were not *precisely* matched, the strength of the hydrogen bond would be that of "normal" hydrogen bonds and differential stabilization of the intermediate and flanking transition states by this mechanism would be impossible.[10,80] Clearly, such objections to the proposal ignored the physical reality that the electronic character of hydrogen bonds and their associated strengths would be a continuous function of ΔpK_a so that the strengths of hydrogen bonds could vary monotonically as ΔpK_a is varied.

5.02.8   PHYSICAL ORGANIC MODELS FOR SHORT STRONG/LOW BARRIER HYDROGEN BONDS

Shan and Herschlag have provided the most persuasive physical organic evidence that significant increases in hydrogen bond strength can, in fact, accompany the large changes in ΔpK_a that are typical for the conversion of substrates to intermediates in active sites.[12]

For a series of substituted salicylates in several solvents, they evaluated whether a Brønsted relationship (Equation (11)) was valid for a large change in ΔpK_a:

$$\log K_{HB} = \beta \times \Delta pK_a + \text{constant} \tag{11}$$

Their analysis was based upon the effect of intramolecular hydrogen bonding on the observed pK_a of the acid, as assessed by the deviations from the intrinsic value estimated in the absence of the intramolecular hydrogen bond (Figure 15).

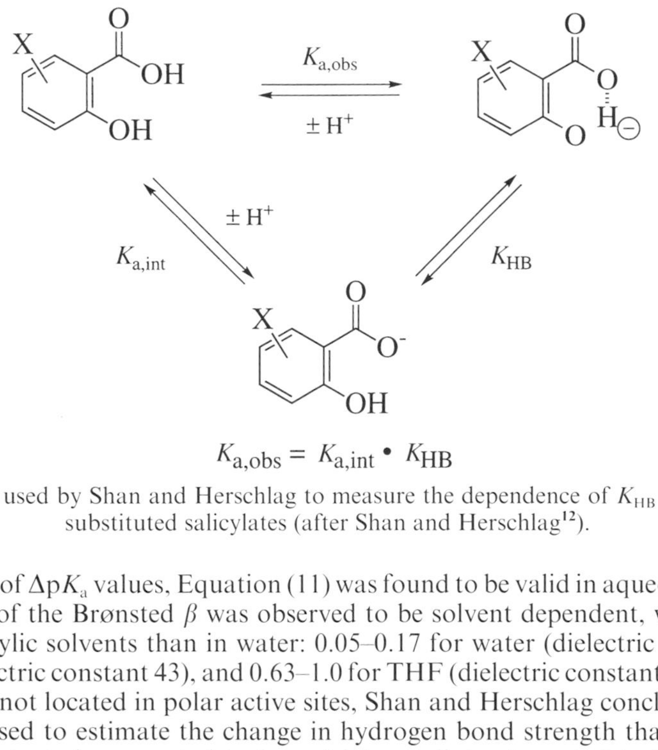

$$K_{a,obs} = K_{a,int} \cdot K_{HB}$$

Figure 15   Approach used by Shan and Herschlag to measure the dependence of K_{HB} on ΔpK_a in a series of substituted salicylates (after Shan and Herschlag[12]).

Over a wide range of ΔpK_a values, Equation (11) was found to be valid in aqueous and nonaqueous solvents. The value of the Brønsted β was observed to be solvent dependent, with the value much larger in nonhydroxylic solvents than in water: 0.05–0.17 for water (dielectric constant 78.5), 0.7–0.9 for DMSO (dielectric constant 43), and 0.63–1.0 for THF (dielectric constant 7). Since disordered water molecules are not located in polar active sites, Shan and Herschlag concluded that a value of 0.7 for β could be used to estimate the change in hydrogen bond strength that would be likely to occur as bound substrate is converted to bound intermediate. As noted in Section 5.02.2.1, the appropriate values for ΔpK_a are ~ 15 for both enolization of carbon acids and addition of nucleophiles to carbonyl groups to form tetrahedral intermediates; from Equation (11), the strengths of hydrogen bonds involving the substrate/intermediate acceptor are predicted to increase by 14 kcal mol^{-1} at 25 °C in both types of reaction (log $K_{HB} = 10.5$), and the rate constant for conversion of the bound to the intermediate complex would increase by a factor of $10^{10.5}$. This calculation could overestimate the effect because it is unlikely that the Brønsted relation is linear for such a large value of ΔpK_a, but it is a valid indication that differential stabilization of the bound intermediate and flanking transition states can be quite large.

In these studies, Shan and Herschlag observed that Equation (11) is valid when $\Delta pK_a = 0$, i.e., no special stabilization is associated with precisely matched proton affinities. Unfortunately, this conclusion was published prior to and independently of their work which demonstrated the application of Equation (11) over a large range of ΔpK_a values, thereby creating confusion in the physical organic and enzymological communities as to the real importance of hydrogen bonding in providing substantial differential stabilization of intermediates.

A second question that has been addressed via physical organic models is that of the strength of hydrogen bonds in nonaqueous solution when the proton affinities of the donors and acceptors are matched. Obviously, the amount of free energy by which a bound intermediate can be stabilized relative to substrate is limited by the absolute strength of the hydrogen bond when the proton affinities are matched (unless the bound substrate is destabilized/strained and this strain can be relieved as the intermediate is formed). Two studies have provided information on this question, those of Schwartz and Drueckhammer[13] and also those of Shan and Herschlag.[10,12]

Schwartz and Drueckhammer quantitated the relative strengths of "normal" and "short strong/low barrier" hydrogen bonds by measuring the positions of the *cis–trans* equilibria of maleic/fumaric acids and their monoanions as a function of solvent (Figure 16).[13] Previous experimental work had established that maleic acid monoanion forms an intramolecular short strong/low barrier hydrogen in aprotic solvents.[82] They observed that while the diacids favored the *trans* (fumaric) isomer irrespective of solvent (H_2O, MeOH, DMSO, and $CHCl_3$), the monoanions favored the *trans* isomer in hydrogen bonding solvents but the *cis* (maleic) isomer in nonhydrogen bonding solvents. By comparing the isomeric distributions for the diacid and the monoanion, they concluded that in $CHCl_3$ the short strong/low barrier hydrogen bond in the monoanion is ~ 4.4 kcal mol^{-1} stronger than normal hydrogen bond in the diacid. However, as pointed out by the authors, the quantitation was difficult and subject to error because in all solvents both the diacid and the dianion greatly favor the *trans* isomer, so any contamination by these due to incomplete or excess titration of the diacid would introduce a potentially large error in the estimated excess hydrogen bond energy associated with the short strong/low barrier hydrogen bond. Subsequent to these studies, the results of *ab initio* calculations confirmed that *in the gas phase* the hydrogen bond in maleic acid monoanion has partial covalent character, i.e., is a short strong/low barrier hydrogen bond, and that the strength of this hydrogen bond relative to that in the diacid is 20.4 kcal mol^{-1}.[83] While the strength of this hydrogen bond is expected to be less due to an increased dielectric constant of the medium, the hydrogen bonded system studied by Schwartz and Drueckhammer is a short strong/low barrier hydrogen bond.

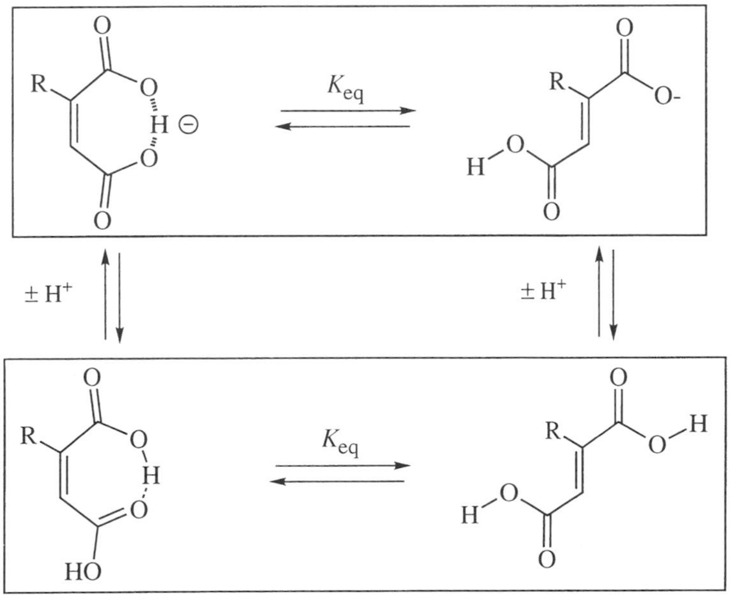

Figure 16 Approach used by Schwartz and Drueckhammer to measure the relative strengths of "normal" and "short strong/low barrier" hydrogen bonds (after Schwartz and Drueckhammer[13]).

An important point for understanding the strength of the hydrogen bond in maleic acid monoanion is that the monoanion is significantly strained:[83] the C(1)–C(2)–C(3) bond angle is 130.2°, i.e., 10.2° greater than that expected for an unstrained double bond. Thus, the hydrogen bond strength as measured or calculated must include a significant contribution from this geometric destabilization.

The previously discussed work of Shan and Herschlag[12] allows the strength of the intramolecular bond in salicylate monoanion to be quantitated. Extrapolation of their data to $\Delta pK_a = 0$ reveals that the maximum strength of the hydrogen bond is 12.2 kcal mol^{-1} (log $K_{HB} = 9$) in DMSO and 3 kcal mol^{-1} in H$_2$O. While these values appear to indicate that only 9 kcal mol^{-1} of differential stabilization of intermediates (relative to the uncatalyzed reaction in H$_2$O) is available by hydrogen bonding, the geometric constraints (a six-membered ring) on this hydrogen bond probably prevent the formation of a short strong/low barrier hydrogen bond; the X-ray structure of the monoanion reveals that the distance between the hydrogen-bonded oxygens is 2.54 Å,[84] considerably longer than the distance that would allow the formation of a short strong/low barrier hydrogen bond.[75] This suspicion is confirmed by the value of the fractionation factor reported by Shan and Herschlag for the unsubstituted salicylate monoanion, i.e., their value of 0.84 ± 0.03 is significantly greater than the value of 0.3 that is considered to be diagnostic for short strong/low barrier hydrogen bonds.

Thus, to date, no appropriate physical organic model has been synthesized and characterized that would allow the strength of a short strong/low barrier hydrogen bond to be measured: (i) the strength of the short strong/low barrier hydrogen bond in maleic acid monoanion as measured by the first pK_a of the diacid includes a (significant?) contribution from distortion, and (ii) the hydrogen bond in salicylate monoanion is geometrically enforced to be "normal" although the proton affinities of the donor and acceptor are similar.

5.02.9   EXPERIMENTAL EVIDENCE FOR STABILIZATION OF INTERMEDIATES BY STRONG HYDROGEN BONDS

Subsequent to the proposal that strong hydrogen bonding can provide significant stabilization of reactive intermediates, direct evidence for the existence of short strong/low barrier hydrogen bonds in several enzyme active sites has been obtained by X-ray crystallographic and/or ^{1}H NMR spectroscopic studies of enzyme–inhibitor complexes. In some cases, kinetic data with mutant enzymes have also been used to estimate the strengths of these hydrogen bonds, and the values so obtained indicate that differential stabilization of a reactive intermediate and its flanking transition states contributes significantly to the rates of the enzyme-catalyzed reactions.

5.02.9.1   Cytidine Deaminase

Cytidine deaminase catalyzes the hydrolysis of the 4-amino group of cytidine (**5**) via a tetrahedral intermediate (**6**) to form uridine. The structure of the complex of the deaminase with zebularine 3,4-hydrate (**7**), a mimic of a tetrahedral intermediate and a competitive inhibitor with a $K_i = 1.2 \times 10^{-12}$ M, revealed a short strong/low barrier hydrogen bond between the OH group of the pyrimidine ring and the carboxylate group of Glu104.[85] Presumably, coordination of the OH group of the inhibitor to the essential active site zinc(II) reduces its pK_a toward the value of the carboxylic group of Glu104. The k_{cat} is reduced by a factor of 10^8 in the E104A mutant, corresponding to an increase in the activation energy by 11 kcal mol^{-1}, thereby providing a quantitative estimate of the importance of the short strong/low barrier hydrogen bond to catalysis.

5.02.9.2   Chymotrypsin

The hydrogen-bonded proton shared by Asp102 and protonated His57 in the catalytic triad in α-chymotrypsin (Section 5.02.5.5; Figure 13) has been studied by both ^{1}H NMR and X-ray crystallography. When His57 is protonated in acidic conditions (pH 3.5), the ^{1}H NMR signal for the proton occurs at a chemical shift of 18.3 ppm in H_2O; in D_2O the ^2H NMR chemical shift of the hydrogen-bonded deuteron is shifted upfield by 1.0 ± 0.4 ppm, as expected for short strong/low barrier hydrogen bonds.[72]

The tetrahedral adducts formed when peptidylfluoromethylketones are complexed with Ser195 (Figure 12) in the active site of α-chymotrypsin have been studied at neutral pH by Abeles, Frey, Ringe, and their co-workers.[71,73,86] The inhibited structure (Figure 12) mimics the tetrahedral intermediate (Figure 13), so His57 is protonated and hydrogen bonded to Asp102. Abeles first reported that the ^1H NMR chemical shift of the hydrogen-bonded proton is 18.7 ppm, and the pK_a of His57 is > 10.5; later studies by Frey refined these measurements, with the chemical shift being 18.9 ppm and the $pK_a \geqslant 12$.[73] The more recent estimate of the pK_a value is elevated by $\geqslant 5$ units from that for histidine in water, indicating that the short strong/low barrier hydrogen bond in the active site is > 7 kcal mol^{-1} stronger than the analogous hydrogen bond in water.[86] This hydrogen bond satisfies the criteria for a short strong/low barrier hydrogen bond and has sufficient strength to contribute significantly to the catalytic rate by indirectly stabilizing the enzyme-bound tetrahedral intermediate.

Although the proton shared by Asp102 and His57 can be observed by ^1H NMR spectroscopy, the hydrogen-bonded protons in the oxyanion hole have not been observed. The carbonyl oxygen of the substrate and the anionic oxygen of the tetrahedral intermediate are located in the oxyanion hole which should be able to provide substantial stabilization of the tetrahedral intermediate.[8] Whether the inability to observe these protons is the result of rapid exchange with solvent is unknown. Mutagenesis of the oxyanion hole in chymotrypsin is impossible, since the hydrogen bond donors are provided by backbone NH groups. However, mutagenesis of the oxyanion hole in the serine protease subtilisin can and has been performed, since it is composed of one or two hydrogen bond donors associated with amino acid side chains: a carboxamide NH of Asn155 and the OH group of Thr220.[87] Mutation of these significantly and additively reduces k_{cat}. The reductions are comparable to those observed for mutagenesis of the Asp-His dyad in trypsin.[88] Thus, both the Asp-His diad and the oxyanion hole are likely to provide significant stabilization of the tetrahedral intermediates in serine proteases.

5.02.9.3 Ketosteroid Isomerase

Ketosteroid isomerase catalyzes the interconversion of α,β- and β,γ-unsaturated ketosteroids via a dienolic intermediate (Figure 17). The reaction is facilitated by Tyr14, an electrophilic catalyst,[89-91] and Asp38, a general basic catalyst.[89] Mildvan and co-workers have studied the interaction of the isomerase with the competitive inhibitor dihydroequilenin (8), a mimic of the intermediate. In the inhibited complex the ^1H NMR chemical shift of the proton shared by Tyr14 and the inhibitor is 18.15 ppm, and the fractionation factor is 0.34;[92] these are characteristic for short strong/low barrier hydrogen bonds. The rates of dissociation of the inhibitor from the wild type isomerase and the Y14F mutant were measured, and from these values the strength of the hydrogen bond in the wild type enzyme was estimated to be $\geqslant 7.6$ kcal mol^{-1}. This value is similar to that deduced ($\geqslant 7.8$ kcal mol^{-1}) from the pK_a for loss of proton that occurs on formation of the isomerase–dihydroequilenin complex, $\leqslant 4.3$.

(8)

5.02.9.4 Citrate Synthase

Citrate synthase catalyzes the Claisen condensation between acetyl CoA and oxaloacetate to yield, after hydrolysis of the thioester bond, citrate and CoA. This reaction probably occurs via the stabilized enolate anion of acetyl CoA. A short hydrogen bond has been observed by X-ray

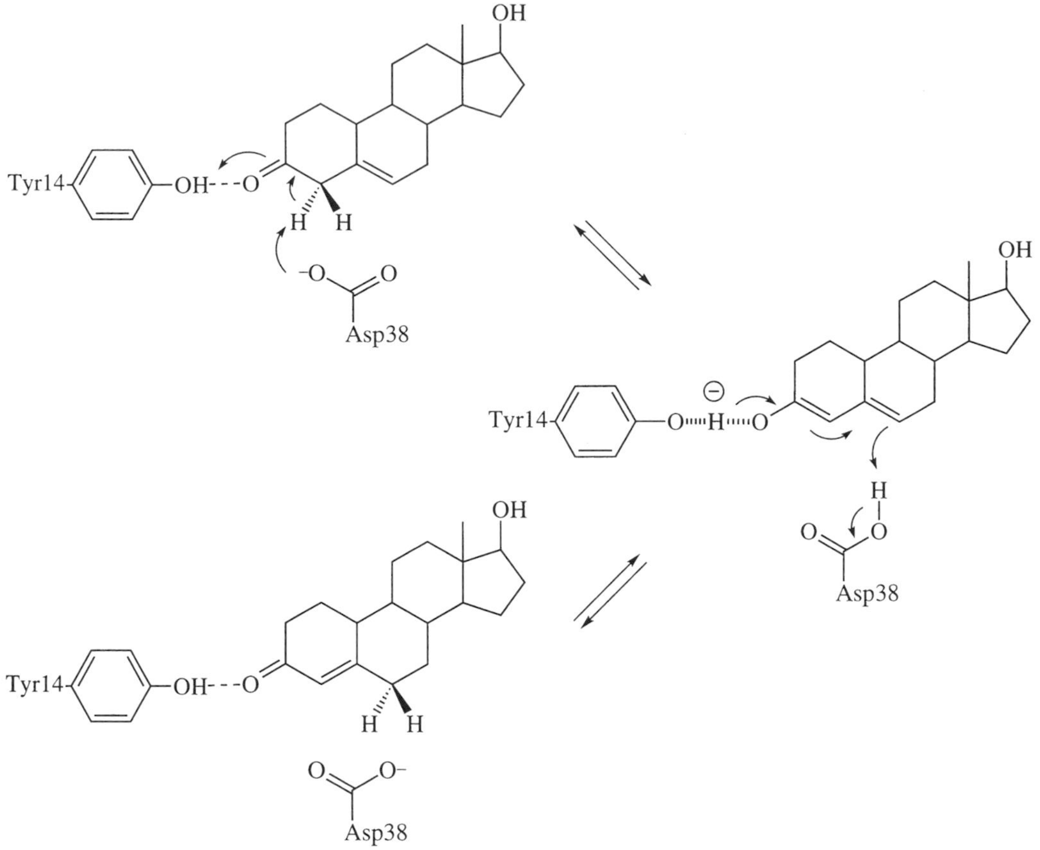

Figure 17   Proposed mechanism for the ketosteroid isomerase-catalyzed reaction.

crystallography in the active site of an inhibited citrate synthase.[93] The carboxylate group of the competitive inhibitor *S*-carboxymethyl CoA, an analogue of the enolate anion of acetyl CoA, is hydrogen bonded to Asp375, the general basic catalyst in the enolization of acetyl CoA. The O···O distance was reported to be <2.4 Å. In contrast, the O···N distance for the complex with an analogous amide-containing inhibitor was ∼2.5 Å. Remington and Drueckhammer and co-workers proposed that the carboxyl inhibitor forms a short strong/low barrier hydrogen bond while the carboxamide inhibitor forms a normal hydrogen bond. However, the K_i for the inhibitors differed by only 18-fold (2 kcal mol^{-1}), although the pK_a values of carboxylic acids and carboxamides differ by about 13 units in aqueous solution. The K_i value for the carboxyl inhibitor was pH dependent, but the lowest pH value used in these experiments, 6.0, exceeds the expected pK_a of either Asp375 or the inhibitor by about 1–2 units. This probably leads to an underestimate of the strength of the hydrogen bond, perhaps by a factor of ∼10, so the strength of the hydrogen bond appears to exceed that of a "normal" hydrogen bond by 3 kcal mol^{-1}.

However, this short hydrogen bond is formed with a *syn* orbital of Asp375 and an *anti* orbital of the carboxyl inhibitor. Since the *syn* orbital of a carboxylate group has a higher proton affinity than the *anti* orbitals,[94] the Brønsted relation in Equation (11) predicts that the strength of this hydrogen bond will not be maximal. Using 0.7 as the value for the Brønsted β and assuming that the difference in proton affinities of the *syn* and *anti* orbitals is a factor of 10^4, the strength of the hydrogen bond observed crystallographically is predicted to be ∼4 kcal mol^{-1} less than the maximal value when the proton affinities are equal, i.e., a short strong/low barrier hydrogen bond between two carboxylic acid/carboxylate groups in an active site could provide as much as 7 kcal mol^{-1} of differential stabilization of a reactive intermediate.

(The reaction catalyzed by citrate synthase does not involve the analogue of such a hydrogen bond since Asp375 is the general base catalyst in the enolization of acetyl CoA. However, the enolate anion of acetyl CoA generated by proton abstraction by Asp375 is likely to be stabilized by a strong hydrogen bond to the neutral His274 in the active site of citrate synthase.)[95]

5.02.10   STABILIZATION OF INTERMEDIATES BY ELECTROSTATIC INTERACTIONS

The original statement of the proposal by Gerlt and Gassman that changes in hydrogen bond strength as the substrate is converted to the intermediate can provide significant stabilization of reactive intermediates was immediately criticized by Guthrie and Kluger.[78] Guthrie and Kluger described an alternative strategy in which electrostatic interactions of anionic reactive intermediates with cationic groups can provide large amounts of stabilization for intermediates in a region of low dielectric constant. In particular, they noted (i) that the active site of mandelate racemase (see Section 5.02.5.3) contains an essential Mg^{2+} (Figure 8); and (ii) that an increase in charge on the carboxylate group could result in as much as 18 kcal mol^{-1} stabilization. Clearly, this amount of stabilization is comparable to or exceeds that which can be provided by short strong/low barrier hydrogen bonds. Although ignored by Guthrie and Kluger, Gerlt and Gassman, in fact, previously had performed a similar analysis of the importance of electrostatic interactions and reached a quantitatively similar conclusion.[58] However, not all active sites contain divalent metal ions that are directly coordinated to the carbonyl/carboxylate oxygens of substrates in enolization reactions, so this electrostatic argument cannot be generally applicable.

As noted in Section 5.02.5.3, both a hydrogen bond donor (Glu317) and an essential Mg^{2+} are coordinated to the substrate carboxylate group in the active site of mandelate racemase. Replacement of the carboxylic acid group of Glu317 with the isosteric but more weakly acidic carboxamide group (E317Q) decreased k_{cat} by a factor of 10^4,[67] suggesting that hydrogen bonding could provide as much as 6 kcal mol^{-1} of differential stabilization of an enolate anion intermediate if the proton affinities of the intermediate and active site hydrogen bond donor were similar. (Note that while the ΔpK_a between the donor and acceptor will decrease irrespective of whether the carboxylic (wild type) or the carboxamide (E317Q) functional group is the hydrogen bond donor, in the latter case the ΔpK_a between the donor and acceptor when the intermediate is present is very large ($\geqslant 10$), so Equation (11) cannot be expected to apply quantitatively. In addition, the E317Q mutant is still a proficient catalyst that tightly binds the transition state/intermediate, suggesting that considerable stabilization of the reactive enolate anion intermediate can occur in this active site.)

The active site of enolase (Section 5.02.5.4) stabilizes an enolate anion intermediate entirely by electrostatic interactions with *two* Mg^{2+} ions. The substrate for enolase, 2-phosphoglycerate (2-PGA), is probably one of the most weakly acidic carbon acids in enzymology, so it is not surprising that electrostatic interactions play an important role in stabilizing the enolate anion intermediate. However, mandelate racemase and enolase are members of the same (enolase) superfamily of enzymes,[96] with most of the substrates for the diverse reactions catalyzed by the members of this superfamily having α-protons that are as weakly acidic like those in 2-PGA. Since the currently available evidence suggests that not all members of the enolase superfamily require two Mg^{2+} ions for catalysis, differential stabilization of the reactive enolate anion intermediate by hydrogen bonding probably can provide stabilization that is comparable to that which is available by purely electrostatic interactions.

5.02.11   SUMMARY

The active sites of enzymes that catalyze reactions involving the formation of oxyanion intermediates, e.g., enolization reactions of carbon acids and nucleophilic substitution of carbonyl compounds, usually contain weakly acidic hydrogen donors that interact directly with the carbonyl oxygen of the substrate. Physical organic models confirm the hypothesis that substantial differential stabilization of the oxyanion intermediates (enolate anions or tetrahedral intermediates) can be accomplished by an increase in hydrogen bond strength as the substrate is converted to the intermediate and the proton affinity of the oxygen increases to become similar to that of the hydrogen bond donor. Structural, spectroscopic, and kinetic studies of several enzymes have provided evidence for the formation of strong hydrogen bonds in the enzyme–intermediate complex. While oxyanion intermediates also can be stabilized by electrostatic interactions with active site cations, substantial stabilization of these intermediates can be accomplished by hydrogen bonding.

5.02.12   REFERENCES

1. A. Thibblin and W. P. Jencks, *J. Am. Chem. Soc.*, 1979, **101**, 4963.
2. W. P. Jencks, *Acc. Chem. Res.*, 1980, **13**, 161.

3. A. Radzicka and R. Wolfenden, *Science*, 1995, **267**, 90.
4. A. R. Fersht, J.-P. Shi, J. Knill-Jones, D. M. Lowe, A. J. Wilkinson, D. M. Blow, P. Brick, P. Carter, M. M. Y. Waye, and G. Winter, *Nature*, 1985, **314**, 235.
5. A. R. Fersht, *Trends Biochem. Sci.*, 1987, **12**, 301.
6. A. R. Fersht, *Biochemistry*, 1988, **27**, 1577.
7. J. A. Gerlt and P. G. Gassman, *J. Am. Chem. Soc.*, 1992, **115**, 11 552.
8. J. A. Gerlt and P. G. Gassman, *Biochemistry*, 1993, **32**, 11 943.
9. W. W. Cleland and M. M. Kreevoy, *Science*, 1994, **264**, 1887.
10. S. Shan, S. Loh, and D. Herschlag, *Science*, 1996, **272**, 97.
11. S. Shan and D. Herschlag, *J. Am. Chem. Soc.*, 1996, **118**, 5515.
12. S. Shan and D. Herschlag, *Proc. Natl. Acad. Sci. USA*, 1996, **93**, 14 474.
13. B. Schwartz and D. G. Drueckhammer, *J. Am. Chem. Soc.*, 1995, **117**, 11 902.
14. A. Bagno, V. Lucchine, and G. Scorrano, *Bull. Chem. Soc. Fr.*, 1987, 563.
15. K. Yates and R. Stewart, *Can. J. Chem.*, 1959, **37**, 664.
16. A. Fischer, G. A. Grigor, J. Packer, and J. Vaughan, *J. Am. Chem. Soc.*, 1961, **83**, 4208.
17. L. A. Flexser, L. P. Hammett, and A. Dingwall, *J. Am. Chem. Soc.*, 1935, **57**, 2103.
18. Y. Chiang, A. J. Kresge, P. Pruszynski, N. P. Schepp, and J. Wirz, *Angew. Chem., Int. Ed. Engl.*, 1990, **29**, 792.
19. J. R. Keefe and A. J. Kresge, in "The Chemistry of Enols," ed. Z. Rappoport, Wiley, New York, 1990, chap. 7.
20. E. M. Arnett, *Prog. Phys. Org. Chem.*, 1963, **1**, 223.
21. J. P. Guthrie, *J. Am. Chem. Soc.*, 1974, **96**, 3608.
22. W. J. Albery and J. R. Knowles, *Biochemistry*, 1976, **15**, 5627.
23. W. J. Albery and J. R. Knowles, *Biochemistry*, 1976, **15**, 5588.
24. J. M. Herlihy, S. G. Maister, W. J. Albery, and J. R. Knowles, *Biochemistry*, 1976, **15**, 5601.
25. S. G. Maister, C. P. Pett, W. J. Albery, and J. R. Knowles, *Biochemistry*, 1976, **15**, 5607.
26. S. J. Fletcher, J. M. Herlihy, W. J. Albery, and J. R. Knowles, *Biochemistry*, 1976, **15**, 5612.
27. P. F. Leadlay, W. J. Albery, and J. R. Knowles, *Biochemistry*, 1976, **15**, 5617.
28. L. M. Fisher, W. J. Albery, and J. R. Knowles, *Biochemistry*, 1976, **15**, 5621.
29. R. T. Raines and J. R. Knowles, *Biochemistry*, 1987, **26**, 7014.
30. W. J. Albery and J. R. Knowles, *Biochemistry*, 1976, **15**, 5631.
31. J. J. Burbaum, R. T. Raines, W. J. Albery, and J. R. Knowles, *Biochemistry*, 1989, **28**, 9293.
32. A. O. Cohen and R. A. Marcus, *J. Phys. Chem.*, 1968, **72**, 4249.
33. R. A. Marcus, *J. Am. Chem. Soc.*, 1969, **91**, 7224.
34. W. J. Albery, *Annu. Rev. Phys. Chem.*, 1980, **31**, 227.
35. G. S. Hammond, *J. Am. Chem. Soc.*, 1955, **77**, 334.
36. J. R. Knowles, *Nature*, 1991, **350**, 121.
37. R. C. Davenport, P. A. Bash, B. A. Seaton, M. Karplus, G. A. Petsko, and D. Ringe, *Biochemistry*, 1991, **30**, 5821.
38. R. T. Raines, E. L. Sutton, D. R. Straus, W. Gilbert, and J. R. Knowles, *Biochemistry*, 1986, **25**, 7142.
39. D. Joseph-McCarthy, L. E. Rost, E. A. Komives, and G. A. Petsko, *Biochemistry*, 1994, **33**, 2824.
40. E. B. Nickbarg, R. C. Davenport, G. A. Petsko, and J. R. Knowles, *Biochemistry*, 1988, **27**, 5948.
41. J. G. Belasco and J. R. Knowles, *Biochemistry*, 1980, **19**, 472.
42. E. A. Komives, L. C. Chang, E. Lolis, R. F. Tilton, G. A. Petsko, and J. R. Knowles, *Biochemistry*, 1991, **30**, 3011.
43. P. J. Lodi and J. R. Knowles, *Biochemistry*, 1991, **30**, 6948.
44. T. C. Bruice and G. L. Schmir, *J. Am. Chem. Soc.*, 1958, **80**, 148.
45. P. J. Lodi and J. R. Knowles, *Biochemistry*, 1993, **32**, 4338.
46. P. C. Babbitt, G. L. Kenyon, B. M. Martin, H. Charest, M. Sylvestre, J. D. Scholten, K.-H. Chang, P.-H. Liang, and D. Dunaway-Mariano, *Biochemistry*, 1992, **31**, 5594.
47. G. Müller-Newman, U. Janssen, and W. Stoffel, *Eur. J. Biochem.*, 1995, **228**, 68.
48. P. C. Babbitt and J. A. Gerlt, *J. Biol. Chem.*, 1997, **272**, 30 591.
49. M. M. Benning, K. L. Taylor, R.-Q. Liu, G. Yang, H. Xiant, G. Wesenberg, D. Dunaway-Mariano, and H. M. Holden, *Biochemistry*, 1996, **35**, 8103.
50. C. K. Engel, M. Mathieu, J. Ph. Zeelen, J. K. Hiltunen, and R. K. Wierenga, *EMBO J.*, 1996, **15**, 5135.
51. G. Yang, P.-H. Liang, and D. Dunaway-Mariano, *Biochemistry*, 1994, **33**, 8527.
52. R. Q. Liu, P.-H. Liang, J. Scholten, and D. Dunaway-Mariano, *J. Am. Chem. Soc.*, 1995, **117**, 5003.
53. K. L. Taylor, R.-Q. Liu, P.-H. Liang, J. Price, D. Dunaway-Mariano, P. J. Tonge, J. Clarkson, and P. R. Carey, *Biochemistry*, 1995, **34**, 13 881.
54. G. Yang, R.-Q. Liu, K. L. Taylor, H. Xiang, J. Price, and D. Dunaway-Mariano, *Biochemistry*, 1996, **35**, 1349.
55. K. L. Taylor, H. Xiang, G. Yang, and D. Dunaway-Mariano, *Biochemistry*, 1997, **36**, 1349.
56. B. J. Bahnson and V. E. Anderson, *Biochemistry*, 1989, **28**, 4173.
57. B. J. Bahnson and V. E. Anderson, *Biochemistry*, 1989, **30**, 5894.
58. J. A. Gerlt and P. G. Gassman, *J. Am. Chem. Soc.*, 1992, **114**, 5928.
59. J. W. Hawes, J. Jaskiewicz, Y. Shimomura, B. Huang, J. Bunting, E. T. Harper, and R. A. Harris, *J. Biol. Chem.*, 1996, **269**, 14 248.
60. G. L. Kenyon, J. A. Gerlt, G. A. Petsko, and J. W. Kozarich, *Acc. Chem. Res.*, 1995, **28**, 178.
61. J. A. Landro, A. T. Kallarakal, S. C. Ransom, J. A. Gerlt, J. W. Kozarich, D. J. Neidhart, G. A. Petsko, and G. L. Kenyon, *Biochemistry*, 1991, **30**, 9264.
62. J. A. Gerlt, J. W. Kozarich, G. L. Kenyon, and P. G. Gassman, *J. Am. Chem. Soc.*, 1991, **113**, 9667.
63. J. A. Landro, J. A. Gerlt, J. W. Kozarich, C. W. Koo, V. J. Shah, G. L. Kenyon, D. J. Neidhart, S. Fujita, J. R. Clifton, and G. A. Petsko, *Biochemistry*, 1994, **33**, 635.
64. A. T. Kallarakal, J. W. Kozarich, J. A. Gerlt, J. R. Clifton, G. A. Petsko, and G. L. Kenyon, *Biochemistry*, 1995, **34**, 2788.
65. J. A. Landro, A. Kallarakal, S. C. Ransom, J. A. Gerlt, J. W. Kozarich, D. J. Neidhart, and G. L. Kenyon, *Biochemistry*, 1991, **30**, 9274.
66. S. L. Schafer, W. C. Barrett, A. T. Kallarakal, B. Mitra, J. W. Kozarich, J. A. Gerlt, J. G. Clifton, G. A. Petsko, and G. L. Kenyon, *Biochemistry*, 1996, **35**, 5662.

67. B. Mitra, A. T. Kallarakal, J. W. Kozarich, J. A. Gerlt, J. R. Clifton, G. A. Petsko, and G. L. Kenyon, *Biochemistry*, 1995, **34**, 2777.
68. S. R. Anderson, V. E. Anderson, and J. R. Knowles, *Biochemistry*, 1994, **33**, 10 545.
69. T. M. Larsen, J. E. Wedekind, I. Rayment, and G. H. Reed, *Biochemistry*, 1996, **35**, 4349.
70. R. R. Poyner, T. Laughlin, G. A. Sowa, and G. H. Reed, *Biochemistry*, 1996, **35**, 1692.
71. K. Brady, A. Z. Wei, D. Ringe, and R. H. Abeles, *Biochemistry*, 1990, **29**, 7600.
72. P. A. Frey, S. A. Whitt, and J. B. Tobin, *Science*, 1994, **264**, 1927.
73. C. S. Cassidy, J. Lin, and P. A. Frey, *Biochemistry*, 1997, **36**, 4576.
74. W. W. Cleland, *Biochemistry*, 1992, **31**, 317.
75. F. Hibbert and J. Emsley, *Adv. Phys. Org. Chem.*, 1990, **26**, 255.
76. P. Gilli, V. Bertolase, V. Ferretti, and G. Gilli, *J. Am. Chem. Soc.*, 1994, **116**, 909.
77. J. A. Gerlt, M. M. Kreevoy, W. W. Cleland, and P. A. Frey, *Chem. Biol.*, 1997, **4**, 259.
78. J. P. Guthrie and R. Kluger, *J. Am. Chem. Soc.*, 1993, **115**, 11 569.
79. A. Warshel, A. Papzyan, and P. A. Kollman, *Science*, 1995, **269**, 102.
80. S. Scheiner and T. Kar, *J. Am. Chem. Soc.*, 1995, **117**, 6970.
81. J. P. Guthrie, *Chem. Biol.*, 1996, **3**, 163.
82. C. L. Perrin, *Science*, 1994, **266**, 1665.
83. M. Garcia-Viloca, A. González-Lafont, and J. M. Lluch, *J. Am. Chem. Soc.*, 1997, **119**, 1081.
84. H. S. Kim and G. A. Jeffrey, *Acta Crystallogr.*, 1971, **B27**, 1123.
85. D. C. Carlow, A. A. Smith, C. C. Yang, A. S. Short, and R. Wolfenden, *Biochemistry*, 1995, **34**, 4220.
86. T. C. Liang and R. H. Abeles, *Biochemistry*, 1987, **26**, 7603.
87. S. Braxton and J. A. Wells, *J. Biol. Chem.*, 1991, **266**, 11 797.
88. C. S. Craik, S. Roczniak, C. Largman, and W. J. Rutter, *Science*, 1987, **237**, 909.
89. A. Kuliopulos, A. S. Mildvan, D. Shortle, and P. Talalay, *Biochemistry*, 1989, **28**, 149.
90. A. Kuliopulos, P. Talalay, and A. S. Mildvan, *Biochemistry*, 1990, **29**, 10 271.
91. L. Xue, P. Talalay, and A. S. Mildvan, *Biochemistry*, 1990, **30**, 10 858.
92. Z. Zhao, C. Abeygunawardana, P. Talalay, and A. S. Mildvan, *Proc. Natl. Acad. Sci. USA*, 1996, **93**, 8220.
93. K. C. Usher, S. J. Remington, D. P. Martin, and D. G. Drueckhammer, *Biochemistry*, 1994, **33**, 7753.
94. R. D. Gandour, *Bioorg. Chem.*, 1981, **10**, 169.
95. A. J. Mulholland and W. G. Richards, *Proteins*, 1997, **27**, 9.
96. P. C. Babbitt, M. S. Hasson, J. E. Wedekind, D. R. J. Palmer, W. C. Barrett, G. H. Reed, I. Rayment, D. Ringe, G. L. Kenyon, and J. A. Gerlt, *Biochemistry*, 1996, **35**, 16 489.

5.03
Keto–Enol Tautomerism in Enzymatic Reactions

CHRISTIAN P. WHITMAN
University of Texas, Austin, TX, USA

5.03.1 INTRODUCTION

In aqueous solution, many carbonyl compounds with a proton on the adjacent carbon exist as an equilibrium mixture of two isomers—the keto form and the enol form (Equation (1)). For many simple carbonyl compounds, there is very little of the enol (or enolate) isomer present and only the keto isomer is detectable.[1] The interconversion of these two isomers is of fundamental importance in many nonenzymatic and enzymatic reactions because carbonyl compounds often react as one isomer or the other. Hence much attention has been focused on the equilibrium between these isomers, the rate of interconversion, and the factors that affect the rates.[2-7]

$$\text{(1)}$$

Enone–dienol tautomerization is observed in the interconversion of unconjugated and conjugated enones and has much in common with keto–enol tautomerization, including the fact that it plays a central role in several nonenzymatic and enzymatic reactions.[2,3] The interconversion of unconjugated and conjugated enones, formally known as a 1,3-allylic isomerization, proceeds through a conjugated enol or a dienol intermediate (Equation (2)). The reaction consists of two parts: the enolization of a β,γ-unsaturated ketone to a dienol (or dienolate) intermediate and the ketonization of the dienol to the α,β-unsaturated isomer.[3] Potentially, all three isomers can exist at equilibrium in aqueous solution.

$$\text{(2)}$$

In general, enols and dienols are thermodynamically unstable and cannot be isolated in sufficient quantities for study.[4] For these reasons, many nonenzymatic and enzymatic reactions involving enols and dienols have been studied in the thermodynamically unfavored direction, that is, in the keto to the enol direction or in the enone to dienol direction. An in-depth discussion of several thermodynamically unstable enols and their role in enzymatic reactions can be found in Chapter 5.02.

In recent years, a variety of stable enols and dienols of biological relevance have been generated or synthesized and their properties studied. Some examples of these isolable compounds include enolpyruvate, generated by pyruvate kinase (PK)[8,9], 3-hydroxy-3,5,7-estratrien-17-one, a substrate analogue for 3-oxo-Δ^5-steroid isomerase,[10,11] and 2-hydroxymuconate, a substrate for 4-oxalo-crotonate tautomerase.[12] These compounds and their corresponding enzymes allow for a comprehensive examination of the solution behavior and also a comparison to the properties observed in the presence of the enzyme. Much of the rationale for such an examination of the solution behavior is to provide a chemical basis for an enzymatic reaction mechanism and to allow a comparison of the free energy profiles so that the enzyme's catalytic ability can be assessed.

This chapter will review first some highlights in the study of biologically relevant enols and dienols followed by a discussion of enzyme-catalyzed keto–enol and enone–dienol tautomerizations. It is well known that keto–enol and enone–dienol tautomerizations in aqueous solution are subject to catalysis by acid and base.[1-3] For this reason, it had been anticipated that many of the corresponding enzyme-catalyzed transformations would involve general acid and general base catalysis.[3] This presumption has been verified by experimentation on several isomerases. Hence this chapter will discuss the experimental basis for the mechanism of selected enzymes that proceed through an enolic or dienolic intermediate, the characteristics of the general acid and general base catalysts, and the prevailing hypotheses about how these catalytic strategies contribute to the overall rate acceleration. As these catalytic strategies unfold, it will be possible to compare and contrast them in order to ascertain whether they change as a function of the varying stabilities of the enolic intermediates.

5.03.2 KETO–ENOL TAUTOMERIZATION IN SOLUTION

Keto–enol tautomerization represents a classical area of study in organic chemistry.[1] The enolization of carbonyl compounds in the presence of acid or base has been studied extensively and the mechanisms and the supporting evidence are reviewed elsewhere.[1-3] The stability of the transition state in either acid- or base-catalyzed enolization or ketonization is influenced by a number of factors resulting in an effect on the rate of enolization and ketonization.[2] Among these factors is stereoelectronic control. Stereoelectronic control is a hypothesis that proton transfer between the oxygen and carbon atoms will occur perpendicular to the plane defined by the sp^2 orbitals of the carbonyl carbon.[2] Stereoelectronic control may also play a role in enzyme-catalyzed reactions in addition to that played by general acid or general base catalysis.

Studies of the ketonization of enols have not been as prolific because they are not as straight-forward. The primary obstacle is the fact that enols are not generally stable in the absence of bulky substituents.[4,5,13] Significant advances in the study of the ketonization of enols resulted when Chiang and Kresge[4] and Capon and Guo[14] developed methodology for the generation of enols in quantities that make their study feasible. Subsequent study of the enols of many simple aldehydes and ketones has provided the rate constants for ketonization, an accurate determination of the keto–enol equilibrium constants, and a determination of pK_a values for both the enols and ketones.[4] The major contribution of these studies is that they indicate which mechanisms are chemically plausible for the enzyme-catalyzed reaction. One example involves the racemization of mandelic acid catalyzed by the enzyme mandelate racemase (Equation (3)).[15] The enol of mandelic acid is believed to be an intermediate in the enzymatic reaction. Studies on the solution chemistry of the enol of mandelic acid indicate that a mechanism in which the enzyme generates the enol and then releases it into solution, where it undergoes spontaneous ketonization, is chemically unreasonable. In order for the enol to be an intermediate in the enzyme-catalyzed reaction, it must remain enzyme bound and be stabilized by nearly 15 kcal mol^{-1}.[16]

$$\text{(3)}$$

In contrast, some enols are surprisingly stable. One such enol is enolpyruvate (Equation (4)).[8,9] The compound has been generated by the action of acid phosphatase on phosphoenolpyruvate in acidic solutions and by the silylation of deprotonated pyruvic acid.[8,9] The latter compound, bis-TMS-enolpyruvate, is stable for several weeks at $-20\,°C$. Kinetic studies of enolpyruvate indicate that it undergoes buffer-catalyzed ketonization.[8,9] These solution studies also provide a partial explanation for the energy released by the hydrolysis of phosphoenolpyruvate (Equation (4)), an intermediate in glycolysis.[4] The energy of phosphoenolpyruvate is derived from a two-step process: the first step involves hydrolysis of phosphoenolpyruvate and the second ketonization of enol-pyruvate. Nearly half of the energy released by the conversion of phosphoenolpyruvate to pyruvate results from the second step.

$$\text{(4)}$$

5.03.3 ENONE–DIENOL TAUTOMERIZATION IN SOLUTION

Enone–dienol tautomerization has been the object of intense study primarily because it is involved in the isomerization of β,γ-unsaturated ketones to their α,β-isomers (Equation (2)). The equilibrium constants for isomerization, the mechanism of isomerization in the presence of acid, base, or amines, and the factors that affect the partitioning of the intermediate dienol are discussed extensively elsewhere.[3] Protonation can occur at either C-α or C-γ of the dienol and the partitioning factor (k_α/k_γ) is a measure of the relative kinetic rates of protonation at these two carbons. Although several factors affect the site of protonation in both the acid- and base-catalyzed mechanisms, steric hindrance to protonation and diene conformation are the only two factors relevant to the following discussion.[3]

As part of their efforts to elucidate the mechanism of the 3-oxo-Δ^5-steroid isomerase-catalyzed reaction, Pollack and co-workers investigated the relationships between the structure and reactivity of several enols, dienols, and trienols.[3,10,11,17,18] Among the compounds studied were a trienol inter-mediate involved in the conversion of 5,7-estradien-3,17-one (**1**) to 4,7-estradien 3,17-one (**3**) (Scheme 1) and a dienol intermediate generated in the conversion of 5-androstene-3,17-dione (**4**) to 4-androstene-3,17-dione (**6**) (Scheme 2).[10,19] The trienol 3-hydroxy-3,5,7-estratrien-17-one (**2**) is isolated as a solid whereas the dienol 3-hydroxy-3,5-androstadien-17-one (**5**) is generated either by the action of acid phosphatase on the corresponding dienol phosphate or by the rapid buffer quenching of the base-treated (**4**).[10,19] The dienol can also be observed spectroscopically ($\lambda_{max} = 256$ nm) in the course of the base-catalyzed isomerization of (**4**) to its α,β-isomer (**6**).[17]

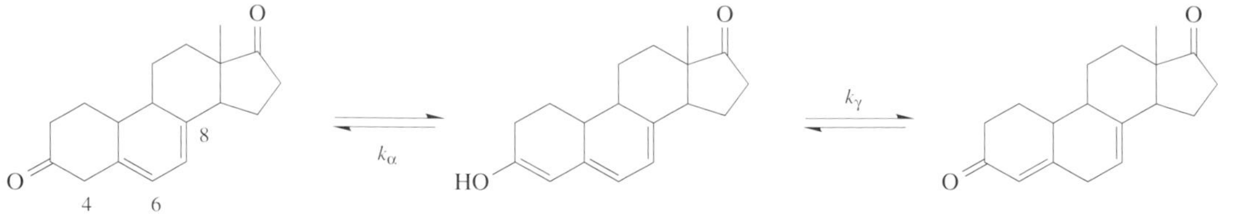

Scheme 1

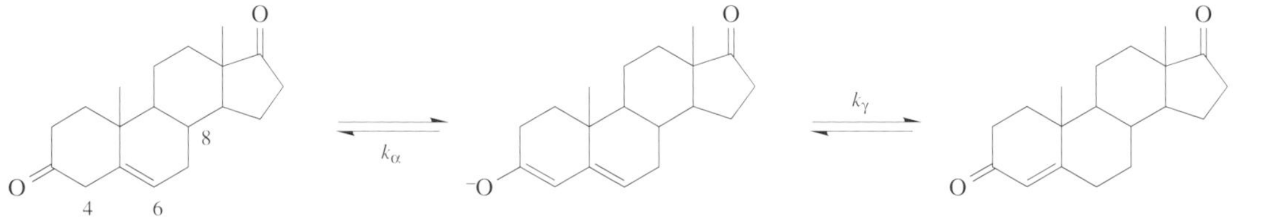

Scheme 2

An extensive kinetic investigation into the ketonization of (**2**) and (**5**) has been performed.[11,18] In aqueous buffer, protonation of (**2**) can occur at either C-4 to generate (**1**), C-6 to generate (**3**), or C-8 to generate 4,6-estradien-3,17-one. Steric hindrance to protonation presumably precludes the fact that protonation is not observed at C-8.[11] In basic solutions, protonation of (**2**) and (**5**) is favored at C-4 (C-α) to generate the β,γ-unsaturated ketones ((**1**) and (**4**), respectively).[18] Facile protonation at the C-α position is typically observed for dienolate ions. For these two compounds, the charge density is not effectively delocalized throughout the π-system of (**2**) or (**5**) because of the twisting of the single bond joining the two double bonds.[11,18] Hence the enzyme, in order to generate its product, the α,β-unsaturated isomer ((**3**) and (**6**), respectively), must overcome the kinetic barrier to protonation at C-6 or C-γ.

The kinetic analysis also allowed for a determination of the pK_a values for (**4**) (12.7) and (**6**) (16.1).[3] These compounds are more acidic than saturated ketones, indicating that the effect of a β,γ-double bond is comparable to that of a phenyl group.[3] The acidifying effect of an α,β-double bond is not as great but still substantial. The implication of this observation for the enzyme-catalyzed reaction is that it suggests that a dienolate intermediate is possible in the reaction because protonation is not required at the oxygen for stabilization.

The pathways involved in the microbial degradation of aromatic compounds are replete with many isolable dienols.[20] A sampling of these dienols and their keto isomers is presented in Schemes 3 and 4. Each of these compounds is a substrate or an intermediate for an enzyme in the pathway. The availability of these compounds and a corresponding enzyme provides an ideal opportunity to study the properties of the intermediate dienols in the absence and presence of an enzyme. In addition, many of these dienols are considerably more stable than the steroidal compounds discussed above (i.e., (**2**) and (**5**)). Although extensive studies have not yet been carried out on the nonenzymatic properties of these dienols, some qualitative observations can be made.

2-Hydroxymuconate (**8**) and 5-(carboxymethyl)-2-hydroxymuconate (**11**) (Scheme 3) are readily synthesized and are stable for several years as solid free acids.[12,21] In aqueous buffer, however, a rapid equilibrium forms between these dienols and their respective β,γ-unsaturated ketones (2-oxo-4-hexenedioate (**7**) and 5-(carboxymethyl)-2-oxo-4-hexenedioate (**10**) before a much slower conversion of the mixture to the α,β-unsaturated isomers (2-oxo-3-hexenedioate (**9**) and 5-(carboxymethyl)-2-oxo-3-hexenedioate (**12**)).[12,21] The composition of the equilibrium mixtures and the partitioning factors for (**8**) and (**11**) are summarized in Table 1. Introduction of a carboxymethyl group at C-5 affects both K_{eq} and the partitioning factor. The carboxymethyl group shows significant inductive destabilization of (**12**) so that the quantity present at equilibrium is less than that of

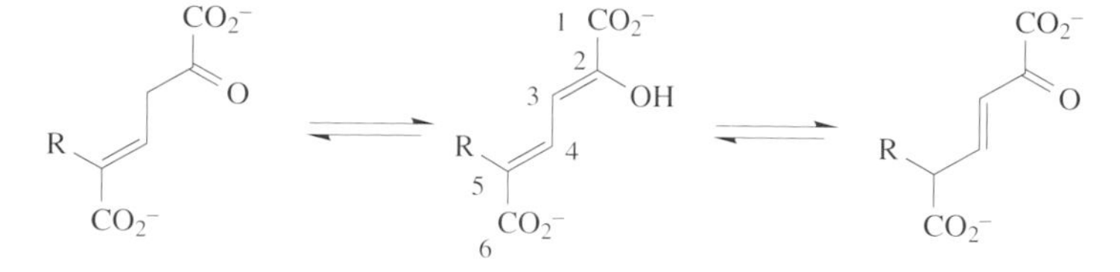

Scheme 3

(7) R = H
(10) R = CH$_2$CO$_2^-$

(8) R = H
(11) R = CH$_2$CO$_2^-$

(9) R = H
(12) R = CH$_2$CO$_2^-$

Scheme 4

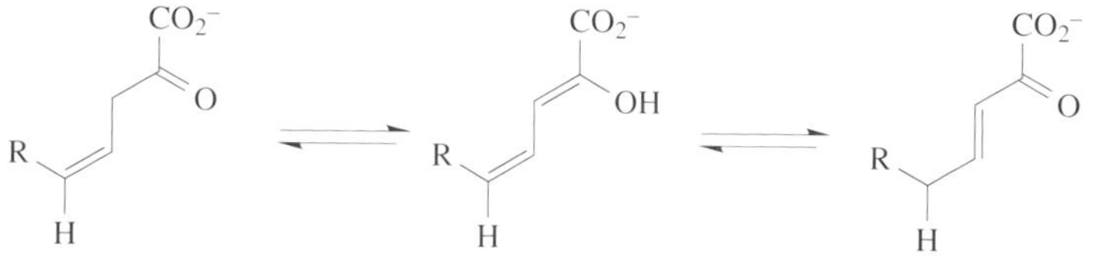

(13) R = CH$_2$CO$_2^-$
(16) R = H

(14) R = CH$_2$CO$_2^-$
(17) R = H

(15) R=CH$_2$CO$_2^-$
(18) R = H

(9).[3,12,21] For both dienols, kinetic protonation is clearly favored at C-3 or C-α. The partitioning factor for (8) is governed by the steric hindrance to protonation at C-5, slight twisting of the bond between the double bonds, and the electronic effect of the C-6 carboxylate group.[12] The larger partitioning factor for (11) is most likely due to two factors: the additional steric bulk at C-5 and a significant amount of twisting of the bond between the double bonds in order to position the two carboxylate groups (C-6 and C-8) at a favorable distance.[21] Finally, the kinetic analysis of the solution behavior of (8) permitted an estimate of the pK_a values for the proton at C-3 of (7) ($\sim$ 12) and for the proton at C-5 of (9) ($\sim$ 13).[22] Again, a dienolate intermediate is a viable entity in the reaction.

Table 1 Composition of equilibrium mixture and partitioning factors (k_α/k_γ) for 2-hydroxymuconate (8) and 5-(carboxymethyl)-2-hydroxymuconate (11).

Compound	Composition (at equilibrium)	k_α/k_γ	Ref.
(8)	8.2% (7); 11.3% (8); 80.5% (9)	$\sim$7–12	12
(11)	16.7% (10); 32.5% (11); 50.8% (12)	$\sim$46	21

These observations provided a starting point for studies of the enzyme-catalyzed ketonization of (8) and (11). For both dienols, the equilibrium constants indicate that the α,β-conjugated isomers are favored thermodynamically, whereas the partitioning factors indicate that the formation of the β,γ-unsaturated ketones is favored kinetically.[12,21] Hence, the enzymes that utilize (8) and (11), 4-oxalocrotonate tautomerase and 5-carboxymethyl-2-hydroxymuconate isomerase (CHMI), respectively, must surmount this kinetic barrier in order to generate an α,β-conjugated isomer which serves as the substrate for the next enzyme in the pathway.

Two other dienols in these pathways subjected to investigation have different equilibrium constants, partitioning factors, and enzymatic properties from those observed for (8) and (11). One compound, 2-hydroxy-2,4-heptadiene-1,7-dioate (14) (Scheme 4) has been isolated as a solid while the second compound, 2-hydroxy-2,4-pentadienoate (17), has been generated as a mixture of (17) and 2-oxo-3-pentenoate (18) by the thermal decomposition of (8).[22–24] With regard to K_{eq}, the α,β-unsaturated isomers (2-oxo-3-heptene-1,7-dioate (15) and (18)) are the predominant species detectable (>95%) at equilibrium.[22,23] In addition, an estimate of the partitioning factors for (14) and (17) suggests that the β,γ-isomers, 2-oxo-4-heptene-1,7-dioate (13) and 2-oxo-4-pentenoate (16), are generated almost exclusively as the concentrations of (15) and (18) cannot be readily measured because of the slow rate of protonation at C-5 (C-γ).[22,23] For example, although (15) is the predominant product at equilibrium, it constitutes only about 10% of the mixture after 8 h. These dienols have properties similar to those observed for 1,3-cyclohexadienol.[25] Finally, in their corresponding enzymatic reactions, (14) and (17) are converted into their β,γ-isomers and not their α,β-isomers, in contrast to the enzymatic reactions utilizing (8) and (11).

5.03.4   UNSTABLE ENOLATE INTERMEDIATES: TRIOSE PHOSPHATE ISOMERASE

5.03.4.1   Identification of the General Base and General Acid Catalysts

Triose phosphate isomerase (TIM) catalyzes the interconversion of dihydroxyacetone (19) and glyceraldehyde 3-phosphate (21) through an enzyme-bound enediolic intermediate (20) (Scheme 5).[26,27] The enzyme plays a central role in glycolysis and its mechanism and energetics have fascinated enzymologists since the 1950s. Affinity labeling with bromohydroxyacetone phosphate and glycidol phosphate, site-directed mutagenesis, and crystallographic studies demonstrate that glutamate-165 acts as the general base catalyst to abstract the *pro-R* hydrogen proton at C-1 of (19).[26] Spectroscopic, site-directed mutagenesis, crystallographic, and pH studies indicate that histidine-95 provides a neutral imidazole side chain that acts as the general acid catalyst.[26,27]

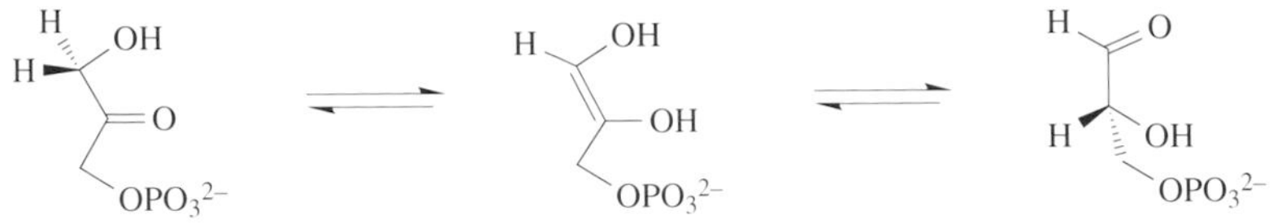

Scheme 5

The TIM reaction illustrates a catalytic strategy used by an enzyme to effect catalysis through the stabilization of an unstable enol intermediate.[7,27] Because the intermediate is not a stable isolable species, its existence and configuration are implicated by indirect evidence typically used to suggest the involvement of an enol intermediate in an enzymatic reaction. The enzyme catalyzes the exchange of the C-1 proton of (19) (or the C-2 proton of (21)) for a deuteron when the reaction is carried out in ^2H_2O. In addition, there is a small but real intramolecular transfer of a proton between C-1 of (19) and C-2 of (21) indicative that a single residue on the enzyme abstracts the proton from either C-1 of (19) or C-2 of (21).[26] Finally, the stereochemical course of the enzyme-catalyzed reaction is *syn*, which is consistent with a single-base mechanism and implies that the enediol intermediate has the *cis* configuration.[26]

The role of Glu-165 was established early in the studies of TIM while the role of His-95 as the general acid catalyst was not clearly determined until several years later. His-95 was first implicated by the observation that it is responsible for the polarization of the carbonyl group of (19) as indicated by a reduction of the IR stretching frequency.[28–30] In addition, crystallographic studies position His-95 within bonding distance of O-1 (2.9 Å) and O-2 (2.8 Å) of 19 such that it can shuttle protons between these positions.[31] Finally, ^{15}N NMR studies show that the imidazole ring is uncharged over the entire pH range of isomerase activity and that a strong hydrogen bond exists between $N^{\epsilon 2}$ of His-95 and the bound inhibitor, phosphoglycolohydroxamate (22) (Scheme 6).[29–31] This inhibitor, a stable analogue of (20), is thought to be the best approximation of the structure of the enzyme in its catalytically competent form.[31,32]

The results of the ^{15}N NMR titration of His-95 indicate that its first pK_a (imidazolium to imidazole) is <4.5, which means it is at least 2 pK_a units from that measured for His-95 in the denatured enzyme.[33] The crystal structure shows that His-95 lies at the amino-terminus of a short α-helix (residues 95–102).[21,31,32] The amino-terminal position is near the positive end of the helix dipole and this proximity accounts for the shift in pK_a value.

5.03.4.2   Stereoelectronic Control

Stereoelectronic control also plays a role in the mechanism of TIM. The orientation of (19) in the active site has been inferred from the binding of (22) in all TIM crystal structures determined to

Scheme 6

date (1997).[32,34–37] These structures show that (**22**) is bound in a coplanar, *cis* conformation. The phosphate ester is in the same plane as the carbonyl group and the hydroxamate hydroxy group. Hence (**19**) is positioned in the active site so that the *pro-R* hydrogen is perpendicular to a plane defined by O-1, C-1, C-2, and O-2 (Scheme 7). The phosphate ester of (**19**) is in the same plane.[35] In this position, the elimination of the phosphate group is not an energetically favored reaction.[38] Glu-165 lies just above the plane. This position of substrate allows for its facile enolization to afford the planar intermediate (**20**) and disfavors the loss of phosphate to afford methylglyoxal.[38,39]

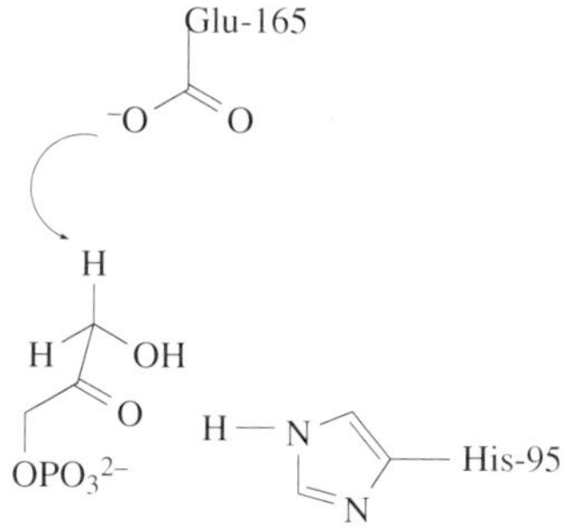

Scheme 7

The contribution of stereoelectronic control to the mechanism of TIM was established partially by an analysis of the structural basis for the loss of activity of the H95N mutant isomerase and the improvement in activity that occurred in the H95N/S96P mutant isomerase.[34,35] The crystal structure of (**22**) bound to the H95N mutant isomerase shows that (**22**) is no longer bound in a coplanar, *cis* conformation.[30,35] The structure implies that the substrate will no longer be bound in a coplanar, *cis* conformation and that the formation of the planar enediolate intermediate in the active site of this mutant will be difficult. This interpretation is consistent with the observation that the H95N mutant isomerase produces three molecules of methylglyoxal for every 10 turnovers to (**19**) or (**21**). The H95N/S96P mutant isomerase regains its ability to bind (**22**) in a coplanar, *cis* conformation so that it is similar to that observed for the wild-type isomerase. As a result, the enediolate intermediate is readily formed. The return of (**19**) to the conformation observed in the wild-type isomerase results in an improvement in catalysis. The precise contribution of stereoelectronic control is difficult to quantify in these mutants because of the role that His-95 plays as a general acid catalyst.[35]

The issue of stereoelectronic control was also addressed by the deletion of four residues (residues 170–173) of a mobile 10-residue loop (residues 168–177) that folds down on bound substrate or bound (**22**).[39] Two residues (Gly-171 and Gly-173) in the deleted segment are believed to be involved in the binding of the phosphate group of substrate through main-chain hydrogen bonds. Deletion of these residues results in an enzyme that catalyzes the elimination of phosphate from either substrate better than it catalyzes the isomerization reaction. Hence it is postulated that these residues either lock the phosphate group in a conformation that disfavors elimination or prevent the release of (**20**) into solution.[39]

5.03.4.3 The Role of General Base–General Acid Catalysis in the Mechanism

A central question about the TIM-catalyzed reaction is how the enzyme generates a sufficiently large quantity of the enediolate intermediate to achieve its rapid rate of reaction using a general base (Glu-165) with a pK_a of ~ 6 to abstract the C-1 hydrogen from (**19**), which has a pK_a of

~ 18–20.[6,7,27,40] One explanation involves the abstraction of the C-1 hydrogen of (19) by Glu-165 in concert with the protonation of the carbonyl group at C-2 by His-95, as explained in Chapter 5.02. Concerted general acid–base catalysis results in an intermediate which is stabilized by a strong, low-barrier hydrogen bond (LBHB).[6,7,27,40] One characteristic of an LBHB is the matched pK_as of the donor (in this case, His-95) and acceptor (*cis*-enediolate (20)).[6,7,27,40] The pK_a of the *cis*-enediolate (20) has been estimated to be ~ 10–11. If the neutral form of His-95 is the general acid catalyst, then its pK_a is ~ 11. Hence the pK_a of the hydroxy group of (20) and that of the general acid catalyst are matched.

The involvement of an LBHB in this reaction has been the subject of much debate. Others have argued that electrostatic stabilization of the enediolate intermediate can provide the necessary energy to enable a rapid reaction.[41,42] These issues are presented and discussed in Chapter 5.02.

5.03.5   METAL-STABILIZED ENOLATE INTERMEDIATES: PYRUVATE KINASE

PK catalyzes the two-step reaction shown in Scheme 8.[9] The first step involves phosphoryl transfer from phosphoenolpyruvate (23) to adenosine diphosphate (ADP) to afford the magnesium-stabilized enolate of pyruvate (enolpyruvate (24)). The second step involves the ketonization of (24) to pyruvate (25) by the addition of a proton to the *2 si* face of (24).[43] Enolpyruvate is a putative intermediate in other reactions, including those catalyzed by phosphoenolpyruvate carboxylase, pyruvate–phosphate dikinase, malic enzyme, and oxaloacetate decarboxylase.[8,44] In addition, vinylogous enolpyruvates are implicated in the two decarboxylation reactions shown in Scheme 9.[45] The decarboxylation of (9) by 4-oxalocrotonate decarboxylase affords (17) as an intermediate and the decarboxylation of (12) by 5-(carboxymethyl)-2-oxo-3-hexene-1,6-dioate decarboxylase generates the isolable (14) as an intermediate.[24,25] The fact that both enzymes require metal for activity suggests that the intermediate dienolates are also metal stabilized.

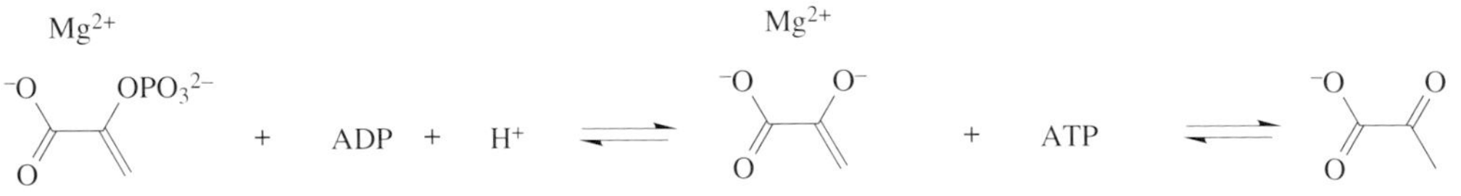

Scheme 8

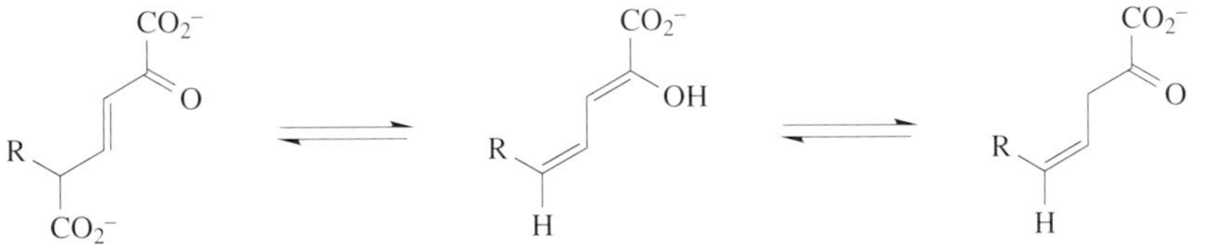

| (9) R = H | (17) R = H | (16) R = H |
| (12) R = CH$_2$CO$_2^-$ | (14) R = CH$_2$CO$_2^-$ | (13) R = CH$_2$CO$_2^-$ |

Scheme 9

The availability of a crystal structure makes the PK-catalyzed reaction an attractive model to study how enzymes such as those listed above make use of metal complexation to generate and ketonize a moderately stable enol.[46,47] Based on model systems, the metal ion is proposed to have a twofold purpose: it decreases the pK_a of the proton at C-3 of (25) by 4–6 pK_a units, which increases the quantity of enolate present for reaction, and it delocalizes the charge density, which stabilizes the enolate species.[44] Unfortunately, the studies of PK are not as advanced as those of other enzymes discussed in this chapter so that little is known about how active site residues lend assistance. However, as these studies progress, a better understanding of the catalytic strategy should emerge.

The identification of (24) as an intermediate in the PK-catalyzed reaction has been demonstrated by several experiments, including the stereospecific ketonization of the exogenously added (24) by the enzyme and the chemical derivatization of the enzyme-bound intermediate.[8,43,48] The rate of ketonization of the exogenously added (24) is estimated to be about 11% of the rate of the overall reaction.[9] The reasons for the apparent kinetic incompetence of this putative intermediate are explained elsewhere.[49]

Two crystal structures of PK shed light on the identity of the general acid catalyst in the PK-catalyzed ketonization of enolpyruvate.[46,47] The first crystal structure of PK, of limited resolution, was solved in the presence of (23) and suggested that Lys-269 acts to protonate the enolate intermediate.[46] This implication is consistent with two observations: exchange data show that three equivalents of tritium are found in pyruvate after a solution of PK in tritiated water is diluted into the components of the assay mixture and pH–rate profiles show that a titratable group with a pK_a of 8.3 is required for activity.[50] No other suitable candidates are located in the area of the bound (23). A second crystal structure, solved in the presence of pyruvate, Mn^{2+}, and K^+, suggests another candidate.[47] In this structure, the amino group of Lys-269 and the hydroxy group of Thr-327 can be placed on either side of the chelated pyruvate.[47] Thr-327 is proposed to act as the general acid catalyst that protonates the double bond of the enolate for one critical reason: its position is consistent with the observed stereospecific protonation of the 2 si face of the enolate. The pK_a of 8.3 can be assigned to Lys-269, but it is proposed that this lysine may interact with the γ-phosphate of adenosine triphosphate.[47]

5.03.6 MODERATELY STABLE DIENOLATE INTERMEDIATES: 3-OXO-Δ^5-STEROID ISOMERASE

5.03.6.1 Identification of the General Base and General Acid Catalysts

3-Oxo-Δ^5-steroid isomerase (known also as Δ^5-3-ketosteroid isomerase or KSI) catalyzes the isomerization of a variety of β,γ-unsaturated 3-oxo steroids (e.g. (4)) (Scheme 2) to their thermodynamically favored conjugated isomers (6) (Scheme 2).[51] The reaction is of particular interest because it is among the fastest known enzyme-catalyzed reactions with a second-order rate constant near the diffusion-controlled rate. KSI provides an example of how an enzyme stabilizes a dienolic intermediate that is moderately stable in solution.

The mechanistic studies of the enzyme prior to 1990 are summarized elsewhere.[51] These studies established that KSI catalyzes a nearly stereospecific intramolecular transfer of the 4β proton of (4) to the 6β position of (6) by a single general base catalyst identified as Asp-38.[51,52] Abstraction of the 4β proton generates the dienolate intermediate (5), which is stabilized by Tyr-14 acting as a general acid catalyst. Ketonization of the dienolate intermediate to afford (6) occurs by protonation at the C-6 position by Asp-38, now acting as a general acid catalyst. Having determined the mechanics of the reaction, much discussion has focused on the nature of the intermediate (dienol vs. dienolate) and how KSI achieves its $\sim 10^{10}$-fold rate enhancement.[19]

5.03.6.2 The Dienolic Intermediate

A dienolic intermediate was initially hypothesized to be involved in the KSI-catalyzed reaction on the basis of solution chemistry, spectroscopic studies, and isotopic labeling studies.[19] Subsequently, three lines of evidence were obtained to support its existence. First, a trienol compound (2), a substrate analogue of the dienol, is converted into (3) by KSI.[10] Second, the dienol compound (5) is partitioned by KSI to a mixture of (4) and (6). Moreover, (5) is processed to (6) faster than (4) is processed to (6), indicating that the dienol is kinetically competent to be an intermediate in the overall reaction.[19] Finally, an intermediate with spectral characteristics of both a dienol and a dienolate species is observed spectroscopically in the reaction of the D38N mutant enzyme with (4). The species is detectable because of the differences in the rates of enolization and ketonization.[53] The enolization of KSI-bound (4) to enzyme-bound (5) is much faster than the subsequent turnover of the enzyme-bound (5) to (6). Consequently, the enzyme–(5) complex has an appreciable lifetime, making its detection and study possible.[53,54]

Whether the intermediate is a dienol or a dienolate has been the subject of many experiments and the topic of much discussion.[55–57] The relatively low pK_a of the C-4 proton of (4) suggests that a dienolate intermediate has sufficient stability to exist at the active site of the enzyme.[3] More compelling evidence implicating a dienolate intermediate comes from the study of the spectral characteristics of 3-amino-1,3,5(10)-estratrien-17β-ol (25) and equilenin (26) (Scheme 10) in the presence of the enzyme.[56] These compounds are analogues of the dienol (5) and can be protonated (25) or deprotonated (26) when bound to KSI. There is no evidence for the protonation of the amine in (25) when bound to KSI. Moreover, the bound species of (26) is anionic.[56] These observations make the inference that the intermediate in the KSI reaction is a dienolate species.

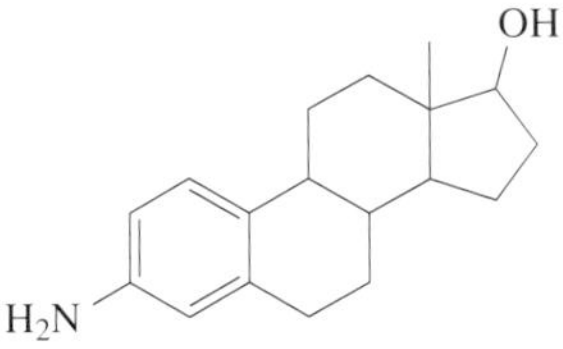

Scheme 10

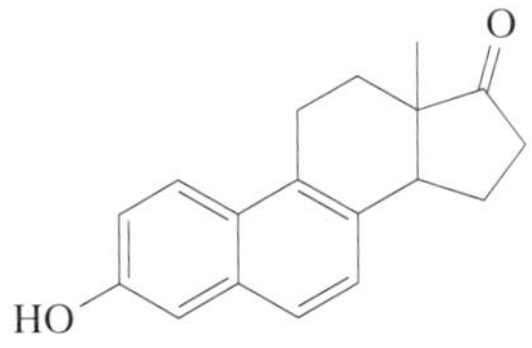

A different perspective on the nature of the intermediate results from a study following the spectral characteristics of three competitive inhibitors of KSI, 19-nortestosterone (27) (Scheme 11), 17β-estradiol (28), and 17β-dihydroequilenin (29), when bound to the enzyme.[57] These compounds are nonreactive analogue intermediates. Each compound undergoes significant polarization upon binding to KSI. The spectral change observed upon the binding of (27) to the enzyme resembles that observed when (27) is exposed to strong acid. The spectral changes observed upon binding of (28) and (29) to the enzyme resemble those of ionized phenolate species in basic solution. Because the full protonation of (27) or the full ionization of (28) or (29) is unfavorable in a solvent-inaccessible active site, the possibility that these observed spectral changes result from strong hydrogen bonding in a nonpolar environment was also investigated. The study of model compounds suggested that such hydrogen bonding between the enzyme and these compounds could account for the observed changes. The conclusion argues against either a dienol or a dienolate intermediate in the KSI-catalyzed reaction and implicates a strong directional hydrogen bond.[57,58]

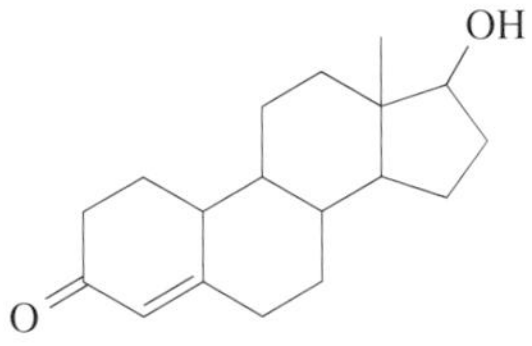

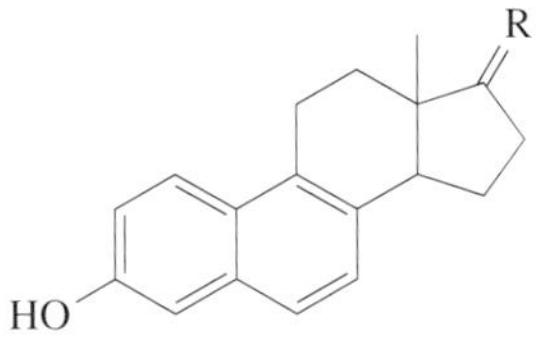

(29) R = α-H, β-OH

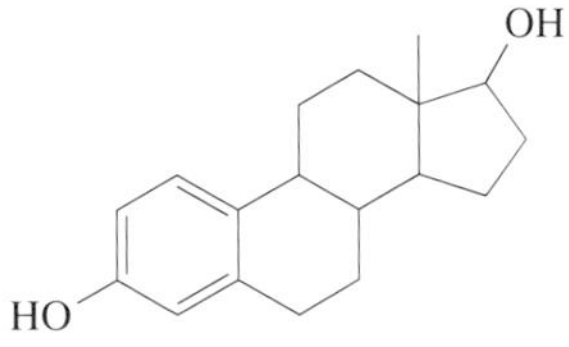

(30) R = O

Scheme 11

5.03.6.3 Does Catalysis Involve a Low-barrier Hydrogen Bond?

The strong directional hydrogen bond proposed to account for the observed spectral characteristics between KSI and bound (27), (28), or (29) hinted that an LBHB forms between Tyr-14 and the 3-carbonyl oxygen of the substrate (4). NMR spectroscopy provides experimental evidence for the existence of LBHBs because the protons show chemical shifts of 17–20 ppm.[40] Accordingly, a highly deshielded NMR signal at 18.15 ppm is observed for a complex of the D38N mutant of KSI and (29), an analogue of the dienolate intermediate.[58] The strong deshielding of the proton results from its almost symmetrical position between two oxygen atoms (the phenolic oxygen of Tyr-14 and the phenolic oxygen of (29)), as shown in Scheme 12. The deshielded proton is proximal to the aromatic protons of both Tyr-14 and (29), as indicated by nuclear Overhauser effect studies.[58] The signal is not observed in a complex of (29) and a mutant isomerase in which Tyr-14 is replaced with phenylalanine. Finally, the bond is not accessible to bulk solvent, as indicated by its exchange behavior. On the basis of these observations, it is concluded that an LBHB exists between Tyr-14 and the dienolate intermediate.[58]

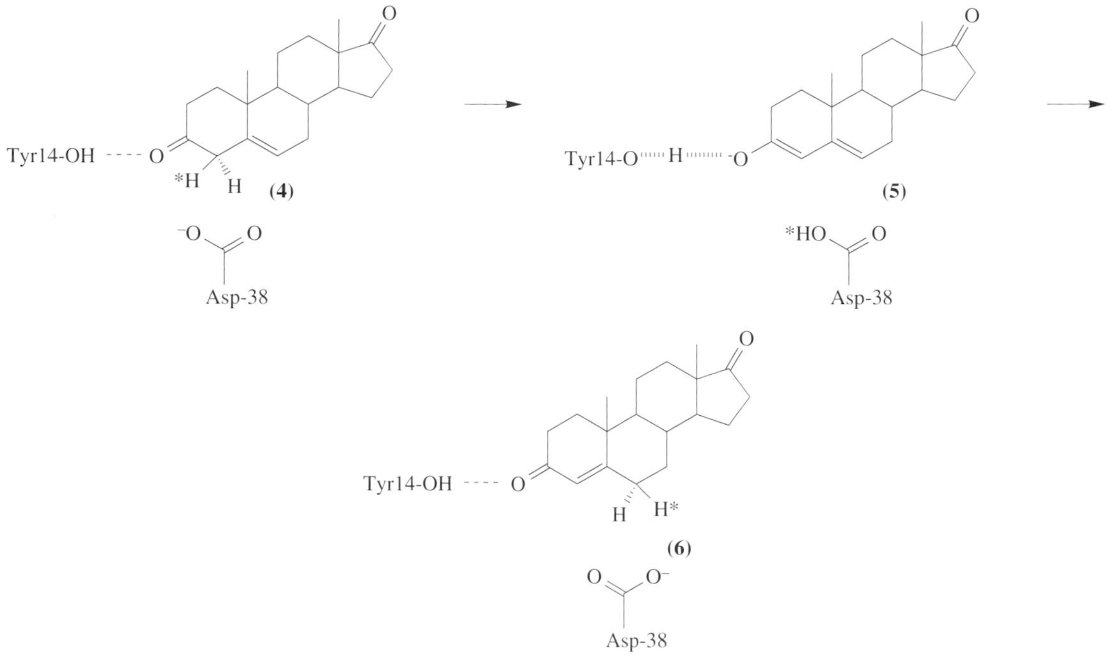

Scheme 12

The participation of an LBHB in the KSI-catalyzed reaction has been challenged. Pollack and co-workers[59] reported a three-dimensional solution structure for KSI and implicated another residue, Asp-99, in the stabilization of (5) and the transition state. When the substrate (4) is docked computationally into the active site of the NMR structure in a position consistent with the roles of Asp-38 and Tyr-14 as well as the observed stereochemistry of the reaction, it is noted that the carboxylate group of Asp-99 is 3 Å from the 3-carbonyl group of (4).[59] This observation prompted a proposal that the carboxylate group of Asp-99 forms a second hydrogen bond to the oxygen of the dienolate intermediate (5) as shown in Scheme 13. Mutagenesis of Asp-99 to alanine results in a 5000-fold reduction in k_{cat} and a 3000-fold reduction in k_{cat}/K_M, consistent with a role for Asp-99. In contrast to a previous proposal that the full rate enhancement of KSI can be attributed to Asp-38 and Tyr-14,[60,61] this model suggests that the rate enhancement results from these two residues in addition to Asp-99 and Phe-101. Phe-101 was implicated in the mechanism in an earlier study.[62] Thus, it is proposed that two hydrogen bonds of moderate strength and the contribution of Phe-101 can stabilize the intermediate dienolate as opposed to a single LBHB.[63,64]

Whether or not an LBHB plays a role in catalysis will require further experimentation, as several issues remain unresolved. For example, the mechanism shown in Scheme 13 is based partially on an observation obtained from a computationally docked substrate which ignores possible conformational changes upon substrate binding. A second issue concerns the origin of the signal at 18.15 ppm attributed to an LBHB. A similar signal observed at 17.4 ppm for a wild-type–(30) complex disappears when Asp-99 is replaced with alanine, suggesting that Asp-99 may play a role in the deshielded signal.[59,65] The structural integrity of the D99A mutant has not been analyzed so that the implication of this observation is not yet clear.

5.03.7 STABLE DIENOL(ATE) INTERMEDIATES: 4-OXALOCROTONATE TAUTOMERASE

5.03.7.1 Background

4-Oxalocrotonate tautomerase (4-OT) was initially reported to catalyze the ketonization of 2-hydroxymuconate (8) to the α,β-unsaturated ketone, 2-oxo-3-hexenedioate (9) (Scheme 3).[20,66] The

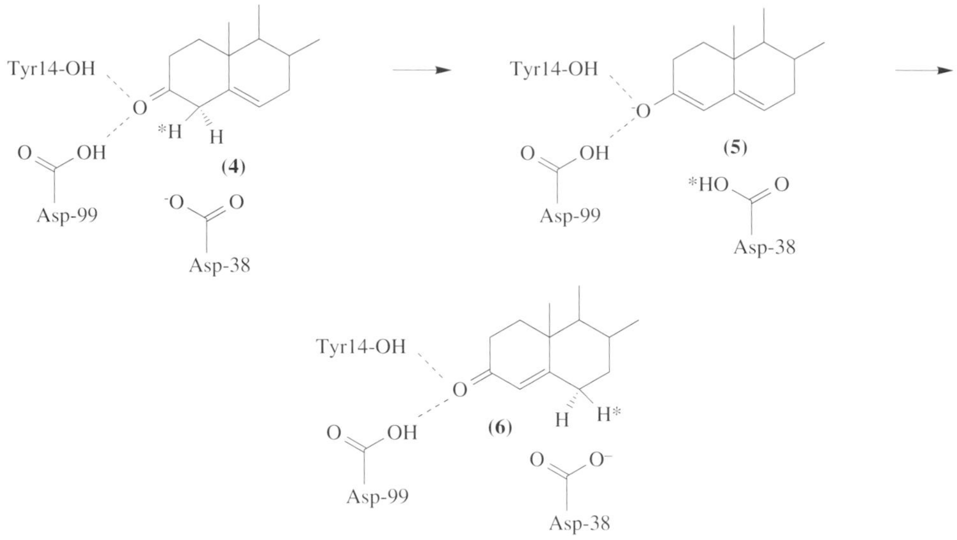

Tyr14-OH

O

O   OH

Asp-99

*H   H

(4)

O   O

O

Asp-38

→

Tyr14-OH

O

O   OH

Asp-99

(5)

*HO   O

O

Asp-38

→

Tyr14-OH

O

O   OH

Asp-99

(6)

H   H*

O   O⁻

Asp-38

Scheme 13

enzyme is expressed as part of a set of inducible enzymes that oxidatively catabolize aromatic hydrocarbons such as toluene (TOL), naphthalene, and phenol to intermediates in the Krebs cycle.[67,68] The entire pathway is plasmid-encoded in toluene-, naphthalene-, and phenol-degrading soil bacteria and enables these organisms to utilize one of these aromatic hydrocarbons as their sole source of carbon and energy.[20,67,68]

Two isozymes of 4-OT have been cloned: one from the TOL plasmid pWW0 in *Pseudomonas putida mt-2* (toluene-degrading) and the other from the plasmid pVI150 harbored in *Pseudomonas* sp. strain CF600 (phenol-degrading).[67,68] Overexpression and purification of these isozymes resulted in the isolation of a hexamer of identical subunits, each containing 62 amino acid residues.[68,69] The two isozymes share 73% sequence homology.[68] The mechanistic studies have been largely carried out on the enzyme from *P. putida mt-2*.

In aqueous buffer, a rapid equilibrium forms between (8) and 2-oxo-4-hexenedioate (7) before a much slower conversion of the mixture to (9).[12] At equilibrium, a mixture of isomers (7)–(9) results, with (9) being the predominant species. The finding that an appreciable concentration of (7) accumulates in buffer raises the question of whether 4-OT acts as an isomerase to convert (7) to (9) through (8) or as a tautomerase that catalyzes the conversion of (8) to (9) (Scheme 3). A straight-forward approach to address this question is complicated by the fact that (7) cannot be synthesized or isolated—it can only be observed in equilibrium with (8).

Kinetic studies following the action of 4-OT on (8) and a mixture of (7) and (8) suggest that the enzyme behaves as an isomerase.[12] The key observation is that the rate of product (9) formation is faster when the enzyme acts on a mixture of (7) and (8) than when the enzyme acts only on (8). In addition, there is no detectable lag time when 4-OT processes (8) to (9). These observations are reflected in the values of k_{cat}/K_M determined for (7) and (8), which show that both reactions are catalyzed near the diffusion-controlled limit although the enzyme shows a preference for (7). The working hypothesis used to explain these kinetic results is that 4-OT converts (7) to (9) through the dienolic intermediate (8).

As an isomerase, 4-OT generates a stable dienolic intermediate (8) from an unstable substrate (7). In this regard, the mechanism for 4-OT may prove to be an interesting contrast to that for either TIM or KSI. Unfortunately, the mechanism for 4-OT is not as well understood as those for TIM and KSI. The identity of the general base catalyst in the 4-OT-catalyzed reaction is secure, but that of the general acid catalyst is not. Further study will be necessary in order to determine how these features contribute to the catalytic power of the enzyme.

5.03.7.2   Stereochemical Studies Implicate an Isomerase Mechanism

Further support for an isomerase mechanism came from stereochemical studies. Because the substrate (7) has not been synthesized or isolated, these studies were performed by utilizing the observation that (8) is an intermediate in the overall reaction and should be partitioned stereospecifically by the enzyme to 3-[^{2}H] (7) and 5-[^{2}H] (9) in ^2H_2O. Using this strategy, it was shown that 4-OT ketonizes (8) to (5S)-5[^{2}H] (9) (Scheme 14).[70] However, it was not possible to do a stereochemical analysis of 3-[^{2}H] (7) because insufficient quantities were available.

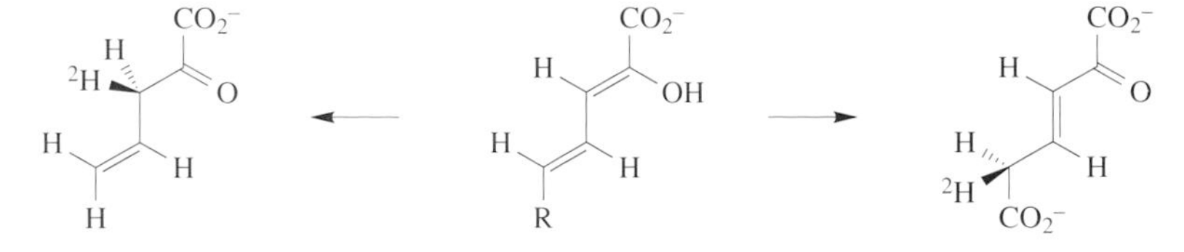

(3R)-[3-^{2}H] (16)   (8) R = CO_2^- (5S)-[5-^{2}H] (9)

(17) R = H

Scheme 14

An alternate strategy was devised based on the observed nonenzymatic and enzymatic properties of 2-hydroxy-2,4-pentadienoate (17) (Scheme 4).[22] In aqueous buffer, (17) ketonizes to the β,γ-unsaturated ketone (16) before a much slower conversion to its α,β-isomer (18). Although 4-OT accelerates both processes, the formation of (16) is significantly faster. This observation, coupled with the fact that (16) accumulates in solution, suggested that the ketonization of (17) to (16) could be used to assign the stereochemical course of 4-OT. Accordingly, [3-^{2}H] (16) was generated by the 4-OT-catalyzed ketonization of (17) in ^2H_2O. The resulting stereochemical analysis indicates that 4-OT ketonizes (17) stereoselectively to (3R)-[3-^{2}H] (16) (Scheme 14).[22] This result and the stereochemical outcome above imply that the isomerization of (7) to (9) is a *syn* process consistent with a one-base mechanism.[22,51]

5.03.7.3 Identification of the General Acid and General Base Catalysts

The identity of the residue responsible for proton transfer remained unknown until a 1.9 Å resolution crystal structure of the 4-OT isozyme from *Pseudomonas* sp. strain CF600 was solved.[71,72] The hexameric structure of the enzyme is arranged as a trimer of dimers. Each dimer consists of two α-helices and four β-strands forming a β-sheet. Each monomer consists of two parallel β-strands and one α-helix. Starting at the N-terminal proline, Pro-1 to Ile-7 form a β-strand. A short loop leads to the α-helix (Asp-13 to Leu-31). Another short loop leads to a second β-strand that begins at Arg-39 and ends at Met-45. The remaining amino acids form the C-terminus loop.[72]

Although the structure was solved in the absence of a bound ligand, Wigley and co-workers[72] identified a region on the surface of each subunit as the active site and postulated the mechanism shown in Scheme 15. The amino-terminal proline acts as the general base and two arginines (Arg-39 and Arg-11) bind the two carboxylate groups of substrate (C-1 and C-6, respectively). Pro-1 and Arg-39 are from the same subunit whereas Arg-11 is from an adjacent subunit. It was further proposed that Arg-39 functions as a general acid catalyst in addition to its role in binding a carboxylate group as shown in Scheme 16. The mechanism assumes that a large conformational change does not occur upon substrate binding.

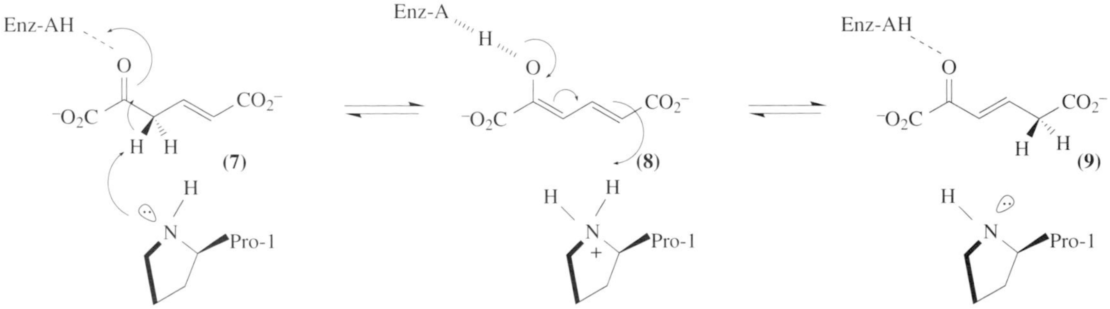

Scheme 15

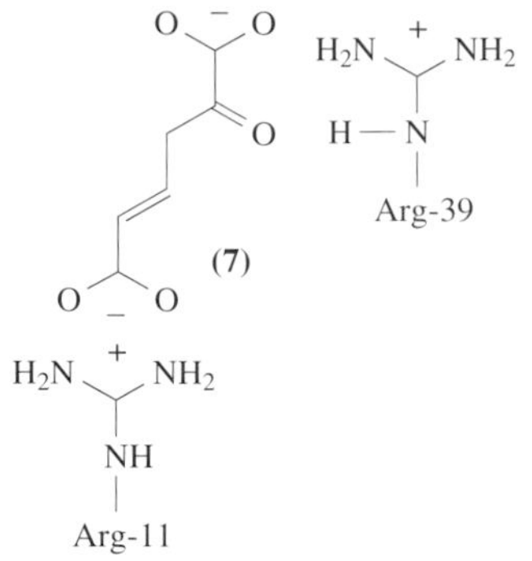

Scheme 16

The role of Pro-1 in the mechanism was confirmed by three observations.[73,74] First, 3-bromo-pyruvate (**31**) (Scheme 17) acts as an active-site-directed irreversible inhibitor which modifies the amino-terminal proline with a stoichiometry of one site per monomer.[73] Second, NMR studies and pH–rate profiles demonstrate that Pro-1 has the correct pK_a value (~ 6.4) to act as the general base as it is in the correct protonation state.[74] Third, the similarity between this pK_a and the pK_a measured in the pH dependence of the inactivation of 4-OT by 3-BP (6.7 ± 0.3) suggests that the same residue is involved in both catalysis and inactivation.[74]

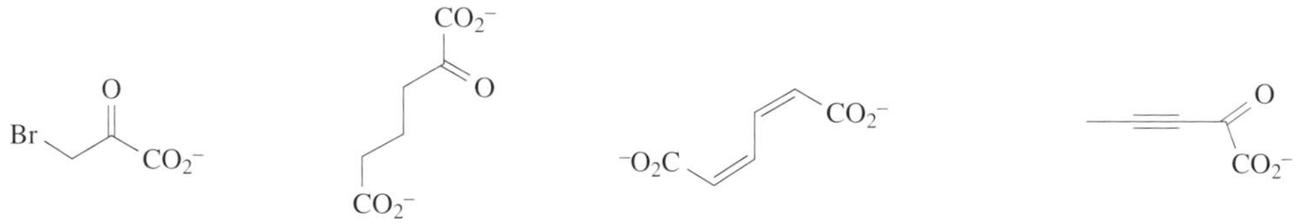

Scheme 17

The pK_a of Pro-1 is perturbed by ~ 3 pK_a units from that measured for a model compound, prolinamide ($pK_a \approx 9.4$). The crystal structure shows that Pro-1 is surrounded by hydrophobic groups and is in proximity to the two arginines. Hence the low dielectric constant of the active site coupled with the nearby electrostatic effects of the arginines may account for the low pK_a.[74]

The pH–rate profiles also implicate an acidic group in the mechanism with a pK_a of 9.0. The identity of this residue is not yet known, but if it is Arg-39, as suggested by the crystal structure, then its pK_a is also lowered by ~ 3.5 pK_a units from that typically observed for an arginine side chain (12.5). The same factors that result in the perturbation of the pK_a of Pro-1 may apply to Arg-39.[74]

The identity of Pro-1 as the general base catalyst in this reaction raises the possibility that a Schiff base may form between Pro-1 and the carbonyl group of (**7**).[75] Although this mechanism cannot be excluded, three experiments suggest that a Schiff-base mechanism is not likely. First, the addition of $NaBH_4$ does not inactivate the enzyme when the reaction is carried out in the presence of a substrate, as would be expected if an imine intermediate is reduced to an amine.[70] Second, there is no evidence for the incorporation of ^{18}O label in product when the reaction is carried out in $H_2^{18}O$ in the presence of substrate. Hydrolysis of the Schiff base in $H_2^{18}O$ should result in the labeling of the product.[45] Finally, there is no NMR spectroscopic evidence for the formation of a Schiff base between 4-OT and either 3-bromopyruvate (**31**) or 2-oxo-1,6-hexanedioate (**32**).[73,76] The latter compound is a partial substrate for the enzyme in that 4-OT catalyzes exchange of the protons at C-3 for deuterons in ^2H_2O.[77]

A role for the two arginine residues in the mechanism was demonstrated by two separate NMR studies. The first study examined the enzyme covalently bonded to (**31**) and a second study examined the enzyme as a complex with a competitive inhibitor, *cis,cis*-muconate (**33**).[76] The chemical shifts for both Arg-11 and Arg-39 change upon the binding of the dicarboxylate (**33**), while the interaction of the monocarboxylate (**31**) with enzyme results only in changes for the chemical shifts assigned to

Arg-39. These combined observations suggest that the 2-oxo acid portion (C-1 and C-2) of substrate interacts with Arg-39, while the C-6 carboxylate group interacts with Arg-11. The absence of an oxygen at C-2 of (**33**) raises a concern about how much of the information gleaned from its interaction with enzyme can be extrapolated to the normal mechanism.

While some of the major features in the mechanism have emerged, the picture of the mechanism is far from complete and many issues remain unresolved including how Arg-39 binds the C-1 carboxylate group of (**7**) and acts as the general acid catalyst. 2-Oxo-3-pentynoate (**34**) has been synthesized and found to be a potent inhibitor of 4-OT. The sole site of modification is Pro-1 and the accumulated evidence suggests that (**34**) is an active-site-directed irreversible inhibitor.[78] A preliminary crystal structure of the enzyme inactivated by this compound has also been obtained.[79] The structure verifies the site of modification as Pro-1 and shows that Pro-1 is attached to C-4 of the species derived from (**34**). More interestingly, however, the crystal structure points to the involvement of Arg-39 as the general acid catalyst and implicates another residue in the binding of the C-1 carboxylate of (**7**). It appears that Arg-61 (within the same dimer as Pro-1 but from the other monomer) binds the C-1 carboxylate group in a bidentate fashion (Scheme 18). These observations suggest that substrate binding induces a conformational change such that the C-terminal loop closes down on the substrate. The contributions of these residues to binding are under investigation.

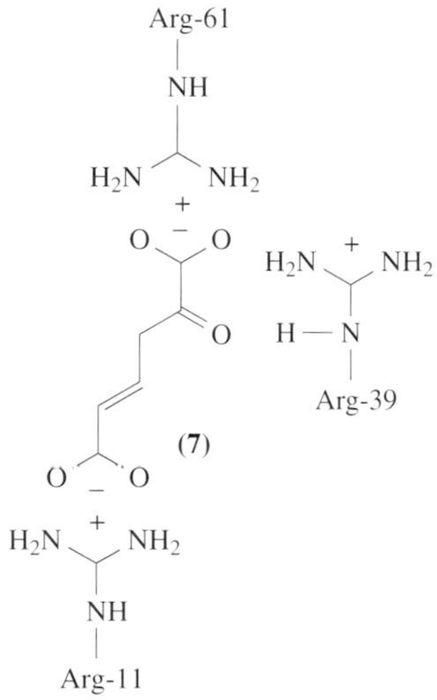

Scheme 18

It remains uncertain how 4-OT utilizes these residues to achieve its impressive catalytic rate. Neither concerted general base–general acid catalysis nor electrostatic stabilization of a dienolate intermediate can be ruled out. The observed pK_a for the general acid catalyst is 9.0. If this pK_a can be assigned to Arg-39, then an LBHB is a possibility in view of the approximately matched pK_as of the putative donor, Arg-39 (~ 9), and the hydroxy group of the acceptor, (**8**) (~ 12).[6,7,22] However, the involvement of a charged, cationic Arg-39 in the mechanism cannot be excluded.[41] Moreover, the low pK_a of the C-3 proton of (**7**) (~ 12) indicates that it is reasonable to have a dienolate species as an intermediate in the reaction. Resolution of these issues requires further experimentation.

5.03.7.4 Construction of Pro-1 Mutants

Proline is unique among the 20 common amino acids because it is a cyclic secondary amino acid. In order to investigate the contributions of the rigidity and the basicity to the mechanism, Pro-1 was changed to Gly and Ala.[77] Thus, the rigid secondary amine was replaced with flexible primary amines. The effect of these P1 mutations is primarily on k_{cat}: there is a 76-fold decrease in k_{cat} for the P1G mutant and a 58-fold decrease for the P1A mutant. The lower basicities of the primary amines in the P1G and P1A mutants compared with the secondary amine of Pro-1 are indicated in the pH–rate profiles by the lower pK_a values of the general base by 0.9 units for Gly-1 (5.3 ± 0.1) and by 0.3 units for Ala-1 (5.9 ± 0.3). Hence, the decrease in basicity would result only in a 7.9- and 2.0-fold decrease in catalysis by the P1G and P1A mutants, respectively. The additional decrease in activity for the P1G and P1A mutants (9.6- and 18-fold, respectively) may reflect the greater flexibility of the primary amines Gly-1 and Ala-1 compared with the secondary amine Pro-1. The flexibility could result in less than optimal positioning of the catalytic base in the P1G and P1A mutants so that it cannot transfer the proton between the C-3 and C-5 positions of (**7**) and (**9**) as efficiently as the wild-type enzyme. The possibility that the decrease in k_{cat} values of the P1G and P1A mutants result from a major structural perturbation is eliminated by several observations, including NMR structural studies (examination of ^{15}N and NH chemical shifts and nuclear Over-hauser effects) of the P1G mutant.[77]

The decreases in the pK_a for the P1G and P1A mutants compared with the Pro-1 enzyme parallel, within experimental error, the differences in the pK_a values for the amino acids themselves, which are 0.8 units for glycine and 0.7 units for alanine vs. proline. Hence it appears that the same factors (environment of low dielectric constant and nearby cationic residues) that lower the pK_a for Pro-1 in the wild-type enzyme have a comparable effect on the pK_a of Gly-1 and Ala-1 in the mutants.

The identity and position of the catalytic base in 4-OT limit the questions that can be addressed for 4-OT using site-directed mutagenesis. One obvious limitation is that alternative secondary amines cannot be introduced by mutagenesis. Hence, the impaired catalysis observed for site-directed mutants will result from a combination of effects as is observed for the P1G and P1A mutants.

The individual contributions of the basicity and flexibility can be ascertained by the synthesis and kinetic analysis of chemical mutants of 4-OT. The total chemical synthesis of 4-OT provides access to these mutants.[80] Accordingly, Pro-1 was replaced with sarcosine, a secondary amine.[81] Assuming that the pK_a of Sar-1 in the chemical mutant is comparable to that observed for the wild-type enzyme, then the five-fold decrease in k_{cat} observed for the Sar-1 mutant may reflect the increased flexibility of the nitrogen as the decrease in activity is comparable to that observed in the P1G and P1A mutants and is ascribed to the increased flexibility. The construction and analysis of additional mutants may provide a better understanding of the role played by the general base and general acid catalysts in the mechanism.

5.03.7.5 5-(Carboxymethyl)-2-hydroxymuconate Isomerase

CHM (**11**) (Scheme 3) is generated in the course of the bacterial catabolism of 4-hydroxy-phenylacetate by the enzymes of the homoprotocatechuate *meta*-fission pathway and is the counter-part to (**8**).[82] The dienol ketonizes chemically in solution and enzymatically by the action of CHMI to (**12**). Kinetic studies on CHMI suggest that it also catalyzes an isomerization reaction and converts (**10**) to (**12**) through the dienolic intermediate (**11**).[24]

The structural resemblance between (**8**) and (**11**) initially suggested that 4-OT and CHMI might be evolutionarily related.[70] Although there is no obvious sequence homology between the two enzymes (with the exception of Pro-1), there are tantalizing pieces of evidence that fostered specu-lation. For example, 4-OT is a hexamer consisting of 62 amino acids per monomer, whereas CHMI is a trimer consisting of 125 amino acids per monomer.[70] In addition, both enzymes process (**8**) to $(5S)$-5-$[^2H]$ (**9**) in ^2H_2O.[70] However, it is apparent that each enzyme clearly favors its own substrate. There is a 46 000-fold decrease in the specificity constant (k_{cat}/K_M) for the interaction of 4-OT with (**11**) and a 10-fold decrease in the specificity constant when CHMI processes (**8**).

When the crystal structures of 4-OT and CHMI were solved, it was found that both the overall folds and the active site regions (including Pro-1) of the two enzymes are nearly superimposable.[72] Hence the mechanistic picture for CHMI is largely inferred from the studies of 4-OT. On this basis, it is postulated that Pro-1 is the general base catalyst and two arginine residues (Arg-40 and Arg-71) from the same subunit bind the C-1 and C-6 carboxylates of substrate, respectively. The same factors that perturb the pK_a of Pro-1 in 4-OT are present in the active site of CHMI, suggesting that the pK_a of Pro-1 in CHMI is likewise perturbed.

The residues involved in the binding of the carboxylate group of the 5-carboxymethyl group of (**10**) are not apparent from the crystal structure.[72] However, the stereochemistry of the reaction indicates that this substituent must protrude out of the active site into solution. This suggests that a conformational change may take place upon the binding of substrate to CHMI in order to position a binding group (or groups) in the vicinity of the protruding carboxylate group. The observation that the C-terminal loop of 4-OT may be involved in the binding of (**34**) has led to speculation that the C-terminal loop of CHMI may also play a role in the binding of the carboxylate group of the 5-carboxymethyl group in addition to the C-1 carboxylate group. These possibilities are being explored.

5.03.7.6 Macrophage Migration Inhibitory Factor

Macrophage migration inhibitory factor (MIF) was first identified in mammals as a cytokine released at the site of an infection by T lymphocytes.[83] As a result, macrophages concentrate at the infection site to carry out antigen processing and phagocytosis. MIF has been found to play a role in other processes such as it is one of the major proteins secreted in response to endotoxin and may be a central factor in septic shock. How MIF exerts many of its physiological roles is not well understood.

When the crystal structure of MIF was solved, an intriguing observation was made: its three-dimensional structure is nearly superimposable on that of CHMI.[84,85] Both proteins are trimers of identical subunits. The MIF subunit consists of 114 amino acids whereas the CHMI subunit is made up of 125 amino acids. These similarities were particularly surprising in view of the fact that there is very little primary sequence homology between the two proteins except that both proteins have an amino-terminal proline. The presence of Pro-1 raised the question of whether MIF catalyzes an enzymatic reaction. Indeed, Rorsman and co-workers[86] showed that MIF catalyzes the enolization of phenylpyruvate (**35**) and the ketonization of (*p*-hydroxyphenyl)pyruvate (**36**) (Scheme 19). This finding led to the conclusion that MIF and phenylpyruvate tautomerase (PPT) are the same protein.[87] Its activity as a cytokine does not appear to be related to the tautomerization reaction, although this question is not entirely resolved.

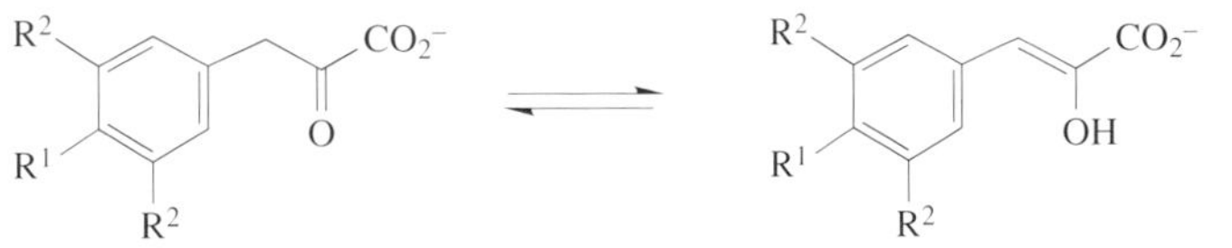

(**35**) $R^1 = H$ $R^2 = H$

(**36**) $R^1 = OH$ $R^2 = H$

(**37**) $R^1 = OH$ $R^2 = I$

Scheme 19

PPT has been studied periodically since the 1950s.[88–90] It utilizes phenylpyruvate (**35**), (*p*-hydroxyphenyl)pyruvate (**36**), and (diiodohydroxyphenyl)pyruvate (**37**) (Scheme 19) as substrates, although the kinetic parameters measured for these compounds suggest that these may not be the physiological substrates.[90] The physiological role for this tautomerization reaction is not known but it has been proposed that one tautomer may be involved in the biosynthesis of thyroxine or in the biosynthesis of alkaloids.[91,92] In view of the structural and mechanistic similarities among 4-OT, CHMI, and MIF/PPT, it is tempting to speculate that MIF/PPT may act as an isomerase and that the intermediate may be an isolable stable enol.[93] In accord with this speculation, it is of interest that (**36**) is primarily the enol isomer as determined by NMR spectroscopy.[87]

Studies by Bucala and co-workers[94] and Whitman and co-workers[95] implicate Pro-1 as the general base catalyst in the tautomerization reaction catalyzed by the recombinant mouse MIF. The pH–rate dependence of the reaction suggests that a basic group with a pK_a of 6.0 ± 0.1 is required for activity. The pK_a of this group, presumably Pro-1, is clearly perturbed. The crystal structure shows that Pro-1 is once again surrounded by a number of aromatic residues resulting in an active site of low dielectric constant. This may account for the low pK_a of Pro-1. If this pK_a value can be assigned to Pro-1, then it is in the correct protonation step to act as the general base catalyst. In addition, the tautomerization reaction is inhibited irreversibly by bromopyruvate (**31**) in a time-dependent pseudo-first order process.[95] The group modified has been identified as Pro-1. Moreover, because the inactivation of MIF by (**31**) shows a similar pH dependence to that observed for the enolization of (**35**), it can be concluded that the same group, Pro-1, is involved in both substrate catalysis and inactivation.

The general acid catalyst is not known, but the active site regions of 4-OT, CHMI, and MIF have been superimposed in order to identify potential candidates. The result did not provide a clear candidate. Stereochemical studies suggest that (**35**) binds in the active site of MIF (assuming Pro-1 is the general base) in the same orientation as do the substrates for 4-OT and CHMI. Retey *et al.*[89] determined that PPT removes the *pro-R* hydrogen of (**35**). It has been shown that 4-OT abstracts the *pro-R* hydrogen of (**16**).[22] On the basis of this stereochemical analysis, it has been inferred that 4-OT abstracts the *pro-R* hydrogen of (**7**) and that CHMI abstracts the *pro-R* hydrogen of (**12**).[24] If the three substrates do bind in the same fashion, then Phe-49 occupies approximately the same position as Arg-39 in 4-OT (or Arg-40 in CHMI) and Lys-32 occupies approximately the same position as Arg-11 in 4-OT (or Arg-71 in CHMI). Because phenylalanine cannot be a general acid catalyst, it seems likely that substrate binding moves another residue into position to play this role. There are four tyrosines (Tyr-36, Tyr-75, Tyr-95, and Tyr-98) in the vicinity of Pro-1. Studies are in progress to identify the general acid catalyst.

5.03.8 CONCLUSIONS

One issue in enzymology that has attracted much attention is how enzymes abstract a weakly acidic hydrogen from a carbon adjacent to a carbonyl group using a weakly basic residue and generate sufficient quantities of an enolic intermediate that are consistent with the observed rapid rates of enzyme-catalyzed reactions.[6,7] Albery and Knowles[96] proposed that enzymes accelerate enolization reactions by equalizing the relative energies of the bound substrate and bound intermediate so that $K_{eq} \approx 1$ on the enzyme. The intermediate is stabilized by the enzyme. A contrasting theory resulted from an analysis of various enzyme mechanisms by Gerlt and Gassman.[6,7] They proposed that the bound intermediate is much less stable than the bound substrate so that $K_{eq} \ll 1$ on the enzyme. Different mechanisms have been proposed to account for how the kinetic and thermodynamic barriers to enolization are lowered.

This chapter presented four different enzymes in which the stability of the enolic or dienolic intermediate varies. KSI and 4-OT illustrate how the mechanistic strategy of an enzyme may change depending upon the relative stability of the intermediate.[97] In their studies of KSI, Hawkinson *et al.*[54] concluded that the enzyme decreases the thermodynamic barrier for the formation of (5) by lowering ΔG° by $\sim 10\ kcal\ mol^{-1}$. In contrast, Stivers *et al.*[74] concluded that 4-OT must decrease both the thermodynamic barrier ($\sim 3.6\ kcal\ mol^{-1}$) and the kinetic barrier ($\sim 4.4\ kcal\ mol^{-1}$) to the enolization of (7) to (8) in order to explain the rate enhancement over the nonenzymatic reaction. The reason for this difference may be due to the ~ 1000-fold greater stability of (8) vs. (5) in aqueous solution.

Our understanding of these enzyme mechanisms has significantly advanced, but it is still not possible to make firm conclusions based on a comparison of the mechanisms used to decrease these barriers. Although in three of these enzymes a general base catalyst has been identified, it is not yet resolved or not yet known whether the general base catalyst acts in a concerted fashion with a general acid catalyst or whether the enolate (or dienolate) intermediate is stabilized electrostatically. Further study of these enzymes will eventually make such comparisons possible.

5.03.9 REFERENCES

1. T. H. Lowry and K. S. Richardson, "Mechanism and Theory in Organic Chemistry," 2nd edn., Harper & Row, New York, 1981, p. 657.
2. R. M. Pollack, *Tetrahedron*, 1989, **45**, 4913.
3. R. M. Pollack, P. L. Bounds, and C. L. Bevins, in "The Chemistry of Enones," eds. S. Patai and Z. Rappoport, Wiley, Chichester, 1989, p. 559.
4. Y. Chiang and A. J. Kresge, *Science*, 1991, **253**, 395.
5. J. R. Keeffe, A. J. Kresge, and N. P. Schepp, *J. Am. Chem. Soc.*, 1990, **112**, 4862.
6. J. A. Gerlt and P. G. Gassman, *J. Am. Chem. Soc.*, 1992, **114**, 5928.
7. J. A. Gerlt and P. G. Gassman, *J. Am. Chem. Soc.*, 1993, **115**, 11 552.
8. J. A. Peliska and M. H. O'Leary, *J. Am. Chem. Soc.*, 1991, **113**, 1841.
9. D. J. Kuo, E. L. O'Connell, and I. A. Rose, *J. Am. Chem. Soc.*, 1979, **101**, 5025.
10. S. Bantia and R. M. Pollack, *J. Am. Chem. Soc.*, 1986, **108**, 3145.
11. G. D. Dzingeleski, S. Bantia, G. Blotny, and R. M. Pollack, *J. Org. Chem.*, 1988, **53**, 1540.
12. C. P. Whitman, B. A. Aird, W. R. Gillespie, and N. J. Stolowich, *J. Am. Chem. Soc.*, 1991, **113**, 3154.
13. Z. Rappoport and S. E. Biali, *Acc. Chem. Res.*, 1988, **21**, 442.
14. B. Capon and B. Guo, *J. Am. Chem. Soc.*, 1988, **110**, 5144.
15. G. L. Kenyon, J. A. Gerlt, G. A. Petsko, and J. W. Kozarich, *Acc. Chem. Res.*, 1995, **28**, 178.
16. Y. Chiang, A. J. Kresge, P. Pruszynski, N. P. Schepp, and J. Wirz, *Angew. Chem., Int. Ed. Engl.*, 1990, **29**, 792.
17. R. M. Pollack, J. P. G. Mack, and S. Eldin, *J. Am. Chem. Soc.*, 1987, **109**, 5048.
18. R. M. Pollack, B. Zeng, J. P. G. Mack, and S. Eldin, *J. Am. Chem. Soc.*, 1989, **111**, 6419.
19. D. C. Hawkinson, T. C. M. Eames, and R. M. Pollack, *Biochemistry*, 1991, **28**, 6756.
20. R. C. Bayly and M. G. Barbour, in "Microbiological Degradation of Organic Compounds," ed. D. T. Gibson, Dekker, New York, 1984, p. 253.
21. G. Hajipour, W. H. Johnson, Jr., P. D. Dauben, N. J. Stolowich, and C. P. Whitman, *J. Am. Chem. Soc.*, 1993, **115**, 3533.
22. H. Lian and C. P. Whitman, *J. Am. Chem. Soc.*, 1993, **115**, 7978.

23. W. H. Johnson, Jr., G. Hajipour, and C. P. Whitman, *J. Am. Chem. Soc.*, 1992, **114**, 11 001.
24. W. H. Johnson, Jr., G. Hajipour, and C. P. Whitman, *J. Am. Chem. Soc.*, 1995, **117**, 8719.
25. G. D. Dzingeleski, G. Blotny, and R. M. Pollack, *J. Org. Chem.*, 1990, **55**, 1019.
26. R. T. Raines, E. L. Sutton, D. R. Straus, W. Gilbert, and J. R. Knowles, *Biochemistry*, 1986, **25**, 7142.
27. J. A. Gerlt, *Curr. Opin. Struct. Biol.*, 1994, **4**, 593.
28. E. B. Nickbarg, R. C. Davenport, G. A. Petsko, and J. R. Knowles, *Biochemistry*, 1988, **27**, 5948.
29. P. J. Lodi and J. R. Knowles, *Biochemistry*, 1991, **30**, 6948.
30. E. A. Komives, L. C. Chang, E. Lolis, R. F. Tilton, G. A. Petsko, and J. R. Knowles, *Biochemistry*, 1991, **30**, 3011.
31. R. C. Davenport, P. A. Bash, B. A. Seaton, M. Karplus, G. A. Petsko, and D. Ringe, *Biochemistry*, 1991, **30**, 5821.
32. Z. Zhang, S. Sugio, E. A. Komives, K. D. Liu, J. R. Knowles, G. A. Petsko, and D. Ringe, *Biochemistry*, 1994, **33**, 2830.
33. P. J. Lodi and J. R. Knowles, *Biochemistry*, 1993, **32**, 4338.
34. E. A. Komives, J. C. Lougheed, K. Liu, S. Sugio, Z. Zhang, G. A. Petsko, and D. Ringe, *Biochemistry*, 1995, **34**, 13 612.
35. E. A. Komives, J. C. Lougheed, Z. Zhang, S. Sugio, N. Narayana, N. H. Xuong, G. A. Petsko, and D. Ringe, *Biochemistry*, 1996, **35**, 15 474.
36. D. Joseph-McCarthy, E. Lolis, E. A. Komives, and G. A. Petsko, *Biochemistry*, 1994, **33**, 2815.
37. P. J. Lodi, L. C. Chang, J. R. Knowles, and E. A. Komives, *Biochemistry*, 1994, **33**, 2809.
38. J. P. Richard, *J. Am. Chem. Soc.*, 1984, **106**, 4926.
39. D. L. Pompliano, A. Peyman, and J. R. Knowles, *Biochemistry*, 1990, **29**, 3186.
40. W. W. Cleland and M. M. Kreevoy, *Science*, 1994, **264**, 1887.
41. J. P. Guthrie and P. Kluger, *J. Am. Chem. Soc.*, 1993, **115**, 11 569.
42. G. Alagona, C. Ghio, and P. A. Kollman, *J. Am. Chem. Soc.*, 1995, **117**, 9855.
43. D. J. Kuo and I. A. Rose, *J. Am. Chem. Soc.*, 1978, **100**, 6288.
44. M. H. O'Leary, *Enzymes*, 1992, **20**, 235.
45. H. Lian and C. P. Whitman, *J. Am. Chem. Soc.*, 1994, **116**, 10 403.
46. H. Muirhead, D. A. Clayden, D. Barford, C. G. Lorimer, L. A. Fothergill-Gilmore, E. Schiltz, and W. Schmitt, *EMBO J.*, 1986, **5**, 475.
47. T. M. Larsen, L. T. Laughlin, H. M. Holden, I. Rayment, and G. H. Reed, *Biochemistry*, 1994, **33**, 6301.
48. S. H. Seeholzer, A. Jaworowski, and I. A. Rose, *Biochemistry*, 1991, **30**, 727.
49. W. W. Cleland, *Biochemistry*, 1990, **29**, 3194.
50. I. A. Rose, D. J. Kuo, J. V. B. Warms, *Biochemistry*, 1991, **30**, 722.
51. J. M. Schwab and B. S. Henderson, *Chem. Rev.*, 1990, **90**, 1203.
52. M. E. Zawrotny, D. C. Hawkinson, G. Blotny, and R. M. Pollack, *Biochemistry*, 1996, **35**, 6438.
53. L. Xue, A. Kuliopulos, A. S. Mildvan, and P. Talalay, *Biochemistry*, 1991, **30**, 4991.
54. D. C. Hawkinson, R. M. Pollack, and N. P. Ambulos, Jr., *Biochemistry*, 1994, **33**, 12 172.
55. D. C. Hawkinson and R. M. Pollack, *Biochemistry*, 1993, **32**, 694.
56. B. Zeng, P. L. Bounds, R. F. Steiner, and R. M. Pollack, *Biochemistry*, 1992, **31**, 1521.
57. Q. Zhao, A. S. Mildvan, and P. Talalay, *Biochemistry*, 1995, **34**, 426.
58. Q. Zhao, C. Abeygunawardana, P. Talalay, and A. S. Mildvan, *Proc. Natl. Acad. Sci. USA*, 1996, **93**, 8220.
59. Z. R. Wu, S. Ebrahimian, M. E. Zawrotny, L. D. Thornburg, G. C. Perez-Alvarado, P. Brothers, R. M. Pollack, and M. F. Summers, *Science*, 1997, **276**, 415.
60. A. Kuliopulos, P. Talalay, and A. S. Mildvan, *Biochemistry*, 1990, **29**, 10 271.
61. L. Xue, P. Talalay, and A. S. Mildvan, *Biochemistry*, 1991, **30**, 10 858.
62. P. N. Brothers, G. Blotny, L. Qi, and R. M. Pollack, *Biochemistry*, 1995, **34**, 15 453.
63. D. C. Hawkinson, T. C. M. Eames, and R. M. Pollack, *Biochemistry*, 1991, **30**, 10 849.
64. M. E. Zawrotny and R. M. Pollack, *Biochemistry*, 1994, **33**, 13 896.
65. T. C. M. Eames, R. M. Pollack, and R. F. Steiner, *Biochemistry*, 1989, **28**, 6269.
66. J. M. Sala-Trepat and W. C. Evans, *Eur. J. Biochem.*, 1971, **20**, 400.
67. S. Harayama, M. Rekik, K.-L. Ngai, and L. N. Ornston, *J. Bacteriol.*, 1989, **171**, 6251.
68. V. Shingler, J. Powlowski, and U. Marklund, *J. Bacteriol.*, 1992, **174**, 711.
69. L. H. Chen, G. L. Kenyon, F. Curtin, S. Harayama, M. E. Bembenek, G. Hajipour, and C. P. Whitman, *J. Biol. Chem.*, 1992, **267**, 17 716.
70. C. P. Whitman, G. Hajipour, R. J. Watson, W. H. Johnson, Jr., M. E. Bembenek, and N. J. Stolowich, *J. Am. Chem. Soc.*, 1992, **114**, 10 104.
71. D. I. Roper, H. S. Subramanya, V. Shingler, and D. B. Wigley, *J. Mol. Biol.*, 1994, **243**, 799.
72. H. S. Subramanya, D. I. Roper, Z. Dauter, E. J. Dodson, G. J. Davies, K. S. Wilson, and D. B. Wigley, *Biochemistry*, 1996, **35**, 792.
73. J. T. Stivers, C. Abeygunawardana, A. S. Mildvan, G. Hajipour, C. P. Whitman, and L. H. Chen, *Biochemistry*, 1996, **35**, 803.
74. J. T. Stivers, C. Abeygunawardana, A. S. Mildvan, G. Hajipour, and C. P Whitman, *Biochemistry*, 1996, **35**, 814.
75. A. Fersht, "Enzyme Structure and Mechanism," 2nd edn., Freeman, San Francisco, 1985, p. 69.
76. J. T. Stivers, C. Abeygunawardana, A. S. Mildvan, and C. P. Whitman, *Biochemistry*, 1996, **35**, 16 036.
77. R. M. Czerwinski, W. H. Johnson, Jr., C. P. Whitman, T. K. Harris, C. Abeygunawardana, and A. S. Mildvan, *Biochemistry*, 1997, **36**, 14 551.
78. W. H. Johnson, Jr., R. M. Czerwinski, M. C. Fitzgerald, and C. P. Whitman, *Biochemistry*, 1997, **36**, 15 724.
79. A. B. Taylor and M. L. Hackert, unpublished results, 1997.
80. M. C. Fitzgerald, I. Chernushevich, K. G. Standing, S. B. H. Kent, and C. P. Whitman, *J. Am. Chem. Soc.*, 1995, **117**, 11 075.
81. M. C. Fitzgerald and C. P. Whitman, unpublished results, 1996.
82. J. R. Jenkins and R. A. Cooper, *J. Bacteriol.*, 1988, **170**, 5317.
83. R. Mitchell, M. Bacher, J. Bernhagen, T. Pushkarskaya, M. F. Seldin, and R. Bucala, *J. Immunol.*, 1995, **154**, 3863.
84. M. Suzuki, H. Sugimoto, A. Nakagawa, I. Tanaka, J. Nishihira, and M. Sakai, *Nature Struct. Biol.*, 1996, **3**, 259.
85. A. G. Murzin, *Curr. Opin. Struct. Biol.*, 1996, **6**, 386.
86. E. Rosengren, R. Bucala, P. Aman, L. Jacobsson, G. Odh, C. N. Metz, and H. Rorsman, *Mol. Med.*, 1996, **2**, 143.

87. E. Rosengren, P. Aman, S. Thelin, C. Hansson, S. Ahlfors, P. Bjork, L. Jacobsson, and H. Rorsman, *FEBS Lett.*, 1997, **417**, 85.
88. W. E. Knox, *Methods Enzymol.*, 1955, **2**, 297.
89. J. Retey, K. Bartl, E. Ripp, and W. E. Hull, *Eur. J. Biochem.*, 1977, **72**, 251.
90. M. C. Pirrung, J. Chen, E. G. Rowley, and A. T. McPhail, *J. Am. Chem. Soc.*, 1993, **115**, 7103.
91. F. Blasi, F. Fragomele, and I. Covelli, *J. Biol. Chem.*, 1969, **244**, 4864.
92. F. Blasi, F. Fragomele, and I. Covelli, *Eur. J. Biochem.*, 1968, **5**, 215.
93. P. C. Babbitt, M. S. Hasson, J. E. Wedekind, D. R. J. Palmer, W. C. Barrett, G. H. Reed, I. Rayment, D. Ringe, G. L. Kenyon, and J. A. Gerlt, *Biochemistry*, 1996, **35**, 16 489.
94. K. Bendrat, Y. Al-Abed, D. J. E. Callaway, T. Peng, C. N. Metz, R. Bucala, *Biochemistry*, 1997, **36**, 15 356.
95. S. L. Stamps, W. H. Johnson, Jr., R. M. Czerwinski, C. P. Whitman, and M. C. Fitzgerald, *Biochemistry*, submitted for publication.
96. W. J. Albery and J. R. Knowles, *Biochemistry*, 1996, **15**, 5631.
97. W. P. Jencks, *Acc. Chem. Res.*, 1980, **13**, 161.

5.04
Nucleophilic Epoxide Openings

RICHARD N. ARMSTRONG

Vanderbilt University, Nashville, TN, USA

5.04.1 INTRODUCTION

5.04.1.1 Basic Chemistry of the Oxirane Ring

The solution chemistry of epoxides is dominated by the strain of the oxirane ring and the ambident character of the substrate. Nucleophilic opening of the epoxide is an exothermic process typically characterized by an early (substrate-like) transition state. Under the neutral or slightly basic conditions encountered in biological systems, nucleophilic substitutions on oxiranes generally follow a classical direct displacement (S_N2) mechanism with attack of the nucleophile at the least hindered carbon and inversion of configuration at the substituted carbon as illustrated in Figure 1. Such displacements occur more readily in protic solvents capable of transferring a proton to the leaving alkoxide. The classical S_N2 mechanism cannot be distinguished from the alternative nucleophilic displacement of an intimate ion pair intermediate.[1]

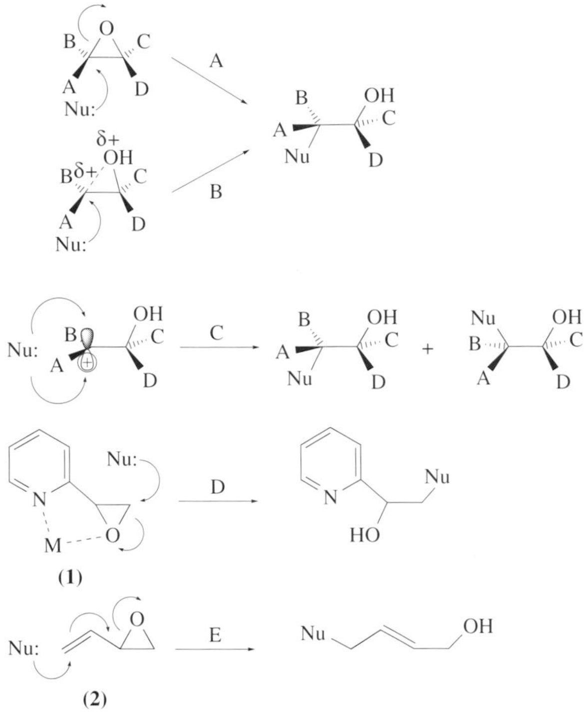

Figure 1   Mechanisms for nucleophilic additions to epoxides: (A) classical S_N2, (B) acid-catalyzed, borderline S_N2, (C) acid-catalyzed, classical S_N1, (D) metal-catalyzed, and (E) conjugate addition.

Acidic conditions catalyze the addition of nucleophiles to epoxides either through a "borderline" S_N2 mechanism or through rate-limiting formation of a carbocation intermediate as in the classical S_N1 process (Figure 1). The former mechanism is likely when the carbocation is not significantly stabilized by neighboring groups while the latter is usually only observed in systems where the *p*-orbital of the carbocation is conjugated with an adjacent π-system. The hallmark of S_N1 chemistry lies in the stereochemical outcome of nucleophilic addition. S_N2 and "borderline" S_N2 reactions invariably result in Walden inversion at the reacting carbon. Mixed stereochemistry is often observed in S_N1 processes resulting from attack of the nucleophile on either side of the planar carbocation. The complete retention of configuration observed in some reactions is taken as evidence for a process involving two inversions and anchimeric assistance of an adjacent functional group. The solution chemistry of epoxides has been extensively reviewed over the years.[2-5]

Metal ions may also accelerate the addition of nucleophiles to epoxides through coordination with the oxygen, particularly if the remainder of the molecule provides additional ligands to the metal. The first well-characterized example of metal ion activation of an epoxide was in nucleophilic additions to 2-pyridyloxirane (**1**) (Figure 1) described by Hanzlik and co-workers where chelation of the metal makes the oxygen a much better leaving group.[6,7]

Oxiranes with adjacent π-systems can undergo ring opening by conjugate addition of a nucleophile at a position remote from the ring. An early example of this type of ring opening was observed in the 1,4-addition of azide ion to 3,4-epoxy-1-butene (**2**) where one of the products formed is 1-azido-5-hydroxy-2-butene.[8] A more biologically relevant reaction of this type is the acid-catalyzed hydrolysis of leukotriene A_4 where the hydroxide is added seven carbons away from where the oxirane bond is cleaved.[9]

5.04.1.2 Enzyme-catalyzed Epoxide Ring Openings

Much of our current knowledge about enzyme-catalyzed nucleophilic additions to epoxides derives from early interest in the metabolism of polycyclic aromatic hydrocarbons. That many of these molecules are highly mutagenic and carcinogenic is directly related to the formation of electrophilic metabolites. Oxidative degradation of this class of compounds is initiated by mono-oxygenases and often results in the formation of arene oxides as metabolic intermediates.[10] Two enzymatic routes for the further metabolism of these electrophilic intermediates involve the addition of oxygen or sulfur nucleophiles to the oxirane ring as illustrated in Figure 2. The enzymatic addition of water is catalyzed by a group of enzymes known as epoxide hydrolases, though, as will be seen later, water is not necessarily the actual nucleophile that adds to the oxirane. The addition of the peptide glutathione (GSH) is an alternative pathway for nucleophilic epoxide opening and is catalyzed by a family of enzymes known as the glutathione transferases. Both of these types of reactions proceed with inversion of configuration at the site of addition.

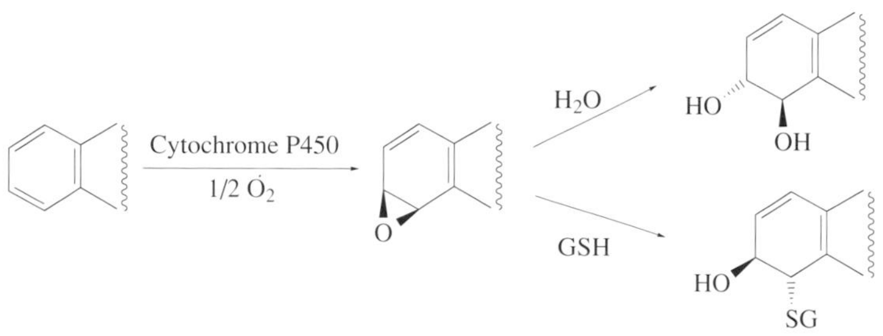

Figure 2 Formation and cleavage of arene oxides by the addition of water catalyzed by epoxide hydrolase and the addition of glutathione (GSH) catalyzed by glutathione transferase.

Many of the examples of solution chemistry briefly summarized in Section 5.04.1.1 have enzymatic counterparts. The canonical epoxide hydrolases and glutathione transferases involved in the metabolism of foreign and endogenous compounds catalyze S_N2 processes. In contrast, the more specialized epoxide hydrolase, leukotriene A_4 hydrolase, catalyzes a remote conjugate addition of water to the substrate. The fosfomycin resistance protein which catalyzes the addition of glutathione to the oxirane ring of the antibiotic fosfomycin is a metalloenzyme with a mechanism somewhat reminiscent of the transition metal-catalyzed opening of 2-pyridyloxirane (Figure 1).[7] This chapter is divided into two major sections illustrating examples of each of these enzymes. Section 5.04.2 deals with the addition of oxygen nucleophiles catalyzed by the epoxide hydrolases while Section 5.04.3 focuses on sulfur nucleophiles and the glutathione transferases.

5.04.2 EPOXIDE HYDROLASES

5.04.2.1 Convergent Families of Epoxide Hydrolases

There are two mechanistically and structurally distinct groups of epoxide hydrolases currently known that represent independent or convergent evolutionary solutions to the problem of catalyzing the nucleophilic addition of water to epoxides. The first group is a subset of the α/β-hydrolase fold superfamily, a diverse collection of enzymes that catalyze a wide variety of hydrolytic reactions by formation of covalent enzyme intermediates.[11] The second group catalyzes the hydration of a very specific substrate, leukotriene A_4.

5.04.2.1.1 *The α/β-hydrolase fold family*

The majority of epoxide hydrolases characterized to date are members of the α/β-hydrolase fold superfamily. This group of enzymes catalyzes the hydrolysis of a number of different substrates including peptides, esters, lactones, haloalkanes, and epoxides. In addition to a common structural framework and active site motif, all members of this superfamily catalyze their respective reactions by initial attack of an active site nucleophile on the substrate to form either an acyl-enzyme or alkyl-enzyme intermediate as illustrated in Figure 3.[11] Depending on the enzyme, the nucleophile may be the hydroxyl group of a serine as in acetylcholine esterase, the sulfhydryl group of a cysteine as in dienelactone hydrolase, or the carboxyl group of an aspartate as in haloalkane dehalogenase and epoxide hydrolase. The hydrolytic reaction is completed by attack of water on the carbonyl group of the intermediate assisted by a histidine and a charge relay carboxylate group. The fundamental difference between the acyl-enzyme and the alkyl-enzyme intermediates is that the oxygen of the water molecule is transferred directly to the product in the former and to the enzyme in the latter (Figure 3). Thus, epoxide hydrolase acts as a covalent shuttle for the oxygen from the solvent pool to the product. It is interesting to note that an epoxide hydrolase that catalyzes the direct attack of water on the oxirane carbon has yet to be demonstrated.

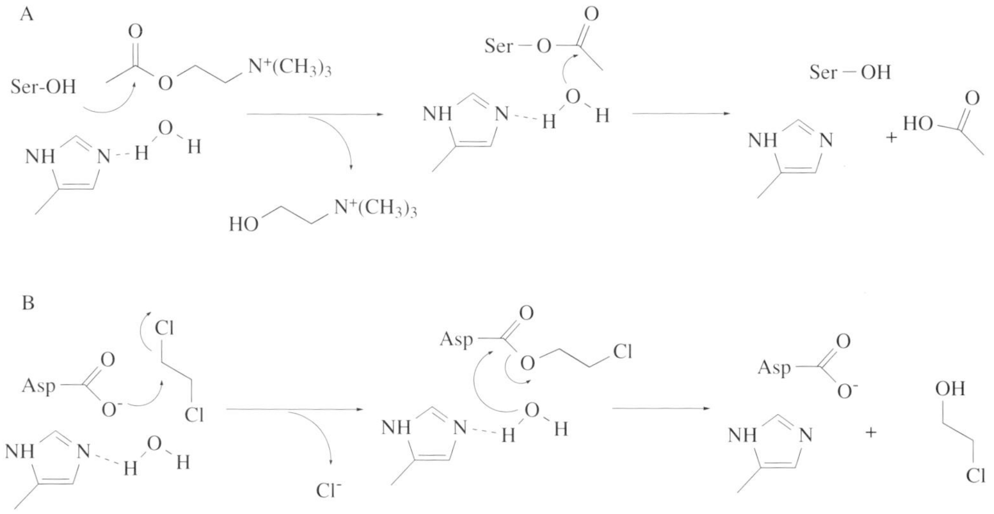

Figure 3   Mechanisms of the α/β-hydrolase fold enzymes: (A) acetylcholine esterase, and (B) haloalkane dehalogenase.

The best characterized epoxide hydrolases are those from vertebrates thought to be involved in the detoxification of foreign and endogenous epoxides. There are two main classes of these enzymes. The first to be discovered was microsomal epoxide hydrolase (mEH), a 455 residue polypeptide that is associated, as are the cytochromes P450, with the endoplasmic reticulum of the cell.[12] The other enzyme thought to be involved in xenobiotic metabolism is soluble epoxide hydrolase (sEH), a dimer of 62 kDa subunits with a substrate preference distinct from that of mEH.[13] In general, mEH prefers mono- or cis-1,2-disubstituted epoxides, whereas sEH is more efficient with trans-1,2-disubstituted substrates. Neither enzyme requires a metal ion or other cofactor for catalysis.

The primary structures of mEH and sEH from a number of vertebrate species have been determined or deduced from cDNA clones.[14–17] Although the overall similarities in amino acid sequence are quite low and initial reports suggested the contrary,[16,17] sEH and mEH are structurally and mechanistically related.[18,19] Statistically significant sequence similarities between sEH, mEH, and haloalkane dehalogenase are evident in the regions near the active site nucleophile, the oxyanion hole, and the general base.[18–20] Although no three-dimensional structure of an epoxide hydrolase has been reported, the vertebrate enzymes are clearly members of the α/β-hydrolase fold superfamily. Similar epoxide hydrolases have been characterized from microorganisms and insects.

5.04.2.1.2   *Leukotriene A₄ hydrolase*

In addition to the epoxide hydrolases involved in xenobiotic metabolism, there are two enzymes that appear to have specific endogenous substrates. One is a distinct microsomal enzyme thought to be specific for cholesterol 5,6-oxide.[21] Little is known about the structure or mechanism of this enzyme. The other is leukotriene A_4 hydrolase, which catalyzes the conjugate addition of water to leukotriene A_4 to give leukotriene B_4. It is a zinc metalloenzyme that appears not to be related to the other known epoxide hydrolases.[18,22] Rather, the enzyme does show significant sequence similarity to known zinc-dependent metallopeptidases. Interestingly enough, the protein also has an arginine aminopeptidase activity, the biological importance of which remains obscure.

5.04.2.2 Catalytic Mechanism of Microsomal and Soluble Epoxide Hydrolases

5.04.2.2.1 Evidence for an alkyl-enzyme intermediate

For almost two decades the chemical mechanism of microsomal epoxide hydrolase was thought to involve direct nucleophilic attack of water on the substrate. This conclusion was based on the observation of inversion of configuration at the oxirane carbon attacked and the involvement of a histidine residue, presumably as a general base, in catalysis.[23-25] The first experimental indication that a covalent alkyl-enzyme intermediate was involved in the reaction came from single-turnover experiments with the microsomal enzyme in ^{18}O water.[20] The isotopic label was found to be incorporated into the enzyme and not the product, as illustrated in Scheme 1. In addition, when the enzyme, labeled with one atom of ^{18}O by a single-turnover in $H_2^{18}O$, was isolated and used in a second single-turnover reaction in $H_2^{16}O$, 0.5 equivalents of ^{18}O were incorporated into the product, a result consistent with the intervention of a carboxyl group in the reaction. Soluble epoxide hydrolase gave similar results in single turnover reactions as well as evidence that the ^{18}O was incorporated into Asp333 of the murine enzyme.[26]

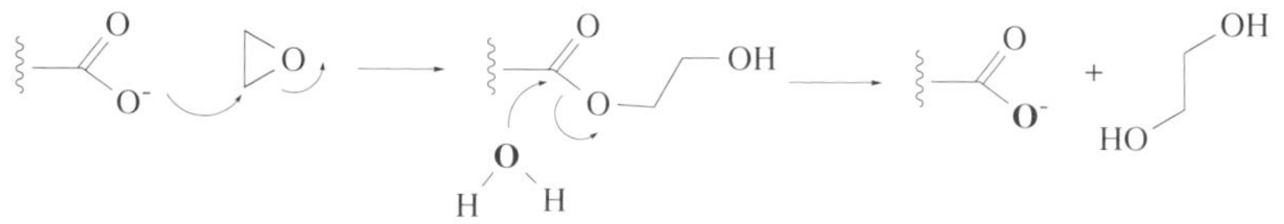

Scheme 1

Additional evidence for an alkyl-enzyme intermediate in the reaction catalyzed by the soluble epoxide hydrolase includes rapid-quench trapping of radiolabeled protein using tritiated juvenile hormone III as the substrate.[27] The extent of radiolabel incorporated increased with decreasing pH due to less rapid turnover of the ester. The most direct evidence for the ester intermediate is the detection of a peptide which includes Asp333 and bears the alkyl group of the substrate.[26]

5.04.2.2.2 Kinetic mechanism of the enzyme

The most salient kinetic issue with respect to a two-step mechanism is the identity of the rate-limiting step in turnover of the enzyme. The rate-limiting step in reactions catalyzed by the micro-somal enzyme with the substrates studied so far is hydrolysis of the alkyl-enzyme intermediate.[28] It is easily imaginable that with an enzyme of such broad substrate selectivity, this need not be true in every instance. Rate-limiting hydrolysis of the ester is a tremendous advantage in dissecting the reaction mechanism, in particular with respect to the specific role of active site residues in catalysis, since each step can be studied in isolation. The kinetics of events in the presteady-state phase of the reaction are used to probe the alkylation of the enzyme.

The presteady-state and steady-state kinetics of the microsomal enzyme have been examined in some detail with the enantiomers of glycidyl-4-nitrobenzoate (**3**).[28] Rapid mixing of the enzyme with an excess of either enantiomer of (**3**) results in a rapid loss of intrinsic protein fluorescence as the enzyme is alkylated followed by a long lag (steady state) and eventual recovery of the original fluorescence. The observed rate constant for the approach to steady state is dependent on the concentration of the substrate and, at least with (2R)-(**3**), is saturable, indicating that reversible formation of a Michaelis complex precedes the alkylation step. The simplest kinetic mechanism consistent with the results is shown in Figure 4. It is clear that the alkylation reaction occurs at a rate that is 10^2–10^3-fold faster than hydrolysis of the intermediate. Perhaps the most interesting observation from the presteady-state kinetics is that the alkylation reaction appears to be reversible. Although the alternative, rapid irreversible alkylation of the enzyme followed by a slower reversible conformational change cannot be entirely ruled out, in either case the alkylation reaction is fast relative to hydrolysis.

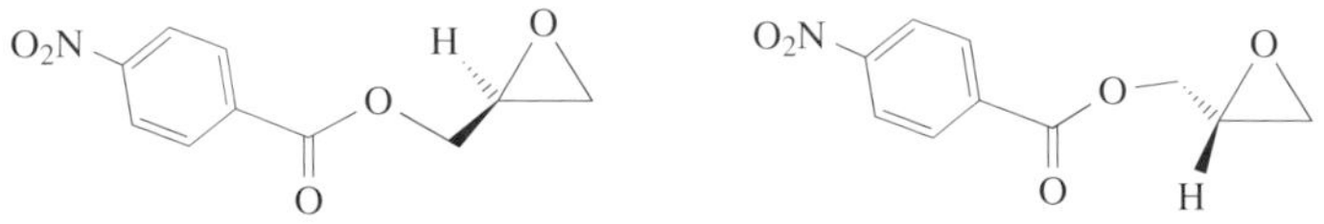

(2S)-(3)   (2R)-(3)

$$S = (2R)\text{-}(3) \quad E + S \underset{4 \text{ s}^{-1}}{\overset{2 \text{ mM}}{\rightleftharpoons}} E{\cdot}S \overset{330 \text{ s}^{-1}}{\rightleftharpoons} E\text{-}I \xrightarrow{0.8 \text{ s}^{-1}} E + P$$

$$S = (2S)\text{-}(3) \quad E + S \underset{0.5 \text{ s}^{-1}}{\overset{\geq 3 \text{ mM}}{\rightleftharpoons}} E{\cdot}S \overset{\geq 250 \text{ s}^{-1}}{\rightleftharpoons} E\text{-}I \xrightarrow{0.07 \text{ s}^{-1}} E + P$$

Figure 4 Minimal kinetic mechanism for microsomal epoxide hydrolase with (2R)- and (2S)-(3).

The enantioselectivity of the microsomal epoxide hydrolase is manifest in both half-reactions but is most pronounced in the hydrolysis of the intermediate. The primary difference between (2R)-(3) and (2S)-(3) in the alkylation reactions is that saturation behavior is observed with the former and not the latter. Interestingly, there is very little kinetic selectivity in the alkylation reaction as judged by k_2/K_s, being $1.7 \times 10^5 \text{ M}^{-1} \text{ s}^{-1}$ with (2R)-(3) and $1.1 \times 10^5 \text{ M}^{-1} \text{ s}^{-1}$ with (2S)-(3). The alkylation reactions with both (2R)-(3) and (2S)-(3) are regiospecific with attack at the least hindered carbon, as is commonly seen in the solution chemistry of epoxides. The largest difference between the two enantiomers is seen in the hydrolysis of the intermediate, where the slower turnover of (2S)-(3) is due to the formation of a more stable alkyl-enzyme.

5.04.2.2.3 Chemistry of formation of the alkyl-enzyme intermediate

The rapid elucidation of many of the active site residues in microsomal and soluble epoxide hydrolase was facilitated by sequence alignments between haloalkane dehalogenase, for which the crystal structure of the free enzyme and the ester intermediate are available. Mapping the sequence of epoxide hydrolases on the crystal structures provided the first indication of the identity of the active site nucleophile and other residues involved in catalysis, as illustrated in Figure 5. Experimental support for the hypothetical active site was first obtained for the soluble enzyme by Hammock and co-workers.[29] The most clearly defined active site residue identified in the sequence alignments is that of the nucleophile, Asp333 and Asp226 in the case of soluble and microsomal enzymes, respectively. Replacement of Asp333 with Asn (D333N) results in an enzyme with a low residual amount of activity which increases over a period of a few days. Thus, the D333N mutant conscripts the hydrolytic machinery normally used to cleave the ester intermediate and catalyzes the hydrolysis of the amide leading to native enzyme.

The identity of Asp226 as the nucleophile in the microsomal enzyme is based on similar evidence. The D226E mutant is inactive and the D226N mutant undergoes a hydrolytic autoactivation with a $t_{1/2}$ of 9.3 days (pH 8, 37 °C).[30] For comparison, the pentapeptide GGNWG, which places an asparagine in the same sequence context as the active site of the enzyme, hydrolyzes under similar conditions but with a $t_{1/2}$ of 60 days. The autoactivation of soluble and microsomal epoxide hydrolase mutants is compelling experimental evidence that Asp333 and Asp226 are the active site nucleophiles.

The tryptophan residue immediately following the nucleophile in haloalkane dehalogenase is known to form part of the chloride ion (leaving group) binding site. That this residue is conserved in both microsomal and soluble epoxide hydrolases suggests that it may also be involved in electrophilic assistance by donating a hydrogen bond from the indole NH to the oxirane oxygen in formation of the ester intermediate. This is not the case with the microsomal enzyme since the

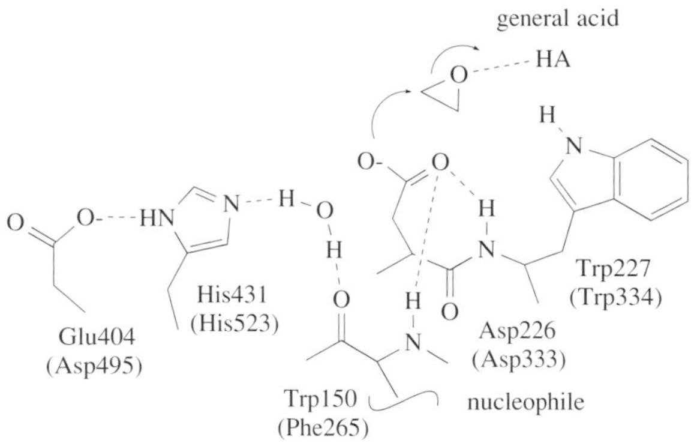

Figure 5 Proposed active site for microsomal and soluble epoxide hydrolase derived from sequence alignments with haloalkane dehalogenase. The residue numbers for the soluble enzyme are shown in parentheses.

W227F mutant is normal with respect to the alkylation half-reaction. Whether other residues may participate in stabilization of the negative charge developing on the oxirane oxygen in the alkylation half-reaction remains an open question.

5.04.2.2.4 *Chemistry of hydrolysis of the alkyl-enzyme intermediate*

Hydrolysis of the alkyl enzyme is dependent on a general-base/charge relay diad to activate the water molecule and an oxyanion hole for stabilization of the tetrahedral intermediate as illustrated in Figure 6. That the histidine about 30 residues from the C-terminus is essential for efficient turnover of the microsomal enzyme was first demonstrated by Bell and Kasper.[25] Sequence alignments with other members of the α/β-hydrolase fold family indicates that H431 serves as the general base in the hydrolysis reaction. In principle, loss of the general base need not impede the formation of the alkyl-enzyme intermediate. Two lines of evidence suggest that this is true. The H431S mutant of the microsomal enzyme is alkylated by (2R)-(**3**) with an apparent second-order rate constant ($k_2/K_s = 2.5 \times 10^5$ M^{-1} s^{-1}), which is similar to that of the native enzyme, but is unable to complete the catalytic cycle.[31] In addition, mutants at the analogous position in the soluble enzyme are catalytically inactive but can be specifically labeled with a single molecule of substrate.[29,32] Similar behavior has been documented with haloalkane dehalogenase.[33]

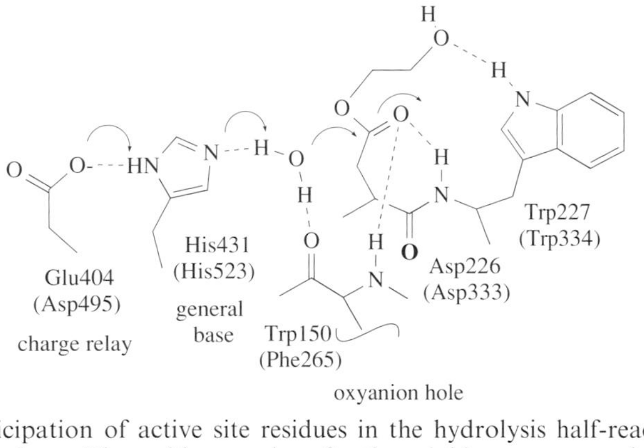

Figure 6 Proposed participation of active site residues in the hydrolysis half-reaction of microsomal and soluble epoxide hydrolases. The residue numbers for the soluble enzyme are shown in parentheses.

The identification of the charge relay residue in the hydrolysis reaction has proven much more problematic due to the low degree of sequence similarity of epoxide hydrolases and other α/β-hydrolase fold enzymes in the relevant part of the primary structure. A low but statistically significant sequence similarity between the soluble epoxide hydrolase and haloalkane dehalogenase suggested that Asp495 of the soluble enzyme was a likely candidate.[18] That the D495H mutant of this enzyme has less than 0.3% of the activity of the native enzyme is certainly consistent with this notion. Single mutants of all the possible carboxylate residues in the relevant region of the microsomal enzyme has revealed that Glu404 is the most likely charge relay residue. A detailed examination of the kinetic properties of the E404Q mutant with (2*R*)-(**3**) as the substrate indicates that the mutant undergoes alkylation at a normal rate $k_2/K_s = 1.5 \times 10^5 \text{ M}^{-1} \text{ s}^{-1}$ but turns over at a rate (0.05 s^{-1}) less than one tenth that of native enzyme.[31] Qualitatively, this is exactly the behavior expected of a charge relay mutant. The decrease in turnover number is much more modest than large (10^4) decreases encountered with charge relay mutants of proteases, perhaps suggesting that other unidentified charge relay interactions function in the epoxide hydrolases. It is also interesting that the soluble enzyme uses an aspartate residue while the microsomal enzyme has conscripted a glutamate.

5.04.2.3   Epoxide Hydrolases in Metabolism and Synthesis

5.04.2.3.1   *Metabolism of xenobiotics*

There is extensive literature on the participation of epoxide hydrolases in the metabolism of foreign compounds, too extensive to be thoroughly reviewed here. In terms of understanding their involvement in xenobiotic metabolism, the most pressing issues are the stereoselectivity and regioselectivity of the enzymes. Extensive information has been gathered regarding the substrate preferences of the microsomal enzyme, particularly toward arene oxides of polycyclic aromatic hydrocarbons.[10,34-39] The regiochemical and stereochemical preferences of the enzymes have led to proposals concerning the topology of the active sites. Although the accuracy of the structural models are unclear, the information does allow some reasonably good predictions of the outcome of metabolism.

The enantioselectivity and regioselectivity of the cytochromes P450 and epoxide hydrolase are crucial in determining the biological consequences of the metabolism of polycyclic aromatic hydrocarbons and certain drugs.[40,41] One well-established example is the metabolism of benzo[*a*]pyrene (**4**) to the highly carcinogenic (7*R*,8*S*,9*S*,10*R*)-7,8-diol-9,10-epoxide as illustrated in Figure 7. In this instance microsomal epoxide hydrolase is regiospecific in the addition of water to the 8-position of (7*R*,8*S*)-benzo[*a*]pyrene-7,8-oxide (**5**), the predominate isomer produced by cytochrome P450, giving the (7*R*,8*R*)-dihydrodiol (**6**), which is stereoselectively oxidized to the diolepoxide (**7**). Of the four possible diastereomeric diolepoxides, the one formed (**7**) is the most potent carcinogen. Its formation is a direct consequence of the stereochemical preferences of cytochrome P450 and epoxide hydrolase.

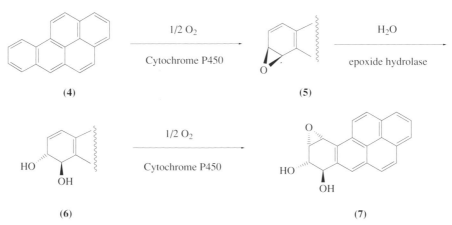

Figure 7   Predominate pathway for the formation of (7*R*,8*S*,9*S*,10*R*)-7,8-dihydroxy-9,10-oxy-7,8,9,10-tetra-
hydrobenzo[*a*]pyrene, (**7**) from benzo[*a*]pyrene, (**4**).[40]

There is also considerable documentation of the stereoselectivity and regioselectivity of the soluble enzyme with numerous substrates.[42] The soluble enzyme appears to play an important role in the metabolism of fatty acid epoxides.[43-45] Although the true substrates for the soluble enzyme are not known, the enzyme may be involved in protection of the cell against oxidative stress.[42]

5.04.2.3.2 *Epoxide hydrolases in organic synthesis*

As the use of enzymes in organic synthesis becomes more commonplace, it is only natural that epoxide hydrolases begin to be considered part of the enzymological armamentarium of the synthetic chemist. The high-degree regioselectivity and stereoselectivity make them useful in resolving chiral epoxides or producing chiral vicinal diols in high enantiomeric excess. Several examples of such resolutions have appeared in the literature.[46-51] These reactions need not be done with purified enzymes but can be accomplished with crude enzyme preparations from a variety of organisms including fungi and bacteria. Moreover, the availability of plasmid-derived expression systems for the enzyme from a number of sources makes purified enzymes from both vertebrate and microbial sources available in large quantities.[25,30,31,52,53] In principle, the expression vectors for these enzymes can be mutated to alter active site structures and give rise to catalysts with altered catalytic specificities. This remains an undeveloped area of research where the possibilities are enormous.

It should also be possible to use epoxide hydrolases in the reverse direction for the stereospecific synthesis of epoxides from diol substrates or other precursors capable of alkylating the carboxyl group. That the formation of the ester from the epoxide is reversible indicates that the enzyme is capable of catalyzing the formation of epoxides once an alkyl-enzyme intermediate is formed. Under the appropriate, perhaps anhydrous, conditions, the condensation of the enzyme with a diol or similar compound of the correct stereochemistry should be possible. Leaving groups other than the hydroxyl group can be contemplated for the alkylation step.

5.04.2.4 Structure and Mechanism of Leukotriene A$_4$ Hydrolase

5.04.2.4.1 *Structure of the enzyme*

The sequences of leukotriene A$_4$ hydrolase provide some insight into the mechanism of action of the enzyme and its relationship to other hydrolytic enzymes. The most striking finding from sequence comparisons with other known proteins is that leukotriene A$_4$ hydrolase resembles several Zn^{2+}-dependent metalloproteinases, an observation consistent with biochemical evidence that the hydrolase activity of the enzyme requires Zn^{2+}.[22,54-56] Furthermore, the enzyme has a very efficient arginine aminopeptidase activity with k_{cat}/K_M values exceeding, in some instances, 10^6 M^{-1} s^{-1}, that also requires Zn^{2+}.[57] Sequence alignments and site-directed mutagenesis have identified the ligands that comprise the Zn^{2+} binding site to be two histidines and a glutamate.[58] Sequence comparisons of the murine enzyme with thermolysin have also suggested catalytic roles for other residues, which are discussed in more detail below. Diffraction-quality crystals of the enzyme complexed with the inhibitor bestatin have been reported though, at the time of writing, the structure solution has not.[59]

5.04.2.4.2 *Mechanism of catalysis*

Leukotriene A$_4$ hydrolase is the only epoxide hydrolase known to catalyze the conjugate addition of water to an epoxide converting leukotriene A$_4$ [(5S)-*trans*-5,6-oxido-7,9-*trans*-11,14-*cis*-eicosatetraenoic acid (**8**)] to leukotriene B$_4$ [(5S,12R)-5,12-dihydroxy-6,14-*cis*-8,10-*trans*-eicosatetraenoic acid (**9**)]. The reaction catalyzed by the enzyme is shown in Figure 8 and differs from the spontaneous hydration only in the stereospecificity of addition and control of the double-bond geometry. As a consequence, some of the most important aspects of the mechanism involve the role of the metal ion in catalysis, how the regiochemistry and stereochemistry of the hydroxy group addition are controlled, and how the intervening double bond geometry is determined. Unfortunately, many of the details of the catalytic mechanism of the enzyme remain obscure.

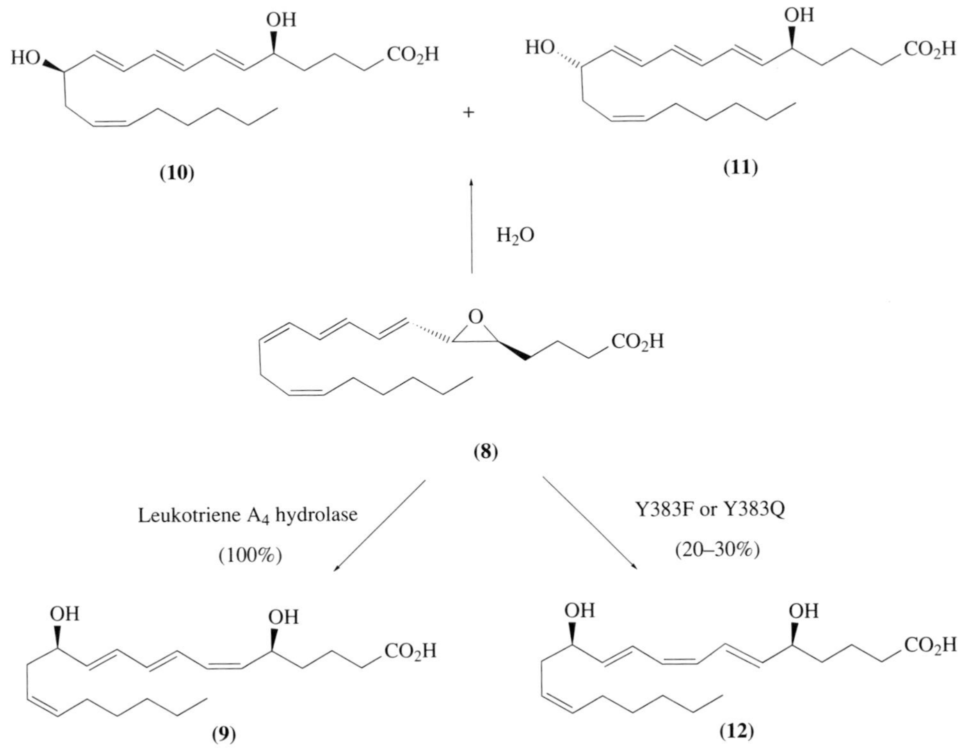

Figure 8 Spontaneous and enzyme-catalyzed hydrations of leukotriene A_4 (**8**).

Although there are several reasonable ideas concerning the mechanism of catalysis, there is a paucity of experimental evidence to support any of the proposals. The metal ion could be involved in more than one aspect of the reactions catalyzed by the enzyme. It could, for example, serve to activate a water molecule for nucleophilic attack in either the peptidase or epoxide hydrolase reaction or it might assist stabilizing the tetrahedral intermediate in peptide hydrolysis or in polarizing the C—O bond of the oxirane ring. The former is the most reasonable hypothesis, particularly with respect to the peptidase activity. Both possibilities are illustrated in Scheme 2 for the epoxide hydrolase reaction. Although the mechanism shown involves direct attack of the water on the substrate, as yet there is no evidence that rules out a covalent intermediate in the reaction similar to those discussed for canonical epoxide hydrolases.

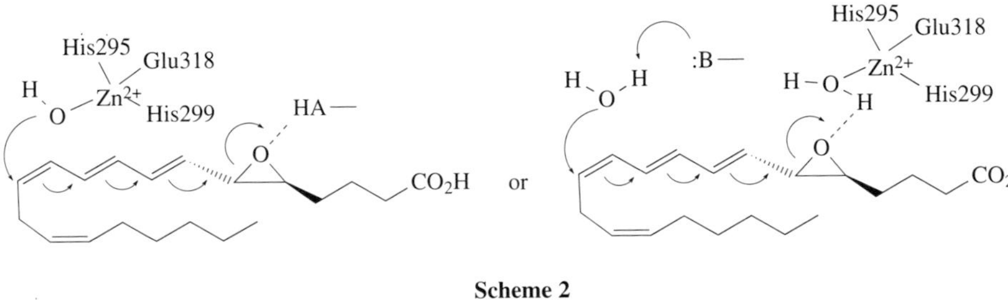

Scheme 2

It is important to consider that the enzyme has two distinct hydrolytic activities that have quite different requirements for transition state stabilization and, therefore, may utilize completely different aspects of the active site. Inasmuch as leukotriene A_4 rapidly decomposes to a mixture of (5S,12R)- and (5S,12S)-5,12-dihydroxy-6,8,10-*trans*-14-*cis*-eicosatetraenoic acid ((**10**) and (**11**)) (Figure 8) the details of stereochemical control, rather than activation of water or the substrate, seem to be the most intriguing issues in understanding the epoxide hydrolase activity. Some progress has been made in this regard.

Several lines of evidence suggest that Tyr383 is involved in catalysis and helps determine the geometry of the double bonds. Leukotriene A$_4$ hydrolase undergoes suicide inactivation by the substrate with a partition ratio of 130 turnovers per inactivation event.[60,61] The inactivation involves covalent modification of Tyr383, though the structure of the adduct is not known. It has been speculated that the inactivation may play some regulatory function. Unlike the native enzyme, the Y383F mutant is not inactivated by the substrate.[61] Although the Y383F and Y383Q mutants of the enzyme still efficiently catalyze the epoxide hydrolase reaction and the majority of the product is leukotriene B$_4$, a significant fraction (20–30%) of the product is the geometric isomer (12), (5S,12R)-5,12-dihydroxy-6,10-*trans*-8,14-*cis*-eicosatetraenoic acid (Figure 8).[62] Thus, one role of the tyrosine residue is to control the stereochemistry of the double bonds during the conjugate addition. Interestingly, the alternative product (12) is a naturally occurring isomer possessing biological activity.[63] Finally, the tyrosine residue is also important for efficient peptide hydrolysis by the enzyme, where it is postulated to act as a proton donor to the leaving amine.[64]

Leukotriene B$_4$ is a potent proinflammatory mediator with leukocyte chemotactic and aggregating properties. The compound directs the migration of neutrophils toward loci of inflammation and stimulates their adhesion to the vascular endothelium. Their involvement in the inflammatory process makes the enzymes participating in their formation targets for the development of new anti-inflammatory agents. A number of inhibitors of leukotriene A$_4$ hydrolase based on the involvement of Zn^{2+} in catalysis have been developed.[65,66] Others have been isolated from natural sources.[67] Further work in this area will surely be stimulated by a structure of the enzyme.

5.04.3 GLUTATHIONE TRANSFERASES

5.04.3.1 Convergent Families of Glutathione Transferases

There appear to be three superfamilies of GSH transferases that have arisen through convergent evolutionary pathways.[68] All are capable of catalyzing the addition of glutathione (13) to epoxides, though the efficiency varies with the substrate structure and the nature of the enzyme. The most extensively studied group are the soluble enzymes that are found in the cytosol of cells in virtually all aerobic organisms. The second superfamily are membrane-bound enzymes that include an enzyme involved in xenobiotic metabolism and the leukotriene C$_4$ synthases.[69] A third family is represented by the fosfomycin resistance proteins, which are the only known metalloglutathione transferases.[70] Each of the superfamilies represent an independent solution to the problem of how to catalyze GSH addition to electrophilic compounds including epoxides.

(13)

5.04.3.1.1 Soluble glutathione transferases

The cytosolic or soluble enzymes derive from a superfamily of genes that encode at least six individual classes of enzymes, five of which are known to be represented in vertebrates. The enzymes are named with respect to the family or class in which they fall (alpha, mu, pi, kappa, sigma, and theta). Members of the soluble enzyme superfamily are all dimers and are related to bacterial stress proteins and elongation factor 1$_\gamma$.[71] The three-dimensional structures of members of five of the families are known and specific features in the active sites of several have been identified as necessary for efficient catalysis toward epoxide substrates as discussed below.

5.04.3.1.2 *Microsomal enzymes and leukotriene C₄ synthase*

There are two general types of membrane-bound glutathione transferases known: one is involved in xenobiotic metabolism and the other is leukotriene C_4 synthase. Neither of the enzymes are related to the soluble GSH transferases discussed above. Microsomal GSH transferase is an integral membrane protein which has been characterized from several sources and appears to be involved in the metabolism of xenobiotics. Moreover, it appears to be related to the other membrane-bound enzyme, leukotriene C_4 synthase, in that the two proteins are about the same size, share a small amount of sequence identity, and are both membrane-bound. Leukotriene C_4 synthase catalyzes the addition of glutathione to the oxirane ring of leukotriene A_4. Although neither of the membrane-bound enzymes are related to the soluble proteins, they do show a remarkable degree of sequence similarity to the 5-lipoxygenase activating protein.

5.04.3.1.3 *Metalloglutathione transferases*

There are two closely related plasmid-encoded bacterial glutathione transferases that are associated with bacterial resistance to the antibiotic fosfomycin, $(1R,2S)$-1,2-epoxypropylphosphonic acid, **(14)**.[70,72] Their primary structures are not related to any of the soluble or microsomal GSH transferases nor do these enzymes catalyze the addition of GSH to any of the normal electrophilic substrates used to assay the soluble enzymes. The enzymes are very specific at catalyzing the nucleophilic opening of the epoxide of fosfomycin. The fosfomycin-resistance proteins are members of a metalloenzyme superfamily that includes glyoxalase I and bacterial dioxygenases.[70]

(14)

5.04.3.2 Structure and Mechanism of the Soluble Enzymes

5.04.3.2.1 *Structure of the enzymes*

There is an enormous amount of structural information about the canonical soluble GSH transferases. The details of the structures have been reviewed extensively and will not be reiterated here except where related to epoxides as substrates.[68,73–75] All of the structures reveal the same basic protein fold, which consists of two domains (Figure 9). The overall fold of the N-terminal domain is identified as a member of the thioredoxin superfamily fold, which includes glutaredoxin and glutathione peroxidase. This glutathione binding domain constitutes about one-third of the protein and consists of a β-α-β-α-β-β-α structural motif that forms a mixed four-strand β-sheet (Figure 9). The C-terminal two-thirds of the protein is an all-α-helical domain with a unique protein fold and a core consisting of a bundle of four helices. Most of the interactions between the protein and GSH that are responsible for binding the substrate are with residues in the N-terminal domain. Among the most important interactions are those with the sulfur of GSH. The alpha, mu, pi, and sigma enzymes have an active site tyrosine residue, the hydroxyl group of which acts as a hydrogen bond donor to the sulfur of the peptide. A seryl hydroxyl group replaces the tyrosine in the class theta enzyme. The net result of the interaction with the sulfur is to lower the pK_a of the bound sulfhydryl group.[77,78]

The structures of product complexes have suggested that many of the interactions between the protein and the xenobiotic substrate are with the C-terminal domain. Some of these interactions appear to be very important in dictating the substrate selectivity of the isoenzymes. One example that is relevant to epoxide ring opening reactions is the structure of a class mu enzyme (the M1-1 isoenzyme from rat) with the products of addition of GSH to the model arene oxide, phenanthrene 9,10-oxide, **(15)** (Scheme 3).[79] One of the structures reveals a hydrogen bonding interaction between the hydroxyl group of Tyr115 in the C-terminal domain and the hydroxyl group in the product as illustrated in Figures 9 and 10. The observation clearly suggests that the tyrosyl hydroxyl group is

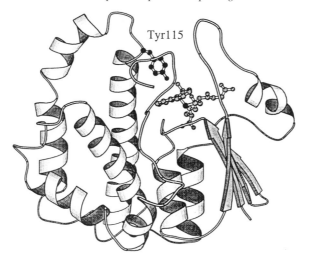

C-terminal domain   N-terminal domain

Figure 9 MOLSCRIPT[76] representation of the structure of one subunit of the GSH transferase M1-1 from rat in complex with (9S,10S)-(**16**) illustrating domain structure and the proximity of the hydroxyl group of Tyr115 (side chain atoms shown in black) to the hydroxyl group of the product. The coordinates are from PDB file 2GST.

in a position to act as a hydrogen bond donor to the oxirane oxygen in the transition state for the nucleophilic addition.[79,80] In fact there is a clear correlation with the presence of a tyrosine at this position in the sequence of other isoenzymes and their ability to catalyze additions to epoxides.[81]

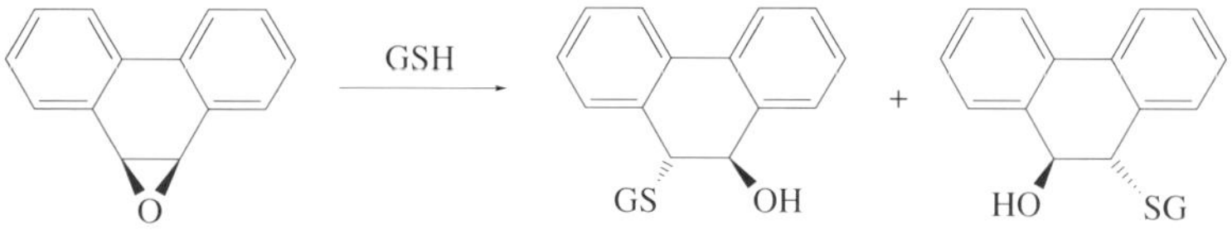

Scheme 3

5.04.3.2.2 *Mechanism of catalysis*

A discussion of the mechanism of nucleophilic additions to epoxides can be broken down into three aspects, including how the enzyme activates the nucleophile, how the enzyme might specifically stabilize the transition state for epoxide opening, and what determines the regioselectivity and stereoselectivity of the addition. With respect to the first issue of how the enzyme enhances the nucleophilicity of the thiol of GSH, it is clear that the enzyme is capable of lowering the pK_a of the thiol to below physiological pH so that the major species in the active site under normal conditions is the thiolate anion. The details of how this occurs has been discussed in detail elsewhere and usually involves the participation of an active site hydroxyl group (e.g., E—O—H $\cdots$ $^-$SG). Moreover, the activation is accomplished in a solvent restricted environment, which should further enhance the reactivity of the anion.

More importantly, some GSH transferases have apparently evolved specific catalytic devices to enhance nucleophilic addition to epoxides. The structure of the class mu M1-1 isoenzyme in complex with (9S,10S)-9-(S-glutathionyl)-10-hydroxy-9,10-dihydrophenanthrene [(9S,10S)-(**16**)], one of the diastereomeric products of the reaction of GSH with phenanthrene 9,10-oxide, indicates that the hydroxyl group of Tyr115 is within hydrogen bonding distance of the 10-hydroxyl group of the bound product and, by implication, is proximal to the oxirane oxygen of the substrate in the transition state. Mutation of Tyr115 to phenylalanine (Y115F) has no significant effect on the catalytic efficiency of the enzyme toward 1-chloro-2,4-dinitrobenzene. The native enzyme is reasonably efficient in the addition of GSH to (**15**) with $k_{cat} = 0.4$ s^{-1}. However, the Y115F mutant is

Figure 10   Expanded view of the active site of the M1-1 GSH transferase illustrating the positions of residues important in the regioselectivity and stereoselectivity of the enzyme. The side chain atoms of Val9, Ile111, and Ser209 are shown in black while the side chain atoms of Tyr115 are shown in light gray.

severely impaired in the same reaction ($k_{cat} = 0.0044$ s^{-1}), evidence that the hydroxyl group of Tyr115 provides electrophilic assistance in the epoxide ring opening, as illustrated in Figure 11. Interestingly, the class mu and pi enzymes, which have a tyrosyl residue in this position, are generally better catalysts with epoxides than are class alpha and sigma enzymes, which do not.[81] Furthermore, the F108Y mutant of the class sigma enzyme is more efficient than the native enzyme towards (**15**).[81] Certain isoenzymes appear to be very efficient with particular epoxide substrates, as is the case with a murine class alpha enzyme toward the (7*R*,8*S*,9*S*,10*R*)-7,8-dihydrozy-9,10-oxy-7,8,9,10-tetrahydrobenzo(*a*)pyrene, (**7**).[82] The structural basis for this efficiency remains to be established.

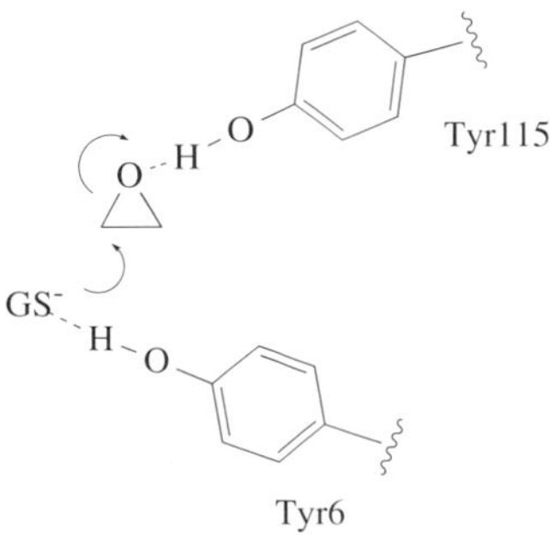

Figure 11   Proposed role of the tyrosine residue in epoxide ring openings catalyzed by class mu and pi GSH transferases.

The stereoselectivity and regioselectivity of the GSH transferases is influenced considerably by the nature of the amino acids in the xenobiotic binding site. For example, the M2-2 isoenzyme from rat gives exclusively (9*S*,10*S*)-(**16**) as the product (Scheme 3), whereas the M1-1 isoenzyme yields roughly an equal mixture of (9*S*,10*S*)-(**16**) and (9*R*,10*R*)-(**16**). The difference in stereoselectivity is principally dictated by three residues, at positions 9, 111, and 209, that line the xenobiotic substrate binding site as illustrated in the expanded view of the active site in Figure 10. The triple mutant V9Y/I111A/S209A of the M1-1 isoenzyme gives 95% (9*S*,10*S*)-(**16**) compared with 43% for M1-1.[83] These structural changes also contribute to the differences observed in the efficiency of GSH transferases with very unstable epoxides such as the potent hepatocarcinogen, aflatoxin-8,9-epoxide.[84,85]

5.04.3.3 Structure and Mechanism of the Microsomal Enzymes

5.04.3.3.1 *Structure of membrane-bound GSH transferases*

Much less is known about the structures of the membrane-bound GSH transferases. The microsomal enzyme involved in xenobiotic metabolism, for which a two-dimensional projection structure derived from electron crystallography at a resolution of 4 Å is available, is the best characterized in this family.[86] The two-dimensional crystals suggest that the enzyme is a trimer in the membrane consisting of a core of six high-density regions, which may well be membrane-spanning α-helices. Unfortunately, it is not possible to draw an envelope defining a single subunit in the trimer nor is it possible to locate the active site. Additional information on the membrane topology of the protein has been derived from a variety of sources including the primary structures, cross-linking experiments, radiation inactivation analysis, and hydrodynamic properties. All of the information is consistent with a trimeric structure of the enzyme.[87,88] The overall topology is best described as an N-terminal in/C-terminal out orientation with five membrane-spanning regions. Although the leukotriene C$_4$ synthases are members of this family, it is unclear how similar their tertiary or quaternary structures are to the xenobiotic metabolizing enzyme.

5.04.3.3.2 *Mechanism of microsomal glutathione transferases*

The catalytic mechanism of the microsomal enzyme is poorly understood. It is reasonable to assume that the enzyme lowers the pK_a of the thiol of bound GSH with devices similar to those found in the cytosolic enzyme. However, recent site-directed mutagenesis experiments suggest that a tyrosyl residue is not involved.[89] Although chemical modification and site-directed mutagenesis experiments have been carried out on the enzyme, they have yet to lead to the unambiguous identification of any residues involved in catalysis.[89,90] The situation is similar with leukotriene C$_4$ synthase where the results of site-directed mutagenesis have been interpreted to implicate Arg51 and Tyr93 in specific roles in catalysis.[91] For example, replacement of Arg51 with Thr or Ile abolishes activity while replacement with Lys or His results in fully active enzyme. The authors suggest that Arg51 is therefore involved in specific protonation of the oxirane. Were this true, substitution of the much smaller His would not be expected to support activity. It seems equally possible that Arg51 serves some crucial structural role. That the Y93F mutant exhibits reduced activity and an altered pH profile perhaps suggests that the hydroxyl group of Y93 may serve to stabilize the thiolate of bound GSH. However, it might just as well serve to protonate the oxirane oxygen. In the absence of supporting information, the conclusions from mutagenesis must be viewed with caution.

The microsomal enzyme involved in xenobiotic metabolism is not particularly efficient in nucleophilic additions to epoxides. In contrast, leukotriene C$_4$ synthase is highly specific in catalyzing the addition of GSH to the oxirane ring of leukotriene A$_4$, as illustrated in Scheme 4. Unlike the hydration of leukotriene A$_4$, the nucleophile is added to one of the oxirane carbons. As expected, the addition occurs with inversion of configuration. The high reactivity of the substrate suggests that the role of the enzyme may be to increase the reactivity of the thiol of GSH as well as to control the regiochemistry of the addition. These issues are likely to remain open until more structural information is available.

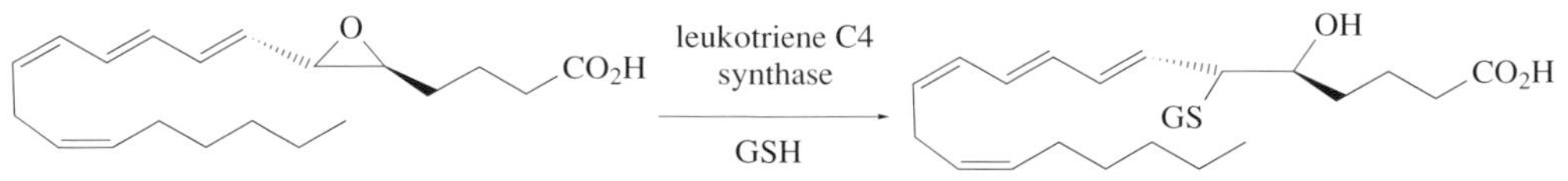

Scheme 4

5.04.3.4   Structure and Mechanism of the Fosfomycin Resistance Protein

5.04.3.4.1   *Mechanism of action of fosfomycin*

Fosfomycin (**14**) is an antibiotic produced by selected strains of *Streptomyces*.[92,93] The functional moiety in the molecule is an oxirane ring that is unusually stable under normal physiologic conditions.[94] The stability of the epoxide can be attributed primarily to the fact that the phosphonate group bears a net negative charge preventing the approach of anionic nucleophiles. Under basic conditions the additions of thiolate anions do not occur at appreciable rates. Addition of thiols occurs only under strongly acidic conditions (e.g., liquid HF, $-40\,°C$).

The mechanism of action of the antibiotic is by alkylation of a cysteine residue (Cys115) in the active site of UDP-*N*-acetylglucosamine-3-enolpyruvyltransferase or MurA, resulting in irreversible inactivation of the enzyme catalyzing the first committed step in peptidoglycan biosynthesis.[94,95] The regiochemistry of this reaction is illustrated in Figure 12. The inactivation occurs only in the presence of the substrate UDP-*N*-acetylglucosamine. The three-dimensional structure of MurA inactivated with the antibiotic reveals that the negative charge on the phosphonate is accommodated by an anion binding site on the protein composed of the side chains of Lys22, Arg120, and Arg397 (Figure 12).[96] In addition, the 3-hydroxyl group of the bound UDP-*N*-acetylglucosamine is within hydrogen bonding distance of the hydroxyl group of the bound drug. However, the presence of this hydroxyl group is not crucial to the inactivation since the enzyme is also effectively inactivated by fosfomycin in the presence of 3-deoxy-UDP-*N*-acetylglucosamine. The latter result suggests that the role of the substrate in the inactivation is to induce the appropriate conformation for alkylation of Cys115.[95]

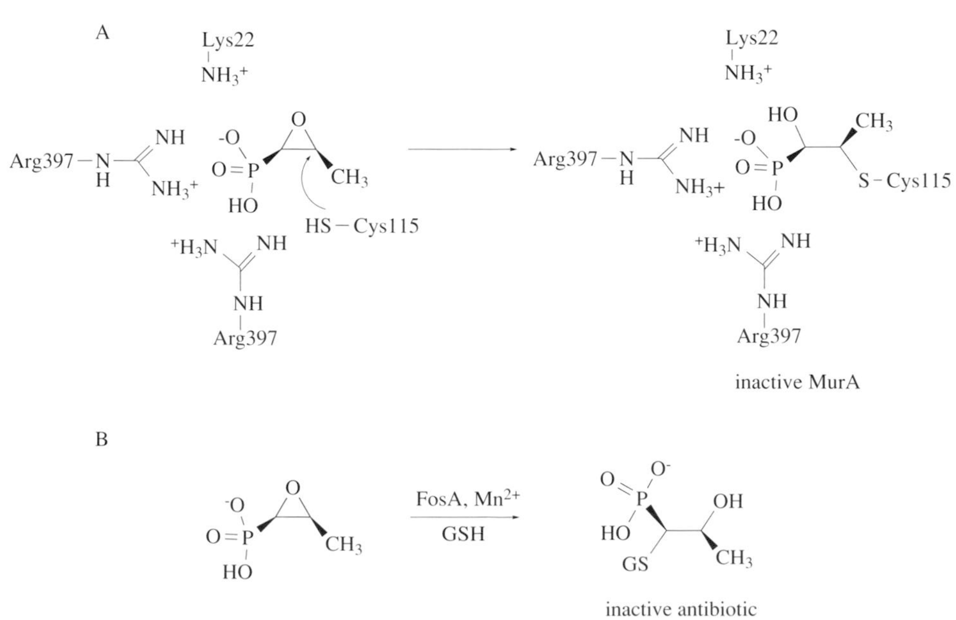

Figure 12   (A) Inactivation of MurA with fosfomycin and (B) the inactivation of fosfomycin by the fosfomycin resistance protein, FosA.

One of the principal clinical problems associated with fosfomycin as an antibiotic is a rapidly acquired resistance to the drug associated with either transport, alterations in the target MurA, or plasmid-encoded resistance proteins. Investigations of the plasmid-borne resistance revealed that the resistance gene encoded a 16 kDa polypeptide (FosA), which catalyzed the addition of glutathione to the antibiotic, as shown in Figure 12, rendering it inactive.[97,98]

5.04.3.4.2 Structure of FosA and FosB

At least two fosfomycin resistance proteins, FosA and FosB, have been identified. Although the *fosB* gene product from *Staphylococcus epidermidis* has not been studied extensively nor been directly demonstrated to be a glutathione transferase, it shares 38% sequence identity with FosA and confers resistance to the antibiotic.[99] The three-dimensional structure of either FosA or FosB has yet to be determined. FosA is known to exist as a homodimer under physiological conditions. Each subunit contains one divalent cation binding site which, in the native state, is occupied by Mn^{2+}.[70,100] The dimer is stable in the absence of metal, indicating that it is not necessary for the integrity of the quaternary structure.

A BLAST[101] search of the protein sequence database revealed that both FosA and FosB shared regions of sequence similarity with two other metalloenzymes, glyoxalase I and the extradiol dioxygenases, whose three-dimensional structures are known.[102,103] Mapping of the sequence similarities of FosA and FosB onto the crystal structures of both enzymes indicates that the conserved regions of the primary structure involve amino acid residues that participate in metal ion binding. Two of the ligands to the metal in FosA are tentatively identified as His7 and Glu113 by analogy with glyoxalase I and 2,3-hidroxybiphenyl-1,2-dioxygenase.[70] Interestingly, glyoxalase I and the dioxygenases have the same $\beta\alpha\beta\beta$ domain topology but differ with respect to the connectivity of the domains.[103] It is tempting to speculate that FosA will have the same basic domain topology.

5.04.3.4.3 Mechanism of FosA

The FosA-catalyzed addition of GSH to the antibiotic occurs with inversion of configuration at the oxirane carbon[104] and with the opposite regiochemistry observed in the alkylation of Cys115 of MurA (Figure 12).[97] The extreme stability of fosfomycin toward nucleophilic attack immediately suggests a role for the metal ion in the catalytic mechanism, either to neutralize the negative charge on the phosphonate or to polarize the C—O bond or both. The paramagnetic properties of Mn^{2+} can be used as a probe of the environment of the metal ion in various substrate and product complexes. EPR spectroscopy and water proton NMR relaxation rates indicate that the Mn^{2+} center interacts strongly with the substrate fosfomycin ($K_d = 17$ μM), weakly with the product ($K_d = 1.1$ mM), and very weakly with GSH. The superhyperfine interactions between the electron spin of enzyme-bound Mn^{2+} and the nuclear quadrapole of $H_2{}^{17}O$ suggests that a minimum of three of the metal coordination sites are occupied by water. A pronounced decrease in the paramagnetic contribution to the water proton NMR relaxation rates upon addition of fosfomycin to the $E \cdot Mn^{2+}$ complex is consistent with the loss of one or more water molecules from the inner coordination sphere of the metal.

A reasonable mechanism for the enzyme based on these observations is illustrated in Figure 13. The chemistry proposed is similar in many respects to that proposed for metal-catalyzed nucleophilic additions to 2-pyridyloxirane (Figure 1).[7] The principal differences are the charge neutralization afforded by the metal in the FosA-catalyzed reaction and the possible specific activation of orientation of the nucleophile by other interactions in the active site.

5.04.3.4.4 Relationship to other metalloenzymes

On further consideration FosA appears to be a member of a superfamily of metalloenzymes that at first glance seem to have little in common with respect to the reactions they catalyze. The enzymes even use different metals and catalyze entirely different reactions. Glyoxalase I is a Zn^{2+}-dependent enzyme that catalyzes a shielded proton transfer to accomplish rearrangement of the thiohemiacetal adduct between GSH and methylglyoxal to *S*-(D-lactoyl)glutathione, while the dioxygenases are Fe^{2+}- or Mn^{2+}-dependent enzymes that catalyze the oxidative cleavage of a carbon–carbon bond

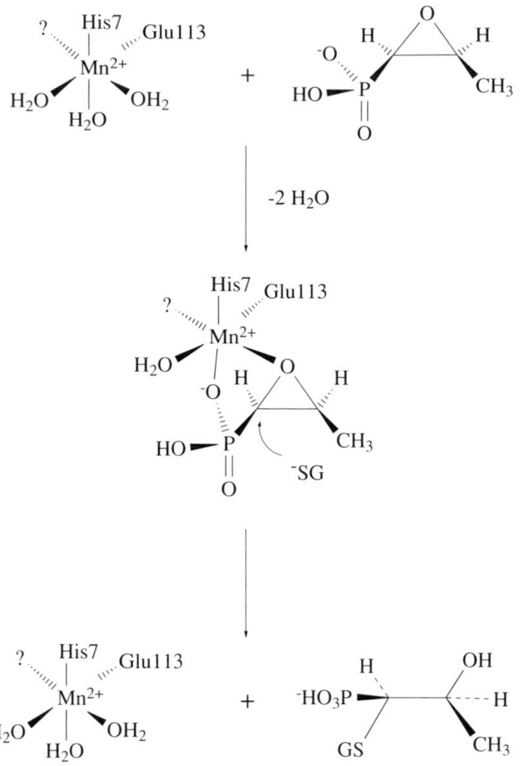

Figure 13 Proposal for the role of the metal in the FosA-catalyzed addition of GSH to fosfomycin.

adjacent to the vicinal hydroxyl groups in catechol-like structures.[105,106] In fact, there is one mechanistic feature common to all of the reactions, the chelation of vicinal oxygen ligands on a substrate or intermediate to promote a reaction. In each instance there are open (solvent occupied) coordination sites on an octahedrally coordinated metal. The mechanistic imperative driving the evolution of this superfamily of metalloenzymes appears to be bidentate coordination to a metal center for electrophilic activation of a substrate or stabilization of an intermediate in the enzyme-catalyzed reactions.

ACKNOWLEDGMENTS

The work cited from this laboratory was funded by Grants from the National Institutes of Health (R01 GM49878, R01 GM30910, R01 AI42756 and P30 ES00133). I express my thanks to all my current and former co-workers whose work is cited here.

5.04.4 REFERENCES

1. P. Y. Bruice, T. C. Bruice, H. Yagi, and D. M. Jerina, *J. Am. Chem. Soc.*, 1976, **98**, 2973.
2. A. S. Rao, S. K. Paknikar, and J. G. Kirtane, *Tetrahedron*, 1983, **39**, 2323.
3. D. M. Jerina, H. Yagi, and J. W. Daly, *Heterocycles*, 1973, **1**, 267.
4. T. C. Bruice and P. Y. Bruice, *Acc. Chem. Res.*, 1976, **9**, 378.
5. A. Hassner (ed.), "Small Ring Heterocycles—Part 3 Oxiranes, Arene Oxides, Oxaziridines, Dioxetanes, Thietanes, Thietes, Thiazetes," Wiley, New York, 1985.
6. R. P. Hanzlik and W. J. Michaely, *J. Chem. Soc., Chem. Commun.*, 1975, 113.
7. R. P. Hanzlik and A. Hamburg, *J. Am. Chem. Soc.*, 1978, **100**, 1745.
8. C. A. VanderWerf, R. Y. Heisler, and W. E. McEwen, *J. Am. Chem. Soc.*, 1954, **76**, 1231.
9. P. Borgeat and B. Samuelsson, *Proc. Natl. Acad. Sci. USA*, 1979, **76**, 3213.
10. R. N. Armstrong, *CRC Crit. Rev. Biochem.*, 1987, **22**, 39.
11. D. L. Ollis, E. Cheah, M. Cygler, B. Dijkstra, F. Frolow, S. M. Franken, M. Harel, S. J. Remington, I. Silman, J. Schrag, J. L. Sussman, K. H. G. Verschueren, and A. Goldman, *Protein Eng.*, 1992, **5**, 197.
12. A. Y. H. Lu, D. Ryan, D. M. Jerina, J. W. Daly, and W. Levin, *J. Biol. Chem.*, 1975, **250**, 8283.
13. K. Ota and B. D. Hammock, *Science*, 1980, **207**, 1479.

14. F. S. Heinemann and J. Ozols, *J. Biol. Chem.*, 1984, **259**, 797.
15. T. D. Porter, T. W. Beck, and C. B. Kasper, *Arch. Biochem. Biophys.*, 1986, **248**, 121.
16. M. Knehr, H. Thomas, M. Arand, T. Gebel, H.-D. Zeller, and F. Oesch, *J. Biol. Chem.*, 1993, **268**, 17623.
17. D. F. Grant, D. H. Storms, and B. D. Hammock, *J. Biol. Chem.*, 1993, **268**, 17628.
18. G. M. Lacourciere and R. N. Armstrong, *Chem. Res. Toxicol.*, 1994, **7**, 121.
19. M. Arand, D. F. Grant, J. K. Beetham, T. Friedberg, F. Oesch, and B. D. Hammock, *FEBS Lett.*, 1994, **338**, 251.
20. G. M. Lacourciere and R. N. Armstrong, *J. Am. Chem. Soc.*, 1993, **115**, 10466.
21. N. T. Nashed, D. P. Michaud, W. Levin, and D. M. Jerina, *Arch. Biochem. Biophys.*, 1985, **241**, 149.
22. J. F. Medina, O. Rådmark, C. D. Funk, and J. Z. Haeggstrom, *Biochem. Biophys. Res. Commun.*, 1991, **176**, 1516.
23. G. C. DuBois, E. Appella, W. Levin, A. Y. H. Lu, and D. M. Jerina, *J. Biol. Chem.*, 1978, **253**, 2932.
24. R. N. Armstrong, W. Levin, and D. M. Jerina, *J. Biol. Chem.*, 1980, **255**, 4698.
25. P. A. Bell and C. B. Kasper, *J. Biol. Chem.*, 1993, **268**, 14011.
26. B. Borhan, A. D. Jones, F. Pinot, D. F. Grant, M. J. Kurth, and B. D. Hammock, *J. Biol. Chem.*, 1995, **270**, 26923.
27. B. D. Hammock, F. Pinot, J. K. Beetham, D. F. Grant, M. E. Arand, and F. Oesch, *Biochem. Biophys. Res. Commun.*, 1994, **198**, 850.
28. H.-F. Tzeng, L. T. Laughlin, S. Lin, and R. N. Armstrong, *J. Am. Chem. Soc.*, 1996, **118**, 9436.
29. F. Pinot, D. F. Grant, J. K. Beetham, A. G. Parker, B. Borhan, S. Landt, A. D. Jones, and B. D. Hammock, *J. Biol. Chem.*, 1995, **270**, 7968.
30. L. T. Laughlin, H.-F. Tzeng, S. Lin, and R. N. Armstrong, *Biochemistry*, 1998, **37**, in press.
31. H.-F. Tzeng, L. T. Laughlin, and R. N. Armstrong, *Biochemistry*, 1998, **37**, in press.
32. M. Arand, H. Wagner, and F. Oesch, *J. Biol. Chem.*, 1996, **171**, 4223.
33. F. Pries, J. Kingma, G. H. Krooshof, C. M. Jeronimus-Stratingh, A. P. Bruins, and D. B. Janssen, *J. Biol. Chem.*, 1995, **270**, 10405.
34. M. Shou, F. J. Gonzalez, and H. V. Gelboin, *Biochemistry*, 1996, **35**, 15807.
35. S. J. Roberts-Thomson, M. E. McManus, C. C. Duke, R. Agnew, and G. M. Holder, *Mol. Pharmacol.*, 1996, **49**, 105.
36. J. D. Adams, Jr., H. Yagi, W. Levin, and D. M. Jerina, *Chemico-Biol. Interact.*, 1995, **95**, 57.
37. J. M. Sayer, H. Yagi, P. J. van Bladeren, W. Levin, and D. M. Jerina, *J. Biol. Chem.*, 1985, **260**, 1630.
38. R. N. Armstrong, B. Kedzierski, W. Levin, and D. M. Jerina, *J. Biol. Chem.*, 1981, **256**, 4726.
39. G. Bellucci, C. Chiappe, and G. Ingrosso, *Chirality*, 1994, **6**, 577.
40. W. Levin, M. K. Buening, A. W. Wood, R. L. Chang, B. Kedzierski, D. R. Thakker, D. R. Boyd, G. S. Gadaginamath, R. N. Armstrong, H. Yagi, J. M. Karle, T. J. Slaga, D. M. Jerina, and A. H. Conney, *J. Biol. Chem.*, 1980, **255**, 9067.
41. G. Bellucci, G. Berti, C. Chiappe, A. Lippi, and F. Marioni, *J. Med. Chem.*, 1987, **30**, 768.
42. L. Meijer and J. W. DePierre, *Chemico-Biol. Interact.*, 1988, **64**, 207.
43. D. C. Zeldin, S. Wei, J. R. Falck, B. D. Hammock, J. R. Snapper, and J. H. Capdevila, *Arch. Biochem. Biophys.*, 1995, **316**, 443.
44. D. C. Zeldin, J. Kobayashi, J. R. Falck, B. S. Winder, B. D. Hammock, J. R. Snapper, and J. H. Capdevila, *J. Biol. Chem.*, 1993, **268**, 6402.
45. M. F. Moghaddam, K. Motoba, B. Borhan, F. Pinot, and B. D. Hammock, *Biochim. Biophys. Acta*, 1996, **1290**, 327.
46. R. V. A. Orva, W. Kroutil, and K. Faber, *Tetrahedron Lett.*, 1997, **38**, 1753.
47. G. Bellucci, C. Chiappe, A. Cordoni, and G. Ingrosso, *Tetrahedron Lett.*, 1996, **37**, 9089.
48. I. V. J. Archer, D. J. Leak, and D. A. Widdowson, *Tetrahedron Lett.*, 1996, **37**, 8819.
49. W. Kroutil, M. Mischitz, P. Plachota, and K. Faber, *Tetrahedron Lett.*, 1996, **37**, 8379.
50. S. Pedragosamoreau, A. Aarchelas, and R. Furstoss, *Tetrahedron Lett.*, 1996, **37**, 3319.
51. S. Pedragosamoreau, A. Archelas, and R. Furstoss, *Tetrahedron Lett.*, 1996, **52**, 4593.
52. G. M. Lacourciere, V. N. Vakharia, C. P. Tan, D. I. Morris, G. H. Edwards, M. Moos, and R. N. Armstrong, *Biochemistry*, 1993, **32**, 2610.
53. R. Rink, M. Fennema, M. Smids, U. Dehmel, and D. B. Janssen, *J. Biol. Chem.*, 1997, **272**, 14650.
54. B. L. Vallee and D. S. Auld, *Proc. Natl. Acad. Sci. USA*, 1990, **87**, 220.
55. J. Z. Haeggstrom, A. Wetterholm, R. Shapiro, B. L. Vallee, and B. Samuelsson, *Biochem. Biophys. Res. Commun.*, 1990, **172**, 965.
56. J. A. Mancini and J. F. Evans, *Eur. J. Biochem.*, 1995, **231**, 65.
57. L. Orning, J. K. Gierse, and F. A. Fitzpatrick, *J. Biol. Chem.*, 1994, **269**, 11269.
58. J. F. Medina, A. Wetterholm, O. Rådmark, R. Shapiro, J. Z. Haeggstrom, B. L. Vallee, and B. Samuelsson, *Proc. Natl. Acad. Sci. USA*, 1991, **88**, 7620.
59. H. Tsuge, H. Ago, M. Aoki, M. Furuno, M. Noma, M. Miyano, M. Minami, T. Izumi, and T. Shimizu, *J. Mol. Biol.*, 1994, **238**, 854.
60. J. F. Evans, D. J. Nathaniel, R. J. Zamboni, and A. W. Ford-Hutchinson, *J. Biol. Chem.*, 1985, **260**, 10966.
61. M. J. Mueller, M. Blomster, U. C. T. Oppermann, H. Jornvall, B. Samuelsson, and J. Z. Haeggstrom, *Proc. Natl. Acad. Sci. USA*, 1996, **93**, 5931.
62. M. J. Mueller, M. B. Andberg, B. Samuelsson, and J. Z. Haeggstrom, *J. Biol. Chem.*, 1996, **271**, 24345.
63. F. Stromberg, M. Hamberg, U. Rosenqvist, S.-E. Dahlen, and J. Z. Haeggstrom, *Eur. J. Biochem.*, 1996, **238**, 599.
64. M. Blomster, A. Wetterholm, M. J. Mueller, and J. Z. Haeggstrom, *Eur. J. Biochem.*, 1995, **231**, 528.
65. W. Yuan, B. Munoz, C.-H. Wong, J. Z. Haeggstrom, A. Wetterholm, and B. Samuelsson, *J. Med. Chem.*, 1993, **36**, 211.
66. J. H. Hogg, I. R. Ollmann, J. Z. Haeggstrom, A. Wetterholm, B. Samuelsson, and C.-H. Wong, *Bioorg. Med. Chem.*, 1995, **10**, 1405.
67. B. L. Parnas, R. C. Durley, E. E. Rhoden, B. F. Kilpatrick, N. Makkar, K. E. Thomas, W. G. Smith, and D. G. Corley, *J. Natural Products*, 1996, **59**, 962.
68. R. N. Armstrong, *Chem. Res. Toxicol.*, 1997, **10**, 2.
69. P. J. Jakobsson, J. A. Mancini, and A. W. Ford-Hutchinson, *J. Biol. Chem.*, 1996, **271**, 22203.
70. B. A. Bernat, L. T. Laughlin, and R. N. Armstrong, *Biochemistry*, 1997, **36**, 3050.
71. E. V. Koonin, A. R. Mushegian, R. L. Tatusov, S. F. Altschul, S. H. Bryant, P. Bork, and A. Valencia, *Protein Sci.*, 1994, **3**, 2045.

72. P. Arca, C. Hardisson, and J. E. Saurez, *Antimicrob. Agents Chemother.*, 1990, **34**, 844.
73. R. N. Armstrong, *Adv. Enzymol. Relat. Areas Mol. Biol.*, 1994, **69**, 1.
74. M. C. J. Wilce and M. W. Parker, *Biochim. Biophys. Acta*, 1994, **1205**, 1.
75. H. Dirr, P. Reinemer, and R. Huber, *Eur. J. Biochem.*, 1994, **220**, 645.
76. J. P. Kraulis, *J. Appl. Crystallogr.*, 1991, **24**, 946.
77. S. Liu, P. Zhang, X. Ji, W. W. Johnson, G. L. Gilliland, and R. N. Armstrong, *J. Biol. Chem.*, 1992, **267**, 4296.
78. X. Ji, P. Zhang, R. N. Armstrong, and G. L. Gilliland, *Biochemistry*, 1992, **31**, 10169.
79. X. Ji, W. W. Johnson, M. A. Sesay, L. Dickert, S. M. Prasad, H. L. Ammon, R. N. Armstrong, and G. L. Gilliland, *Biochemistry*, 1994, **33**, 1043.
80. W. W. Johnson, S. Liu, X. Ji, G. L. Gilliland, and R. N. Armstrong, *J. Biol. Chem.*, 1993, **268**, 11508.
81. X. Ji, E. C. von Rosenvinge, W. W. Johnson, S. I. Tomarev, J. Piatigorsky, R. N. Armstrong, and G. L. Gilliland, *Biochemistry*, 1995, **34**, 5317.
82. X. Hu, S. K. Srivastava, H. Xia, Y. C. Awasthi, and S. V. Singh, *J. Biol. Chem.*, 1996, **271**, 32684.
83. S. Shan and R. N. Armstrong, *J. Biol. Chem.*, 1994, **269**, 32373.
84. K. D. Raney, D. J. Meyer, B. Ketterer, T. M. Harris, and F. P. Guengerich, *Chem. Res. Toxicol.*, 1992, **5**, 470.
85. W. W. Johnson, Y.-F. Ueng, M. Widersten, B. Mannervik, J. D. Hayes, P. J. Sherratt, B. Ketterer, and F. P. Guengerich, *Biochemistry*, 1997, **36**, 3056.
86. H. Hebert, I. Schmidt-Krey, and R. Morgenstern, *EMBO J.*, 1995, **14**, 3864.
87. C. Andersson, E. Mosialou, R. Weinander, H. Herbert, and R. Morgenstern, *Biochim. Biophys. Acta*, 1994, **1204**, 298.
88. T. D. Boyer, D. A. Vessey, and E. Kempner, *J. Biol. Chem.*, 1986, **261**, 16963.
89. R. Weinander, L. Ekstrom, C. Andersson, H. Raza, T. Bergman, and R. Morgenstern, *J. Biol. Chem.*, 1997, **272**, 8871.
90. C. Andersson and R. Morgenstern, *Biochem. J.*, 1990, **272**, 479.
91. B. K. Lam, J. F. Penrose, K. Xu, M. H. Baldasaro, and K. F. Austen, *J. Biol. Chem.*, 1997, **272**, 13923.
92. D. Hendlin, E. O. Stapley, M. Jackson, H. Wallick, A. K. Miller, F. J. Wolf, T. W. Miller, L. Chaiet, F. M. Kahan, E. L. Flotz, H. B. Woodruff, J. M. Mata, S. Hernandez, and S. Mochales, *Science*, 1969, **166**, 122.
93. B. G. Christensen, W. J. Leanza, T. R. Beattie, A. A. Patchett, B. H. Arison, R. E. Ormond, F. A. Kuehl, G. Albers-Schonberg, and O. Jardetzky, *Science*, 1969, **166**, 123.
94. F. M. Kahan, J. S. Kahan, P. J. Cassidy, and H. Kroop, *Ann. N.Y. Acad. Sci.*, 1974, **235**, 364.
95. J. L. Marquardt, E. D. Brown, W. S. Lane, T. M. Haley, Y. Ichskawa, C.-H. Wong, and C. T. Walsh, *Biochemistry*, 1994, **33**, 10646.
96. T. Skarzynski, A. Mistry, A. Wonacott, S. E. Hutchinson, V. A. Kelly, and K. Duncan, *Structure*, 1996, **4**, 1465.
97. P. Arca, M. Rico, A. F. Brana, C. J. Villar, C. Hardisson, and J. E. Suarez, *Antimicrob. Agents Chemother.*, 1988, **32**, 1552.
98. P. Arca, C. Hardisson, and J. E. Suarez, *Antimicrob. Agents Chemother.*, 1990, **34**, 844.
99. R. Zilhao and P. Courvalin, *FEMS Microbiol. Lett.*, 1990, **68**, 267.
100. L. T. Laughlin, B. A. Bernat, and R. N. Armstrong, unpublished results.
101. S. F. Altschul, W. Gish, W. Miller, E. W. Myers, and D. J. Lipman, *J. Mol. Biol.*, 1990, **215**, 403.
102. S. Han, L. D. Eltis, K. N. Timmis, S. W. Muchmore, and J. T. Bolin, *Science*, 1995, **270**, 976.
103. A. D. Cameron, B. Olin, M. Ridderstrom, B. Mannervik, and T. A. Jones, *EMBO J.*, 1997, **16**, 3386.
104. B. A. Bernat, L. T. Laughlin, and R. N. Armstrong, unpublished results.
105. J. A. Landro, E. J. Brush, and J. W. Kozarich, *Biochemistry*, 1992, **31**, 6069.
106. L. Shu, Y.-M. Chiou, A. M. Orville, M. A. Miller, J. D. Lipscomb, and L. Que, *Biochemistry*, 1995, **34**, 6649.

5.05
Deamination of Nucleosides and Nucleotides and Related Reactions

VERN L. SCHRAMM
Albert Einstein College of Medicine of Yeshiva University, New York, USA

and

CAREY K. BAGDASSARIAN
College of William and Mary, Williamsburg, VA, USA

5.05.1   INTRODUCTION TO DEAMINATIONS OF PURINES AND PYRIMIDINES

Throughout this chapter "deamination" implies replacement of an amino group by a hydroxyl group. Deaminations of purines and pyrimidines are aromatic substitutions, which differ from nucleophilic substitutions in saturated compounds. Thus, back-side nucleophilic displacements (S_N2) are prohibited by the back lobe of the sp^2 orbital located within the confines of the aromatic ring. Also, S_N1 mechanisms, in which the amino group is protonated and lost in a heterolytic event, are unlikely because of the low pK_a for an exocyclic amino group in a conjugated heterocycle and the highly unfavorable energetics to form an aromatic cation (e.g., 298 kcal mol^{-1} to dissociate benzene to H$^-$ and a phenyl cation). The chemical solution to this problem, summarized in Figure 1, is nucleophilic aromatic substitution (S_NAr) by addition–elimination reactions.[1,2] Attack of a solvent or enzyme-directed nucleophile to a vacant p^* orbital permits bonding to the ring without displacement or ionization of ring substituents (see Figure 1). The unstable Meisenheimer intermediate is formed following the highest energy barrier, which is the attack of the nucleophile. In a subsequent step, the leaving group amine is activated by protonation with loss from the unstable intermediate. The lifetime of the intermediate defines the mechanism of the substitution. If its lifetime is in the order of a bond vibration, it becomes a transition state rather than an intermediate.[3]

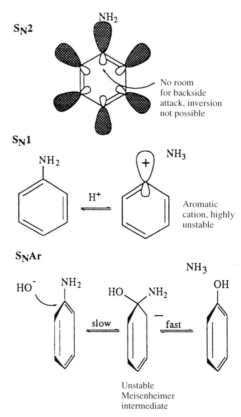

Figure 1   The nature of aromatic nucleophilic substitutions. The upper panel indicates the planar sp^2 orbitals with the back lobes oriented within the plane of the aromatic ring, preventing S_N2 attack. The middle panel demonstrates the unfavorable dissociation of aniline to form the benzyl cation. The lower panel illustrates nucleophilic aromatic substitution of an amino group by hydroxide ion. The sp^3 intermediate is unstable since the ring has lost conjugation which is restored with the loss of NH$_3$.

Direct confirmation of the atomic lifetimes in an S_NAr mechanism has been obtained by femtosecond dissociation and mass spectroscopic observation of the gas phase reaction intermediates for the conversion of iodobenzene to chlorobenzene. The chloroiodobenzene intermediate is formed transiently and decomposes with a large negative free energy change to chlorobenzene. The time scale for decomposition of the intermediate is 880 fs, more than an order of magnitude slower than C—Cl or C—I bond vibrations. These results confirm the existence of a short-lived intermediate in aromatic nucleophilic substitutions.[3] This reaction was defined in the gas phase, where stabilization of intermediates or the transition state cannot occur from the environment.

The same fundamental mechanistic questions pertain to the enzymatic deaminations to be discussed here. For enzymatic deaminations, several additional problems must be addressed. The poorly nucleophilic water molecule must be activated to attack the amino-substituted aromatic carbon. Protonation of the departing amino group is required to permit loss of neutral NH_3. The lifetime of the intermediate may differ between the environment of the catalytic site and that found in gas phase studies, thus both stepwise and concerted mechanisms are conceivable for enzymatic catalysis. The protein environment which leads to large rate enhancements by transition-state stabilization must be defined. Finally, the sum of the enzyme-enforced chemical steps must be rapid to facilitate clearance of the catalytic site, thus permitting the large catalytic turnover numbers found in many of the purine and pyrimidine deaminases.

The results of mechanistic and structural investigations have provided answers to most of these questions, at least for one example among the purine and pyrimidine deaminases. Water activation usually occurs by formation of a metal-bound hydroxy ion at the catalytic site, and the leaving group protonation occurs from a nearby enzymatic acid or from bulk solvent after the leaving amino group rehybridizes. The intermediate Meisenheimer complex has a finite lifetime at the catalytic site. Studies from X-ray crystallography have indicated the protein contacts to substrate and transition-state analogues. The nature of the enzyme-bound transition states are also being defined by kinetic isotope effects and associated computational chemistry.

Advances in RNA transcription and posttranscriptional regulation have revealed a novel biological function for aromatic deaminations. A growing list of RNA molecules are deaminated to change the genetic coding information from that in the DNA. This unexpected upheaval of the central dogma of molecular biology raises new problems of macromolecular recognition and catalysis. These systems are poorly understood, but are recognized to be essential in shaping the protein machinery which provides cellular function.

5.05.2 DEAMINATION OF PURINES

5.05.2.1 Adenosine Deaminase

5.05.2.1.1 *Biological functions*

Adenosine deaminase is widely distributed in both prokaryotes and eukaryotes. The enzyme catalyzes the hydrolytic deamination of adenosine and 2'-deoxyadenosine to form inosine or 2'-deoxyinosine and NH_3 (Scheme 1). The enzyme is essential for biological turnover and salvage of purines, since deamination is an essential step for the subsequent C—N bond scission of the nucleosides. In mammals, this occurs by purine nucleoside phosphorylase which catalyzes phosphorolysis of inosine or guanosine but does not act on adenosine. The biological significance of this deamination is revealed in human adenosine deaminase deficiency, where the failure to deaminate 2'-deoxyadenosine results in the accumulation of dATP in the progenitor cells for B- and T-cell development of the immune system. Elevated dATP inhibits ribonucleotide reductase, and thus DNA synthesis.[4] The disorder prevents clonal expansion in response to an immune challenge, and the resulting phenotype is the severe combined immunodeficiency syndrome.[5] This disorder is

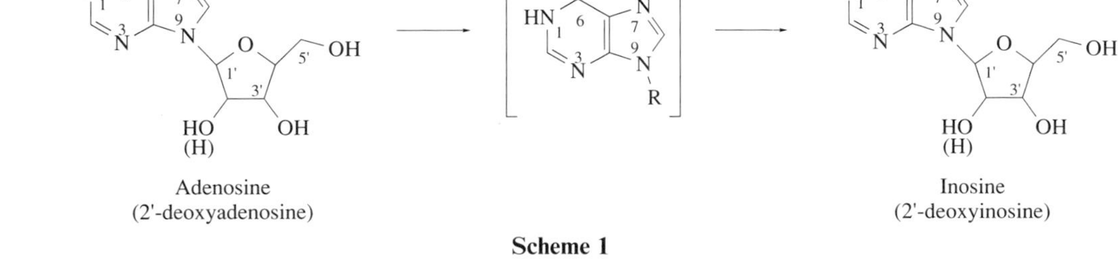

Adenosine
(2'-deoxyadenosine)

Inosine
(2'-deoxyinosine)

Scheme 1

fatal if not treated by bone marrow transplantation or adenosine deaminase replacement therapy. Attempts to treat the disorder led to the first experiments in human gene replacement therapy.[6] The deficiency is now being successfully treated with enzyme therapy using adenosine deaminase covalently modified with polyethylene glycol to prolong its lifetime in the blood.[7]

Transgenic mice deficient in adenosine deaminase are reported to die *in utero* with liver cell and intestinal degeneration.[8] Tissue-specific expression of the enzyme in placenta and the gastrointestinal tract results in partial rescue of the lethal phenotype and results in mice with profound disturbances in purine metabolism, similar to those found in human adenosine deaminase deficiency.[9]

5.05.2.1.2 *Kinetic properties*

(i) *Substrates and kinetic constants*

Adenosine deaminase was first described by Schmidt in studies of ammonia metabolism in muscle.[10] The enzyme commonly used for kinetic and mechanistic studies is that from mammalian (calf) intestine, where it is present in relatively large amounts.[11] The enzyme specifically attacks the 6-position of the purine ring and is capable of displacing a variety of leaving groups including F, Cl, Br, I, NHMe, NH_2NH_2, and OMe.[12-14] The K_m values for adenosine, 2′-deoxyadenosine, and 3′-deoxyadenosine are nearly equivalent at 25–35 μM, while that for 2′,3′-dideoxyadenosine is 100 μM. The catalytic rates for these four substrates are similar at 190, 176, 99 and 234 s^{-1} at 25 °C. In contrast, 5′-deoxyadenosine and adenine give K_m values of 450 μM and 150 μM with catalytic rates of only 0.05 s^{-1} and 0.003 s^{-1}, respectively. Thus the k_{cat}/K_m is 3.3×10^{-6} for adenine relative to adenosine. Ribose binds poorly to the enzyme, demonstrating a necessary covalent link between the ribosyl and adenine moieties for efficient catalysis. The kinetic results are consistent with the observed biological function of adenosine and 2′-deoxyadenosine deamination. The correlation of activity with the 5′-hydroxy establishes that this contact is important for the formation of the transition-state ensemble.[15] Based on the x-ray crystallographic results, a zinc–histidine bridge is formed between the attacking hydroxy group and the 5′-hydroxy of ribose, establishing a structural link between the remote ribosyl group and the energy of transition-state stabilization.[16]

Adenosine binding to form the Michaelis complex is followed by a rapid catalytic step.[17,18] The rate of product formation from the Michaelis complex is more rapid than dissociation of the complex to free enzyme and substrate. This commitment to catalysis is the result of a high affinity for adenosine with dissociation constants in the 1–10 μM range, and the rapid catalytic turnover approaching 250 s^{-1}.

Substrate specificity for adenosine deaminases is isozyme-specific, since the adenosine deaminase isolated from *Aspergillus oryzae* shows nearly identical k_{cat}/K_m values for both adenosine and 5′-deoxyadenosine. The mammalian and *Aspergillus* enzymes are similar in their specificity for C-6 substituents and both enzymes are zinc metalloproteins. Inhibition by the transition-state inhibitor 2′-deoxycoformycin is characterized by a K_i^* (the equilibrium dissociation constant) of 2.7×10^{-9} M for the *Aspergillus* enzyme compared to that of 2×10^{-12} M for the mammalian enzyme. The results indicate that the 5′-hydroxy "bridge" is not present in the *Aspergillus* isozyme, however, direct structural data to define this arrangement are not yet available.[19]

Purine nucleoside analogues including formycin and neplanocin A (Figure 2) with amino or other accepted substituents at the 6-position are also substrates for adenosine deaminases.[20] This activity is undesirable when analogues susceptible to adenosine deaminase are used *in vivo*. To avoid deamination, biologically active analogues resistant to adenosine deaminase such as 2-chloro-2′-deoxyadenosine and 2-fluoro-neplanocin A have been developed. Alternatively, therapeutic compounds which are substrates have been co-administered with transition-state inhibitors of adenosine deaminase.[20,21] In other cases, the activity of adenosine deaminase has been used to convert prodrugs into the active form—an example being the conversion of 2-amino-6-methoxypurine arabinoside to guanine arabinoside by hydrolysis of the methoxy group.[22]

(ii) *Inhibitors*

During the isolation and characterization of the natural product formycin, it was observed that a compound isolated from a culture filtrate of *Actinomycetes* had the ability to prevent adenosine deaminase from deaminating formycin, and the material was therefore named coformycin.[23-25] It

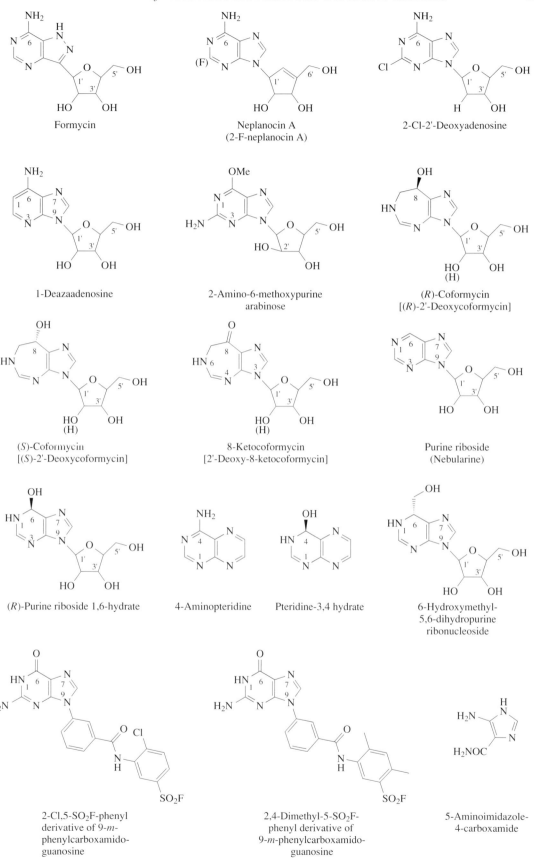

Figure 2 Substrates, substrate analogues, and inhibitors of adenosine deaminase and guanine deaminase.

was shown to have an (R)-secondary alcohol at the 8-position of the 7-membered ring (Figure 2). The K_i value for coformycin was reported to be 6.5×10^{-8} M with adenosine as substrate, the most powerful adenosine deaminase inhibitor then known for the enzyme. More complete inhibition studies established that the inhibition by coformycin and the closely related 2′-deoxycoformycin was characterized by both rapidly reversible and slow-onset phases, to give an equilibrium binding constant of 2×10^{-12} M.[26] This dissociation constant is readily interpreted as the result of transition-state interactions, since the (S)-isomer of 2′-deoxycoformycin and 8-ketocoformycin bind with dissociation constants similar to the K_m for adenosine, while the (R)-isomer of 2′-deoxycoformycin binds 1.3×10^7 times tighter, an energetic difference of 9.9 kcal mol^{-1}.[27] Based on the transition-state stabilization theory for enzymatic reactions, the transition-state binding energy predicted for the enzyme–transition-state complex is 10^{-16} M; therefore, the interaction of the (R)-coformycins is experiencing a substantial fraction of the theoretical maximum binding energy.[28–30] The structures of (R)-coformycin and the transition state reveal the similarity at the reaction center (Scheme 1).

Binding of purine riboside at the catalytic site of adenosine deaminase results in an enzymatic hydration to form the enzyme-bound 1,6-hydrate (Figures 2 and 3). The structure has been confirmed by comparing the NMR shift observed with [6-^{13}C]purine riboside as a result of the sp^2 to sp^3 hybridization change at C-6 to that observed for model compounds. The chemical shift of the inhibitor bound to the enzyme at C-6 is -73.1 ppm, while protonation of purine riboside at N-1 results in chemical shifts of -7.9 ppm at C-6.[14] The structure of the bound inhibitor has also been confirmed by crystallography (see below) to be in the (R)-stereochemistry at C-6.[16] A crystallographic proposal that bound inosine forms an enzyme hydrate 6-diol could not be supported by subsequent NMR studies.[31,32] The close similarity of the purine riboside 1,6-hydrate to the actual transition state is reflected in the equilibrium dissociation constant value of 3×10^{-13} M, a close approach to the theoretical binding limit of 10^{-16} M for the ideal transition-state analogue.[33]

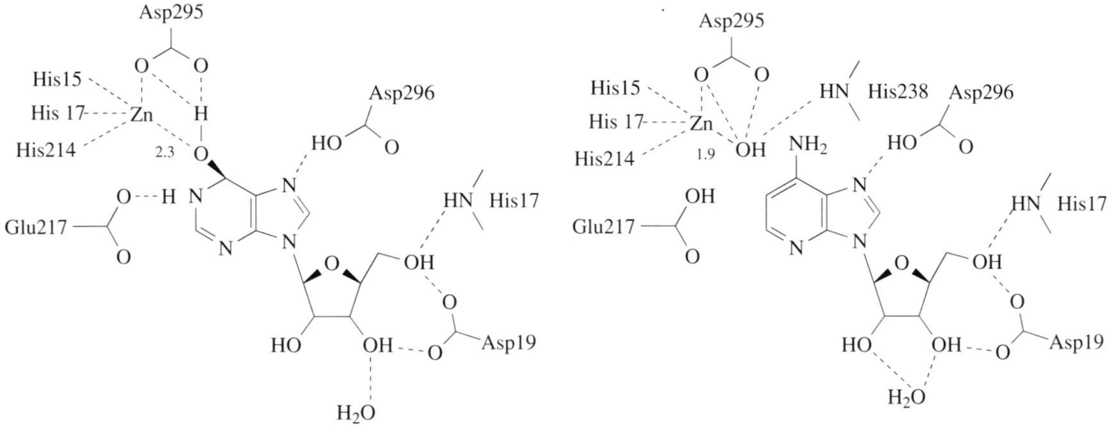

Figure 3   Catalytic site contacts between mouse adenosine deaminase and the 1,6-hydrate of purine riboside (left) or 1-deazadenosine (right). Only contacts of 3.0 Å or less are shown.

The source of the binding energy for the 2′-deoxycoformycin and purine riboside 1,6-hydrate complexes with adenosine deaminase has been dissected in insightful studies from Wolfenden's laboratory.[15,33,34] Deleting specific atoms or groups from transition-state inhibitors followed by reconstitution of the complete inhibitor using distinct molecular fragments was shown to decrease binding energy from that of the intact inhibitor by 7–12 kcal mol^{-1}. The structural integrity of the intact inhibitor is required for the full binding energy, implicating entropic and direct co-operative binding phenomena to ensure the appropriate hydrogen bonds are made to attain the full binding potential.

Small changes in hydrogen bond lengths and angles result in large energetic effects. It is, therefore, not surprising that the combination of molecular fragments from an inhibitor do not equal the binding energy of the intact molecule. Jencks has termed the ligand-induced interaction of additional catalytic site contacts the "Circe effect", and the description of "a transition state in pieces" for adenosine deaminase can be interpreted as an example of the effect.[35] Wolfenden has described the thermodynamics of these interactions as overcoming the entropic components of the bimolecular reaction and thus reducing the equilibrium constant to achieve the transition state.

The deficiency of B- and T-cells observed in human adenosine deaminase deficiency has led to the use of adenosine deaminase inhibitors to reduce populations of these cells in proliferative disorders. This application has also drawn interest to the development of additional tight binding inhibitors, some of which are shown in Figure 2.[36-38] For example the 6-hydroxymethyl-5,6-dihydropurine ribonucleoside mimics the transition state by the loss of conjugation from the purine ring, protonation of N-1 and the presence of the alcohol to occupy the enzymatic site for H_2O (Scheme 1). Few inhibitors are as efficient as the (*R*)-coformycins or bind as tightly as the intermediate analogue, the 1,6-hydrate of purine riboside.

(iii) Reaction mechanism

Reaction rates of 6-substituted substrates with different leaving group ability can be used to indicate if hydroxy attack or leaving group departure is rate-limiting in the adenosine deaminase mechanism. Early studies with the enzyme found k_{cat} ratios of 4:2:1 for 6-Cl, 6-Br, and 6-I purine ribosides, consistent with rate-determining addition of the attacking water hydroxy.[12] Other studies gave a ratio of seven for the relative k_{cat} values with 6-F- to 6-Cl purine ribosides using the calf intestinal enzyme.[14] Although this F/Cl reactivity ratio is slightly smaller than usually found for chemical reactivity in aromatic nucleophilic substitutions, it was considered to be consistent with an early transition state with nucleophilic addition as the largest barrier on the reaction coordinate.[13,14]

Kinetic isotope effects in D_2O are inverse for adenosine deaminase, with the reaction rate being 1.3-fold faster in the presence of D_2O than in H_2O.[17] This unusual isotope effect occurs when the equilibrium constant for deuterium transfer from a donor to acceptor is more favorable than for protium, and this transfer occurs in an equilibrium step prior to achieving the transition state.[39] Proton transfer from a sulfhydryl to a nitrogen or oxygen acceptor would be expected to result in an inverse solvent deuterium isotope effect. An early mechanistic proposal for adenosine deaminase was for a catalytic site Cys nucleophile to form the sp^3 intermediate with the subsequent attack of the water nucleophile displacing the enzyme. The inverse solvent deuterium isotope effects occur on ionization of the sulfhydryl and were consistent with this double displacement mechanism.[17,39] However, this mechanism was revised on the basis of the tight-binding of the 6-hydroxymethyl-5,6-dihydropurine ribonucleoside which is characterized by an sp^3 center with an alcohol group at the reaction center and the observation that the enzyme catalyzed the hydration of pteridine and the dehydration of pteridine hydrate at approximately the same rate as it catalyzed the deamination of 4-aminopteridine.[40] It was suggested that the transition state is reached before a tetrahedral intermediate is fully developed, which was an early and insightful prediction for purine deaminases. With information from the crystal structures of bound inhibitors, it was discovered that the zinc-activated H_2O acted directly without a prior covalent enzyme-substrate intermediate.[16]

5.05.2.1.3 Protein structure

The X-ray crystal structure of adenosine deaminase from mouse revealed a monomeric α-β TIM barrel. The catalytic site ligand is in a pocket which is closed by the flaps of helix and loop regions. The presence of tightly bound zinc at the bottom of a deep catalytic site cavity permitted a new interpretation of the earlier mechanistic observations.[16] Zinc had not been previously detected in adenosine deaminase, since its tight binding prevented removal by chelating agents and added metal had no effect because zinc copurifies with the enzyme, the hallmark of a metalloprotein. No Cys residues were found within 5 Å of the catalytic site, thus the proposal for Cys adducts or proton donation from Cys could be ruled out.[16,17]

(i) Catalytic site structure

Amino acids in contact with substrate analogue and transition-state complexes bound in the catalytic site of mouse adenosine deaminase are shown schematically in Figure 3. The catalytic site water molecule is in contact with the Zn^{2+} when the substrate analogue 1-deazaadenosine is bound, and is bound as the Zn^{2+}-hydroxide on the basis of the 1.9 Å zinc-oxygen bond.[41] Attack on C-6 cannot proceed without the assistance at N-1, whose protonation by Glu217 assists in the loss of the π-conjugated ring system. This structure represents the Michaelis complex as 1-deazaadenosine is a competitive inhibitor which binds with affinity approximately equivalent to that of the substrate. The interaction of adenosine deaminase with 1-deazaadenosine is characterized by six hydrogen bonds <3.0 Å. In contrast, with purine riboside bound, a full carbon–oxygen bond is made at C-6 where the oxygen is now covalently bound to the purine C-6 and is 2.3 Å from the Zn^{2+}. This complex is an "amino-less" mimic of the Meisenheimer intermediate which forms in the normal deamination reaction coordinate. Since there is no appropriate leaving group, the enzyme stabilizes this inherently unstable hydration complex because of its resemblance to the transition state. The major transition state features include protonation of N-1, the pro(R)-sp^3-secondary alcohol at C-6, and the loss of the conjugated ring system. Interactions with Glu217 and the $Zn–H_2O$ contacts permit a binding energy approaching that of the transition state.

Nine hydrogen bonds at <3.0 Å define the interaction between adenosine deaminase and the hydrate of purine riboside.[16] An assumption of modest energies of -4 kcal mol^{-1} per bond results in a sum of -36 kcal mol^{-1} binding energy, adequate to overcome entropic, conformational, and covalent forces required to convert substrates to the transition-state complex. The connection between the ribosyl 5′-hydroxy and catalytic efficiency is linked to His17, since this group contacts both the 5′-hydroxy and the Zn^{2+} atom which ionizes the attacking water molecule. The poor substrate activity with 5′-deoxyadenosine (see above) can be explained by a small shift in the His17—Zn—H_2O—C-6 distance changing the pK_a of bound H_2O and the O—C-6 distance. Both would have negative effects on catalysis. The equal efficiency of mammalian adenosine deaminases with nucleosides or 2′-deoxynucleosides arises from the lack of strong enzymatic interactions at the 2′-position.

The largest energetic barrier for adenosine deamination is formation of the sp^3 transition state, which requires loss of the conjugated bond system. Protonation of N-1 by Glu217 has been proposed to result in a strong hydrogen bond, and has been proposed to be a "low barrier hydrogen bond" which gives rise to the inverse solvent kinetic isotope effect.[16,42] This proposal has not been substantiated by independent NMR spectroscopic evidence for the low barrier hydrogen bond, despite experimental attempts.[43] Isotope-edited difference Raman spectroscopy has assigned a 4 to 10 kcal mol^{-1} energy for this bond.[44]

(ii) Mutational analysis

Confirmation of the functional assignment of the catalytic site residues in adenosine deaminase has come from active-site mutagenesis, kinetic, and crystallographic analysis. Replacement of Asp295Glu gave an enzyme in which the zinc-bound H_2O is replaced by the Glu carboxylate. Catalytic function is lost, but substrates bind normally. Asp296Ala or Asn result in decreased substrate affinity and decreased but measurable catalytic activity, consistent with a role in forming the Michaelis complex in addition to electrostatic stabilization of the transition state.[45] Mutation of Glu217 results in an enzyme which binds substrates with near normal affinity but has lost most catalytic capacity and the ability to hydrate purine riboside. The apparent binding affinity for purine riboside is the same in native and Glu217 mutants, despite the failure to hydrate.[43] This revealing finding is consistent with the chemical energy for hydration arising from the increased favorable contacts as the transition-state complex is achieved. The intrinsic binding energy is therefore converted to the chemical energy of hydration as the binding of the purine riboside proceeds from the Michaelis complex to the hydrate. Changes in Asp295 improve the ability to form the Michaelis complex, but decrease catalytic rates to about 10^{-3} of normal.[45] The bound purine riboside hydrate is still formed, demonstrating that the $Zn–H_2O–His238$ interactions are capable of activating the water nucleophile. Mutation of His238 to Ala or Glu resulted in catalytic impairment of 10^{-3} to 10^{-6} with a 20-fold increased or unchanged affinity for substrate. The His238Ala enzyme was also able to hydrate purine riboside, therefore, in the absence of His238, Asp295 is capable of water activation. The results indicate that His238 and Asp295 co-operate in water activation and play a role in electrostatic stabilization of the transition state.[46]

In a random-selection mutational investigation, growth of cultured cells under selection pressure by 2′-deoxycoformycin yielded a family of adenosine deaminase mutations resistant to 2′-deoxycoformycin. Genetic analysis revealed that all of the mutations which were analyzed were remote from the catalytic site.[47] This intriguing result suggests that global conformational changes are involved in transition-state inhibitor binding. Studies of this kind will provide interesting mutants for the analysis of the protein conformational changes with respect to catalysis. However, this class of mutant protein has not yet been widely exploited.

(iii)   Human mutations

Knowledge of the catalytic mechanism of adenosine deaminase prejudices the view that mutations leading to human deficiency syndromes may cluster at the catalytic site. The distribution of the first 25 missense human mutants reveals that this is not the case. Only two of these mutations are in the nine hydrogen bonding amino acids which interact with the transition-state complex, and the majority of mutations are remote from the catalytic site.[48–50] Some mutations are predicted to influence folding, others mark the protein for rapid degradation, and some may interfere with the adenosine deaminase binding proteins which are of unknown function in humans.[51] Others may prevent conformational changes which are essential to reach the transition state. The random nature of genetic mutations probes every aspect of the biological function of a catalyst, only one of which is the direct amino acid contacts which promote the chemical reaction. Based on the mutational distribution with human adenosine deaminase deficiencies, much remains to be learned about the interactions between conformational and chemical functions.

5.05.2.1.4   *Chemical mechanism*

(i)   Activation of H_2O and NH_2

The enzyme-bound zinc converts zinc-bound water to zinc hydroxide, based on the crystallographically-determined Zn—O distance of 1.9 Å in the presence of 1-deazaadenosine, a competitive substrate analogue.[41] Chemical model studies with zinc chelates have indicated a pK_a value near 7 for zinc-bound H_2O.[52,53] The pK_a of the complex in adenosine deaminase is likely to be well below 7, since the crystallographic result gives the expected Zn—O distance for a hydroxide (1.9 Å) in crystals grown and analyzed at pH 4.2.[41] In the extensively characterized zinc proteases, the structural arrangement is: base–water–Zn–electrophile, in which the base accepts the water proton and the electrophile is proposed to hydrogen bond to the scissile carbonyl of the substrate, promoting attack of the zinc/base activated water.[54] The same arrangement is present in adenosine deaminase without the same connectivity of the amino acids. Asp295 bridges the zinc and the purine-bonded hydroxy proton in the structure with purine riboside. His238 may accept or facilitate transfer of the proton from the water, and the nearby Glu217 protonates N-1 to facilitate loss of aromatic ring structure as the water attacks at C-6.[16,41]

The broad leaving group specificity for adenosine deaminase, even with groups having larger molecular volumes than NH_3, implies that few or no specific interactions are involved in leaving group activation. In crystal structures no proton donor for the departing NH_2 is apparent. Proton donation by His238 has been proposed but it is 4.5 Å from the position of the proposed departing NH_2. As C-6 rehybridizes to sp^3, the amino group rehybridizes from sp^2 to sp^3 and a lone pair becomes available for protonation. Prior to departure, the group accepts a proton since NH_2^- is not readily formed as a leaving group. Chemical instability of the C-6-amino, C-6-hydroxy substituted intermediate leads to C-6-amino bond loss, driven by the energetically favorable formation of the aromatic product.

(ii)   Transition-state structure

The transition state for enzyme-catalyzed S_NAr reactions refers to the highest energy structure on the reaction coordinate. From chemical analogy, this is expected to be the enforced attack of the water nucleophile for the reaction catalyzed by adenosine deaminase. The transition-state barrier for decomposition of the putative intermediate is likely to be modest relative to that for hydroxy attack and requires minimal enzymatic assistance. Indirect information for the transition-state structure of adenosine deaminase is provided from the crystal structure with purine riboside, from transition-state inhibitors, and from site-directed mutagenesis. Direct information has been obtained from kinetic isotope effects for the ^{15}N-leaving group and from D_2O solvent studies. The current body of experimental evidence for transition-state structure is incomplete, but obviously includes those features represented in the coformycins and purine riboside hydrate. These tight-binding analogues resemble the unstable intermediate since they are fully rehybridized to sp^3 at the position equivalent to C-6 in the substrate. This qualifies the inhibitors as reaction intermediate analogues. However, it is unlikely that the enzyme exerts large forces to stabilize the bound intermediate, since this would provide an anticatalytic effect. The tight binding of the intermediate analogues indicates that the transition state for hydroxylation is late, that it resembles the sp^3-hybridized intermediate, and that it does not require the presence of the C-6-amino group for tight-binding interactions. Although the transition state is not fully characterized, it sufficiently resembles the inhibitors to ensure that attack of the hydroxide is well advanced, and that departure of the leaving group follows the rate limiting hydroxy attack according to chemical reactivity series studies, and that N-1 is protonated. In this reaction mechanism, the amino group is a spectator which is not required for nucleophilic attack of the hydroxide, and which departs with minimal assistance from the enzyme.

5.05.2.2   AMP Deaminase

5.05.2.2.1   *Biological functions*

The chemical deamination catalyzed by AMP deaminase uses AMP as the sole physiologically relevant substrate but is otherwise identical to that of adenosine deaminase. However the proteins and their biological functions are distinct. AMP deaminases are tetramers of subunits each of ~ 100 kDa.[55–57] The large, allosterically controlled enzyme consists of substrate sites, allosteric activator sites for ATP, and allosteric inhibitor sites for GTP and PO_4.[58,59] AMP deamination is thought to regulate the size of the cellular adenine nucleotide pool and the production of Krebs-cycle intermediates through the purine nucleotide cycle.[60,61] In muscle, it is also responsible for the production of ammonia for neutralization of lactic acid produced in anaerobic glycolysis.[62] Human deficiency in the muscle isozyme of AMP deaminase results in a form of muscle weakness with no NH_3 release from skeletal muscle upon exercise.[63,64] AMP deaminase deficiencies are relatively common, found in 2–3% of all muscle biopsies.[64] Several inborn errors have been identified and include an early stop codon, at base 34 of the open reading frame. In another mutation of the liver isozyme, the allosteric inhibition by GTP is decreased, resulting in uric acid overproduction. Thus continual removal of adenine nucleotides by AMP deaminase requires increased *de novo* purine biosynthesis and results in the phenotype of gout from uric acid overproduction. Some deficiencies of the erythrocyte isozymes of AMP deaminase have no known phenotype.[65] The enzyme is likely to have as yet unknown functions in muscle contraction because of its stoichiometric and highly localized association with the myofibrils in the region of the A band, with 2 moles of enzyme per mole of myosin in muscle.[66,67] In heart, the production of superoxide free radical ions during heart attacks results from the oxidation of hypoxanthine, and is linked to the degradation of the adenine nucleotide pool, which requires AMP deaminase.[68] An application of AMP deaminase inhibitors in heart attacks has been proposed to prevent degradation of the adenylate pool and to reduce the production of free radicals during ischemia and in the reperfusion phase of an ischemic event.[69]

(i)   Allosteric regulation

Adenosine 5′-phosphate deaminase was first characterized by Schmidt in 1928 in studies of muscle metabolism and known as "Schmidt's deaminase".[70] Its kinetic properties were more completely characterized by work carried out in the 1950s and 1960s.[71–74] The complex nature of the kinetic response to AMP was recognized by 1967.[75] The yeast enzyme was one of the first AMP deaminases to be sequenced and overexpressed and the kinetic properties have been extensively characterized.[56,58] The yeast enzyme shows 51% identity with that from rat muscle, with the major regions of identity being in the C-terminal catalytic site region.

AMP deaminase subunits contain one catalytic site for AMP and two regulatory sites which bind ATP, GTP, or PO_4.[58] The catalytic sites bind AMP weakly with an $S_{0.5}$ of 1.3 mM and a Hill coefficient of 2.1. Allosteric activation by ATP causes loss of the co-operativity and increased affinity to 0.2 mM.[76] The interaction with ATP has a K_d of 6 μM and the binding is not co-operative. Both GTP and PO_4 are allosteric inhibitors, reversing the activating effect of ATP with kinetic inhibition constants of 48 μM and 350 μM, respectively. In direct binding studies, each enzyme subunit was shown to contain one site for coformycin 5′-phosphate, a transition-state analogue, and two regulatory sites per subunit at which ATP, GTP, and PO_4 can bind in a mutual competition. This unusual regulatory feature has also been reported for muscle AMP deaminase.[77]

The co-operative kinetics of the substrate saturation curve arise from differences in the kinetic rate constants for the tetrameric enzyme subunits as the individual catalytic sites fill.[58] Thus the turnover number for the first site is less than that of the second and continues until all sites are filled. When filled, the turnover number is 1200 s^{-1} at each catalytic site. This mechanism is antithetical to the usual explanations for allosteric effects, in which co-operative binding interactions are invoked.[78,79] In yeast AMP deaminase and AMP nucleosidase (an unrelated allosteric enzyme) where the co-operative saturation curves have been directly analyzed by comparative binding and kinetic experiments, both enzymes had equivalent substrate binding at sequential sites, but increased catalytic efficiency as the adjacent sites fill.[58,80]

Regulation of AMP deaminase activity in cells depends on cellular levels of AMP, ATP, GTP, and PO_4. The three regulatory molecules bind more tightly to the regulatory sites than their cellular concentrations, thus the regulatory sites are always filled by the competing mixture of activating ATP and inhibiting PO_4 and GTP ligands. When the concentration of ATP dominates, the enzyme is active and when PO_4 and/or GTP dominate, the enzyme is less active. The detailed role of the enzyme in metabolism remains unknown. Yeast cells in which the enzyme has been genetically disrupted are capable of growth, but at a diminished rate.[56] The human muscle weakness which occurs in muscle AMP deaminase deficiency establishes a role in contraction, but the link with contractile elements and AMP regulation remains to be resolved.[64]

(ii) Distribution and isozymes

Distinct isozymes of AMP deaminase in humans include muscle (M), liver (L), and two erythrocyte isoforms, E1 and E2, which are also found in heart.[47] Genetic and immunologic analyses have established that the enzymes have distinct amino termini and highly conserved carboxy termini.[81-83] The isozymes show distinct kinetic properties, since the allosteric activation of human M-form by ATP can be replaced by high potassium concentrations, unlike the L-form. Kinetic studies with purified rabbit heart AMP deaminase indicate that regulatory controls on this enzyme are stringent, since the k_{cat} is more than an order of magnitude slower than that for yeast or muscle enzymes.[68] The human isoforms are similar to the yeast enzyme in the C-terminal portions but are distinct in the N-terminal. Mutational and truncation studies and chimeric molecules of AMP deaminases have confirmed that the C-terminal portion of the protein contains the catalytic domain.

5.05.2.2.2 Chemical mechanism and transition-state structure

(i) Catalytic site structure

Alignment of the yeast and rat AMP deaminases with mouse adenosine deaminase has revealed a striking identity of nearly all of the amino acids known to be in contact with the zinc, nucleophilic H_2O and bound purine riboside.[84] One significant difference in the conserved catalytic site amino acids is the HisLeuAsp at position 17–19 of adenosine deaminase which corresponds to HisLysAsp at the corresponding 424–426 of yeast AMP deaminase (Figure 4). This region in adenosine deaminase interacts with the 5′-hydroxy of adenosine, and in AMP deaminase, the Lys is proposed to ion pair with the negative charge of the 5′-phosphate. The similarity of the catalytic site alignment has led to assumptions that the enzymes are catalytically similar, however, the tighter binding of substrate and high commitment to catalysis for adenosine deaminase is in contrast to the kinetic properties of AMP deaminase.[84] In addition, the enzymes are highly specific for their respective substrates. Many of the same residues are also conserved in rat muscle AMP deaminase, with His to Gly substitution in one of the zinc ligands, a conservative Lys to Arg replacement for ion-pairing to the 5′-phosphate, and an Asp to Gln change for hydrogen bonding to the ribosyl (Figure 4).

(ii) Kinetic isotope effects

Kinetic and pH studies established that the reaction mechanism for yeast AMP deaminase is rapid equilibrium, with AMP binding in near-equilibrium with the enzyme.[84] Under these conditions, experimentally measured substrate kinetic isotope effects are intrinsic, and the results can be used

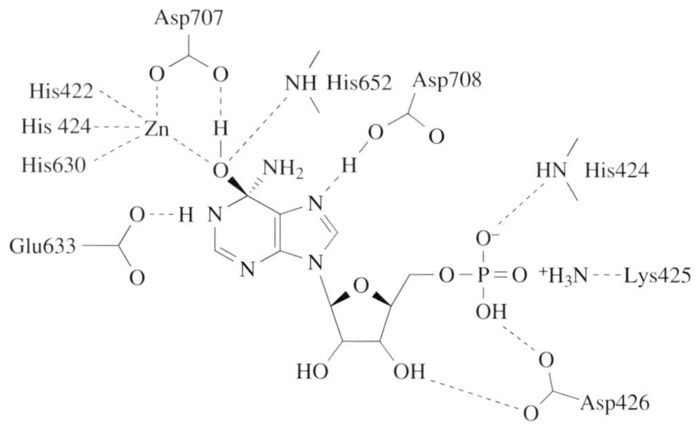

Asp707

His422

His 424

His630

Glu633

NH His652   Asp708

HN   His424

O—P=O ^+H_3N---Lys425

Asp426

HO   OH

	15	17	19		214	217		238		295	296
Mouse Adenosine deaminase	•••HisValHisLeuAsp•••195 AA•••HisxxGlu•••20 AA•••His•••56 AA•••AspAsp•••										

	422	424	426		630	633		652		707	708
Yeast AMP deaminase	•••HisAlaHisLysAsp•••204 AA•••HisxxGlu•••18 AA•••His•••54 AA•••AspAsp•••										

	363	365	367		572	575		594		649	650
Rat AMP deaminase	•••HisAlaGlyArgGln•••204 AA•••HisxxGlu•••18 AA•••His•••54 AA•••AspAsp•••										

Figure 4 The proposed catalytic site contacts for AMP at the transition state for yeast AMP deaminase. The sequence alignments for catalytic site elements are compared for mouse adenosine deaminase (352 amino acids), yeast AMP deaminase (810 amino acids), and rat muscle AMP deaminase (748 amino acids). The three enzymes contain 1 mol Zn per mol enzyme. The Lys425 of yeast AMP deaminase and Arg at the equivalent position for rat AMP deaminase are proposed to provide the specificity for the 5′-phosphate by ion pairing. This position corresponds to a neutral Leu in the adenosine deaminase enzyme.

to interpret transition-state structure.[85] Kinetic isotope effects were measured with the results shown in Figure 5.[86] Solvent D_2O isotope effects for V_{max}/K_m are inverse, with the reaction being more rapid in D_2O than H_2O. The isotope effect varied from 0.71 with AMP as substrate to 0.33 with the slow substrate 8-Br-AMP. Inverse solvent deuterium isotope effects indicate that a proton donor–acceptor pair (or pairs) which promote catalysis prefers deuterium to hydrogen.[87] The result precludes a proton being transferred at the transition state, which is expected to give a large normal isotope effect (as large as 6). The inverse solvent isotope effects were interpreted as proton transfers at Zn^{2+}-liganded oxygens, each of which is expected to give an inverse effect of ~ 0.7. The ^{15}N-isotope effect of 1.014 is indicative of a small change in the vibrational environment, relative to a maximum value of 1.04 for an additional full bond at the transition state.[88] The larger isotope effect would occur if the amino group were protonated prior to reaching the transition state. The ^{14}C-6 isotope effect is diagnostic of the extent of bond rehybridization at C-6 in the transition state. The isotope effects were used in normal mode vibrational analysis to build transition-state models of the reaction consistent with all of the experimentally measured kinetic isotope effects.[89]

(iii) Transition-state structure

Three transition-state and reaction coordinate models were considered for analysis of the kinetic isotope effects: (i) attack of the zinc hydroxide prior to amine protonation; (ii) loss of the protonated NH_2 (NH_3) from the Meisenheimer complex; and (iii) a reaction coordinate where the attack of the hydroxy and the loss of the NH_3 are equal energetic barriers in the reaction coordinate, thus isotope effects would be a composite of these two transition states. The third reaction coordinate pattern is one commonly found for catalytically efficient enzymes, where internal energetic barriers are approximately equal for all steps of the reaction. The family of experimental kinetic isotope effects was consistent only with a transition-state structure in which the Pauling bond order to the attacking hydroxide nucleophile is near 0.8 and the leaving group NH_2 is fully bonded at a near-sp^3 center (see Figure 5).[86] Proton transfer to N-1 occurs prior to reaching the transition state to assist formation of the sp^3 center. At the transition state, the AMP has not yet reached the Meisenheimer intermediate. Steps past the transition state include the formation of the sp^3 intermediate, protonation of NH_2 to form NH_3, and the rapid loss of NH_3. All steps subsequent to transition-state

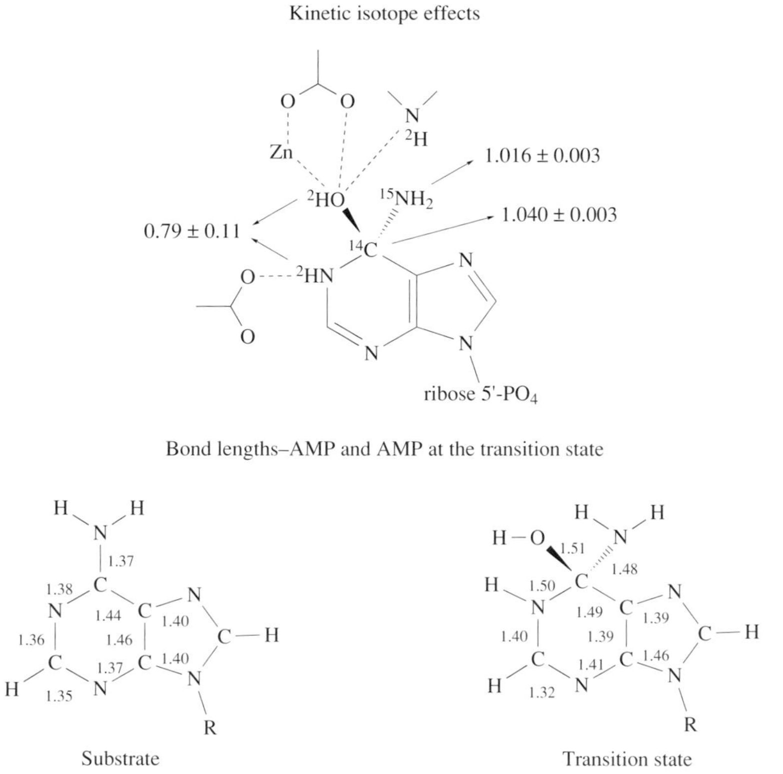

Kinetic isotope effects

Bond lengths–AMP and AMP at the transition state

Substrate

Transition state

Figure 5 Kinetic isotope effects for the hydrolysis of AMP by yeast AMP deaminase and the transition-state geometry. The values shown above are the ratio (k_{cat}/K_m natural abundance)/(k_{cat}/K_m isotope label) determined from measuring the indicated kinetic isotope effects. The structures below show the bond lengths in angstroms for the adenine portions of AMP and the transition-state structure stabilized by AMP deaminase.

formation are fast, since formation of the transition state is the rate-limiting step. *In vacuo*, the chloroiodobenzene Meisenheimer intermediate has a lifetime of 880 fsec,[3] thus in the active site of AMP deaminase, a lifetime for 6-hydroxy-6-aminopurine ribose 5′-phosphate up to 6 orders of magnitude longer will still accommodate the experimentally determined turnover number of 1200 s^{-1} for AMP deaminase.

(iv) Transition-state inhibitors

Soon after the discovery of coformycin and 2′-deoxycoformycin it was found that the 5′-phosphates of these inhibitors were powerful inhibitors of AMP deaminases.[90] (*R*)- and (*S*)-2′-deoxy-coformycin, (*R*)-coformycin, and the corresponding 5′-monophosphates have been characterized as inhibitors of yeast AMP deaminase.[91] The equilibrium dissociation constants ranged from 4.2×10^{-3} M to 10^{-11} M. The presence of the 2′-hydroxy, the 5′-phosphate, and the (*R*)-stereochemistry at the reaction center are necessary for the optimum binding of inhibitors. Slow-onset inhibition with the best inhibitors conforms to the mechanism shown in Scheme 2, where E = enzyme, I = inhibitor, K_i is an equilibrium binding constant for a weakly-bound enzyme–inhibitor complex, and k_5 and k_6 are the rate constants for formation and relaxation of a tightly bound enzyme inhibitor complex, EI*. Only the combination of E and I is rapidly reversible. For a family of coformycin inhibitors,

$$\text{E} + \text{I} \xrightleftharpoons{K_i} \text{E·I} \underset{k_6}{\overset{k_5}{\rightleftharpoons}} \text{E·I*}$$

$$K_i^* = \frac{[\text{E}][\text{I}]}{[\text{E·I}] + [\text{E·I*}]} = \frac{K_i k_6}{k_5 + k_6}$$

Scheme 2

K_i varied by a factor of 3000 and the overall inhibition constant (K_i^*) varied by a factor of 2×10^5. The rate constant k_6 varied by a factor of 650, while k_5 was similar for all slow-onset inhibitors. The results argue against inhibitor-induced transition-state conformations for the slow-onset inhibitors with this enzyme and suggest that the enzyme attains the transition-state configuration at a low rate (0.04 s^{-1}) independent of the inhibitor structure. The predictive effects of these provocative studies for the kinetics of protein conformational changes need to be further explored with additional inhibitors and in other enzyme systems.

(v) Molecular electrostatic potential surfaces

Transition-state theory predicts that the tight-binding transition-state inhibitor (*R*)-coformycin 5′-phosphate will resemble the altered AMP molecule at the transition state. AMP deaminase is one of the few enzymes for which direct experimental data exists for transition-state structure and where powerful transition-state inhibitors are available. This combination permits a direct comparison of the features which are important in transition-state inhibitor recognition. The comparison has been made by *ab initio* calculations for the purine portion of the substrate, the transition state, and for the equivalent portion of (*R*)-coformycin 5′-PO$_4$. The molecular electrostatic potentials were then extracted at an electron density equivalent to the van der Waals surfaces for comparison.[92] These structures (Figure 6) provide visual confirmation that the electrostatic potential surfaces are closely related for the transition-state inhibitor and the transition state, and that both differ considerably from the substrate. This difference permits the differential stabilization of transition-state structure relative to the substrate and provides the catalytic potential for AMP deaminase.

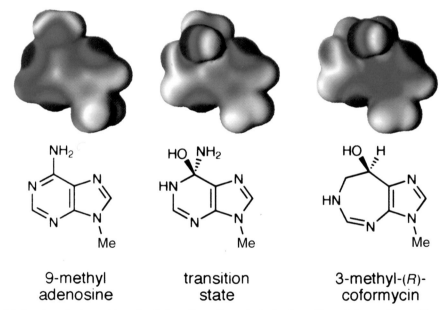

Figure 6 Molecular electrostatic potential surfaces for substrate, transition state, and transition-state inhibitor of AMP deaminase. The molecules are truncated by replacing the ribose 5′-phosphate by a methyl group since the ribose 5′-phosphate is unchanged by the reaction. The surface shown approximates to the van der Waals surface. Regions which are electron deficient (partial positive charge) are shown in red. Regions which are electron rich (partial negative charge) are shown in blue. Areas which are near neutrality are yellow, and green represents areas which are slightly electron rich. The stick figures below the electrostatic potential surfaces show the chemical structures on the same scale and in the same geometry as shown above.

5.05.2.2.3 *Transition-state similarity measures for adenosine and AMP deaminases*

Transition-state theory for enzyme-catalyzed reactions proposes that molecules which are similar to the transition state should bind tightly to the enzyme.[30] The availability of experimental transition-state information for the AMP deaminase reaction, and by extension, the adenosine deaminase reaction, makes it possible to test this proposition. Molecules most similar to the experimentally determined transition-state structure would be expected to bind the most tightly. Molecular similarity measures have been used to compare the substrate, transition state, reactant-state inhibitors, and transition-state inhibitors for the AMP and adenosine deaminase reactions.[93,94] The basis for ranking the molecules for similarity to the transition state is the distribution of the molecular electrostatic potential at the electron density near the van der Waals surface of the molecules. Substrates and inhibitors are compared to the transition state by orienting the molecules so that the van der Waals surfaces are maximally coincident. At this geometry, the similarity is based on the relative spatial distribution of the electrostatic potential for the test molecule as compared to that of the transition state. When the similarity measure was applied to transition-state inhibitors for AMP and adenosine deaminases, a strong correlation between the binding free energies and the similarity of the transition-state inhibitor to the transition state was found (Figure 7). The results support the validity of the kinetic isotope effect method for transition-state determination for the deaminases. The results also indicate that the more closely the inhibitors resemble the transition states in their molecular electrostatic potential surface interactions, the more tightly they bind to the target enzymes. The transition states for AMP and adenosine deaminases are similar, based on the similar interaction energies for the transition-state analogues.

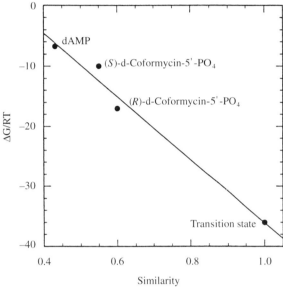

Figure 7 Correlation of the binding free energy for substrate, transition state, and inhibitors of AMP deaminase with the degree of similarity to the transition state. The line is fixed by the binding energy of the substrate (left ordinate) and the energy of transition-state binding based on the extent of rate acceleration relative to nonenzymic catalysis. The inhibitors are placed on the curve according to their similarity and the energy of binding based on inhibition studies. Free energies are for the 2'-deoxy-molecules.

5.05.2.3 Other Purine Deaminases

The group of purine deaminases in this section are not as well characterized as adenosine and AMP deaminases. The lack of an important or recognized metabolic significance applies to adenine deaminase and guanine deaminase. The newly discovered RNA deaminases have generated vast interest, but their recent discovery limits the current mechanistic and structural information. However, the advent of genomic DNA sequencing now makes these enzymes readily available, and they should be better understood in the near future.

5.05.2.3.1 Adenine deaminase

Adenine deaminase has been reported to be present in some bacteria and lower eukaryotes, but there is no evidence for the enzyme in mammalian tissues.[95] The genes encoding for adenine deaminase have been cloned or sequenced from *Saccharomyces cerevisiae* and *Bacillus subtilis*.[96,97]

There is no evidence for the enzyme in *Escherichia coli*. Only partial purifications of the protein have been reported, and there are no definitive kinetic or structural characterizations. Reports of adenine deaminase from the genomic or mRNA–cDNA sequencing projects now provide new sources for investigating this enzyme.

Adenine deaminase is the only enzyme in *B. subtilis* capable of deaminating adenylates, thus it plays a unique role. The gene encodes a 65 kDa protein located at 130° on the *B. subtilis* chromosomal map.[98] The predicted amino acids at 42–61 include the conserved sequence found in dihydroorotase and related enzymes and is thought to be a metal ion binding site. Protozoan parasites lack *de novo* purine biosynthesis and several are known to express adenine deaminase. For example, *Tritrichomonas foetus* uses adenine deaminase to form hypoxanthine which is then salvaged by hypoxanthine phosphoribosyl transferase.[99] The organism has no adenine phosphoribosyl transferase to salvage adenine directly, thus the enzyme is essential for utilization of adenine. In humans, adenine is formed in the phosphorolysis of 5′-thiomethyl adenosine. The pathway for recycling of adenine in mammals is via adenine phosphoribosyl transferase, therefore, no adenine deaminase is necessary.

5.05.2.3.2   *Guanine deaminase*

Pathways for degradation of guanylates involve formation of free guanine in the purine nucleoside phosphorylase reaction followed by guanine deamination to form xanthine. Xanthine oxidase oxidizes xanthine to uric acid, the end product of guanine-containing nucleotides. Guanine deaminase is widely distributed in mammals with the interesting exception of swine, which excrete guanine directly into the urine.

The human liver enzyme has been purified and is a dimer of 59 kDa subunits. The K_m for guanine is 15 μM and the k_{cat} is 20 s^{-1}.[100] Similar properties were reported for a homogeneous preparation from rat brain.[101] The rat brain enzyme was immunologically identical to that from rat liver. An inhibitor which binds with 50 μM affinity is 5-aminoimidazole-4-carboxamide (AICA, a precursor of purine *de novo* synthesis), and 9-(*p*-carbetoxyphenyl)guanine binds with 5 μM affinity to the pig brain enzyme.[101,102] A family of active-site directed tight-binding and irreversible inhibitors has been synthesized for guanine deaminase based on 9-phenylguanine derivatives and containing chemically reactive groups such as sulfonylfluoride.[103] The 2-chloro-5-sulfonylfluorophenyl derivative of 9-*m*-phenylcarboxamido-guanosine binds to guanine deaminase with an $I_{0.5}$ of 35 nM followed by the time-dependent inactivation of the enzyme. The similar 2,4-dimethyl-5-sulfonylfluorophenyl derivative is a 42 nM inhibitor but does not cause covalent inactivation (see Figure 2). A potential application of guanine deaminase inhibitors is to prevent the enzyme from deaminating carcinostatic agents such as 8-azaguanine which is a substrate for the enzyme. Imidazolylthiocarbamide and quinazolone inhibitors have also been explored for this purpose.[104,105] Despite the availability of purified enzyme, the genes, and powerful inhibitors, the reaction mechanism remains obscure.

5.05.2.3.3   *RNA adenylate deaminases*

The focus on the structural and catalytic properties of RNA during the 1990s has led to the remarkable discovery that base-specific posttranscriptional modifications occur in both purines and pyrimidines of RNA, and in mRNA as well as tRNA. The RNA adenylate deaminases will be discussed here and the RNA cytidine deaminases are discussed in Sections 5.05.3.3 and 5.05.3.4.

The RNA adenylate deaminases are widely-distributed 100–140 kDa proteins which contain multiple double-stranded RNA binding domains, a nuclear targeting signal, and an adenosine deaminase motif, all in the same peptide.[106] The proteins are not inhibited by (*R*)- or (*S*)-coformycin, thus the catalytic site structure and the transition-state structure is expected to differ from that of adenosine or AMP deaminases.[107] Conversion of A to I in double-stranded RNA weakens the interaction by 2 kcal mol^{-1}, thus the enzyme was first discovered as a dsRNA "unwinding protein".[108] The reaction was characterized by the use of [U-^{13}C]adenosine and [^{18}O-H$_2$O] to demonstrate that the 6-oxygen comes from water and the base is not exchanged during the conversion from an adenosine site to an inosine site.[109]

The enzymes have no significant activity on adenosine. The interferon-inducible, double-stranded RNA-specific adenosine deaminase is a 1226 amino acid enzyme with three motifs common to double-stranded RNA-binding proteins. The postulated adenosine deaminase motif is near the

C-terminal, and mutation of the conserved CysHisAlaGlu sequence present in adenosine deaminases destroyed enzymatic activity, but did not affect RNA-binding activity.[110] Likewise, mutation of the putative zinc-binding amino acids His910, Cys966, and Cys1036 abolished activity. The wide distribution of the double-stranded RNA adenosine deaminases (dsRAD) and its nanomolar affinity for z-DNA[111] have led to the proposal that it interacts with the DNA strand following passage of RNA polymerase.[112] In this region, the DNA is more likely to have the z-structure, and double-stranded RNA is present as a consequence of transcription. The proposal suggests that mRNA is then edited to deaminate specific adenosines to inosines, which causes the double-stranded RNA to unfold. The change of adenosine to inosine results in the site being translated as a guanosine site, since inosines are read as guanosines in translation. As anticipated from its substrate specificity, antibody immunochemistry experiments have localized the HeLa cell enzyme to the nucleus, and have demonstrated the widespread distribution of the enzyme in the brain.

One of the dsRNA deaminase functions occurs in editing the mRNA for the glutamate receptor subunit, and has been investigated in neuroblastoma cells. The change is classified as a Q/R edit, since changing one specific residue from A to I changes a glutamine to an arginine in the GluR-B ion channel, altering its properties to make it more permeable to calcium ions. The modification requires both a specific dsRNA adenosine deaminase and an accessory protein. The mechanism for this reaction and the role of the second accessory protein are unknown.[113,114] In the hepatitis D virus RNA, an adenosine is converted to an inosine, resulting in an A to G change in the antigenomic DNA, to give rise to the edited sequence which is ultimately expressed in viral proteins.[115]

The wide distribution of this newly described RNA editing protein and the multiple functions already attributed to the reaction suggest that dsRNA editing will have a rich biological and mechanistic life. Based on the limited sequence homology and mutational information available, it is proposed that the enzyme catalytic site, including the proposed ligands for zinc, is more like the cytidine deaminase catalytic site than that of adenosine deaminase. The large size of the catalytic protein, the need for accessory proteins, and the need for double-stranded RNA substrates makes this a complicated and interesting system for further exploration in purine deamination reactions.

5.05.3 DEAMINATION OF PYRIMIDINES

5.05.3.1 Cytidine Deaminase

5.05.3.1.1 Biological function

Cytidine deaminase catalyzes the hydrolytic deamination of cytidine into uridine and ammonia with a 4×10^{11}-fold rate enhancement.[28] Uridine is converted to the base uracil, and the ultimate catabolism of cytidine yields the amino acid β-alanine and malonyl-CoA which can enter into fatty acid biosynthesis. Alternatively, uridine can be metabolized to all of the pyrimidine triphosphate nucleic acid precursors.[116] The focus of the following is an enzyme from *E. coli* for which extensive biochemical and crystallographic information exists. Attention will be drawn to enzymes from other sources to highlight the present discussion. The large number of transition-state analogues that inhibit cytidine deaminase makes it an attractive system for study, both experimentally and computationally.

5.05.3.1.2 Kinetic properties

After the discovery of cytidine deaminase activity in extracts of *E. coli*, purification was achieved by Wang *et al.*,[117] and Cohen and Wolfenden.[118] Molecular weight determinations yielded a value of 54 kDa or higher, and SDS-polyacrylamide gel electrophoresis gave a single band with a molecular weight about half these values.[118–122] From the crystal structure, the enzyme was confirmed to be dimeric with identical subunits of 31 540 Da each.[123] For the substrate cytidine (Figure 8(a)) K_m is reported in the range 5×10^{-5} to 2.22×10^{-4} M, and a representative V_{max} is 233 μmol min^{-1} mg^{-1}, with $k_{cat} = 137$ s^{-1}.[32,118,120,122,124] The uncatalyzed analogous deamination of cytidine to uridine (Figure 8(a)) has a rate of 3.2×10^{-10} s^{-1} at 25 °C, for an enzymatic rate enhancement of 4×10^{11}.[28] The *E. coli* enzyme is not co-operatively dependent on substrate concentration, and it is not allosterically regulated by any ribonucleosides which serve as competitive inhibitors.[119,121] In contrast, cytidine deaminase from yeast is allosterically inhibited by both cytidine 5′-PO$_4$ and excess cytidine, and the sheep liver enzyme has allosteric sites for deoxy-thymidine triphosphate and deoxy-uridine triphosphate.[125,126] For comparison, for mouse kidney cytidine deaminase, $K_m = 2.1 \times 10^{-4}$ M for cytidine; for sheep liver enzyme, $K_m = 1.6 \times 10^{-4}$ M; for yeast enzyme, K_m is reported as 2.5×10^{-4} M; and the human liver deaminase has $K_m = 9.2 \times 10^{-6}$ M.[127,128]

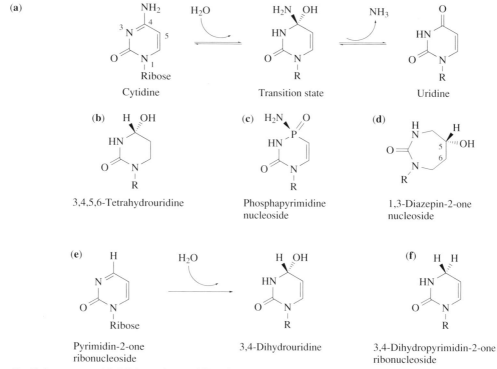

Figure 8   Substrates and inhibitors for cytidine deaminase. The transition state in (a) is proposed to have 0.8 bond order to the hydroxyl.

E. coli deaminase also catalyzes the hydrolytic deamination of 5,6-dihydrocytidine into 5,6-dihydrouridine. This alternative substrate has k_{cat} approximately 1/9 that for the reaction with cytidine, and $K_m = 1.13 \times 10^{-4}$ M for the dihydrosubstrate.[124] For 5,6-dihydrocytidine the enzyme enhances the reaction rate 1.4×10^5-fold.[122] Deamination of the saturated 5,6-dihydrocytidine suggests that there is no nucleophilic addition across the 5,6-double bond, either by an enzymatic group or water, during catalysis of cytidine.[124] Water addition across the 3,4-double bond during hydrolytic deamination of cytidine (or of 5,6-dihydrocytidine) by cytidine deaminase is implicated and the chemistry is analogous to that of adenosine and AMP deaminase catalysis. Figure 8(a) shows the mechanism with a schematic of the anticipated near-tetrahedral reaction transition state for cytidine (and for the 5,6-dihydrosubstrate reaction if the 5,6-bond is saturated). A potent inhibitor of cytidine deaminase, 3,4,5,6-tetrahydrouridine (Figure 8(b)), features a tetrahedral C-4 with an (*R*)-hydroxy group and a protonated nitrogen at N-3. These moieties mimic those of the proposed transition state of the deamination reaction, and this inhibitor is found to be a transition-state analogue, most closely related to the transition state of the 5,6-dihydro-substrate.[129]

Methylation of the amine on cytidine gives a substrate with k_{cat}/K_m three orders of magnitude less than for cytidine, while dimethylation inactivates the substrate, probably for steric reasons.[118,130] 2'-Deoxycytidine is a better substrate for the enzyme than is cytidine: k_{cat}/K_m for the former is 11 times that for the latter.[120]

Certain cytidine derivatives—5-azacytidine, cytosine arabinoside, and FIAC (2'-fluoro-5-iodo-1-β-D-arabinofuranosylcytosine)—are antineoplastic and antiviral agents but undergo degradation by nucleosidase deaminase in mammalian tissues.[131–135] 5-Azacytidine is a substrate for human liver cytidine deaminase with $K_m = 5.8 \times 10^{-5}$ M and a relative $k_{cat} = 0.17$ (compared with cytidine for which $K_m = 9.2 \times 10^{-6}$ M and $k_{cat} = 1$).[128] Human leukemia cells metabolize 5-azacytidine with $K_m = 4.3 \times 10^{-3}$ M, while $K_m = 2.2 \times 10^{-5}$ M for cytidine.[133] Additionally, 2',3'-dideoxycytidine, a selective inhibitor of HIV reverse transcriptase, is degraded by cytidine deaminase.[123] For these agents to work with efficiency, it becomes necessary to inhibit cytidine deaminase. *In vitro* and *in vivo* studies have established that 3,4,5,6-tetrahydrouridine (Figure 9(b)) is a potent inhibitor.[128,132,136–138] This and other inhibitors are now discussed.

5.05.3.1.3 *Transition-state inhibitors*

At the transition state for an enzymatically catalyzed reaction, substrate is typically bound to the enzyme 10^{10}–10^{15} times tighter than in the Michaelis complex.[30] The dissociation constant for the transition state K_{TX} can be calculated as: $K_{TX} = (k_{non}/k_{cat})K_m$, where k_{non} is the rate constant for the uncatalyzed reaction, K_m is the Michaelis constant for the substrate, and k_{cat} is the rate of the catalyzed reaction.[129] Explicitly, K_{TX} is the equilibrium constant for the (hypothetical) dissociation of the transition-state complex from the enzyme into solution. For deamination by cytidine deaminase, $K_{TX} \approx 10^{-16}$ M using the kinetic constants quoted above. The alternative substrate 5,6-dihydro-cytidine gives $K_{TX} \approx 10^{-9}$ M. The transition state for cytidine is more tightly bound because enzymatic catalysis enhances the rate (over that in solution) of the cytidine reaction by a greater factor than it does the dihydrocytidine reaction. If a chemically stable transition-state analogue resembles electrostatically and geometrically the transition state, then an extremely tight-binding inhibitor of the enzyme will have been created. For the case of cytidine deaminase, it is advantageous to mimic the transition state of cytidine rather than that of 5,6-dihydrocytidine (see Figure 8(a)).

A feature of many transition-state inhibitors is that they exhibit slow onset of inhibition[139–141] when compared with substrate-analog inhibitors. It appears that the conformation of the ground-state enzyme is such that it can bind rapidly and reversibly substrate or substrate analogues but cannot readily accommodate into the active site a molecule that resembles the altered geometry of the transition state. For transition-state analogues an initial weak interaction between analogue and enzyme (defined by K_d) is followed by a slow conformational change in the enzyme to maximize interactions with the ligand, leading to the final tight-binding inhibition. The dissociation constant for the overall process is defined as K_{i*}, without decomposition into the two-step process (Scheme 2). These enzymatic conformational changes are probably related to conformational changes which occur during the catalytic process.

3,4,5,6-Tetrahydrouridine was found to be a slow onset, tight-binding inhibitor of human liver cytidine deaminase with $K_i^* = 2.9 \times 10^{-8}$ M.[128] The tetrahedral configuration at C-4 mimics that of the transition state (Figures 8(a) and 8(b)), as does the protonated N-3. Direct measurements by ligand techniques of the kinetics of enzyme–inhibitor binding and release gave $K_i^* = 4.5 \times 10^{-8}$ M.[137] From steady-state measurements of the *E. coli* enzyme, tetrahydrouridine showed an inhibition constant of $K_i^* = 2.4 \times 10^{-7}$ M.[118] Kinetic parameters revealed that the inhibitor for the bacterial enzyme shows slow onset in a two-step process: initial, rapid complex formation and subsequent slow conformational rearrangement.[139] Since $K_{TX} \approx 10^{-9}$ M for the enzymatic transition state derived from 5,6-dihydrocytidine, the structure for which tetrahydrouridine is a close analogue, and recalling that $\Delta G = RT \ln K_i^*$, 3,4,5,6-tetrahydrouridine captures 74% of the binding free energy of the transition state–enzyme complex. In contrast, the substrate 5,6-dihydrocytidine achieves only 44% of the transition state's binding energy. For a penetrating discussion, see Frick *et al.*[122]

A phosphapyrimidine nucleoside (Figure 8(c)) is another potent inhibitor of *E. coli* cytidine deaminase.[120,139,142] This compound binds to the enzyme without covalent modification as determined by UV-difference spectroscopy, and the binding is fully reversible with dissociation of the intact inhibitor from the enzyme. The analysis is complicated by the fact that the inhibitor is hydrolyzed in solution to a species that binds 30 000 times less tightly than the parent compound. Analysis at pH 6, where the inhibitor is relatively stable in solution, revealed that it was released from the enzyme in active form. The K_i^* is 0.9×10^{-9} M for the phosphapyrimidine nucleoside and it exhibits slow onset of inhibition with $k_{on} = 8300$ M^{-1}s^{-1} and $k_{off} = 7.8 \times 10^{-6}$ s^{-1} for the overall reaction. Because the molecule is not covalently bound to the enzyme, but shows slow and tight-binding behavior, and has structural similarity to the transition state of the catalyzed reaction, phos-phapyrimidine nucleoside is considered a transition-state analogue.[139]

The products of the reactions for cytidine and 5,6-dihydrocytidine—uridine and 5,6-dihydro-uridine (Figure 8(a))—are competitive inhibitors with $K_i = 2.5 \times 10^{-3}$ M and $K_i = 3.4 \times 10^{-3}$ M, respectively.[118] Note that C-4 of each is sp^2-hybridized, while C-4 of the transition state is tetrahedral, facilitating the release of products. Although N-3 is protonated in both products and in the respective transition states, the C-4 hybridization dominates the tight-binding at the transition state.

A series of inhibitors based on the parent structure 1,3-diazepin-2-one nucleoside (Figure 8(d)) was synthesized to inhibit mammalian cytidine deaminase.[138,143,144] Note the similarity to (*R*)-coformycin, an inhibitor for the adenosine deaminase reaction. However, the stereochemistry, as deduced from the crystal structure of enzyme–transition-state analogue complexes, at the hydroxy-bearing carbon is enantiomeric to that of (*R*)-coformycin for maximal interaction with the enzyme. For the human liver enzyme, $K_i = 4 \times 10^{-8}$ M for the 1,3-diazepin-2-one nucleoside inhibitor having the correct stereochemistry, while $K_i = 9 \times 10^{-7}$ M for the enantiomer.[138] Removal of the hydroxyl group from this inhibitor and introduction of a 5,6-double bond gives a particularly tight binding inhibitor ($K_i = 2.5 \times 10^{-8}$ M, human liver enzyme), though without slow-onset inhibition.[143] Perhaps the lack of a hydroxy moiety permits a rapid entry into the active site.

Pyrimidin-2-one ribonucleoside (Figure 8(e)) and its 5-fluoro-analogue are inhibitors of mammalian and yeast cytidine deaminase, and subsequently were found to inhibit the *E. coli* enzyme with nominal inhibition constants of $K_i = 3.0 \times 10^{-7}$ M and $K_i = 3.5 \times 10^{-8}$ M, respectively.[122,145] If these two competitive inhibitors were bound to the active cleft as the covalent hydrates 3,4-dihyrouridine and the analogous fluoro-compound—by H_2O addition across the 3,4-double bond as shown in Figure 8(e)—they would be extremely tight-binding, since 3,4-hydration is energetically unfavorable in solution. Through analysis of model compounds, the equilibrium constant for formation of the water adduct of pyrimidin-2-one ribonucleoside was calculated as $K = 4.7 \times 10^{-6}$ (unit water activity).[122] A thermodynamic cycle gave for the dissociation constant of the hydrated inhibitor 3,4-dihydrouridine $K_i = 1.2 \times 10^{-12}$ M. A similar treatment for the fluoro-derivative yields $K_i = 3.9 \times 10^{-11}$ M.[146] Evidence that 3,4-dihydrouridine and the flouro-compound are actually bound to the enzyme in covalently hydrated forms was provided through spectroscopic and NMR studies.[122,146] 3,4-Dihydrouridine differs from the proposed transition state simply by substitution of a hydrogen for the -NH_2 group of the latter—it is not surprising that this transition-state analogue is extraordinarily tight-binding (see Figure 8). The crystal structures of the two pyrimidin-2-one ribonucleosides complexed with the *E. coli* enzyme confirmed them to be bound as the covalent water adducts, as discussed below. Hydrated pyrimidin-2-one ribonucleoside captures 75% of the binding free energy of the transition state for cytidine deamination, while the substrate achieves only about 25%. If the hydroxyl group on the bound ribonucleoside is replaced by hydrogen, giving 3,4-dihydropyrimidin-2-one ribonucleoside (Figure 8(f)), $K_i = 3.0 \times 10^{-5}$ M. The latter inhibitor is less tightly bound than its parent transition-state analogue by 10.1 kcal mol^{-1}, underscoring the importance of the hydroxy for transition-state stabilization by the enzyme. Since pyrimidin-2-one ribonucleoside is not appreciably hydrated in solution, it binds to the enzyme as a substrate analogue—the covalent hydration occurs at the active site. The binding of pyrimidine ribonucleosides to cytidine deaminase is analogous to the interaction of purine riboside with adenosine deaminase.

5.05.3.1.4 *Protein structure*

(i) *Protein fold*

The structures of *E. coli* cytidine deaminase complexed with several inhibitors have been determined crystallographically. The strategy has been to elucidate active-site-substrate modifications during catalysis by complexing the enzyme variously with substrate analogues, transition-state analogues, and with product.[123,147–150] Except at the active site, the global structure of the enzyme is virtually identical for any choice of ligand. The protein fold discussed below is based upon the enzyme-5-fluoropyrimidin-2-one ribonucleoside complex at 2.3 Å resolution.[123,147]

The crystal structure establishes that 5-fluoropyrimidin-2-one ribonucleoside (fluoro-derivative of the structure in Figure 8(e)) binds as the covalently hydrated transition-state analogue 5-fluoro-3,4-dihydrouridine. The same is true for the unfluorinated inhibitor.[148] *E. coli* cytidine deaminase is dimeric, with the monomers related through a twofold rotation through the molecular dyad. Each monomer consists of three distinct domains. The first, unrelated to the active site, is 48 residues. The other two domains, differentiated as N- or C-terminal, are themselves rotationally related, with the axis perpendicular to that of the entire molecule. The enzyme is effectively tetrameric in the crystal. Cytidine deaminases from both *B. subtilis* and humans are homotetramers, though there is otherwise little structural homology between the *E. coli* enzyme and other proteins.[151,152] The N-terminal domain of a monomer contains the active site, while the structurally related C-terminal domain features a nonfunctional active site. Within the dimeric interface, the functional active site of one monomer is completed through residues from the nonfunctional site of the other, a configuration that renders the bound inhibitor virtually solvent inaccessible.

A novel structure, which defines the location of the active site, is found in the N-terminal domain. The third β-strand of a three-stranded antiparallel sheet gives way abruptly to an α-helix, which in turn yields a fourth β-strand, and then a second α-helix. This second helix turns into the final β-strand of the motif. The two helices, nearly parallel with a 30° angle between them, feature N-termini in close spatial proximity forming a crevice. The dipole moments of these helices are focused together and close to a coordinated zinc ion and the hydroxy oxygen of the bound inhibitor. A comparison of the bacterial enzyme to other deaminases and to the structure of adenosine deaminase is given by Betts *et al.*[123] and Carter.[147]

(ii) Catalytic-site residues and coordinated zinc

The catalytic site has been described with reference to the bound transition-state analogues 5-fluoro-3,4-dihydrouridine or 3,4-dihydrouridine.[123,148] The interactions of these inhibitors with the active site are nearly identical. As the hydration of these inhibitors takes place in the active cleft, the process is analogous to the first step in hydrolytic deamination of substrate cytidine. Because the inhibitors have a hydrogen atom replacing the amino leaving group of the substrate, they are chemically stable as the bound tetrahedral transition-state-like structures.

The catalytic site contains a tetrahedrally-coordinated zinc ion: two cysteine thiolate ligands (Cys129 and Cys132), a histidine (His102), and a water molecule complex the zinc. By comparison, the adenosine deaminase zinc is penta-coordinated without Cys interactions. The water, zinc-activated to a hydroxy ion, adds to C-4 of the bound inhibitor or substrate forming the tetrahedral configuration. Figure 9a sketches these interactions, as well as the remaining hydrogen bonds, for the transition state. For the inhibitor -H replaces -NH$_2$ on C-4 and the hydrogen bond to the backbone carbonyl oxygen of Thr127 is suppressed. Note that the hydroxy moiety, in addition to being coordinated to the zinc, is hydrogen bonded to Cys129 and Glu104: these three interactions are responsible for transition-state stabilization during catalysis. The proton on N-3 is hydrogen bonded to the crucial Glu104. Mutation of Glu104 into alanine reduced the enzymatic activity by eight orders of magnitude.[153,154] Likewise, mutations of the amino acids complexed to zinc had severe consequences for catalysis.[155] One of the hydrogen bonds to the ribose ring originates from Ala631 in the nonfunctional active site of the other monomer across the dimeric interface. Because the 2'-OH group is not complexed with the enzyme, 2'-deoxycytidine is a good substrate.

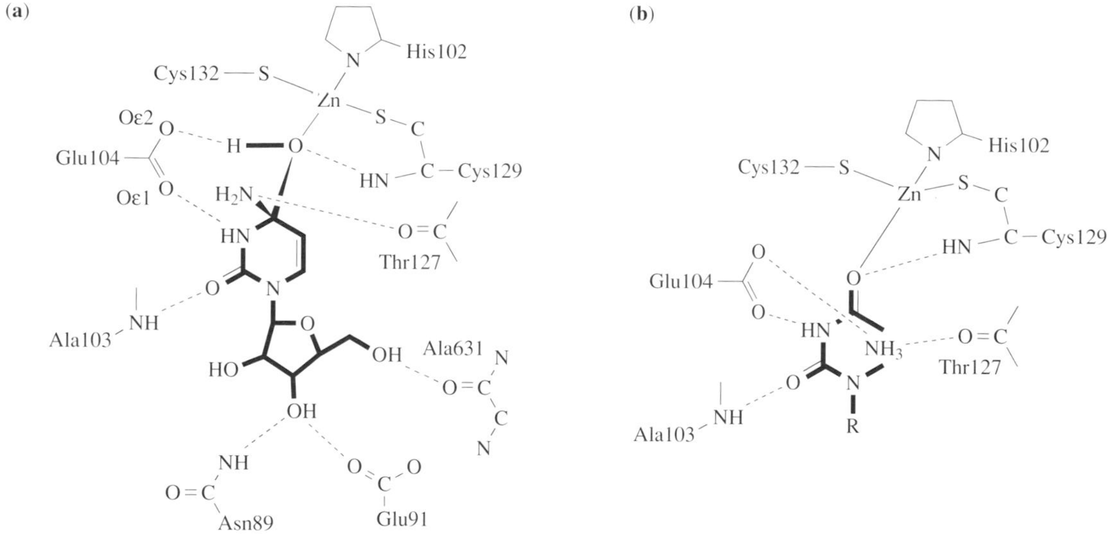

Figure 9 Active-site interactions for (a) the transition state of cytidine deaminase and (b) uridine deduced from the crystallographic studies summarized in the text.

Inhibitor–enzyme contacts are augmented by interactions of the hydrophobic pyrimidine edge with three aromatic side chains: two, from Phe71 and Tyr126, are in the active site, the third, Phe565, originates in the nonfunctional site. Phe233 is positioned underneath the ribose ring. The pocket into which the leaving group must fit is small, explaining why 4-*N*-methylcytidine is a poorer substrate than is cytidine.

(iii) Structure of bound inhibitors

Cytidine deaminase has been additionally crystallized in complex with the substrate analogue 3-deazacytidine, 3,4-dihydropyrimidin-2-one ribonucleoside (Figure 8(f)), and with the product uridine (Figure 8(a)).[148-150] The sequence of bound inhibitors—3-deazacytidine, 3,4-dihydropyrimidin-2-one ribonucleoside, 3,4-dihydrouridine, and uridine—approximates the reaction coordinate from substrate to transition state to product. Since a carbon atom replaces N-3 of cytidine, 3-deazacytidine cannot undergo addition at the 3,4-double bond. When complexed in the active site, this inhibitor traps the zinc-bound water molecule without steric interference. The amino group is positioned in the leaving group pocket and hydrogen bonded to the backbone of Thr127. There is no hydrogen bond to C-3. Though 3,4-dihydropyrimidin-2-one ribonucleoside is like a substrate analogue in its lack of a hydroxy group, it is tetrahedral at C-4 and has a protonated N-3 like the transition state. This inhibitor also traps the zinc-bound water in the active site. Here, not enough volume is available to accommodate both the inhibitor and the trapped water, leading to unfavorable van der Waals interactions in the vicinity of C-4. Otherwise, the inhibitor enjoys active-site interactions similar to those described for the transition-state analogues. The sequence 3-deazacytidine, 3,4-dihydropyrimidin-2-one ribonucleoside, 3,4-dihydrouridine shows progressive movement of the pyrimidine ring towards the activated water, suggesting that during catalysis the amino group remains in the binding site for the leaving group. Uridine, the product, is found to bind as the 4-ketopyrimidine, as predicted on energetic grounds (see Figure 9(b)).[32,150] In the crystal structure a water or ammonia molecule is found in the leaving group pocket, hydrogen bonded to the backbone carbonyl oxygen of Thr127. Here, an ammonia molecule is sketched in the figure. The keto-oxygen—derived from the hydroxy group at C-4 of the transition state—is not hydrogen bonded to Glu104, though O-4 still interacts with zinc and Cys129. Instead, the carboxy oxygen of Glu104 that was associated with the hydroxy group is now pivoted to form a hydrogen bond with the water or ammonia molecule in the leaving-group pocket. This crucial residue, after stabilizing the hydroxy group in the transition state, shuttles a proton from that moiety to the leaving group, fostering product formation. Finally, the N-1—C-1′ glycosidic bond is distorted by 19° so that dissociation of the product is energetically favored.

5.05.3.1.5 Mechanism

The microscopic mechanism for hydrolytic deamination of cytidine is described by Betts *et al.*[123] and Carter and co-workers.[147-150] In the substrate-free enzyme, via proton transfer to the carboxylate oxygen $O\varepsilon1$ of Glu104, the zinc-bound water becomes activated to hydroxide ion (see Figure 9(a) for placement of $O\varepsilon1$ and $O\varepsilon2$). When substrate enters the cleft, the hydrogen of $O\varepsilon1$ binds to N-3, with increasing transfer to the substrate as the reaction proceeds. Attack at C-4 by the zinc-bound hydroxide forms the first tetrahedral intermediate as sketched in Figure 9(a), with a hydrogen bond from hydroxide to $O\varepsilon2$ of Glu104. The involved proton is abstracted from the zinc-bound hydroxide ion by $O\varepsilon2$ giving a second tetradehral enolate O-4 alkoxide intermediate. Finally, Glu104 $O\varepsilon2$ reorients, donating the proton to the leaving group and completing the enol-to-keto tautomerization into uridine. The path of the proton shuttle can be inferred from Figures 9(a) and (b).

5.05.3.1.6 Transition-state similarity calculations for cytidine deaminase: application of neural networks

Computational neural networks, consisting of simple units or neurons that operate in parallel, are trained to learn the features of input or training patterns through adjustment of the interactions or weights connecting the neurons.[156,157] Each pattern in the training set—in this case the molecular electrostatic potential surfaces of the inhibitors, substrates, and transition states—is associated with an output target, the binding free energy. During training the weights are updated through many iterations until the network learns to recognize all the training molecules by calculating the binding free energy of each with minimum of error. Once this is accomplished, the network is used to predict the free energy of binding of a newly introduced inhibitor.

The powerful pattern recognition capabilities of neural networks were explored for the large number of substrates, transition states, and inhibitors for cytidine deaminase.[158] Molecular electrostatic potential surfaces were prepared and oriented as described in Section 5.05.2.2.3 for the molecules of the AMP and adenosine deaminase systems. A representative set of points describing the entire surface electrostatic potentials of a molecule constitutes the input for that molecule. All molecules shown in Figure 8, except for phosphapyrimidine ribonucleoside, 1,3-diazepin-2-one nucleoside, and pyrimidin-2-one ribonucleoside, were used in the training set, which included also the 5-flouro-derivative of 3,4-dihydrouridine and molecules of the 5,6-dihydrocytidine reaction. Neural networks were trained variously with either nine, seven, or five molecules and used to predict the inhibition constants for any of the molecules not present in the respective training set as well as the binding free energy of the transition state. When fewer training molecules are used, a range of weak and tight-binding inhibitors must be represented for subsequent successful predictions: otherwise, the choice for inhibitor inclusion in the training set could be random. The method was found to be both accurate and robust, with an average error of 4% for the nine-molecule training set and 11% for the seven- and five-molecule sets. The transition-state prediction was (in units of $\Delta G/RT$) within 3 units of the total -36 units of transition-state binding free energy (Table 1).[158]

Table 1   Neural net prediction of binding energy for cytidine deaminase ligands.[a]

Ligand	$\Delta G/RT$ (experimental)	$\Delta G/RT$ predicted (neural network)
Transition-state structure of cytidine	-36	-34
Hydrated 5-fluoro-pyrimidin-2-one riboside	-29	-32
Hydrated pyrimidin-2-one riboside	-27	-27
Transition state for 5,6-dihydrocytidine	-21	-21
3,4,5,6-Tetrahydrouridine	-16	-15
3,4-Dihydrozebularine	-10	-10
Cytidine	-9.9	-9.3
5,6-Dihydrocytidine	-9.1	-9.7
Uridine	-6.0	-5.8
5,6-Dihydrouridine	-5.7	-6.0

[a] The training set for each predicted $\Delta G/RT$ was the molecular electrostatic potential surface of the other nine molecules.[94] The algorithm used this data base to predict the $\Delta G/RT$ for the indicated molecule. The method provides predictive power for substrates, substrate analogues, and transition-state inhibitors.[158]

5.05.3.2   Other Pyrimidine Deaminases

5.05.3.2.1   *Cytosine deaminase*

Cytosine deaminase, present in bacteria and fungi, but not in mammalian cells, catalyzes the hydrolytic deamination of the pyrimidine base cytosine (cytidine without the ribose moiety) to uracil.[159] This product, when converted to uridine and then to uridine monophosphate, serves as a precursor for nucleic acid synthesis. *E. coli* cytosine deaminase has $K_m = 0.22$ mM for cytosine, with $k_{cat} = 185$ s^{-1}.[160] The enzyme requires a divalent metal ion for catalysis, and iron(II) can be removed from it, leaving an apoenzyme with $<5\%$ of the native catalytic activity. The apoenzyme can be reconstituted with addition of divalent iron, manganese, cobalt, or zinc. Although the K_m values for the resulting active enzymes are similar, reconstitution with iron gives the most active form, while zinc addition yields an enzyme with $k_{cat} = 32$ s^{-1}. Copper and zinc ions inhibit the fully functional deaminase, and the enzyme contains two metal binding sites. Inhibition with copper does not involve displacement of the catalytically essential metal.

Much of the interest in cytosine deaminase stems from its ability to convert 5-fluorocytosine, a relatively nontoxic compound or prodrug, into the highly cytotoxic 5-fluorouracil. Cytosine deaminase has been conjugated to monoclonal antibodies targeted to antigen on the cell surfaces of solid mammalian tumors.[161] These conjugates are thus delivered specifically to the tumor cell population, and oral administration of 5-fluorocytosine leads to localized active drug production at the tumor. Gene therapy strategies introduce (through retroviral vectors, for example) the cytosine deaminase gene into mammalian tumor cells.[162,163] In the presence of the prodrug, these cells commit metabolic suicide. A new delivery system employs genetically-engineered anaerobic bacteria to express cytosine deaminase while growing in the hypoxic regions of tumors.[164]

5.05.3.2.2   *dCMP deaminase*

Deoxycytidylate (dCMP) deaminase, catalyzing the formation of dUMP (deoxyuridine monophosphate) from dCMP (deoxycytidine monophosphate), has been isolated from a number of sources including human cells and T4 bacteriophage-infected *E. coli*. Since it is allosterically activated by 2′-deoxycytidine triphosphate (dCTP) and inhibited by deoxythymidine triphosphate (dTTP), both end products of the metabolic pathway, a central role in deoxyribonucleotide synthesis is played by the enzyme.[165] Cells deficient in dCMP deaminase show an imbalance in dCTP and dTTP pools, leading to increased mutations during DNA replication.[165] dUMP is converted to dTMP by thymidylate synthase, an enzyme interacting strongly with dCMP deaminase in the multienzyme complex responsible for deoxynucleotide biosynthesis.[166,167]

Human dCMP deaminase, with 178 amino acids and a zinc-binding domain per subunit, is hexameric.[165,168] The K_m value for dCMP is 58 μM, with $k_{cat} = 960$ s^{-1}. Tetrahydro-dUMP inhibits the (chick embryo) deaminase with $K_i = 10^{-8}$ M. The human enzyme is inhibited by 2′-β-D-deoxyribose-pyrimidin-2-one 5′-phosphate with $K_i = 1.2 \times 10^{-8}$ M. As the modes of action of dCMP and cytidine deaminase are thought to be the same, this inhibitor is perhaps covalently complexed to dCMP deaminase since it is the 5′-nucleotide, 2′-deoxy-analogue of pyrimidin-2-one ribonucleoside (Figure 8(e)).[33,169] T4-phage dCMP deaminase is hexameric as well but with two zinc-binding sites per monomer.[170,171] One of the zinc atoms, apparently bound to two cysteines and a histidine residue as in cytidine deaminase, is essential for catalysis, while the other is complexed to a unique zinc-binding motif. The allosteric inhibitor dTTP forms a photolabile covalent bond with Phe112, and if this residue is mutated to alanine the enzyme loses regulation by dCTP and dTTP.

5.05.3.2.3   *dCTP deaminase*

dCTP deaminase activity was initially detected in PBS$_1$-phage infected *B. subtilis*.[172] The enzyme catalyzes the deamination of dCTP to deoxyuridine triphosphate (dUTP). In *E. coli*, *Salmonella typhimurium*, and *Rickettsia prowazekii*, the agent of epidemic typhus, dCTP deaminase activity followed sequentially by the degradation of dUTP to dUMP provides most of the substrate for thymidylate synthase.[173–175] *Salmonella* dCTP deaminase requires magnesium for activity (or manganese, cobalt, or calcium). The true substrate of the reaction is Mg·dCTP; the divalent metal ion does not bind as a ligand to the enzyme. dCTP deaminase, highly specific for dCTP, shows a sigmoidal saturation curve for the substrate, and dTTP and dUTP are powerful inhibitors. However, dUTP probably does not function *in vivo* as a negative effector since dUTPase prevents its accumulation. The gene for the *E. coli* enzyme has been cloned, and it yields the 21 kDa polypeptide monomer of the tetrameric enzyme.[173]

5.05.3.3   Mammalian mRNA Cytidine Deaminase

RNA editing refers to processes that alter the informational content of RNA molecules through means other than splicing, 5′- and 3′-end formation, or hypermodification of nucleotide bases. The most extensively studied system is the mammalian apolipoprotein (apo) B messenger RNA editing complex, and the following discussion stems from reviews from the mid-1990s.[176–179] ApoB mRNA editing is a posttranscriptional event occurring within the nuclear compartment whereby a cytidine to uridine (C → U) modification at nucleotide 6666 of the mRNA converts a glutamine (CAA) into a stop codon (UAA). The unedited mRNA is translated into the full length apoB-100 protein of 512 kDa, while the editing process yields the smaller apoB-48 (241 kDa). Both these proteins are involved in lipoprotein metabolism, and, even though apoB-48 corresponds to 48% of apoB-100, they perform different functions. A metabolic switch from production of one protein in favor of the other can provide homeostatic control. ApoB-100 is a component of very low density (VLDL), intermediate density (IDL), and low density lipoproteins (LDL). It is necessary for the transport of endogenously synthesized triglyceride and cholesterol, and the overproduction of these lipoproteins results in a condition strongly correlated with atherosclerosis. ApoB-48 is a component of chylomicrons, secreted by the enterocytes of the small intestine for delivery of dietary triglycerides to peripheral tissues.

The C → U editing event at nucleotide 6666 of apoB mRNA proceeds through hydrolytic deamination at C-4 of the cytidine residue. The catalytic subunit, called apobec-1, of the apoB mRNA editing system has been identified and cloned from rat, human, rabbit, and mouse sources. Apobec-1 does not by itself edit apoB mRNA but confers editing activity to, for example, chick intestinal extracts which do not have intrinsic editing capability. Catalytic activity is inhibited by the zinc-specific chelating agent *o*-phenanthroline. Apobec-1 from rat is 27 kDa and its amino acid sequence is homologous to those of cytidine deaminase from *E. coli*, *B. subtilis*, yeast, and humans. Most importantly, there is strong homology around the active site residues as identified for *E. coli* cytidine deaminase (see Section 5.05.3.1.4). In apobec-1, with analogy to bacterial cytidine deaminase, His61, Cys93, and Cys96 are postulated to coordinate a zinc atom. Zinc activates a water molecule to hydroxide ion in the hydrolytic deamination of cytidine 6666 into a uridine residue. Glu63 serves as the proton shuttle during catalysis, in accord with the role of Glu104 in *E. coli* cytidine deaminase. The mutation of Cys93 and 96 into Ala residues, either individually or in concert, abolish zinc binding, while His61 conversion to Arg or Ser and Glu63 replacement by Gln preserve zinc-binding capability. This is the case for the analogous residues in the *E. coli* enzyme. However, these His61 and Glu63 mutations abolish catalytic activity entirely. It is postulated that the zinc-binding and catalytic mechanism for rat apobec-1 is similar to that for *E. coli* cytidine deaminase. Similar results are reported for rabbit apobec-1. A leucine-rich region at the carboxy-terminus of apobec-1 is also found in the *E. coli* enzyme. In rat, human, and rabbit apobec-1 the presence of two proline residues destroy the classic coiled coil structure of a leucine-zipper. The leucine-rich carboxy terminus may be involved in the dimerization of apobec-1 or in interaction with other proteins. Mutation of four leucine residues in the leucine-rich motif results in diminished editing activity *in vitro*, presumably through disruption of dimerization or of the association of apobec-1 with other necessary proteins: the cytidine deaminase activity is unaffected.

Purified apobec-1 binds to apoB mRNA, although the association shows low specificity. The binding is mediated by amino acid residues in the cytidine deaminase catalytic site. For example, Cys93 conversion to either Ser or Ala affects both RNA and zinc binding, while identical alterations to Cys96 do not greatly diminish RNA-binding capabilities. Therefore, even though the active site can be damaged, RNA binding might not be affected. His61 mutations into Arg, Ser, or Ala, while affecting zinc coordination only moderately, are deleterious to RNA binding. Mutation to the proton shuttle Glu63 also destroys apobec-1's ability to bind RNA. Phe66 and Phe87, encompassed in the active site, are important to the binding process as well. Deletion of 14 amino acids from the amino terminus and five from the carboxy terminus has shown that these domains may also interact with RNA.

Apobec-1 is part of a macromolecular editing complex or editosome which is targeted to apoB mRNA. Most alterations to a crucial sequence (5'-UGAUCAGUAUA-3') downstream from the edited cytidine reduce or destroy editing. Insertion of this sequence into a heterologous mRNA induces cytidine deamination upstream from it. Apobec-1 itself binds to an AU-rich region (5'-UAUAUU-3') downstream and overlapping with the above sequence. Apobec-1, when immobilized on an affinity column, binds proteins from rat hepatoma cell extracts of 145, 87, 75, 66, 61, and 50 kDa along with others ranging from 35 to 45 kDa.[180] Some of these auxiliary proteins must comprise the editosome and when complexed with apobec-1 confer to it editing activity; apobec-1 cannot catalyze the C → U transition alone. Included in the associated proteins is the protein that recognizes the 11 nucleotide "mooring" sequence above, previously identified as a 60 kDa species which interacts with the sequence subset 5'-UGAU-3'. From baboon kidney, a 65 kDa complementing protein that activates apobec-1 was isolated in the absence of apoB mRNA.[181] The complementing protein interacts directly with the catalytic subunit without involvement of its leucine-rich carboxy terminus. The simplest model of the editosome complex involves only the catalytic apobec-1 subunit and its complementing protein which recognizes and binds to the mooring sequence.

5.05.3.4 Plant Mitochondrial mRNA Cytidine Deaminase

Editing of mitochondrial mRNA in higher plants was first reported in 1989, and the following discussion is based on reviews from the late 1980s and the 1990s.[182–187] RNA editing is a post-transcriptional modification of the nucleotide sequence of mRNA in the mitochondrial compartment. The editing event is site specific, involving the conversion of genomically encoded cytidine residues into uridines. In contrast to the mammalian editing system described above, here multiple cytidine sites are targeted for editing. Less frequently, the reverse process, U → C, is encountered. Because most of the editing occurs within mRNA regions encoding for protein, the amino acid sequence of the resulting polypeptide is different from that predicted from the parent mitochondrial DNA. Editing events can also introduce initiation and termination codons in mRNA, while a small number of these C → U modifications are not reflected in the protein sequence. Posttranscriptional modification of mRNA—occurring after DNA has been transcribed into mRNA—makes it virtually impossible to predict a protein sequence from the corresponding gene. The net effect of the editing process is an increase in sequence similarity between plant mitochondrial proteins and their nonplant analogues. For example, editing creates amino acid residues in wheat or potato mRNAs (that code for subunit 9 of ATP synthase) which are already present in the analogous positions of the DNA sequence of the gene from other plant, fungi, and animal sources. Importantly, the discovery of mRNA editing dispels the idea that plant mitochondrial DNA is interpreted through a nonstandard genetic code. Based on the sequence of the cytochrome oxidase subunit II gene (*coxII*), it was surmised that an essential tryptophan was coded by the CGG triplet which usually specifies arginine: however, CGG → UGG editing at the level of mRNA gives the standard codon for tryptophan. Finally, these edits often give functional protein. In all plants other than wheat, the *coxII* gene codes for a Cys in position 228 which binds copper and is essential for protein activity. In wheat, a CGU (Arg) to UGU (Cys) mRNA edit precedes translation into protein. Introduction of unedited mRNA into the mitochondria of transgenic plants leads to nonfunctional proteins and cytoplasmic male sterility.

The editing process is site specific; if a fully edited mRNA is incubated with mitochondrial lysate, the remaining cytidine residues are not converted into uridines. There is no clear understanding of the determinants of this specificity. No conserved mRNA sequence has been discovered which can unambiguously specify an editing site, although the nucleotide distribution in the regions flanking editing sites differs from the mean nucleotide frequency. Similarly, no secondary or tertiary mRNA structures have been correlated to editing events, and guide RNAs (small RNA molecules complementary to unedited mRNA and dictating the information for an editing site) have not been uncovered.

C → U conversion can proceed in one of four ways: (i) site-specific hydrolytic deamination or transamination of cytidine to uridine (analogous to the deamination processes described above); (ii) a transglycosylation reaction in which the cytosine base is exchanged with uracil via a break and reformation of the glycosyl bond; (iii) replacement of cytidine nucleotide with uridine, involving cleavage and religation of the phosphodiester backbone of the mRNA; and (iv) modification of cytosine by attachment of the amino acid lysine at position 2, giving lysidine which is read as uridine by most enzymes. Labeling experiments with [α-^{32}P]CTP have demonstrated that the phosphodiester linkages are unaffected during editing, thereby eliminating the possibility of nucleotide substitutions. The labeled CTP can be administered to intact mitochondria, with radiolabel showing on the UMP residues that result after hydrolysis of mRNA transcripts. Alternatively, labeled *in vitro* mRNA strands can be incubated with mitochondrial lysate, showing that the editing process is not coupled with transcription. Furthermore, the possibility of lysidine conversion has been unambiguously eliminated. Radiolabels on the cytosine base are preserved in the editing process—if transglycosylation were involved, there would be no label on the resulting uracil base. In plant mitochondrial mRNA editing, the simplest mechanism consistent with these results is hydrolytic deamination at C-4 of the cytosine base. For analogy, in animal RNA editing systems the deamination is catalyzed by a protein having significant homology to prokaryotic and eukaryotic cytidine deaminases. Plant mitochondrial editing of the ATP synthase mRNA involves association of an editosome complex with mRNA at the editing site. The editosome is composed of RNA binding proteins, perhaps involved in the recognition process, the cytidine deaminase catalytic subunit, and several unknown factors. The reverse U → C modification probably proceeds through a different mechanism.

mRNA editing has been discovered in chloroplasts and in mitochondrial tRNA as well.[188–190] The editing of tRNA usually occurs in the double-stranded stem of the folded precursor molecule and is essential for proper processing of tRNA precursors. For example, a cytidine to uridine edit in potato mitochondrial phenylalanine tRNA precursor corrects a C·A mismatched base pair and alters the secondary structure of the precursor, rendering it susceptible to efficient recognition and processing by Rnase P. Other systems that feature RNA editing through deamination are summarized by Blanc *et al.*[187]

5.05.4 CONCLUSIONS

Deaminations of purines and pyrimidines use protein-bound metals to activate a water to the metal-hydroxy. Protonation of the purine or pyrimidine base assists in deconjugating the energetically favored aromatic ring structure. The transition states are reached just before the *sp³* center is formed, and deamination follows rapidly. Knowledge of the chemistry and transition-state structure, assisted by powerful inhibitors from natural product chemistry, have permitted the synthesis of tightly-bound inhibitors for the deaminases. Newly discovered members of the deaminases edit RNA and raise new questions of substrate specificity, chemical mechanism, and biological function. These enzymes are essential to the full understanding of the genetic code.

5.05.5 REFERENCES

1. F. A. Carey and R. F. Sundberg, "Advanced Organic Chemistry Part A: Structure and Mechanism," 3rd edn., Plenum, New York, 1990, p. 579.
2. F. A. Carey and R. F. Sundberg, "Advanced Organic Chemistry Part B: Reactions and Synthesis," 3rd edn., Plenum, New York, 1990, p. 588.
3. D. Zhong, S. Ahmad, P. Y. Chang, and A. H. Zewail, *J. Am. Chem. Soc.*, 1997, **119**, 2305.
4. M. S. Coleman, J. Donofrio, J. J. Hutton, L. Hahn, A. Daoud, B. Lampkin, and J. Dyminski, *J. Biol. Chem.*, 1978, **253**, 1619.
5. E. R. Giblett, J. E. Anderson, F. Cohen, B. Pollard, and H. J. Meuwissen, *Lancet*, 1972, **2**, 1067.
6. C. A. Mullen, K. Snitzer, K. W. Culver, R. A. Morgan, W. F. Anderson, and R. M. Blaese, *Hum. Gene Ther.*, 1996, **7**, 1123.
7. M. S. Hershfield, *Clin. Immunol. Immunopathol.*, 1995, **76**, S228.
8. A. A. Migchielsen, M. L. Breuer, M. A. van Roon, H. te Riele, C. Zurcher, F. Ossendorp, S. Toutain, M. S. Hershfield, A. Berns, and D. Valerio, *Nature Gen.*, 1995, **10**, 279.
9. M. R. Blackburn, S. K. Datta, M. Wakamiya, B. S. Vartabedian, and R. E. Kellems, *J. Biol. Chem.*, 1996, **271**, 15 203.
10. G. Schmidt, *Z. Physiol. Chem.*, 1932, **208**, 185.
11. K. A. Mohamedali, O. M. Guicherit, R. E. Kellems, and F. B. Rudolph, *J. Biol. Chem.*, 1993, **268**, 23 728.
12. B. M. Chassy and R. J. Suhadolnik, *J. Biol. Chem.*, 1967, **242**, 3655.
13. R. Wolfenden, *Biochemistry*, 1968, **7**, 2409.
14. L. C. Kurz and C. Frieden, *Biochemistry*, 1987, **26**, 8450.
15. W. M. Kati, S. A. Acheson, and R. Wolfenden, *Biochemistry*, 1992, **31**, 7356.
16. D. K. Wilson, F. B. Rudolph, and F. A. Quiocho, *Science*, 1991, **252**, 1278.
17. P. M. Weiss, P. F. Cook, J. D. Hermes, and W. W. Cleland, *Biochemistry*, 1987, **26**, 7378.
18. L. C. Kurz, L. Moix, M. C. Riley, and C. Frieden, *Biochemistry*, 1992, **31**, 39.
19. J. Grosshans and R. Wolfenden, *Biochim. Biophys. Acta*, 1992, **285**, 28.
20. S. Shuto, T. Obara, H. Itoh, Y. Kosugi, Y. Saito, M. Toriya, S. Yaginuma, S. Shigeta, and A. Matsuda, *Chem. Pharm. Bull.*, 1994, **42**, 1688.
21. E. Beutler and D. A. Carson, *Blood Cells*, 1993, **19**, 559.
22. C. U. Lambe, D. R. Averett, M. Paff, J. E. Reardon, J. G. Wilson, and T. A. Krenitsky, *Cancer Res.*, 1995, **55**, 3352.
23. T. Sawa, Y. Fukagawa, I. Homma, T, Takeuchi, and H. Umezawa, *J. Antibiot.*, 1967, **201**, 227.
24. N. Nakamura, G. Koyama, Y. Ititaka, M. Ohno, N. Yagisawa, S. Kondo, K. Maeda, and H. Umezam, *J. Am. Chem. Soc.*, 1974, **96**, 4327.
25. P. K. W. Woo, H. W. Dion, S. M. Lange, and L. F. Dahl, *J. Heterocycl. Chem.*, 1974, **11**, 641.
26. R. P. Agarwal, T. Spector, and R. E. Parks, Jr., *Biochem. Pharmacol.*, 1977, **26**, 359.
27. V. L. Schramm and D. C. Baker, *Biochemistry*, 1985, **24**, 641.
28. L. Frick, J. P. MacNeela, and R. Wolfenden, *Bioorg. Chem.*, 1987, **15**, 100.
29. R. Wolfenden, *Nature (London)*, 1969, **223**, 704.
30. R. Wolfenden, *Acc. Chem. Res.*, 1972, **5**, 10.
31. D. K. Wilson and F. A. Quiocho, *Biochemistry*, 1994, 33.
32. P. Shih and R. Wolfenden, *Biochemistry*, 1996, **35**, 4697.
33. W. M. Kati and R. Wolfenden, *Biochemistry*, 1989, **28**, 7919.
34. W. Jones and R. Wolfenden, *J. Am. Chem. Soc.*, 1986, **108**, 7444.
35. W. P. Jencks, *Adv. Enzymol.*, 1975, **43**, 219.
36. S. Y. Li and J. D. Stoeckler, *J. Med. Chem.*, 1994, **37**, 3844.
37. G. Cristalli, A. Eleuteri, R. Volpini, S. Vittori, E. Camaioni, and G. Lupidi, *J. Med. Chem.*, 1994, **37**, 201.
38. C. J. Carrera, A. Saven, and L. D. Piro, *Hemat. Oncol. Clin. N. Am.*, 1994, **8**, 357.
39. K. B. J. Schowen, in "Transition States of Biochemical Processes," eds. R. D. Gandour and R. L. Schowen, Plenum, New York, 1978, p. 225.
40. B. Evans and R. Wolfenden, *Biochemistry*, 1973, **12**, 392.
41. D. K. Wilson and F. A. Quiocho, *Biochemistry*, 1993, **32**, 1689.
42. W. W. Cleland, *Biochemistry*, 1992, **31**, 317.
43. K. A. Mohamedali, L. C. Kurz, and F. B. Rudolph, *Biochemistry*, 1996, **35**, 1672.
44. H. Deng, L. C. Kurz, F. B. Rudolph, R. Callender, *Biochemistry*, 1998, **37**, 4968.
45. V. Sideraki, K. A. Mohamedali, D. K. Wilson, Z. Chang, R. E. Kellems, F. A. Quiocho, and F. B. Rudolph, *Biochemistry*, 1996, **35**, 7862.
46. V. Sideraki, D. K. Wilson, L. C. Kurz, F. A. Quiocho, and F. B. Rudolph, *Biochemistry*, 1996, **35**, 15 019.
47. M. M. Ibrahim, I. T. Weber, and T. B. Knudsen, *Biochem. Biophys. Res. Commun.*, 1995, **209**, 407.
48. M. S. Hershfield and B. S. Mitchell, in "The Metabolic and Molecular Bases of Inherited Disease," 7th edn., eds. C. R. Scriver, A. L. Beaudet, W. S. Sly, and D. Valle, McGraw Hill, New York, 1995, p. 1725.
49. R. Hirschhorn, *Clin. Immunol. Immunopathol.*, 1995, **76**, S219.

50. M. L. Markert, *Immunodeficiency*, 1994, **5**, 141.
51. W. P. Schrader, C. A. West, U. H. Rudofsky, and W. A. Samsonoff, *J. Histochem. Cytochem.*, 1994, **42**, 775.
52. J. Groves and J. R. Olson, *Inorg. Chem.*, *J. Am. Chem. Soc.*, 1985, **24**, 2715.
53. T. H. Fife and T. J. Przystus, *J. Am. Chem. Soc.*, 1986, **108**, 4631.
54. D. W. Christianson and W. N. Lipsomb, in "Mechanistic Principles of Enzyme Activity," eds. J. F. Liebman and A. Greenberg, VCH Publishers, New York, 1988, p. 1.
55. C. J. Coffee and W. A. Kofke, *J. Biol. Chem.*, 1975, **250**, 6653.
56. S. L. Meyer, K. L. Kvalnes-Krick, and V. L. Schramm, *Biochemistry*, 1989, **28**, 8734.
57. R. L. Sabina, R. Marquetant, N. M. Desai, K. Kaletha, and E. W. Holmes, *J. Biol. Chem.*, 1987, **262**, 12 397.
58. D. J. Merkler and V. L. Schramm, *J. Biol. Chem.*, 1990, **265**, 4420.
59. B. Ashby and C. Frieden, *J. Biol. Chem.*, 1978, **253**, 8728.
60. J. M. Lowenstein and M. N. Goodman, *Fed. Proc.*, 1978, **37**, 2308.
61. J. J. Aragon and J. M. Lowenstein, *Eur. J. Biochem.*, 1980, **110**, 371.
62. M. N. Goodman and J. M. Lowenstein, *J. Biol. Chem.*, 1977, **252**, 5054.
63. A. Katz, K. Sahlin, and J. Henriksson, *Am. J. Physiol.*, 1986, **250**, C834.
64. R. L. Sabina and E. W. Holmes, in "The Metabolic and Molecular Basis of Inherited Disease," 7th edn., eds. C. R. Scriver, A. L. Beaudet, W. S. Sly, and D. Valle, McGraw Hill, New York, 1995, p. 1725.
65. B. Norman, B. Glenmark, and E. Jansson, *Pharm., World Sci.*, 1994, **16**, 55.
66. B. Ashby, C. Frieden, and R. Bischoff, *J. Cell. Biol.*, 1979, **81**, 361.
67. B. Ashby and C. Frieden, *J. Biol. Chem.*, 1977, **252**, 1869.
68. J. K. Thakkar, D. R. Janero, C. Yarwood, H. Sharif, and D. R. Hreniuk, *Biochem. J.*, 1993, **280**, 335.
69. Y. Xia, G. Khatchikian, and J. L. Zweier, *J. Biol. Chem.*, 1996, **271**, 10 096.
70. C. L. Zielke and C. H. Suelter, "The Enzymes," 3rd edn, 1971, **4**, 47.
71. Y.-P. Lee, *J. Biol. Chem.*, 1957, **227**, 987.
72. Y.-P. Lee, *J. Biol. Chem.*, 1957, **227**, 993.
73. Y.-P. Lee, *J. Biol. Chem.*, 1957, **227**, 999.
74. D. H. Turng and J. F. Turner, *Biochem. J.*, 1961, **79**, 143.
75. K. L. Smiley and C. H. Suelter, *J. Biol. Chem.*, 1967, **242**, 1980.
76. D. J. Merkler, A. S. Wali, J. Taylor, and V. L. Schramm, *J. Biol. Chem.*, 1989, **264**, 21 422.
77. R. Wolfenden and Y. Tomozawa, *Biochemistry*, 1970, **9**, 3400.
78. J. Monod, J. Wyman, and J.-P. Changeux, *J. Mol. Biol.*, 1965, **12**, 88.
79. D. E. Koshland, G. Nemethy, and D. Filmer, *Biochemistry*, 1966, **5**, 365.
80. V. L. Schramm, *J. Biol. Chem.*, 1976, **251**, 3417.
81. M. T. Bausch-Jurken and R. L. Sabina, *Arch. Biochem. Biophys.*, 1995, **321**, 372.
82. R. Eddy, S. Sait, and R. L. Sabina, *Biochim. Biophys. Acta*, 1996, **1308**, 122.
83. F. Van den Bergh and R. L. Sabina, *Biochem. J.*, 1995, **312**, 401.
84. D. J. Merkler and V. L. Schramm, *Biochemistry*, 1993, **32**, 5792.
85. V. L. Schramm, B. A. Horenstein, and P. C. Kline, *J. Biol. Chem.*, 1994, **269**, 18 259.
86. D. J. Merkler, P. C. Kline, P. Weiss, and V. L. Schramm, *Biochemistry*, 1993, **32**, 12 993.
87. D. M. Quinn and L. D. Sutton, in "Enzyme Mechanism from Isotope Effects," ed. P. F. Cook, CTC Press, Boca Raton, 1991, p. 72.
88. L. Melander and W. J. Saunders, Jr., "Reaction Rates of Isotopic Molecules," Wiley, New York, 1980, p. 4.
89. L. B. Sims and D. E. Lewis, in "Isotopes in Organic Chemistry," eds. E. Buncel and C. C. Lees, Elsevier, New York, 1984, vol. 6, p. 161.
90. C. Frieden, H. R. Gilbert, W. H. Miller, and R. L. Miller, *Biochem. Biophys. Res. Commun.*, 1979, **91**, 278.
91. D. J. Merkler, M. Brenowitz, and V. L. Schramm, *Biochemistry*, 1990, **29**, 8358.
92. P. C. Kline and V. L. Schramm, *J. Biol. Chem.*, 1994, **269**, 22 385.
93. C. K. Bagdassarian, B. B. Braunheim, V. L. Schramm, and S. D. Schwartz, *Int. J. Quantum Chem.*, 1996, **60**, 73.
94. C. K. Bagdassarian, V. L. Schramm, and S. D. Schwartz, *J. Am. Chem. Soc.*, 1996, **118**, 8825.
95. H. A. Simmonds, A. S. Sahota, and K. J. Van Acker, in "The Metabolic and Molecular Basis of Inherited Disease," 7th edn., C. R. Scriver, A. L. Beaudet, W. S. Sly, and D. Vallea, McGraw Hill, New York, 1995, p. 1707.
96. R. Borriss, S. Porwollik, and R. Schroeter, *Microbiology*, 1996, **142**, 3027.
97. M. C. Deeley, *J. Bacteriol.*, 1992, **174**, 3102.
98. P. Nygaard, P. Duckert, and H. H. Saxild, *J. Bacteriol.*, 1996, **178**, 846.
99. C. C. Wang, R. Verham, A. Rice, and S. Tzeng, *Mol. Biochem. Parasitol.*, 1983, **8**, 325.
100. N. K. Gupta and M. D. Glantz, *Arch. Biochem. Biophys.*, 1985, **236**, 266.
101. S. Miyamoto, H. Ogawa, H. Shirake, and H. Nakagawa, *J. Biochem.*, 1982, **91**, 167.
102. A. Lucacchini, U. Montali, and D. Seganini, *Ital. J. Biochem.*, 1977, **26**, 27.
103. B. R. Baker and H. U. Siebenneick, *J. Med. Chem.*, 1971, **14**, 802.
104. A. K. Saxona, S. Ahmad, K. Shanker, K. P. Bhargava, and K. Kishor, *Pharmazie*, 1980, **35**, 16.
105. A. K. Saxona, S. Ahmad, K. Shanker, and K. Kishor, *Pharmacol. Res. Commun.*, 1984, **16**, 243.
106. U. Kim and K. Nishikura, *Semin. Cell Biol.*, 1993, **4**, 285.
107. U. Kim., T. L. Garner, T. Sanford, D. Speicher, J. M. Murray, and K. Nishikura, *J. Biol. Chem.*, 1994, **269**, 13 480.
108. B. L. Bass and H. Weintraub, *Cell*, 1987, **48**, 607.
109. A. G. Polson, P. F. Crain, C. Pomerantz, J. A. McCloskey, and B. L. Bass, *Biochemistry*, 1991, **30**, 11 507.
110. F. Lai, R. Drakas, and K. Nishikura, *J. Biol. Chem.*, 1995, **270**, 17 098.
111. U. Kim, T. L. Garner, T. Sanford, D. Speicher, and J. M. Murray, *J. Biol. Chem.*, 1994, **269**, 13 480.
112. A. Herbert and A. Rich, *J. Biol. Chem.*, 1996, **271**, 11 595.
113. G. A. Dabiri, F. Lai, R. A. Drakas, and K. Nishikura, *EMBO Journal*, 1996, **15**, 34.
114. S. Maas, T. Melchery, A. Herb, P. H. Seeburg, W. Keller, S. Krause, M. Higuchi, and M. A. O'Connell, *J. Biol. Chem.*, 1996, **271**, 12 221.
115. J. L. Casey and J. L. Gerin, *J. Virol.*, 1995, **69**, 7593.
116. D. Voet and J. D. Voet, "Biochemistry," 2nd edn, Wiley, New York, 1995, p. 795.

117. T. P. Wang, H. Z. Sable, and J. O. Lampen, *J. Biol. Chem.*, 1950, **184**, 17.
118. R. M. Cohen and R. Wolfenden, *J. Biol. Chem.*, 1971, **246**, 7561.
119. H. Hosono and S. Kuno, *J. Biochem.*, 1973, **74**, 797.
120. G. W. Ashley and P. A. Bartlett, *J. Biol. Chem.*, 1984, **259**, 13 615.
121. A. Vita, A. Amici, T. Cacciamani, M. Lanciotti, and G. Magni, *Biochemistry*, 1985, **24**, 6020.
122. L. Frick, C. Yang, V. E. Marquez, and R. Wolfenden, *Biochemistry*, 1989, **28**, 9423.
123. L. Betts, S. Xiang, S. A. Short, R. Wolfenden, and C. W. Carter, Jr., *J. Mol. Biol.*, 1994, **235**, 635.
124. B. E. Evans, G. N. Mitchell, and R. Wolfenden, *Biochemistry*, 1975, **14**, 621.
125. P. L. Ipata, G. Cercignani, and E. Balestreri, *Biochemistry*, 1970, **9**, 3390.
126. G. B. Wisdom and B. A. Orsi, *Eur. J. Biochem.*, 1969, **7**, 223.
127. W. A. Creasey, *J. Biol. Chem.*, 1963, **238**, 1772.
128. D. F. Wentworth and R. Wolfenden, *Biochemistry*, 1975, **14**, 5099.
129. A. Radzicka and R. Wolfenden, *Methods Enzymol.*, 1995, **249**, 284.
130. I. Wempen, R. Duschinsky, L. Kaplan, and J. J. Fox, *J. Am. Chem. Soc.*, 1961, **83**, 4755.
131. G. W. Camiener and C. G. Smith, *Biochem. Pharmacol.*, 1965, **14**, 1405.
132. G. W. Camiener, *Biochem. Pharmacol.*, 1968, **17**, 1981.
133. B. A. Chabner, J. C. Drake, and D. G. Jones, *Biochem. Pharmacol.*, 1973, **22**, 2763.
134. T.-C. Chou, A. Feinberg, A. J. Grant, P. Vidal, U. Reichman, K. A. Watanabe, J. Fox, and F. S. Philips, *Cancer Res.*, 1981, **41**, 3336.
135. B. Chandrasekaran, R. L. Capizzi, T. E. Kute, T. Morgan, and J. Dimling, *Cancer Res.*, 1989, **49**, 3259.
136. A. R. Hanze, *J. Am. Chem. Soc.*, 1967, **89**, 6720.
137. R. G. Stoller, C. E. Myers, and B. A. Chabner, *Biochem. Pharmacol.*, 1978, **27**, 53.
138. V. E. Marquez, P. S. Liu, J. A. Kelley, J. S. Driscoll, and J. J. McCormack, *J. Med. Chem.*, 1980, **23**, 713.
139. G. W. Ashley and P. A. Bartlett, *J. Biol. Chem.*, 1984, **259**, 13 621.
140. C. Frieden, L. C. Kurz, and H. R. Gilbert, *Biochemistry*, 1980, **19**, 5303.
141. P. A. Bartlett and C. K. Marlowe, *Biochemistry*, 1983, **22**, 4618.
142. G. W. Ashley and P. A. Bartlett, *Biochem. Biophys. Res. Commun.*, 1982, **108**, 1467.
143. P. S. Liu, V. E. Marquez, J. S. Driscoll, R. W. Fuller, and J. J. McCormack, *J. Med. Chem.*, 1981, **24**, 662.
144. C.-H. Kim, V. E. Marquez, D. T. Mao, D. R. Haines, and J. J. McCormack, *J. Med. Chem.*, 1986, **29**, 1374.
145. J. J. McCormack, V. E. Marquez, P. S. Liu, D. T. Vistica, and J. S. Driscoll, *Biochem. Pharmacol.*, 1980, **29**, 830.
146. D. C. Carlow, S. A. Short, and R. Wolfenden, *Biochemistry*, 1996, **35**, 948.
147. C. W. Carter, Jr., *Biochimie*, 1995, **77**, 92.
148. S. Xiang, S. A. Short, R. Wolfenden, and C. W. Carter, Jr., *Biochemistry*, 1995, **34**, 4516.
149. S. Xiang, S. A. Short, R. Wolfenden, and C. W. Carter, Jr., *Biochemistry*, 1996, **35**, 1335.
150. S. Xiang, S. A. Short, R. Wolfenden, and C. W. Carter, Jr., *Biochemistry*, 1997, **36**, 4768.
151. B.-H. Song and J. Neuhard, *Mol. Gen. Genet.*, 1989, **216**, 462.
152. S. Vincenzetti, A. Cambi, J. Neuhard, E. Garattini, and A. Vita, *Protein Expr. Purif.*, 1996, **8**, 247.
153. C. Yang, D. Carlow, R. Wolfenden, and S. A. Short, *Biochemistry*, 1992, **31**, 4168.
154. D. C. Carlow, A. A. Smith, C. C. Yang, S. A. Short, and R. Wolfenden, *Biochemistry*, 1995, **34**, 4220.
155. A. A. Smith, D. C. Carlow, R. Wolfenden, and S. A. Short, *Biochemistry*, 1994, **33**, 6468.
156. J. Zupan and J. Gasteiger, "Neural Networks for Chemists," VCH, Weinheim, 1993, p. 119.
157. M. Wagener, J. Sadowskij, and J. Gasteiger, *J. Am. Chem. Soc.*, 1995, **117**, 7769.
158. B. B. Braunheim, C. K. Bagdassarian, V. L. Schramm, and S. D. Schwartz, 1998, submitted for publication.
159. K. Wei and B. E. Huber, *J. Biol. Chem.*, 1996, **271**, 3812.
160. D. J. T. Porter and E. A. Austin, *J. Biol. Chem.*, 1993, **268**, 24 005.
161. P. M. Wallace and P. D. Senter, *Methods Find. Exp. Clin. Pharmacol.*, 1994, **16**, 505.
162. M. P. Deonarain, R. A. Spooner, and A. A. Epenetos, *Gene Ther.*, 1995, **2**, 235.
163. C. A. Mullen, *Pharmocol. Ther.*, 1994, **63**, 199.
164. M. E. Fox, M. J. Lemmon, M. L. Mauchline, T. O. Davis, A. J. Giaccia, N. P. Minton, and J. M. Brown, *Gene Ther.*, 1996, **3**, 173.
165. G. F. Maley, A. P. Lobo, and F. Maley, *Biochim. Biophys. Acta*, 1993, **1162**, 161.
166. C. K. Mathews, *Prog. Nucleic Acid Res. Mol. Biol.*, 1993, **44**, 167.
167. K. M. McGaughey, L. J. Wheeler, J. T. Moore, G. F. Maley, F. Maley, and C. K. Mathews, *J. Biol. Chem.*, 1996, **271**, 23 037.
168. K. X. B. Weiner, R. S. Weiner, F. Maley, and G. F. Maley, *J. Biol. Chem.*, 1993, **268**, 12 983.
169. W. M. Kati and R. Wolfenden, *Science*, 1989, **243**, 1591.
170. J. T. Moore, R. E. Silversmith, G. F. Maley, and F. Maley, *J. Biol. Chem.*, 1993, **268**, 2288.
171. J. T. Moore, J. M. Ciesla, L-M. Changchien, G. F. Maley, and F. Maley, *Biochemistry*, 1994, **33**, 2104.
172. F. Tomita and I. Takahashi, *Biochim. Biophys. Acta*, 1969, **179**, 18.
173. L. Wang and B. Weiss, *J. Bacteriol.*, 1992, **174**, 5647.
174. C. F. Beck, A. R. Eisenhardt, and J. Neuhard, *J. Biol. Chem.*, 1975, **250**, 609.
175. R. R. Speed and H. H. Winkler, *J. Bacteriol.*, 1991, **173**, 4902.
176. J. Scott, N. Navaratnam, S. Bhattacharya, and J. R. Morrison, *Curr. Opin. Lipidol.*, 1994, **5**, 87.
177. L. Chan, *Biochimie*, 1995, **77**, 75.
178. J. Scott, *Cell*, 1995, **81**, 833.
179. R. Benne, *Curr. Opin. Genet. Develop.*, 1996, **6**, 221.
180. Y. Yang and H. C. Smith, *Biochem. Biophys. Res. Commun.*, 1996, **218**, 797.
181. A. Mehta, S. Banerjee, and D. M. Driscoll, *J. Biol. Chem.*, 1996, **271**, 28 294.
182. P. S. Covello and M. W. Gray, *Nature*, 1989, **341**, 662.
183. J. M. Gualberto, L. Lamattina, G. Bonnard, J.-H. Weil, and J.-M. Grienenberger, *Nature*, 1989, **341**, 660.
184. R. Hiesel, B. Wissinger, W. Schuster, and A. Brennicke, *Science*, 1989, **246**, 1632.
185. M. W. Gray and P. S. Covello, *FASEB J.*, 1993, **7**, 64.
186. W. Yu, T. Fester, H. Bock, and W. Schuster, *Biochimie*, 1995, **77**, 79.

187. V. Blanc, X. Jordana, S. Litvak, A. Araya, *Biochimie*, 1996, **78**, 511.
188. B. Hoch, R. M. Maier, K. Appel, G. L. Igloi, and H. Kössel, *Nature*, 1991, **353**, 178.
189. L. Maréchal-Drouard, A. Cosset, C. Remacle, D. Ramamonjisoa, and A. Dietrich, *Mol. Cell. Biol.*, 1996, **16**, 3504.
190. A. Marchfelder, A. Brennicke, and S. Binder, *J. Biol. Chem.*, 1996, **271**, 1898.

5.06
Ester Hydrolysis

DANIEL M. QUINN, ROHIT MEDHEKAR and NATHAN R. BAKER
University of Iowa, Iowa City, IA, USA

5.06.1 INTRODUCTION

Ester hydrolysis is an ubiquitous reaction, both in the organic chemical and biocatalytic realms. Biocatalytic ester hydrolysis is involved in numerous physiological functions, ranging from the destruction of neurotransmitters to the absorption of dietary nutrients and to the detoxification of xenobiotics.[1] Before describing the structural and mechanistic motifs that have evolved for the enzymatic hydrolysis of esters, it is instructive to describe the basic mechanisms by which non-enzymatic ester hydrolysis proceeds.

5.06.1.1 Mechanistic Diversity in Ester Hydrolysis[2–4]

There are two limiting stepwise mechanisms for the hydrolysis of esters, as outlined in Scheme 1. For esters that possess strongly basic leaving groups, a stepwise addition–elimination mechanism operates that transits a tetrahedral intermediate that is higher in energy than reactants or products by 60 kJ mol^{-1} or more. For esters whose leaving groups are weakly basic (i.e., the conjugate acid of the leaving group is a strong acid), an elimination–addition mechanism operates that transits a high-energy acylium ion intermediate that is rapidly trapped by solvent. Lying between these conceptual extremes is a wide array of possible concerted mechanisms in which there are various combinations of total bond order in the developing H_2O–carbonyl carbon and breaking RO–carbonyl carbon bonds. Because many biotic esters have strongly basic alkoxy leaving groups, it is likely that enzyme catalytic power has evolved under the selective pressure to stabilize high-energy tetrahedral intermediates and the transition states for their formation and decomposition. Therefore, this chapter will explore structural motifs that enzymes might utilize to stabilize such structures. However, there are also biotic esters that possess less strongly basic phenoxy leaving groups, and therefore a concerted mechanism for their enzyme-catalyzed hydrolysis is worth considering. Not shown explicitly in Scheme 1 are the roles played by general acids (e.g., proton transfer to the leaving group), general bases (e.g., proton removal from the attacking water), of intervening nucleophiles (e.g., in a mechanism that involves an intermediate transacylation stage), or of electrophilic catalysts (e.g., metal ions that polarize the substrate carbonyl by Lewis acid–base interaction). These catalytic themes will be considered for particular classes of esterolytic enzymes in this chapter.

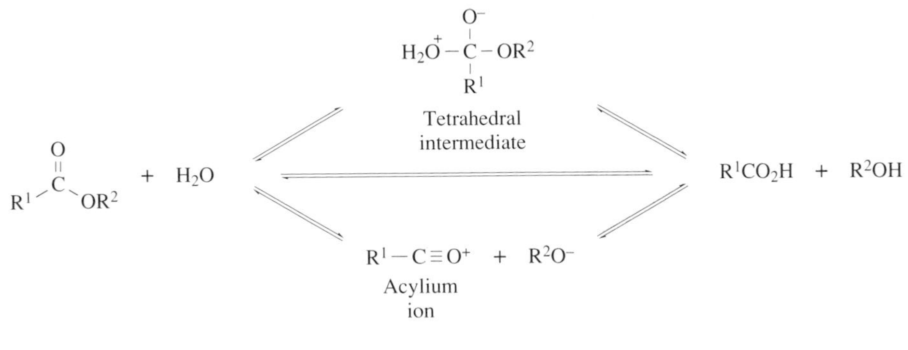

$$H_2\overset{+}{O} - \underset{R^1}{\overset{O^-}{\underset{|}{\overset{|}{C}}}} - OR^2$$

Tetrahedral
intermediate

$$\underset{R^1\diagdown \overset{O}{C}\diagup OR^2}{} \ + \ H_2O \ \rightleftharpoons \ R^1CO_2H \ + \ R^2OH$$

$$R^1 - C \equiv O^+ \ + \ R^2O^-$$

Acylium
ion

Scheme 1

The philosophy that will guide description herein of enzymatic ester hydrolysis is to provide a mechanistic picture that is similar in detail to that which is customary for well-characterized organic reactions. Consequently, no attempt will be made to provide an encyclopedic listing of the enzymes that are known to catalyze ester hydrolysis. Rather, the pertinent major enzyme families will be introduced, and for representative enzymes from each family a reasonably detailed view of the chemical and physical events that comprise the catalytic mechanisms will be developed. A multifarious approach will be pursued. Particular, though not exclusive, attention will be paid to enzymes for which crystal structures are available from the Brookhaven Protein Data Bank. The traditional staple of the organic chemist, arrow pushing, will be used to provide intuitively satisfying mechanistic outlines. However, where relevant, mechanistic elements that are informed by the tools of the computational chemist, such as quantum mechanics or molecular dynamics, will be described. And finally, the results of site-directed mutagenesis experiments will be cited that illuminate the relationship of structure and function in esterolytic biocatalysis. It is the combination of approaches such as these that offers the best hope for understanding the specificity and catalytic power of enzymes in general, and esterases in particular.

5.06.1.2 Families of Ester Hydrolyzing Enzymes

There are three major families of enzymes that have evolved to catalyze the hydrolysis of esters. One family is the acid lipases, which function in the lysosomes of cells or in the gastric juice of the stomach, and which have maximal activities in the approximate pH range 2 to 5. The second family is the phospholipases A_2, which are maximally active at neutral pH and most of which utilize Ca^{2+} as a Lewis acid in catalysis. The third family is the esterases of the α/β hydrolase fold supergene family. This diverse group of enzymes is comprised largely of serine esterases that utilize Ser-His-Asp (or Glu) catalytic triads. However, one member of the family utilizes the thiol function of cysteine as its active site nucleophile, and two members utilize unusual and surprising catalytic triads. Enzymes from each of the three major families will be discussed throughout the ensuing segments of this chapter.

5.06.1.3 Homogeneous Versus Heterogeneous Enzyme Catalysis

Esterases usually catalyze the hydrolysis of substrates that are unimolecularly dispersed. Such reactions are examples of homogeneous catalysis, and, in the absence of substrate and/or product activation and/or inhibition, are empirically described by the Michaelis–Menten equation:

$$v_i = \frac{V_{max}[A]}{K_m + [A]} \tag{1}$$

However, for lipases, which are esterases that catalyze the hydrolysis of trenchantly insoluble lipid ester substrates, the substrates are organized in supramolecular complexes, such as micelles, lipid bilayers, or lipoproteins. For such supramolecular substrates, binding of the enzyme to the lipid–water interface either precedes or is concomitant with binding of a substrate monomer in the active site of the enzyme. Therefore, molecular recognition by lipases involves not only the traditionally acknowledged interaction with substrate monomers, but also the less well characterized recognition by the enzyme of the lipid–water interface.

Heterogeneous lipolytic catalysis can be viewed as a logical extension of the Michealis–Menten mechanism,[5] as outlined in Scheme 2. In this mechanism, binding of the enzyme to the lipid–water interface is required for subsequent substrate binding to the active site and catalytic turnover. The corresponding kinetics equation predicts that the reaction velocity increases in a manner that is mathematically equivalent to the hyperbolic Michealis–Menten equation:

$$v_i = \frac{V_{max}^{app}\left(X_A \dfrac{I_A}{V}\right)}{K_m^{app} + \left(X_A \dfrac{I_A}{V}\right)} \tag{2}$$

In this equation X_A, I_A, and V are the substrate concentration in molecules (interface area)$^{-1}$, the total interface area, and the reaction volume, respectively. X_A and I_A may alternately be expressed in units of mole fraction and total moles of interface components, respectively. Saturation kinetics are observed not because the enzyme active site is increasingly bound by substrate as the apparent substrate concentration $X_A I_A/V$ is increased, but rather because the enzyme is increasingly bound to the interface as I_A increases. The substrate concentration X_A in the interface does not vary because it is a feature of the composition of the interface.

The hyperbolic parameters K_m^{app} and V_{max}^{app} are also more complex than the corresponding parameters of homogeneous Michealis–Menten kinetics:[5]

$$V_{max}^{app} = \frac{k_{cat}^*[E]_T X_A}{K_m^* + X_A} \tag{3}$$

$$K_m^{app} = \frac{(k_d/k_p)K_m^* X_A}{K_m^* + X_A} \tag{4}$$

Equations (3) and (4) show that both of the apparent Michaelis–Menten parameters are hyperbolic functions of the interfacial substrate concentration X_A. V_{max}^{app} contains not only the interfacial catalytic rate constant k_{cat}^* ($=k_3$ under initial velocity conditions), but also the fractional saturation

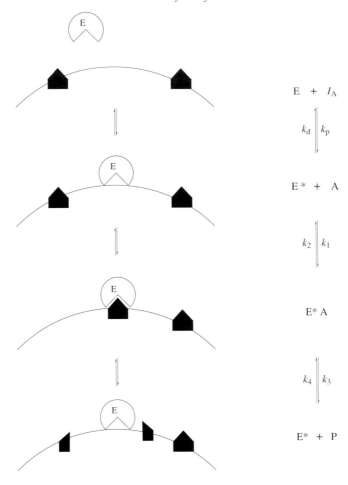

E + I_A

$k_d \parallel k_p$

E* + A

$k_2 \parallel k_1$

E* A

$k_4 \parallel k_3$

E* + P

E = enzyme in solution   E* = enzyme at interface
A, P = interfacial substrate and product
E*A = interfacial Michaelis complex I_A = interfacial area

Scheme 2

of the active sites of the interface-bound enzyme by substrate, $X_A/(K_m^* + X_A)$. K_m^{app} likewise depends on the fractional saturation of the interface-bound enzyme, but also contains the dissociation constant of the enzyme–interface complex, k_d/k_p, and the interfacial Michaelis constant $K_m^* = (k_2 + k_3)/k_1$. Therefore, saturation kinetics for the interface-bound enzyme can only be effected, and k_{cat}^*, K_m^*, and k_d/k_p determined, if one can systematically vary the interfacial substrate concentration X_A. This cannot be done for some substrate-containing complexes, such as cell membranes or lipoproteins, of lipolytic esterases. Fortunately, interfacial saturation kinetics can be assessed with model supramolecular substrates of particular lipases, and examples thereof will be discussed in appropriate later sections of this chapter.

The catalytic power of esterases and other enzymes has been evolutionarily optimized under the selective pressure of specific stabilization of transition states of the chemical steps in the relevant mechanisms. However, for enzymes that function by the mechanism of Scheme 2, the interfacial binding event must also be optimized for efficient catalytic function. An efficient lipase will both bind specifically to its cognate interface and catalyze interfacial substrate turnover in a highly processive manner, i.e., the enzyme will catalyze turnover at high substrate/enzyme stoichiometries before dissociating from the interface. Such a processive mechanism insures that the enzyme does not spend a high time fraction in the bulk medium, where it is inactive as a lipase. Therefore, for lipases, the structural bases of both interfacial recognition and transition state stabilization will be explored.

5.06.2 PHOSPHOLIPASE A$_2$

5.06.2.1 Catalytic Mechanism

Calcium-dependent phospholipase A$_2$ (PLA2) is perhaps the best characterized of lipolytic enzymes.[6-8] Though calcium-independent PLA2s have also been studied,[9] attention is focused herein on the better characterized calcium-dependent enzymes. Secretory PLA2s are found in numerous species, including the digestive juices and inflammatory exudates of humans and other mammals, and the venoms of snakes and insects. The enzyme catalyzes the stereospecific and regiospecific hydrolysis of the *sn*-2 ester bond of phospholipid substrates by the mechanism outlined in Scheme 3. Important elements in the catalytic mechanism are the His-Asp diad and an enzyme-bound Ca^{2+}, which respectively serve as a general acid–base catalyst in the formation and breakdown of the putative tetrahedral intermediate and as a Lewis acid to stabilize the tetrahedral intermediate. In inflammatory response, this reaction releases arachidonic acid, which is subsequently converted to effectors of cellular function, such as thromboxanes and prostaglandins. Consequently, PLA2 catalysis is of considerable interest, both from the biochemical and biomedical standpoints.

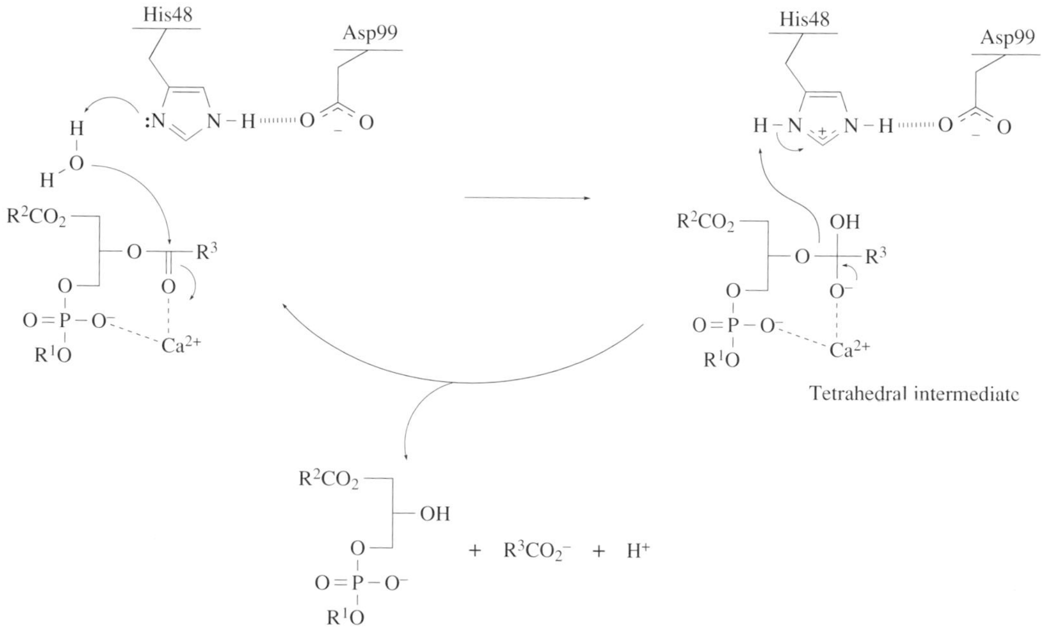

Scheme 3

5.06.2.2 Enzyme Structure

An appreciation of the catalytic power and substrate specificity of PLA2 begins with the structure of the enzyme. Accordingly Figure 1 shows the structure of bovine pancreatic PLA2.[10,11] The enzyme has a molecular mass of 14 kDa, consists of six helical segments and a two-stranded β-sheet connected by short loops, and is cross-linked by seven disulfide bonds. The active site His48-Asp99 diad is situated ∼13 Å from the protein surface in an approximately cylindrical channel. Particular residues on the surface of the enzyme that surrounds the opening of the active site channel are thought to comprise the interfacial recognition surface, as discussed later.

Figure 2 shows a closer view of the interactions between active site residues and the transition state analogue inhibitor L-1-(*O*-octyl-2-(heptylphosphonyl))-*sn*-glycero-3-phosphoethanolamine;[12] similar complexes have been described for PLA2s from the venoms of the Chinese cobra[13] (*Naja naja atra*) and the honey bee[14] (*Apis meliflora*) and from inflammatory exudate.[15] In the uncomplexed enzyme, the active site Ca^{2+} is seven-coordinated in a pentagonal bipyramidal geometry, with the axial ligands provided by a water molecule and the peptide carbonyl oxygen of Tyr28, and the equatorial ligands provided by Asp49 (both carboxylate oxygens), a water molecule, and the peptide carbonyl oxygens of Gly30 and Gly32.[16] When the transition state analogue binds, a Ca^{2+} water

C–terminus

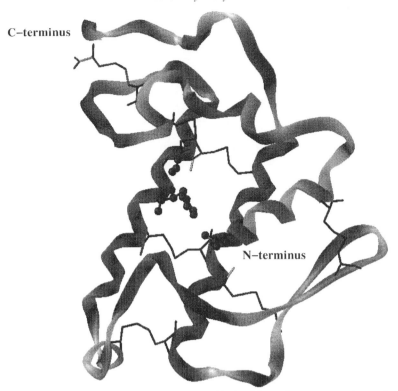

N–terminus

Figure 1 Structure of bovine pancreatic PLA2. The peptide backbone is represented as a shaded ribbon, and the positions of seven disulfide bonds are indicated by black capped sticks.

ligand is displaced by an oxygen of the phosphorylethanolamine headgroup. The second Ca^{2+} water ligand is displaced by a nonbridging oxygen of the *sn*-2-phosphonyl function, which also is H-bonded to the peptide NH of Gly30. The *sn*-2-phosphonyl function mimics the tetrahedral intermediate of the catalytic mechanism (cf. Scheme 3), and therefore Ca^{2+} and the peptide NH of Gly30 comprise the oxyanion hole that stabilizes the tetrahedral intermediate and the transition states for the formation and decomposition of the tetrahedral intermediate. The second nonbridging oxygen of the *sn*-2-phosphonyl function has displaced the water molecule that is H-bonded to δN of His48 in the uncomplexed enzyme and that is thought to be the nucleophilic water in the catalytic mechanism (cf. Scheme 3). Therefore, this complex provides a view of the PLA2-substrate interactions that are important for catalytic function of the enzyme.

5.06.2.3 Structure–Function Relationships

5.06.2.3.1 *Characterization of interfacial enzyme kinetics*

Physiological PLA2 catalysis is an example of interfacial biocatalytic ester hydrolysis. Therefore, in order to probe the structure–function relationships of PLA2, and to cogently design PLA2 inhibitors that have therapeutic potential, it is necessary to characterize the interfacial Michaelis–Menten kinetics of the enzyme. Two approaches have been developed to achieve this aim: surface dilution kinetics, and kinetics in the scooting mode. Each approach is outlined below.

(i) *Surface dilution kinetics*

The surface dilution kinetics method for characterizing the interfacial kinetics of PLA2 catalysis has been extensively championed by Dennis and co-workers.[17] In this method, the phospholipid substrate is formulated in mixed micelles with a detergent, such as Triton X100, that does not inhibit the enzyme. Initial velocities are determined as a function of bulk concentration of the substrate (i.e., [A]), in which case the total interface area is being varied. Initial velocities are also determined

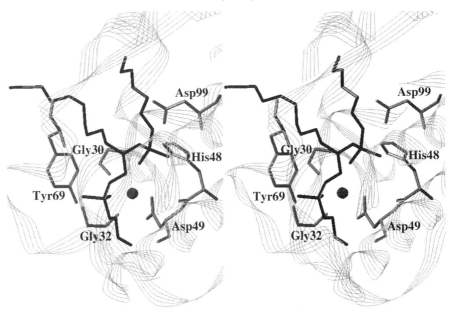

Figure 2 Crossed stereoview of the complex of bovine pancreatic PLA2 with a phosphonate transition state analogue. The active site Ca^{2+} and the inhibitor are represented as a sphere and black capped sticks, respectively.

as a function of the mole fraction of the phospholipid substrate in the mixed micelles. The resulting data are analyzed according to Equation (5):

$$v_i = \frac{V^*_{max}X_A[A]}{K_d K^*_m + K^*_m[A] + X_A[A]} \qquad (5)$$

This equation is an expansion of Equation (2) according to Equations (3) and (4), albeit with $K_d = k_d/k_p$, $V^*_{max} = k^*_{cat}[E]_T$, and with [A] replacing I_A/V, and is identical *pro forma* to the equation for equilibrium ordered bisubstrate enzyme kinetics. However, the first binding event in this case is binding of the enzyme to the interface, which occurs in saturable fashion when the interface area per unit volume, and hence [A], is increased. Accordingly, the dependence of $1/v_i$ on $1/[A]$ is linear:

$$\frac{1}{v_i} = \frac{K^*_m + X_A}{V^*_{max}X_A} + \frac{K_d K^*_m}{V^*_{max}X_A} \frac{1}{[A]} \qquad (6)$$

The intercept of this equation is a linear function of $1/X_A$, where X_A is the mole fraction of the phospholipid substrate in the mixed micelles. Therefore, if measurements of v_i as a function of [A] are conducted at various X_A, then an intercept replot allows one to determine V^*_{max} and K^*_m. The slope of Equation (6) is also a linear function of $1/X_A$, and consequently a slope replot gives $K_d K^*_m/V^*_{max}$, from which one calculates K_d, the dissociation constant of the enzyme from the micelle interface.

Surface dilution kinetics have been utilized by Hendrickson and Dennis[18] to analyze the mechanism of cobra venom (*Naja naja naja*) PLA2-catalyzed hydrolysis of synthetic, chiral dithioester analogues of didecanoyl phosphatidylethanolamine (PE) and didecanoyl phosphatidylcholine (PC). The kinetics were consistent with a "dual phospholipid model" in which the enzyme binds first to the micelle interface, concomitant with binding to an interfacial phospholipid molecule. The interfacial enzyme then binds an additional phospholipid molecule prior to catalytic turnover. Though mixed micelles with these substrates had similar values of K_d (0.1–0.2 mM) and K^*_m (0.1 mol fraction), V^*_{max} for the PC substrate was five times that for the PE substrate. Further analysis provided an explanation for the difference in catalytic turnover of the two types of substrates. When the concentration of the PE substrate was fixed and the TX100 concentration varied, an approach that varies X_A, hyperbolic saturation kinetics were observed, as predicted by Equation (5). However, sigmoidal kinetics were observed in similar experiments with the PC substrate, an observation that suggests that substrate activation is occurring via secondary binding of additional PC molecules to the interfacial enzyme. PC and lysoPC ligands serve as activators of PE turnover, which is consistent with this substrate activation mechanism.[19]

(ii) Kinetics in the scooting mode

Pancreatic PLA2 binds tightly to vesicles that contain anionic phospholipids, such as dimyristoyl phosphatidylmethanol (PM), but weakly and reversibly to uncharged vesicles of PC.[20] In addition, exchange of anionic phospholipids among vesicles is a slow process. These features of anionic phospholipid vesicles have been exploited by Jain and collaborators to construct an interfacial kinetics system for PLA2 that is highly processive. That is, the enzyme binds irreversibly to the vesicle and remains bound while all of the substrate in the outer leaflet of the vesicle is hydrolyzed, a situation that Jain et al. call kinetics in the "scooting mode" (reviewed in references 1 and 21). In the contrasting situation of kinetics in the "hopping mode," such as operates in PLA2-catalyzed hydrolysis of zwitterionic phospholipids in uncharged vesicles or micelles, the enzyme binds transiently and reversibly to the interface. Scooting mode systems offer various advantages over hopping mode systems for characterizing interfacial lipolytic enzyme function in general and PLA2 reaction dynamics in particular. In a hopping kinetics system, intervesicle or intermicelle exchange of enzyme or substrate could be the rate-limiting step, and therefore the effect of substrate or enzyme structure on reactivity would be masked. In a hopping kinetics system, inhibitors might block activity by preventing enzyme binding to the interface, an occurrence that complicates the interpretation of the inhibition mechanism. Because scooting systems are not beset with these problems, they are particularly useful for structure–function investigations of the catalytic mechanisms and inhibition of lipolytic enzymes.

The various equations that are used to analyze the kinetics of PLA2 catalysis in the scooting mode are described herein. For more detailed discussions of the derivations and uses of these equations, the reader is referred to reviews.[1,21] Under scooting conditions, the enzyme does not dissociate from the interface, which in Equation (2) corresponds to the condition $X_A I_A / V \gg K_m^{app}$. Also, when the interface contains only substrate, $X_A = 1$. Therefore, Equation (2) reduces to the following equation, wherein the rate is expressed per enzyme molecule:

$$v_0 = V_{max}^{app} = \frac{k_{cat}^*}{K_m^* + 1}. \tag{7}$$

This equation shows that, like enzyme kinetics in homogeneous solution, scooting kinetics at the lipid–water interface are described by the Michaelis–Menten equation, save that the units of K_m^* and v_0 are mole fraction and s^{-1}, respectively. However, unlike in homogeneous enzyme kinetics, in scooting systems it is generally not possible to vary the mole fraction of substrate without also varying the physical properties of the interface. This complication is circumvented by fitting the reaction time course to the integrated form of the interfacial Michaelis–Menten equation:

$$k_i t = \left(\frac{k_i N_A}{v_0} - 1\right)\frac{P}{P_\infty} - \ln\left(1 - \frac{P}{P_\infty}\right) \tag{8}$$

P and P_∞ are the product concentrations at times t and ∞, respectively, and the rate constant k_i is given by the following equation:

$$k_i = \frac{k_{cat}^*}{N_A K_m^*[1 + (1/K_P^*)]}. \tag{9}$$

In Equations (8) and (9) N_A and K_P^* are the number of substrate molecules in the outer leaflet of the vesicle and the dissociation constant of the interfacial E^*P complex, respectively.

The values of k_i and v_0 that are determined from fitting time course data to Equation (8) are further factored into their constituent fundamental kinetic constants by utilizing competitive and irreversible inhibitors. Jain et al.[22] have determined the dissociation constants of complexes of PLA2 and reversible competitive inhibitors (K_I^* values) by ligand protection experiments, wherein increasing concentrations of the reversible inhibitor progressively block irreversible inactivation of the micelle-bound enzyme by reagents that alkylate His48 of the active site. The same reversible competitive inhibitor is then used to inhibit the hydrolysis rate in a scooting vesicle system. In this case the ratio of the initial velocity in the absence (v_0^0) to that in the presence (v_0^I) of a competitive inhibitor is given by Equation 10:

$$\frac{v_0^0}{v_0^1} = 1 + \frac{1+(1/K_{\mathrm{I}}^*)}{1+(1/K_{\mathrm{m}}^*)}\frac{X_1}{1-X_1} \qquad (10)$$

Consequently, a plot of v_0^0/v_0^1 versus $X_1/(1-X_1)$ is a straight line, from the slope of which K_{m}^* is calculated by using the value of K_{I}^* determined in ligand protection experiments. Once K_{m}^* is known, k_{cat}^* is calculated according to Equation (7). An independent check of the value of K_{m}^* can be had from the ratio $k_{\mathrm{i}}N_{\mathrm{A}}/v_0 = [1+(1/K_{\mathrm{m}}^*)]/[1+(1/K_{\mathrm{P}}^*)]$, determined by fitting time course data to Equation (8). Examples of the use of scooting vesicles for characterizing structure–function relationships for PLA2 are discussed below.

5.06.2.3.2 *Molecular recognition of lipid interfaces*

The active site opening of pancreatic PLA2 is ringed by an annulus of residues that are thought to comprise the interfacial recognition domain[10,23] of the enzyme. The locations of these residues are displayed in Figure 3. The interfacial recognition domain contains clusters of residues near the amino (Ala1, Leu2, Trp3, Asn6) and carboxyl (Lys116, Asn117, Asp119, Lys121, Lys122) termini and from two intermediate regions of the sequence (Glu17, Leu19, Leu20, Asn23, Asn24, Leu31; and Lys56, Val65, Asn67, Tyr69, Thr70, Asn72).

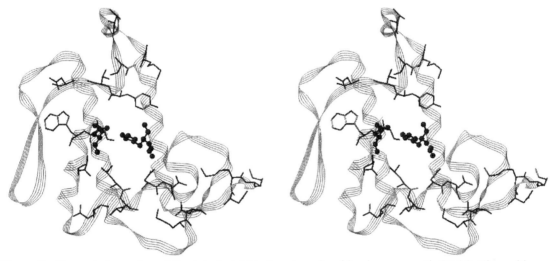

Figure 3 Crossed stereoview of the interfacial binding domain of bovine pancreatic PLA2. The residues of the interfacial binding domain are represented as black capped sticks, and the Asp99-His48-H$_2$O triad of the active site as black balls and sticks. Trp3 near the amino terminus is just to the left of center, and the carboxyl terminus is in the lower right.

Generally, the effects are modest of altering the identities of residues of the putative interfacial binding domain of bovine pancreatic PLA2 by site-specific mutagenesis. Nonetheless, particular results that are informative or surprising are summarized herein. Tomasselli *et al.*[24] proposed that substrate-level acylation of Lys56 leads to dimerization of the enzyme and a large enhancement of catalytic activity. However, two lines of evidence refute this model: (i) PLA2 is fully active in a scooting kinetics system when each vesicle contains at most one enzyme monomer;[25] (ii) conversion of Lys56 to Met, Ileu, Phe, Asn, or Thr gives enzymes that are more active than wild-type, this despite the fact that all but the Thr mutant are incapable of being acylated by the substrate.[26]

Electrostatics likely play a major role in absorption of PLA2 to the interface,[27] and in subsequent interfacial catalysis. Jain and collaborators[28] have dissected the effects of negative charge at the interface by analysis of mutants of Lys53 and Lys56 of bovine pancreatic PLA2. When each of these residues is changed to Met, the resulting mutant enzymes have $V_{\mathrm{max}}^{\mathrm{app}}$ and $K_{\mathrm{m}}^{\mathrm{app}}$ values for hydrolysis of micellar diheptanoyl-PC that are an order of magnitude larger than and about half that of the wild-type enzyme, respectively. However, these differences are eliminated in an anionic micelle system (dioctanoyl-PM), in a scooting kinetics system (dimyristoyl-PM), and by increasing the concentration of NaCl in the reaction medium from 0.1 M to 4 M. Detailed analyses of scooting and micellar systems showed that the mutations and increasing concentrations of NaCl did not affect K_{m}^* for substrate turnover or K_{I}^* for interaction with active-site ligands, but did affect the binding of the enzyme to PC interfaces (i.e., K_d was increased), and consequent interfacial catalytic turnover (i.e., k_{cat}^* was increased). Therefore, the charged residues Lys53 and Lys56 play two roles, both electrostatic in origin: one is to enhance binding to anionic phospholipid interfaces, while the other is to enhance catalytic turnover at the interface by an allosteric mechanism that is mediated by charge neutralization of Lys53 and Lys56 on interaction with anionic phospholipids. The role of Lys56 in recognition of anionic interfaces is also supported by the work of Cho and colleagues,[29] who found that mutation to Glu reduces by ∼100-fold the binding affinity of the enzyme to polymerized vesicles of anionic phospholipids.

There are two classes of low molecular weight (14 kD), Ca^{2+}-dependent PLA2s.[7] The class I enzymes include those from mammalian pancreas (such as bovine pancreatic PLA2) and the snake venoms of the genera elapidae and hydrophodiae. Class II enzymes include a mammalian nonpancreatic PLA2 and enzymes from the snake venoms of the genera crotalidae and viperidae. Class II enzymes possess interfacial recognition domains whose charged residue topologies are distinct from that of class I enzymes. For class II enzymes mutagenesis experiments underscore the importance of electrostatics for recognition of anionic phospholipid interfaces. Cho and colleagues[30] have shown that the Lys7Glu/Lys10Glu double mutant of the PLA2 from the venom of *Agkistrodon piscivorus piscivorus* (App-PLA2) binds about 500-fold more weakly than does the wild-type enzyme to polymerized vesicles of the anionic phospholipid 1,2-bis[12-(lipoyloxy)-dodecanoyl]-*sn*-glycero-3-phosphoglycerol. The effect of the double mutant corresponds to a binding energy difference of 3.8 kcal mol^{-1}, which is about half of the energy difference for binding of the enzyme to polymerized vesicles of anionic versus uncharged phospholipids. Moreover, analysis of the two corresponding single mutants indicates that Lys7 and Lys10 make additive contributions to the energy of interfacial recognition. Cho, Gelb, and colleagues[31] utilized the same polymerized vesicle system to assess by site-directed mutagenesis the roles of patches of cationic residues on interfacial recognition by human secretory group IIa PLA2. They found that, as for App-PLA2, a patch near the amino terminus that consists of Arg7, Lys10, and Lys16 makes a sizable contribution to catalysis and interfacial recognition. When these three residues were converted as a set to Glu residues, k_{cat}^*/K_m^* was reduced by 15-fold, and K_d, the dissociation constant of the enzyme from the interface, was increased 300-fold. Additional important contributions were found for patches in the middle of the sequence (Lys74, Lys87, Arg92) and near the carboxyl terminus (Lys124, Arg127). These investigations underscore the fact that, even though the charged residue topologies of the interfacial recognition domains of class I and class II PLA2s differ, positive residues play similar roles in interfacial recognition.

Trp3 of bovine and porcine pancreatic PLA2 has long been recognized as contributing to interfacial recognition, particularly because Trp3 fluorescence serves as a reporter of enzyme binding to phospholipid micelles and vesicles.[32] Liu *et al.*[33] probed the function of Trp by mutagenesis and kinetic analysis of phospholipid micelles and scooting vesicles. When compared to the wild-type enzyme, the Trp3Ala mutant showed a 7-fold reduction in apparent k_{cat} and a 5.5-fold increase in apparent K_m for hydrolysis of micellar dioctanoyl-PC. These results suggest that replacement of Trp3 with Ala impairs the binding of the enzyme to the interface and may also affect catalytic turnover. Analysis of the Trp3Ala mutant by scooting kinetics in the dimyristoyl-PM system showed that k_{cat}^* is only 38% less than that of the wild-type enzyme, while K_m^* is increased ∼5-fold. Therefore, one of the effects of mutation is to perturb the E* to E*A step, i.e., interfacial binding of substrate in the active site (cf. Scheme 2). Moreover, the wild-type enzyme only hydrolyzes that amount of substrate in the outer leaflet of the scooting vesicles, in concert with the fact that the enzyme binds irreversibly to the interface. However, the Trp3Ala mutant hops from vesicle to vesicle and consequently hydrolyzes the available substrate in the outer leaflet of all vesicles in the assay. Therefore, mutation also impairs binding to the vesicle interface.

The solution-phase NMR structures of porcine pancreatic PLA2, either free or bound to micelles of dodecylphosphocholine,[34,35] provide insights into the conformational changes that underpin activation of the enzyme at interfaces.[21] Two flexible regions of PLA2 become more ordered when the enzyme binds to the interface: the amino terminus, including residues Ala1, Leu2 and Trp3; and a surface loop that contains Tyr69. However, the amino terminus adopts the rigid α-helical conformation found in the crystal structure only when the substrate analogue (*R*)-1-octyl-2-(*N*-dodecanoylamino)-2-deoxyglycero-3-phosphoglycol is bound in the active site of the micelle-bound enzyme. These results indicate that successive conformational changes occur when the enzyme binds to the interface and subsequently binds substrate monomers in its active site. Site-specific mutagenesis experiments that are described in the next section are consistent with these observations, and illuminate specific roles played by Tyr69 and residues near the amino terminus.

5.06.2.3.3 *Molecular recognition of phospholipid monomers*

Figure 2 and the accompanying discussion described the interaction of the active site of bovine pancreatic PLA2 with a transition state analogue inhibitor that mimics the tetrahedral intermediate of the reaction mechanism. Figure 4 shows additional interactions of the alkyl chains of the transition state analogue that mimic the *sn*-1 and *sn*-2 acyl chains of phospholipid substrates with aromatic and hydrophobic amino acids that line the active site channel.[12] This complex and similar complexes with human secretory group IIa PLA2[15] and PLA2 from Chinese cobra[13] (*Naja naja atra*) and honey bee[14] venoms indicate that acyl-mimetic chains of the transition state analogue project through the active site channel to the surface of the enzyme. Therefore, PLA2 should interact with at least the eight proximal carbons of the acyl chains of substrates. The chain-length dependence for inhibition of cobra venom (*Naja naja naja*) PLA2 by amide analogues of PC substrates, reported by Yu and Dennis,[36] is consistent with this idea. For 1-(hexylthio)-2-acylamido-1,2-dideoxy-*sn*-glycero-3-phosphocholine analogues that have acyl chains from 1 to 12 carbons in length, critical micelle concentration (cmc) values were sufficiently high that inhibition of the enzyme was observed in homogeneous solution. In this situation, inhibitor potency, plotted as $\log(IC_{50})$, correlated linearly with chain length for analogues that have 4–10 carbons in the 2-acylamido chain, but leveled off at $\sim 200~\mu$M and ~ 300 nM, respectively, for the shorter and longer-chain analogues. These results suggested that the enzyme interacts with the proximal 9 or 10 carbons of the *sn*-2 acyl chain of phospholipid substrates, which is supported by molecular modeling of the interaction of dimyristoyl phosphatidylethanolamine with the enzyme.[8] Moreover, since the bound inhibitor is in equilibrium with the inhibitor in homogeneous aqueous solution, the correlation of inhibitor potency with acylamido chain length underscores the importance of hydrophobic interaction in acyl chain molecular recognition by PLA2. Yu and Dennis[36] also characterized the effect on the enzyme of analogues that contained a hexadecylthio function at the *sn*-1 position and acylamido groups from 2 to 20 carbons in length at the *sn*-2 position. These compounds were assayed as inhibitors in mixed micelles with synthetic dithioester analogues of dihexanoyl and didecanoyl phosphatidylcholines. In this surface dilution kinetics system, inhibitor potency varied little with the length of the acylamido chain, as one expects when the inhibitor is extracted from a hydrophobic micelle into the hydrophobic active site channel of PLA2.

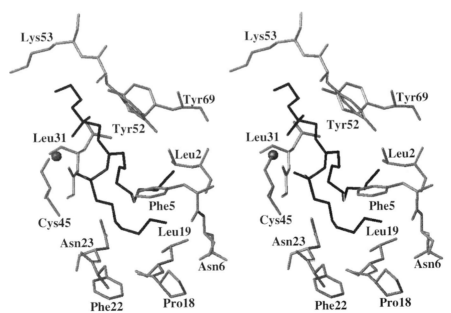

Figure 4 Crossed stereoview of fatty acyl interactions in the complex of bovine pancreatic PLA2 and a phosphonate transition state analogue. The active site Ca²⁺ and inhibitor are represented as a sphere and black capped sticks, respectively. Residues that interact with the *sn*-1 alkyl chain include Leu2, Trp3 (not shown) Leu31 and Tyr69; residues that interact with the *sn*-2 alkyl chain include Phe5, Ile9 (not shown), Pro18, Leu19, Phe22, Asn23, Phe106 (not shown), and the Cys29–Cys45 disulfide bond.

Figure 4 shows that Leu2 and Phe5 of bovine pancreatic PLA2 make close interactions with the heptylphosphonyl function of the bound transition state analogue inhibitor.[12] Site-specific mutagenesis, in conjunction with kinetic analysis in a scooting vesicle system,[33] illuminates the function of these two residues in molecular recognition. When Phe5 was replaced by Val, the resulting mutant enzyme was not impaired in binding to the interface of dimyristoyl phosphatidylmethanol scooting vesicles, and substrate and inhibitor binding to the active site at the interface were unaffected. However, v_0 (cf. Equation (7)) was reduced 165-fold, this on the strength of a 190-fold reduction in k_{cat}^*. These results indicate that the function of Phe5 is to specifically stabilize the transition state by interacting with the *sn*-2 acyl chain. Leu2, on the other hand, plays a critical role in substrate specificity. When the Leu2Ala mutant is compared to the wild-type enzyme, the apparent k_{cat} for hydrolysis of dioctanoyl-PC is unchanged, and v_0 and k_{cat}^* for hydrolysis of dimyristoyl-PM are only reduced by 3-fold and 4-fold, respectively. In the Leu2Trp mutant, the apparent k_{cat} for dioctanoyl-PC is again unaffected, but v_0 for dimyristoyl-PM is decreased by a factor of 33. These results indicate that Leu2 provides binding interactions that enable PLA2 to effectively hydrolyze long-chain phospholipid substrates. The crystal structure of the complex of the substrate analogue 2-dodecanoylamino-1-hexanol phosphoglycol with porcine pancreatic PLA2 supports this conclusion, in that Leu2 makes a close contact with C_{12} of the *sn*-2 acylamino chain of the inhibitor.[37]

Site-specific mutagenesis and experiments with substrate analogues suggest that Tyr69 of porcine pancreatic PLA2 is involved both in transition state stabilization and stereospecificity of bovine pancreatic PLA2.[1,38] In class II PLA2s, Lys replaces Tyr69. Replacing Tyr69 with Lys or Phe resulted in comparable reductions in V_{max} for hydrolysis of dioctanoyl-PC, 8.7-fold or 4.5-fold, respectively, and in V_{max}/K_m, 25-fold or 12-fold, respectively. The Lys mutant retained the L-stereospecificity of the native enzyme for the hydrolysis of the *sn*-2 ester function of phospholipid substrates, but the Phe mutant showed partial loss of L-stereospecificity. By using thionophospholipids that are stereogenic at phosphorus by virtue of a sulfur-for-oxygen substitution, the >130-fold $(R_p)/(S_p)$ headgroup stereospecificity of the native enzyme was found to be reduced to 25-fold in the Phe mutant. Phosphonolipids, which have reduced fatty acyl to phosphorus distances, are hydrolyzed less efficiently than phospholipids by both native and mutant enzymes, and with partial loss of L-stereospecificity. These results suggest that a Tyr or Lys sidechain at position 69 participates in transition state stabilization by hydrogen bonding to the phosphodiester function of the phospholipid headgroup. Moreover, this interaction results in precise positioning of the substrate in the active site, and in the absence of this interaction the loss of this effect is manifested by reduction in stereospecificity for hydrolysis of the *sn*-2 ester function.

A considerable literature describes various aspects of the structure–function relationships of PLA2, as probed by site-specific mutagenesis. Additional examples of such work have been reviewed.[1]

5.06.3   ESTERASES OF THE α/β HYDROLASE FOLD FAMILY

In 1992 Ollis *et al.*[39] compared the crystal structures of five hydrolytic enzymes of diverse catalytic function: *Torpedo californica* acetylcholinesterase, *Geotrichum candidum* lipase, dienelactone hydrolase from *Pseudomonas* sp. B13, wheat carboxypeptidase II, and haloalkane dehalogenase from *Xanthobacter autotrophicus*. This comparison introduced a fundamental new enzyme topology, which was named the α/β hydrolase fold. The secondary structural elements that are common to the catalytic cores of the five enzymes include eight β-strands that form a superhelically twisted β-sheet. The β-strands are connected by loops and α-helices, and six of the helices are conserved secondary structural elements. Four of the enzymes have an identical topology[40] of the eight central β-strands, as shown in Figure 5: +1, +2, −1x, +2x, (+1x)₃. In wheat carboxypeptidase II a β-hairpin loop is inserted between β-strands 7 and 8, so that the topology is +1, +2, −1x, +2x, +1x, +1x, +3x, −1, −1. Also shown in Figure 5 are the positions of the six conserved α-helices. The residues of the active site triad are situated on loops that separate β-strands and α-helices. The active site nucleophile, which in all esterases of the α/β fold family save dienelactone hydrolase is a serine residue, is the central residue in a sharp γ-turn between strand 5 and helix C (cf. Figure 5). The acid constituent of the triad, either Asp or Glu, is situated on the loop that follows strand 7. The His constituent lies near the end of a turn that follows the end of strand 8. Subsequent comparisons of conserved elements that are involved in tertiary structure[41] and of the primary sequences of 70 proteins,[42] some of which are devoid of catalytic activity, greatly broadened the range of proteins that belong to the α/β hydrolase fold supergene family.[1] The ESTHER database, which can be accessed on the Internet at www.ensam.inra.fr/cholinesterase, currently contains primary sequences and additional information on more than 320 enzymes and structural proteins that belong to this family. Nearly 90% of the entries in the database are esterases that belong to three subfamilies: lipase (L-family), carboxylesterase/cholinesterase (C-family), hormone-sensitive lipase (H-family). Moreover, the database contains links to more than 110 structures of 32 enzymes of the α/β hydrolase family. Therefore, the α/β hydrolase fold family has rapidly developed into one of the most extensive of protein supergene families.

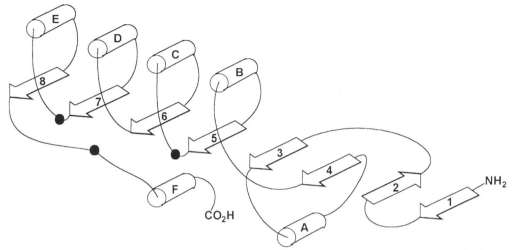

Figure 5 Conserved secondary structural elements and topology of the of the α/β hydrolase fold. α-Helices and β-strands are represented as cylinders and arrows, respectively, and the locations of the residues of the active site triad are shown as black circles.

5.06.3.1 Functional Diversity

The esterases of the α/β hydrolase fold supergene family catalyze a wide range of reactions,[42,43] as outlined in Table 1. Though the list of biocatalytic functions in this table is by no means exhaustive, it serves to illustrate the diverse functions that the enzymes serve. Among them are the acetyl-cholinesterase-catalyzed hydrolysis of the neurotransmitter acetylcholine,[1,43–48] the catabolic function of lipases in dietary lipid absorption and lipoprotein metabolism,[49,50] and the detoxification of xenobiotic esters by liver enzymes.[42] This diversity of functions renders esterases of the α/β hydrolase fold family useful reagents in *ex vivo* biocatalytic synthetic transformations, as well as therapeutic targets in various disease states. A more extensive delineation of the biocatalytic functions of esterases of the α/β hydrolase fold family is provided in the ESTHER database alluded to in the preceding paragraph.

5.06.3.2 Cholinesterases

There are two cholinesterases in mammals, acetylcholinesterase (AChE) and butyrylcholinesterase (BuChE), while AChE is also found in insects, reptiles, and marine species.[48] The function of AChE is the hydrolytic destruction of the excitatory neurotransmitter acetylcholine (ACh) at nerve–nerve synapses and neuromuscular junctions.[1,43–48] Consequently, this enzyme has been the focus not only of an extensive biotechnology industry (i.e., carbamate and organophosphorus pesticides for the farm and garden), but regrettably also of chemical warfare. Moreover, AChE is the sole current target of FDA-approved drugs for palliative treatment of Alzheimer's disease. Therefore, AChE catalysis has broad implications for human health, national security, and agribusiness. AChE has a sharp specificity for substrates that have small acyl groups, and has highest activity on esters of acetic acid. BuChE, on the other hand, shows comparable activities on acetyl, propanoyl, butanoyl, and benzoyl substrates. The structural basis of the distinct acyl specificities of AChE and BuChE will be described later in this chapter. Unlike AChE, the physiological function of BuChE is not clearly defined. The highest BuChE activity is found in the bloodstream, where the enzyme may function as a broad spectrum esterase for hydrolytic degradation of xenobiotics.

Table 1 Biocatalytic functions of esterases of the α/β hydrolase fold family.

Enzyme class	Examples	EC number	Function
Cholinesterases	acetylcholinesterase	3.1.1.7	hydrolysis of acetylcholine at synapses in the nervous system
	butyrylcholinesterase	3.1.1.8	unknown; perhaps xenobiotic ester hydrolysis in bloodstream
Lipases	pancreatic lipase	3.1.1.3	hydrolysis and absorption into bloodstream of alimentary lipid esters
	cholesterol esterase	3.1.1.13	hydrolysis and absorption into bloodstream of alimentary lipid esters; lipoprotein metabolism
	lipoprotein lipase	3.1.1.34	triacylglycerol-rich lipoprotein metabolism; adipose uptake of fatty acids
	Geotrichum candidum lipase	3.1.1.3	yeast lipid metabolism
	Pseudomonas glumae lipase	3.1.1.3	bacterial lipid metabolism
	hormone-sensitive lipase	3.1.1.3	hydrolytic lipid mobilization in mammalian cells
Esterases	*Fusarium solani pisi* cutinase	3.1.1.–	plant infection: hydrolysis of cutin polymer coating of leaves
	hepatic esterases	3.1.1.1	hydrolysis of xenobiotic esters in the liver
	dienelactone hydrolase	3.1.1.45	bacterial hydrolysis of dienelactones

5.06.3.2.1 Enzyme structure

A hallmark of AChE function is the very high catalytic power of the enzyme,[44,45] a subject that will be detailed in the following section. The situation of the active site at the bottom of a narrow, 20 Å deep gorge in the X-ray structure of *T-californica* AChE[51] therefore came as a surprise. As Figure 6 shows, AChE contains five important functional loci:[1,43,47] the catalytic triad, the oxyanion hole, the acyl binding locus, the quaternary ammonium binding locus of the active site, and the peripheral anionic binding site. Figure 6(a) shows the interactions of the active site triad, Ser203(*200*)-His447(*440*)-Glu334(*327*), and the oxyanion hole with the transition state analogue inhibitor *m*-(*N,N,N*-trimethylammonio)-trifluoroacetophenone (TMTFA).[52] The numbers in normal print and italics after each amino acid denote the sequence positions in mammalian and *T. californica* AChEs, respectively. The catalytic triad is reminiscent of similar triads in the active sites of serine proteases,[53] and functions as a general acid–base proton transfer network in the catalytic mechanism. The three-pronged oxyanion hole of AChE stabilizes oxyanionic tetrahedral intermediates and transition states, as do the less elaborate two-pronged oxyanion holes of serine proteases. Figure 6(b) shows the interactions of TMTFA with the quaternary ammonium binding locus of the active site. The Me_3N^+ function of TMTFA makes close contacts with Trp86(*84*), Phe337(*330*), Glu202(*199*), and three water molecules, and is involved in an ion–dipole interaction with Tyr133(*130*). These groups provide a polar and counterionic environment for the cationic function of choline substrates and cholinergic ligands. Figure 6(c) shows the constituents of the acyl binding locus of the active site. The CF_3 group of TMTFA makes close contacts with five residues: Gly122(*119*), Trp236(*233*), Phe295(*288*), Phe297(*290*) and Phe338(*331*). This highly circumscribed concave binding site readily accommodates the alkyl portions of acetyl and propanoyl acyl functions, but sterically excludes butanoyl and benzoyl groups. Though the X-ray structure of BuChE has yet to be reported, an homology model of the enzyme has been constructed by using the structure of *T. californica* AChE as a template.[54,55] Among other differences, the BuChE model possesses a more expansive acyl binding locus in which Phe295(*288*) and Phe297(*290*) of AChE are replaced by Leu and Val, respectively. Site-directed mutagenesis supports the idea that residues in these two locations determine the relative acyl specificities of AChE and BuChE. When the two Phe residues of *T. californica* or human AChE are replaced by smaller aliphatic amino acids, the resulting mutant enzymes display the broad acyl specificity of BuChE.[54,56–58]

Figure 6(d) shows the relative locations in AChE of the peripheral anionic site and the catalytic triad of the active site, and Scheme 4 shows various ligands that bind to the peripheral site.[43,47,48] Among the ligands in the scheme are cationic inhibitors propidium and d-tubocurarine, which bind exclusively to the peripheral site. These ligands probably inhibit AChE by providing a steric blockade to substrate entry into and product release from the active site during catalytic turnover.[59] Fasciculin-2 (FAS2), a 61 residue peptide from the venom of mambas, binds to the peripheral site[60] and inhibits

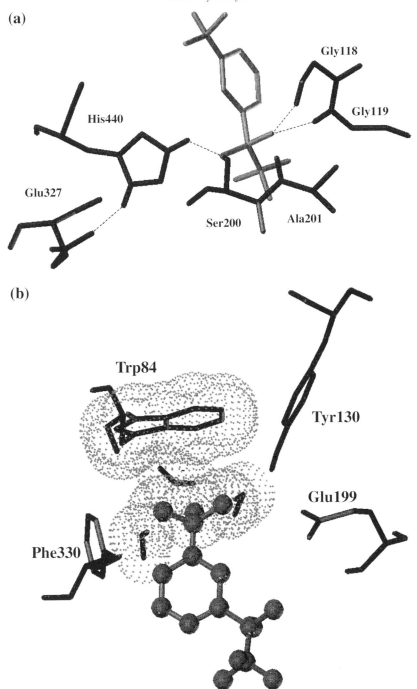

Figure 6 Functional loci in *Torpedo californica* AChE. (a) Interactions of the transition state analogue TMTFA (gray capped sticks) with the active site triad (Ser200-His440-Glu327) and oxyanion hole (Gly118-Gly119-Ala201). H-bonds are denoted as dotted lines. (b) Interactions of TMTFA (balls and sticks) with residues of the quaternary ammonium binding site. The close interaction of the quaternary ammonium function of TMTFA with Trp84 is shown as dot surfaces. (c) Interactions of TMTFA with residues of the acyl binding locus. (d) Crossed stereoview of the relative locations of the active site triad (space-filled) and the peripheral anionic site (balls and sticks; Tyr70, Asp72, Tyr121, Glu278, and Trp279).

AChE both by providing a steric blockade[61] and by allosterically affecting the conformation of the active site,[62] in particular that of Trp86(*84*). The X-ray structure of the complex of fasciculin with mouse AChE shows the contacts that the inhibitor makes with residues of the peripheral site and that access of ligands to the active site is blocked in the complex.[63] Certain rod-shaped ligands, such

(c)

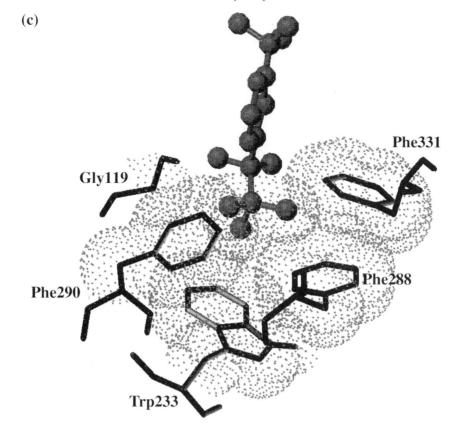

(d)

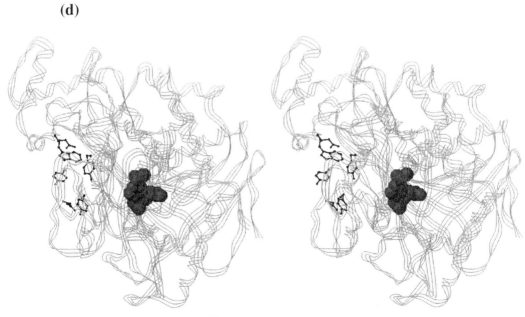

Figure 6   (continued)

as the bisquaternary decamethonium and the Alzheimer's disease drug aricept[64] of Scheme 4, span both the active and peripheral sites. Substrate inhibition at high substrate concentrations is a common feature of AChE-catalyzed hydrolysis of ACh and the surrogate chromogenic substrate acetylthiocholine (ATCh, cf. Scheme 4). Site-directed mutagenesis supports the notion that the substrate inhibition binding site is the peripheral anionic locus.[47,48,65]

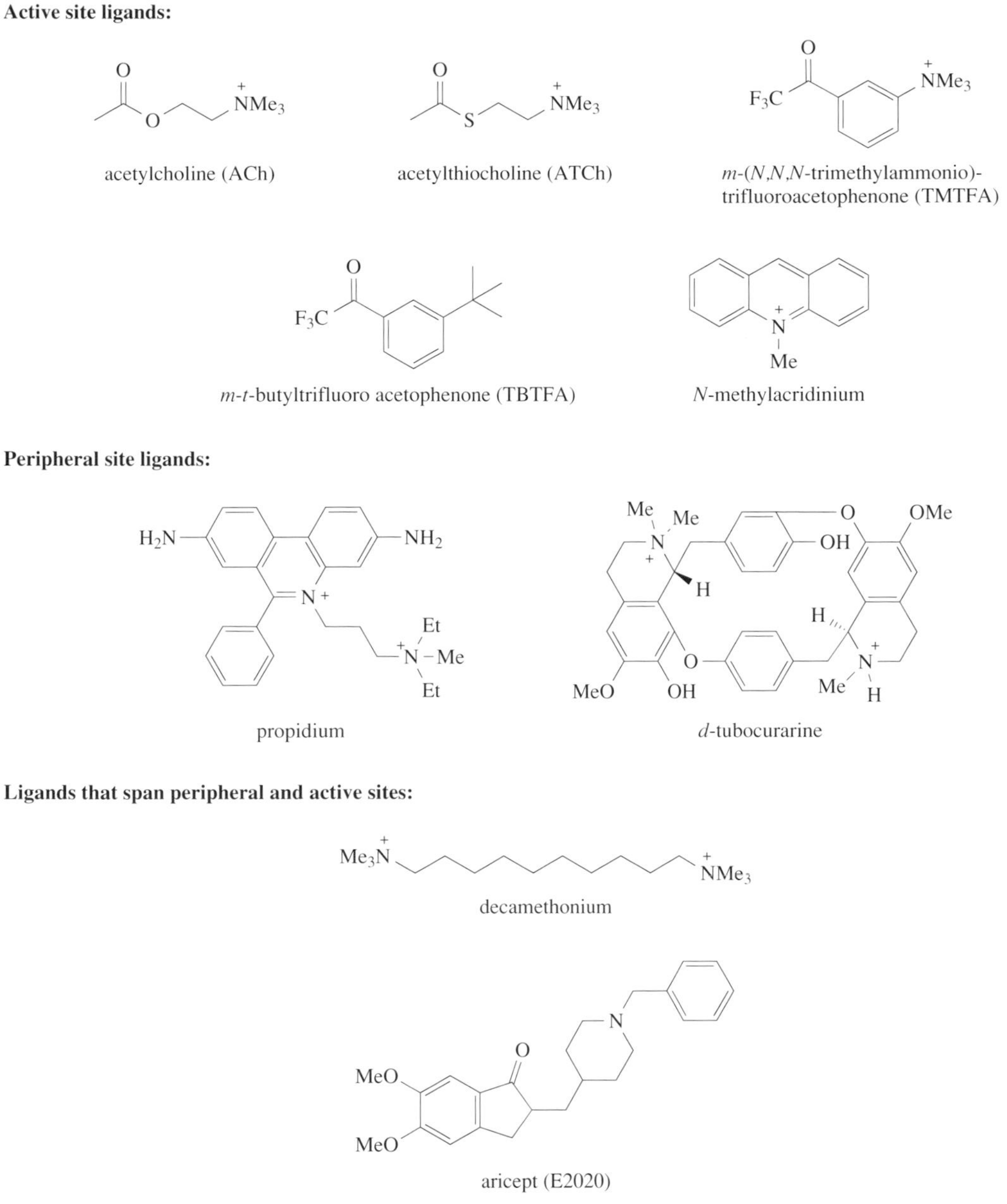

Active site ligands:

acetylcholine (ACh)

acetylthiocholine (ATCh)

m-(*N*,*N*,*N*-trimethylammonio)-
trifluoroacetophenone (TMTFA)

m-*t*-butyltrifluoro acetophenone (TBTFA)

N-methylacridinium

Peripheral site ligands:

propidium

d-tubocurarine

Ligands that span peripheral and active sites:

decamethonium

aricept (E2020)

Scheme 4

5.06.3.2.2 *Catalytic mechanism*

Nearly all of the enzymes of the α/β hydrolase fold supergene family are serine esterases that utilize chemical mechanisms similar to that of AChE in Scheme 5. The reaction occurs in successive acylation and deacylation stages, each of which involves high-energy tetrahedral intermediates. The AChE-TMTFA complex of Figure 6 is a structural analogue of the tetrahedral intermediate in the acylation stage of catalysis. The first direct evidence for the acylenzyme mechanism for AChE catalysis was provided by Froede and Wilson,[66] who trapped the radiolabeled acetylenzyme intermediate in *Electrophorus electricus* AChE-catalyzed hydrolysis of ACh and ATCh by rapid flow-quench techniques. Their results demonstrated that for both substrates the acylation and deacylation stages of catalysis contribute comparably to rate limitation of k_{cat}.

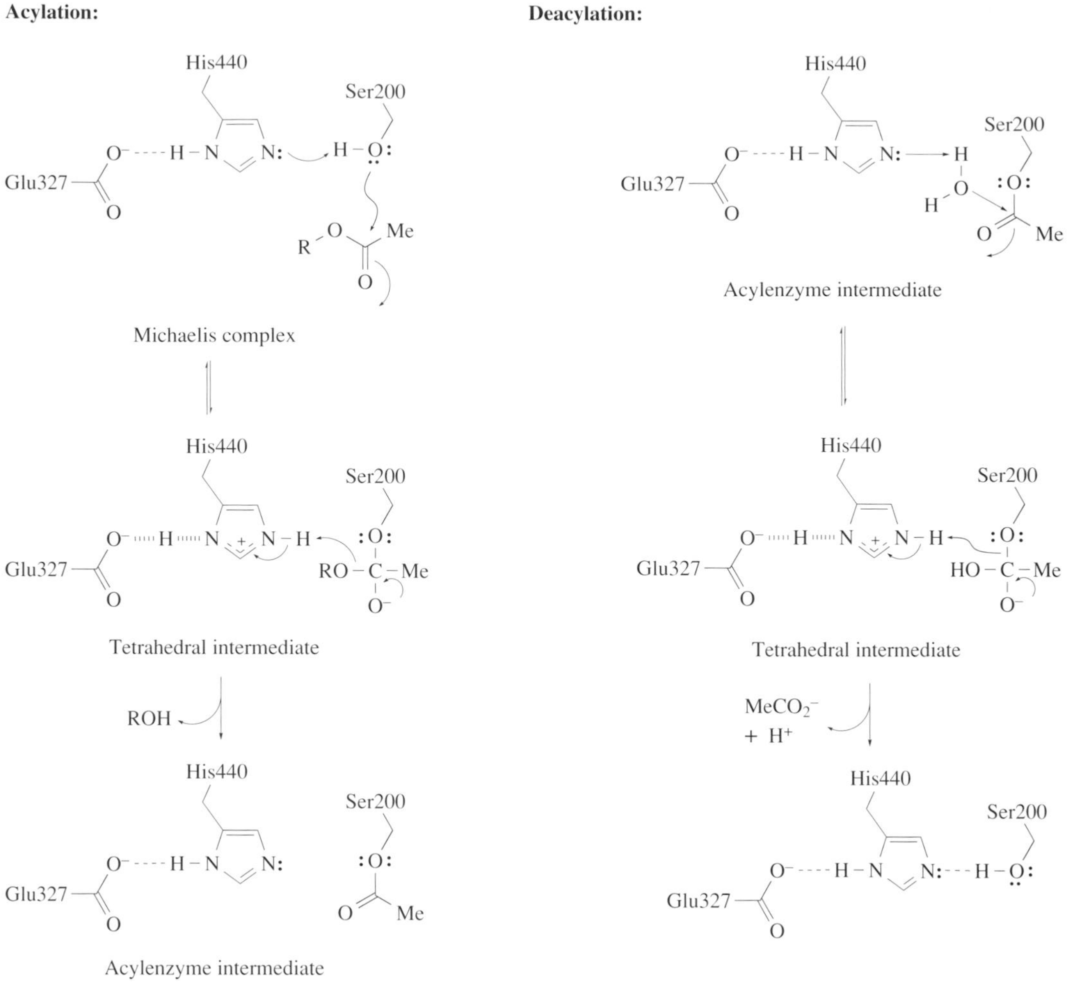

Acylation: **Deacylation:**

His440   His440
Ser200   Ser200

Glu327   Glu327

Michaelis complex   Acylenzyme intermediate

His440   His440
Ser200   Ser200

Glu327   Glu327

Tetrahedral intermediate   Tetrahedral intermediate

ROH $MeCO_2^-$ + H^+

His440   His440
Ser200   Ser200

Glu327   Glu327

Acylenzyme intermediate

Scheme 5

(i) Catalytic power

The kinetic mechanism that corresponds to the acylenzyme mechanism of Scheme 5 is outlined in Scheme 6. The corresponding Michaelis–Menten rate constants are given by Equations (11) and (12). As mentioned earlier, AChE operates at very high catalytic efficiency.[44,45] For example, the turnover number k_{cat} is $\sim 10^4\,s^{-1}$ at 25°C. Because k_3 (acylation) and k_5 (deacylation) are comparably rate limiting, $k_3 \sim k_5 \sim 2 \times 10^4\,s^{-1}$, which corresponds to an average lifetime of $\sim 50\,\mu s$ for the acetylenzyme intermediate. Moreover, k_{cat} for AChE-catalyzed hydrolysis of ACh is $\sim 10^{13}$-fold faster than spontaneous hydrolysis at pH = 7,[52] which corresponds to 74 kJ mol^{-1} of free energy of stabilization of the transition state. The second-order rate constant k_E ($\equiv k_{cat}/K_m$) contains contributions from successive substrate binding (k_1) and acylation catalysis (k_3, cf. Scheme 6). Values of k_E are $\sim 10^8\,M^{-1}\,s^{-1}$ at an ionic strength of ~ 0.2, but approach values $> 10^9\,M^{-1}\,s^{-1}$ as the ionic strength approaches zero.[67] The magnitude of k_E is comparable to the speed limit for biological catalysis estimated by Eigen and Hammes,[68] and to the diffusion-controlled Stern–Volmer quenching of electronic excited states.[69] These comparisons strongly suggest that under bimolecular conditions AChE-catalyzed hydrolyses of ACh and ATCh are at least partly diffusion-controlled reactions. The structural and dynamic bases for the very high catalytic efficiency of AChE are discussed in further detail in the next two sections.

$$k_{cat} = \frac{V_{max}}{[E]_T} = \frac{k_3 k_5}{k_3 + k_5} \tag{11}$$

$$k_E \equiv \frac{k_{cat}}{K_m} = \frac{k_1 k_3}{k_2 + k_3} \tag{12}$$

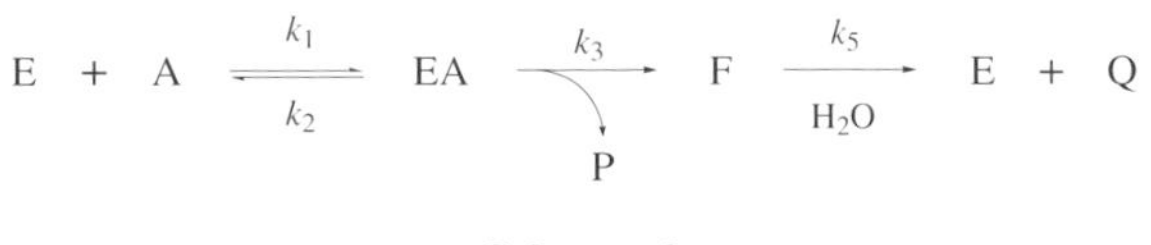

$$E + A \; \underset{k_2}{\overset{k_1}{\rightleftharpoons}} \; EA \; \overset{k_3}{\longrightarrow} \; F \; \underset{H_2O}{\overset{k_5}{\longrightarrow}} \; E + Q$$
$$\searrow P$$

Scheme 6

(ii) Function of the catalytic triad

The chemical mechanism of Scheme 5 shows that Ser203(*200*) and His447(*440*) of the catalytic triad respectively function as nucleophilic and general acid–base catalytic elements. As in the serine proteases, however, the inclusive function of Glu334(*327*) of the triad is not clear. Two limiting functions for the carboxylate groups of serine enzyme triads can be envisioned. (a) Transition states for the formation and decomposition of tetrahedral intermediates are stabilized by electrostatic interaction between the carboxylate and the incipient imidazolium cation.[70] This interaction is thought to orient the imidazole function of histidine for efficient acid–base catalysis, a poor choice when one considers the weak r^{-1} dependence of electrostatic interactions. (b) As does the imidazole group of histidine, the carboxylate functions as a general acid–base catalyst in a mechanism that has long been called the "charge-relay" mechanism.[71] This mechanism avoids the formation of an imidazolium–carboxylate ion pair in active sites of modest polarity, but has been criticized because imidazolium to carboxylate proton transfer is thermodynamically unfavorable.[72]

A third mechanism avoids the pitfalls of these two limiting mechanisms, though again not without controversy. In this mechanism the transition state is stabilized by a short, strong H-bond between the incipient imidazolium cation and the carboxylate of Glu or Asp. This mechanism was first proposed by Elrod *et al.*[73] to rationalize the quadratic dependence of k_{cat} on n (n = atom fraction of deuterium in mixed buffers of H_2O and D_2O) for trypsin-catalyzed hydrolysis of a tripeptide *p*-nitroanilide substrate. The short, strong His-Asp H-bond was referred to as a "solvation catalytic" proton bridge, and is well-precedented in nonenzymic acyl transfer reactions. Frey and co-workers[74] have identified this H-bond in α-chymotrypsin at low pH as a low-barrier H-bond (LBHB) on the basis of its downfield chemical shift at 18.3 ppm. Furthermore, in trifluoroketone transition state analogue complexes, the chemical shift of the His-Asp H-bond is further downfield at ∼19 ppm.[75] This downfield shift in tetrahedral transition state analogue complexes is interpreted as indicating that in the tetrahedral intermediate and flanking transition states the His-Asp LBHB is stronger than in the reactant ES complex, and therefore would contribute to catalytic acceleration by selective stabilization of the transition state and tetrahedral intermediate.

The data in Table 2 compare His-Glu (or Asp) H-bond distances for various serine enzymes. All H-bond distances were measured in the active site triads of enzyme structures that were downloaded from the Brookhaven Protein Data Bank. The most notable feature of the data in Table 2 is that H-bond distances for AChE and for the serine proteases chymotrypsin, trypsin, elastase, subtilisin, and α-lytic protease are palpably shorter than 2.9 Å, the sum of the van der Waals radii of N–O H-bond donor–acceptor pairs.[76] However, for all lipases in the table the corresponding distances exceed 2.8 Å and average about 2.9 Å. These comparisons suggest that, if the carboxylate residue of the catalytic triad serves the same function in all serine enzymes, a conformational adjustment is required to fully arm the lipase triad for catalysis. It is tempting to speculate that such a conformational change is a component of the mechanism of interfacial activation.

The short H-bond distances in AChE and serine proteases raise the possibility that the bridging proton lies in a potential whose barrier to proton transfer lies below the vibrational zero-point energy level of the asymmetric stretch. This possibility was assessed by using the model of the AChE catalytic triad that is shown in Figure 7. This model was constructed by extracting the coordinates of the triad from the X-ray structure of the *T. californica* AChE-TMTFA complex,[52] and by subsequently truncating the triad as shown in the figure. The remaining atoms of the protein were represented at their respective coordinates as point charges. The proton that bridges His447(*440*) and Glu334(*327*) was placed at various positions on a linear transit between Nδ of His and Oε1 of Glu. Electronic energies were then calculated for each nuclear geometry by utilizing the 6–311++G** basis set in the quantum mechanics package GAUSSIAN 94;[77] correlation energies were included by using Becke–Perdew–Wang density functionals.[78] Figure 8 shows a plot of the dependence of the calculated energy on the proton position in the His-Glu H-bond, as well as a fit of the data to a fourth-order polynomial. The Schrödinger equation was also solved for this quartic vibrational potential,[79] and the corresponding square modulus of the vibrational wave function (i.e., Ψ^2) and the zero-point energy are included in the figure. Since Ψ^2 provides the probability of the

Table 2 His-Glu (or Asp) H-bond distances in serine enzymes.

Enzyme	H-bond distance (Å)[a]	Comments
Acetylcholinesterase	2.73 ± 0.18	9 structures; free E and EI complexes
	2.52	native unliganded AChE
	2.66	AChE-TMTFA complex
	2.54 ± 0.12	2 structures of complex of AChE with the inhibitor edrophonium
Trypsin	2.70 ± 0.07	4 structures of unliganded native enzyme
	2.61 ± 0.05	4 structures of trypsinogen
α-Chymotrypsin	2.63 ± 0.04	4 distances from 2 dimer structures
γ-Chymotrypsin	2.67 ± 0.09	5 structures of unliganded native enzyme
	2.63	complex with N-acetyl-L-Phe trifluoromethyl ketone
	2.49	complex with N-acetyl-L-Leu-L-Phe trifluoromethyl ketone
	2.81 ± 0.04	2 aged phosphoryl adducts
Elastase	2.64 ± 0.06	2 structures of free native enzyme
	2.84 ± 0.08	8 structures of enzyme–peptide anilide complexes
Subtilisin	2.57 ± 0.18	12 structures; native and mutant enzymes and EI complexes
	2.59 ± 0.21	5 structures of free native enzyme
α-Lytic protease	2.77 ± 0.07	2 structures of wild-type enzyme
	2.84 ± 0.15	4 structures of mutant enzymes
	2.73 ± 0.05	7 structures; complexes of wild-type enzyme with peptide boronic acids
Lipases	2.88 ± 0.10	average distance from 23 unliganded lipase structures (range of distances; specific examples follow)
Pancreatic lipase	2.88	native free enzyme structure
	2.86 ± 0.06	2 structures of lipase–colipase complex inhibited with C_{11} phosphonate
Cholesterol esterase	2.81	native bovine pancreatic enzyme
	2.84	enzyme from *Candida cylindracea*
Candida antarctica lipase B	2.97 ± 0.09	4 structures of free wild-type enzyme
Cutinase	2.92 ± 0.09	43 structures; wild-type and mutant enzymes and EI complexes
	2.84 ± 0.02	3 free wild-type enzyme structures
	2.92 ± 0.08	7 structures of phosphylated EI complexes

[a]Uncertainties are ± one standard deviation of the mean for H-bond distances that are averaged from three or more structures. When distances are averaged for two structures, uncertainties are ± one-half of the range of the two measurements.

proton position as a function of the N–H coordinate, Figure 8 shows that the proton is highly delocalized and has maximum probability of being found about equidistant from Nδ of His447(*440*) and Oε1 of Glu334(*327*). Though there is a small classical potential energy barrier in the plot of Figure 8, the zero-point energy is well above the top of the barrier and therefore the His-Glu H-bond in the active site triad of AChE is an LBHB. One of the potential effects of this LBHB is to delocalize the negative charge of Glu334(*327*) through the H-bond into the imidazole ring of His447(*440*). This should increase the basicity of His447(*440*) and in turn increase the nucleophilicity of Ser203(*200*), which is H-bonded to Nε of His447(*440*). This possibility is currently being addressed by additional computational studies.

(iii) Role of electrostatics

The considerable catalytic power of AChE is manifested in k_E values that approach or exceed the diffusion-controlled limit.[44,45] For example, $k_E = 4.2 \times 10^9$ M^{-1} s^{-1} at zero ionic strength for *E. electricus* AChE-catalyzed hydrolysis of ATCh.[67] This value exceeds the estimate of Eigen and Hammes[68] for the maximum bimolecular rate constant of enzyme-catalyzed reactions, and is comparable to the calculated rate constant for diffusion-controlled quenching of electronic excited states, $k_q = 6.5 \times 10^9$ M^{-1} s^{-1}.[69] However, the value of k_E at zero ionic strength does not provide the upper limit for the rate constant for bimolecular association of charged substrates with the active site. This is so because the AChE active site binds the fully extended *tt* conformation of ACh or ATCh in which the dihedral angles for the N$^+$–C–C–X (X = O or S) and C–C–X–C fragments are ~180°.[80,81] This conformation only comprises about 13% of the equilibrium population of conformations of ACh.[82] Scheme 7 can be used to estimate the intrinsic bimolecular binding rate constant for the fully extended conformation of ACh or ATCh as $k_E^C = k_E/K_C$, where k_E and K_C are respectively the measured value of k_{cat}/K_m and the equilibrium constant for conformational transition to the fully extended form of the substrate. Table 3 contains biomolecular rate constants for binding

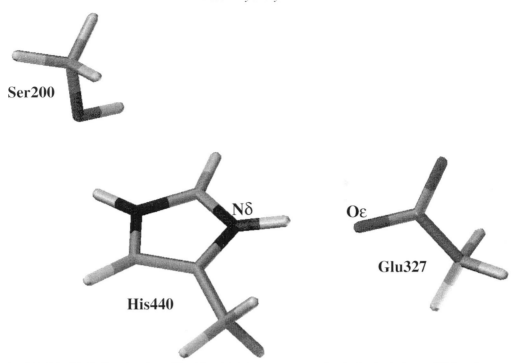

Figure 7 Model of AChE active site utilized in *ab initio* quantum mechanical calculations of H-bond structure and energetics. His440 is in the protonated histidinium form.

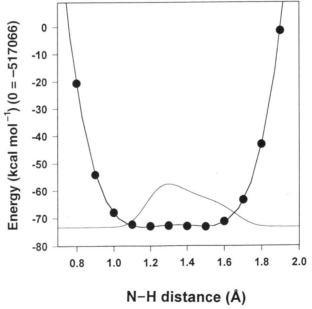

Figure 8 Dependence of electronic energy (dark circles) on N–H distance for the H-bond that bridges $N\delta$ of His440 and $O\varepsilon$ of Glu327 in the active site of AChE, calculated for the model in Figure 7. Overlaid on the plot is the square modulus of the wave function of the zero-point vibrational energy level, which represents the probability of localization of the proton.

of various ligands to AChE. Three features of the data in this table are noteworthy: (a) binding rate constants at zero ionic strength for the conformationally constrained, cationic ligands TMTFA[83,84] and *N*-methylacridinium[67] are $> 10^{10}$ M^{-1} s^{-1}; (b) k_E^C for ATCh at zero ionic strength[67] is comparable to the binding rate constants for conformationally constrained ligands. This similarity supports the procedure described herein for calculating the intrinsic binding rate constant for the extended *tt* conformation of choline substrates. (c) The cationic inhibitor TMTFA binds 5000- and 430-fold faster to *E. electrophorus*[84] and mouse[83] AChEs, respectively, than does the isosteric neutral inhibitor TBTFA.

$$E + A_s \xrightleftharpoons{K_C} E + A_{tt} \xrightarrow{k_E^C} E + P + Q$$

Scheme 7

Table 3 Rate constants for binding of ligands and substrates to AChE.

Compound[a]	Rate constant ($M^{-1} s^{-1}$)	Comments	Ref.
N-Methylacridinium	8.4×10^8	rate constant for inhibitor binding at $\mu = 0.12$ and 25 °C	67
	1.1×10^{10}	rate constant for inhibitor binding extrapolated to $\mu = 0$ and 25 °C	67
Acetylcholine (ACh)	1.6×10^8	k_E measured at 25 °C and $\mu = 0.1$	44,45
	1.2×10^9	k_E^C calculated from preceding k_E, as described in text	
Acetylthiocholine (ATCh)	4.2×10^9	k_E at 25 °C, extrapolated to $\mu = 0$	67
	3.2×10^{10}	k_E^C calculated from preceding k_E, as described in text	
m-(*N,N,N*-Trimethylammonio)-trifluoroacetophenone (TMTFA)	1.8×10^{10}	rate constant for inhibitor binding to *E. electrophorus* AChE, extrapolated to $\mu = 0$	84
	1.6×10^{10}	rate constant for inhibitor binding to mouse AChE, extrapolated to $\mu = 0$	83
m-*t*-Butyltrifluoroacetophenone (TBTFA)	3.6×10^6	rate constant for inhibitor binding to *E. electrophorus* AChE, extrapolated to $\mu = 0$	84
	3.7×10^7	rate constant for inhibitor binding to mouse AChE, extrapolated to $\mu = 0$	83

[a]See structure of active site ligands in Scheme 4.

The means by which AChE effects the very large bimolecular binding rate constants and discrimination for cationic ligands was first suggested by the electrostatics calculations of Ripoll *et al.*[85] AChE not only has a high net negative charge at neutral pH, but also possesses a nonsymmetric charge distribution such that there is a preponderance of negatively charged amino acid residues in the hemisphere of the protein that contains the active site gorge. This charge distribution is associated with a high dipole moment, ~ 1700 debyes,[86] whose vector is directly aligned with the active site gorge, as illustrated in Figure 9. Consequently, it has been suggested that the electrical field of AChE accelerates the delivery of substrate to the active site. This model, as well as the cationic versus neutral ligand discrimination, are supported by Brownian dynamics simulations of ligand binding to the enzyme.[86,87]

Mutagenesis experiments show that the negative potential that arises from Glu and Asp residues on the surface of the enzyme adjacent to the opening of the active site gorge does not make a large contribution to the rate of cationic ligand binding. For example, when seven negative charges of human AChE are neutralized by mutagenesis, the negative component of the electrical potential is compressed to the surface of the enzyme, but k_E for ATCh hydrolysis is only reduced by a factor of 2.5.[88] Similar neutralization of six negative charges in mouse AChE only slowed binding of the cationic active site ligand TMTFA about five-fold, but slowed binding of the peripheral site ligand FAS2 by a factor of 260.[83] These rather diminutive effects on active site ligand binding are accounted for by electrostatics calculations, which show that neutralization of surface charges has very little effect on the gradient of the electrical potential (i.e., the electric field) in the vicinity of the active site.[86] Moreover, rate constants that were calculated by Brownian dynamics simulations for cationic ligand binding to wild type and mutant mouse AChEs correlate well with the experimental measurements.[83]

Neutralization by mutation of anionic active center residues has larger effects on cationic ligand binding to the active site.[83,89] In mouse AChE,[83] Asp74(*72*)Asn and Glu450(*443*)Gln mutations reduce k_{on} for binding of the cationic inhibitor TMTFA by factors of 25 and 8, respectively, while in the Glu202(*199*)Gln mutant k_{on} is reduced by less than two-fold. On the other hand, these changes either have no effect on or modestly increase k_{on} for the neutral analogue TBTFA. The distinct effect of neutralizing active center anionic amino acids on binding of the cationic versus the neutral transition state analogue strongly suggests that the residues promote active center cationic ligand

Figure 9 Representation of the negative electrostatic potential, contoured as a mesh surface at $-1kT/e$, for *Torpedo californica* AChE. The polypeptide backbone is shown as a worm. The dipole moment vector originates near Ser200 and tracks through the active site gorge of the enzyme. This figure was constructed by using the electrostatics modeling program GRASP.[90]

binding by electrostatic guidance. The additive effect of multiple active site mutations supports this idea. In the Asp74(*72*)Asn/Glu202(*199*)Gln/Glu450(*443*)Gln triple mutant of mouse AChE k_E for ATCh is reduced by 1.4×10^4, mostly on the strength of a 400-fold increase in K_m, while the reduction in k_{on} for TMTFA binding is $\geqslant 70$-fold.[83] However, the role of these residues in electrostatic guidance of cationic ligand binding to the active site has been challenged, particularly for Asp74(*72*).[89] In the Asp74(*72*)Asn mutant of human AChE k_E for ATCh turnover and k_{on} for TMTFA binding are reduced by factor of 6 and 36, respectively. However, changing Asp74(*72*) to Glu or Gly has about the same effect on k_E, while k_{on} for TMTFA is nearly the same for the Asp74(*72*)Asn and Asp74(*72*)Lys mutants. These results have been interpreted as indicating that the negative charge of Asp74(*72*) does not have a large effect on the association rates of charged ligands. It seems likely that single mutations of negatively charged residues will generally not have very large effects on the rates of cationic ligand binding, because the steering effect of the electric field of AChE results from the overall charge distribution of the protein, as suggested by the additive effect of multiple mutations of mouse AChE alluded to above.[83] Further experimentation should show whether electrostatic guidance is a feature of catalysis by AChEs from certain species, or is a universal feature of AChE function.

(iv) Structural basis of molecular recognition

The structural basis of the acyl specificities of AChE and BuChE were described earlier. However, additional interactions with constituents of the active site gorge play a decisive role in conferring on these enzymes their specificities for choline substrates and other organic cations. Nearly half of the surface area of the active site gorge of AChE is provided by aromatic amino acid residues.[51] The function of these residues is to provide short-range electrostatic interactions with the quaternary ammonium group of ligands. Two types of structure–activity relationships support this contention. In the first type, the affinities of TMTFA and a series of *meta*-substituted analogues were measured for AChEs from *T. californica* and *E. electrophorus*.[91] Multiple linear free-energy relationships were observed that indicated that electron-withdrawing substituents stabilize the tetrahedral hemiketal adduct with Ser203(*200*), and that molecular recognition in the quaternary ammonium binding locus depended on the molar refractivity of the substituents. Molar refractivity is a measure of the partial molar volume and polarizability of substituents, and accounts for 10^5-fold of the 10^7-fold range of substituent-dependent inhibitor affinities. Therefore, molar refractivity accounts for 29 kJ mol^{-1} of interaction free energy in the quaternary ammonium binding locus of AChE, and molecular recognition in this locus provides almost half of the 74 kJ mol^{-1} free-energy of stabilization of the transition state (see Section 5.06.3.2.2(i)). These results were interpreted as indicating that dispersion interactions are important in molecular recognition in the quaternary ammonium binding locus, and show that this binding locus accounts for almost half of the catalytic power of AChE. Hydrophobic effects, on the other hand, are not important in molecular recognition of quaternary ammonium ligands, since the affinities of the series of transition state analogues are not well correlated by the hydrophobic constants of the substituents.

The second type of structure–activity relationship that probes the roles of amino acid residues in the quaternary ammonium binding locus (cf. Figure 6(b)) is site-specific mutagenesis. Shafferman and colleagues[92] have shown that k_E values for human AChE-catalyzed hydrolysis of ATCh are reduced by 3000-fold and 6000-fold in the Trp86(*84*)Ala and Trp86(*84*)Glu mutants, respectively, while a much smaller 10-fold activity reduction is seen in the Trp86(*84*)Phe mutant. For the isosteric substrate, 3,3-dimethylbutyl acetate, mutation effects were modest; the largest effect was a nine-fold reduction in k_E in the Trp86(*84*)Glu mutant. These effects were interpreted in terms of a model in which the primary role of Trp86(*84*) is to provide a site for cation-π interactions with the quaternary ammonium group of active site ligands. Dougherty and co-workers[93] have characterized the cation-π effect by quantum chemical calculations and studies of the binding of organic cation guests to synthetic, aromatic hosts. They find that the energies of cation-π interactions depend approximately on r^{-n} ($n < 2$), and therefore are of a shorter range than interactions that are described by the electrostatic law, which depend on r^{-1}. Moreover, the cation-π effect arises from the interaction of cations with the negative component of the quadrupole moment of the aromatic ring. The comparable inactivities of the Trp86(*84*)Glu and Trp86(*84*)Ala mutants shows that an anionic counterion cannot replace the effect of the aromatic quadrupole. This accords well with the fact that only two anionic residues (Asp74(*72*), Glu202(*199*)) line the surface of the active site gorge, while aromatic residues provide nearly half of the gorge surface. Mutagenesis experiments show that additional residues, in particular Glu202(*199*) and Tyr337(*330*) (in *T. californica* AChE the corresponding residue is Phe330), contribute to molecular recognition in the quaternary ammonium binding locus, though not as decisively as Trp86(*84*).[52] The cumulative effect of mutations of residues in the quaternary ammonium binding locus is in reasonable quantitative agreement with the 10^5-fold substituent effect determined by ligand structure–activity studies, as described above.

Harel *et al.*[52] overlaid the TMTFA-AChE complex on a model of the tetrahedral intermediate in the acylation stage of AChE-catalyzed hydrolysis of ACh.[51] This overlay showed that the quaternary ammonium functions of TMTFA and the tetrahedral intermediate nearly coincide, and in both cases project against the aromatic fused-ring system of Trp86(*84*). This binding geometry is as expected for interaction of the cationic quaternary ammonium function and the π-electron density of Trp86(*84*), and provides an explanation for the preference of AChE for the extended *tt* conformation of ACh or ATCh.[80–82] In the extended *tt* conformation the stabilizing cation-π interaction is maximized, and steric interaction between Ser203(*200*) and the ethano bridge or the quaternary ammonium function of the substrate is minimized. Substrate β-deuterium kinetic isotope effect experiments have indicated that the transition step for the nucleophilic attack step in the acylation stage of catalysis structurally resembles the tetrahedral intermediate.[94] Therefore, the views provided by the TMTFA complex[52] and the tetrahedral intermediate model[51] are germane to the mechanism and power of AChE catalysis.

5.06.3.3 Lipases

A large number of esterases of the α/β hydrolase fold family are lipases that have as their physiological functions the catalyzed hydrolysis of water-insoluble lipid ester bonds of substrates that are contained in supramolecular assemblies, such as lipoproteins, micelles, membranes, or intracellular lipid inclusions (cf. Table 1). A surprising array of structural variations on the basic α/β hydrolase fold and associated mechanistic nuances have evolved to catalyze lipid hydrolysis by enzymes in this family. Illustrative examples are discussed below that underscore this structural and mechanistic diversity.

5.06.3.3.1   Enzyme structure

The first high-resolution X-ray structures that were reported for lipases were those of human pancreatic lipase[95] and a fungal lipase from *Rhizomucor miehei*.[96] In both of these enzymes, access to the Ser-His-Asp active site triad is occluded by surface loops, as shown in Figure 10 for the fungal lipase, and the oxyanion hole is distorted from the geometry that it adopts in covalent complexes with active site ligands. These features suggest that interfacial activation in these enzymes must involve movement of the occluding surface loops and concomitant shifts in active site architecture. However, the two enzymes have evolved distinct themes for substrate access to the active site, as discussed in the next section. Crystal structures of bovine pancreatic cholesterol esterase (CEase) have been reported by two different research groups,[97,98] and comparative features of these structures suggest yet another mode of regulation of access to the active site. Chen *et al.*[97] grew crystals of the enzyme in space group C2 in the absence of detergent and determined the structure to 1.6 Å resolution. In their structure the active site is covered by a surface loop, consisting of residues 116–125, and the hydrophobic carboxyl terminus of the enzyme is inserted into the active site and distorts the architecture of the oxyanion hole. In contrast, Wang *et al.*[98] crystallized the enzyme in the space group $P3_12_1$ in the presence of the detergent zwittergent 3–12, and determined the structures of the apoenzyme and the apoenzyme–taurocholate complex to 2.8 Å resolution. In their apoenzyme structure the carboxyl terminus has been displaced from the active site, while in the apoenzyme–taurocholate complex the bile salt cofactor interacts with the 116–125 loop and displaces the loop to allow access of substrate to the active site. The structure of guinea pig pancreatic lipase contains an unoccluded, preformed active site,[99] and therefore represents yet another distinct mode of regulation of lipid substrate access to the active site. The structure of the bacterial lipase cutinase also shows an open access to a preformed active site.[100]

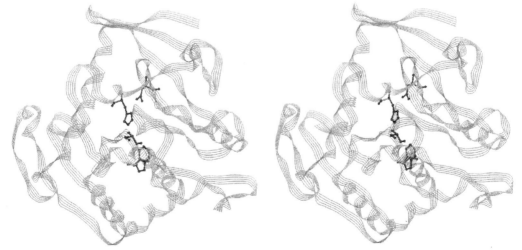

Figure 10   Stereoview of the closed form of *Rhizomucor miehei* lipase. Access of lipid substrates to the active site is blocked by a helical lid that consists of residues 85–92. The active site triad, Ser144-His257-Asp203, and Trp88 of the lid are shown as balls and sticks.

5.06.3.3.2   Catalytic mechanism

Like cholinesterases, lipases of the α/β hydrolase fold family utilize the general acylenzyme mechanism that is outlined in Scheme 5. For mammalian pancreatic cholesterol esterase, the acylenzyme intermediate has been trapped with alternate nucleophiles, such as short-chain alcohols and hydroxylamine.[101,102] In trapping experiments, the acylenzyme intermediate partitions between reaction with water or with the alternate nucleophile, as outlined in Scheme 8, and the corresponding effect of nucleophile concentration on the kinetics of the reaction is described by Equations (13) and (14).

$$V_{max} = k_{cat}[E]_T = \frac{k_3(k_5 + k_7[Nu])}{k_3 + k_5 + k_7[Nu]}[E]_T \tag{13}$$

$$K_m = \frac{k_2 + k_3}{k_1} \frac{k_5 + k_7[Nu]}{k_3 + k_5 + k_7[Nu]} \tag{14}$$

These equations show that nucleophilic trapping is manifested by an increase in V_{max} (activation) and a corresponding increase in K_m, and therefore the parameter V_{max}/K_m is unaffected. A series of parallel lines is observed when kinetic data are presented as Lineweaver–Burk plots, and hence activation follows a ping-pong mechanism, as expected for a transacylation from the active site Ser to the alternate nucleophile. Nucleophilic trapping experiments can also be used to determine whether acylation or deacylation is rate limiting for k_{cat}. If deacylation is solely rate limiting (i.e., k_5 and $k_7[Nu]$ are $\ll k_3$), then V_{max} and K_m in Equations (13) and (14) are linear functions of [Nu] and $V_{max} = k_5[E]_T$ in the absence of the alternate nucleophile. This is the observed behavior when porcine pancreatic CEase-catalyzed hydrolyses of fatty acyl p-nitrophenyl esters are run in the presence of NH_2OH.[102] If acylation is solely rate-limiting (i.e., $k_3 \ll k_5$), then $V_{max} = k_3[E]_T$ and $K_m = (k_3 + k_2)/k_1$, and therefore added nucleophile has no effect on the Michaelis–Menten kinetic parameters. However, in this case product analysis will detect the transacylation product. If acylation and deacylation are comparably rate-limiting, then Equations (13) and (14) predict that K_m and V_{max} are nonlinear, hyperbolic functions of [Nu]. In this case the rate constants k_3, k_5, and k_7 can be determined by least-squares analysis of the dependence of V_{max} on [Nu].

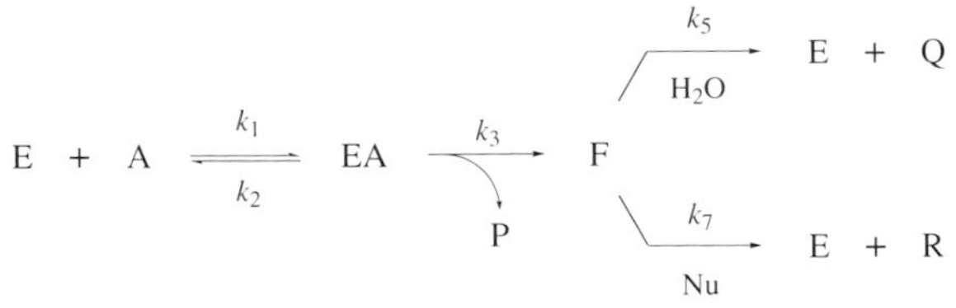

Nu = nucleophile   R = transacylation product

Scheme 8

Also like the cholinesterases, lipases of the α/β hydrolase fold family utilize catalytic triads for general acid–base catalysis and oxyanion holes for stabilization of tetrahedral intermediates and transition states. However, unlike the cholinesterases, not all lipases utilize tripartite oxyanion holes. For example, human[103] and equine[104] pancreatic lipases utilize the peptide NH functions of Phe77 and Leu153 in a bipartite oxyanion hole. In lipases that utilize tripartite oxyanion holes, two H-bond donors are contributed by peptide NH functions and a third donor is the γOH group of a Ser or Thr sidechain. For example, in the crystal structure of the covalent adduct in which Ser120 of the active site of cutinase is diethylphosphorylated, Ser42 contributes the OH and peptide donors for interaction with the phosphoryl oxygen.[100] When Ser42 is replaced by Ala by site-specific mutagenesis the three-dimensional structure of the enzyme is not significantly altered, but enzyme activity is reduced by a factor of 450,[105] which indicates that γOH of Ser42 provides 16 kJ mol^{-1} of transition state stabilization. This estimate of transition state stabilization by an amino acid side chain H-bond donor is similar to that determined by mutagenesis for Asn155 of the oxyanion hole of the serine protease subtilisin.[106,107]

(i) Interfacial recognition and substrate access to the active site

For lipases of the α/β hydrolase fold family, the specific regions of the enzyme that are involved in interfacial recognition have not yet been defined. However, X-ray structures of ligand complexes suggest the means by which certain lipases bind to lipid–water interfaces, and coincidentally how substrate access to the active site is regulated. Pancreatic lipase requires the presence of a cofactor, the 93-residue protein colipase, for full activity.[108] In the unliganded enzyme, the Ω-disulfide loop

Cys237–Cys261 occludes entry to the active site of ligands that are contained in the lipid interface or in the bulk medium. In this closed form of the enzyme, Trp252 of the loop interacts with Phe77, Leu153, and Tyr114 of the catalytic core of pancreatic lipase.[1,108] In the absence of lipids or lipid analogues, colipase interacts with pancreatic lipase only on the predominately β-stranded C-terminal domain of the enzyme, and active site access remains occluded by the Ω-disulfide loop. However, in a noncovalent adduct between the active site and didodecanoyl-PC, two major conformational changes have occurred.[109] One is that the Ω-disulfide loop has been displaced from the surface of the enzyme by a hinge-like movement that establishes a second locus of interaction with colipase, as shown in Figure 11. Therefore, colipase appears to stabilize the open form of pancreatic lipase that is required for access of substrates to the active site. A thermodynamically equivalent interpretation is that the presence of lipid substrate in the active site stabilizes the lipase–colipase interaction, as evidenced by the fact that interaction affinity increases in the presence of lipid interfaces.[1,108] The second major conformational change[109] involves the loop that consists of residues 76–85, Phe77 of which interacts with Trp252 of the Ω-disulfide loop in the closed form of the enzyme. In the open form Phe77 folds back on the catalytic core of the enzyme, where it contributes its peptide NH function to the now properly aligned bipartite oxyanion hole. Therefore, the conformation changes that accompany lipid substrate binding in the active site and colipase interaction with the Ω-disulfide loop reshape the geometry of the active site so that complimentarities to the geometries of the tetrahedral intermediates and flanking transition states of the catalytic mechanism are optimized. This elegant model provides a picture, at atomic resolution, of the mechanism of interfacial activation of pancreatic lipase. It is possible that the homologous enzyme lipoprotein lipase, which requires the small protein cofactor apoCII for full activity,[1,108,110] operates by a similar interfacial activation mechanism. However, crystal structures of the lipoprotein lipase-apoCII system have not yet been reported.

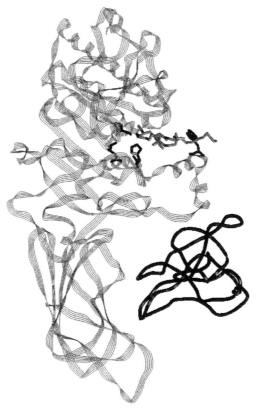

Figure 11 Structure of the ternary complex of human pancreatic lipase (five-stranded ribbon), porcine colipase (black tube), and didodecanoyl-PC (gray capped sticks). The active site triad, Ser152-His263-Asp176, and Trp252 of the Ω-disulfide loop Cys237-Cys261 are shown as black capped sticks.

As discussed earlier (see Section 5.06.3.3.1), binding of lipid substrates likely dislodges the hydrophobic carboxyl terminus from the active site of cholesterol esterase (CEase), and binding of the cofactor taurocholate repositions the loop 116–125 to promote access to the active site.[97,98] In addition, CEase is known to preferentially bind to interfaces that have a net negative charge,[1,111] such as those of anionic mixed bile salt micelles in the intestines. Homology modeling of the structures of human and rat pancreatic CEases suggests the molecular origins of this interfacial recognition.[111] As for AChE,[85,86] the asymmetric distribution of charged residues in CEase is associated with a net dipole moment, the positive pole of which probably interacts with negative interfaces and orients the active site for interaction with lipid substrates in the interface. Gratifyingly, the dipole moment vector emerges from the enzyme surface near the sequence segment Lys61-Lys62-Arg63; this segment is the proposed site for interaction of CEase with the anionic glycosaminoglycan polymer heparin.[111] However, the electrostatic model described herein suggests that CEase utilizes an ultrastructural, topological motif for interfacial recognition, rather than a locus on the enzyme that involves just a few amino acid sidechains.

Cygler and co-workers have described the structures of a fungal lipase from *Candida rugosa* that alternately crystallizes in closed or open forms.[1] In the closed form of the enzyme the active site is occluded by an Ω-disulfide loop that consists of residues Cys60–Cys97. The residues on the exterior surface of this loop are primarily hydrophilic, while those on the interior portion are hydrophobic and overlay the active site triad, Ser209-His449-Glu341, that lies near the bottom of a hydrophobic active site cavity. The observation of crystal structures of both closed and open forms of the enzyme suggests that the two forms may be in equilibrium in solution. The open form of the enzyme is also observed in complexes in which Ser209 is covalently modified by lipid-mimetic transition state analogues. Therefore, it is likely that interaction of the enzyme with the amphipathic lipid–water interface triggers the conformational change of the Ω-disulfide loop that subsequently provides access of substrates in the interface to the active site. This mechanism of conformational equilibrium between closed and open forms and of substrate-triggered conformational change is likely shared by other lipases,[1,43,108] including those from *Rhizomucor miehei* (cf. Figure 10 for closed structure of the enzyme) and *Pseudomonas cepacia*,[112] and is probably a general mechanism for interfacial recognition/activation for lipases whose activities are not affected by cofactors.

(ii) Molecular recognition of lipid monomers

The results of X-ray crystal structure determinations also provide detailed information on the origins of molecular recognition of lipid monomers by the extended active sites of lipases of the α/β hydrolase fold supergene family. In the covalent complex formed when Ser209 of *C. rugosa* lipase reacts with hexadecanesulfonyl chloride,[1,113] one of the sulfonyl oxygens is situated in the oxyanion hole, therein making H-bonds with the peptide NH groups of Gly123, Gly124, and Ala210. The alkyl chain of the inhibitor occupies a long and narrow cavity in the interior of the enzyme, as illustrated in Figure 12. This cavity is lined predominately with hydrophobic and aromatic amino acid side chains, and enforces *gauche* conformations on the alkyl chain of the inhibitor at C_7–C_8, C_{10}–C_{11}, and C_{11}–C_{12}. These *gauche* conformations wrap the alkyl chain around Leu302 and enforce an overall conformation in which the proximal and distal portions of the alkyl chain are approximately perpendicular. The geometry of this kinked internal cavity may explain the preference of *C. rugosa* lipase for lipid substrates that contain oleic fatty acyl chains. At high inhibitor concentrations the enzyme is covalently linked at Nε of His449 of the active site triad to another hexadecanesulfonyl moiety, which projects into the bulk medium through the opening of the active site gorge. Based on these observations, Grochulski *et al.*[113] proposed that *C. rugosa* lipase binds to a triacylglycerol substrate monomer that is in a tuning fork conformation, with the *sn*-2 acyl chain projecting into the internal cavity illustrated in Figure 12, and with the *sn*-1 and *sn*-3 acyl chains protruding through the active site opening into the interface of the lipid-containing supramolecular complex.

The structure of the complex of pancreatic lipase with didodecanoyl-PC[109] that is displayed in Figure 11 provides an accounting of the interactions utilized by the enzyme to accommodate lipid monomers in its active site. As in catalyses by PLA2 (see Section 5.06.2.3.3) and *C. rugosa* lipase,[1,113] pancreatic lipase employs primarily hydrophobic and aromatic amino acid side chains to interact with the acyl chains of the substrate. The *sn*-2 acyl chain of the bound phosphatidylcholine monomer mimics the *sn*-2 chain of a triacylglycerol substrate, and interacts with the side chains of Ile78 and the helix formed by residues 251–259. Recall that in the closed form of the enzyme this helix is a constituent of the Ω-disulfide loop Cys237–Cys261 that overlays the active site and that Trp252 of the loop interacts with Phe77. In the open, active form, Phe77 contributes its peptide NH to the oxyanion hole. In the structure of Figure 11 Trp252 makes closest contact with C_{10} of the *sn*-2 acyl chain. Since Trp252 in the complex is near the surface of the enzyme, this model suggests that, like

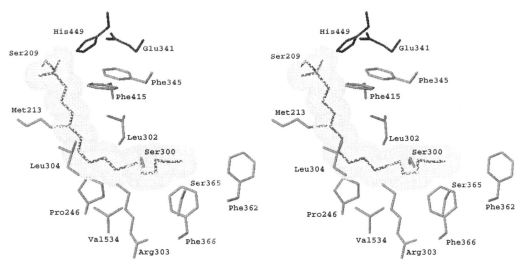

Figure 12 Crossed stereoview of fatty acyl recognition in the extended active site of *Candida rugosa* lipase. The alkyl chain of the covalently bound hexadecanesulfonyl inhibitor is represented as a dot surface, and the active site triad (Ser209-His449-Glu341) and residues that interact with the alkyl chain as capped sticks.

the other lipases discussed previously, pancreatic lipase extracts a portion of the substrate from the interface, with the remaining portions of the substrate embedded in the interface.

Pancreatic lipase also interacts with the proximal portion of the *sn*-1 acyl chain. The *sn*-1 chain of the bound phospholipid, which mimics the triacylglycerol chain that contains the scissile ester bond, interacts with hydrophobic and aromatic amino acids, among them Phe77, Tyr114, Pro180, Leu202, Ile209, and Phe215. Interestingly, in the crystal structure of the lipase–colipase complex in which Ser152 is covalently attached to an undecanoylphosphonate moiety,[1,108] the alkyl chain of the inhibitor utilizes essentially the same set of interactions as does the *sn*-1 acyl chain in the complex of Figure 11. Moreover, in both the undecanoylphosphonate and phosphatidylcholine complexes, C_{10} of the respective alkyl and *sn*-1 acyl chains is at the enzyme surface. Therefore, the enzyme extracts from the interface about the proximal halves of the *sn*-1 and *sn*-2 acyl chains. This molecular recognition strategy is a reasonable expression of the evolution of lipase catalytic power, since the enzyme does not pay the full free-energy cost of total extraction of trenchantly insoluble, highly hydrophobic substrate monomers from the supramolecular lipid-containing complex.

5.06.3.3.3 *Hormone-sensitive lipase*

Hormone-sensitive lipase (HSL) catalyzes the hydrolysis of acylglycerols and cholesteryl esters that are contained in cellular lipid inclusions.[114] The primary physiological function of the enzyme is the mobilization of fatty acids from their triacylglycerol stores in adipose cells, a process that provides fuel for energy metabolism in mammals. HSL also mobilizes free cholesterol from cholesteryl ester stores for subsequent steroidogenesis. The enzyme is divided roughly into regulatory and catalytic domains, and is activated by cAMP-dependent phosphorylation at Ser552 of the regulatory domain when catecholamines bind to β-adrenergic receptors. Activation of HSL is blocked in two ways: (i) cAMP-independent phosphorylation of Ser554 does not affect activity, but prevents phosphorylation of Ser552; (ii) dephosphorylation of the Ser552 adduct accompanies binding of insulin to the insulin receptor, thereby accounting for the antilipolytic activity of the hormone.

HSL and homologous proteins comprise the H-family of the α/β hydrolase fold supergene family, but show virtually no sequence homology with proteins of the C- and L-families. However, there are several reasons for including HSL in the α/β hydrolase fold supergene family. Cloning of the enzyme from adipose tissue and subsequent mutagenesis experiments (discussed in reference 114) indicate that HSL is a serine esterase, with the active site nucleophile (Ser424) a component of a Gly-Asp-Ser-Ala-Gly lipase consensus sequence. Contreras *et al.*[114] used the PredictProtein PHD program to generate predictions of secondary structural elements of the catalytic domain of HSL. They noted extensive homology at the secondary structural level between HSL and other α/β hydrolases, such as AChE, *C. rugosa* lipase (CRL) and *G. candidum* lipase. Consequently, they used the crystal structure of CRL as a template to construct a three-dimensional model of HSL that spans residues 327 to 748 of the primary sequence. Their model confirmed that Ser424 and His723 are constituents of the active site triad, and revealed that the third member of the triad is Asp693. Secondary structural elements that are shared with CRL include β-strands 2 through 8 and nine α-helices. The regulatory domain, which spans residues 454 through 681 of the primary sequence and intervenes β-strands 6 and 7, is an insert in the catalytic domain and was not included in the model. The X-ray structure of an H-family enzyme, the bacterial brefeldin A esterase, has been solved, but as of the writing of this chapter there was a hold on the requisite coordinates at the Brookhaven Protein Data Bank.

5.06.3.4 Thioesterases

The crystal structure of the myristoyl-ACP-specific thioesterase from *Vibrio harveyi* was reported by Lawson *et al.* in 1994,[115] and reveals several interesting features of the enzyme. Though this is the only three-dimensional structure currently available for a thioesterase, the similarity in tertiary fold between the thioesterase and haloalkane dehalogenase[116] from *Xanthobacter autotrophicus* definitively places the enzyme in the α/β hydrolase fold family.[39] The nucleophilic serine of the Ser114-His241-Asp211 triad is in a strained conformation in a sharp γ-turn that intervenes β-strand 5 and α-helix C, a structural feature that is common in serine esterases and lipases and is referred to as the "nucleophilic elbow."[39] However, the sequence about Ser114 does not conform to the Gly-X-Ser-X-Gly consensus lipase/esterase sequence, since the Gly residues in the flanking amino terminal and carboxyl terminal directions are replaced by larger amino acids, Ala and Ser, respectively. This comprised the first observation of divergence from the consensus active site sequence among enzymes of the α/β hydrolase fold family. Moreover, the oxyanion hole is not fully formed in the native enzyme, since only one peptide NH, that of Leu115, is well positioned to serve as an H-bond donor to stabilize oxyanionic tetrahedral intermediates and quasitetrahedral transition states. Lawson *et al.*[115] suggested that a mobile loop that consists of residues 168–171 and that is adjacent to the substrate binding site may change conformation on substrate binding to provide a second H-bond donor for the oxyanion hole. This idea recalls the conformation changes that form the oxyanion hole in the active sites of lipases and that contribute to interfacial activation.[108]

A detailed kinetic study by Li *et al.*[117] demonstrated that, as expected, the myrisotyl-ACP-specific thioesterase from *V. harveyi* utilizes the acylenzyme catalytic mechanism that is outlined in Scheme 5. They showed that the stoichiometry of acylenzyme formation on interaction of the enzyme with myristoyl-ACP was 0.7 mol per mol of enzyme, and obtained a similar result from the amplitude of the pre-steady-state burst of *p*-nitrophenol when *p*-nitrophenyl myristate was the substrate. These observations show that, for both the physiological and the artificial substrates, acylation and deacylation contribute comparably to rate limitation under conditions of substrate saturation. Quantitative analyses of the acylenzyme kinetics of the thioesterases from *Vibrio harveyi* and rat mammary gland, and the effects of site-specific mutations on these kinetics, have been reviewed.[1]

5.06.3.5 Dienelactone Hydrolase

Dienelactone hydrolase (DLH) from *Pseudomonas* sp. *B13* catalyzes the hydrolysis of the cyclic ester function of 4-carboxymethylenebut-2-en-4-olide to maleyl acetate,[118,119] as outlined in Scheme 9. The enzyme is the third in the β-ketoadipate pathway of bacterial and fungal degradation of aromatic substrates. Also shown in Scheme 9 are alternate substrates and inhibitors of DLH. Though a member of the α/β hydrolase fold supergene family, DLH utilizes an unprecedented active site triad and a novel mechanism to effect catalysis of ester bond hydrolysis. These are discussed in the two sections that follow.

5.06.3.5.1 *Active site structure*

The active site triad of DLH consists of Cys123, His202, and Asp171,[118,119] and consequently is a hybrid of the Ser-His-Asp (or Glu) triads of serine hydrolases and the Cys-His-Asn triads of cysteinyl

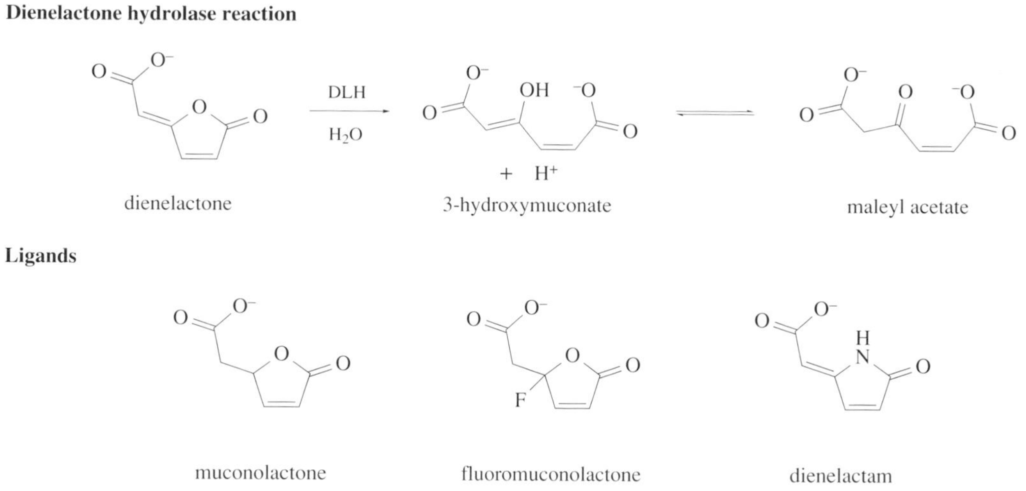

Dienelactone hydrolase reaction

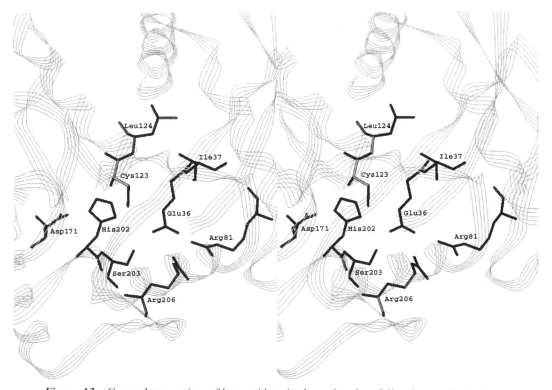

dienelactone   3-hydroxymuconate   maleyl acetate

Ligands

muconolactone   fluoromuconolactone   dienelactam

Scheme 9

proteases. Figure 13 shows the location of key residues in and near the active site of the enzyme. In addition to the triad, six residues are important to the function of DLH. Among these are Arg81, Arg206, and Ser203, which interact with the carboxylate group of the substrate, and Glu36, which in the free enzyme forms an ion pair with Arg206. His202 and Asp171 form an additional ion pair in the free enzyme, but Cys123 is not interacting with His202. Rather, Cys123, as the neutral thiol, forms a hydrogen bond with Glu36 in the interior of the enzyme. Changes in these amino acid side chain interactions of the free enzyme are observed when active site ligands bind, such as the dienelactam inhibitor of Scheme 9. These changes provide the basis for a novel substrate-induced activation mechanism, which is outlined in the following section.

Figure 13 Crossed stereoview of key residues in the active site of dienelactone hydrolase.

Ester Hydrolysis

5.06.3.5.2 *Catalytic mechanism*

Like that of other esterases of the α/β hydrolase family, the catalytic mechanism of DLH occurs in successive acylation and deacylation stages that respectively involve the formation and breakdown of tetrahedral intermediates.[118,119] Aside from these features, however, the formal similarity ends. Unlike the other esterases, which form oxyester acylenzyme intermediates with the active site serine, DLH forms a thioester acylenzyme intermediate with the thiolate form of Cys123. Therefore, in the acylation stage of catalysis, His202 of DLH does not serve as it does in serine hydrolases as a general base catalyst for the formation of the tetrahedral intermediate. Rather, the histidinium form of His202 stabilizes the thiolate form of Cys123, a role that is like that of the active site His in cysteinyl protease catalysis. However, as mentioned earlier, Cys123 in the resting enzyme is probably in the neutral thiol form and is H-bonded to Glu36 in the interior of the enzyme. When the substrate binds to the active site, Arg206 forms an ion pair with the carboxylate function of the substrate, which in turn weakens the Arg206–Glu36 ion pair.[118] This enhances the basicity of Glu36, which moves toward and deprotonates the thiol of Cys123. The resulting thiolate anion moves from the enzyme interior into the active site, where it interacts with His202 and nucleophilically attacks the lactone ring of the substrate. In this way, binding of the substrate is linked to events which activate the enzyme and initiate the acylation stage of catalysis.[118]

Scheme 10 outlines a chemical mechanism proposed by Ollis and colleagues[119] for DLH catalysis. In this mechanism, the tetrahedral intermediate formed in the acylation stage of catalysis collapses to the acylenzyme intermediate by expulsion of the enolate oxygen, albeit not assisted by general acid catalysis by the histidinium form of His202 of the triad. Support for this mechanistic assignment is that the dienelactone and fluoromuconolactone of Scheme 9 are substrates for DLH, but the dienelactam and muconolactone are not. Consequently, the enzyme does not hydrolyze substrates with basic leaving groups, which suggests that general acid catalysis is not a feature of transition state stabilization for conversion of the tetrahedral intermediate to the acylenzyme. Moreover, in the crystal structure of the DLH complex with the dienelactam inhibitor, the lactam nitrogen is not close enough to His202 for a strong H-bond to form. This model is consistent with the inability of DLH to hydrolyze substrates with basic leaving groups, whereas serine esterases of the α/β hydrolase fold family, such as AChE,[44] are capable of doing so. An additional unusual feature of the mechanism is the use of the incipient enolate oxygen of the acylenzyme intermediate as a general base catalyst in the deacylation stage of catalysis. Ollis *et al.*[119] supported this idea by building a model of the enolate acylenzyme intermediate. They found that in this model a water molecule could not be situated between His202 and the thioester function without distorting the active site and weakening the ion pair interactions between the substrate carboxyl and Arg81 and Arg206. However, a model in which a water molecule is situated for deprotonation by the enolate oxygen and accompanying attack on the thioester function was not hampered by these geometrical problems.

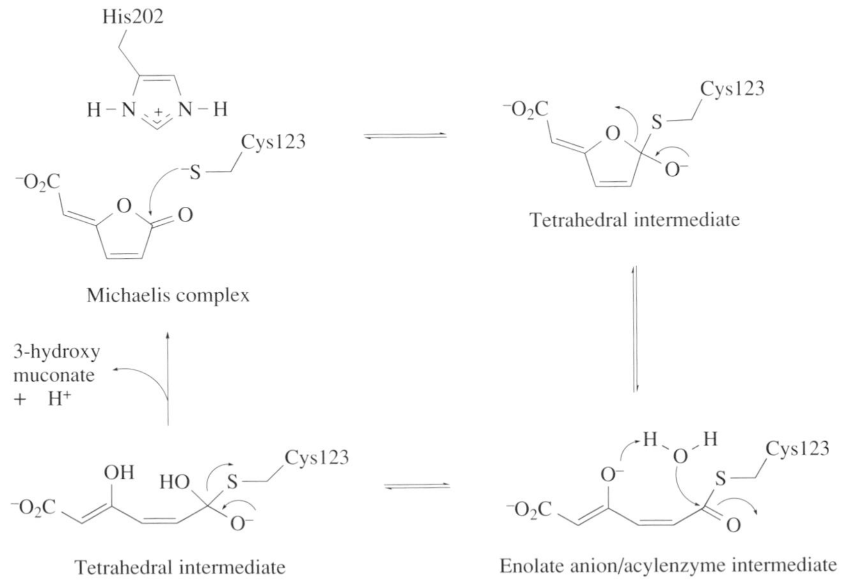

Scheme 10

5.06.3.6 Esterases with Unusual Catalytic Triads

Dienelactone hydrolase is not the only esterase whose active site does not fit the Ser-His-Asp (or Glu) catalytic triad paradigm of the serine proteases and serine esterases.[1] The structures of two lipases, one from *Pseudomonas glumae*[120] and another from *Streptomyces scabies*,[121] also reveal novel active site motifs. When Asp263 of the catalytic triad of *P. glumae* lipase is replaced by Glu and Ala, the mutant enzymes retain 45% and 22%, respectively, of the activity of the wild-type enzyme.[120] It is conceivable that Glu288, which is adjacent to Asp263, may compensate for the loss of the function of Asp263 in the mutant enzymes, which in turn implies that Glu288 and Asp263 are redundant carboxylates in an active site tetrad. In the triad of *S. scabies* esterase the active site triad consists of Ser14, His283, and Trp280,[121] and the main chain carbonyl oxygen of the Trp residue is in a position that is equivalent to the carboxylate oxygen that is H-bonded to His in the more prevalent triads of serine esterases and lipases.

5.06.4 ACID LIPASES

Acid lipases are enzymes that have maximal activities at acidic pH values and catalyze the hydrolysis of the ester bonds of triacylglycerols and/or cholesteryl esters.[122,123] Two acid lipases, lingual lipase and gastric lipase, initiate the hydrolysis of dietary triacylglycerols in the stomach. Lingual lipase is secreted by the serous glands in the back of the tongue, whereas gastric lipase is synthesized and secreted by the chief cells of the fundic mucosa of the stomach. The partial hydrolysis of triacylglycerols that is effected by these enzymes is completed in the duodenum by pancreatic lipase and cholesterol esterase. The coordinated operation of these four enzymes leads to complete hydrolysis of dietary lipid esters and their subsequent absorption into the bloodstream. Lysosomal acid lipase, on the other hand, catalyzes the intralysosomal hydrolysis of triacylglycerols and cholesteryl esters that are taken up by the low density lipoprotein receptor system. The enzyme is important for cellular cholesterol regulation, since the free cholesterol that it liberates inhibits HMG-CoA reductase, the rate limiting enzyme in cholesterol biosynthesis. Rare genetic deficiencies of lysosomal acid lipase activity lead to accumulation of cellular lipid esters and consequent biomedical complications, such as premature atherosclerosis.

5.06.4.1 Gene Structure and Sequence Similarities

Comparisons of gene structures and amino acid sequences, themselves deduced from cDNA sequences, suggest that rat lingual lipase (RLL), human gastric lipase (HGL), and human lysosomal acid lipase (HLAL) are members of an acid lipase gene family.[122] The genes of all three enzymes are composed of 10 exons interrupted by nine introns, though the length of genomic DNA occupied by the three genes varies considerably (HGL, 14 kb; RLL, 18.7 kb; HLAL, 38.8 kb). The exon sizes and intron phase classes of RLL and HGL are identical, which suggests that RLL is the rat gastric lipase expressed in the serous glands of the tongue. Human gastric lipase consists of 379 residues and has appreciable homology with rat lingual lipase (377 amino acids, 76% identity) and human lysosomal acid lipase/cholesteryl ester hydrolase (378 amino acids, 58% identity). None of these enzymes, however, has significant sequence homology with the lipases and esterases of the α/β hydrolase supergene family, which were discussed in the preceding section of this chapter.

5.06.4.2 Catalytic Mechanism

The enzymes of the acid lipase family likely catalyze the hydrolysis of ester bonds of acylglycerols and cholesteryl esters via an acylenzyme mechanism,[123] such as that utilized by enzymes of the α/β hydrolase family and outlined in Scheme 5. Chemical modification studies suggest that the enzymes are serine esterases, but loss of activity is also noted in the presence of sulfhydryl-modifying reagents. The serine esterase mechanism is supported by a comparative site-specific mutagenesis study of HLAL and HGL.[123] Both HLAL and HGL contain two lipase consensus sequences -Gly-X-Ser-X-

Gly- (X = variable amino acid), one centered on Ser99 and one on Ser153, that contain putative active site nucleophiles. Replacement of Ser153 in HLAL with Thr produced a mutant enzyme that had no activity against cholesteryl oleate, triolein, or tributyrin substrates. However, replacement of Ser99 by Thr gave a mutant HLAL that retained ~30–70% of the activity of the wild-type enzyme. Parallel mutagenesis experiments in HGL showed that Ser153 was necessary for activity against triolein or tributyrin, but Ser99 was not. Therefore, Ser153 is implicated as the nucleophile in the catalytic triads of acid lipases.

Eight Asp residues that are conserved in the three acid lipases were replaced by Gly in a series of single mutants of HLAL.[123] Three of the mutants (Asp89Gly, Asp124Gly, and Asp257Gly) retained significant catalytic activity toward both cholesteryl oleate and triolein substrates, and therefore cannot contribute the carboxylate residue to the catalytic triad. Three additional mutants (Asp93Gly, Asp130Gly, and Asp328Gly) retained significant activity toward triolein but were catalytically inactive toward cholesteryl oleate, establishing a differential role for these residues in the two lipid ester hydrolyses for which HLAL is a catalyst. However, substitution of either Asp324 or Asp331 by Gly abolished activity toward both substrates, and therefore one of these two residues is a likely component of the active site triad. This issue was settled by constructing a series of single mutants in which Asp324, Asp328, or Asp331 of HGL is replaced by Gly. Of these, the Asp328Gly and Asp331Gly mutants were almost as active as wild-type enzyme toward both triolein and tributyrin substrates, whereas the Asp324Gly mutant was devoid of activity. Therefore, comparative mutagenesis in HLAL and HGL suggests that Asp324 is the acid component of the catalytic triad.

Six His residues are conserved in HLAL, HGL, and RLL, and each of these was replaced by Gln in a series of single mutants.[123] Substitutions of His262, His298, and His345 in HLAL gave mutant enzymes that retained significant catalytic activity, while replacements of the remaining three conserved residues His65, His274, and His353 gave enzymes that were catalytically inactive. Replacement of His274 in HGL by Gln gave a mutant enzyme that retained significant catalytic activity, and therefore either His65 or His353 was suggested to be the general acid/base element of the catalytic triad. Of these residues, His353 is favored because its sequence position gives an ordering of triad residues, that is, Ser153-Asp324-His353, that is like those found in numerous additional lipases and esterases.

Therefore, the acid lipases, like the esterases and lipases of the α/β hydrolase fold family and the peptidases of the serine protease families, have arrived by convergent evolution on the catalytic triad theme of transition state stabilization. However, the operation of active site triads in acid lipases raises a number of questions. In serine proteases and esterases, the pK_a of the imidazole side chain of the active site His falls in the range 6–7.5.[44,53] What structural features of the active sites of acid lipases lower the pK_a of the His to such a degree that the enzymes have high activities at pH values as low as 2? Is His353 the general acid/base component of a traditional triad? Or does the protonated form of His353 serve solely as a general acid to assist the departure of the leaving group from the tetrahedral intermediate (cf. Schemes 1 and 5) or as an H-bond donor in the oxyanion hole? Do the acid lipases share topological features with esterases and lipases of the α/β hydrolase fold family, or do they possess a unique fold that defines a new supergene family? Answers to these and other questions must await the report of the first X-ray structure of an acid lipase.

5.06.5 CONCLUSIONS

Progress proceeds at a dizzying pace on identification of the structural motifs that are utilized by esterases and lipases to effect molecular recognition of the transition state and of the biological milieu in which their substrates are imbedded. This progress is driven by several factors, among them the increasing frequency with which relevant enzyme structures are deposited in the Brookhaven Protein Data Bank, the ready availability and extensiveness of on-line databases, and the conservative use of site-specific mutagenesis to illuminate the catalytic and structural roles of individual amino acids. High-level computational methods are increasingly being utilized to address detailed issues of catalytic function, such as the role of long-range electrostatics or the nature of H-bonding in the active site. Because computational power is exponentially increasing at an ever decreasing cost, computational biochemistry is likely to revolutionize our grasp of enzyme structure–function relationships. Consequently, one can reasonably expect that our understanding of the structures and functions of lipases and esterases, and their exploitations for biomedical or biotechnological purposes, will also increase.

5.06.6 REFERENCES

1. D. M. Quinn and S. R. Feaster, in "Comprehensive Biological Catalysis," ed. M. Sinnott, Academic Press, London, 1998, vol. 1, p. 455.
2. W. P. Jencks, "Catalysis in Chemistry and Enzymology," McGraw-Hill, New York, 1969, p. 508.
3. A. Williams, *Acc. Chem. Res.*, 1989, **22**, 387.
4. R. A. McClelland and L. J. Santry, *Acc. Chem. Res.*, 1983, **16**, 394.
5. R. Verger and G. H. deHaas, *Annu. Rev. Biophys. Bioeng.*, 1976, **5**, 7.
6. E. A. Dennis, *The Enzymes*, 1983, **16**, 307.
7. F. F. Davidson and E. A. Dennis, *J. Mol. Evol.*, 1990, **31**, 228.
8. E. A. Dennis, *J. Biol. Chem.*, 1994, **269**, 13 057.
9. E. J. Ackermann and E. A. Dennis, *Biochim. Biophys. Acta*, 1995, **1259**, 125.
10. B. W. Dijkstra, J. Drenth, and K. H. Kalk, *Nature*, 1981, **289**, 604.
11. B. W. Dijkstra, K. H. Kalk, W. G. J. Hol, and J. Drenth, *J. Mol. Biol.*, 1981, **147**, 97.
12. K. Sekar, A. Kumar, X. Liu, M.-D. Tsai, M. H. Gelb, and M. Sundaralingam, *Acta Crystallogr. D*, in press.
13. S. P. White, D. L. Scott, Z. Otwinowski, M. H. Gelb, and P. B. Sigler, *Science*, 1990, **250**, 1560.
14. D. L. Scott, Z. Otwinowski, M. H. Gelb, and P. B. Sigler, *Science*, 1990, **250**, 1563.
15. D. L. Scott, S. P. White, J. L. Browning, J. J. Rosa, M. H. Gelb, and P. B. Sigler, *Science*, 1991, **254**, 1007.
16. K. Sekar, S. Eswaramoorthy, M. K. Jain, and M. Sundaralingam, *Biochemistry*, 1997, **36**, 14 186.
17. G. M. Carman, R. A. Deems, and E. A. Dennis, *J. Biol. Chem.*, 1995, **270**, 18 711.
18. H. S. Hendrickson and E. A. Dennis, *J. Biol. Chem.*, 1984, **259**, 5734.
19. H. S. Hendrickson and E. A. Dennis, *J. Biol. Chem.*, 1984, **259**, 5740.
20. M. K. Jain, J. Rogers, D. V. Jahagirdar, J. F. Marecek, and F. Ramiriz, *Biochim. Biophys. Acta*, 1986, **860**, 435.
21. M. H. Gelb, M. K. Jain, A. M. Hanel, and O. G. Berg, *Annu. Rev. Biochem.*, 1995, **64**, 653.
22. M. K. Jain, B.-Z. Yu, J. Rogers, G. N. Ranadive, and O. G. Berg, *Biochemistry*, 1991, **30**, 7306.
23. F. Ramirez and M. K. Jain, *Proteins: Struct. Funct., Genet.*, 1991, **9**, 229.
24. A. G. Tomasselli, J. Hui, J. Fisher, H. Zürcher-Neely, I. M. Reardon, E. Oriaku, F. J. Kézdy, and R. L. Heinrikson, *J. Biol. Chem.*, 1989, **264**, 10 041.
25. M. K. Jain, G. Ranadive, B.-Z. Yu, and H. M. Verheij, *Biochemistry*, 1991, **30**, 7330.
26. J. P. Noel, C. A. Bingman, T. Deng, C. M. Dupureur, K. J. Hamilton, R.-T. Jiang, J.-G. Kwak, C. Sekharudu, M. Sundaralingam, and M. D. Tsai, *Biochemistry*, 1991, **30**, 11 801.
27. D. L. Scott, A. M. Mandel, P. B. Sigler, and B. Honig, *Biophys. J.*, 1994, **67**, 493.
28. J. Rogers, B.-Z. Yu, M.-D. Tsai, O. G. Berg, and M. K. Jain, *Biochemistry*, 1998, **37**, 9549.
29. R. Dua, S.-K. Yu, and W. Cho, *J. Biol. Chem.*, 1995, **270**, 263.
30. S. K. Han, E. T. Yoon, D. L. Scott, P. B. Sigler, and W. Cho, *J. Biol. Chem.*, 1997, **272**, 3573.
31. Y. Snitko, R. S. Koduri, S. K. Han, B. J. Molini, D. C. Wilton, M. H. Gelb, and W. Cho, *Biochemistry*, 1998, **37**, in press.
32. C. M. Dupureur, B.-Z. Yu, M. K. Jain, J. P. Noel, T. Deng, Y. Li, I.-J. Byeon, and M.-D. Tsai, *Biochemistry*, 1992, **31**, 6402.
33. X. Liu, H. Zhu, B. Huang, J. Rogers, B.-Z. Yu, A. Kumar, M. K. Jain, M. Sundaralingam, M. D. Tsai, *Biochemistry*, 1995, **34**, 7322.
34. N. Dekker, A. R. Peters, A. J. Slotboom, R. Beelens, R. Kaptein, and G. deHaas, *Biochemistry*, 1991, **30**, 3135.
35. A. R. Peters, N. Dekker, L. van den Berg, R. Boelens, R. Kaptein, A. J. Slotboom, and G. H. de Haas, *Biochemistry*, 1992, **31**, 10 024.
36. L. Yu and E. A. Dennis, *J. Am. Chem. Soc.*, 1992, **114**, 8757.
37. M. M. G. M. Thunnissen, E. Ab, K. H. Kalk, J. Drenth, B. W. Dijkstra, O. P. Kuipers, R. Dijkman, G. H. de Haas, and H. M. Verheij, *Nature*, 1990, **347**, 689.
38. O. P. Kuipers, N. Dekker, H. M. Verheij, and G. H. de Haas, *Biochemistry*, 1990, **29**, 6094.
39. D. L. Ollis, E. Cheah, M. Cygler, B. Dijkstra, F. Frolow, S. M. Franken, M. Harel, S. J. Remington, I. Silman, J. Schrag, J. L. Sussman, K. H. G. Verschueren, and A. Goldman, *Protein Eng.*, 1992, **5**, 197.
40. J. S. Richardson, *Adv. Prot. Chem.*, 1981, **34**, 167.
41. M. Cygler, J. D. Schrag, J. L. Sussman, M. Harel, I. Silman, M. K. Gentry, and B. P. Doctor, *Protein Sci.*, 1993, **2**, 366.
42. M. K. Gentry and B. P. Doctor, in "Enzymes of the Cholinesterase Family," eds. D. M. Quinn, A. S. Balasubramanian, B. P. Doctor, and P. Taylor, Plenum Press, New York, 1995, 493.
43. D. M. Quinn, in "Comprehensive Toxicology," ed. F. P. Guengerich, Elsevier Science, New York, 1997, vol. 3, p 243.
44. D. M. Quinn, *Chem. Rev.*, 1987, **87**, 955.
45. T. L. Rosenberry, *Adv. Enzymol. Relat. Areas Mol. Biol.*, 1975, **43**, 103.
46. H. C. Froede and I. B. Wilson, *The Enzymes*, 1971, **5**, 87.
47. P. Taylor and Z. Radić, *Annu. Rev. Pharmacol. Toxicol.*, 1994, **34**, 281.
48. J. Massoulié, L. Pezzementi, S. Bon, E. Krejci, and F.-M. Vallette, *Prog. Neurobiol.*, 1993, **41**, 31.
49. E. A. Rudd and H. L. Brockman, in "Lipases," eds. B. Borgström and H. L. Brockman, Elsevier, Amsterdam, 1984, 185.
50. L. C. Smith and H. J. Pownall, in "Lipases," eds. B. Borgström and H. L. Brockman, Elsevier, Amsterdam, 1984, 263.
51. J. L. Sussman, M. Harel, F. Frolow, C. Oefner, A. Goldman, L. Toker, and I. Silman, *Science*, 1991, **253**, 872.
52. M. Harel, D. M. Quinn, H. K. Nair, I. Silman, and J. L. Sussman, *J. Am. Chem. Soc.*, 1996, **118**, 2340.
53. L. Polgar, in "Hydrolytic Enzymes," eds. A. Neuberger and K. Brocklehurst, Elsevier, Amsterdam, 1984, 159.
54. M. Harel, J. Sussman, E. Krejci, S. Bon, P. Chanal, J. Massoulié, and I. Silman, *Proc. Natl. Acad. Sci. USA*, 1992, **89**, 10 827.
55. C. B. Millard and C. A. Broomfield, *Biochem. Biophys. Res. Commun.*, 1992, **189**, 1280.
56. I. Silman, E. Krejci, N. Duval, S. Bon, P. Chanal, M. Harel, J. Sussman, and J. Massoulié, in "Multidisciplinary Approaches to Cholinesterase Functions," eds. A. Shafferman and B. Velan, Plenum Press, New York, 1992, 177.
57. D. C. Vellom, Z. Radić, Y. Li, N. A. Pickering, S. Camp, and P. Taylor, *Biochemistry*, 1993, **32**, 12.

58. A. Ordentlich, D. Barak, C. Kronman, Y. Flashner, M. Leitner, Y. Segall, N. Ariel, S. Cohen, B. Velan, and A. Shafferman, *J. Biol. Chem.*, 1993, **268**, 17 083.
59. T. Szegletes, W. D. Mallendar, and T. L. Rosenberry, *Biochemistry*, 1998, **37**, 4206.
60. Z. Radić, R. Duran, D. C. Vellom, Y. Li, C. Cervenansky, and P. Taylor, *J. Biol. Chem.*, 1994, **269**, 11 233.
61. T. L. Rosenberry, C.-R. Rabl, and E. Neumann, *Biochemistry*, 1996, **35**, 685.
62. Z. Radić, D. M. Quinn, D. C. Vellom, S. Camp, and P. Taylor, *J. Biol. Chem.*, 1995, **270**, 20 391.
63. Y. Bourne, P. Taylor, and P. Marchot, *Cell*, 1995, **83**, 503.
64. A. Inoue, T. Kawai, M. Wakita, Y. Iimura, H. Sugimoto, and Y. Kawakami, *J. Med. Chem.*, 1996, **39**, 4460.
65. A. Shafferman, B. Velan, A. Ordentlich, C. Kronman, H. Grosfeld, M. Leitner, Y. Flashner, S. Cohen, D. Barak, and N. Ariel, *EMBO J.*, 1992, **11**, 3561.
66. H. C. Froede and I. B. Wilson, *J. Biol. Chem.*, 1984, **259**, 11 010.
67. H.-J. Nolte, T. L. Rosenberry, and E. Neumann, *Biochemistry*, 1980, **19**, 3705.
68. M. Eigen and G. G. Hammes, *Adv. Enzymol. Relat. Subj. Biochem.*, 1963, **25**, 1.
69. T. H. Lowry and K. S. Richardson, "Mechanism and Theory in Organic Chemistry," Harper & Row, New York, 1987, 3rd edn., p. 994.
70. A. Warshel, G. Naray-Szabo, F. Sussman, and J.-K. Hwang, *Biochemistry*, 1989, **28**, 3629.
71. D. M. Blow, J. J. Birktoft, and B. S. Hartley, *Nature*, 1969, **221**, 337.
72. W. W. Bachovchin and J. D. Roberts, *J. Am. Chem. Soc.*, 1978, **100**, 8041.
73. J. P. Elrod, J. L. Hogg, D. M. Quinn, K. S. Venkatasubban, and R. L. Schowen, *J. Am. Chem. Soc.*, 1980, **102**, 3917.
74. P. A. Frey, S. A. Whitt, and J. B. Tobin, *Science*, 1994, **264**, 1927.
75. C. S. Cassidy, J. Lin, and P. A. Frey, *Biochemistry*, 1997, **36**, 4576.
76. F. Hibbert and J. Emsley, *Adv. Phys. Org. Chem.*, 1990, **26**, 255.
77. M. J. Frisch, G. W. Trucks, H. B. Schlegel, *et al.*, GAUSSIAN 94 (Revision D.4), 1995, Gaussian, Inc., Pittsburgh, PA.
78. J. P. Perdew and Y. Wang, *Phys. Rev. B*, 1986, **33**, 8822.
79. P. S. Herman, Ph.D. Thesis, University of Iowa, 1989.
80. E. E. Smissman and G. R. Parker, *J. Med. Chem.*, 1973, **16**, 23.
81. K. B. Schowen, E. E. Smissman, and W. F. Stephen, *J. Med. Chem.*, 1975, **18**, 292.
82. B. S. Zhorov, N. N. Shestakova, and E. V. Rozengart, *Quant. Struct.-Act. Relat.*, 1991, **10**, 205.
83. Z. Radić, P. D. Kirchoff, D. M. Quinn, J. A. McCammon, and P. Taylor, *J. Biol. Chem.*, 1997, **272**, 23 265.
84. S. Malany, N. Baker, M. Verweyst, R. Medhekar, D. M. Quinn, B. Velan, C. Kronman, and A. Shafferman, *Chem.-Biol. Interact.*, in press.
85. D. R. Ripoll, C. H. Faerman, P. H. Axelsen, I. Silman, and J. L. Sussman, *Proc. Natl. Acad. Sci. USA*, 1993, **90**, 5128.
86. J. Antosiewicz, J. A. McCammon, S. T. Wlodek, and M. K. Gilson, *Biochemistry*, 1995, **34**, 4211.
87. H.-X. Zhou, J. M. Briggs, and J. A. McCammon, *J. Am. Chem. Soc.*, 1996, **118**, 13 069.
88. A. Shafferman, A. Ordentlich, D. Barak, A. Kronman, R. Ber, T. Bino, N. Ariel, R. Osman, and B. Velan, *EMBO J.*, 1994, **13**, 3448.
89. A. Shafferman, A. Ordentlich, D. Barak, C. Kronman, N. Ariel, and B. Velan, in "Structure and Function of Cholinesterases and Related Proteins," eds. B. P. Doctor, D. M. Quinn, R. L. Rotundo, P. Taylor, and M. K. Gentry, Plenum Press, New York, in press.
90. A. Nicholls, K. Sharp, and B. Honig, *Proteins*, 1991, **11**, 281.
91. H. K. Nair, J. Seravalli, T. Arbuckle, and D. M. Quinn, *Biochemistry*, 1994, **33**, 8566.
92. A. Shafferman, A. Ordentlich, D. Barak, C. Kronman, N. Ariel, M. Leitner, Y. Segall, A. Bromberg, S. Reuveny, D. Marcus, T. Bino, A. Lazar, S. Cohen, and B. Velan, in "Enzymes of the Cholinesterase Family," eds. D. M. Quinn, A. S. Balasubramanian, B. P. Doctor, and P. Taylor, Plenum Press, New York, 1995, 189.
93. J. C. Ma and D. A. Dougherty, *Chem. Rev.*, 1997, **97**, 1303.
94. S. Malany, R. S. Sikorski, J. Seravalli, R. Medhekar, D. M. Quinn, Z. Radić, P. Taylor, B. Velan, C. Kronman, and A. Shafferman, in "Structure and Function of Cholinesterases and Related Proteins," eds. B. P. Doctor, D. M. Quinn, R. L. Rotundo, P. Taylor, and M. K. Gentry, Plenum Press, New York, in press.
95. F. K. Winkler, A. D'Arcy, and W. Hunziker, *Nature*, 1990, **343**, 771.
96. L. Brady, A. M. Brzozowski, Z. S. Derewenda, E. Dodson, G. Dodson, S. Tolley, J. P. Turkenburg, L. Christiansen, B. Huge-Jensen, L. Norskov, L. Thim, and U. Menge, *Nature*, 1990, **343**, 767.
97. J. C.-H. Chen, L. J. W. Miercke, J. Krucinski, J. R. Starr, G. Saenz, X. Wang, C. A. Spilburg, L. G Lange, J. L. Ellsworth, and R. M. Stroud, *Biochemistry*, 1998, **37**, 5107.
98. X. Wang, C.-S. Wang, J. Tang, F. Dyda, and X. C. Zhang, *Structure*, 1997, **5**, 1209.
99. A. Hjorth, F. Carrière, C. Cudrey, H. Wöldike, E. Boel, D. M. Lawson, F. Ferrato, C. Cambillau, G. G. Dodson, L. Thim, and R. Verger, *Biochemistry*, 1993, **32**, 4702.
100. C. Martinez, A. Nicolas, H. van Tilbeurgh, M.-P. Egloff, C. Cudrey, R. Verger, and C. Cambillau, *Biochemistry*, 1994, **33**, 83.
101. J. S. Stout, L. D. Sutton, and D. M. Quinn, *Biochim. Biophys. Acta*, 1985, **837**, 6.
102. L. D. Sutton, J. S. Stout, and D. M. Quinn, *J. Am. Chem. Soc.*, 1990, **112**, 8398.
103. M.-P. Egloff, F. Marguet, G. Buono, R. Verger, C. Cambillau, and H. van Tilbeurgh, *Biochemistry*, 1995, **34**, 2751.
104. Y. Bourne, C. Martinez, B. Kerfelic, D. Lombardo, C. Chapus, and C. Cambillau, *J. Mol. Biol.*, 1994, **238**, 709.
105. A. Nicolas, M. Egmond, C. T. Verrips, J. de Vlieg, S. Longhi, C. Cambillau, and C. Martinez, *Biochemistry*, 1996, **35**, 398.
106. P. Bryan, M. W. Pantoliano, S. G. Quill, H. Hsiao, and T. Poulos, *Proc. Natl. Acad. Sci. USA*, 1986, **83**, 3743.
107. J. A. Wells, B. C. Cunningham, T. P. Graycar, and D. A. Estell, *Phil. Trans. R. Soc. London A*, 1986, **317**, 415.
108. Z. S. Derewenda, in "Advances in Protein Chemistry," ed. V. N. Schumaker, Academic Press, San Diego, CA, 1994, vol. 45, p 1.
109. H. van Tilbeurgh, M.-P. Egloff, C. Martinez, N. Rugani, R. Verger, and C. Cambillau, *Nature*, 1993, **362**, 814.
110. H. van Tilbeurgh, A. Roussel, J. M. Lalouel, and C. Cambillau, *J. Biol. Chem.*, 1994, **269**, 4626.
111. S. R. Feaster, D. M. Quinn, and B. L. Barnett, *Protein Sci.*, 1997, **6**, 73.
112. K. K. Kim, H. K. Song, D. H. Shin, K. Y. Hwang, and S. W. Suh, *Structure*, 1997, **5**, 173.

113. P. Grochulski, F. Bouthillier, R. J. Kazlauskas, A. N. Serreqi, J. D. Schrag, E. Ziomek, and M. Cygler, *Biochemistry*, 1994, **33**, 3494.
114. J. A. Contreras, M. Karlsson, T. Østerlund, H. Laurell, A. Svensson, and C. Holm, *J. Biol. Chem.*, 1996, **271**, 31 426.
115. D. M. Lawson, U. Derewenda, L. Serre, S. Ferri, R. Szittner, Y. Wei, E. A. Meighen, and Z. S. Derewenda, *Biochemistry*, 1994, **33**, 9382.
116. S. M. Franken, H. J. Rozeboom, K. H. Kalk, and B. W. Dijkstra, *EMBO J.*, 1991, **10**, 1297.
117. J. Li, R. Szittner, Z. S. Derewenda, and E. A. Meighen, *Biochemistry*, 1996, **35**, 9967.
118. E. Cheah, C. Austin, G. W. Ashley, and D. Ollis, *Protein Eng.*, 1993, **6**, 575.
119. E. Cheah, G. W. Ashley, J. Gary, and D. Ollis, *Proteins*, 1993, **16**, 64.
120. M. E. M. Noble, A. Cleasby, L. N. Johnson, M. R. Egmond, and L. G. J. Frenken, *FEBS Lett.*, 1993, **331**, 123.
121. Y. Wei, J. L. Schottel, U. Derewenda, L. Swenson, S. Patkar, and Z. S. Derewenda, *Nat. Struct. Biol.*, 1995, **2**, 218.
122. P. Lohse, P. Lohse, S. Chahrokh-Zadeh, and D. Seidel, *J. Lipid Res.*, 1997, **38**, 880.
123. P. Lohse, S. Chahrokh-Zadeh, P. Lohse, and D. Seidel, *J. Lipid Res.*, 1997, **38**, 892.

5.07

Chemistry and Enzymology of Phosphatases

THEODORE S. WIDLANSKI and WILLIAM TAYLOR
Indiana University, Bloomington, IN, USA

5.07.1 INTRODUCTION

This chapter focuses on the mechanisms of enzyme-catalyzed phosphate ester hydrolysis, with an emphasis on developments in our understanding of these enzymes and the techniques used to study them. Phosphatases form a large class of structurally and mechanistically disparate enzymes that catalyze the hydrolysis of phosphate monoesters (Equation (1)). Hundreds of phosphatases have been identified and the total number of human phosphatases has been estimated to exceed one thousand, or greater than 1% of the genome.[1,2] Given the large number of phosphatases that have been studied, these efforts represent, in a microcosm, many of the important problems that mechanistic enzymologists have struggled with over the years. The chapter is therefore written to function as an appropriate resource for those teaching or wishing to learn about the enzymology and chemistry of phosphatases. The authors have gone to some pains to present a cohesive treatment of the mechanistic tools that have been employed in the study of phosphatases, in addition to presenting recent developments at the time of writing. There is an introductory section describing global relationships among differing types of phosphatases. However, this chapter will not discuss structural questions in great detail, since these concerns have been reviewed elsewhere.[3–6]

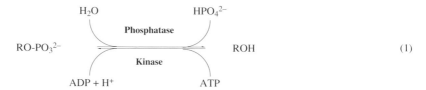

$$RO\text{-}PO_3^{2-} \qquad\qquad\qquad ROH \qquad\qquad (1)$$

Phosphate esters and the enzymes that metabolize them play a remarkable number of important and diverse roles in biochemistry. RNA/DNA repair and synthesis, signal transduction, phospholipid metabolism, energy storage, transcription control, protein activation/deactivation, cell transformation, and a host of other biological activities all rely on the enzyme-catalyzed cleavage of phosphate esters. In addition, phosphatases play an essential role in the virulence and/or life cycle of a number of pathogenic organisms such as *Yersinia* (*Y. Pestis* is the causative agent of the bubonic plague), *Salmonella*, *Variola* viruses (smallpox), and *Vaccinia* viruses.[7] Our emerging understanding of the role of phosphatases as important biological regulatory agents has provided an impetus for the isolation and characterization of many new enzymes. Crystal structures for a number of molecules, representative of most classes of phosphatases, are now available.[3–6,8]

5.07.2   CLASSIFICATION OF PHOSPHATASES

Phosphatases are generally classified based on substrate specificity and/or mechanism of action. A third, historically popular classification scheme is based (sometimes erroneously) on pH optimum for catalytic activity; ergo the classification of these enzymes as "acid" or "alkaline" phosphatases. Phosphatase substrates range from small phosphorylated metabolites such as phospholipids, carbohydrate, and cyclitol phosphates, up to large phosphorylated proteins such as glycogen synthase. Based on their substrate specificity, phosphatases can be divided into three major groups (Figure 1). (i) Nonspecific—a mechanistically broad class of enzymes including both alkaline and acid phosphatases. These enzymes generally catalyze the hydrolysis of almost any unhindered phosphate ester and usually employ an active site nucleophile. (ii) Phosphoprotein specific—enzymes that prefer phosphoproteins or phosphopeptides as substrates. The largest subclasses to date are the protein serine/threonine phosphatases (PPases), protein tyrosine phosphatases (PTPases), and dual-specificity phosphatases (capable of hydrolyzing both phosphotyrosine and phosphoserine/threonine-containing substrates). (iii) Small molecule specific—enzymes that are specific for one (or a related group of structurally similar) substrate(s).

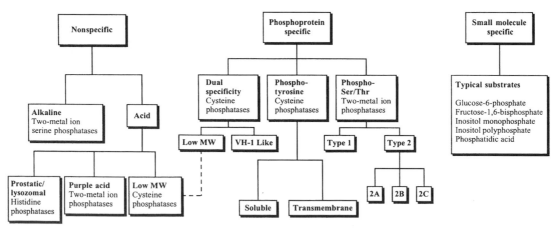

Figure 1   Classification of phosphatases according to specificity and mechanistic type.

As new phosphatases are discovered and existing enzymes further characterized, this classification scheme will have to be amended. For example, because the catalytic mechanism employed by phosphatidic acid phosphohydrolase is unknown, there is not enough information to permit its classification. The presence of phosphohistidine, phosphoarginine, and phospholysine residues within proteins suggests that phosphatases (as well as kinases) specific for these transformations exist, but relatively little is known about these enzymes.[9,10] Finally there is still some uncertainty about the true nature of the *in vivo* substrates for a number of these enzymes, and classification based on *in vitro* substrate specificity may sometimes be erroneous.

Classification of phosphatases according to notional mechanistic criteria leads to two major groupings: enzymes that utilize an active site nucleophile as the initial phosphoryl group acceptor; and enzymes that transfer the phosphoryl group immediately to water. Virtually all known non-specific enzymes (purple acid phosphatase being a notable exception) catalyze phosphoryl group transfer to an active site nucleophile. The identity of the nucleophile may be used to group these enzymes into separate classes such as histidine, cysteine, or serine phosphatases. A second criterion for subgrouping considers whether the enzyme utilizes active site metal ions. Many phosphatases utilize a two-metal ion dyad to bind phosphate esters and catalyze their subsequent hydrolysis. Indeed, this metal ion motif seems to be a ubiquitous theme in phosphoryl group transfer biochemistry, as it is also employed by phosphodiesterases (e.g., nucleases) and phosphotriesterases. Three-metal ion phosphodiesterases are also quite common.

Two-metal ion phosphatases display a great deal of heterogeneity with respect to structure, sequence, metal identity, metal ligation, active site residues, and even mechanism. For example, alkaline phosphatase utilizes a third active site metal (an Mg^{II} that does not directly contact the phosphate ester) as well as a nucleophilic serine residue. In all other cases that have been examined, the two-metal ion phosphatases appear to catalyze direct transfer of the phosphoryl group to water. The occurrence of one or more Zn^{II} ions in the active site of a two-metal ion phosphatase is fairly frequent. However, many other metal ions such as Mn^{II}, Fe^{II}, and Fe^{III} are found in the active sites of these enzymes.[11] It is also clear that other divalent ions, such as Co^{II} in the case of purple acid phosphatase, will support catalysis.[12] For many phosphatases, the identity of the biologically relevant metal ions has not yet been firmly established.

5.07.3 MECHANISTIC CONCERNS

Perhaps the single most important advance in our ability to understand phosphate ester cleavage and the mechanisms used by phosphatases was the development (in the 1970s and 1980s) of appropriate stereochemical probes of this reaction. This advance, in conjunction with cloning, site-directed mutagenesis, and X-ray crystallographic refinements, revealed many of the essential details of phosphatase chemistry. Despite this wealth of knowledge, the full mechanism of catalysis is not known unless there is an understanding of both the nature of transition states and the pathway leading to transition state formation. Moreover, if an enzyme's structure and mechanism of action are understood, the design of specific tight-binding inactivators should be facilitated. The use of sensitive isotope effects to probe the transition states of phosphoryl group transfer,[13] the development of new design motifs for phosphatase inhibitors,[14] and the emergence of new methods for trapping and observing phosphoenzyme intermediates are interesting developments in chemically (and mechanistically) oriented research on phosphatases. However, it is clear that isotope effects, though informative, may sometimes provide ambiguous answers about the nature of transition states of enzyme-catalyzed phosphoryl group transfer reactions. In addition, the design of phosphatase inhibitors, although rapidly developing, is still in its infancy. This chapter will focus on these new developments in research, as well as summarizing pre-existing methodology.

For many hydrolytic enzymes such as proteases, phosphatases, glycosidases, and sulfatases, it is possible to compare the enzyme-catalyzed reaction with a variety of nonenzymatic processes. Such comparisons may underscore the fundamental mechanistic requirements for a specific reaction, or lead to a better understanding of the special properties and advantages of enzyme catalysis. Any detailed discussion of the chemistry of phosphatases must therefore start with a review of the nonenzymatic reaction.

5.07.3.1 Nonenzymatic Phosphate Ester Hydrolysis

While carbon and sulfate esters are relatively reactive molecules, both phosphate diesters and monoesters (particularly dianions) display a high degree of kinetic stability.[15] The stability of the

DNA phosphodiester backbone is of critical importance for the transmission of genetic information to new generations. Nevertheless, phosphate mono-, di-, and triesters are all susceptible to hydrolysis under a wide variety of conditions, including specific or general acid/base-catalyzed hydrolysis, as well as metal ion-catalyzed hydrolysis.

A continuum of mechanistic possibilities for a nucleophilic substitution reaction at phosphorus may be adequately shown in a More–O'Ferrall–Jencks diagram (Figure 2). It is clear that, as in carbon substitution reactions, there are two mechanistic extremes: a dissociative pathway via a metaphosphate intermediate (akin to an S_N1 substitution via a carbocation), or an associative pathway. However, unlike an S_N2 reaction at carbon, the associative cleavage of a phosphate ester may in principle proceed either via a concerted displacement of the leaving group, or via a stepwise, nonconcerted reaction with the intermediacy of a pentavalent phosphorane.

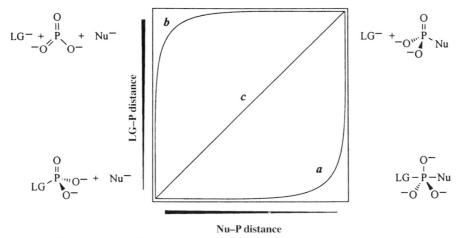

Figure 2 More–O'Ferrall–Jencks diagram for nucleophilic cleavage of phosphate monoester dianions. Path *a* is fully associative, while path *b* is fully dissociative. Path *c* is intermediary between these two extremes. Nu is a nucleophile, LG is a leaving group.

The stability of a number of known organophosphorus(V) compounds[16] may lead one to consider the stepwise associative pathway as an attractive possibility. However, a consideration of the kinetic requirements for this pathway immediately casts doubt on its viability. Consider the cleavage of *p*-nitrophenyl phosphate under basic conditions. A notional free energy profile for the stepwise phosphorane pathway is shown in Figure 3. *p*-Nitrophenoxide is a much better leaving group than hydroxide. Therefore, the intermediate phosphorane should partition almost entirely toward product, rather than return to starting material. This leads to irreversible formation of the phosphorane, a situation that is most easily accounted for if formation of the phosphorus(V) species is the rate-determining step for the reaction. The base-catalyzed cleavage of phosphate aryl esters, however, displays high β-leaving group values,[17] which would only occur if the leaving group alcohol departed during the rate-determining step. On this basis it appears that the stepwise associative pathway for cleavage of aryl phosphate dianions is unlikely.

A more promising alternative is the metaphosphate pathway originally suggested by Westheimer and Bunton.[18,19] This mechanism adequately accounts for the high β-leaving group values often observed during this reaction, and predicts that phosphoryl group transfer should show little discrimination between various nucleophiles. Indeed, β-nucleophile values for these reactions are often quite low,[20] and mixtures of simple alcohols and water give statistical product distributions.[21,22] The three-phase test devised by Rebeck also provides indirect evidence that metaphosphate may be an intermediate in a variety of reactions.[23] However, there are some inconsistencies between the observed data and that predicted for a free metaphosphate intermediate. For example, there is some discrimination between various types of alcohols, particularly hindered ones.[17]

With the advent of chiral phosphorus technology (Section 5.07.3.2.2(ii)) developed independently by Knowles and Lowe, a more definitive test of the nature of this reaction became available.[24,25] The simple prediction was that phosphate ester cleavage via a metaphosphate intermediate, which is planar and symmetrical, would lead to racemization at phosphorus. An in-line displacement, whether stepwise or concerted, would lead to inversion of stereochemistry. Somewhat surprisingly, the reaction of chiral phosphate esters with simple unhindered alcohols proceeds with inversion.[26,27] This result is inconsistent with the intermediacy of free metaphosphate in solution. At the same time however, it was clear that a stepwise phosphorane mechanism was also unlikely. A resolution

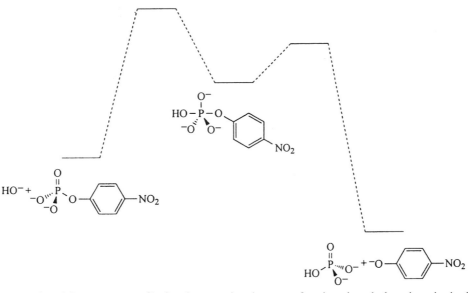

Figure 3 Notional free energy profile for the stepwise cleavage of *p*-nitrophenyl phosphate by hydroxide.

of this dilemma came with the suggestion that the reaction proceeds via the preassociation of a nucleophile.[28,29] The transition state of this reaction is "exploded" with very little bond making and a great deal of bond breaking (Figure 4). The transition state is therefore metaphosphate-like. However, the incoming nucleophile and departing leaving group must still be *anti* to each other, thus explaining the observed stereochemistry. With such a mechanism, very weak nucleophiles might well compete poorly with much better ones. Indeed, this controversy was resolved very nicely when it was shown that cleavage of chiral phosphate esters in *t*-butanol leads to racemic products.[30] Evidently, *t*-butanol in acetonitrile is too poor a nucleophile to follow the preassociative pathway.

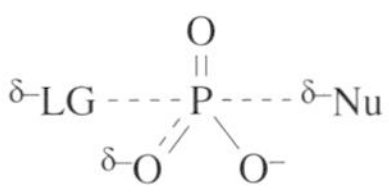

$$\delta\text{-LG}----\overset{\displaystyle\overset{O}{\|}}{\underset{\delta\text{-}O \quad O^-}{P}}----\delta\text{-Nu}$$

Figure 4 Proposed transition state structure for a preassociative nucleophilic attack on a phosphate monoester dianion. Note the high degree of bond breaking and low degree of bond making.

The above results are further supported by studies that have determined the heavy atom isotope effects associated with phosphoryl transfer. These experiments, as well as results for both enzyme-catalyzed and solution hydrolysis of phosphate esters, are described in Section 5.07.3.2.3.

5.07.3.2 Enzyme-catalyzed Phosphate Ester Hydrolysis

It is convenient to break up enzyme-catalyzed reaction mechanisms into three parts: kinetic, catalytic, and chemical. The kinetic mechanism of the enzymatic process covers reaction rates, order of steps, and the relative timing of those steps. Knowledge of the catalytic mechanism requires that we identify which protein functional groups are engaged in the catalytic process and what their roles are. The chemical mechanism defines the electron flow during the reaction and the nature of enzyme-bound intermediates and transition states.

For phosphatases, there are three simple questions that provide a wealth of information about kinetic, chemical, and catalytic mechanism. First, is a covalent intermediate formed during catalysis? Second, what is the nature of such an intermediate? Third, does the enzyme utilize metals for catalysis? Because the vast majority of phosphatases fall into recognizable mechanistic groups (*vide supra*), answering these three questions allows us to classify a phosphatase and come up with a reasonable hypothesis for its mode of action. However, unambiguously determining the existence and nature of putative phosphoenzyme intermediates is often a remarkably taxing process. As a result, such efforts occupy a central position in mechanistic studies of phosphatases.

5.07.3.2.1 Nature of phosphoenzyme intermediates

There are three forms of covalent enzyme intermediates used by phosphatases: phosphoserine, phosphohistidine, and phosphocysteine (Figure 5). These are not the only phosphorylated amino acids that occur in nature. Phosphoarginine, phospholysine, and mixed anhydrides with the carboxy group of various amino acids (aminoacyl adenylates for example) are all species that appear either as reactive intermediates or as constituents of modified proteins. In addition, acyl phosphates (e.g., acetyl phosphate) and guanidino phosphates (e.g., creatine phosphate) are well-known phosphagens. To date, there is no definitive evidence for the intermediacy of any of these latter species in enzyme-catalyzed phosphate ester hydrolysis. However, given the precedent from other biochemical phosphoryl group transfer reactions, it would not be surprising if such phosphatases were discovered.

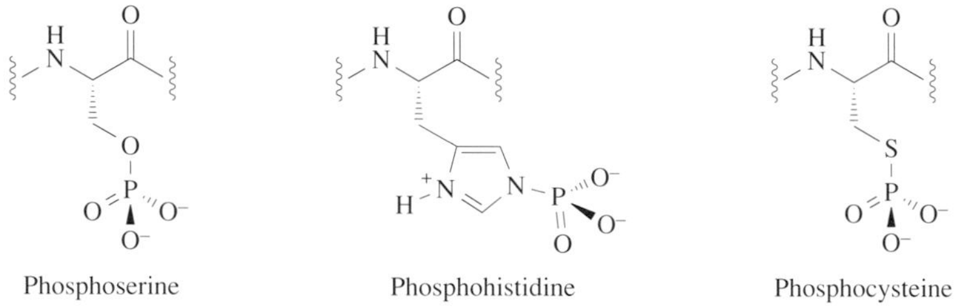

Phosphoserine   Phosphohistidine   Phosphocysteine

Figure 5   Typical phosphoenzyme intermediates.

Enzymes utilizing a phosphoserine intermediate are best exemplified by alkaline phosphatases. Phosphohistidine-containing enzymes include prostatic acid phosphatase, the highly homologous lysozomal acid phosphatase, and glucose-6-phosphate phosphatase. Cysteine phosphatases include PTPases of both high and low molecular weight forms, as well as dual-specificity phosphoprotein phosphatases. These cysteine phosphatases have varying degrees of sequence homology, although all contain a phosphate binding loop with an essential cysteine residue. Interestingly, one cannot convert a cysteine phosphatase to a serine phosphatase. Despite the close structural homology between these two amino acids, cysteine phosphatases lose all catalytic activity upon mutation of the active site cysteine to a serine.[31,32] Clearly, the nature of the active site nucleophile is crucial for these enzymes. In contrast, alkaline phosphatase retains a substantial amount of activity upon mutagenesis of the active site serine to an alanine.[33] This is one of a very few examples of a mutated enzyme that can use a water molecule in place of the active site nucleophile.

5.07.3.2.2 Detection of phosphoenzyme intermediates

There are a number of direct and indirect methods that can be used to test for the formation of a phosphoenzyme intermediate during enzyme-catalyzed phosphate ester hydrolysis. These methods include: direct observation of the phosphoenzyme; stereochemical probes; ^{18}O wash-in into inorganic phosphate; transfer of the phosphoryl group to other nucleophiles (such as low molecular weight alcohols); and a variety of kinetic criteria, such as burst kinetics or the absence of leaving group effects. Of these tests, the direct observation of a kinetically competent phosphoenzyme and clear inversion (no phosphoenzyme) or retention (phosphoenzyme) of stereochemistry at phosphorus are the only unambiguous results. Negative results, such as the inability to trap a phosphoenzyme, cannot be construed as definitive proof that there is no phosphoenzyme intermediate. Indeed, even positive results must be interpreted cautiously. The interpretation of these results depends largely on the relative magnitude of the microscopic rate constants shown in Scheme 1.

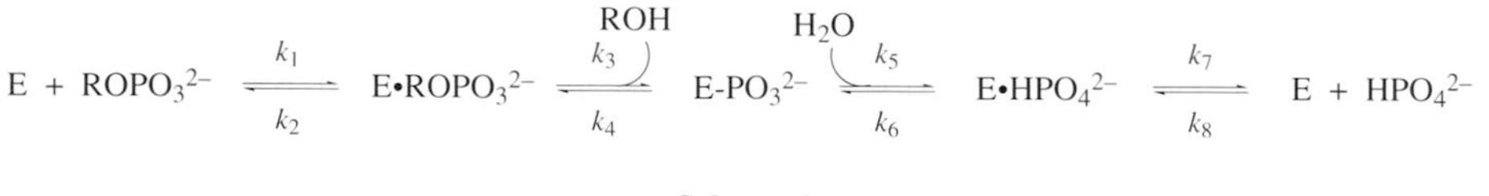

Scheme 1

The importance of understanding this kinetic scheme cannot be overstated, since the relative magnitudes of the rate constants will dictate which experimental approaches are feasible. The kinetic constant k_1 represents the rate of substrate binding and k_2 is the rate of substrate release, k_3 is the rate of enzyme phosphorylation from substrate, and k_4 is the rate of phosphoryl transfer to product alcohol. Under normal conditions, when the concentration of product alcohol is low, the phosphoenzyme species is trapped by water and phosphorylation of the enzyme is effectively irreversible. k_5 is the rate of phosphoryl group transfer to water, and for many phosphatases is the rate-determining step for the overall transformation. k_7 is the rate of release of inorganic phosphate.

(i) Direct observation of phosphoenzyme intermediates

In favorable cases it may be possible to isolate and/or utilize spectroscopic techniques to directly observe phosphoenzyme intermediates. This requires the rate of enzyme phosphorylation (k_3) to be substantially faster than dephosphorylation (k_5). In addition, k_3 is usually an irreversible step, because the concentration of ROH is low and k_4[E-P][ROH] is negligible. Under these conditions, the phosphoenzyme intermediate is the major enzyme species in solution and is therefore amenable to detection and isolation. Under such conditions, one can expect to observe burst kinetics and no leaving group effect.

Quench experiments are usually carried out by mixing the enzyme (at sufficiently high concentration to ensure easy detection of any phosphoenzyme that is formed) with sufficient substrate to assure saturation, and then quenching the enzymatic reaction with a chemical agent (i.e., acid or base) or some other denaturant (such as sodium dodecyl sulfate). The choice of quenching agent depends on the particulars of the enzymes involved, and on the chemical nature of the phosphoenzyme. For example, phosphocysteine is relatively acid stable, whereas phosphohistidine is acid labile. When simply isolating a phosphoenzyme, it is not necessary to use rapid quench techniques, providing that the enzyme turns over slowly enough to ensure that the substrate concentration is still high when the quench is performed.

Phosphoenzyme intermediates have been identified and quantitated by use of ^{32}P labeled substrates followed by electrophoresis,[34] ^{31}P NMR,[35] and mass spectrometry.[36] The use of radioactive substrates is by far the most common method for identification of a phosphoenzyme. The major drawbacks of this method are that information about the nature of the phosphoenzyme linkage is not accessible without proteolysis and sequencing of the isolated peptide. This requires that the phosphoenzyme be stable to somewhat lengthy handling. Under these conditions, one must exercise due caution with respect to the possibility of intramolecular migration of the phosphoryl group to a different amino acid side chain. In addition, synthesis of ^{32}P labeled substrates may not be routine, especially if the substrate is a complex one such as a phosphopeptide. In such cases, identification of the phosphoenzyme by ^{31}P NMR has proven to be a useful tool. This method has the advantage that chemical shift information can give a fairly good idea of the nature of the phosphoenzyme intermediate without resorting to proteolysis. In the future, this may turn out to be a particularly useful method for studying labile phosphoenzyme intermediates.

A most promising development in this area is the emergence of mass spectrometry techniques. As these methods continue to improve, it seems likely that mass spectrometry will become a routine technique for studying covalent enzyme–substrate adducts.[37] Mass spectrometry has already been used to detect a β-lactamase acyl enzyme adduct and (coupled with rapid quench techniques) measure the rate of formation and decay of this intermediate.[38] Matrix assisted laser desorption (MALDI) time of flight mass spectrometry has been used to demonstrate the formation of a phosphoenzyme adduct for the dual-specificity phosphatase Stp1.[36] Typical results are shown in Figure 6.

(ii) Stereochemistry

Simple phosphate esters can be rendered chiral by use of the three oxygen isotopes at the peripheral oxygen positions, or by replacing one of these oxygens with sulfur (Figure 7). With such an ester in hand, it is possible to evaluate the stereochemical course of a P—O bond cleavage reaction. In dissecting the stereochemistry of phosphoryl group transfer reactions, it is important to make the distinction between the stereochemical course of an individual transfer and the overall stereochemical course of the reaction. In principle, in an individual transfer event, the addition of a nucleophile to chiral phosphoryl groups can give products with inversion, racemization, or retention of stereochemistry. Inversion would result from a direct in-line displacement of the leaving group.

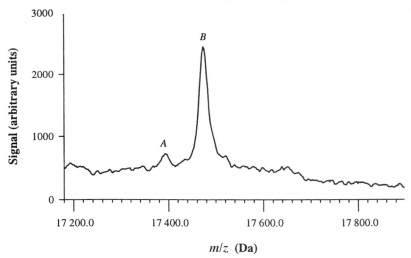

Figure 6   MALDI time-of-flight mass spectrum of the dual-specificity phosphatase Stp1 after acid quenching of the enzyme under steady-state conditions. Peak *A* is the native enzyme. Peak *B* is the trapped phosphoenzyme. The measured mass difference is 79.1 ± 1.6 Da.

Racemization would imply that the nucleophile has equal access to either side of the phosphoryl group or that a rapidly rotating, long-lived metaphosphate intermediate is formed. This situation is unlikely to occur in enzyme-catalyzed reactions. Indeed, no phosphatase has ever been reported to catalyze phosphate ester hydrolysis with concomitant racemization at phosphorus.

$$\left[\begin{array}{c} {}^{18}O \\ \| \\ {}^{17}O^{,,,}P^{\diagup}OR \\ {}^{16}O \end{array}\right]^{2-} \qquad \left[\begin{array}{c} {}^{18}O \\ \| \\ S^{,,,}P^{\diagup}OR \\ {}^{16}O \end{array}\right]^{2-}$$

Figure 7   The two types of substrates used for chiral phosphate analyses.

The last possibility is an "adjacent" mechanism, in which the leaving group is not positioned to adopt an apical position in a trigonal bipyramidal intermediate. In this case, there must first be a pseudorotation to exchange apical and equatorial positions so that collapse of the phosphorane intermediate leads to expulsion of the appropriate leaving group (Scheme 2). The overall process results in retention of stereochemistry at phosphorus. While this mechanism is required for the hydrolysis of certain types of cyclic phosphate esters, there is no evidence that phosphatases employ adjacent displacement mechanisms. Indeed, there is abundant evidence from X-ray crystallographic studies that the enzymatic nucleophile is always *anti* to the departing group. Accordingly, the reasonable assumption is made that for phosphatases, all single displacements occur with inversion.

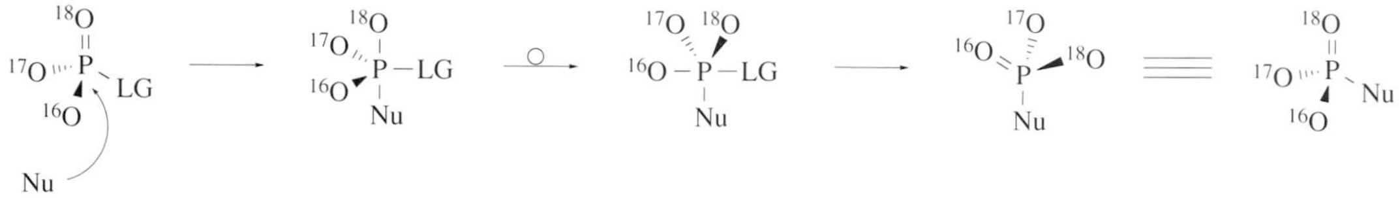

Scheme 2

Given that all single enzyme-catalyzed phosphoryl transfers take place with inversion, the overall stereochemical course of the phosphatase reaction is simply a function of how many transfers there are. If there are an odd number, inversion will be observed. An even number of displacements results in retention. It is straightforward to see how this methodology differentiates between enzymes that use a covalent intermediate (two transfers, or net retention of stereochemistry) and those that do not (a single transfer, or inversion).

There are a number of related methods for performing chiral phosphorus analyses. In a typical method, chiral phosphate esters are synthesized from a chiral phosphochloridate, followed by deblocking (Scheme 3). The chloridate is obtained by reaction of ephedrine with $POCl_3$, followed by separation of the two diastereomers by flash chromatography. To analyze the stereochemistry of phosphoryl group transfer using [${}^{16}O$, ${}^{17}O$, ${}^{18}O$]phosphate esters, the enzyme must catalyze a transphosphorylation reaction. Usually the reaction is run in the presence of a chiral diol acceptor such as butanediol. The product ester is cyclized, methylated, and analyzed by ^{31}P NMR. In those cases where the enzyme will not catalyze a transphosphorylation event, a chiral thiophosphate ester or anhydride must be used.

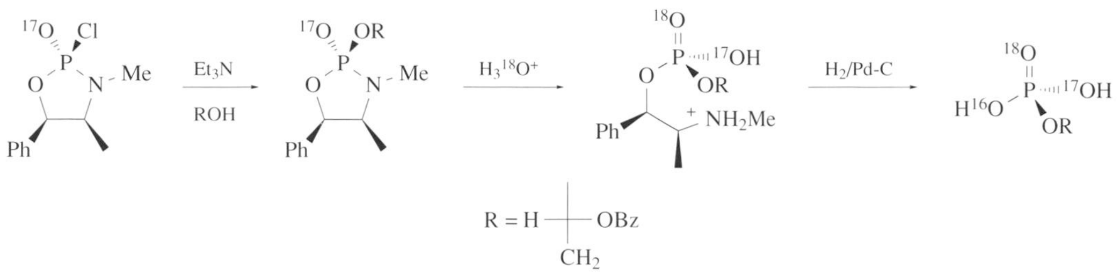

Scheme 3

Unfortunately, chiral phosphorus analyses cannot be routinely performed in most laboratories, and are generally unavailable for enzymes that use relatively complex substrates (such as phosphopeptides) owing to the difficulty of synthesizing the requisite labeled substrates. Nevertheless, this method has played an important role in defining some particularly elusive mechanisms. For example, based on kinetic criteria, it was thought that purple acid phosphatase employed an active site nucleophile.[39,40] However, chiral phosphorus analysis revealed that the stereochemical course of phosphoryl group transfer is clean inversion, implying a direct transfer to water.[41] The X-ray structure of this enzyme shows that it is a prototypical member of the two-metal ion phosphatases, all of which are thought to utilize water as the phosphoryl group acceptor.

(iii) Kinetic studies

Any putative enzyme mechanism must account for the observed kinetics of the enzyme-catalyzed reaction. As such, kinetic studies can be very useful for discriminating between mechanistic alternatives. It should always be understood that kinetic data cannot prove a mechanism, only disprove it, or lead us to new mechanistic hypotheses. In many cases it is far simpler to carry out routine kinetic analyses than to examine the stereochemistry of hydrolysis or to isolate a phosphoenzyme intermediate. For phosphatases that form phosphoenzyme intermediates, dephosphorylation of the enzyme is generally the rate-limiting step (Scheme 1). Under these conditions, the rate of enzyme turnover (k_{cat}) will be insensitive to the nature of the leaving group (Brønsted plots give β-leaving group values close to zero). This type of experiment works well for nonspecific phosphatases or phosphatases that handle simple aromatic substrates (i.e., PTPases). Another simple test for a phosphoenzyme intermediate is transfer of the phosphoryl group to an appropriate nucleophilic acceptor. If dephosphorylation of the enzyme is rate limiting, then the addition of a potent nucleophile may also enhance the rate of enzyme turnover. Indeed, it is possible to gain information about the dephosphorylation step by varying the nature of the nucleophile and observing changes in the reaction rate.[42] Finally, the enzyme should catalyze the incorporation of ^{18}O-labeled water into inorganic phosphate by a simple reversal of the dephosphorylation step.

If bond breaking to the leaving group is significantly faster than a subsequent step, the enzyme will show burst kinetics. The plot of product alcohol release vs. time will not extrapolate to zero concentration at zero time, and the difference will be stoichiometric with the amount of enzyme used. Such observations are usually consistent with the formation of a phosphoenzyme intermediate whose breakdown is rate limiting. There are therefore four relatively simple tests for the formation of a phosphoenzyme intermediate: small β-leaving group values (with nonspecific enzymes), trapping of the intermediate by nucleophiles, incorporation of ^{18}O-labeled water into inorganic phosphate, and burst kinetics.

Even if breakdown of a phosphoenzyme intermediate is rate limiting, it is possible to probe earlier steps in the reaction pathway. This can be done by studying the enzyme under nonsaturating or so-called V/K conditions. Under these conditions, rate measurements give information up to and including the first irreversible step. Since loss of the alcohol leaving group is generally irreversible, these measurements can give important information about steps that occur up to and including enzyme phosphorylation. Examples of this include a number of elegant studies on alkaline phosphatase. For example, it has been determined that alkaline phosphatase-catalyzed hydrolysis of a series of aryl esters under V/K conditions gives small negative β-values.[43] There are two reasonable interpretations for this observation. First, the leaving group may be protonated or bound to an active metal ion, in which case the leaving group bears little negative charge in the transition state. Alternatively, some step prior to loss of the leaving group may be partly or even fully irreversible.

To address these possibilities, Jenck's group studied the alkaline phosphatase-catalyzed hydrolysis of a series of N-phosphorylated pyridinium salts. These compounds already bear a positive charge, and should therefore neither become protonated nor coordinate to a metal ion prior to P—O cleavage. Somewhat surprisingly, these compounds also give β-values close to zero, suggesting that there is an irreversible enzymatic step prior to cleavage of the P—N bond.[44] In such a case, how does one go about probing the nature of P—O cleavage? One interesting approach is that provided by Herschlag's group, who measured the alkaline phosphatase-catalyzed hydrolysis of a series of substituted aryl thiophosphate esters.[45] These compounds are such poor substrates for the enzyme that the chemical step of enzyme phosphorylation becomes rate limiting and all prior steps are reversible. With these substrates, a β-value of close to -1 is observed, indicating that stabilization of the charge on the leaving group is important in spite of the probable ligation of the leaving group to an active site zinc ion.

The case is somewhat different for cysteine phosphatases. The rate-limiting step for these enzymes is dephosphorylation of the phosphocysteine intermediate. As expected, k_{cat} values for the hydrolysis of a series of substituted aromatic phosphate esters catalyzed by small tyrosine phosphatase (Stp1) gives a β-leaving group value close to zero.[46] The β-leaving group value for V/K, however, is about -0.4. In this case, protonation of the leaving group by the active site acid diminishes the need to stabilize the charge on the leaving group, resulting in a modest β value. Mutagenesis of the active site general acid results in a mutant enzyme that gives biphasic Brønsted plots with β-values close to -1 for very good leaving groups.

5.07.3.2.3 *Isotope effects on phosphate ester hydrolysis*

The transition states for hydrolysis of phosphate esters may, in principle, be dissociative and metaphosphate-like; associative and concerted ("S_N2 like"); or associative via the intermediacy of a pentacovalent phosphorane. These possibilities may be examined by measuring the heavy atom isotope effects associated with phosphate ester bond cleavage. The results obtained in this way dovetail nicely with data from classical chemical investigations (Sections 5.07.3.1 and 5.07.3.2), giving a clear picture of the solution reaction and a foundation for interpreting the enzyme-catalyzed reaction. This section gives an overview of how isotope effects on phosphate monoester hydrolysis are measured, summarizes the results for nonenzyme-catalyzed reactions, and compares these results to those obtained with several phosphatases.

(i) *Determination of isotope effects*

Compared to a deuterium isotope effect, the primary isotope effect resulting from the replacement of ^{16}O with ^{18}O will always be very modest, owing to the small difference in mass between these atoms. Secondary isotope effects arising from this substitution are even smaller, and until recently measurement of these effects on phosphate ester hydrolysis was effectively precluded.

Isotope ratio mass spectrometry, which can detect isotope ratios with errors as small as 0.0001, permits the measurement of $^{12}C/^{13}C$, $^{14}N/^{15}N$, and $^{16}O/^{18}O$ ratios with great accuracy.[47] Although isotope ratio mass spectrometers are designed to operate with natural abundance isotope levels, they are limited in that they can only be used to detect CO_2 and N_2. If the isotopically labeled atom cannot be converted easily into one of these two forms, as is the case with $^{16}O/^{18}O$ phosphate esters, then one must use a remote label to detect the desired isotope effect. The Cleland group[47] has pioneered the use of this method in probing the transition state structure of phosphate ester hydrolysis. Analyses of this type measure the distribution of heavy atoms during the course of a reaction, and reveals isotopic discrimination that occurs up to and including the first irreversible step.

Remote labeling involves the use of a "reporter" group in the molecule of interest that can ultimately be transformed into CO_2 or N_2. In the case of p-nitrophenoxide (p-NPP), the nitro group fulfills this role, as it is easily converted to N_2. Two versions of the molecule must be synthesized:

one that has ^{15}N in the nitro group as well as ^{18}O in the phosphate (either bridging, for measurement of primary effects, or nonbridging, for measurement of secondary effects), and one that has only ^{14}N in the nitro group, but has a natural abundance ^{16}O/^{18}O ratio in the phosphate group. These two substrates are then mixed to restore the natural abundance ratio of ^{14}N/^{15}N (Scheme 4). The isotope effects measured using this substrate mixture are the product of the ^{16}O/^{18}O effect and the ^{14}N/^{15}N effect. Measurement of the isotope effects with natural abundance in both positions gives the isolated ^{14}N/^{15}N effect, and from these two results the ^{16}O/^{18}O effect can be calculated.

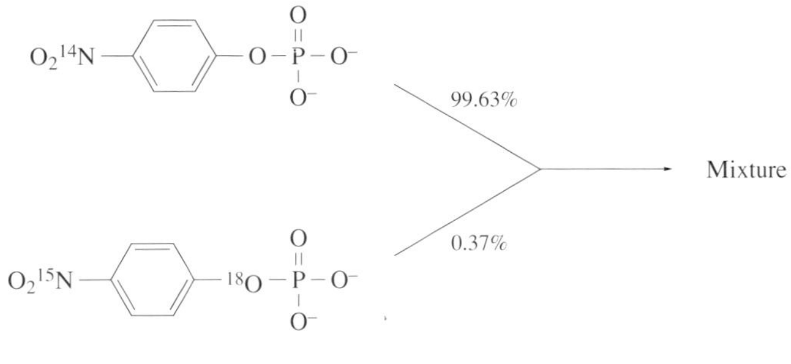

Scheme 4

(ii) Solution hydrolysis of p-NPP

The primary ^{18}O, secondary ^{18}O, and secondary ^{15}N isotope effects on the hydrolysis of p-NPP have been determined for both the monoanion and the dianion. These results are shown in Table 1. The primary ($^{18}k_{bridge}$) effects indicate the amount of P—O cleavage in the transition state, while the secondary ($^{18}k_{nonbridge}$) effects reflect changes in bond order to the peripheral oxygens. The ^{15}k effects are a measure of the charge development on the leaving group oxygen. This effect arises from the quinoid resonance form of p-nitrophenoxide (Figure 8), which makes a more significant contribution to the anion than it does to either the substrate or the protonated phenol. As the nitrogen is less tightly bonded in this resonance form, transition states with a large amount of charge development lead to a normal isotope effect.

Table 1 Heavy atom isotope effects on the hydrolysis of p-NPP.

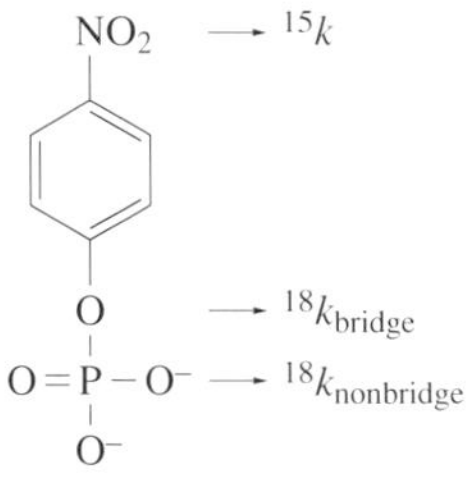

Ionization state	^{15}k	$^{18}k_{bridge}$	$^{18}k_{nonbridge}$	Ref.
Monoanion	1.0005 ± 0.0002	1.0106 ± 0.0003	1.0224 ± 0.0005	48
Dianion	1.0034 ± 0.0002	1.0230 ± 0.0005	0.9993 ± 0.0007	48

Figure 8 Resonance structures for p-nitrophenoxide.

The primary ^{18}O isotope effect for nonenzymatic cleavage of the p-nitrophenylphosphate dianion is greater than 2%. This provides evidence that the reaction does not proceed via a pentavalent phosphorane intermediate, as the irreversible formation of such a species would suppress any intrinsic isotope effect on bond cleavage. Brønsted correlations indicate almost complete bond breaking in the transition state for this reaction, and departure of the leaving group as the anionic species. This suggests that the observed ^{18}O$_{bridge}$ isotope effect for hydrolysis of the dianion reflects that of a fully dissociative transition state. The primary ^{18}O isotope effect on hydrolysis of the monoanion is substantially smaller, probably because protonation of the leaving group oxygen (inverse effect) partially masks the normal effect produced by P—O cleavage. This protonation is most likely mediated by one or more water molecules (Figure 9).

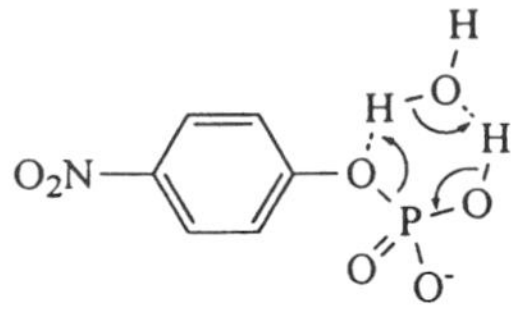

Figure 9   Water-mediated proton transfer mechanism for the dissociative cleavage of phosphate monoester monoanions.

The large ^{15}N effect on cleavage of the dianion reflects the highly charged nature of the leaving group. Remarkably, the value of this isotope effect is larger than that measured for deprotonation of *p*-nitrophenol.[13] One potential explanation is that charge repulsion from the negatively charged PO_3 group causes increased delocalization of charge from the departing phenolic oxygen into the benzene ring. The small but measurable ^{15}N effect on hydrolysis of the monoanion supports the hypothesis that the leaving group oxygen is protonated in the transition state, although this transfer lags slightly behind bond cleavage.

Within experimental error, there is no secondary ^{18}O isotope effect for dianion hydrolysis. This result is in conflict with the traditional view of metaphosphate, in which the P—O$_{nonbridge}$ bond order approximates 5/3 (Structure A, Figure 10). A transition state of this nature should lead to a significant inverse $^{18}k_{nonbridge}$ isotope effect. However, this simple picture of bonding in metaphosphate may be inaccurate. An alternative is that metaphosphate is positively charged on phosphorus, and is more accurately described by structures B and C, Figure 10.[49,50] Hydrolysis of the monoanion gives a normal $^{18}k_{nonbridge}$ effect, due to the primary isotope effect associated with transfer of a proton from a peripheral oxygen to the leaving group oxygen.

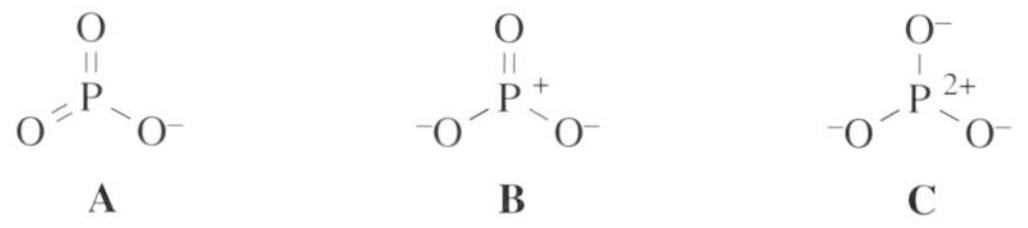

Figure 10   Metaphosphate resonance structures.

The results described here are consistent with a dissociative transition state for the nonenzymatic hydrolysis of both the dianion and monoanion of *p*-NPP. The values obtained in these studies provide a valuable guide for interpreting the isotope effects on enzyme-catalyzed phosphoryl transfer.

(iii) Results with phosphatases

Kinetic isotope effects using *p*-nitrophenylphosphate as substrate have been determined for four classes of phosphatases to date: high MW PTPases (Yop51, PTP1);[51] low MW PTPases (Stp1);[52] dual-specificity phosphatases (*Vaccinia* H-1 related (VHR));[53] and S/T PPases (calcineurin)[54] (Table 2). There are no observed isotope effects on alkaline phosphatase-catalyzed hydrolysis because the isotopically sensitive step occurs after an irreversible step.[48]

Table 2   Heavy atom isotope effects on the hydrolysis of *p*-NPP.

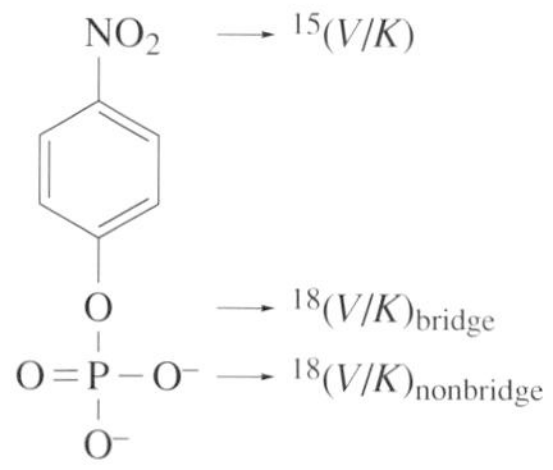

Phosphatase	$^{15}(V/K)$	$^{18}(V/K)_{bridge}$	$^{18}(V/K)_{nonbridge}$	Ref.
Yop51	0.9999 ± 0.0003	1.0152 ± 0.0006	0.9998 ± 0.0013	51
PTP1	1.0001 ± 0.0002	1.0142 ± 0.0004	0.9981 ± 0.0015	51
Stp1	1.0007 ± 0.0001	1.0160 ± 0.0005	1.0018 ± 0.0003	52
VHR	0.9999 ± 0.0004	1.0118 ± 0.0020	1.0003 ± 0.0003	53
PP2b	1.0014 ± 0.0001	1.0115 ± 0.0012	0.9942 ± 0.0007	54

All of these enzymes (with the exception of alkaline phosphatase) display a normal $^{18}(V/K)_{bridge}$ isotope effect, with almost all values intermediary between that obtained for the dianion in solution and the monoanion in solution. This suggests a large amount of P—O cleavage, and is indicative of a dissociative transition state. In comparison, the $^{18}O_{bridge}$ isotope effect for O,O-diethyl-4-nitrophenylphosphate, which proceeds via an associative transition state, is 1.0060.[55]

With the exception of calcineurin, all of these enzymes utilize an active site aspartic acid residue to protonate the leaving group oxygen. Protonation is predicted to lead to low $^{15}(V/K)$ isotope effects, as there is essentially no charge development on this oxygen in the transition state. Indeed, the largest observed $^{15}(V/K)$ isotope effect is 1.0007, only slightly larger than that for the monoanion in solution. Were the bridging oxygen to depart as the phenoxide, one would expect a value closer to that observed for hydrolysis of the p-NPP dianion in solution.

Mutation of the catalytic acid to asparagine or alanine leads to a dramatic increase in $^{15}(V/K)$ values, consistent with this residue's role in aiding the expulsion of the alcohol leaving group via protonation. This mutation also leads to an increase in $^{18}(V/K)_{bridge}$ effects. In the wild-type enzymes, the observed $^{18}(V/K)_{bridge}$ effects are the product of P—O cleavage (normal effect) and protonation by aspartic acid (inverse effect). Removal of the inverse isotope effect gives $^{18}(V/K)_{bridge}$ values even larger than that for the dianion solution reaction, indicating almost complete P—O cleavage in the transition state for the enzyme-catalyzed reaction. The size of these effects also suggests that the intrinsic isotope effects are fully expressed in the mutant enzymes, and most likely in the native enzymes as well.

Calcineurin displays a much larger $^{15}(V/K)$ isotope effect than those observed with the phosphatases discussed above, though still smaller than the effect for hydrolysis of the p-NPP dianion in solution. This suggests partial neutralization of the developing charge on the leaving group oxygen, possibly through interaction with a metal-bound water, or from an active site histidine or metal ion. In contrast to PTPases and dual-specificity PTPases, however, leaving group stabilization is thought to lag behind P—O cleavage. An alternate explanation is that the observed isotope effects are not the intrinsic isotope effects for catalysis, and that values for both $^{18}(V/K)_{bridge}$ and $^{15}(V/K)$ are suppressed by a nonisotopically sensitive event. Further investigations, perhaps with site-directed mutants, are needed to resolve this issue.

The $^{18}(V/K)_{nonbridge}$ isotope effects for Yop51, PTP1, and VHR are all similar to that for the p-NPP dianion in solution: within experimental error, there is no isotope effect. The $^{18}(V/K)_{nonbridge}$ effects with Stp1 are slightly normal, indicative of a transition state with more associative character, although the $^{18}(V/K)_{bridge}$ and $^{15}(V/K)$ effects suggest that P—O bond cleavage is largely rate limiting. Calcineurin gives the most puzzling $^{18}(V/K)_{nonbridge}$ isotope effects, with a considerable inverse value observed. If one considers the transition state for p-NPP dianion hydrolysis (metaphosphate-like) to be a mechanistic extreme, then the $^{18}(V/K)_{nonbridge}$ value for calcineurin cannot indicate a more dissociative transition state. Perhaps the interaction between active-site metals and peripheral phosphate oxygens leads to the observed inverse isotope effect. In any event, it is important to remember that p-NPP is a very poor substrate for this enzyme, and that results with this compound may not be indicative of the behavior of more physiologically relevant substrates.

The above results have been interpreted as indicating a dissociative transition state for enzyme-catalyzed phosphoryl transfer, given the large normal primary ^{18}O effects and the small or absent secondary ^{18}O effects. These data are clearly inconsistent with a synchronous concerted mechanism (S_N2-like), which should give a relatively small normal value for the primary ^{18}O effect. Therefore, this mechanism can be excluded. There remains, however, another mechanistic possibility, the formation of a pentacovalent intermediate whose breakdown is at least partially rate limiting. The isotope effects expected for a pathway of this type are consistent with those described above. Rate-limiting cleavage of the phosphate ester linkage should give a sizable $^{18}(V/K)_{bridge}$ effect, while concomitant protonation of the departing phenol should reduce the $^{15}(V/K)$ effect. The $^{18}(V/K)_{nonbridge}$ effect should be the product of the inverse effect on intermediate formation and the normal effect on intermediate decomposition. The magnitude of this effect, and whether it should be overall inverse or normal, is difficult to predict.

In contrast to the solution hydrolysis of p-NPP, the existence of a pentacoordinate intermediate in an enzyme-catalyzed reaction is not precluded on the basis of leaving group ability. In the solution reaction, the formation of such an intermediate would of necessity be slower than decomposition, and no isotope effects would be observed. The enzyme-catalyzed reaction has no such requirement, as positioning and movement of active-site residues may define a kinetic pathway in which breakdown of the intermediate is rate limiting.

Crystal structures of enzyme–inhibitor complexes may also provide evidence for the existence of a pentacoordinate covalent enzyme species. Phosphatases co-crystallized with vanadate often show a trigonal bipyramidal coordination sphere with continuous electron density between the active site nucleophile and vanadium.[56] A structure of the Ras·RasGAP·GDP·Mg^{2+}·AlF_3 complex suggests the existence of a trigonal bipyramidal intermediate (or transition state) here as well.[57] However, a dissociative transition state may be considered trigonal bipyramidal, albeit with elongated apical substituents. How well these structures represent the true pathway for phosphate ester hydrolysis is a matter of speculation. What is clear, however, is that based on the available data the existence of a phosphorane intermediate cannot be excluded.

5.07.4 INDIVIDUAL PHOSPHATASES

In the following sections, representative examples of several classes of phosphatases will be discussed. The selected enzymes have been the subject of several mechanistic, mutagenesis, and chemical labeling studies. There are X-ray crystal structures of these enzymes, and this information validates many of the initial hypotheses concerning the mechanisms of catalysis these enzymes employ. In some cases, however, most notably those of the more recently discovered PTPases and the Ser/Thr phosphatases, the crystal structures have been instrumental in guiding mechanistic thinking. In any event, understanding of these enzymes is constantly evolving, and the ideas presented in these sections must surely be viewed as provisionary.

Phosphatases are an example of the successful evolution of differing catalytic strategies for catalysis of the same reaction, phosphate ester hydrolysis. However, certain key chemical components appear to be universally employed by all phosphatases. Phosphate monoesters are intrinsically resistant to nucleophilic attack because they are (largely) dianions at neutral pH. To be effective catalysts, all phosphatases must somehow enhance the electrophilicity of the phosphate ester substrate. These enzymes must also be capable of activating a nucleophile, stabilizing a pentacoordinate intermediate or transition state, and facilitating the departure of a leaving group. A comparison between PTPases and Ser/Thr phosphatases reveals interesting similarities in how these challenges are met. PTPases effectively neutralize the phosphate ester by providing strong hydrogen bonds to the phosphate group from main chain amides on the phosphate binding loop. Active site arginine residues aid in both binding the negatively charged phosphate and in stabilizing the transition state for phosphoryl transfer. The Ser/Thr phosphatases accomplish the same goals via the use of a metal dyad and a pair of active site arginines. Furthermore, the local positive charge of the active sites in both types of enzymes facilitates the deprotonation of active site nucleophiles, the formation of phosphorane intermediates (or transition states), and the departure of a leaving group. In PTPases, the latter is accomplished by the use of an active-site general acid. The Ser/Thr phosphatases are thought to use a Lewis acid. Thus, the two classes of enzymes utilize similar global strategies for catalyzing phosphate ester hydrolysis, although the specific catalytic mechanisms employed are quite different.

When thinking about the catalytic mechanisms employed by phosphatases, it is useful to keep in mind a mental inventory of the elements necessary for successful catalysis. How does the enzyme neutralize the charge of the ester? How does it activate the nucleophile? How does it stabilize transition states and/or intermediates? In the following sections the specifics of these questions will be addressed.

5.07.4.1 Nonspecific Phosphatases

There are four families of nonspecific phosphatases. Three of these are nucleophilic: serine phosphatases (alkaline phosphatase), histidine phosphatases (prostatic and lysozomal acid phosphatases), and low molecular weight cysteine acid phosphatases. The low molecular weight cysteine phosphatases bear substantial structural similarity to a variety of phosphotyrosyl and dual specificity phosphoprotein phosphatases. The true biological substrates for these enzymes are not known. For this reason, the low molecular weight enzymes have been grouped with the cysteine phosphoprotein phosphatases. The last class of nonspecific enzymes are the purple acid phosphatases (so called because the Zn^{II}/Fe^{III} version of the enzyme has a charge transfer band that gives it a purple color). This group of enzymes is unique among the nonspecific phosphatases in that an active site nucleophile is not employed. However, these enzymes pay a price for this in being the poorest catalysts of the whole group.

5.07.4.1.1 Serine phosphatase (alkaline phosphatase)

Alkaline phosphatase is probably the most thoroughly studied and understood of all phosphatases. The enzyme is ubiquitous in both prokaryotes and eukaryotes and is almost completely nonspecific, catalyzing the hydrolysis of almost any phosphate monoester dianion. In eukaryotes, multiple isoforms of the enzyme are found in a wide variety of tissues. The role of these enzymes is not known, however, eukaryotic enzymes are typically more active than their bacterial counterparts.[58] K_m and k_{cat} values for simple phosphate esters are in the range 1–250 µM and 30–150 s^{-1}, respectively.

Bacterial alkaline phosphatase was the first phosphatase to have its crystal structure determined.[59,60] This is unfortunate in some respects, since alkaline phosphatase is rather unique and is not a representative example of most metallophosphatases. All alkaline phosphatases are homodimeric enzymes that utilize a catalytic zinc(II) dyad, a magnesium(II) ion, a nucleophilic active site serine, and an active site arginine residue. Alkaline phosphatase is the only metallophosphatase that is known to use a nucleophilic active site residue, and the only phosphatase that utilizes three metal ions in the active site. It is also the only phosphatase to use serine as a nucleophile. Remarkably, the crystal structure of the bacterial enzyme reveals the complete absence of any active site amino acids that might function as either general acids or general bases. Evidently, the serine residue is coordinated to an active site zinc as the alkoxide. Substrate binds to the two active site zinc ions (Figure 11) with the leaving group *anti* to the incoming serine nucleophile. The phosphate ester is further activated by binding directly to the active site arginine and to the magnesium and several other residues via the intermediacy of water-mediated hydrogen bonds. Mutagenesis of the arginine residue results in both a higher K_m and k_{cat}, suggesting that this residue plays a significant but noncritical role in both binding and catalysis.[61,62]

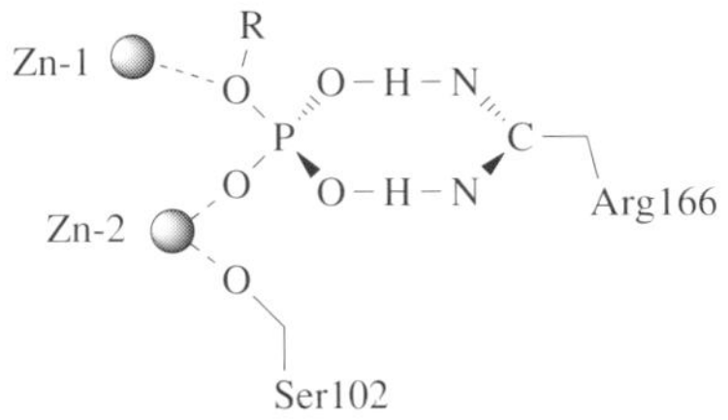

Figure 11 Orientation of the phosphate monoester in the active site of alkaline phosphatase.

Addition of the active site serine to the phosphate ester leads to the formation of a phosphorane-like transition state or intermediate, in which the charge on the departing alkoxide is stabilized by coordination to one of the zinc ions. Expulsion of the leaving group then yields a covalently bound phosphoserine group. Cleavage of this species occurs by the microscopic reversal of its formation, with the exception that hydroxide takes the place of the ester alcohol.

The rate-limiting step for this process depends on the pH of the solution. At acidic pH, cleavage of the phosphoserine group is rate limiting. At higher pH, the release of the product, inorganic phosphate, is rate limiting. Inorganic phosphate is also an exceptionally good inhibitor of this enzyme, with a K_i of about 5 µM for the bacterial enzyme. A number of mutant alkaline phosphatases with enhanced k_{cat} values have been reported.[63] In most of these mutants, enhancement of k_{cat} probably results from reduced affinity of the enzyme for inorganic phosphate, leading to faster product release and a higher turnover rate. A bacterial alkaline phosphatase with a D153A mutation shows a 6.3-fold increase in k_{cat}, a 13.7-fold increase in V/K, and a 159-fold increase in K_i for inorganic phosphate. This residue forms a hydrogen bond to a water molecule in direct contact with the active site magnesium.

Examination of the active site of the enzyme shows immediately why the enzyme binds phosphate so tightly, yet is unable to discriminate between various phosphate esters. The active site is a shallow declivity that packs three divalent ions and an arginine (presumably protonated) into close proximity. There is sufficient room for the phosphate portion of the phosphate ester to bind. However, the ester portion of the substrate must point away from the enzyme. Due to the shallowness of the binding pocket, it is not possible for the pendant ester group to profitably contact the enzyme. The high degree of positive charge in the active site of this enzyme is a general hallmark of phosphatases and is critical for substrate binding and activation. The use of an arginine residue to facilitate binding and catalysis is also a common feature of many phosphatases.

Because of their short lifetime, it is normally very difficult to obtain crystallographic information about covalent enzyme intermediates. In the case of bacterial alkaline phosphatase, however, there are two crystal structures of enzymes with covalently bound phosphate. In the first structure, obtained by Wyckoff's group, substitution of cadmium for both active site zinc ions results in an enzyme in which the covalent phosphoserine is quite stable and amenable to X-ray crystallography.[64] The Kantrowitz group determined the X-ray crystal structure of a mutant enzyme with a covalently bound phosphate.[65] Mutation of His331 to Gln (His331 is one of the ligands for Zn-1) gives an enzyme with only slightly altered kinetic constants (50% reduction in k_{cat} and no change in K_m in the presence of added Zn). Evidently, perturbation of the ligand sphere of Zn-1, the metal thought to activate the incoming water molecule, stabilizes the phosphoenzyme sufficiently so that crystals containing a covalently bound phosphate may be obtained. A water (or hydroxide) is also found bound to Zn-1 in a position *anti* to the P—O bond of the phosphoserine intermediate, suitably positioned for an in-line displacement.

5.07.4.1.2 Purple acid phosphatase

Purple acid phosphatases are a large family of enzymes found in both plants and animals. The best characterized of these enzymes are the kidney bean enzyme, which uses an Fe^{III}–Zn^{II} metal dyad, the porcine enzyme (known as uteroferrin) and the bovine spleen enzyme. The latter two utilize an Fe^{III}–Fe^{II} metal dyad. The promiscuity of these metallophosphatases with respect to metal ion usage is quite remarkable. The current case illustrates how relatively homologous enzymes from different sources may utilize different metals. Indeed, it is possible to replace the active site Zn^{II} of the kidney bean enzyme with either Fe^{II} or Co^{II} and still maintain high activity.[12] This situation is also the case with phosphoserine/threonine phosphatases, which, depending on the specifics of the enzyme, seem to have varied predilections for active site metal ions. Though there is little sequence homology between purple acid phosphatase and the protein Ser/Thr phosphatases, a substantial degree of similarity in the three-dimensional architecture and metal binding motifs of these enzymes exists.

It was reported that this enzyme utilizes an active site nucleophile for catalysis.[39,40] This judgment was based largely on kinetic evidence. However, subsequent stereochemical experiments revealed that the enzyme catalyzes a direct transfer of the phosphoryl group to water.[41] The X-ray crystal structure of the kidney bean enzyme shows a conserved water or hydroxide bound to the active site Fe^{III} (Figure 12). This is the probable phosphoryl group acceptor. There are three active site histidines that may be catalytically important: His202, His295, and His296 are all in position to make hydrogen bond contacts to a bound phosphate ester. In particular, His296 seems well positioned to function as a general acid in protonating the leaving group alcohol. Interestingly, this is one of only a few phosphatases that does not seem to utilize an active site arginine residue for binding and catalysis. This may be one of the reasons why this enzyme has a relatively poor affinity for phosphate esters. The K_m of the kidney bean enzyme for phosphate esters (ca. 700 μM) is over an order of magnitude greater than that observed for typical nonspecific phosphatases. Alternatively, it is possible that we simply have not yet discovered the true *in vivo* substrate for this enzyme.

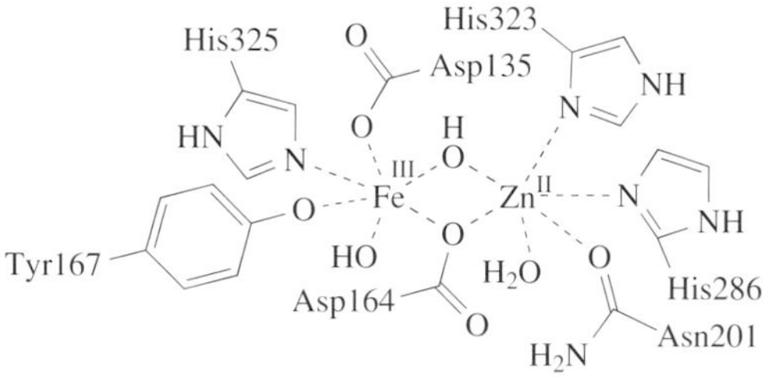

Figure 12   Metal dyad in the active site of purple acid phosphatase.

5.07.4.1.3 Histidine phosphatases

Prostatic acid phosphatase is a member of a large class of nonspecific phosphatases that are fairly homologous and contain a nucleophilic active site histidine. Other prominent members of this class include lysozomal acid phosphatases and bacterial acid phosphatases, among others. The choice of nomenclature for these enzymes is somewhat unfortunate, since several of these enzymes exhibit broad pH optima for k_{cat}. This misnomer is a direct result of the use of *p*-nitrophenyl phosphate as a test substrate. *p*-Nitrophenyl phosphate has a fairly low pK_a (5.4).[66] Because these enzymes utilize the monoanionic form of the ester, which is present at very low concentrations at a pH of 7, the enzyme activity at this pH is often underestimated. A hallmark of these enzymes is their potent inhibition by L-(+)-tartrate.[67] Interestingly, another dicarboxylate, endothall, is a good inhibitor of calcineurin,[68] a metallophosphatase with little apparent similarity to the histidine phosphatases.

A variety of active site-directed agents inactivate prostatic acid phosphatase with concomitant modification of the nucleophilic histidine residue. Similar studies have implicated arginine and an aspartate or glutamate as essential residues.[69–71] The crystal structure of rat acid phosphatase confirms the essential roles of these amino acids. There are two active site histidine residues, three arginines, and an essential aspartate that are all positioned to interact with the phosphate ester. An X-ray crystal structure of the enzyme with bound vanadate (Figure 13) reveals the probable roles of these residues.[56] His12 is covalently bound to the vanadate, which has a trigonal bipyramidal geometry, probably mimicking a high-energy intermediate or transition state. Asp258 donates a hydrogen bound to an apical oxygen that is *anti* to His12. This aspartate probably functions as an acid to protonate the leaving group alcohol, and as a base to deprotonate an attacking water molecule. (A similar acid–base motif is used by the phosphotyrosine phosphatases.) The three arginine residues all interact with the equatorial oxygens of the vanadate. These residues play an important role in binding and activating the phosphate ester toward nucleophilic attack. Similarly, a protonated histidine (His257) donates a hydrogen bond to an equatorial oxygen of the bound vanadate. The basis of this enzyme's selectivity for phosphate ester monoanions is unclear from this structure. However, it is possible that His257 is found as the neutral species in the active form of the enzyme, and functions as a hydrogen bond acceptor from a bound phosphate ester monoanion.

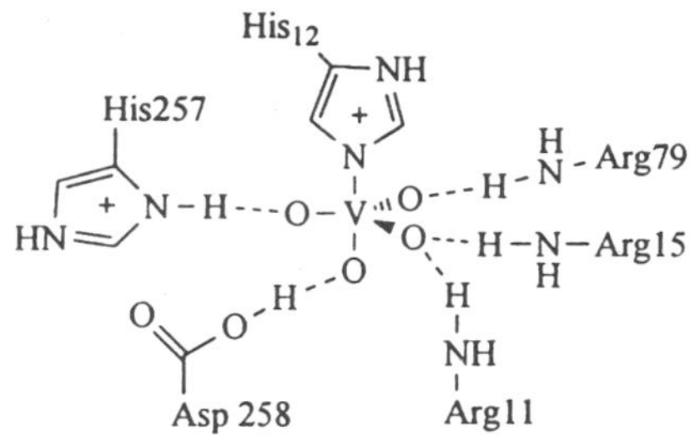

Figure 13 Binding of vanadate to the active site of rat acid phosphatase.

Although prostatic acid phosphatase is a fairly nonspecific enzyme *in vitro*, there is substantial circumstantial evidence that this protein plays a role in regulating levels of protein phosphorylation, particularly on tyrosine.[72–74] A variety of tyrosine phosphorylated proteins serve as exceptionally good substrates for this enzyme, with K_ms in the low nanomolar range. These substrates bind so tightly that the rate-limiting step of the reaction changes from cleavage of the phosphoenzyme to release of the dephosphorylated protein.[75] This would suggest a potential role for this enzyme in the etiology of prostate cancer.

5.07.4.2 Phosphoprotein Phosphatases

Although phosphoserine/threonine phosphatases have long been known, until recently our understanding of the molecular structure and catalytic mechanism of these enzymes was very poor. Our knowledge of PTPases was even more rudimentary, since the recognition of these enzymes as a unique family is a fairly recent development. Despite this, we now have a very good understanding of how PTPases function, as well as a rapidly growing appreciation of the molecular architecture and mechanism of the phosphoserine/threonine phosphatases. This development was due in part to the rapid cloning and subsequent site-directed mutagenesis of a number of phosphatases (particularly of PTPases) in these two families of enzymes. These studies were followed by representative X-ray crystal structures of both PTPases and phosphoserine/threonine phosphatases. Although the use of these crystal structures for the *de novo* design of phosphatase inhibitors has yet to be accomplished, these X-ray structures have been very useful for rationalizing the function of known inhibitors. In addition, X-ray structures of inhibitor–enzyme complexes have played an important role in verifying and testing the mechanistic evidence gathered through site-directed mutagenesis (*vide infra*).

5.07.4.2.1   *Protein tyrosine phosphatases*

All known phosphotyrosine-specific phosphatases share substantial homology in the catalytic domain. (It should be borne in mind however, that this fact does not mean that all PTPases that exist in nature are homologous or use the same catalytic motif.) This homology includes a conserved phosphate binding loop that contains an active site cysteine. At physiological pH, this cysteine usually exists as the anion (pK_a ca. 4.5 in Yop51, the enzyme from *Yersinia enterocolitica*[76]), and is flanked by a neighboring histidine, reminiscent of the Cys–His catalytic dyad found in cysteine proteases. There is strong evidence that the hydrolysis of phosphomonoesters catalyzed by these enzymes proceeds through a phosphoenzyme intermediate and that the enzyme nucleophile is the active site cysteine. Yop51 catalyzes the hydrolysis of a variety of aromatic phosphates at approximately the same rate, regardless of the nature of the leaving group. Although K_ms may vary substantially, k_{cat} for all of these reactions is close to 1200 s^{-1}.[77] This suggests that these enzyme-catalyzed reactions proceed through a common rate-determining step, most likely dephosphorylation of the enzyme. Treatment of a recombinant rat brain PTPase with ^{32}P labeled *p*-nitrophenyl phosphate results in the incorporation of radioactivity into the enzyme. Mutant proteins containing a serine in the place of the active site cysteine are incapable of incorporating label and are also catalytically inactive, although substrate binding is unaffected.[31] Treatment of a variety of PTPases with iodoacetate leads to covalent modification of this highly conserved cysteine residue and concurrent loss of activity.[76]

In an elegant study from the Walsh and Anderson laboratories, the formation and breakdown of the phosphorylated enzyme was followed directly. Using a ^{32}P labeled synthetic phosphopeptide as substrate, rapid quench kinetics reveals that a burst of phosphorylation of the leukocyte antigen related PTPase (LAR) occurs on the millisecond timescale, followed by a slower hydrolysis. Phosphorus-31 NMR of the quenched phosphoenzyme intermediate shows a resonance whose chemical shift is consistent with a phosphocysteine residue.[35]

Although the Cys–His dyad found in PTPases is evocative of cysteine proteases, it is clear that there is a constellation of other active residues that contribute greatly to the catalytic process. Extensive mutagenesis studies on Yop51 carried out by Dixon's group provide evidence for the involvement of a variety of conserved residues. These residues include His402, Glu290, and Asp356, as well as a number of conserved arginine residues.[76,78] The histidine may play a role in altering the pK_a of the active site cysteine, since site-directed mutagenesis of this residue shifts the pH optimum for catalysis and has an affect on the optimal pK_a for alkylation of this cysteine by iodoacetate.[76] Conversion of the Glu to Gln and the Asp to Asu results in substantial diminution of enzyme activity. Most significantly, the wild type enzyme displays a bell-shaped pH/rate profile, consistent with the involvement of both an active site general base and a general acid. The Asp–Asn mutant shows a pH/rate profile consistent with the involvement of only a single base. The Glu–Gln mutant shows the opposite pH/rate profile, consistent with dependence on only an active site acid. This data suggests that the active site Glu functions as a general base and the active site Asp functions as a general acid.[78] As is common with many phosphatases, the enzyme contains active site arginine residues, in this case two. The role of these residues is probably to provide both substrate binding energy and to stabilize the pentacoordinate phosphorane intermediate (or transition state) that forms during the reaction.[79]

The first available crystal structure of a PTPase, PTP1B, provided direct confirmation of the existence of the cysteine residue positioned to function as an active site nucleophile.[80] However, in this structure, neither the essential glutamate nor aspartate residue appeared to be in position to play a role in the enzyme-catalyzed reaction. This ambiguity was resolved upon obtaining a crystal structure of Yop51 with the inhibitor tungstate bound at the active site.[81] Upon binding of tungstate, the essential aspartate, which is located on a flexible loop, closes down on the active site so that it is in position to function as a general acid to assist in departure of the leaving group. (Presumably, this residue then functions as a general base to facilitate dephosphorylation of the enzyme in a subsequent step.) This study confirms the importance of the active site aspartate and reveals that the essential glutamate is not directly involved in catalysis, though it plays a role in ion pairing to an essential arginine residue. This study underscores the importance of using enzyme–inhibitor (and in available cases, enzyme–substrate) complexes to yield a mechanistically relevant crystal structure. In addition, it highlights both the strengths and limitations of site-directed mutagenesis as a mechanistic tool.

A subsequent X-ray structure of a complex of the catalytically inactive PTP1B (containing a Cys to Ser mutation) with a high-affinity phosphopeptide substrate (DADEpYL-NH$_2$), also shows the expected positioning of the active site aspartate (Asp181).[82] In addition, this structure reveals interesting details about peptide binding. Phosphotyrosine-containing peptides with N-terminal acidic residues are the preferred substrates for PTPases. This crystal structure reveals some of the molecular interactions responsible for this specificity. In particular, an arginine residue (Arg47) is positioned to form a salt bridge to acidic residues at position P-1 and P-2 of the peptide substrate. Moreover, while there are a number of hydrogen-bonding interactions between the peptide backbone of the substrate and the enzyme, it is clear from this structure that most of the peptide backbone does not contact the enzyme directly. This suggests that the use of peptidomimetics is likely to be very successful for the generation of biologically active inhibitors. An interesting pioneering study in the development of PTPase inhibitors was the screening of a peptide library containing a cinnamic acid unit as the phosphotyrosine mimic.[83] The best inhibitor is potent in the low nanomolar range against PTP1B. The inhibitor consists of a cinnamic acid linked to Gly-Glu-Glu via an amide linkage between a *p*-carboxy group and the amino group of glycine. Although no crystal structure of this inhibitor bound to PTP1B is as yet available, the similarity of this inhibitor to the preferred peptide substrate (acidic residues at P-1 and P-2) leaves little doubt as to the molecular basis for the potent inhibition.

5.07.4.2.2 *Protein Ser/Thr phosphatases*

These enzymes are customarily divided into two categories based on biochemical criteria such as substrate specificity and response to various inhibitors.[84] Type 1 enzymes (PP1) specifically dephosphorylate the β-subunit of phosphorylase kinase and are inhibited by low molecular weight proteins known as inhibitor-1 (I1) and inhibitor-2 (I2). Type 2 enzymes, including 2A, 2B (also referred to as calcineurin), and 2C, prefer the α-subunit of phosphorylase kinase as a substrate and are insensitive to I1 and I2. However, grouping the type 2 enzymes together is somewhat misleading, since the phosphatases, PP1, 2A, and 2B actually have highly homologous catalytic domains while 2C does not.

All serine/threonine phosphatases are metalloenzymes. Calcineurin contains a zinc–iron dyad, whereas PP1 appears to utilize two manganese ions.[85] Thus, despite the fair sequence homology between these enzymes, the metal dyad is quite different. For many of the enzymes in this class, the identity of the active site metals is unknown. The crystal structures of calcineurin and the catalytic subunit of PP1 reveal that the metal ion binding motif (central β-α-β-α-β motif) found in these enzymes is similar (despite the lack of sequence homology) to the motif found in purple acid phosphatase.[86,87] This two-metal ion binding motif also appears in other phosphatases including inositol monophosphatase and fructose-1,6-bisphosphatase. The enzyme also contains the signature sequence DXH(X)$_n$GDXXD(X)$_n$GNHD/E (n ~25) that is found in a variety of phosphatases including Ser/Thr PPases.[88] In addition, this enzyme contains two catalytically important arginine residues at the active site. This also appears to be a common feature of all the Ser/Thr PPases.

Based on the structural information for Ser/Thr PPases, and by analogy to other metallophosphatases, reasonable roles for the active site metals include binding of the phosphate monoester, activation of a nucleophile, and/or stabilizing the transition state of phosphate ester cleavage. The likeliest candidate for a nucleophile is a metal-bound water or hydroxide. Alternatively, the conserved His125 (coupled with Asp95) could act as the nucleophile, or as a general acid to promote the departure of the leaving serine residue.

5.07.4.3 Small-molecule Specific Phosphatases

A number of phosphorylated molecules such as inositol phosphates and phosphatidic acid play important roles in signal transduction. Not surprisingly, there are specific phosphatases and kinases that operate to control the cellular concentrations of these molecules. In addition, there are specific phosphatases associated with carbohydrate metabolism (e.g., glucose 6-phosphatase and fructose-1,6-bisphosphatase). Undoubtedly, there are more small-molecule specific phosphatases of these types that have yet to be discovered. In the cases where the biochemistry of these enzymes is understood, they fit the general patterns that have already been described. For example, inositol monophosphatase and fructose-1,6-bisphosphatase are metallophosphatases that utilize a metal dyad, while glucose 6-phosphatase and fructose-2,6-bisphosphatase appear to be histidine phosphatases. In general, these enzymes have received less scrutiny than either the phosphoprotein phosphatases or the nonspecific phosphatases. This will probably change as specific biological roles for these enzymes become clear. The biochemistry of inositol monophosphatase may be of particular interest since it has been postulated that this enzyme is the target of lithium ion treatment for depression.[89]

Though there is limited sequence homology between fructose-1,6-bisphosphatase and inositol monophosphatase, they both share a similar layered β-α-β-α-β structure. The overall three-dimensional structures of these enzymes is also quite similar.[90] The metal ion binding site of inositol polyphosphatase may also be similar to these enzymes.[91] Although not yet definitively established, it appears likely that these three enzymes all use a metal-bound water or hydroxide as the phosphoryl group acceptor. Given the structural similarity between these otherwise unrelated enzymes, it seems likely that this motif will be found in other metallophosphatases as well.

In contrast to these metalloenzymes, both glucose-6-phosphatase and fructose-2,6-bisphosphatase utilize an active site histidine nucleophile. An X-ray structure of the phosphoenzyme intermediate obtained in the fructose-2,6-bisphosphatase catalyzed reaction has been obtained by flash freezing techniques.[92] The use of a histidine nucleophile may turn out to be a fairly common motif for specific small-molecule phosphatases. A database search has identified a novel sequence motif KXXXXXXRP-(X$_{12-54}$)-PSGH-(X$_{31-54}$)-SRXXXXXHXXXD that is found in a variety of lipid phosphatases, mammalian glucose-6-phosphatase, and a variety of nonspecific bacterial acid phosphatases.[93] Note the similarity of the conserved residues to those that are important in prostatic acid phosphatase.

5.07.5 PHOSPHATASE INHIBITORS

Many phosphatases interact largely, or even solely, with the phosphoryl portion of the phosphate ester. For example, alkaline phosphatase binds inorganic phosphate more tightly than it does phosphate monoesters. Indeed, inorganic oxoanions such as vanadate, arsenate, tungstate, and molybdate are potent nonselective inhibitors of almost all phosphatases.[94,95] Unfortunately, compounds such as these afford little inspiration for the design of new inhibitor motifs that might ultimately serve as the basis for drug design. Despite this generally unsatisfactory state of affairs, some inroads into the development of phosphatase inhibitors have been made over the last few years. Table 3 outlines some newly discovered motifs for phosphatase inhibition. It is anticipated that the introduction of these motifs into combinatorial libraries of small molecules or peptidomimetics will yield inhibitors of even higher potency and selectivity. Moreover, a number of these inhibitor types are sufficiently developed to function as practical and useful research tools.

5.07.5.1 Competitive Inhibitors

Approximately half of all phosphatases, as well as a substantial number of phosphodiesterases and triesterases, contain active site metal ions that are essential for catalysis. Compounds that incorporate an extra metal binding ligand have proven to be very effective inhibitors of metalloproteases such as angiotensin converting enzyme.[103] Surprisingly, no one had designed phosphatase inhibitors that utilized this same strategy, although the naturally occurring phosphatase inhibitor endothall may be an example of such a molecule.[68] A number of very simple compounds that inhibit metallophosphatases are shown below ((8)–(15)).[96] All of these compounds bear a pendant group that is capable of ligating to an active site metal, except for the unsubstituted phosphonate (15) which functions as a control. These compounds display a great range of potencies. For example, the aromatic thiol (11) is the most potent small-molecule inhibitor of alkaline phosphatase that is known, with a K_i of approximately 200 nM. However, compounds such as (12) and (14) are relatively modest inhibitors, though still better than (15).

A crystal structure of (12) bound to alkaline phosphatase has been determined.[104] This structure, represented in Figure 14, clearly shows that the pendant group of the inhibitor reaches around to engage in an extra contact to an active site zinc. In a sense, this inhibitor mimics the binding of a high-energy phosphorane intermediate or transition state, as these species probably make three zinc ion contacts, while the bound substrate makes but two.

This simple notion for inhibitor design will probably be exploitable for the design of both potent and selective phosphatase and phosphodiesterase inhibitors. Given the ubiquitous nature of alkaline phosphatase, inhibitors such as (11) should be quite useful as a tool for studying phenomena associated with protein/small-molecule phosphorylation. In this regard, it is worth noting that many

Table 3 Phosphatase inhibitors.

Inhibitor motif	Enzyme target(s)	Specific example	Inhibitor type	Ref.
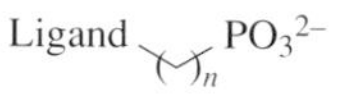 Ligand—(—)$_n$—PO$_3^{2-}$	Metallophosphatases	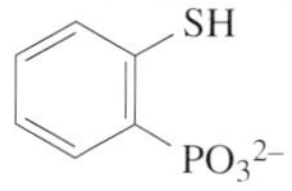 (1)	Competitive	96
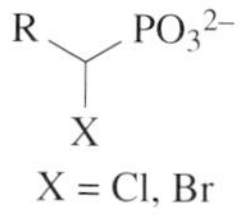 R—CH(X)—PO$_3^{2-}$, X = Cl, Br	PTPases	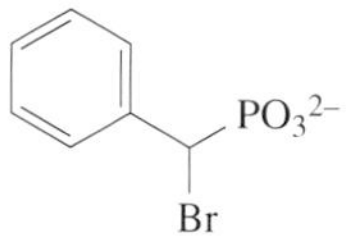 (2)	Affinity reagent	97
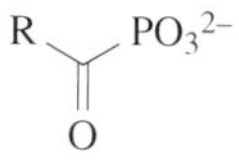 R—C(O)—PO$_3^{2-}$	PTPases, Possibly others	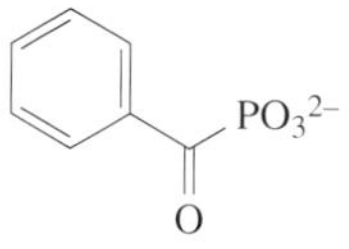 (3)	Affinity reagent	98
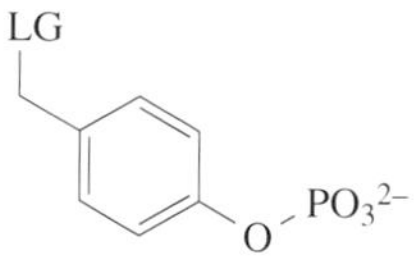 LG–CH$_2$–C$_6$H$_4$–O–PO$_3^{2-}$	Prostatic Acid Phosphatase PTPases	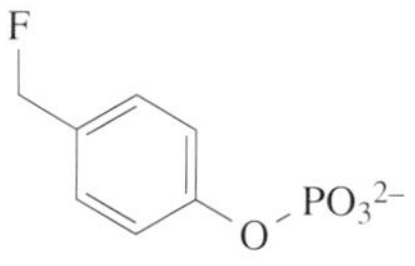 (4)	Mechanism-based	99 100
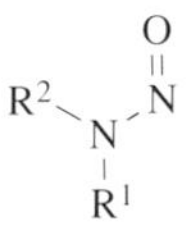 R^2–N(N=O)–R^1	PTPases Possibly others	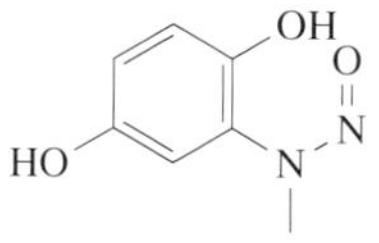 (5)	Unknown	101
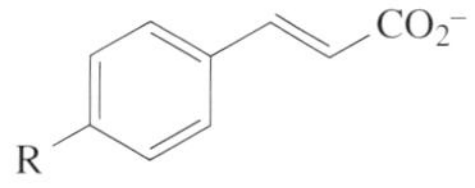 R–C$_6$H$_4$–CH=CH–CO$_2^-$	PTPases	R = Glu containing peptide (6)	Competitive	83
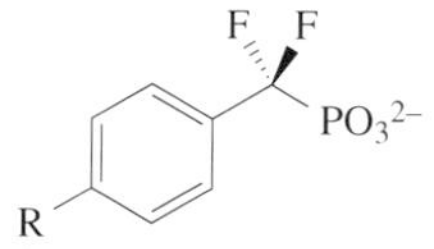 R–C$_6$H$_4$–CF$_2$–PO$_3^{2-}$	PTPases	R = Glu containing peptide (7)	Competitive	102

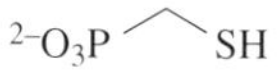

$^{2-}$O$_3$P—CH$_2$—SH (8)

$^{2-}$O$_3$P—CH$_2$CH$_2$—SH (9)

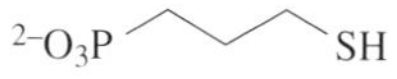

$^{2-}$O$_3$P—CH$_2$CH$_2$CH$_2$—SH (10)

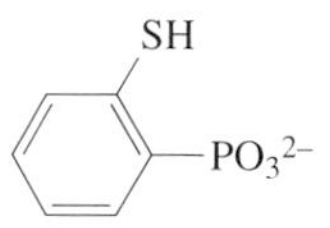

(11)

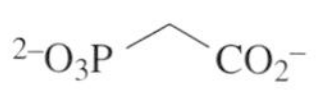

$^{2-}$O$_3$P—CH$_2$—CO$_2^-$ (12)

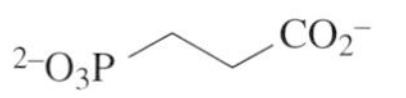

$^{2-}$O$_3$P—CH$_2$CH$_2$—CO$_2^-$ (13)

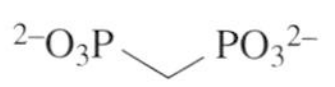

$^{2-}$O$_3$P—CH$_2$—PO$_3^{2-}$ (14)

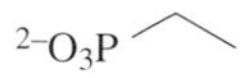

$^{2-}$O$_3$P—CH$_2$CH$_3$ (15)

of these compounds exhibit surprising degrees of selectivity. For example, while (11) is a very potent inhibitor of alkaline phosphatase, it is a fairly poor inhibitor of purple acid phosphatase, another phosphatase that utilizes a zinc-containing metal dyad.

Compounds (6) and (7) (Table 3) represent two different competitive inhibitor types for phosphotyrosine phosphatases. Inhibitor (6) is based on a cinnamic acid motif that was appended to

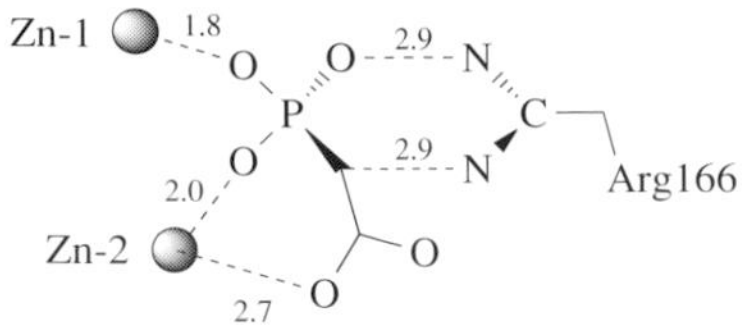

Figure 14 Binding of phosphonoacetic acid to the active site of alkaline phosphatase. Distances shown are in angstroms.

various peptide-like pieces using combinatorial approaches.[83] The result of this exercise was the development of some fairly potent PTPase inhibitors that show varying selectivity even among closely related phosphatases. Compound (7) is based on a difluorophosphonate motif. Inclusion of a phosphonate analogue of phosphotyrosine into a suitable peptide yields inhibitors that can be quite effective.[102]

A small inhibitor library based on an oxazole pharmacophore was also reported.[105] These compounds show marked selectivity between PTPases and dual-specificity enzymes.

5.07.5.2 Affinity Reagents

Two newly developed classes of affinity reagents are the α-halophosphonates[97] and the α-ketophosphonates.[98] To date, little is known about the selectivity and mechanism of action of the α-ketophosphonates, although they appear to be potent in the low micromolar range and show time-dependent inactivation when tested against PTPases. The α-halophosphonates have been studied more extensively. These inhibitors appear to be highly specific for PTPases, and do not react with strong nucleophiles such as azide, hydroxylamine, or thiolates. The compounds are easily prepared in three steps from the corresponding aldehydes. Given their selectivity and ease of preparation, these inhibitors show great promise for use in developing a new generation of PTPase inhibitors.

5.07.5.3 Mechanism-based Inhibitors

Currently, there exists only one general type of mechanism-based inhibitor of a phosphatase.[99,100] Compound (4) is an example of such an inhibitor. These inhibitors are potent against both prostatic acid phosphatase and phosphotyrosine phosphatases. Scheme 5 shows the probable mechanism by which these compounds inactivate phosphatases. Cleavage of the phosphate ester leads to the rapid elimination of fluoride ion to give a quinone methide. This alkylating agent must react with an active site nucleophile prior to its release from the active site. Although some very potent phosphatase inhibitors have been developed using this motif, it seems unlikely that this methodology will be useful for the development of therapeutic agents because of the ubiquitous presence of nonselective phosphatases that would hydrolyze these inhibitors before they have an opportunity to reach their site of action. However, these compounds do represent a useful strategy for the pursuit of phosphodiesterase inhibitors, and a ribonuclease A inhibitor has been developed using this motif.[106]

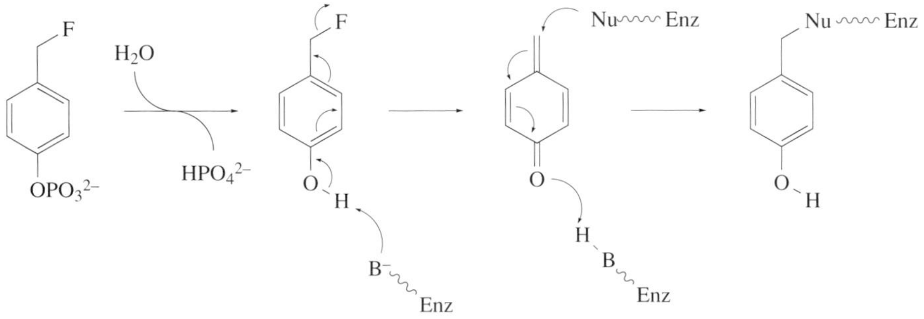

Scheme 5

5.07.6 CONCLUSION

Since the late 1980s phosphatases have emerged as important players in complex biological phenomena. Molecular and structural biology have combined to yield a wealth of new information about this diverse group of enzymes. However, we still know very little about a substantial number of these enzymes and this field will remain a fertile research area for some time to come. In addition, the role of phosphatases in regulating cell growth and metabolism suggests that a number of these enzymes may be clinical targets. Ultimately, it will be chemists who answer these types of questions by designing potent and specific phosphatase inhibitors. Given the relative paucity of results in this area so far, such an endeavor will require both a thorough understanding of the biochemistry of phosphatases and a good degree of ingenuity.

5.07.7 REFERENCES

1. D. A. Pot and J. E. Dixon, *Biochim. Biophys. Acta*, 1992, **1136**, 35.
2. H. Charbonneau and N. K. Tonks, *Annu. Rev. Cell Biol.*, 1992, **8**, 463.
3. E. B. Fauman and M. A. Saper, *TIBS*, 1996, **21**, 413.
4. D. Barford, *TIBS*, 1996, **21**, 407.
5. D. E. Wilcox, *Chem. Rev.*, 1996, **96**, 2435.
6. N. Sträter, W. N. Lipscomb, T. Klabunde, and B. Krebs, *Angew. Chem., Int. Ed. Engl.*, 1996, **35**, 2024.
7. M. Barinaga, *Science*, 1996, **272**, 1261.
8. G. Schneider, Y. Lindqvist, and P. Vihko, *EMBO J.*, 1993, **12**, 2609.
9. H. Ohmori, M. Kuba, and A. Kumon, *J. Biol. Chem.*, 1993, **268**, 7625.
10. K. Yokoyama, H. Ohmori, and A. Kumon, *J. Biochem.*, 1993, **113**, 236.
11. F. Rusnak, L. Yu, and P. Mertz, *SBIC*, 1996, **1**, 388.
12. J. L. Beck, M. J. McArthur, J. De Jersey, and B. Zerner, *Inorg. Chim. Acta*, 1988, **153**, 39.
13. W. W. Cleland and A. C. Hengge, *FASEB J.*, 1995, **9**, 1585.
14. T. S. Widlanski, J. K. Myers, B. Stec, K. M. Holtz, and E. R. Kantrowitz, *Chem. Biol.*, 1997, **4**, 489.
15. F. H. Westheimer, *Science*, 1987, **235**, 1173.
16. N. N. Greenwood and A. Earnshaw, "Chemistry of the Elements," Pergamon, New York, 1984.
17. A. J. Kirby and A. G. Varvoglis, *J. Am. Chem. Soc.*, 1967, **89**, 415.
18. W. W. Butcher and F. H. Westheimer, *J. Am. Chem. Soc.*, 1955, **77**, 2420.
19. D. W. C. Barnard, C. A. Bunton, D. R. Llewellyn, K. G. Oldham, B. L. Silver, and C. A. Vernon, *Chem. Ind. (London)*, 1955, 760.
20. N. Bourne and A. Williams, *J. Am. Chem. Soc.*, 1984, **106**, 7591.
21. F. Ramirez and J. F. Marecek, *Pure Appl. Chem.*, 1980, **52**, 1021.
22. F. Ramirez, J. F. Marecek, and S. S. Yemul, *Tetrahedron Lett.*, 1982, **23**, 1515.
23. J. Rebek, Jr., F. Caviña, and C. Navarro, *J. Am. Chem. Soc.*, 1978, **100**, 8113.
24. S. L. Buchwald, D. E. Hansen, A. Hassett, and J. R. Knowles, *Methods Enzymol.*, 1982, **87**, 279.
25. G. Lowe, *Acc. Chem. Res.*, 1983, **16**, 244.
26. S. L. Buchwald and J. R. Knowles, *J. Am. Chem Soc.*, 1982, **104**, 1438.
27. S. L. Buchwald, J. M. Friedman, and J. R. Knowles, *J. Am. Chem. Soc.*, 1984, **106**, 4911.
28. W. P. Jencks, *Acc. Chem. Res.*, 1980, **13**, 161.
29. W. P. Jencks, *Chem. Soc. Rev.*, 1981, **10**, 345.
30. J. M. Friedman, S. Freeman, and J. R. Knowles, *J. Am. Chem. Soc.*, 1988, **110**, 1268.
31. K. Guan and J. E. Dixon, *J. Biol. Chem.*, 1991, **266**, 17026.
32. Z.-Y. Zhang and L. Wu, *Biochemistry*, 1997, **36**, 1362.
33. J. E. Butler-Ransohoff, S. E. Rokita, D. A. Kendall, J. A. Bunzon, K. S. Carano, E. T. Kaiser, and A. R. Matlin, *J. Org. Chem.*, 1992, **57**, 142.
34. K. Guan and J. E. Dixon, *J. Biol. Chem.*, 1991, **266**, 17026.
35. H. Cho, R. Krishnaraj, E. Kitas, W. Bannwarth, C. T. Walsh, and K. S. Anderson, *J. Am. Chem. Soc.*, 1992, **114**, 7296.
36. C. T. Houston, W. P. Taylor, Z.-Y. Zhang, J. P. Reilly, and T. S. Widlanski, unpublished results.
37. D. B. Northrop and F. B. Simpson, *Bioorg. Med. Chem.*, 1997, **5**, 641.
38. I. Saves, O. Burletschiltz, L. Maveyraud, J. P. Samama, J. C. Prome, and J. M. Masson, *Biochemistry*, 1995, **34**, 11660.
39. J. B. Vincent, M. W. Crowder, and B. A. Averill, *Biochemistry*, 1992, **31**, 3033.
40. J. B. Vincent, M. W. Crowder, and B. A. Averill, *J. Biol. Chem.*, 1991, **266**, 17737.
41. E. G. Mueller, M. W. Crowder, B. A. Averill, and J. R. Knowles, *J. Am. Chem. Soc.*, 1993, **115**, 2974.
42. Y. Zhao and Z.-Y. Zhang, *Biochemistry*, 1996, **35**, 11797.
43. A. D. Hall and A. Williams, *Biochemistry*, 1986, **25**, 4784.
44. B. I. Labow, D. Herschlag, and W. P. Jencks, *Biochemistry*, 1993, **32**, 8737.
45. F. Hollfelder and D. Herschlag, *Biochemistry*, 1995, **34**, 12255.
46. L. Wu and Z.-Y. Zhang, *Biochemistry*, 1996, **35**, 5426.
47. W. W. Cleland, *Methods Enzymol.*, 1995, **249**, 341.
48. A. C. Hengge, W. A. Edens, and H. Elsing, *J. Am. Chem. Soc.*, 1994, **116**, 5045.
49. A. Rajca, J. Rice, A. Streitweiser and H. Schaefer, *J. Am. Chem. Soc.*, 1987, **109**, 4189.
50. H. Horn and R. Ahlrichs, *J. Am. Chem. Soc.*, 1990, **112**, 2121.
51. A. C. Hengge, G. A. Sowa, L. Wu, and Z.-Y. Zhang, *Biochemistry*, 1995, **34**, 13982.
52. A. C. Hengge, Y. Zhao, L. Wu, and Z.-Y. Zhang, *Biochemistry*, 1997, **36**, 7928.

53. A. C. Hengge, J. M. Denu, and J. E. Dixon, *Biochemistry*, 1996, **35**, 7084.
54. A. C. Hengge and B. L. Martin, *Biochemistry*, 1997, **36**, 10 185.
55. S. R. Caldwell, F. M. Raushel, P. M. Weiss, and W. W. Cleland, *Biochemistry*, 1991, **30**, 7444.
56. Y. Lindqvist, G. Schneider, and P. Vinko, *Eur. J. Biochem.*, 1994, **221**, 139.
57. K. Scheffzek, M. R. Ahmadian, W. Kabsch, L. Wiesmüller, A. Lautwein, F. Schmitz, and A. Wittinghofer, *Science*, 1997, **277**, 333.
58. J. E. Murphy and E. R. Kantrowitz, *Mol. Microbiol.* 1994, **12**, 351.
59. J. M. Sowadski, M. D. Handschumacher, H. M. K. Murthy, B. A. Foster, and H. W. Wyckoff, *J. Mol. Biol.*, 1985, **186**, 417.
60. H. W. Wyckoff, *Am. Soc. Microbiol.* 1987, 118.
61. J. E. Butler-Ransohoff, D. A. Kendall, and E. T. Kaiser, *Proc. Natl. Acad. Sci. USA*, 1988, **85**, 4276.
62. A. Chaidaroglou and E. R. Kantrowitz, *Protein Eng.*, 1989, **3**, 127.
63. E. R. Kantrowitz, *Phosphate Microorg.*, 1994, 319.
64. E. E. Kim and H. W. Wyckoff, *J. Mol. Biol.*, 1991, **218**, 449.
65. J. E. Murphy, B. Stec, L. Ma, and E. R. Kantrowitz, *Nature Struct. Biol.*, 1997, **4**, 618.
66. F. Millich and E. L. Hayes, Jr., *J. Am. Chem. Soc.*, 1964, **89**, 2914.
67. R. L. Van Etten and M. S. Saini, *Clin. Chem.*, 1978, **24**, 1525.
68. J. H. Tatlock, M. Angelica Linton, X. J. Hou, C. R. Kissinger, L. A. Pelletier, R. E. Showalter, A. Tempczyk, and J. E. Villafranca, *Bioorg. Med. Chem. Lett.*, 1997, **7**, 1007.
69. O. Bodansky, *Adv. Clin. Chem.*, 1972, **15**, 43.
70. R. L. van Etten, *Ann. N Y Acad. Sci.*, 1982, **390**, 27.
71. R. L. van Etten, R. Davidson, P. Stevis, H. MacArthur, and D. Moore, *J. Biol. Chem.*, 1991, **266**, 2313.
72. M. Lin, C. Lee, and G. M. Clinton, *Mol. Cell. Biol.*, 1986, **6**, 4753.
73. J. Le Goff, P. Martin, and J. Raynaud, *Endocrinology*, 1988, **123**, 1693.
74. L. Nguyen, A. Chapdelaine, and S. Chevalier, *Clin. Chem.*, 1990, **36**, 1450.
75. M. Lin and G. M. Clinton, *Adv. Prot. Phosphatases*, 1987, **4**, 199.
76. Z.-Y. Zhang and J. E. Dixon, *Biochemistry*, 1993, **32**, 9340.
77. Z.-Y. Zhang, A. M. Thieme-Sefler, D. Maclean, D. J. McNamara, E. M. Dobrusin, T. K. Sawyer, and J. E. Dixon, *Proc. Natl. Acad. Sci. USA*, 1993, **90**, 4446.
78. Z.-Y. Zhang, Y. Wang, and J. E. Dixon, *Proc. Natl. Acad. Sci. USA*, 1994, **91**, 1624.
79. Z.-Y. Zhang, Y. Wang, Li Wu, E. B. Fauman, J. A. Stuckey, H. L. Schubert, M. A. Saper, and J. E. Dixon, *Biochemistry*, 1994, **33**, 15 266.
80. D. Barford, A. J. Flint, and N. K. Tonks, *Science*, 1994, **263**, 1397.
81. J. A. Stuckey, H. L. Schubert, E. B. Fauman, Z.-Y. Zhang, J. E. Dixon, and M. A. Saper, *Nature*, 1994, **370**, 571.
82. Z. Jia, D. Barford, A. J. Flint, and N. K. Tonks, *Science*, 1995, **268**, 1754.
83. E. J. Moran, S. Sarshar, J. F. Cargill, M. M. Shahbaz, A. Lio, A. M. M. Mjalli, and R. W. Armstrong, *J. Am. Chem. Soc.*, 1995, **117**, 10 787.
84. P. Cohen, *Annu. Rev. Biochem.*, 1989, **58**, 453.
85. Y. Lian, A. Haddy, and F. Rusnak, *J. Am. Chem. Soc.*, 1995, **117**, 10 147.
86. J. Goldberg, H. Huang, Y. Kwon, P. Greengard, A. C. Nairn, and J. Kuriyan, *Nature*, 1995, **376**, 745.
87. J. P. Griffith and M. A. Navia, *Cell*, 1995, **82**, 507.
88. E. V. Koonin, *Protein Sci.*, 1994, **3**, 356.
89. S. J. Pollack, J. R. Atack, M. R. Knowles, G. McAllister, C. I. Ragan, R. Baker, S. R. Fletcher, L. L. Iversen, and H. B. Broughton, *Proc. Natl. Acad. Sci. USA*, 1994, **91**, 5766.
90. Y. Zhang, J. Liang, and W. N. Lipscomb, *Biochem Biophys. Res. Commun.*, 1993, **190**, 1080.
91. A. F. Neuwald, J. D. York, and P. W. Majerus, *FEBS Lett.*, 1991, **294**, 16.
92. Y.-H. Lee, T. W. Olson, C. M. Ogata, D. G. Levitt, L. J. Banaszak, and A. J. Lange, *Nature Struct. Biol.*, 1997, **4**, 615.
93. J. Stuckey and G. M. Carman, *Protein Sci.*, 1997, **6**, 469.
94. D. C. Crans, C. M. Simone, R. C. Holz, and L. Que, Jr., *Biochemistry*, 1992, **31**, 11 731.
95. A. K. Saha, D. C. Crans, M. T. Pope, C. M. Simone, and R. H. Glew, *J. Biol. Chem.*, 1991, **266**, 3511.
96. J. K. Myers, S. M. Antonelli, and T. S. Widlanski, *J. Am. Chem. Soc.*, 1997, **119**, 3163.
97. W. P. Taylor, Z.-Y. Zhang, and T. S. Widlanski, *Bioorg. Med. Chem.*, 1996, **4**, 1515.
98. T. S. Widlanski, W. P. Taylor, and J. Roestamadji, unpublished results.
99. J. K. Myers and T. S. Widlanski, *Science*, 1993, **262**, 1451.
100. Q. Wang, U. Dechert, F. Jirik, and S. G. Withers, *Biochem. Biophys. Res. Commun.*, 1994, **200**, 577.
101. M. Imoto, H. Kakeya, T. Sawa, C. Hayashi, M. Hamada, T. Takeuchi, and K. Umezawa, *J. Antibiot.*, 1993, **46**, 1342.
102. B. Ye and T. R. Burke, *Tetrahedron*, 1996, **52**, 9963.
103. D. W. Cushman, H. S. Cheung, E. F. Sabo, and M. A. Ondetti, *Biochemistry*, 1977, **16**, 5484.
104. T. S. Widlanski, J. K. Myers, B. Stec, K. M. Holtz, and E. R. Kantrowitz, *Biochemistry*, submitted for publication.
105. R. L. Rice, J. M. Rusnak, F. Yokokawa, S. Yokokawa, D. J. Messner, A. L. Boynton, P. Wipf, and J. S. Lazo, *Biochemistry*, 1997, **36**, 15 965.
106. J. K. Stowell, T. S. Widlanski, T. G. Kutateladze, and R. T. Raines, *J. Org. Chem.*, 1995, **60**, 6930.

5.08
Mechanistic Investigations of Ribonucleotide Reductases

STUART LICHT and JOANNE STUBBE
Massachusetts Institute of Technology, Cambridge, MA, USA

5.08.1   INTRODUCTION

Ribonucleotide reductases (RNRs) catalyze the conversion of nucleotides to deoxynucleotides (Equation (1)), the monomeric precursors required for DNA biosynthesis. These enzymes are of interest for a variety of reasons and have consequently been the focus of a number of reviews. Reductases have been proposed to provide a link between the RNA and the DNA world.[1] Their central role in nucleotide metabolism has made them the successful target for design of antitumor and antiviral agents.[2–4] Their inducibility by DNA damaging agents suggests that they may play a role in cell cycle regulation.[5] The assembly of the essential diferric •Y cofactor of one class of reductases has provided a paradigm for studying and understanding posttranslational modification by metal/oxygen-based chemistry and biochemistry. Finally, the observation that reductases possess protein radicals[6] has attracted the attention of those interested in understanding how enzymes harness the chemical reactivity of free radicals to execute very difficult chemistry in an exquisitely controlled fashion.

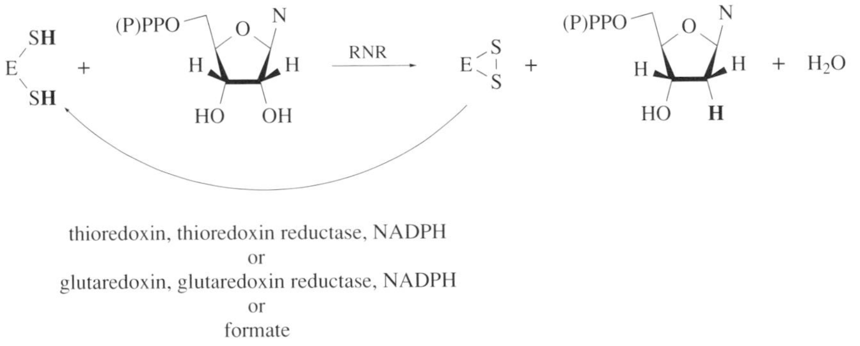

$$
\text{(1)}
$$

thioredoxin, thioredoxin reductase, NADPH
or
glutaredoxin, glutaredoxin reductase, NADPH
or
formate

This chapter will focus specifically on the last topic. A detailed chemical mechanism for nucleotide reduction will be presented. The chemical and biochemical evidence in support of each step will be described. Model reactions providing precedent for each step will be presented and discussed. In some cases, alternative mechanisms for a given step will be presented, and the alternatives evaluated based on biochemical evidence. General principles of free radical-based enzymatic transformations learned from these studies will be put forward with the goal of facilitating future studies on less well-understood enzymatic systems of similar complexity.

Four classes of RNRs have been isolated and characterized (Figure 1), all of which catalyze the same reaction (Equation (1)), although some reductases use nucleoside diphosphates as substrates while others use triphosphates. The class I enzymes are composed of two homodimeric subunits (α2 = R1 and β2 = R2) (Figure 2). The R2 subunit contains a diferric cluster, •Y cofactor, essential for reduction. While it is still controversial, there is approximately one •Y and two diferric clusters per R2. The R1 subunit is the business end of this reductase, and will quite likely have a very similar tertiary fold and active site in all four classes of reductases. It possesses three cysteine residues within the active site, all of which are essential for reduction. The structure of R1 reveals that these three cysteines are within 6 Å of each other.[7,8] Two additional cysteines in class I and II RNRs, invisible in the structure of the *Escherichia coli* R1 because the C-terminus of the R1 peptide is disordered, are required for *in vivo* reduction, shuttling reducing equivalents into the active site by the physiological thioredoxin or glutaredoxin reducing systems.[9,10]

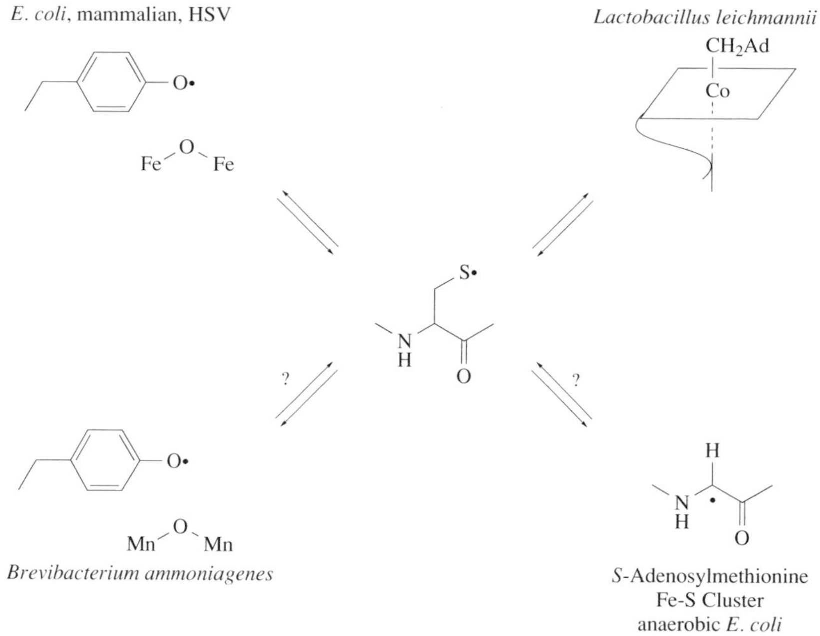

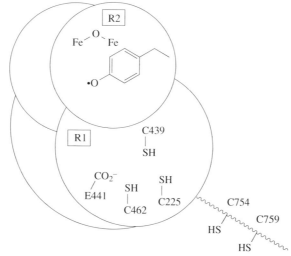

Figure 1 Metallocofactors used by ribonucleotide reductases.

Figure 2 The class I reductase from *E. coli*. The R2 subunit contains the diferric cluster and the •Y. The R1 subunit contains the substrate binding site, the thiol that becomes the putative thiyl radical (C439), the cysteines that provide reducing equivalents to the nucleotide (C462 and C225), the cysteines that shuttle reducing equivalents from an external reducing system (C754 and C759), and a glutamate proposed to act as a general base (E441).

The class II enzymes require adenosylcobalamin (AdoCbl) as a cofactor. Three of the members of this class are composed of monomers,[11–14] while the fourth has an $\alpha2\beta2$ subunit composition.[15] Despite the availability of sequences for all of these proteins and the X-ray structure of the adenosylcobalamin-requiring methylmalonyl-CoA mutase,[16] an AdoCbl binding motif has not been identified.

The class III reductases are composed of two homodimeric subunits ($\alpha2\beta2$). The cofactor-generating subunit, $\beta2$, bears sequence homology to the pyruvate formate lyase activating enzyme.[17,18] Data suggest that both of these enzymes possess a single iron–sulfur cluster at their subunit interface and bind *S*-adenosylmethionine (SAM). The iron–sulfur cluster in the presence of SAM generates a glycyl radical on the $\alpha2$ subunit, in a sequence context similar to the glycyl radical generated on pyruvate formate lyase.[17,19–21] During glycyl radical generation, SAM is converted to 5′-deoxyadenosine (5′-dA) and methionine. The possible analogy between this unprecedented chemistry and the reactivity of AdoCbl has not gone unnoticed.[18,22] This enzyme differs from the class I and II reductases in that formate, rather than thioredoxin or glutaredoxin, provides the reducing equivalents for nucleotide reduction and, in the process, it is oxidized to CO_2. It is probably not a coincidence that formate is also one of the products of pyruvate formate lyase.

Based on growth requirements, the loss of enzymatic activity in the presence of hydroxyurea and the UV-visible spectrum of the purified protein, the class IV reductases are proposed to contain a dinuclear manganese cluster and a •Y.[23,24] EPR studies support the latter proposal.[25] Few details on this protein, however, have emerged.

Despite the apparent diversity in cofactor requirement, subunit organization, and primary sequence, a unifying mechanistic theme for RNRs has emerged. The domains or subunits in which nucleotide reduction occurs in each class of RNR are predicted to have similar secondary and tertiary structures and to catalyze the reduction process through similar radical-based chemical mechanisms. What is unique to each class is the metallocofactor that functions as the radical chain initiator of the reduction process. Each cofactor has been proposed to generate a thiyl radical in the active site of the subunit in which reduction occurs and this thiyl radical is proposed to initiate the radical-dependent reduction process. In the first half of this review, the postulated mechanisms by which each cofactor generates a thiyl radical will be presented and discussed. In the second half of the chapter, a detailed mechanism for the radical-mediated nucleotide reduction process is presented. Data in support of this mechanism, including detailed studies with a variety of mechanism-based nucleotide inhibitors of RNR and studies of site-directed mutants of RNR, which must be accommodated by any proposed alternative mechanisms, will be presented and discussed.

5.08.1.1   Outline of the Proposed Mechanisms for Thiyl Radical Formation, Nucleotide Reduction, and Mechanism-based Inactivation by Substrate Analogues

What is known about the generation of the putative essential thiyl radical for RNRs (Figure 1) will be discussed in detail for the the class I *E. coli* and class II *L. leichmannii* enzymes. In the case of the class I reductases, the •Y on the R2 subunit, formed during the assembly of the diferric cluster,[26–31] is proposed[32–35] to initiate a chain of proton-coupled electron transfers between amino acid residues (Figure 3) that results in a net intersubunit electron transfer and formation of a thiyl radical on the R1 subunit. The pathway for this process has been proposed based on the structures of the R1 and R2 subunits and on biochemical studies, which suggest a mode of docking of these subunits. In addition, sequences of 40 class I RNRs are available and all of the indicated residues are conserved.[36] Finally, studies from the Sjöberg, Thelander, and Gräslund laboratories have shown that mutation of any of these residues destroys the ability of the enzymes to catalyze nucleotide reduction.[32–35] A distance of ∼35 Å is proposed between the Y122 radical on R2 and the thiol of C439 on R1. The mechanism of communication between these subunits is a major focus of effort in many laboratories. The thermodynamics and kinetics of •Y reduction will be discussed, as will the possible significance of proton transfer.

The class II RNRs utilize AdoCbl as a radical chain initiator. The *L. leichmannii* reductase catalyzes homolysis of the carbon–cobalt bond of AdoCbl. The unpaired spin generated at the 5′ position of the axial adenosine ligand of the cofactor is proposed to generate an active site thiyl radical on residue C408. Thus, AdoCbl has been proposed to serve as the functional equivalent of the R2 subunit in the class I reductase (Figure 4).

The major difference between these two classes, as will be described in detail subsequently, is that the cofactor is directly involved in hydrogen atom abstraction from cysteine 408, thus requiring the cofactor to bind in a region close to the active site cysteines. The factors influencing catalysis of carbon–cobalt bond homolysis, the question of whether carbon–cobalt bond homolysis and thiyl radical formation are concerted or stepwise, and the issue of whether the carbon–cobalt bond is reformed after each nucleotide reduction event will be addressed.

The proposed mechanism for nucleotide reduction[37,38] is shown in Scheme 1. In the first step, the metallocofactor generates the essential thiyl radical, the detailed mechanism of which is cofactor-dependent. Once the thiyl radical is generated, the mechanisms of all RNRs are proposed to become congruent. The thiyl radical is proposed to initiate catalysis by abstraction of the 3′ hydrogen of the nucleotide substrate to generate a 3′-nucleotide radical intermediate. This intermediate is proposed to rapidly lose its 2′ hydroxy group as water, and the hydrogen of its 3′-hydroxy group to glutamate, forming a 3′-keto-2′-deoxynucleotide radical. Reduction is proposed to proceed via a single electron transfer from one of the cysteines located on the α face of the nucleotide concomitant

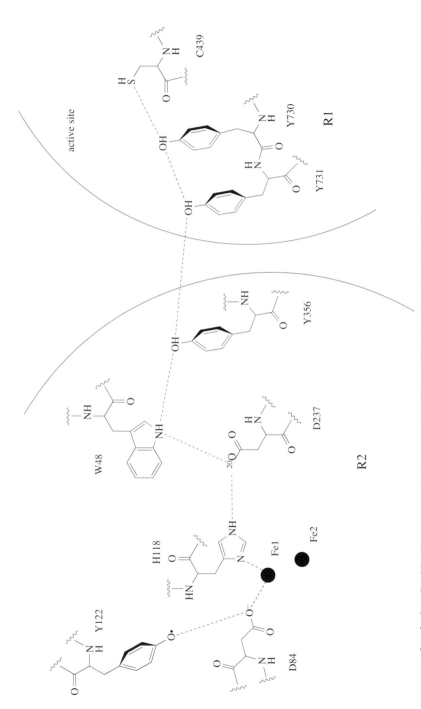

Figure 3 Residues proposed to be involved in electron transfer between the R2 and R1 subunits of the *E. coli* RNR. Y122 of R2 is the stable •Y. C439 of R1 is proposed to form the catalytically essential thiyl radical. Other residues shown represent a proposed electron/proton transfer pathway between the two proteins. Adapted from ref. 34.

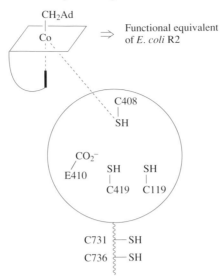

Figure 4   The class II reductase from *L. leichmannii*. Shown are the cysteine believed to be oxidized to a thiyl radical (C408), the cysteines proposed to provide reducing equivalents to the nucleotide (C119 and C419), the cysteines proposed to shuttle reducing equivalents from an external reducing system (C731 and C736), and a glutamate proposed to act as a general base (E410).

with rapid proton transfer to generate a 3′-keto-2′-deoxynucleotide intermediate and a disulfide radical anion. A second one-electron reduction from the resulting disulfide radical is proposed to generate a 3′-ketodeoxynucleotide radical and a disulfide, and regenerate the anionic form of glutamate. Reabstraction of hydrogen atom from the cysteine on the β face of the nucleotide by the 3′-deoxynucleotide radical completes the first turnover and regenerates the thiyl radical. In order for multiple turnovers to take place, the disulfide must be rereduced. Thermodynamic and kinetic considerations related to each of the proposed steps will be described, with a special emphasis on the influence of protonation state on the stability of radical intermediates. Alternative mechanisms proposed by others will also be examined.

Since the initial report of Eckstein, Thelander, and co-workers in 1976[39] that the *E. coli* RNR could be inactivated by 2′-chloro and 2′-azido-2′-deoxynucleotides, these and similar compounds have played an important role in understanding the catalytic capabilities of the RNRs. Any mechanism proposed for the nucleotide reduction process must also explain the unusual chemistry observed with 2′-substituted nucleotide analogues. Over the years a mechanistic paradigm (Scheme 2) has unfolded to explain the observed inactivation of ribonucleotide reductases by 2′-substituted-2′-deoxynucleotides.[38] Almost all of these nucleotides are mechanism-based inhibitors in which, as for the normal substrate, the reaction is initiated by abstraction of the 3′ hydrogen by the active site thiyl radical. In addition, in the case of the 2′-chloro- and fluoro-derivatives, inactivation is also accompanied by deoxynucleotide production. To explain the difference in the results observed with these inhibitors in comparison with the normal substrate, the 2′ substituent is proposed to depart without the requirement for protonation of the leaving group. Thus, the same 3′-keto-2′-deoxynucleotide radical is present in the active site as is observed during the normal reduction process, but both thiols on the α face of the nucleotide are proposed to remain protonated. The 3′-keto-deoxynucleotide radical can then be reduced via hydrogen atom transfer, not electron transfer as proposed with the normal substrate. Reduction from the β face regenerates the thiyl radical, allowing active cofactor to be regenerated as well. Reduction from the α face forms a disulfide radical. In both cases, the product is the 3′-ketodeoxynucleotide, which has been identified by trapping with sodium borohydride.[40,41] Appropriate labeling studies using [2′ or 3′-³H]-labeled nucleotide analogues has provided strong support for this proposal. Reduction from the top face is tantamount to a 1,2 hydrogen shift and is markedly similar to reactions catalyzed by enzymes using AdoCbl to mediate rearrangement reactions.

Regardless of which thiol provides the reducing equivalents, a 3′-ketodeoxynucleotide is generated. This compound is released from the enzyme active site into solution, where its chemical instability results in release of base (N), loss of inorganic pyrophosphate or tripolyphosphate, and generation of 2-methylene-3-2*H*-furanone. This sugar analogue is highly activated toward nucleophilic attack, and can alkylate and inactivate the R1 subunit or its equivalent. Remarkably,

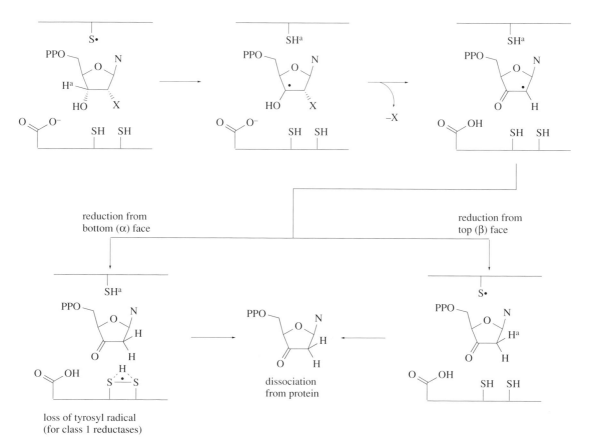

Scheme 1

Scheme 2

the normal nucleotide substrate can become a mechanism-based inhibitor when either of the cysteines in the active site are mutated to serines, disrupting the normal reduction process, or when the cysteines are oxidized to a disulfide.[9,10,42-44] In addition, in studies with R1 mutants of the conserved glutamate within the active site, the normal substrate again acts as a mechanism-based inhibitor.[45,46] Similarities and differences between chemical model systems for the inactivation processes and the inactivation chemistry observed in enzymatic systems will be described. Whatever model is proposed, it must be able to account for a wealth of biochemical data that has been accumulated with these mechanism-based inhibitors.

5.08.2 GENERATION OF THIYL RADICALS AT AN RNR ACTIVE SITE BY ADENOSYLCOBALAMIN

5.08.2.1 Evidence for a Thiyl Radical

The cofactor whose role in catalysis is best understood is AdoCbl in the class II RNRs. The RNR from *L. leichmannii* (RTPR) has been studied most extensively of the class II enzymes. Evidence will be presented that homolytic bond cleavage of the carbon–cobalt bond of AdoCbl allows abstraction of a hydrogen atom active site cysteine, C408 (Scheme 3). This system has provided the only direct spectroscopic evidence for thiyl radicals in RNRs to date. However, the biochemical and chemical similarities between this enzyme and the *E. coli* class I RNR strongly suggest that thiyl radicals are intermediates in this enzyme, as well.

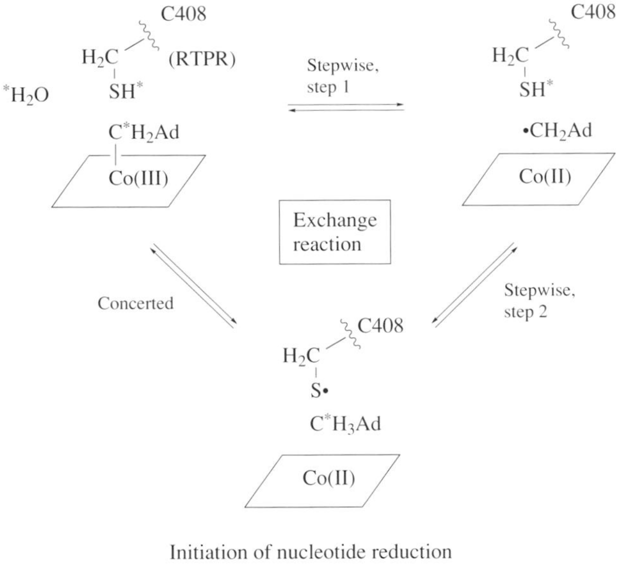

Scheme 3

A key to understanding the function of the cofactor using presteady state stopped flow (SF) UV-vis spectroscopy and rapid freeze quench (RFQ) EPR spectroscopy came with the discovery of a second reaction catalyzed by this enzyme. It was reported in the 1960s by two groups that if [5′-³H]-AdoCbl was incubated with RTPR, allosteric effector, and reductant the 5′ hydrogens of AdoCbl could exchange with solvent.[47-49] A proposed mechanism for this process is shown in Scheme 3. The proposal is that the cofactor generates, in a stepwise or a concerted fashion, a thiyl radical, 5′-dA and cob(II)alamin and that this process is reversible. Based on the relatively fast rate of this process in comparison to the nucleotide-reduction process, it has been argued in detail that this exchange reaction offers a glimpse of the mechanism by which the cofactor generates the thiyl radical, and hence reproduces the key first step in the nucleotide-reduction process.[50]

Historically, the first evidence that the *L. leichmannii* RTPR catalyzes the cleavage of the carbon–cobalt bond of AdoCbl came from the characterization of this exchange reaction. Incubation of

RTPR with [5'-³H]-AdoCbl labeled nonspecifically in both the 5'-pro-R and pro-S positions was found to cause exchange of all of the tritium in the cofactor with the solvent, indicating that the two 5' methylene hydrogens become equivalent in the course of this reaction.[51] This observation suggests that the enzyme catalyzes formation of 5'-dA and cob(II)alamin from AdoCbl.

The hypothesis that RTPR catalyzes the formation of cob(II)alamin was confirmed in the 1970s by spectroscopic characterization of intermediates formed during the exchange reaction (and nucleotide reduction).[52,53] SF UV-vis spectroscopy showed that under the conditions of the exchange reaction, using dithiothreitol (DTT) as a reductant, RTPR catalyzes the conversion of AdoCbl into cob(II)alamin with an apparent first-order rate constant of ~ 40 s^{-1}.[52] Carbon–cobalt bond cleavage is thus kinetically competent to be on the pathway for nucleotide reduction, which occurs with a rate constant of 1.5 s^{-1}. RFQ EPR experiments revealed that the intermediate formed on carbon–cobalt bond homolysis is unique and distinct from cob(II)alamin bound to the enzyme in the presence of 5'-dA when turnover is not taking place. In the latter case, the spectrum exhibits a $g_\perp$ of 2.23, a $g_\parallel$ of 2.0, and a cobalt nuclear hyperfine interaction of 110 G.[54] In the former case, the freeze-quenched intermediate has a g-value of 2.12 and a cobalt hyperfine of 54 G.[50,53] This intermediate appears with a rate constant ~ 40 s^{-1}, the same as the rate constant for cob(II)alamin formation measured by SF.[50,53] The EPR signal was originally hypothesized to result from a cob(II)alamin/5'-dA• radical pair. However, the repetition of the experiment with [5'-²H₂]- and [5'-¹³C]-AdoCbl did not alter the EPR spectrum of this intermediate, as would have been predicted for an intermediate with unpaired spin density on the 5' position of 5'-dA. Experiments to establish the identity of this intermediate provide strong evidence that it is an enzyme-based thiyl radical exchange-coupled to cob(II)alamin, and that this thiyl radical mediates both nucleotide reduction and the exchange of the 5' hydrogens of AdoCbl.

Evidence that the species that gives rise to the observed EPR signal is a thiyl radical comes from several different types of experiments. Cloning and overexpression of RTPR[12] has allowed a number of mechanistically informative site-directed mutagenesis studies. The X-ray structure of the R1 subunit of the *E. coli* RNR, in conjunction with many biochemical studies, strongly suggested that C439 of R1 is the cysteine that is converted to a thiyl radical via the •Y. Based on a short, statistically insignificant, sequence homology with the *E. coli* RNR, C408 was predicted to act as the corresponding thiyl radical in RTPR (Figure 3).

L. leichmannii RTPR: TNPC408GEISLA
E. coli RDPR: SNLC439LEIAP

When this residue was mutated to a serine, the mutant enzyme was unable to catalyze either nucleotide reduction or the exchange reaction, consistent with the proposed role for C408.[10] To test this hypothesis further, RTPR was prepared in which all of the cysteines contained deuterium at their β positions ([β-²H₂]-cysteine), and this labeled RTPR was used in RFQ EPR experiments. This isotopic substitution resulted in a pronounced narrowing of the cobalt hyperfine features of the EPR spectrum of the intermediate, establishing its predicted identity.[50] Computer simulation of the observed EPR signals, at both 9 and 35 GHz, further confirmed its identification, and allowed an estimate of the distance between the thiyl radical and cob(II)alamin to be 5.5–6 Å.[55]

Although the experiments described above were carried out in the absence of substrate, experiments in the presence of substrate indicate that an enzyme-based thiyl radical with very similar spectroscopic properties is also generated in a kinetically competent fashion. When reaction mixtures containing RTPR, substrate, allosteric effector, and TR (thioredoxin)/TRR (thioredoxin reductase)/NADPH are freeze-quenched during turnover, the EPR-active intermediate trapped is similar in lineshape and effective g-value to the intermediate trapped in the exchange reaction. It also exhibits cobalt hyperfine narrowing on substitution of the β hydrogens of cysteine residues with deuterium.[50] This species, generated with k_{obs} of >200 s^{-1}, is kinetically competent to be involved in turnover. These results support our contention that the exchange reaction provides an excellent model for how AdoCbl acts as a radical chain initiator.

While direct spectroscopic observation of a thiyl radical intermediate has only been established with the *L. leichmannii* class II RNR (although similar results are observed with the *Thermophilia acidophilus* RNR[56]), the striking biochemical similarities between this enzyme and *E. coli* class I RNR strongly imply that a thiyl radical intermediate is also formed with this RNR during nucleotide reduction. Both enzymes react with 2'-chloro-2'-deoxynucleotide mechanism-based inactivators, forming inorganic pyrophosphate (PP), tripolyphosphate (PPP), base, and 2-methylene-3-(2H)-furanone which reacts with the protein to inactivate the enzyme and form a chromophore at 320 nm (Scheme 2).[40,41,57–59] In both enzymes, the phenotypes of five conserved cysteine to serine site-directed mutants, including the cysteine proposed to form the thiyl radical in the *L. leichmanni* reductase (C408), are strikingly similar, consistent with a commonality of mechanism.[9,10,42,43,60,61] What is known mechanistically about the normal reduction process is also very similar in both RNRs. In both cases, the 2′ hydroxyl group of the nucleotide is replaced, with retention of configuration, by a solvent-derived hydrogen during the normal reduction process.[62,63] Isotope effects on 3′ C—H bond cleavage are comparable in the two enzymes,[64,65] consistent with a common mechanism for 3′ hydrogen abstraction. The allosteric regulation patterns for both enzymes are similar.[66–69] Finally, both enzymes react with 2′-difluoro-2′-deoxynucleotides (F_2CDP or F_2CTP, respectively) and 2′-fluoromethylene cytidine 5′-di or triphosphates (FMCDP or FMCTP) to form new radical species shown in one case to be substrate-derived.[45] Since the mechanisms of nucleotide reduction appear to be similar in the two enzymes, and the cysteine residues involved in the mechanism appear to be performing the same functions in both enzymes, it is likely that the thiyl radical intermediate will be present in the *E. coli* RNR as well as the *L. leichmannii*.

5.08.2.2 Catalysis of Carbon–Cobalt Bond Cleavage

One of the critical questions in the mechanism of RTPR is how the enzyme catalyzes homolytic carbon–cobalt bond cleavage, and how this bond homolysis leads to formation of a thiyl radical. Answering this question must begin with a measurement of the bond dissociation energy of the carbon–cobalt bond in AdoCbl. Studies from the Finke[70,71] and Halpern[72] laboratories have established that this bond dissociation energy is ~ 30 kcal mol^{-1} in neutral aqueous solution, with the rate constant determined to be 10^{-9} s^{-1} at room temperature.[73] Activation parameters for this reaction have also been determined, both in neutral aqueous solution ($\Delta H\ddagger = 31.8 \pm 0.7$ kcal mol^{-1}, $\Delta S\ddagger = 6.8 \pm 1.0$ cal mol^{-1} K^{-1})[73] and in ethylene glycol, where viscosity makes cage effects more important[74] ($\Delta H\ddagger = 34.5 \pm 0.8$ kcal mol^{-1}, $\Delta S\ddagger = 14 \pm 1.0$ cal mol^{-1} K^{-1}).[71]

The observation of a rate constant for cob(II)alamin formation of the order of 200 s^{-1} for RTPR (under turnover conditions) requires a rate acceleration of $\sim 10^{11}$ over the uncatalyzed homolysis,[73] which corresponds to a $\Delta G\ddagger$ of 16 kcal mol^{-1}. There is also evidence that the thermodynamics of the carbon–cobalt bond cleavage reaction are highly perturbed when the cofactor is bound to RTPR. Up to 50% of bound cobalamin is in the form of cob(II)alamin under the conditions of the exchange reaction,[52] suggesting an equilibrium constant on the order of unity ($\Delta G \sim 0$) between AdoCbl and its homolysis products. This perturbation of the equilibrium thermodynamics of carbon–cobalt bond homolysis is likely to be required for catalysis. The activation parameters for carbon–cobalt bond homolysis in solution indicate that the transition state is not much higher in energy than the products, as expected for a highly endergonic reaction. Thus, lowering the activation barrier by ~ 16 kcal mol^{-1} would not be possible by lowering the energy of the transition state alone. Enzymatic destabilization of the reactant ground states and/or stabilization of the product is also required.

5.08.2.3 Chemical Models of the Role of Steric Strain and Basicity of the Axial Ligand in Acceleration of Carbon–Cobalt Bond Homolysis

As outlined below, model systems have inspired a variety of possibilities for enzymatic mechanisms of catalysis of carbon–cobalt bond homolysis. There is no general consensus on which, if any, of these mechanisms are actually used. One mechanism for the modulation of the energetics of carbon–cobalt bond cleavage has been proposed to be through the use of steric effects of the *trans* axial ligand of AdoCbl.[75–78] Studying heterolytic dealkylations of alkylcobalamins, Grate and Schrauzer observed that the presence of the nitrogen base coordinating to cobalt increased the dealkylation rate.[79] They suggested that the axial nitrogen ligand causes an upward bending of the corrin, applying steric strain to the axial alkyl substituent of cobalt and accelerating bond cleavage. Although this study examined heterolytic cleavage, the same principle would apply to homolytic bond cleavage. For cobaloximes, studies on the effects on carbon–cobalt bond strength of bulky phosphine axial ligands support this hypothesis,[80] as does the increased carbon–cobalt bond strength observed for benzylcobalt complexes of a porphyrin compared with the analogous cobaloxime, which is less rigid.[81]

A second mechanism suggests that the basicity of the axial ligand can affect the homolysis rate, with lower basicity making the carbon–cobalt bond more labile. The bond dissociation enthalpies of a series of cobaloximes with substituted pyridines or imidazole as *trans*-axial ligands were found to increase with increasing basicity of the *trans*-axial ligand. These results are consistent with electron donation by the ligand stabilizing the cobalt(III) oxidation state of the intact cobaloxime over the cobalt(II) state resulting from homolytic bond cleavage.[82]

Work on the lability of neopentylcobalamin (a derivative of AdoCbl in which a neopentyl group replaces the 5′-deoxyadenosyl moiety) supports the importance of both steric and electronic factors. Neopentylcobalamin undergoes uncatalyzed carbon–cobalt bond cleavage up to 10^6 times faster than does AdoCbl.[83] In addition, the base-on form of neopentylcobalamin is $\sim 10^3$ times more reactive than the base-off form.[84,85]

Other studies on this system emphasize the effect of the entropy due to mobility of the acetamide side chains on the energetics of carbon–cobalt bond homolysis.[83,85–88] While the activation enthalpies are approximately the same for the base-on and base-off forms, the activation entropies differ by ~ 10 cal mol^{-1} K^{-1} in aqueous solution[86] and ~ 20 cal mol^{-1} K^{-1} in ethylene glycol,[83] indicating that the destabilization is entropic. This observation led to the suggestion that steric interactions between the neopentyl group and one or more of the acetamide side chains cause neopentylcobalamin to be entropically destabilized relative to the products of carbon–cobalt bond homolysis.[85] In support of this hypothesis, epimerization at C13 of the corrin, which puts another acetamide side chain in a position to interact with the neopentyl group, increases the entropy of activation of carbon–cobalt bond homolysis.[87]

Structural studies have also been used to analyze the factors that contribute to carbon–cobalt bond strength. X-ray crystallographic studies are consistent with a role for the *trans*-axial ligand in governing the length and, by implication, the strength of the carbon–cobalt bond.[77] Comparison of the structure of a cobalamin with imidazole and cyanide as axial ligands with that of cyanocobalamin (dimethylbenzimidazole and cyanide as axial ligands) showed that the bulkier dimethyl-benzimidazole causes a greater "upward" distortion of the corrin ring than does imidazole (an angle between the "northern" and "southern" halves of the corrin ring of $18 \pm 0.3°$ for the cobalamin containing dimethylbenzimidazole vs. $11.3 \pm 0.2°$ for the cobalamin containing imidazole).[89]

However, resonance Raman experiments indicate that the carbon–cobalt bond stretching frequency is the same for the base-on and base-off forms of AdoCbl, suggesting that the *trans* ligand does not alter the bond strength of the carbon–cobalt bond itself in cobalamins.[90] Nonetheless, the base-off form of AdoCbl has a bond dissociation energy that is ~ 5 kcal mol^{-1} larger than the base-on form (which is also in contrast to what would be predicted from the electronic effects outlined above).[91] One explanation for these results is that the base affects the stability of cob(II)alamin without having a large effect on the stability of AdoCbl. Thus, while model systems have demonstrated how the energetics of a carbon–cobalt bond might be modulated, the specific factors that determine the bond dissociation energy for the physiologically relevant cobalamins remain an active area of investigation.

5.08.2.4 Molecule-induced Homolysis as a Model for Acceleration of Carbon–Cobalt Bond Cleavage

In the case of RTPR, the chemistry may be distinct from the model studies due to the involvement of a protein radical. In the enzymatic reaction, we have proposed that carbon–cobalt bond cleavage occurs in a concerted fashion with thiyl radical formation (Scheme 3).[50] This process would be less endergonic than carbon–cobalt bond homolysis alone. Using 30 kcal mol^{-1} as the homolytic bond dissociation enthalpy of the carbon–cobalt bond of AdoCbl,[73] 88–91 kcal mol^{-1} as the homolytic bond dissociation enthalpy of the RS—H bond,[92,93] and 98 kcal mol^{-1} as the homolytic bond dissociation enthalpy of the C—H bond,[94] the enthalpy for a concerted reaction can be estimated to be 15–20 kcal mol^{-1} as opposed to 30 kcal mol^{-1} for the carbon–cobalt bond homolysis alone.

This concerted reaction may be analogous to the class of radical reactions known as molecule-induced homolyses, in which homolytic cleavage of a nonradical species is accelerated by interaction with another nonradical species.[95] The radical chain halogenation of styrene by *t*-butyl hypochlorite procedes rapidly and exothermically in the dark, even though the bond dissociation energy of the oxygen–chlorine bond in *t*-butyl hypochlorite is ~ 40 kcal mol^{-1}.[96] Observing 1,2-dichloro-phenylethane among the products of this reaction, Walling *et al.* hypothesized that radical initiation requires a concerted O—Cl bond homolysis and addition of chloro radical to styrene (Scheme 4(a)).

Since the latter reaction is estimated to be exothermic by 49 kcal mol^{-1} the concerted reaction would be expected to be exothermic.[96] A similar molecule-induced homolysis has been reported for the radical addition of iodine to styrene (Scheme 4(b)), in which formation of radical species is 10^6 times faster than the homolysis of iodine in the absence of styrene.[97]

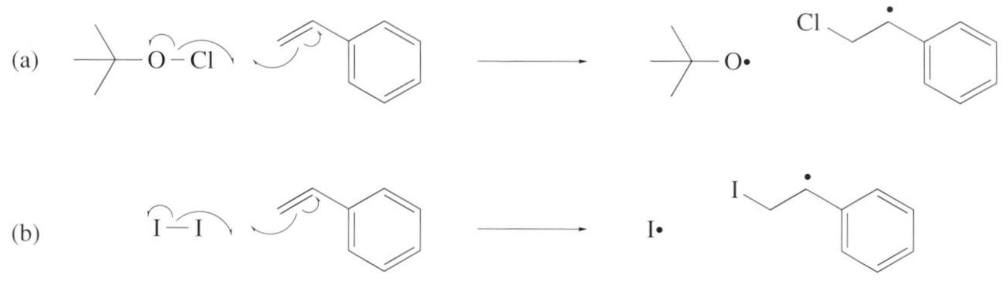

Scheme 4

Although concerted reactions can be favorable from an enthalpic standpoint, they are generally entropically less favorable than the corresponding stepwise reactions due to the formation of a more ordered transition state. In an enzymatic system, the binding energy of AdoCbl would be expected to compensate for the entropy cost of preorganization of the reacting species at the enzyme active site.[98] In this regard, it may be significant that the concentration dependence of the observed rate constants for cob(II)alamin formation under both exchange reaction and turnover conditions, in combination with the failure to detect binding of AdoCbl to enzyme in equilibrium binding experiments on the catalytically inactive C408S mutant, suggest that the K_d for binding of AdoCbl to RTPR is >100 μM.[99] Given the number of binding determinants on AdoCbl, a much lower K_d should be possible and might have been expected. Perhaps unfavorable entropic factors associated with preorganization of the enzyme and the cofactor contribute to this relatively high K_d.

In nonenzymatic systems, a favorable enthalpy change of a reaction can facilitate concerted reactions even when the entropy change of the reaction is very negative. The oxidative addition of methane to rhodium(II) porphyrin complexes (Figure 5) is another example of how molecule-assisted homolysis can allow rapid homolytic cleavage of strong bonds. Wayland *et al.* have shown that this reaction is second order in rhodium complex, exhibits a sizable deuterium isotope effect ($k_{CH4}/k_{CD4} = 8.6$), and has a relatively small activation enthalpy ($\Delta H^{\ddagger} = 7.1 \pm 1.0$ kcal mol^{-1}) and a large negative activation entropy ($\Delta S^{\ddagger} = -39 \pm 5$ kcal mol^{-1} K^{-1}). A concerted reaction with a four-centered transition state was proposed to account for these observations (Figure 5(b)). In this mechanism, C—H bond breaking occurs in concert with Rh—H and Rh—Me bond formation.[100] The oxidative addition of methane to a diporphyrin dirhodium complex with the porphyrin units tethered by a diether spacer occurred $\sim 10^2$-fold faster than the termolecular reaction, consistent with preorganization of the rhodium units reducing the entropy cost of the four-centered transition state (Figure 5(c)).[100] The thermodynamics of these model reactions are roughly analogous to the thermodynamics expected for concerted carbon–cobalt bond cleavage and thiyl radical formation. In the case of the rhodium complexes, the formation of two weaker bonds (Rh—H and Rh—CH$_3$, both ~ 60 kcal mol^{-1}) "pays for" the homolytic cleavage of a stronger bond (Me—H, 105 kcal mol^{-1}).[100] In the case of a concerted carbon–cobalt bond cleavage, the formation of a stronger bond (the CH$_2$—H bond of 5'-dA, ~ 100 kcal mol^{-1}) partially "pays for" the homolytic cleavage of two weaker bonds (Co—C, ~ 30 kcal mol^{-1} and S—H, ~ 90 kcal mol^{-1}).

5.08.2.5 Structural Information on Cobalamin Enzymes and its Relevance to the Problem of Carbon–Cobalt Bond Cleavage

As summarized above, data from model systems suggest that the identity and conformation of the *trans* axial ligand of cobalamin bound to an enzyme could affect the carbon–cobalt bond strength, although the precise causes of *trans* effects in model systems are incompletely understood. One of the reasons attention has focused on the *trans* axial ligand is the unusual 5,6-dimethylbenzimidazole (DMB) moiety that serves this purpose in AdoCbl. This ligand might affect the structure and reactivity of the cofactor through both the steric effects mentioned previously[89] (Section 5.08.2.3) and electronic effects. The pK_a of the DMB moiety in AdoCbl has been measured

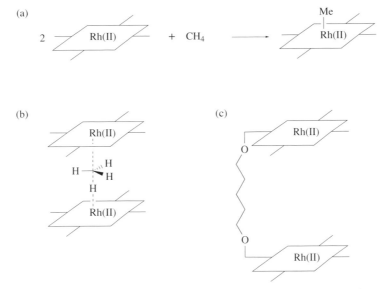

Figure 5 (a) Oxidative addition of methane to rhodium porphyrins. (b) Four-centered transition state for the oxidative addition of methane. (c) Covalently linked porphyrin rhodium dimer.

to be 3.7,[101] compared with a pK_a of 7 for imidazole, consistent with different σ donor properties for these two ligands.

Crystallographic and EPR studies of the cobalamin-containing enzymes methionine synthase (which catalyzes carbon–cobalt bond heterolysis)[102] and methylmalonyl-CoA mutase (which catalyzes carbon–cobalt bond homolysis)[16,103] indicate that the *trans* axial ligand is likely to be very important for establishing the reactivity of the cofactor, albeit in a surprising way. In both cases, the DMB base is not coordinated to cobalt in the enzyme-bound cobalamin, and a histidine residue from the enzyme serves as the *trans* axial ligand in its place.

In the case of methylmalonyl-CoA mutase, the cobalamin in the crystal is a mixture of cobalt(II) and cobalt(III) forms, and no electron density corresponding to the 5′-deoxyadenosyl moiety is observed.[16] These observations suggest that the crystal structure is close to the structure the enzyme assumes after carbon–cobalt bond cleavage. The crystallographic data have been interpreted as indicating a long histidine–cobalt bond (2.5 Å vs. 1.95–2.2 Å in free cobalamins).[16] It was hypothesized that this unusual Co—N bond length promotes carbon–cobalt bond homolysis by stabilizing the cobalt(II) oxidation state of the homolysis products over the cobalt(III) AdoCbl state. This view is at odds with Spiro and Banerjee's interpretation of their resonance Raman data, which posits that for cobalamins, coordination of benzimidazole alters the stability of cob(II)alamin (the homolysis product) rather than AdoCbl.[90] In addition, the existing crystallographic data do not rule out a role for steric interactions in catalysis. Unfavorable steric interactions between the enzyme and AdoCbl might be relieved after homolysis, resulting in the observed structure, which is consistent with a flat, rather than a "flexed," corrin ring. The mechanism of labilization of the carbon–cobalt bond by this enzyme thus remains an interesting and controversial question. Nonetheless, the structural data rule out the possibility that the unusual DMB cofactor, long postulated to be the mediator of steric weakening of the carbon–cobalt bond at enzyme active sites, interacts directly with either cobalt or the corrin ring to weaken the carbon–cobalt bond in methylmalonyl-CoA mutase-bound AdoCbl.

While the methylmalonyl-CoA mutase structure has been extremely thought-provoking for the field of cobalamin biochemistry, the *L. leichmannii* RTPR is likely to activate the carbon–cobalt bond for homolytic cleavage by a different mechanism. EPR studies on unlabeled (^{14}N in DMB) cob(II)alamin bound to [U-^{15}N]-RTPR in the presence of an allosteric effector and a mechanism-based inactivator, the triphosphate of 2′-methylene-2′-deoxycytidine, have shown ^{14}N hyperfine interaction with cobalt. These results demonstrate that the enzyme does not provide the *trans* axial nitrogen ligand and strongly suggest that the DMB is bound to cobalt.[104] Thus, this enzyme may use the added steric bulk of the DMB for catalysis in ways that appear to be unnecessary in methylmalonyl-CoA mutase. Further speculation must await determination of a three-dimensional structure of RTPR.

5.08.2.6 Mechanistic Studies on the *L. leichmannii* RNR Provide Evidence for a Concerted Mechanism in which Carbon–Cobalt Bond Homolysis is Entropy-driven

To address the question of how RTPR accomplishes catalysis of carbon–cobalt bond homolysis, the kinetics and thermodynamics of carbon–cobalt bond homolysis/thiyl radical formation have been studied for the exchange reaction. Efforts were initially focused on establishing if this reaction proceeds in a stepwise or a concerted fashion. One piece of evidence favoring the concerted model is the inability of C408S RTPR to catalyze carbon–cobalt bond homolysis.[105]

To make a distinction between the two mechanisms, kinetic and equilibrium isotope effects on cob(II)alamin formation were measured using unlabeled AdoCbl in D_2O, [5′-$^2H_2$]-AdoCbl in H_2O, and with [5′-$^2H_2$]-AdoCbl in D_2O.[106] Comparison of the k_{obs} in these experiments with that measured with unlabeled AdoCbl in H_2O revealed k_H/k_D of 1.7, 1.6, and 2.7, respectively. An equilibrium isotope effect was also measured. Twice as much cob(II)alamin was generated with [5′-$^2H_2$]-AdoCbl in D_2O than with AdoCbl in H_2O. The kinetic isotope effects on k_{obs}, which contains contributions from both forward and reverse rate constants, can be ascribed to primary isotope effects on both abstraction of hydrogen from thiol to form the thiyl radical and abstraction of hydrogen from 5′-dA to regenerate the thiol of C408. The equilibrium isotope effect is consistent with the low fractionation factor (0.4–0.5) expected for thiols.[107] Because the heavier isotope will fractionate onto the carbon center of 5′-dA in preference to the cysteine thiol, deuteration of the thiol group drives the reaction toward formation of thiyl radical and cob(II)alamin.[106]

Quantitative modeling of the kinetics was required to make a distinction between the two mechanisms. Kinetic simulations were carried out taking into account a variety of constraints imposed by factors such as the fractionation factor associated with thiols and the statistical factors expected for hydrogen abstraction from mixed isotopomers of 5′-dA. In addition, global analysis[108] of the isotope effect data was performed, placing no constraints on possible rate constants. Both of these approaches favored the concerted mechanism, and thus provided the framework for interpreting the temperature-dependent studies on this process.

The kinetics of cob(II)alamin formation in the exchange reaction have also been measured as a function of [AdoCbl] and temperature. The [AdoCbl] dependence of the k_{obs} allowed assignment of microscopic rate constants both for carbon–cobalt bond homolysis/thiyl radical formation and for carbon–cobalt bond reformation/thiol regeneration. The variation of these microscopic rate constants with temperature provided the activation enthalpies and entropies ($\Delta H\ddagger$ and $\Delta S\ddagger$) for both forward and reverse steps. A similar analysis of the amount of cob(II)alamin formed at equilibrium as a function of [AdoCbl] and temperature provided an independent check on the enthalpy and entropy of this reaction (ΔH and ΔS).

The results of these studies provide further evidence that the enzyme makes the reactant (AdoCbl-bound) state and the product (cob(II)alamin/5′-dA/thiyl radical) state approximately equal in energy. This dramatic thermodynamic perturbation, in addition to transition-state stabilization, is required to account for the large rate acceleration observed on this process, as discussed above. The ΔH measured is consistent with the calculated net ΔH for nonenzymatic carbon–cobalt bond homolysis, S—H bond homolysis, and C—H bond formation. This observation suggests that weakening of the carbon–cobalt bond by strain or electronic effects, proposed from model studies, is not the predominant factor in the observed thermodynamic perturbation. The free energy of the enzymatic reaction (ΔG) is dominated by a large $T\Delta S$ term. Thus, entropic effects, rather than enthalpic effects, appear to be most important in determining the thermodynamics of carbon–cobalt bond homolysis/thiyl radical formation.

The $\Delta H\ddagger$ is measured to be ~ 45 kcal mol^{-1}, which is likely to represent contributions from carbon–cobalt bond cleavage, thiyl radical formation, protein conformational changes, and solvent reorganization. This large $\Delta H\ddagger$ is significant in that, like ΔH, its magnitude suggests that the enzyme does not function by weakening the carbon–cobalt bond. As observed for ΔG, $\Delta G\ddagger$ is dominated by a favorable $T\Delta S\ddagger$, which brings $\Delta G\ddagger$ to ~ 15 kcal mol^{-1}. While the source of this large entropic term is not known, solvent release and protein conformational entropy are likely to be critical in the entropy changes that drive catalysis.

Comparable experiments in the presence of substrate will be difficult to interpret, due to the complexity of the kinetics.[52] However, neither the rate of carbon–cobalt bond homolysis (~ 200 s^{-1} vs. 40 s^{-1}), nor the amount of cob(II)alamin in the steady state are dramatically perturbed in the presence of substrate relative to the exchange reaction conditions. Thus, large entropic effects will likely play a major role in the turnover process as well. While the mechanisms for thiyl radical formation will be different for enzymes containing different metallocofactors, these studies on the *L. leichmannii* RTPR show that a metallocofactor at an enzyme active site can effect rapid conversion of a cysteine residue to a thiyl radical.

5.08.2.7 Carbon–Cobalt Bond Reformation Follows Each Turnover

The exchange reaction, which has been so useful as an avenue for studying the mechanism and energetics of carbon–cobalt bond homolysis, has also shed light on the issue of whether the carbon–cobalt bond is reformed after every turnover during the normal reduction process (Scheme 3). Since it requires thiyl radical abstraction of tritium from 5′-dA, the rate of tritium release to the solvent provides a lower limit for the rate of carbon–cobalt bond reformation.

To investigate the chain length of the nucleotide reduction process, rapid acid quench experiments were carried out with RTPR, [5′-³H]-AdoCbl, effector, and [¹⁴C]-NTP. The relative rates of formation of tritiated water and [¹⁴C]-dNTP are indicative of the chain length for nucleotide reduction. In one limiting case, where every dNTP is accompanied by reformation of the carbon–cobalt bond, the chain length would be 1. However, due to the kinetic isotope effect on this process and the statistical effect (associated with the three equivalent hydrogens in the methyl group of 5′-dA), not every carbon–cobalt bond reformation will result in tritiated water release. Assuming a selection effect against tritium abstraction of 10 and a statistical effect of 2, a chain length of 1 should result in 1/20 of an equivalent of tritiated water released for every turnover. In the case of multiple turnovers, the number should be much lower.[106]

The rapid acid quench experiments showed that after 300 ms, 0.6 equivalent of [2-¹⁴C]-dATP is formed and 0.06 equivalent of tritiated water is released. No lag phase was evident in the formation of either product. Furthermore, the ratio of ³H₂O to dATP increased as a function of time. These results indicate that most of the tritiated water is released after nucleotide reduction and suggest that reformation of the carbon–cobalt bond occurs after every turnover. Additional data supporting this model come from steady-state kinetic analysis of dNTP formation when [RTPR] ≫ [AdoCbl].[106] The short chain length can be easily rationalized chemically as a method of protecting the reactive thiyl radical. If carbon–cobalt bond reformation did not follow each turnover, the thiyl radical would be present during dissociation of the product and binding of another molecule of substrate. It would thus be exposed to the solvent, rendering it vulnerable to reaction with oxygen or other reactive species in solution. The cob(II)alamin under these conditions would also have an enhanced probability of oxidation. Knowledge of the chain length of the reaction is also critical to defining the rate acceleration mediated by RTPR[75]. If, for example, one carbon–cobalt homolysis event catalyzed formation of many dNTPs, then the actual rate acceleration catalyzed by RTPR could be much less than the 10^{11} predicted based on k_{obs}. Thus the chain length of approximately 1 suggests that 10^{11} is the actual rate acceleration.

This section has described evidence for the existence of a thiyl radical intermediate in nucleotide reduction by the *L. leichmannii* class II RNR, and offered hypotheses as to how this radical might be formed rapidly with complete control. In the next section, formation of a thiyl radical in the *E. coli* class I reductase will be discussed.

5.08.3 PROPOSED ROLE OF PROTEIN-BASED TYROSYL RADICAL IN CATALYSIS AND MODEL STUDIES IN SUPPORT OF THIS ROLE

5.08.3.1 Tyrosyl Radicals are Proposed to Generate Active Site Thiyl Radicals

While there is extensive biochemical and structural data available on the *E. coli* RNR, and the •Y was shown to be essential for the nucleotide reduction process in the early 1970s,[109] the function of the •Y in catalysis has eluded experimental verification. What is amazing about the Class I RNRs is that their small subunit (R2) contains this remarkably stable •Y (located on residue 122 in *E. coli* RNR, and having a $t_{1/2}$ of 4 days at 4 °C),[26] yet it is this radical that is proposed to serve as the initial electron acceptor in a series of coupled electron and proton transfer reactions that lead to the generation of putative thiyl radical at the active site of the large subunit (R1) of the enzyme (Figure 3).

Analysis of the structures of R2 and R1, in conjunction with extensive biochemical analysis of this system, has allowed Uhlin and Eklund to dock the two subunits together, with the resulting structure suggesting that the distance between the Y122 on R2 and the active site C439 of R1 is ∼35 Å (Figure 3).[7,8,110] The R2 environment must allow generation of this •Y from Y122 by the diferrous form of R2 in the presence of O_2 and reductant. It further must allow stabilization of this radical, which would normally have a $t_{1/2}$ in solution of 10 ms,[111] so that it is not quenched adventitiously by external reductants or amino acid residues surrounding the cofactor site. In addition, this radical, in the presence of the second subunit, R1, and substrate, nucleoside diphosphate (NDP), must react with other protein residues to accomplish a net electron transfer over a very long distance to generate a single specific thiyl radical. As described in detail (Figure 3), a specific electron transfer pathway comprised of conserved amino acid residues with extensive hydrogen bonding interactions has been proposed based on the structures of R1 and R2.[34,112,113] Site-directed mutagenesis experiments on R2[114,115] and R1[34,116] have shown that mutation of any of these residues causes reduction of enzymatic activity to the level of contaminating wild-type activity in all the mutant preparations.

In order to understand the unique role that the protein plays in the reactivity of the R2 •Y, it is necessary to understand the thermodynamics and kinetics of the reversible oxidation and reduction of Y to •Y. This can be studied most easily using tyrosine and tyrosine-containing model peptides. How the R2 environment can modulate these properties may then be examined based on the structure of R2 and a variety of R2 mutants.

5.08.3.2   Thermodynamics and Kinetics of Tyrosyl Radical Formation: Dependence on Protonation State

The diferrous iron center of R2 catalyzes the oxidation of the Y122 to a neutral tyrosyl radical, •Y122, in the presence of oxygen and reductant. In contrast to many reports in the literature, it is clear from thermodynamic arguments that no tyrosyl radicals of biological importance are in the protonated state (that is, exist as cation radicals), since the pK_a of these species are of the order of −2.[117,118] Both ENDOR[119] and resonance Raman spectroscopy[120] provide direct support for this conclusion. The detailed analysis of the mechanism by which the diferrous form of R2 generates the •Y is beyond the scope of this chapter. However, analysis of the effects of pH on the kinetics and thermodynamics of tyrosine oxidation can furnish insight into how the protein environment could influence the redox properties of Y122.

Pulse radiolysis experiments have established that oxidation of a neutral tyrosine to a •Y by an azide radical occurs with a rate constant of $\sim 10^8$ M^{-1} s^{-1},[121,122] while deprotonated tyrosine is oxidized 10 times faster.[121] Modulation of the pK_a of the phenol of the tyrosine residue is thus one way that the protein can affect the rate of hydrogen atom or electron transfer. Even if the electron transfer step is not rate limiting, such a modulation of rates might be important to ensure the specificity of such a process; for example, if a tyrosine residue is close to a tryptophan residue, formation of the tryptophan radical might be kinetically favored if the tyrosine residue was protonated, but less favored if the tyrosine residue was deprotonated.

The protonation state of a tyrosine residue is also important in determining the thermodynamics of the oxidation of tyrosine. Klapper and co-workers have used cyclic voltammetry and pulse radiolysis methods[122] to obtain a midpoint potential of 0.93 V for oxidation of tyrosine to a •Y at pH 7.[123,124] Furthermore, only modest differences in pK_a and oxidation potential between the free amino acid and the amino acid incorporated into di- or tripeptides have been reported.[124] The issue of whether the protein environment perturbs this potential thus remains to be elucidated. Nonetheless, these model studies suggest that modulation of the protonation states of tyrosine could be an important mechanism for maintaining specificity of proton and electron transfer within a protein.

5.08.3.3   Spectroscopic Approaches to Determining the Hydrogen Bonding State of the R2 Tyrosyl Radical

While Y122 is a neutral radical in the active form of the cofactor of R2, it has been proposed that the ability of this residue to mediate oxidation of an adjacent amino acid side chain via hydrogen atom transfer (Figure 3) might be related to the hydrogen-bonding environment of this residue.[125] EPR spectroscopy and related spectroscopies (ENDOR, ESEEM (electron spin echo envelope modulation)) have provided a way of looking selectively at •Y and the distribution of its spin density. Accordingly, these techniques have provided a probe of the structural and electronic properties of tyrosyl radicals in class I RNRs from a variety of sources.

ENDOR spectroscopy in D_2O,[119] EPR spectroscopy of R2 incorporating [^{17}O]-labeled tyrosine,[126] and high-field EPR spectroscopy (139.5 GHz[127] and 245 GHz[128]) all provide evidence that the •Y in *E. coli* R2 is not hydrogen bonded to another protein residue or to bulk water. In the latter cases,[127,128] the *g*-anisotropy, especially in *g*1, has been proposed to be an important indicator of hydrogen bonding status. Interestingly, the X-ray structure of diferrous R2 shows that Y122 is hydrogen-bonded to D84, a ligand to its proximal iron,[7] but that after cluster assembly and •Y formation, this proton, of necessity, is no longer present.[129] It could in fact be transferred to the putative water (or disordered hydroxide) of intermediate X in the assembly process, generating the water observed on the iron adjacent to the •Y in the resting state of the protein.[110,130] This same water molecule might also provide a proton to the phenoxy oxygen of Y122 on reduction of the •Y during turnover (Figure 3). Tyrosyl radicals in other class I RNRs have also been examined using EPR. The details of these studies have been summarized.[131] Surprisingly, the spin density in almost all tyrosyl radicals appears to be the same regardless of their environment. While it was initially thought that the *E. coli* •Y would serve as the prototype of all class I tyrosyl radicals, this may not be the case. Recent high-field EPR experiments on the mouse RNR have revealed a *g*-anisotropy consistent with a hydrogen bond to the •Y.[132] This interpretation needs to be confirmed by other spectroscopic methods, however. While the hydrogen bonding of the •Y in the resting enzyme appears to vary from species to species, the change in hydrogen bonding of this residue during electron transfer remains unknown and is most probably of importance in electron transfer. Unfortunately, experiments to test this hypothesis are very difficult in biological systems. The next section describes model systems for the coupling of proton and electron motion.

5.08.3.4 Model Studies on Proton-coupled Electron Transfer

Model studies are required to gain a basic understanding of how coupling of proton motion to electron transfer might operate. Nocera and co-workers have reported several systems for studying proton-coupled electron transfer. One is based on formation of a hydrogen-bonded interface between an electron donor (a zinc porphyrin) and an electron acceptor (3,4-dinitrobenzoic acid), both bearing carboxylic acid moieties (Figure 6(a)).[133] Transient absorption spectroscopy shows that the rate constant for forward electron transfer is $\sim 5 \times 10^{10}$ s^{-1}. Thus, electron transfer in this system is fast, and comparable in rate to electron transfer in a similar system in which a zinc porphyrin is covalently linked to a quinone acceptor.[134] The observation of kinetic isotope effects on forward and back electron transfer ($k_H/k_D = 1.7$ and 1.6, respectively) when the acid groups are deuterated demonstrates that proton motion accompanies electron transfer. This study suggests that while the dicarboxylic acid bridge is the medium for electron transfer, it is not a static medium; protons in the bridge rearrange in response to the transfer of an electron.

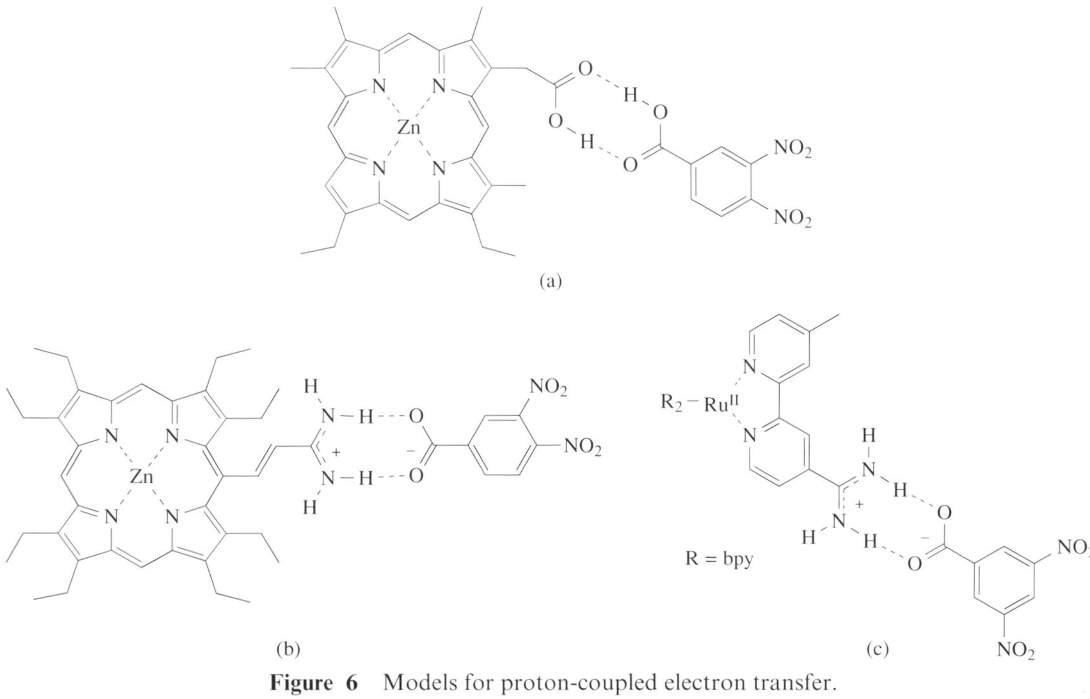

Figure 6 Models for proton-coupled electron transfer.

In order to investigate whether changes in charge and polarity associated with proton motion affect the rate of electron transfer, Nocera and co-workers devised systems containing donors and acceptors linked by an asymmetric hydrogen-bonded interface, an amidinium-carboxylate salt bridge.[135,136] One such system is a zinc porphyrin donor linked to 3,4-dinitrobenzoic acid through an amidinium-carboxylate salt bridge (Figure 6(b)).[135] Measurement of the excited state lifetime of the zinc porphyrin when complexed to the acceptor gave a rate constant of $\sim 8 \times 10^8$ s^{-1}, slower than that observed for the system in which donor and acceptor were linked through a dicarboxylic acid bridge. Similarly, a Ru-(bpy)$_3$$^{+2}$ donor linked to a 3,5-dinitrobenzene acceptor through an amidinium-carboxylate salt bridge (Figure 6(c)) exhibited an electron transfer rate constant $\sim$ 2-fold smaller than the analogous system containing a dicarboxylic acid bridge, even though the driving force for the latter system is 0.07 V smaller.[136] Again, a kinetic isotope effect was observed on substitution of exchangeable hydrogens for deuterium ($k_H/k_D = 1.34$). These observations suggest that proton-coupled electron transfer also occurs in systems with an asymmetric hydrogen-bonded bridge, but that it is less efficient than in symmetric systems. Nocera and co-workers suggest that the charge rearrangement that accompanies proton motion in the asymmetric systems gives rise to solvent reorganization, which manifests itself in a slower electron transfer rate (additional Franck–Condon factors).

These model studies are consistent with the hypothesis that electron transfer in proteins can be mediated through hydrogen bonds.[137] Furthermore, they suggest that the rate of electron transfer through a hydrogen bond will depend on whether proton motion can be coupled to electron motion without large reorganization of solvent or of neighboring amino acid residues. Perhaps hydrogen bonding to amino acid residues involved in electron transfer not only modulates the driving force of electron transfer reactions, but also minimizes charge rearrangements that would increase the reorganization energy of electron transfer.

5.08.3.5 Thermodynamic and Kinetic Studies of Formation and Reduction of the R2 Tyrosyl Radical

The reduction potential of the •Y122 of *E. coli* R2 has been estimated to be 1.0 ± 0.1 V.[138] The inability of redox mediators to equilibrate with the radical in a reasonable amount of time has made precise determination of this number impossible. However, these results are similar to the model studies (~ 1.0 V at pH 7),[123,124] suggesting that the protein has not dramatically perturbed the reduction potential, at least in the *E. coli* R2.

While the thermodynamic properties of •Y122 appear to be relatively unperturbed by the protein environment of R2, this is not the case for its chemical reactivity. The lifetime of days for •Y122 in the protein in comparison with milliseconds for •Y in solution is consistent with the buried and hydrophobic environment of •Y122 established crystallographically.[129] The detailed structural basis for this kinetic stability has been the focus of much effort.[139–141]

Evidence for the role of the protein in maintaining the kinetic stability of •Y122 comes from experiments using hydroxyurea (HU), a known reductant of the •Y. The •Y of R2 reacts with HU at least 10 times faster in the presence of R1, ATP (an allosteric effector), and CDP (a substrate) or dCDP (a product) than it does in the presence of R1 alone, suggesting that the conformation of the holoenzyme governs its kinetic accessibility.[139] Analogous experiments with mutant R1s in which the three active site cysteines that interact directly with the substrate are mutated to alanines (C225A, C462A, and C439; Figure 2) demonstrated that these mutations had differing effects on the stability of the •Y. While the C225 and C462 R1 mutants had effects on the reactivity of the R2 •Y comparable to that of the wild-type, the C439 mutant effectively abolished the increase in reactivity observed with the wild-type R1, increasing the rate of •Y loss by only $\sim$ 1.4-fold over the rate in the presence of bovine serum albumin, ADP, and dCDP. This effect is not due to a simple inability of C439A R1 to bind to R2, as this mutant functions as an inhibitor of ribonucleotide reduction with a low μM K_i.[61]

These results are consistent with a specific role for the active site cysteines in maintaining a holoenzyme conformation that modulates the ability of HU to participate in a chain of hydrogen atom abstractions that ultimately results in quenching of the •Y. This interpretation assumes a hydrogen atom transfer mechanism for quenching by HU, a point that has not been experimentally determined. Regardless of the detailed mechanism, the ability of subtle alterations in the R1 subunit to affect the HU reactivity of the R2 subunit suggests that the conformation of R1 effects electron transfer/proton transfer within R2 and between R1 and R2.

5.08.3.6   Attempts at Direct Observation of Electron Transfer Between R1 and R2

One area that requires further effort is obtaining direct evidence for kinetically competent electron transfer within R2 and between R1 and R2. Efforts to observe reduction of the •Y during substrate (NDP) reduction by means of RFQ EPR and SF UV-vis spectroscopy have been unsuccessful.[37] However, the inability to observe •Y reduction is not at all inconsistent with this step being on the catalytic pathway. One way in which •Y reduction might escape direct detection is if a step preceding its reduction were rate-limiting, and all the steps subsequent to that step were relatively fast. In that case, the only observable species would contain the •Y, even under presteady-state conditions. A conformational change, perhaps triggered by substrate binding, that is required for intersubunit electron transfer could be such an early rate-limiting step.

The need for further investigation of the hypothesis that intersubunit electron transfer is essential for nucleotide reduction is emphasized by a report from Cooperman and co-workers that the Y177F M2 (the mouse homologue of R2), is able to support nucleotide reduction.[142] They report that Y177F M2 catalyzes nucleotide reduction with a turnover number ~0.5% that of the wild-type enzyme. Control experiments show that any amount of •Y177 that may be present is insufficient to account for the enzymatic activity. The observation that a nonapeptide corresponding to the C-terminus of M2 (which is essential for interaction with M1) inhibits this activity, but the C-terminal peptide of *E. coli* R2 does not is also consistent with the assignment of the enzymatic activity to M2 rather than an RNR in the host strain. Cooperman and co-workers ascribe this activity to a putative transient radical species generated during the assembly and disassembly of the required diferric •Y cofactor. In contrast to the *E. coli* R2, the iron center and tryosyl radical in M2 can be reduced by DTT, present in their reaction mixtures as a reductant to generate deoxynucleotides, and can be reoxidized in the presence of O_2. Transient intermediates in the reoxidation, such as the Fe^{3+}/Fe^{4+} intermediate X in the assembly of the active cofactor,[28,29,143,144] are capable of generating alternative protein radicals[32,141,145] which could eventually generate the required thiyl radical on R1. Interestingly, HU also inhibits the observed enzymatic activity, although it is unknown whether this inhibition is due to reduction of a transient radical or to reaction of HU with the iron center.[35,146] The quantification of the number of equivalents of O_2 and ferrous iron consumed relative to the amount of dNDP produced would establish the validity of this intriguing model. It is of interest to note that Bollinger *et al.*, when studying the assembly of the cofactor of Y122F R2 from *E. coli*, observed a species (10% of the protein) that had a UV/vis spectrum identical to a •Y.[27] It was suggested at that time that it might be associated with Y356 (Figure 3), one of the residues on the putative electron transfer pathway. However, no group has been able to associate this transient absorption feature with an EPR-active species.[31] A similar experiment should be carried out with Y177F M2. If a transient intermediate is observed, then a presteady-state three-syringe experiment could be attempted. Rapid mixing of R1 and NDP with R2 at the time of maximum formation of the transient •Y should lead to dNDP formation.

5.08.3.7   Indirect Evidence for Electron Transfer and its Pathway

Despite much effort from Sjöberg and her colleagues,[32–35] there is no direct evidence to support the proposed electron transfer pathway between C439 on R1 and Y122 on R2 (Figure 3). Their effort to test this model has focused on the generation of a mutant of each of the residues in the proposed pathway. While all of the mutants are "inactive," no buildup of any transient radical has been observed. The X-ray structure of the R1 double mutant Y730,731F in the absence of R2 suggests no major disruptions of R1 that would lead to loss of activity.[34] Unfortunately, the overall phenotype of inactivity is difficult to interpret. Several similar experiments have been carried out with Y730F and Y731F mutants of R1. The model suggests that for Y730F R1 in the presence of substrate, allosteric effector and R2, reduction of •Y122 R2 might be detected concomitant with transient generation of •Y on Y731F of R1, assuming similar oxidation potentials for these two residues. Disappointingly, no such transient radical is observed using RFQ EPR methods. The loss of activity in this mutant has been rationalized in terms of the disruption of the putative H-bonding network, proposed to be crucial in this process.[34] This is a case where unnatural amino acids, with the same H-bonding network and minimal structural perturbation but altered oxidation perturbation might allow buildup of a radical in the second subunit, concomitant with loss in the •Y.

While no direct evidence for the required reversible electron-transfer pathway has been obtained, several indirect methods demonstrating irreversible electron transfer have provided strong support for reduction of the •Y122 on R2 and concomitant oxidation of a nucleotide substrate analogue or an amino acid residue on R1. The most compelling evidence has resulted from the study of the interaction of 2′-vinylfluoro-2′-deoxycytidine 5′-diphosphate (VFCDP) with *E. coli* RNR.[45] As shown in Scheme 5, VFCDP is a stoichiometric mechanism-based inhibitor of RNR. Inactivation of the enzyme results from destruction of the •Y on the R2 subunit and covalent stoichiometric labeling, most probably of E441 (Figure 2) of R1. Loss of the •Y is accompanied by formation of a new radical which isotopic labeling studies have established is allylic and nucleotide-derived.[147] Thus, reduction of •Y122 on R2 results in formation of a second radical in the active site of R1. These results establish that long-range electron transfer is feasible, but say nothing about the pathway for this process.

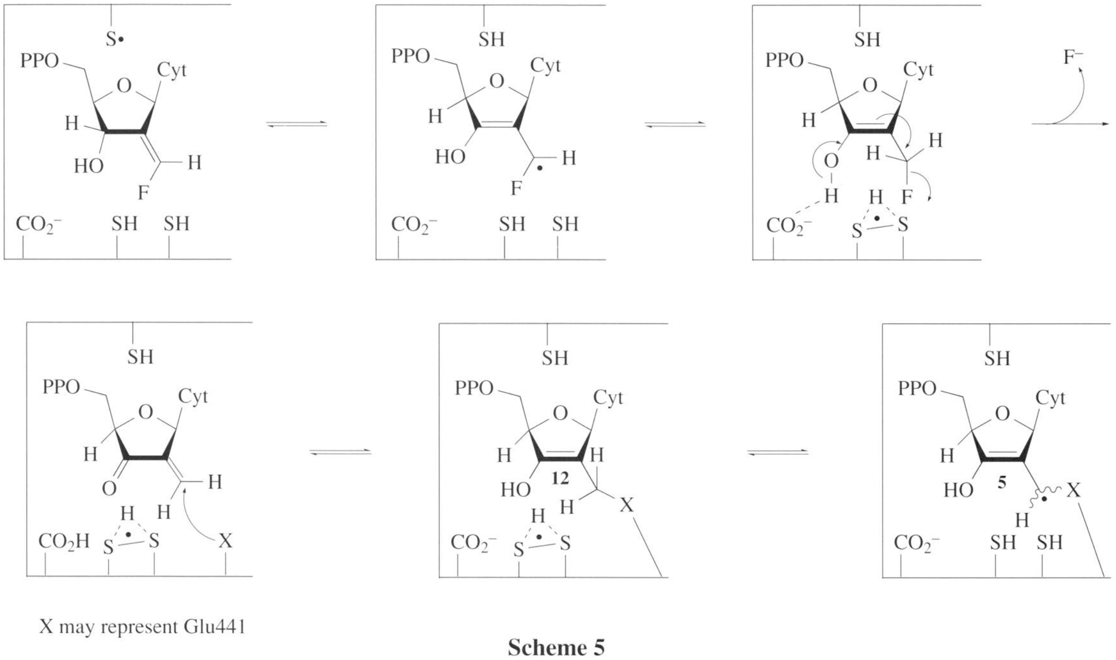

X may represent Glu441

Scheme 5

A second example of electron transfer between R1 and R2 comes from studies of the stoichiometric mechanism-based inhibitor 2′-azido-2′-deoxyuridine diphosphate (N₃UDP) (Scheme 6). Inactivation by this compound results from destruction of the essential •Y122 on R2.[147,148] Loss of this radical is accompanied by formation of a second radical.[148,149] Nitrogen gas release, uracil and pyrophosphate formation, and, eventually, alkylation of R1 with 2-methylene-3-(2*H*)-furanone also occur in the course of mechanism-based inactivation.[147,148] Studies using [¹⁵N]-N₃UDP and [β-²H-cysteine]- RDPR have shown that a new nitrogen-centered radical derived from the azido group of the inhibitor is generated (~0.4 equiv. of new radical formed per equiv. of •Y122 lost)[148,150,151] and that this radical is covalently attached to a cysteine of R1.[152] Site-directed mutagenesis studies suggest that C225 (Figure 2) is the modified residue.[150,152] Furthermore, studies of the inactivation of RDPR with [3′-²H]-N₃UDP reveal an isotope effect on loss of the •Y122.[150] While the chemistry is obviously complex, the data strongly suggest that the radical on R2 can be transferred to R1 via the nucleotide.

Finally, two R1 active mutants, C225S R1 and E441Q R1, provide additional evidence for electron transfer between the subunits. When C225S R1 is incubated with UDP and R2, the •Y122 on R2 is lost and the normal substrate is converted to a mechanism-based inhibitor.[43] Loss of this radical on R2 results in polypeptide cleavage on R1 (Equation (2)). The unusual products (a C-terminal peptide with a carboxamide terminus and a formylated N-terminal peptide) of this cleavage between S224 and S225 suggest that the radical has been transferred from one subunit to the second.[44] Using [3′-²H]-UDP, V_{max} and V/K isotope effects of ≈ 2.0 have been detected on loss of the •Y.[43] These effects are likely to arise from a coupled pair of reactions: reduction of the •Y to form a thiyl radical at C439 of R1, and abstraction of the 3′ hydrogen by the thiyl radical. Studies with E441Q R1, CDP, and R2 also result in the conversion of the normal substrate into a mechanism-based inhibitor.[45,46] Under one set of conditions, the •Y on R2 is reduced and several new radicals, thought to be nucleotide based, are detected. While the details of these systems remain to be elucidated, both suggest that once again electron transfer is occurring between the two subunits.

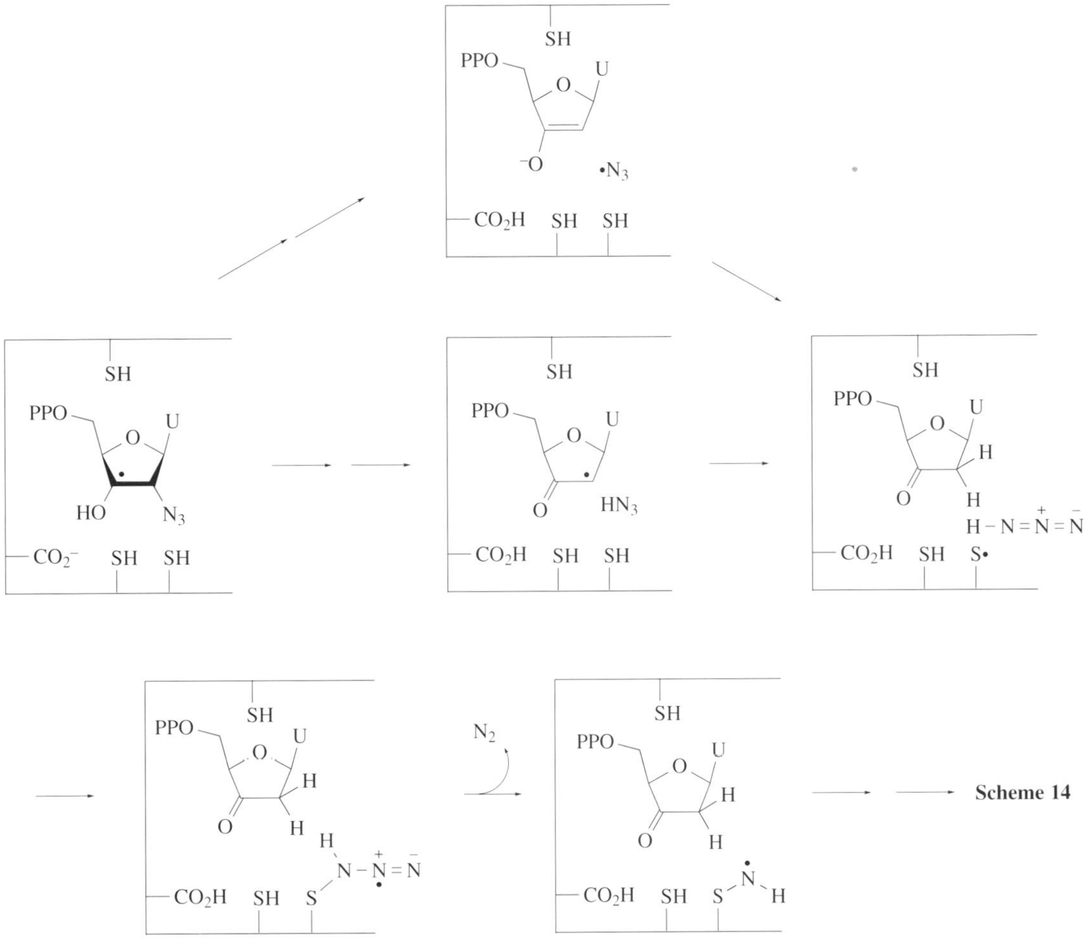

Scheme 6

$$C225S\ R1 \xrightarrow[\text{CDP}]{O_2} \text{\~\~\~SGVRTPTRQFSS}_{224}-NH_2 \qquad \underset{H}{\overset{O}{\|}}{C} V_{226}LIECGDSLDSINAT\text{\~\~\~} \qquad (2)$$

5.08.4 GENERATION OF THE THIYL RADICAL ON R1 FROM THE TYROSYL RADICAL ON R2

5.08.4.1 Precedent for the Formation of Thiyl Radicals from Phenoxy Radicals

Reactions of thiols with phenoxy radicals have previously been reported to produce thiyl radicals and disulfide radical anions derived from these thiyl radicals.[153,154] Moldéus and co-workers have used spin trap reagents to obtain evidence for generation of thiyl radicals in a system containing glutathione or cysteine and an oxidizing system consisting of acetaminophen, horseradish peroxidase, and hydrogen peroxide (at pH 8.0).[153] In rapid-flow EPR experiments, Mason and co-workers found that the same system (at pH 7.5) produces the disulfide radical anion of glutathione or cysteine.[154] They propose that the phenoxy radical generated via the oxidation of acetaminophen by horseradish peroxidase reacts with glutathione to form a thiyl radical, which in turn can react with another molecule of glutathione (in the thiolate form) to form the disulfide radical anion. These model studies show that a phenoxy radical can generate a thiyl radical, although the yield of this process was not reported. In addition, in the model system, in contrast to the reductase system, the reaction could be driven to the right uniquely by the experimental conditions, such as the presence of a spin trap. The proposed mechanism for RNR requires that a reasonable amount of thiyl radical be formed without a very exergonic step immediately following the generation of this radical. Nevertheless, this model system illustrates how an oxidizing system might generate a thiyl radical through the intermediacy of a phenoxy radical, in analogy with the diferric cluster assembly of R2 generating a •Y that functions to form a thiyl radical.

Pulse radiolytic experiments also provide support for the formation of thiyl radicals from tyrosyl radicals. Prütz *et al.* have obtained evidence for this reaction in several different contexts. They demonstrated that the presence of thiols inhibits a tyrosine dimerization proposed to occur through •Y intermediates, suggesting that the •Y was reduced by hydrogen atom abstraction from the thiol.[111] They also generated a •Y on the Trp-Tyr dipeptide, and showed that the presence of glutathione (2 mM) at pH 8.1 accelerated the decay of the •Y, allowing estimation of a rate constant of $\sim 2 \times 10^5$ M^{-1} s^{-1} for this process.[155] A rate constant for the reverse reaction, abstraction of a hydrogen from tyrosine by a thiyl radical, was measured to be 5.8×10^6 M^{-1} s^{-1} for the Gly-Tyr dipeptide and glutathione at pH 8.1. These studies suggest that the equilibrium between •Y and thiyl radical lies toward the former, but not so far that generation of thiyl radical from •Y is thermodynamically unfeasible. At pH 8.1, a significant fraction of the glutathione is deprotonated, allowing for the possibility of a thermodynamically favorable single electron transfer between a •Y and the thiolate.

5.08.4.2 Thermodynamics of Thiyl Radical Formation

Critical to the proposed mechanism for the *E. coli* RNR is the •Y-mediated thiyl radical formation. Model studies described above suggest that this is, in fact, possible. The midpoint potential for the tyrosine/•Y couple in neutral solution has been measured as -0.93 V.[123] However, this potential is dependent on the protonation state of the tyrosine. The tyrosinate is clearly more easy to oxidize than neutral tyrosine (at pH 11, the oxidation potential is -0.7 V). The midpoint potential for oxidation of protonated β-thioethanol or penicillamine to a thiyl radical has been measured to be -1.33 V at pH 7,[156] while that for one-electron oxidation of glutathione is -0.91 V at pH 5.0.[157] Cysteine might be expected to have a similar oxidation potential, and hence its oxidation via tyrosine would be thermodynamically unfavorable. One-electron oxidation of cysteinate, on the other hand, has a midpoint potential of -0.73 V.[156] Thus, depending on the protonation state of the C439 and Y122, the equilibrium amounts of thiyl radical and •Y could vary substantially.

5.08.4.3 Kinetics of Thiyl Radical Formation

Model systems indicate that the abstraction of hydrogen from a thiol by many different radical species occurs at diffusion-controlled rates. Phosphite radicals abstract a hydrogen atom from penicillamine with a rate constant of 3×10^8 M^{-1} s^{-1}.[158] The thiyl radical derived from β-mercaptoethanol can abstract a hydrogen atom from dithiothreitol with a rate constant of 1.7×10^7 M^{-1} s^{-1}.[159] Carbon-centered radicals derived from ethylene glycol and D-ribose abstract hydrogen atoms form dithiothreitol with rate constants of 2.6×10^7 M^{-1} s^{-1} and 9×10^7 M^{-1} s^{-1}, respectively.[160] Formation of thiyl radicals by electron transfer has also been reported to be rapid, with the OH• adduct of deoxyguanosine reacting with cysteine with a rate constant of 8.4×10^8 M^{-1} s^{-1} at pH 9.5 (where cysteine will be predominantly in the thiolate form).[161] Thus, while structural data suggest that C439 will be protonated in R1 in the resting state,[8] thiyl radical formation would be expected to be kinetically facile by either hydrogen atom abstraction or by electron transfer, if the cysteine thiol is transiently deprotonated. Thus, the success of the oxidation might be governed by proton-mediated electron transfer, with Y730 being one, of potentially many (Figure 3), important residues.

5.08.5 A GLYCYL RADICAL MAY GENERATE AN ACTIVE SITE THIYL RADICAL IN THE ANAEROBIC *E. COLI* RNR

The anaerobic *E. coli* RNR is composed of two homodimeric subunits ($\alpha 2\beta 2$). The $\beta 2$ subunit is an activating enzyme that binds *S*-adenosylmethionine (SAM)[162,163], flavodoxin[164], and flavodoxin reductase. It contains a single [4Fe–4S] cluster at the dimer interface.[165,166] This activating enzyme catalyzes formation of an essential glycyl radical on the $\alpha 2$ homodimeric subunit that is required for nucleotide reduction.[165] The details of the mechanism of this reaction remain to be determined. One of many possibilities being considered is that SAM is reductively cleaved to form methionine and 5′-deoxyadenosyl radical (5′-dA•), which abstracts a hydrogen from glycine. Pyruvate formate lyase activating enzyme is the prototype for the $\beta 2$ chemistry catalyzed by RNR. It is required to generate the glycyl radical of pyruvate formate lyase (PFL) that is required for formation of acetyl-CoA and formate from pyruvate and coenzyme A.[18] SAM is converted to 5′-dA and methionine during this reaction,[167] and studies have shown that using [α-^{2}H] glycine labeled PFL, deuterium is found in the 5′ position of 5′-dA.[168] Determination of the thermodynamics of the reductive cleavage of SAM will be an important step in evaluating the mechanism of this process and the roles of the enzyme and the iron sulfur cluster in catalyzing the formation of glycyl radical.

Using the mechanistic information about the *E. coli* and *L. leichmannii* reductases as a guide, it was postulated that the role of the glycyl radical in the anaerobic *E. coli* reductase is to generate a thiyl radical that initiates nucleotide reduction by abstracting the 3′-hydrogen of the substrate.[38] Evidence that the protein radical that initiates nucleotide reduction is not the glycyl radical is provided by a study in which nucleotide reduction was studied in D_2O.[169] When the deoxynucleotide product was isolated and examined by NMR spectroscopy, in addition to deuterium located in the pro-R position at C-2′, 1–2% deuteration was also observed at the 3′ position. This important result demonstrates that the 3′-hydrogen atom is abstracted during the course of the reaction. Moreover, it suggests that the protein radical that abstracts the 3′-hydrogen, and presumably also returns it, can exchange with the solvent. Although the glycyl radical in pyuvate formate lyase has been shown by EPR to exchange with solvent,[170,171] the rate of this process is 10^{-5} the normal turnover rate. Furthermore, the glycyl radical in the anaerobic *E. coli* reductase does not exchange with solvent.[172,173] Thus, the observation of deuterium incorporation from solvent suggests that the glycyl radical does not abstract hydrogen from the 3′ position of the substrate, and is consistent with the hypothesis that a thiyl radical serves that purpose.

Model studies have provided information about the thermodynamics and kinetics of the abstraction of hydrogen from a thiol by a glycyl radical. The reduction potential for the alanyl radical (unpaired spin on the α carbon)/alanine couple has been estimated to be 1.22 V,[174] based on the reduction potential of ethylamine radical[175] and the estimated effect of an α carboxyl group on the C—H bond energy[176] using Benson's group additivity rules.[177] Using the reduction potential of 1.33 V measured for the thiyl radical/thiol couple,[156] an equilibrium constant of 200 was estimated for the equilibrium between a thiyl radical and an alanyl radical.[178] The corresponding value for a glycyl radical is likely to be similar, although the inductive effect of the methyl group will probably make alanine easier to oxidize than glycine. In the same paper, the rate constant for abstraction of hydrogen from the α position of glycine by a cysteine thiyl radical was determined to be $3.2 \times 10^5 \, M^{-1} \, s^{-1}$.[178] The thermodynamics and kinetics of this model reaction indicate that hydrogen abstraction by the putative enzyme-based thiyl radical to regenerate the glycyl radical should be both thermodynamically favorable and rapid. However, this model system also suggests that hydrogen abstraction by a glycyl radical to form a thiyl radical would be thermodynamically unfavorable. For the mechanism to be feasible, the enzymatic system must alter the energetics of the redox chemistry to make both the forward and reverse reactions reasonably facile.

Theoretical quantum mechanical studies predict that a planar, extended conformation will be highly favored for the glycyl radical,[179] making it difficult for the enzyme to modulate the reactivity of the glycyl radical through a conformational change (e.g., making the glycyl radical more reactive by enforcing a conformation in which captodative stabilization was minimal). It is thus more plausible that the enzyme modulates the reactivity of the thiyl radical. As discussed in Section 5.08.4.1, the protonation state of a cysteine residue has a large effect on the redox potential, with the reduction potential for the thiyl radical/thiolate couple being 0.73 V, compared with 1.33 V for the thiyl radical/thiol couple. While full deprotonation of the putative active site cysteine would preclude hydrogen atom abstraction, hydrogen bonding that weakened the S—H bond might shift the redox potential of this residue enough that formation of the thiyl radical would be thermodynamically favorable, or at least energetically neutral.

5.08.6   THE MECHANISM OF NUCLEOTIDE REDUCTION: ABSTRACTION OF THE 3′ HYDROGEN BY A THIYL RADICAL

5.08.6.1   Thermodynamics of Hydrogen Abstraction by Thiyl Radicals

The model postulated in Scheme 1 is that the thiyl radical abstracts a hydrogen atom from the 3′ position of the nucleotide to initiate the nucleotide reduction process. The initial question that has been raised by many critics of this model is whether this reaction is thermodynamically favorable. The homolytic bond dissociation energy of an S—H bond (derived from heats of formation) is 88–91 kcal mol^{-1},[92,93] while the bond dissociation energy of the hydrogen of a deoxyribose sugar is ~ 91 kcal mol^{-1}.[180] Thus, the ΔH of this reaction allows for the possibility of a favorable ΔG.

Experimentally, however, equilibria in solution greatly favor thiyl radicals over carbon-centered radicals. Asmus and co-workers measured the forward and reverse rate constants for the reaction of penacillamine thiyl radicals with 2-propanol to be $1.4 \pm 0.3 \times 10^4$ M^{-1} s^{-1} and $1.2 \pm 0.3 \times 10^8$ M^{-1} s^{-1}, respectively.[181] This corresponds to an equilibrium constant of 10^{-4}, or a ΔG° of 5 kcal mol^{-1}. For abstraction of hydrogen from dimethyltetrahydrofuran by the thiyl radical, the analogous equilibrium constant is $\sim 10^{-5}$,[182] while for hydrogen atom abstraction from HCO$_2^-$, the equilibrium constant is $\sim 5 \times 10^{-4}$.[183]

Deprotonation of the 3′-hydroxy of the substrate is one way to perturb the energetics of 3′-hydrogen atom abstraction. Theoretical studies of Evans and coworkers have shown that deprotonation of the hydroxy group of methanol lowers the C—H bond dissociation energy by ~ 15 kcal mol^{-1}.[184] Although full deprotonation of the 3′ hydroxy of the substrate is unlikely to occur at the enzyme active site, this study suggests that hydrogen bonding of the 3′ hydroxy (for example to the 2′ hydroxy) could lower the bond dissociation energy of the 3′ C—H enough that hydrogen abstraction by a thiyl radical would be exergonic.

Hydrogen bonding from groups within the active site of the enzyme could also contribute to the net energetics of this step. While the 3′ hydrogen of the substrate would not participate in hydrogen bonding, the thiol resulting from this abstraction could function as a hydrogen bond donor. While the primary role of enzymes is usually thought to be lowering the transition state energy of a given process, in a multi-step mechanism, equalizing the relative ground state energies of intermediates can also make a significant contribution to catalysis.[185]

5.08.6.2   Kinetics of Hydrogen Abstraction by Thiyl Radicals

As noted above, the rates of hydrogen abstraction by thiyl radicals are typically quite fast. The thiyl radicals of cysteine, glutathione, and penicillamine abstract hydrogen from 2-propanol with rate constants of $\sim 10^4$ M^{-1} s^{-1}.[180,181] With even moderate effective concentrations of thiyl radical at the active site, therefore, these rate constants would allow for rates of hydrogen atom abstraction greatly in excess of the overall rate constant for turnover to produce deoxynucleotide (~ 5 s^{-1}).

The most reasonable mechanism to accommodate an unfavorable equilibrium is the coupling of this equilibrium to a subsequent fast, irreversible step in the mechanism. This type of reaction with thiyl radicals has ample precedent in the literature through the work of Huyser and Kellogg,[186] which in fact provided the experimental model for this mechanistic proposal. As will be discussed in detail in Section 5.08.7.1, the loss of water from the 2′ position of the nucleotide, proposed to follow 3′ hydrogen abstraction (Scheme 1) is fast and irreversible, and model studies suggest that the coupling of this step to the abstraction of the 3′ hydrogen could easily give rise to kinetically competent formation of the 2′ ketyl radical intermediate even if the equilibrium between thiyl radical and nucleotide radical is unfavorable.

5.08.6.3   Evidence for 3′ Hydrogen Atom Abstraction in the Enzymatic Reaction

Strong evidence for 3′ hydrogen abstraction comes from studies of kinetic isotope effects on the reduction of 3′-tritium and deuterium labeled nucleotides. When ribonucleoside diphosphate reductase (RDPR) reduces [3′-^{3}H]-NDPs, tritium isotope effects on V/K are typically of the order of 2, with values ranging from 1.4 to 4.7 depending on the pH and the allosteric effector.[64,187] Deuterium substitution at the 3′-position of the substrate does not produce an isotope effect on V_{max}, but has the expected value for a V/K isotope effect based on the tritium isotope effect and the Swain–Schaad relation. Substitution of tritium at other positions on the sugar of the nucleotide does not give rise to selection effects. In addition, a small but significant amount of tritium ($\sim 0.5\%$ of total tritium at 50% conversion) is released to the solvent on incubation of [3′-^{3}H]-NDP with RDPR. This tritium washout is pH dependent as well, with slightly more tritium being volatilized at more basic pH. These results require that hydrogen be abstracted from the 3′ position of the nucleotide, that this hydrogen abstraction take place before the first irreversible step, and that the hydrogen abstraction not be the overall rate-limiting step in nucleotide reduction.

The magnitude of the $^T(V/K)$ is relatively small, but a comparison of the $^T(V/K)$ isotope effect with the $^D(V/K)$ isotope effect shows that these effects are consistent with a primary kinetic isotope effect on C—H bond cleavage. Using the $^T(V/K)$ and $^D(V/K)$ isotope effects to calculate the intrinsic deuterium isotope effect on C—H bond cleavage by the method of Northrup (Equation (3)),[188] where ^Dk is the intrinsic deuterium isotope effect, gives a value of ~ 5 for k_H/k_D, a reasonable primary isotope effect. The intrinsic deuterium isotope effect in turn allows calculation of the sum of the forward and reverse commitment factors (Equation (4)).[188] In this case, the forward commitment represents the tendency of the thiyl radical to abstract the 3′ hydrogen, rather than to partition back to the previous intermediate, while the reverse commitment is the tendency of the 3′ nucleotide radical to reabstract a hydrogen from C439 to regenerate the thiyl radical as opposed to undergoing dehydration. The sum of the commitment factors is calculated to be ~ 3. Assuming that the dehydration that follows the hydrogen atom abstraction is fast and irreversible (Section 5.08.7), the reverse commitment will be zero, so the value of ~ 3 would represent the forward commitment. This analysis suggests that the observed V/K isotope effects are small because the commitment to catalysis is high, not because the intrinsic isotope effect is low.

$$\frac{^DV/K-1}{^TV/K-1} \;=\; \frac{^Dk-1}{^Dk^{1.442}-1} \qquad\qquad (3)$$

$$^D(V/K)-1 \;=\; \frac{^Dk-1}{1+C_f+C_r} \qquad\qquad (4)$$

Kinetic isotope effects have also been observed with mechanism-based inhibitors labeled with deuterium or tritium at the 3′ position. During inactivation of RDPR by [3′-^{3}H]-N$_3$UDP, ~ 0.2 equiv. of tritiated water is released per equivalent of enzyme inactivated.[189] This indicates that 3′ C—H bond cleavage is required for this reaction. The partitioning between tritiated water release and inactivation corresponds to a fivefold selection effect against abstraction of tritium. As mentioned previously, with [3′-^{2}H]-N$_3$UDP, an isotope effect is observed on loss of the •Y.[150] The kinetics of •Y loss are multiphasic, and only the fast phase is isotope sensitive. Analyzing only the fast phase, a $^DV_{max}$ isotope effect of 1.5 and a $^D(V/K)$ isotope effect of 2.2 were calculated. The kinetics of inactivation are also isotope sensitive, although the multiphasic nature of these kinetics complicates quantitative interpretation. These observations are consistent with abstraction of hydrogen being partially rate-limiting in inactivation by N$_3$UDP.

Site-directed mutagenesis of C225 (Figure 2) in the *E. coli* class I R1 to a serine produces a mutant enzyme which catalyzes its own cleavage into two polypeptides when incubated with R2 and a nucleotide substrate (Equation (2)).[42–44] This self-inactivation is proposed to occur through the intermediacy of a substrate-based radical. Consistent with this hypothesis, this self-inactivation exhibits kinetic isotope effects similar to those observed with N$_3$UDP. Using [3′-^{3}H]-UDP, the rate of tritium washout to solvent was compared with the rate of uracil release to calculate a selection effect of 3.2 against tritium abstraction.[43] The •Y loss that accompanies self-inactivation also exhibits a kinetic isotope effect when [3′-^{2}H]-UDP is used as the substrate: ^DV and $^D(V/K)$ are 2.0 and 2.0, respectively.[43] These observations are consistent with hydrogen atom abstraction being partially rate-limiting in self-inactivation as well as inactivation by N$_3$UDP.

It appears that while hydrogen atom abstraction is not rate-limiting in substrate reduction by wild-type RDPR, it is partially rate-limiting in mechanism-based inactivation of wild-type and mutant enzymes. One explanation for this is that the inactivation reactions bypass a slow step in the reduction chemistry that occurs after 3′ hydrogen atom abstraction. The proposed reduction of the 3′ keto intermediate by an enzyme-based disulfide radical anion might be that slow step, as model reactions suggest it is a difficult transformation (Section 5.08.8).

5.08.7　LOSS OF THE 2′ HYDROXY GROUP

5.08.7.1　Model Studies on Dehydration of Ethylene Glycol

Following abstraction of the 3′ hydrogen, the nucleotide substrate is proposed to lose the 2′ hydroxy as water. The original proposal focused on a cation radical intermediate generated by this dehydration, although based on model systems, both cation radical and anion radical, as outlined subsequently, could feasibly account for the data. It is still felt that a cation radical mechanism can best accommodate all of the available data on the enzymatic system, although the possibility of a concerted mechanism has also been considered.[38,190] Rapid deprotonation of the 3′-hydroxy by the E441 would yield a ketyl radical (Scheme 1). One-electron oxidation of ethylene glycol and derivatives, an excellent model for the 2′, 3′, *cis* diol of the ribose of nucleotides, provided the basis for the authors' original working hypothesis for this transformation on the enzyme.

EPR studies on the one-electron oxidation of ethylene glycol (Scheme 7) have shown that two radical species can be produced.[191,192] When ethylene glycol is oxidized in the absence of acid using •OH produced from photolysis of H_2O_2, a 24-line spectrum arising from the ethylene glycol alkyl radical is observed. In the presence of acid (pH = 2.5),[193] however, a spectrum consisting of three broad lines is observed in place of the 24-line spectrum and assigned as the aldehyde radical, based on the g-value of 2.0046. This g-value is significantly higher than the values of 2.0024 typically observed for 2-hydroxyalkyl radicals and is close to the g-value of 2.00443 observed for the ketyl radical derived from acetone. A structure in which the oxygen formed a three-membered ring was ruled out based on the inequivalency of hyperfine couplings.[192] Similar results were obtained when

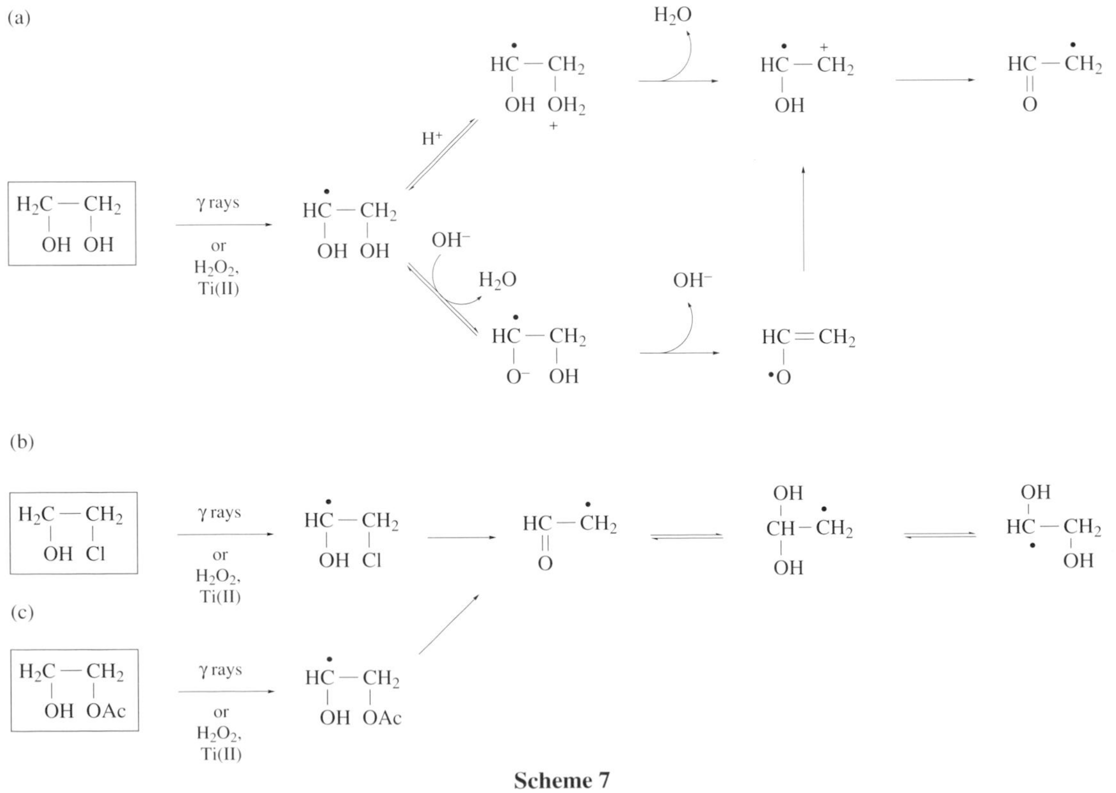

Scheme 7

ethylene glycol was oxidized by H_2O_2/titanous ion in experiments monitored by rapid flow EPR spectroscopy.[191] These results are consistent with a mechanism in which a 2-hydroxyalkyl radical is generated initially, then undergoes acid-catalyzed dehydration to form the formylmethyl radical (Scheme 7).

Rate constants for hydrogen atom abstraction and dehydration have been measured by taking advantage of the ability of the hydroxy radical formed in the Fenton reaction to initiate the one-electron oxidation of ethylene glycol.[194] The stoichiometry of H_2O_2 consumption (mols of H_2O_2 consumed/mol of Fe^{2+} oxidized) depends on the partitioning between two pathways of dehydration of an intermediate radical, one requiring Fe^{3+} and regenerating Fe^{2+}, the other requiring H^+ (Scheme 8). The chain length of the reaction thus depends on the concentrations of acid and Fe^{3+} present. Quantitative analysis of the acid and Fe^{3+} dependence of the partitioning allows estimation of the rate constant for dehydration to form the radical cation. At pH 1.3, this rate constant is estimated to be 1.3×10^8 s^{-1}. This large rate constant suggests that this reaction could occur very rapidly at the active site of RDPR if the enzyme could effect this acid catalysis.

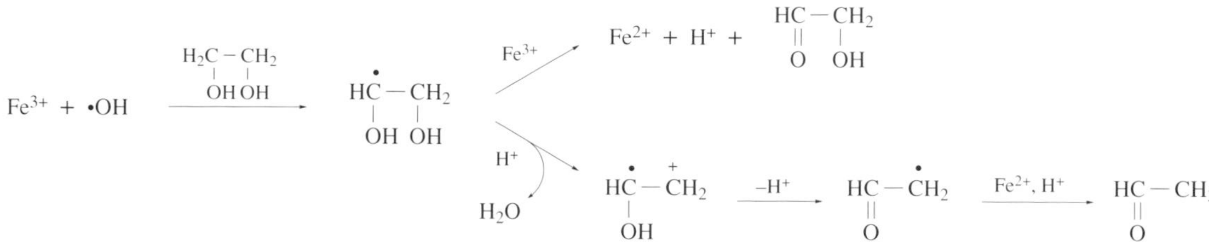

$$Fe^{3+} + \cdot OH \xrightarrow{\underset{OH\,OH}{\overset{H_2C-CH_2}{|\quad|}}} \underset{OH\ OH}{\overset{H\dot{C}-CH_2}{|\quad\ |}}$$

(Scheme 8: the 1,2-dihydroxyethyl radical partitions into two pathways — an Fe^{3+} pathway giving $Fe^{2+} + H^+ + \underset{O\ \ OH}{\overset{HC-CH_2}{||\ \ |}}$, and an H^+ / H_2O pathway giving the radical cation $\underset{OH}{\overset{H\dot{C}-\overset{+}{C}H_2}{|}} \xrightarrow{-H^+} \underset{O}{\overset{H\dot{C}-CH_2}{||}} \xrightarrow{Fe^{2+},\,H^+} \underset{O}{\overset{HC-CH_3}{||}}$)

Scheme 8

Pulse radiolytic studies have also been used to characterize these reactions and to assign rate constants. The observation of acetaldehyde and derivatives after pulse radiolysis of unbuffered aqueous ethylene glycol solutions[195] supports the sequence of reactions in Scheme 7. Elimination of water from sugar radicals has also been invoked to explain why deoxy products and malondialdehyde are formed on irradiation of ribose and other pentoses, especially under alkaline conditions.[196] Kinetic analysis using optical absorption and polarographic measurements of radical formation during the pulse radiolysis of aqueous ethylene glycol solutions[197] shows that from pH 3 to pH 7, only the 1,2-dihydroxyethyl radical is formed. This radical decays with second-order kinetics $(1.1 \times 10^6$ M^{-1} s$^{-1})$, consistent with dimerization of the radical leading to this decay. At pH 1.4, a fast, first-order decay is observed in addition to a slower second-order decay. Fitting the observed kinetics to a mechanism in which a rapid equilibrium protonation is followed by dehydration of the protonated 1,2-dihydroxyethyl radical (Scheme 7) yields an equilibrium constant of 0.18 M^{-1} for the protonation equilibrium ($pK_a = 0.74$) and a rate constant for dehydration of 8.6×10^5 s^{-1}. This rate constant is lower than that obtained by analysis of the stoichiometry of oxidant consumption,[194] but much larger than the rate constant for enzymatic turnover. The unimolecular dehydration step in the acid-catalyzed dehydration of the radical derived from erythritol has also been reported to proceed with a rate constant of 3.5×10^5 s^{-1}, suggesting that ethylene glycol is a good model system for the kinetics of this reaction in sugars and nucleotides.[198]

The dehydration of the 1,2-dihydroxyethyl radical can also be base-catalyzed.[197,199] Pulse radiolytic studies show that OH^- reacts with the neutral 1,2-dihydroxyethyl radical to form the radical anion (simple deprotonation is unlikely, given that the pK_a of this radical is likely to be greater than 10).[197] The rate constant for elimination of OH^- has also been studied by pulse radiolysis, and measured to be 3×10^6 s^{-1}.[199] In contrast, the rate constant for the uncatalyzed dehydration (i.e., in the absence of acid or base) has been measured to be $\sim 10^4$ s^{-1}.[198]

The reactions observed with ethylene glycol also take place with nucleoside radicals, as shown by a model study from Giese's laboratory.[200] Photolysis of a ribonucleoside derivatized with a selenol ester at the 3' position generates a 3' nucleoside radical specifically (Scheme 9), thus providing the closest model yet available for the enzymatic system. Rate constants for the dehydration were measured by competition kinetics. Release of base from the nucleoside was used as a measure of the dehydration reaction in this experiment, as the dehydration leads to a 3'-ketodeoxynucleoside which undergoes loss of the base under the experimental conditions.[201,202] Direct reduction of the 3' nucleoside radical by tributyltin hydride present in the reaction mixture causes formation of the ribo-or xylo-nucleoside, so that comparison of the amounts of nucleoside product with the amount of base release allows determination of the rate of elimination relative to the rate of hydrogen atom abstraction. With this method, a rate constant of 2×10^3 s^{-1} was measured for the dehydration in the absence of buffer, while in the presence of 0.1 M triethylammonium acetate (pH 7), the rate constant was 1.5×10^6 s^{-1}.

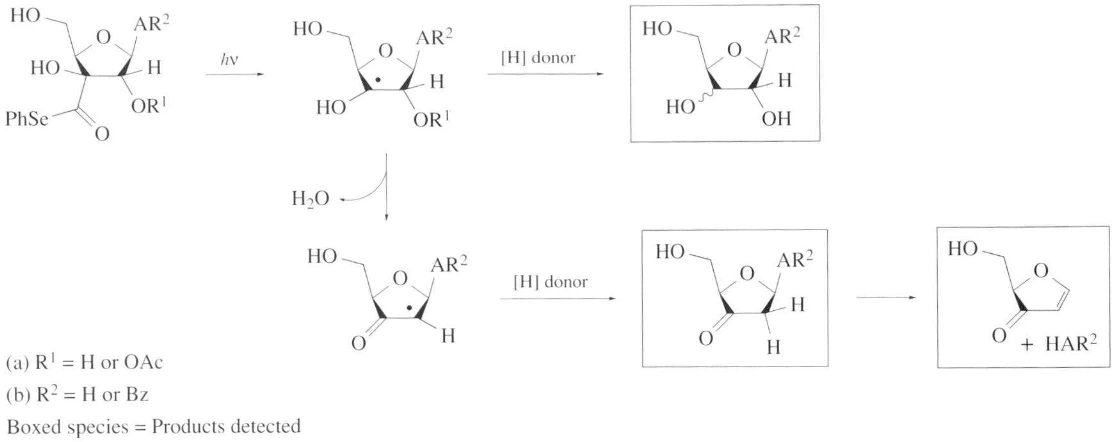

(a) R^1 = H or OAc

(b) R^2 = H or Bz

Boxed species = Products detected

Scheme 9

Lenz and Giese also examined the pH dependence of the rate of dehydration.[200] The rate was found to increase from pH 5 to 7 in 0.1 M phosphate buffer, but to decrease with decreasing pH in the pH range of 3–6 with citrate buffer. These results were interpreted to favor base catalysis over acid catalysis for the dehydration in this model system (Scheme 10). This model reaction was proposed to support a radical anion mechanism for RNRs. In this mechanism, E441 (a conserved residue in both class I and II reductases)[45,46] was proposed to act as a base, generating the radical anion leading to the putative ketyl radical. This interpretation is consistent with the three-dimensional structure of R1, which places E441 in the active site, close to where the 3′ hydroxy of the substrate is proposed to bind.[8,200] However, the earlier studies with ethylene glycol suggest that pH <2 would be required to observe acid catalysis, as the pK_a for the hydroxy group of the radical has been determined to be 1.4[198] (although an earlier measurement puts this value at 0.74[197]). It is thus anticipated that if the studies of Lenz and Giese were extended over a greater pH range (to 1.5), specific acid catalysis would also be observed for this nucleoside model system.

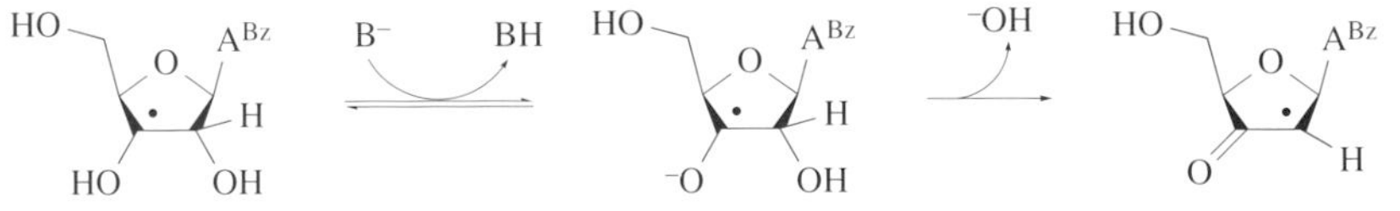

Scheme 10

Both the ethylene glycol model system and the nucleoside model system suggest that acid or base catalysis can accelerate the dehydration by up to 1000-fold relative to the uncatalyzed reaction. As has been noted previously,[37,200] acceleration of this step may be critical in driving the reaction, given the unfavorable equilibrium for the previous proposed step, abstraction of the 3′ hydrogen by a thiyl radical. In fact, model studies now allow a prediction of the rate constants of the hydrogen atom abstraction and radical dehydration steps in the enzymatic mechanism. Experiments from the Asmus and von Sonntag laboratories[181,182] have demonstrated that while the equilibrium between thiyl radical and carbon-centered radical greatly favors the thiyl radical (K_{eq} of 10^{-4}–10^{-5}), the rate of approach to equilibrium (i.e., the sum of the forward and reverse rates) is large, since the forward rate constant is $\sim 10^4$ M^{-1} s^{-1}). From the model studies on ethylene glycol[194,195,197,198] and the nucleoside model of Giese and Lenz,[200] the rate constant for acid- or base-catalyzed dehydration of the 3′-nucleotide radical can be estimated to be $\sim 10^6$ s^{-1}. If the enzyme active site enforces a reasonably high effective concentration ($\geqslant 1$ M) for the thiyl radical and the 3′ nucleotide radical, the rate of approach to equilibrium for the hydrogen abstraction step will be greater than the rate of the dehydration step by a factor of at least 100. In this case, the observed rate of formation of the 3′-ketodeoxynucleotide radical at the enzyme active site will be equal to the product of the equilibrium constant for the hydrogen atom abstraction step and the rate constant for the dehydration step ($K_{eq}k_{dehyd}$). For the acid- or base-catalyzed dehydration, then, the observed rate should be ~ 10–100 s^{-1}, fast enough for this sequence of reactions to be kinetically competent for ribonucleotide reduction. Since model studies[198,200] predict that the uncatalyzed dehydration of the 3′ nucleotide radical would be ~ 1000-fold slower than the catalyzed reaction, enzymatic catalysis of this step would be required for the hydrogen abstraction/dehydration sequence to be kinetically competent unless the enzyme acted to perturb the thermodynamics of the hydrogen abstraction equilibrium (Section 5.08.6).

Both the acid- and base-catalyzed mechanisms that have been proposed for the enzyme would require sizable perturbations in pK_a of substrate and/or enzyme moieties involved in catalysis. Using a pK_a of 4.3 as the value for a glutamate carboxylate in solution,[203] and a pK_a of 9.8 for the 3′ hydroxy of the 3′ radical derived from 2′-deoxyribose[204] (the pK_a for ribose will be lower, due to hydrogen bonding from the 2′ hydroxy), the pK_a of the glutamate and the radical would have to be perturbed by a total of ~ 6 pK units for the glutamate (E441) to deprotonate the 3′ hydroxy effectively. In order for a cysteine (C225 or C462) to protonate the 2′ hydroxy of the 3′ nucleotide radical, the enzyme would have to perturb the pK_a of the cysteine and the radical by a total of 4 pK units, taking a pK_a of 8.3 for the cysteine thiol in solution.[203] Shifts of up to 4 pK units have been reported for a number of enzymatic systems, and a perturbation of the pK_a of histidine in the active site of serine proteases may be as large as 5.5.[205,206]

However, site-directed mutagenesis experiments offer indications that the mechanism of catalysis of dehydration might be more complex than a single enzyme residue acting as a general acid or a general base. If dehydration were required to drive the hydrogen atom abstraction to completion and either C225 or C462 were absolutely required as a proton donor for this dehydration, mutation of the critical cysteine to a serine would be expected to prevent both dehydration and hydrogen atom abstraction. The C225S and C462S RDPR mutants (and the C119S and C419S mutant RTPRs[10]) are in fact able to catalyze both 3′ hydrogen atom abstraction and formation of the 3′-ketodeoxynucleotide intermediate, as judged by release of the base from the sugar, formation of a 320 nm absorbing species on the enzyme, and loss of •Y.[9,42,43,61] If, on the other hand, E441 was absolutely required as a proton acceptor, the E441Q mutant would also be expected to be inactive with respect to hydrogen atom abstraction and dehydration of the radical. However, preliminary studies[45] and later studies by Sjöberg and co-workers[46] indicate that this mutant is able to catalyze release of base from nucleotide substrates, meaning that it can catalyze hydrogen atom abstraction and formation of the 3′-ketodeoxynucleotide.

Several explanations might reconcile the model studies with the phenotypes of site-directed mutants. First, both general acid and general base catalysis might be in effect, so that no single mutation could abolish catalysis. Alternatively, as discussed in Section 5.08.6, the equilibrium between thiyl radical and carbon-centered radical might be perturbed on the enzyme, making catalysis of the dehydration step less important for the overall reaction. Finally, Zipse has suggested that deprotonation of the 3′ hydroxy and elimination of water might be concerted.[190] This suggestion is based on *ab initio* calculations of the interaction of the *trans*-butene radical cation with water, which indicate that the addition of water to this radical cation is unfavorable. Zipse interprets these calculations to suggest that the analogous proposed intermediate in the ribonucleotide reductase mechanism (Scheme 1) will be relatively high in energy and that a concerted formation of the relatively stable 3′ ketyl radical is likely. Although Zipse invokes a hydrogen bond acceptor that can interact with both the 3′ hydroxy group and a protonated 2′ hydroxy simultaneously, a concerted mechanism might also occur with two different hydrogen bond acceptors acting simultaneously, or with the 3′ hydroxy acting as a hydrogen bond acceptor for the 2′ hydroxy.

Working out the sequence of proton transfers that ribonucleotide reductases actually use to effect the hydrogen atom abstraction/water elimination sequence will be extremely challenging. The combination of model reactions, quantum chemical simulations, and mechanistic studies on wild-type and mutant enzymes allows several conclusions to be drawn. First, this sequence of reactions could be kinetically competent for ribonucleotide reduction. Second, deprotonation of the 3′ hydroxy is likely to be important for this sequence of reactions to be kinetically competent for turnover, although it is unclear whether this would take place before hydrogen atom abstraction (in order to render this process less thermodynamically unfavorable[184]) or after (to provide base catalysis[200]). Third, a single residue furnishing general acid or general base catalysis is unlikely to account for the acceleration of this sequence of reactions by the enzyme. Studies on mechanism-based inhibitors, described in the next section, have begun to address the proton transfer question as it relates to the protonation state of the 2′ hydroxy and the enzymatic group(s) that could be involved in this protonation.

5.08.7.2   Model Systems for Reactions of RDPR with Mechanism-based Inhibitors

Replacement of one of the hydroxy groups of ethylene glycol with a good leaving group allows the loss of the leaving group from the radical without acid or base catalysis. When 2-chloroethanol is oxidized (titanous ion/hydrogen peroxide), the formylmethyl radical is observed (Scheme 7).[191,193] The alkyl radical derived from ethylene glycol and its gem diol regioisomer (Scheme 7(b)) are also detected under these conditions. This radical is assigned the structure •CH$_2$CH(OH)$_2$ based on its g-value of 2.0025, which indicates that oxygen is not bound to the atom bearing the unpaired spin, and its hyperfine coupling constants are similar to the radical •CH$_2$CH(OH)(OMe).[193] These products could result from hydration of the protonated formylmethyl radical that would be the direct product of chloride loss from the 2-chloroethanol alkyl radical. These results are consistent with the loss of chloride to give the formylmethyl radical occurring without acid catalysis, in contrast to loss of water from ethylene glycol, which would require acid or base catalysis. Significantly, the radical •CH(OH)CH$_2$Cl was not detected, although a small amount of •CHClCH$_2$OH was observed.[193] The failure to detect •CH(OH)CH$_2$Cl suggests that this species may eliminate chloride too rapidly to be observed.

Similarly, when unbuffered aqueous solutions of acetic acid 2-hydroxyethyl ester (Scheme 7(c)) are subjected to pulse radiolysis, acetic acid is eliminated with a rate constant estimated to be $5 \times 10^5 - 5 \times 10^6$ s^{-1}.[207] This estimated range of rate constants for acetate loss is significant for understanding the mechanism of action of 2′-deoxy-2′-halonucleotide mechanism-based inhibitors of RNRs. It shows that when the hydroxyl in ethylene glycol is replaced by a good leaving group (in this case, an acetate group), elimination does not require acid or base catalysis to be rapid ($\sim 10^6$ s^{-1}). This contrasts with the elimination of water, which requires acid or base catalysis to be accelerated to this extent.[198,200] As discussed in Section 5.08.7, a rate constant of $\sim 10^6$ s^{-1} for the dehydration step in ribonucleotide reduction may be required to drive the 3′ hydrogen atom abstraction.

The principle that elimination of acetic acid from a 3′ nucleoside radical can be rapid even in the absence of acid or base catalysis is illustrated more directly in experiments from Giese's laboratory on a ribonucleoside derivatized with a selenol ester at the 3′ position and an acetate group at the 2′ position (Scheme 9).[200] Photolysis of this compound (an analogue of the model for the normal reduction process discussed in Section 5.08.7)[200] led to rapid elimination of acetic acid even in the absence of acid or base. This conclusion was drawn from the observation that in the presence of tributyltin hydride as a radical scavenger, the products observed from the photolysis were derived entirely from elimination of acetic acid (e.g., acetic acid, the 3′-ketonucleoside), with no products from abstraction of hydrogen from tributyltin hydride observed. A small amount ($\sim 10\%$) of nucleoside product derived from hydrogen abstraction was observed when butyl thiol was used as the radical trap. Although the lack of products derived from hydrogen atom abstraction prevents a detailed kinetic analysis using the competition methods described above, examination of the one set of conditions published by Lenz and Giese that did afford a measurable amount of hydrogen atom abstraction product (see ref. 200, Table 2, entry 4) suggests that the rate constant for elimination of acetic acid is $\sim 10^5$ s^{-1}. This model system thus provides strong support for the idea that a good leaving group at the 2′ position of a nucleotide can obviate the need for acid or base catalysis for elimination.

Model systems of Robins and co-workers offer another perspective on the elimination of leaving groups from the 2′ position of mechanism-based inactivators of RNRs. Robins has proposed that 2′-azido and -chloro substituents, among others, will be expelled as radicals, rather than anions.[193,208,209] In order to generate radicals at the 3′ position of 2′-substituted nucleosides, Robins and co-workers treated protected 3′-*O*-(phenoxythiocarbonyl)-2′-substituted nucleosides (Scheme 11) with radical initiators (either tributyltin hydride/AIBN or triphenyl-silane/dibenzoyl peroxide) in refluxing toluene. The 2′-iodo, chloro, methylthio- and azido-substituted compounds reacted to give the 2′,3′-didehydro-2′,3′-dideoxynucleoside elimination products. In contrast, the 2′-fluoro, -mesyloxy, and -tosyloxy derivatives reacted under the same initiation conditions to give the 3′-deoxy-2′-substituted nucleosides. This difference in reactivity was interpreted as arising from differences in homolytic bond dissociation energy within this series of compounds. While the homolytic bond dissociation energies for the carbon–iodine, carbon–chlorine, and carbon–azide bonds are low, allowing facile loss of iodo-, chloro-, and azido-radicals, the 2′-fluoro, -mesyloxy, and -tosyloxy derivatives would be expected to eliminate the 2′ substituent as an anion.[210,211] Thus, the observation that the iodo-, chloro-, and azido-substituted nucleosides underwent elimination, while the fluoro-, mesyloxy-, and tosyloxy-derivatives did not is consistent with elimination proceeding through a radical mechanism. A key difficulty in applying these results to the mechanism of ribonucleotide reductase action on the di- or triphosphates of 2′-halo- or azido-substituted inactivators is that the model compounds studied lack the crucial 3′ hydroxyl group. The observation of base catalysis for elimination of water from ethylene glycol,[198] the elimination of the 2′-hydroxyl when a radical is generated at the 3′ position of a protected nucleoside,[200] and the failure of monohydric alcohols to undergo facile dehydration on titanous ion oxidation[212] all suggest that the 3′ hydroxyl is essential to the chemical reactivity of the 2′ hydroxy.[37]

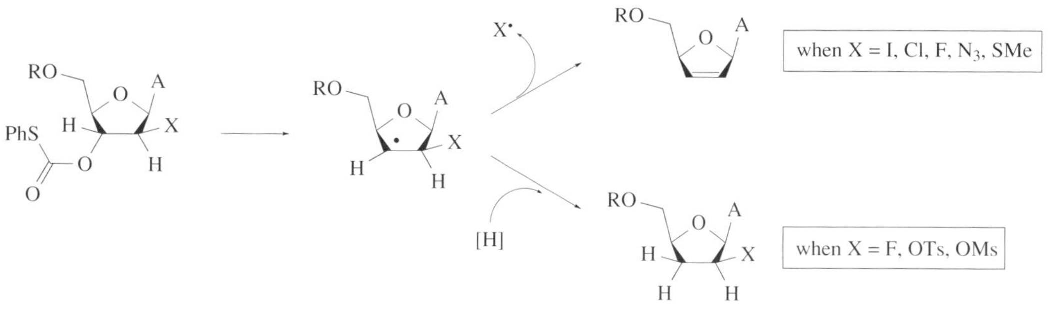

Scheme 11

Work from the Robins group introduces a model system which addresses this concern. Treatment of homoadenosine and homouridine derivatives bearing nitro groups at the 6′ position with tri-butyltin hydride and the radical initiator AIBN produces an alkoxy radical at the 6′ position[209] (Scheme 12). This radical is proposed to react with the 3′ hydrogen of the nucleoside to form a 3′ radical.[209] If the 2′ position has a chloro- or tosylate-substituent, its elimination followed by β elimination of uracil results in formation of the α,β unsaturated ketone (Scheme 12).

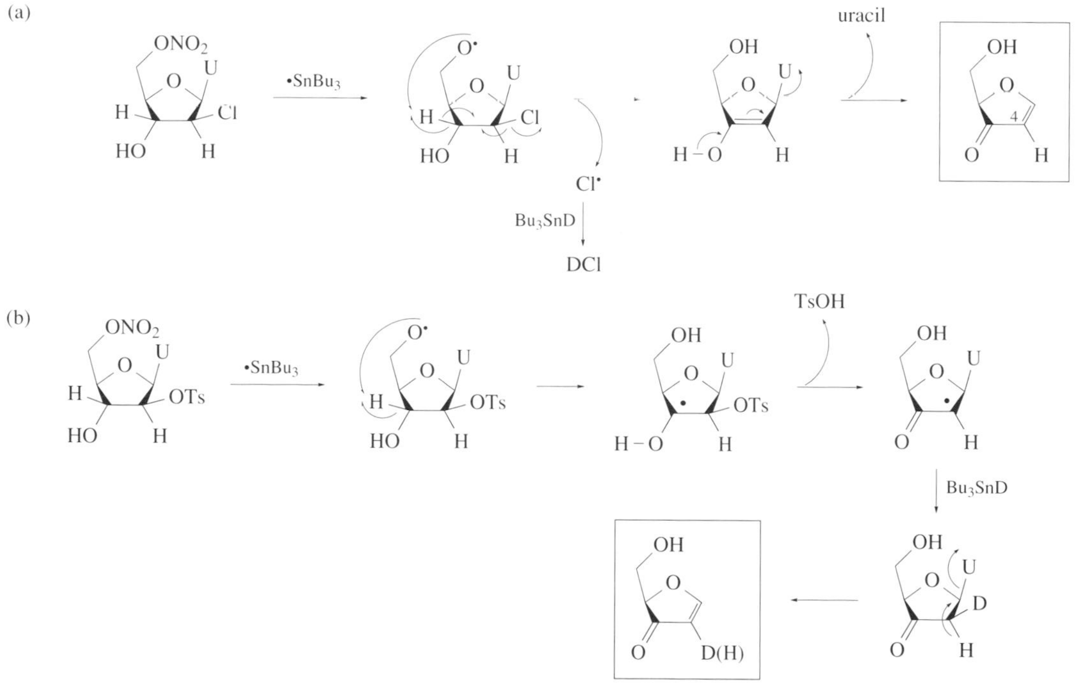

Scheme 12

In order to determine whether the 2′-chloro substituent is lost as chloride anion or as a chloro radical, Robins and co-workers treated 2′-chloro-2′-deoxy-6′-nitrohomouridine with tributyltin deuteride and AIBN (Scheme 12). Deuterium incorporation at C-4 of the final product was not observed, consistent with the hypothesis that the 2′-chloro substituent is lost as chloro radical. These results thus support the proposal that no long-lived nucleoside-based radical intermediate exists subsequent to hydrogen atom abstraction from the 3′ position.[209] If a radical intermediate were present, it would be reduced by tributyltin deuteride. In addition, no 2′-chloro-2′-deoxy-6′-

homouridine was isolated from this reaction,[216] indicating that abstraction of hydrogen from tributyltin deuteride by the putative 3' radical is slow compared with elimination of chlorine.

In contrast, treatment of 2'-*O*-tosyl-6'-nitrohomoadenosine with tributyltin deuteride and AIBN led to ~30% incorporation of deuterium at the C-4 position of the product (Scheme 12(b)), consistent with a mechanism that includes loss of tosylate anion and formation of a 3'-keto-2'-deoxynucleoside radical that can abstract a deuterium from tributyltin deuteride (selective transfer of deuterium to the α face accounts for the 30% incorporation).[209] These results were interpreted to suggest that RNR interaction with 2'-chloro-2'-deoxynucleotides might proceed by loss of Cl• rather than Cl⁻ as proposed in the 1980s.[41,59,217] The mechanistic possibility of Cl• loss was considered in the original formulation of a hypothesis to account for the observations with the enzyme. As outlined below, however, the studies with the enzymatic system cannot be accommodated by loss of Cl•. Thus, while the photolysis of 2'-chloro-2'-deoxy-6'-nitrohomouridine provides a chemically feasible model for mechanism-based inactivation of ribonucleotide reductases, it appears that the enzymatic reaction proceeds by a different mechanism.

5.08.7.3 Mechanism-based Inhibition of Ribonucleotide Reductases by 2'-Halo-substituted Nucleotides

In 1976, Thelander *et al.* reported that incubation of *E. coli* RDPR with ClUDP resulted in its inactivation and that this process was accompanied by chloride ion and uracil release.[147] Since these important early studies, this reaction has been studied in detail using [5'-³H]-, [β-³²P]-, [3'-³H]-, [2'-³H]- and [¹⁴C]-labeled ClUDP and ClUTPs with both class I and II RNRs, respectively. Studies of the products of these reactions have provided much information about catalytic capabilities of RNRs, and have allowed formulation of a detailed mechanism for this process. Any mechanism for this process must provide a reasonable explanation for why these compounds function as inactivators and, under conditions in which the enzymes are protected against inactivation, substrates. The mechanistic model that explains available data is presented in Scheme 2.

Evidence is presented that has led to formulation of this model, in which Cl⁻, rather than Cl•, is lost through a cation radical, rather than an anion radical. Incubation of [3'-³H]-ClND(T)Ps with RNRs leads to release of ³H₂O to the solvent, suggesting that inactivation is mechanism-based, beginning with abstraction of hydrogen from the 3' position of the inactivator, as is the case with the normal substrate (Scheme 1).[41,57,59] As summarized in Scheme 13, this inactivation is accompanied by initial formation of a 3'-ketodeoxynucleotide which dissociates from the enzyme and decomposes into nucleic acid base, pyrophosphate (or tripoly-phosphate), and 2-methylene-3(2H)-furanone.[41] The latter species alkylates the R1 subunit (or its equivalent in other RNRs) in the presence of the thioredoxin reducing system. However, if DTT is used as reductant, this furanone is trapped at the exocyclic methylene, leading to protection against inactivation.[41,217] The ability to effect multiple turnovers results in the ability to detect production of deoxynucleotides.

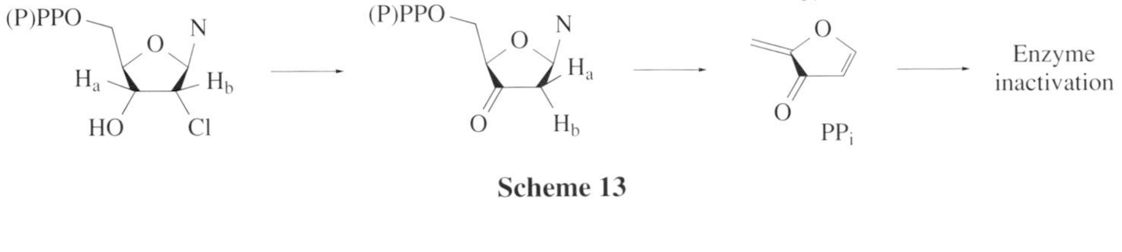

Scheme 13

A working model presented by the authors (compare Schemes 1 and 2) predicts that the 3'-nucleoside radical is a common intermediate in the inactivation and reduction mechanisms and that partitioning between these two processes occurs through the 3'-ketyl radical intermediate. The requirement for protonation of the leaving group was proposed to govern whether two cysteines or a cysteine and a cysteinate were present in this form of the enzyme. If both dNDP and inactivation products are generated through a common intermediate subsequent to 3' hydrogen atom abstraction, they both could exhibit the same isotope effect on this process. Unfortunately, it has not been possible to demonstrate this. In a double label experiment using [3'-³H]- and [β-³²P]-ClUDP, a V/K effect of <1.17 was determined.[57] However, a similar experiment could not be carried out on dNDP produced in this same experiment because the amount generated is too low for accurate dual label analysis. This number is less than that measured ($k_H/k_T = 3.3$) using [3'-³H]-UDP.[187] This might be the expected result, as V/K isotope effects measure the chemistry up to and through the first irreversible step. In the authors' proposed mechanism this step would be Cl^- vs. H_2O release. Based on model studies, Cl^- release would be expected to be faster, not requiring acid or base catalysis, resulting in a larger commitment to catalysis and hence a lower isotope effect.

In support of this proposal, studies with a variety of 2'-ara or ribo halogenated nucleotides (F, Cl, Br, I) have been carried out.[217] ara-ClATP and ara-BrATP inactivate the *L. leichmannii* reductase through a 3'-ketodeoxynucleotide at rates comparable to the ribo isomers. These results suggest that interaction of the leaving group with the enzyme is not important for inactivation (that is, general acid catalysis is not required). An analogous experiment using ara-ATP (which has a hydroxy group at the 2' position) results in no inactivation, demonstrating the requirement for protonation of the leaving group in elimination of H_2O. (Similar experiments with the *E. coli* RNR and the diphosphate forms of these inactivators provide no evidence for halide release, suggesting that this enzyme does not tolerate the increased steric bulk on the β face of the nucleotide.) These observations must be accommodated by any proposed mechanism. Finally, it is also important to note that 2'-fluoro-2'-deoxynucleotides (FUD(T)P) are also substrates and inactivators. It is unlikely that $F\bullet$ would be involved in either process.

If one makes the assumption that a 3'-nucleotide radical is a common intermediate in the reactions of the various 2'-halo-substituted inactivators, the dependence on the pK_a of the leaving group of the observed partitioning between turnover and inactivation can be interpreted as evidence for a mechanism in which the 2' halogen substituents are eliminated as halide ion, rather than as the radical. In experiments performed on the *L. leichmannii* RTPR, the enzyme was found to catalyze the production of ~ 1 equiv. of dUTP per 1.5 equiv. of fluoride released when incubated with FUTP in the presence of 3 mM DTT at pH 7.8.[217] In contrast, incubation of RTPR with ClUTP results in 1 equiv. of dUTP per 220 turnovers (at pH 7.4 in the presence of 3 mM DTT).[217] The partitioning thus correlates with the pK_a of the leaving group. When the leaving group is hydroxy ($pK_a = 16$), all turnovers result in deoxynucleotide formation. When the leaving group is fluoride ($pK_a = 3.2$), both reduction and inactivation occur, with reduction predominating. When the leaving group is chloride ($pK_a = -7$), however, inactivation predominates over reduction. The simplest interpretation of this data, assuming a commonality of mechanism, is that at least one of the cysteines proposed to interact with the α face of the substrate (C119 and C419 for the *L. leichmannii* enzyme, C225 and C462 for the *E. coli* enzyme) must be deprotonated for reduction to occur. If the leaving group is not protonated, deprotonating the appropriate thiol in the active site, the 3'-keto-2'-deoxynucleotide radical intermediate will be reduced by hydrogen atom transfer either from the β face (C439) or the α face (C225 or C462), forming a 3'-keto-2'-deoxynucleotide (Scheme 2). C225 and C462 will thus be in the form of either a dithiol or a neutral disulfide radical. In either case, C225/C462 will not be in the disulfide radical anion state proposed to be required for reduction of the 3'-keto-2'-deoxynucleotide (Section 5.08.8.3). Thus, the 3'-keto-2'-deoxynucleotide will not be reduced rapidly, and dissociation of this intermediate from the active site to form the furanone species in solution competes effectively with reduction.

Supporting this hypothesis is the observation that the partitioning of ClUTP between reduction and inactivation is sensitive to the pH when DTT is used as the reductant. At pH 5.5, the partitioning is 200 turnovers per reduction, while at pH 8.3, reduction occurs once in every 70 turnovers.[217] The partitioning is also sensitive to the concentration of exogenous reductant; changing the DTT concentration from 3 mM to 30 mM increases the ratio of reduction events to turnovers from 1 in 220 to 1 in 120.[217] The DTT dependence suggests that when the redox environment does not allow the α face cysteines to be fully reduced, the rate of reduction is diminished, and furanone formation competes more effectively. Taken together, these data support the hypothesis that halide ion is eliminated from the 2' position, rather than halo radical.

The strongest evidence against halo radical elimination mechanism comes from the modes of inactivation that are observed with 2'-halo-substituted inactivators. The observation that the $\bullet Y$ in the *E. coli* class I RNR is not reduced during inactivation[147] requires that one of the intermediates be able to reabstract a hydrogen from C439, regenerating the thiyl radical which can, in turn, regenerate the $\bullet Y$. The 3'-ketyl radical intermediate could perform this function if Cl^- is eliminated. However, if $Cl\bullet$ is eliminated, then no nucleotide radical species would be available to abstract a hydrogen from C439 unless $Cl\bullet$ could migrate from the α face of the nucleotide to the β face. This would be an unlikely scenario given the steric constraints of the enzyme active site. Thus, elimination of $Cl\bullet$ would trap C439 in the thiol form, preventing reformation of the R2 $\bullet Y$, contrary to experimental observations.

Experiments defining the fate of the 3' hydrogen in the course of inactivation by ClUDP (ClUTP) provide unambiguous support for loss of halide ion. The radiolabeled 2'-deoxy-3'-ketoUDP formed on incubation of RDPR with [3'-^{3}H]-ClUDP was trapped using sodium borohydride. Chemical analysis showed that the tritium originally at the 3' position had migrated to the 2' position on the β face.[41] This result requires a 3' to 2' shift of hydrogen and can be explained as abstraction of the 3' hydrogen from ClUDP by the thiyl radical (C439), followed by elimination of chloride ion to form the 3'-ketyl radical and reabstraction of the hydrogen on C439 (Scheme 2) by the 3'-ketyl radical to place the hydrogen originally at the 3' position at the 2' position. This result is not consistent with chloro radical formation, as chloro radical would have to migrate to the β face of the nucleotide, abstract tritium from C439, and deliver it to the 2' position of the intermediate. Thus, in the enzyme-catalyzed reaction, it is unlikely that the mechanism of C—X bond cleavage is homolytic, in contrast to the model studies of Robins.[208,209]

5.08.7.4 Summary

Both simple model systems based on generation of radicals from ethylene glycol[194,195,197,199] and more complex models that allow generation of radical at specific positions on nucleosides[200,208,209] have provided insight into how RNRs catalyze elimination of the 2' substituent of nucleotide substrates and inactivators. Both base catalysis (deprotonation of the 3'-hydroxy) and acid catalysis (protonation of the 2'-hydroxy) may be involved, although the enzymatic groups involved and the timing of protonation and deprotonation remain to be determined.

The rates of dehydration measured for both ethylene glycol-based and nucleoside-based systems suggest that the acid- or base-catalyzed dehydration is fast enough to drive the potentially unfavorable hydrogen atom abstraction that precedes it. The enzyme may employ some combination of stabilization of the carbon-centered radical (to make the hydrogen atom abstraction less thermodynamically unfavorable) and acceleration of dehydration to make this sequence of reactions kinetically competent for nucleotide reduction.

Model systems predict that elimination of 2'-halo substituents in the course of mechanism-based inactivation of ribonucleotide reductases does not require acid or base catalysis, and might occur either through elimination of halide ion or halo radical. A variety of biochemical studies establish loss of halide ion from C-2'. The differences between the reactions with mechanism-based inactivators and reduction of the normal substrate have been rationalized as arising from a difference in protonation state of the cysteines interacting with the α face of the substrate. Deprotonation of these cysteines is proposed to be critical for nucleotide reduction to proceed normally. The next section will discuss model chemistry which supports the hypothesis that deprotonation of active site thiols is required for nucleotide reduction.

5.08.8 REDUCTION OF THE 3'-KETO-2'-DEOXYNUCLEOTIDE RADICAL BY TWO SINGLE ELECTRON TRANSFERS AND FORMATION OF DEOXYNUCLEOTIDE PRODUCT

5.08.8.1 One-electron Reduction of the 3'-Keto-2'-deoxynucleotide Radical: Model Studies

The next step proposed in the RNR mechanism is the one-electron reduction of the ketyl radical by a pair of cysteine residues at the active site, forming the enolate anion of the 3'-keto-2'-deoxynucleotide and a protonated disulfide radical. The protonated disulfide radical should rapidly protonate the enolate, forming a 3'-keto-2'-deoxynucleotide and a disulfide radical anion (Scheme 1). Once again, model systems have provided a basis for understanding what reactions might be thermodynamically favorable and kinetically competent for nucleotide reduction. As has been the case for many of the proposed reactions in the RNR mechanism, model studies suggest that the protonation state of enzyme side chains is a critical determinant of enzymatic reactivity.

Pulse radiolytic studies from the von Sonntag laboratory have allowed determination of rates of reduction of formylmethyl radicals by thiolate anions (Equation (5)).[160]

$$\overset{\bullet}{\underset{\underset{O}{\|}}{HC}}-CH_2 + RS^- \longrightarrow \overset{H}{\underset{-O}{\diagdown}}C=CH_2 + RS^{\bullet} \qquad (5)$$

The rate of reduction by dithiothreitol of the formylmethyl radical derived from ethylene glycol was measured to be $1.2 \times 10^8 \ M^{-1} \ s^{-1}$ at pH 10, and $3.5 \times 10^8 \ M^{-1} \ s^{-1}$ at pH 11.1. In contrast, the rate constant at pH 8.4 (below the first pK_a of dithiothreitol) is $\leqslant 10^7 \ M^{-1} \ s^{-1}$.[218] These results have been interpreted by von Sonntag and co-workers as indicating that the dithiothreitol thiolate anion (or, at pH 11.1, the even more reactive dianion) is responsible for the reduction, which would thus occur by electron transfer rather than hydrogen atom transfer. Evidence for electron transfer from thiols to organic radicals has also been obtained in studies of the repair of OH• adducts of dGMP by monothiols.[161] The pH dependence of this reaction indicated that the thiolate, rather than the thiol, is the reactive species, suggesting that the reaction proceeds by electron transfer rather than hydrogen atom transfer.

These results are in accord with the hypothesis that the protonation state of the "bottom face" cysteines determine the partitioning between normal turnover and inactivation, as discussed in the previous section. If one of these cysteines can be deprotonated, it can reduce the 3′-keto-2′-deoxynucleotide radical by electron transfer (followed by rapid proton transfer), by analogy with the reduction of formylmethyl radical. If the leaving group at the 2′ position is not a strong enough base to deprotonate one of these cysteines, the reduction will be slow, and the pathway leading to inactivation will begin to predominate.

The initial product of an electron transfer between a cysteine thiolate adjacent to a cysteine thiol and the 3′-keto-2′-deoxynucleotide radical will be the enolate anion and a neutral, protonated disulfide radical. However, the pK_a of a neutral disulfide radical is 5.2.[159] The enolate should therefore be rapidly protonated, resulting in the formation of 3′-keto-2′-deoxynucleotide and disulfide radical anion. Protonation of an enolate by protonated disulfide radical from the α face would also account for the observation that the 2′-hydroxy group of the substrate, in all classes of RNRs, is replaced by a solvent-derived hydrogen atom with retention of configuration.[62,63]

These model studies suggest that this reduction will not be rate-limiting for turnover. With a reasonable effective concentration for the reducing thiolate at the active site (>1 M), the rate of this first electron transfer should be $>10^8$ s^{-1}, much faster than the overall rate of turnover.

5.08.8.2   Evidence from Mechanism-based Inhibitors for Reduction from the α Face of the Substrate

As previously discussed, N$_3$UDP is a stoichiometric inactivator of class I RNRs in which 3′ hydrogen atom abstraction initiates the inactivation process (Scheme 6). Formation of hydrazoic acid could proceed by one of several mechanisms. Based on model studies with 2-azido ethanol and the titanous ion/H$_2$O$_2$ oxidizing system, azide radical could be released subsequent to hydrogen atom abstraction. Thus, one possibility is that azide radical could be reduced in the enzyme active site by a cysteine residue to generate a thiyl radical and hydrazoic acid.[219] Alternatively, hydrazoic acid could be generated directly, through a mechanism analogous to that proposed for the substrate reduction.[152] The 3′-keto-2′-deoxynucleotide radical is proposed to abstract a hydrogen from a cysteine at the α face of the nucleotide, forming a thiyl radical.[150,152] Addition of the resulting thiyl radical to hydrazoic acid (which, despite its pK_a of ~ 5.7, must remain protonated for reaction to occur), followed by loss of N$_2$,[220] is hypothesized to produce a nitrogen-centered radical, which is then proposed to react with the 3′-keto-2′-deoxynucleotide to form a radical that can be trapped and observed by EPR. Detailed analysis of the EPR spectra of this radical at 9 GHz and 140 GHz has allowed full assignment of all three principal g-values and partial assignment of the components of the nitrogen hyperfine tensor.[152] This information, in conjunction with EPR spectroscopy of the nitrogen-centered radical generated using β-deuterium labeled cysteine-R1[152] and the use of ESEEM data on selectively deuterated variants of the inhibitor,[150] allowed the proposal that the nitrogen-centered radical has one of the structures shown in Scheme 14. The proposed initiation of this chemistry by a nucleotide-based radical abstracting a hydrogen from a thiol (C225)[150,152] (Scheme 6) is analogous to the one-electron reduction of an α-keto radical proposed to take place in the normal catalytic mechanism, with the exception that a hydrogen atom, rather than an electron, is transferred. However, the alternative mechanism (Scheme 6) in which azide radical is lost initially, cannot be ruled out as an initial event required for the inactivation process. In fact, this pathway is favored based on model chemistry discussed above.

Direct experimental discrimination between the proposed electron transfer/proton transfer sequence and a hydrogen atom transfer from a "bottom face" cysteine is likely to be quite difficult. Nevertheless, the combination of model studies and the reactivity of different 2′-halo-substituted mechanism-based inhibitors provides substantial indirect evidence for this hypothesis. The model studies also suggest that this step will not be rate limiting for turnover.

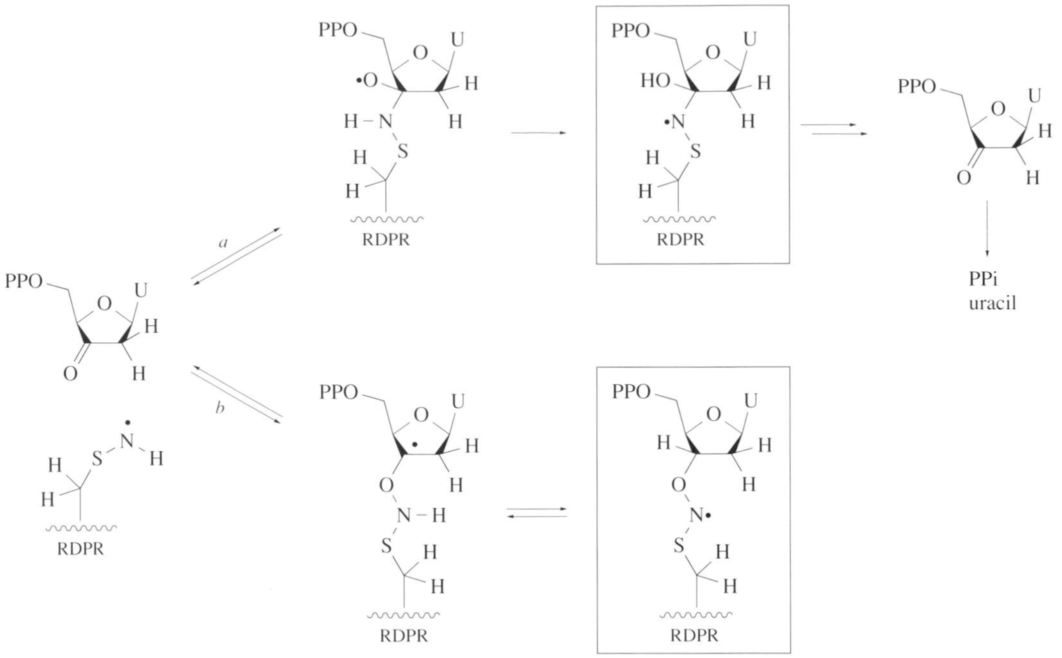

Scheme 14

5.08.8.3 One-electron Reduction of 3′-Keto-2′-deoxynucleotides: Model Studies

Protonation is also predicted to be critical for the second proposed electron transfer, from disulfide radical anion to the 3′-keto-2′-deoxynucleotide (Scheme 1). The reverse reaction, reduction of a disulfide to a disulfide radical anion by a ketyl radical, is highly exergonic. The one-electron reduction potential of dithiothreitol is -1.6 V, while the one-electron reduction potential of acetone is -2 V.[221,222] Using pulse radiolysis, the reduction of oxidized DTT by the α-hydroxyalkyl radical anion derived from 2-propanol was shown by von Sonntag and co-workers to proceed with a rate constant of 4×10^8 M^{-1} s^{-1}.[223] These results suggest that the proposed electron transfer would not be observed if the product were the ketyl radical anion.

However, as Giese has noted,[200] the reduction of a ketone to a protonated ketyl radical (i.e., an α-hydroxyalkyl radical) would be thermodynamically favorable. Using the reduction potential of acetone and the pK_a of the neutral α-hydroxyalkyl radical of 2-propanol, Schwarz and Dodson calculated the reduction potential for conversion of acetone and a proton to a protonated ketyl radical to be -1.4 V.[222] These results suggest that a disulfide radical anion can reduce a ketone, as long as the product ketyl radical is protonated. Steady-state radiolytic studies of von Sonntag and co-workers,[223] in combination with the work of Schwarz and Dodson,[222] allow an estimate of the rate constant for this process. The overall rate constant for reduction of oxidized DTT by neutral alcohol radicals is estimated to be ~ 100 s^{-1}.[223] The reduction of acetone by a disulfide radical anion is predicted to have an equilibrium constant of ~ 2400, based on the difference of 0.2 V between the reduction potentials of disulfide radical anion and acetone.[222] The rate constant for reduction of acetone to a neutral radical by disulfide radical anion can thus be calculated to be $\sim 2 \times 10^5$ M^{-1} s^{-1}.

These results suggest that a glutamate in the active site of the enzyme is necessary to accomplish the second electron transfer. Giese has suggested that E441 of R1 serves this purpose.[200] This is consistent with the phenotype of the E441Q mutant R1, which catalyzes base elimination preferentially over nucleotide reduction.[45,46] If E441 is required to donate a proton to allow this reduction to take place, mutating this glutamate to a glutamine will allow the chemistry to proceed through the formation of the 3′-keto-2′-deoxynucleotide, but prevent the reduction of that intermediate, allowing elimination of base and polyphosphate. The phenotype of E441D R1, which produces deoxynucleotide, albeit at a rate much lower than wild-type R1, is also consistent with this model.

5.08.9 ABSTRACTION OF HYDROGEN FROM A CYSTEINE THIOL BY THE 3'-DEOXYNUCLEOTIDE RADICAL: COMPLETION OF NUCLEOTIDE REDUCTION AND REGENERATION OF THE THIYL RADICAL

The final step in the proposed mechanism is the reabstraction of hydrogen by the 3'-deoxynucleotide radical to complete nucleotide reduction and regenerate the thiyl radical for another catalytic cycle. Extensive model studies have been carried out on abstraction of hydrogen from alcohol radicals by thiols.[224] The rate constant for abstraction of hydrogen from β-thio ethanol by the α-hydroxyalkyl radical derived from 2-propanol is $\sim 5 \times 10^8 \, M^{-1} \, s^{-1}$.[225] The hydrogen reabstraction step in the RNR mechanism is thus expected to be rapid. Finally, while it has been shown in the case of the class II RNR that reformation of the starting form of the cofactor (AdoCbl, in that case) is regenerated on every turnover, in the case of the class I enzymes, the number of turnovers effected per thiyl radical formation has not been established.

5.08.10 SUMMARY AND CONCLUSIONS

Mechanistic studies on enzymatic ribonucleotide reduction and on small-molecule model systems for the steps in this reaction support the mechanism in Scheme 1. In the enzymatic systems, both the proposed thiyl radical (in the presence of physiological substrates) and nucleotide-based radicals (derived from mechanism-based inhibitors) have been observed spectroscopically. Kinetic and product analysis studies with physiological substrates and mechanism-based inhibitors support the proposed mechanism. Model studies have established that the proposed mechanism is plausible from a kinetic and thermodynamic standpoint; evidence from model systems suggests that each proposed step or sequence of steps could occur in a kinetically competent fashion in the enzymatic mechanism.

These mechanistic studies have provided insights into how enzymes effect catalysis and control reactivity. One aspect of enzymatic catalysis that the RNRs illustrate well is the ability of enzymes to couple a thermodynamically unfavorable step to a subsequent thermodyamically favorable step, thus assisting in the cleavage of strong covalent bonds. In the case of the *L. leichmannii* reductase, the homolytic cleavage of the carbon–cobalt bond of AdoCbl (uphill by $\sim 30 \, kcal \, mol^{-1}$) is coupled to the thermodynamically favorable formation of a thiyl radical from 5'-dA• by virtue of both reactions occurring in a single concerted step. For both *E. coli* and *L. leichmannii* reductases, coupling of the abstraction of the 3' hydrogen by the thiyl radical to the rapid and thermodynamically favorable dehydration of a 3'-nucleotide radical may be essential for allowing the hydrogen atom abstraction to occur and nucleotide reduction to proceed. The ribonucleotide reductases thus provide an example of how acceleration of a sequence of single elementary steps may be selected for in evolution, in addition to acceleration of each step individually.

The RNRs also illustrate the importance of protonation states in biological redox chemistry. Model systems suggest that generation of a thiyl radical from a •Y in the *E. coli* reductase requires dynamic regulation of the protonation states of the protein residues involved in electron transfer. Model reaction and the reactions of mechanism-based inactivators also show that redox reactions between the substrate and the active site thiols of the enzyme are highly dependent on the protonation states of the substrate and of active site residues.

The necessity for control of active site residue protonation states also has implications for the design of mechanism-based inhibitors of ribonucleotide reductase and other redox enzymes. Many of the mechanism-based inhibitors of ribonucleotide reductases work in part by preventing the deprotonation of the cysteines that interact with the α face of the nucleotide, thus interfering with their ability to deliver reducing equivalents. These inhibitors show that the pK_a of a leaving group can be as important to the mechanism of inactivation as its rate of expulsion.

Much remains to be learned about the mechanisms of RNRs, particularly with regard to spectroscopic characterization of intermediates in intersubunit electron transfer (for the *E. coli* enzyme) and of substrate-based radicals in the normal catalytic mechanism. Isotopic substitutions or other perturbations in conditions may allow these intermediates to exist long enough to be trapped and characterized. If so, spectroscopic techniques such as high-field EPR, ENDOR, and ESEEM should furnish important information about the structures, electronic properties, and protonation states of these intermediates. This information, in turn, will give further insight into how these fascinating enzymes control radical reactivity.

5.08.11 REFERENCES

1. P. Reichard, *Science*, 1993, **260**, 1773.
2. M. J. Robins, M. C. Samano, and V. Samano, *Nucleosides Nucleotides*, 1995, **14**, 485.
3. J. L. Abbruzzese and W. Plunkett, *J. Clin. Oncol.*, 1991, **9**, 491.
4. L. W. Hertel, J. S. Kroin, C. S. Grossman, G. B. Grindey, A. F. Dorr, A. M. V. Storniolo, W. Plunkett, V. Gandhi, and P. Huang, *ACS Symp. Ser.*, 1996, **639**, 265.
5. S. J. Elledge, Z. Zhou, J. B. Allen, and T. A. Navas, *Bioessays*, 1993, **15**, 333.
6. J. Stubbe, *Annu. Rev. Biochem.*, 1989, **58**, 257.
7. D. T. Logan, X.-D. Su, A. Åberg, K. Regnström, J. Hajdu, H. Eklund, and P. Nordlund, *Structure*, 1996, **4**, 1053.
8. U. Uhlin and H. Eklund, *Nature*, 1994, **370**, 533.
9. S. S. Mao, T. P. Holler, G. X. Yu, J. M. Bollinger, S. Booker, M. I. Johnston, and J. Stubbe, *Biochemistry*, 1992, **31**, 9733.
10. S. Booker, S. Licht, J. Broderick, and J. Stubbe, *Biochemistry*, 1994, **33**, 12 676.
11. A. Jordon, E. Torrents, C. Jeanthon, R. Eliasson, U. Hellman, C. Wenstedt, J. Barbe, I. Gibert, and P. Reichard, *Proc. Natl. Acad. Sci. USA*, 1997, **94**, 13 487.
12. S. Booker and J. Stubbe, *Proc. Natl. Acad. Sci. USA*, 1993, **90**, 8352.
13. D. Panagou, M. D. Orr, J. R. Dunstone, and R. L. Blakley, *Biochemistry*, 1972, **11**, 2378.
14. A. Tauer and S. A. Benner, *Proc. Natl. Acad. Sci. USA*, 1997, **94**, 53.
15. P. K. Tsai and H. P. C. Hogenkamp, *J. Biol. Chem.*, 1980, **255**, 1273.
16. F. Mancia, N. H. Keep, A. Nakagawa, P. F. Leadlay, S. McSweeney, B. Rasmussen, P. Bösecke, O. Diatt, and P. R. Evans, *Structure*, 1996, **4**, 339.
17. X. Sun, R. Eliasson, E. Pontis, J. Andersson, G. Buist, B.-M. Sjöberg, and P. Reichard, *J. Biol. Chem.*, 1995, **270**, 2443.
18. K. K. Wong and J. W. Kozarich, in "Metal Ions in Biological Systems," eds. H. Sigel and A. Sigel, Marcel Dekker, New York, 1994, vol. 30, p. 279.
19. W. Rödel, W. Plaga, R. Frank, and J. Knappe, *Eur. J. Biochem.*, 1988, **177**, 153.
20. J. Tomaschewski and W. Rüger, *Nucleic Acids Res.*, 1987, **15**, 3632.
21. R. D. Fleischmann, M. D. Adams, O. White, R. A. Clayuton, E. W. Kirkness, A. R. Kerlavage, C. J. Bult, J.-F. Tomb, B. A. Dougherty, J. M. Merrick, *et al. Science*, 1995, **269**, 496.
22. J. Knappe and T. Schmitt, *Biochem. Biophys. Res. Commun.*, 1976, **71**, 1110.
23. A. Willing, H. Follmann, and G. Auling, *Eur. J. Biochem.*, 1988, **178**, 603.
24. G. Auling and H. Follmann, in "Metal Ions in Biological Systems," eds. H. Sigel and A. Sigel, Marcel Dekker, New York, 1994, vol. 30.
25. U. Griepenburg, G. Lassmann, and G. Auling, *Free Rad. Res.*, 1996, **26**, 473.
26. C. L. Atkin, L. Thelander, P. Reichard, and G. Lang, *J. Biol. Chem.*, 1973, **248**, 7464.
27. J. M. Bollinger Jr., D. E. Edmonson, B. H. Huynh, J. Filley, J. R. Norton, and J. Stubbe, *Science*, 1991, **253**, 292.
28. J. M. Bollinger Jr., J. Stubbe, B. H. Huynh, and D. E. Edmondson, *J. Am. Chem. Soc.*, 1991, **113**, 6289.
29. J. M. Bollinger Jr., W. H. Tong, N. Ravi, B. H. Huynh, D. E. Edmondson, and J. Stubbe, *J. Am. Chem. Soc.*, 1994, **116**, 8024.
30. N. Ravi, J. M. Bollinger Jr., W. H. Tong, N. Ravi, B. H. Huynh, D. E. Edmondson, and J. Stubbe, *J. Am. Chem. Soc.*, 1994, **116**, 8007.
31. W. H. Tong, S. Chen, S. G. Lloyd, D. E. Edmondson, B. H. Huynh, and J. Stubbe, *J. Am. Chem. Soc.*, 1996, **118**, 2107.
32. M. Sahlin, G. Lassmann, S. Pötsch, B.-M. Sjöberg, and A. Gräslund, *J. Biol. Chem.*, 1995, **270**, 12 361.
33. Å. Larsson, I. Climent, P. Nordlund, M. Sahlin, and B.-M. Sjöberg, *Eur. J. Biochem.*, 1996, **237**, 58.
34. M. Ekberg, M. Sahlin, M. Eriksson, and B.-M. Sjöberg, *J. Biol. Chem.*, 1996, **271**, 20 655.
35. S. Nyholm, L. Thelander, and A. Gräslund, *Biochemistry*, 1993, **32**, 11 569.
36. W. A. van der Donk and J. Stubbe, *Chem. Rev.*, 1998, **98**, 705.
37. J. Stubbe, *Adv. Enzymol. Relat. Areas Mol. Biol.*, 1990, **63**, 349.
38. J. Stubbe and W. A. van der Donk, *Chem. Biol.*, 1995, **2**, 793.
39. L. Thelander, B. Larsson, J. Hobbs, and F. Eckstein, *J. Biol. Chem.*, 1976, **251**, 1398.
40. G. W. Ashley, G. Harris, and J. Stubbe, *Biochemistry*, 1988, **27**, 4305.
41. M. A. Ator and J. Stubbe, *Biochemistry*, 1985, **24**, 7214.
42. S. S. Mao, M. I. Johnston, J. M. Bollinger, and J. Stubbe, *Proc. Natl. Acad. Sci. USA*, 1989, **86**, 1485.
43. S. S. Mao, T. P. Holler, J. M. Bollinger, G. X. Yu, M. I. Johnston, and J. Stubbe, *Biochemistry*, 1992, **31**, 9744.
44. W. A. van der Donk, C. Zeng, K. Biemann, J. Stubbe, A. Hanlon, and J. E. Kyte, *Biochemistry*, 1996, **35**, 10 058.
45. W. A. van der Donk, G. Yu, D. J. Silva, J. Stubbe, J. R. McCarthy, E. T. Jarvi, D. P. Matthews, R. J. Resvick, and E. Wagner, *Biochemistry*, 1996, **35**, 8381.
46. A. L. Persson, M. Eriksson, B. Katterle, S. Potsch, M. Sahlin, and B. M. Sjöberg, *J. Biol. Chem.*, 1997, **272**, 31 533.
47. W. S. Beck, R. H. Abeles, and W. G. Robinson, *Biochem. Biophys. Research Commun.*, 1966, **25**, 421.
48. R. H. Abeles and W. S. Beck, *J. Biol. Chem.*, 1967, **242**, 3589.
49. H. P. C. Hogenkamp, R. K. Ghambeer, C. Brownson, R. L. Blakley, and E. Vitols, *J. Biol. Chem.*, 1968, **243**, 799.
50. S. Licht, G. J. Gerfen, and J. Stubbe, *Science*, 1996, **271**, 477.
51. P. A. Frey, S. S. Kerwar, and R. H. Abeles, *Biochem. Biophys. Res. Commun.*, 1967, **26**, 873.
52. Y. Tamao and R. L. Blakley, *Biochemistry*, 1973, **12**, 24.
53. W. H. Orme-Johnson, H. Beinert, and R. L. Blakley, *J. Biol. Chem.*, 1974, **249**, 2338.
54. J. A. Hamilton, R. Yamada, R. L. Blakley, H. P. C. Hogenkamp, F. D. Looney, and M. E. Winfield, *Biochemistry*, 1971, **10**, 347.
55. G. J. Gerfen, S. Licht, J.-P. Willems, B. M. Hoffman, and J. Stubbe, *J. Am. Chem. Soc.*, 1996, **118**, 8192.
56. J. Wu and J. Stubbe, unpublished results.
57. G. Harris, M. Ator, and J. Stubbe, *Biochemistry*, 1984, **23**, 5214.
58. J. Stubbe and J. W. Kozarich, *J. Am. Chem. Soc.*, 1980, **102**, 2505.
59. J. Stubbe, G. Smith, and R. L. Blakley, *J. Biol. Chem.*, 1983, **258**, 1619.

60. A. Aberg, S. Hahne, M. Karlsson, A. Larsson, M. Örmo, A. Ahgren, and B. M. Sjöberg, *J. Biol. Chem.*, 1989, **264**, 12 249.
61. S. S. Mao, G. X. Yu, D. Chalfoun, and J. Stubbe, *Biochemistry*, 1992, **31**, 9752.
62. T. J. Batterham, R. K. Ghambeer, R. L. Blakley, and C. Brownson, *Biochemistry*, 1967, **6**, 1202.
63. S. David and J. Eustache, *Carbohydr. Res.*, 1971, **20**, 319.
64. J. Stubbe and D. Ackles, *J. Biol. Chem.*, 1980, **255**, 8027.
65. J. Stubbe, D. Ackles, R. Segal, and R. L. Blakley, *J. Biol. Chem.*, 1981, **256**, 4843.
66. W. S. Beck, M. Goulian, A. Larsson, and P. Reichard, *J. Biol. Chem.*, 1966, **241**, 2177.
67. W. S. Beck, *J. Biol. Chem.*, 1967, **242**, 3148.
68. N. C. Brown and P. Reichard, *J. Mol. Biol.*, 1969, **46**, 39.
69. S. Eriksson and B. M. Sjöberg, in "Allosteric Enzymes," ed. G. Hervé, CRC Press, Boca Raton, FL, 1989, p. 189.
70. R. G. Finke and B. P. Hay, *Inorg. Chem.*, 1984, **23**, 3041.
71. B. P. Hay and R. G. Finke, *Polyhedron*, 1988, **7**, 1469.
72. J. Halpern, S.-H. Kim, and T. W. Leung, *J. Am. Chem. Soc.*, 1984, **106**, 8317.
73. B. P. Hay and R. G. Finke, *J. Am. Chem. Soc.*, 1986, **108**, 4820.
74. T. W. Koenig, B. P. Hay, and R. G. Finke, *Polyhedron*, 1988, **7**, 1499.
75. J. Halpern, *Science*, 1985, **227**, 869.
76. J. M. Pratt, in "B12," ed. D. Dolphin, Wiley, New York, 1982, vol. 1, p. 325.
77. L. Randaccio, N. Bresciani Pahor, E. Zangrando, and L. G. Marzilli, *Chem. Soc. Rev.*, 1989, **18**, 225.
78. R. Banerjee, *Chem. Biol.*, 1997, **4**, 175.
79. J. H. Grate and G. N. Schrauzer, *J. Am. Chem. Soc.*, 1979, **101**, 4601.
80. F. T. T. Ng, G. L. Rempel, and J. Halpern, *Inorg. Chim. Acta*, 1983, **77**, L65.
81. M. K. Geno and J. Halpern, *J. Am. Chem. Soc.*, 1987, **109**, 1238.
82. F. T. T. Ng, G. L. Rempel, and J. Halpern, *J. Am. Chem. Soc.*, 1982, **104**, 621.
83. M. D. Waddington and R. G. Finke, *J. Am. Chem. Soc.*, 1993, **115**, 4629.
84. S. M. Chemaly and J. M. Pratt, *J. Chem. Soc., Dalton Trans.*, 1980, 2274.
85. K. L. Brown and H. B. Brooks, *Inorg. Chem.*, 1991, **30**, 3420.
86. S.-H. Kim, H. L. Chen, N. Feilchenfeld, and J. Halpern, *J. Am. Chem. Soc.*, 1988, **110**, 3120.
87. K. L. Brown, X. Zou, and D. R. Evans, *Inorg. Chem.*, 1994, **33**, 5713.
88. K. L. Brown, S. Cheng, and H. M. Marques, *Inorg. Chem.*, 1995, **34**, 3038.
89. B. Kräutler, R. Konrat, E. Stupperich, G. Färber, K. Gruber, and C. Kratky, *Inorg. Chem.*, 1994, **33**, 4128.
90. S. Dong, R. Padmakumar, R. Banerjee, and T. G. Spiro, *J. Am. Chem. Soc.*, 1996, **118**, 9182.
91. B. P. Hay and R. G. Finke, *J. Am. Chem. Soc.*, 1987, **109**, 8012.
92. S. W. Benson, *Chem. Rev.*, 1978, **78**, 23.
93. D. F. McMillen and D. M. Golden, *Ann. Rev. Phys. Chem.*, 1982, **33**, 493.
94. D. R. Lide, "CRC Handbook of Chemistry and Physics," 77th edn., CRC Press, Boca Raton, FL, 1996/1997.
95. W. A. Pryor, "Free Radicals," McGraw-Hill, New York, 1966.
96. C. Walling, L. Heaton, and D. D. Tanner, *J. Am. Chem. Soc.*, 1965, **87**, 1715.
97. G. Fraenkel and P. D. Bartlett, *J. Am. Chem. Soc.*, 1959, **81**, 5582.
98. W. P. Jencks, "Catalysis in Chemistry and Enzymology," McGraw-Hill, New York, 1969.
99. S. Licht and J. Stubbe, unpublished results.
100. B. B. Wayland, S. Ba, and A. E. Sherry, *J. Am. Chem. Soc.*, 1991, **113**, 5305.
101. K. L. Brown and J. M. Hakimi, *J. Am. Chem. Soc.*, 1984, **106**, 7894.
102. C. L. Drennan, S. Huang, J. T. Drummond, R. Matthews, and M. L. Ludwig, *Science*, 1994, **266**, 1669.
103. R. Padmakumar, S. Taoka, R. Padmakumar, and R. Banerjee, *J. Am. Chem. Soc.*, 1995, **117**, 7033.
104. C. Lawrence, S. Licht, and J. Stubbe, unpublished results.
105. S. Booker, Ph.D. Thesis, Massachusetts Institute of Technology, 1994.
106. S. Licht, Ph.D. Thesis, Massachusetts Institute of Technology, 1998.
107. K. B. Schowen and R. L. Schowen, *Methods Enzymol.*, 1982, **87**, 501.
108. P. Kuzmic, *Anal. Biochem.*, 1996, **237**, 260.
109. A. Ehrenberg and P. Reichard, *J. Biol. Chem.*, 1972, **247**, 3485.
110. P. Nordlund and H. Eklund, *J. Mol. Biol.*, 1993, **232**, 123.
111. W. A. Prütz, J. Butler, and E. J. Land, *Int. J. Radiat. Biol.*, 1983, **44**, 183.
112. B.-M. Sjöberg, *Structure*, 1994, **2**, 793.
113. B. Katterle, M. Sahlin, P. P. Schmidt, S. Potsch, D. T. Logan, A. Gräslund, and B. M. Sjöberg, *J. Biol. Chem.*, 1997, **272**, 10 414.
114. I. Climent, B.-M. Sjöberg, and C. Y. Huang, *Biochemistry*, 1992, **31**, 4801.
115. U. Rova, K. Goodtzova, R. Ingemarson, G. Behravan, A. Gräslund, and L. Thelander, *Biochemistry*, 1995, **34**, 4267.
116. D. J. Silva, J. Stubbe, V. Samano, and M. J. Robins, *Biochemistry*, 1998, **37**, 5528.
117. W. T. Dixon and D. Murphy, *J. Chem. Soc., Faraday Trans. 2*, 1975, **72**, 1221.
118. D. M. Holton and D. Murphy, *J. Chem. Soc., Faraday Trans. 2*, 1979, **75**, 1637.
119. C. J. Bender, M. Sahlin, G. T. Babcock, B. A. Barry, T. K. Chandrashekar, S. P. Salowe, J. Stubbe, B. Lindström, L. Petterson, A. Ehrenberg, and B.-M. Sjöberg, *J. Am. Chem. Soc.*, 1989, **111**, 8076.
120. G. Backes, V. L. Davidson, F. Huitema, J. A. Duine, and J. Sanders-Loehr, *Biochemistry*, 1991, **30**, 9201.
121. E. J. Land and W. A. Prütz, *Int. J. Radiat. Biol.*, 1979, **36**, 75.
122. M. R. DeFilippis, C. P. Murthy, M. Faraggi, and M. H. Klapper, *Biochemistry*, 1989, **28**, 4847.
123. A. Harriman, *J. Phys. Chem.*, 1987, **91**, 6102.
124. M. R. DeFilippis, C. P. Murthy, F. Broitman, D. Weinraub, M. Faraggi, and M. H. Klapper, *J. Phys. Chem.*, 1991, **95**, 3416.
125. G. T. Babcock, M. Espe, C. Hoganson, N. Lydakis-Simantiris, J. McCracken, W. Shi, S. Styring, C. Tommos, and K. Warncke, *Acta Chem. Scand.*, 1997, **51**, 533.
126. C. W. Hoganson, M. Sahlin, B.-M. Sjöberg, and G. T. Babcock, *J. Am. Chem. Soc.*, 1996, **118**, 4672.
127. G. J. Gerfen, B. F. Bellew, S. Un, J. M. Bollinger Jr., J. Stubbe, R. G. Griffin, and D. J. Singel, *J. Am. Chem. Soc.*, 1993, **115**, 6420.

128. S. Un, M. Atta, M. Fontecave, and A. W. Rutherford, *J. Am. Chem. Soc.*, 1995, **117**, 10 713.
129. P. Nordlund, B.-M. Sjöberg, and H. Eklund, *Nature*, 1990, **345**, 593.
130. J. P. Willems, H. I. Lee, D. Burdi, D. E. Doan, J. Stubbe, and B. M. Hoffman, *J. Am. Chem. Soc.*, 1997, **119**, 9816.
131. A. Gräslund and M. Sahlin, *Annu. Rev. Biophys. Biomolec. Struct.*, 1996, **25**, 259.
132. P. P. Schmidt, K. K. Andersson, A.-L. Barra, L. Thelander, and A. Gräslund, *J. Biol. Chem.*, 1996, **271**, 23 615.
133. C. Turro, C. K. Chang, G. E. Leroi, R. I. Cukier, and D. G. Nocera, *J. Am. Chem. Soc.*, 1992, **114**, 4013.
134. M. R. Wasielewski, M. P. Niemczyk, and W. A. Svec, *J. Am. Chem. Soc.*, 1985, **107**, 1080.
135. J. P. Kirby, N. A. van Dantzig, C. K. Chang, and D. G. Nocera, *Tet. Lett.*, 1995, **36**, 3477.
136. J. A. Roberts, J. P. Kirby, and D. G. Nocera, *J. Am. Chem. Soc.*, 1995, **117**, 8051.
137. R. Langen, J. L. Colón, D. R. Casimiro, T. B. Karpishin, J. R. Winkler, and H. B. Gray, *J. Biol. Inorg. Chem.*, 1996, **1**, 221.
138. K. E. Silva, T. E. Elgren, J. L. Que, and M. T. Stankovich, *Biochemistry*, 1995, **34**, 14 093.
139. M. Karlsson, M. Sahlin, and B. M. Sjöberg, *J. Biol. Chem.*, 1992, 12 622.
140. M. Ormo, K. Regnstrom, Z. Wang, L. J. Que, M. Sahlin, and B. M. Sjoberg, *J. Biol. Chem.*, 1995, **270**, 6570.
141. M. Örmo, F. deMare, K. Regnstrom, A. Aberg, M. Sahlin, J. Ling, T. M. Loehr, J. Sanders-Loehr, and B. M. Sjöberg, *J. Biol Chem*, 1992, **267**, 8711.
142. M. A. Henriksen, B. S. Cooperman, J. S. Salem, L.-S. Li, and H. Rubin, *J. Am. Chem. Soc.*, 1994, **116**, 9773.
143. B. E. Sturgeon, D. Burdi, S. Chen, B.-H. Huynh, D. E. Edmonson, J. Stube, and B. M. Hoffman, *J. Am. Chem. Soc.*, 1996, **118**, 7551.
144. J. M. Bollinger Jr., W. H. Tong, N. Ravi, B. H. Huynh, D. E. Edmondson, and J. Stubbe, *J. Am. Chem. Soc.*, 1994, **116**, 8015.
145. M. Sahlin, G. Lassmann, S. Pötsch, A. Slaby, B.-M. Sjöberg, and A. Gräslund, *J. Biol. Chem.*, 1994, **269**, 11 699.
146. G. A. McClarty, A. K. Chan, B. K. Choy, and J. A. Wright, *J. Biol. Chem.*, 1990, **265**, 7539.
147. W. van der Donk, G. J. Gerfen, and J. Stubbe, *J. Am. Chem. Soc.*, 1998, **120**, 4252.
148. M. Ator, S. P. Salowe, J. Stubbe, M. H. Emptage, and M. J. Robins, *J. Am. Chem. Soc.*, 1984, **106**, 1886.
149. B.-M. Sjöberg, A. Gräslund, and F. Eckstein, *J. Biol. Chem.*, 1983, **258**, 8060.
150. S. P. Salowe, J. M. Bollinger Jr., M. Ator, J. Stubbe, J. McCracken, J. Peisach, M. C. Samano, and M. J. Robins, *Biochemistry*, 1993, **32**, 12 749.
151. A. Sancar, *Biochemistry*, 1994, **33**, 2.
152. W. A. van der Donk, J. Stubbe, G. J. Gerfen, B. F. Bellew, and R. G. Griffin, *J. Am. Chem. Soc.*, 1995, **117**, 8908.
153. D. Ross, E. Albano, U. Nilsson, and P. Moldéus, *Biochem. Biophys. Res. Commun.*, 1984, **125**, 109.
154. D. N. R. Rao, V. Fisher, and R. Mason, *J. Biol. Chem.*, 1990, **265**, 844.
155. W. Prütz, J. Butler, E. J. Land, and A. J. Swallow, *Int. J. Radiat. Biol.*, 1989, **55**, 539.
156. P. S. Surdhar and D. A. Armstrong, *J. Phys. Chem.*, 1987, **91**, 6532.
157. M. Tamba and P. O'Neill, *J. Chem. Soc. Perkin Trans. II*, 1991, 1681.
158. K. Schäfer and K. D. Asmus, *J. Phys. Chem.*, 1981, **85**, 852.
159. M. S. Akhlaq and C. von Sonntag, *Z. Naturforsch.*, 1987, **42c**, 134.
160. M. S. Akhlaq, S. Al-Baghdadi, and C. von Sonntag, *Carb. Res.*, 1987, **164**, 71.
161. P. O'Neill, *Radiat. Res.*, 1983, **96**, 198.
162. R. Eliasson, M. Fontecave, H. Jörnvall, M. Krook, E. Pontis, and P. Reichard, *Proc. Natl. Acad. Sci. USA*, 1990, **87**, 3314.
163. J. Harder, R. Eliasson, E. Pontis, M. Ballinger, and P. Reichard, *J. Biol. Chem.*, 1992, **267**, 25 548.
164. V. Bianchi, R. Eliasson, M. Fontecave, E. Mulliez, D. M. Hoover, R. G. Matthews, and P. Reichard, *Biochem. Biophys. Res. Commun.*, 1993, **197**, 792.
165. X. Sun, J. Harder, H. Jörnvall, B. M. Sjöberg, and P. Reichard, *Proc. Natl. Acad. Sci. USA*, 1993, **90**, 577.
166. S. Ollagnier, E. Mulliez, J. Gaillard, R. Eliasson, M. Fontecave, and P. Reichard, *J. Biol. Chem.*, 1996, **271**, 9410.
167. J. Knappe, F. A. Neugebauer, H. P. Blaschkowski, and M. Gänzler, *Proc. Natl. Acad. Sci. USA*, 1984, **81**, 1332.
168. M. Frey, M. Rothe, A. F. V. Wagner, and J. Knappe, *J. Biol. Chem.*, 1994, **269**, 12 432.
169. R. Eliasson, P. Reichard, E. Mulliez, S. Ollagnier, M. Fontecave, E. Liepinsh, and G. Otting, *Biochem. Biophys. Res. Commun.*, 1995, **214**, 28.
170. V. Unkrig, F. A. Neugebauer, and J. Knappe, *Eur. J. Biochem.*, 1989, **184**, 723.
171. C. V. Parast, K. K. Wong, S. A. Lewisch, J. W. Kozarich, J. Peisach, and R. S. Magliozzo, *Biochemistry*, 1995, **34**, 2393.
172. E. Mulliez, M. Fontecave, J. Gaillard, and P. Reichard, *J. Biol. Chem.*, 1993, **268**, 2296.
173. X. Sun, S. Ollagnier, P. P. Schmidt, M. Atta, E. Mulliez, L. Lepape, R. Eliasson, A. Gräslund, M. Fontecave, P. Reichard, and B.-M. Sjöberg, *J. Biol. Chem.*, 1996, **271**, 6827.
174. R. Zhao, J. Lind, G. Merényi, and T. E. Eriksen, *J. Am. Chem. Soc.*, 1994, **116**, 12 010.
175. D. A. Armstrong, A. Rauk, and D. Yu, *J. Am. Chem. Soc.*, 1993, **115**, 666.
176. G. Merenyi and J. Lind, *J. Am. Chem. Soc.*, 1994, **116**, 7872.
177. S. W. Benson, "Thermochemical Kinetics," Wiley, New York, 1976.
178. R. Zhao, J. Lind, G. Merenyi, and T. E. Eriksen, *J. Am. Chem. Soc.*, 1994, **116**, 12 010.
179. V. Barone, C. Adamo, A. Grand, F. Jolibois, Y. Brunel, and R. Subra, *J. Am. Chem. Soc.*, 1995, **177**, 12 618.
180. C. Schöneich, M. Bonifacic, U. Dillinger, and K.-D. Asmus, in "Sulfur-Centered Reactive Intermediates in Chemistry and Biology," eds. C. Chatgilialoglu and K.-D. Asmus, Plenum Press, New York, 1990, vol. 197, p. 367.
181. C. Schöneich, M. Bonifacic, and K.-D. Asmus, *Free Rad. Res. Commun.*, 1989, **6**, 393.
182. M. S. Akhlaq, H.-P. Schuchmann, and C. von Sonntag, *Int. J. Radiat. Biology*, 1987, **51**, 91.
183. P. S. Surdhar, S. P. Mezyk, and D. A. Armstrong, *J. Phys. Chem.*, 1989, **93**, 3360.
184. M. L. Steigerwald, W. A. Goddard III, and D. A. Evans, *J. Am. Chem. Soc.*, 1979, **101**, 1994.
185. W. J. Albery and J. R. Knowles, *Biochemistry*, 1976, **15**, 5627.
186. E. S. Huyser and R. M. Kellogg, *J. Org. Chem.*, 1966, **31**, 3366.
187. J. Stubbe, J. Ator, and T. Krenitsky, *J. Biol. Chem.*, 1983, **258**, 1625.
188. D. B. Northrup, *Methods Enzymol.*, 1982, **87**, 607.
189. S. P. Salowe, M. Ator, and J. Stubbe, *Biochemistry*, 1987, **26**, 3408.

190. H. Zipse, *J. Am. Chem. Soc.*, 1995, **117**, 11 798.
191. A. L. Buley, R. O. C. Norman, and R. J. Pritchett, *J. Chem. Soc. B*, 1966, 849.
192. R. Livingston and H. Zeldes, *J. Am. Chem. Soc.*, 1966, **88**, 4333.
193. B. C. Gilbert, J. P. Larkin, and R. O. Norman, *J. Chem. Soc. Perkin II*, 1971, 794.
194. C. Walling, *Acc. Chem. Res.*, 1975, **8**, 125.
195. C. von Sonntag and E. Thoms, *Z. Naturforsch.*, 1970, **25b**, 1405.
196. H. Scherz, *Radiat. Res.*, 1970, **43**, 12.
197. K. M. Bansal, M. Grätzel, A. Henglein, and E. Janata, *J. Phys. Chem.*, 1973, **77**, 16.
198. S. Steenken, M. J. Davies, and B. C. Gilbert, *J. Chem. Soc. Perkin II*, 1986, 1003.
199. S. Steenken, *J. Phys. Chem.*, 1979, **83**, 595.
200. R. Lenz and B. Giese, *J. Am. Chem. Soc.*, 1997, **119**, 2784.
201. R. W. Binkley, D. G. Hehemann, and W. W. Binkley, *J. Org. Chem.*, 1978, **43**, 2573.
202. F. Hansske, D. Madej, and M. J. Robins, *Tetrahedron*, 1984, **40**, 125.
203. E. P. Serjeant and B. Demprey, "Ionization Constants of Organic Acids in Aqueous Solution," Pergamon, Oxford, 1979, p. 989.
204. E. Hayon and M. Simic, *Acc. Chem. Res.*, 1973, **7**, 114.
205. K. Brady and R. H. Abeles, *Biochemistry*, 1990, **29**, 7600.
206. A. R. Fersht, "Enzyme Structure and Mechanism," W. H. Freeman, New York, 1985.
207. T. Matsushige, G. Koltzenburg, and D. Schulte-Frohlinde, *Ber. Bun. Gesell.*, 1975, **79**, 657.
208. M. J. Robins, S. F. Wnuk, A. E. Hernandez-Thirring, and M. C. Samano, *J. Am. Chem. Soc.*, 1996, **118**, 11 341.
209. M. J. Robins, Z. Guo, and S. F. Wnuk, *J. Am. Chem. Soc.*, 1997, **119**, 3637.
210. D. P. Curran, *Synthesis*, 1988, 417.
211. D. P. Curran, *Synthesis*, 1988, 489.
212. W. T. Dixon and R. O. C. Norman, *J. Chem. Soc.*, 1963, 3119.
213. J. C. Lopez, R. Alonso, and B. Fraser-Reid, *J. Am. Chem. Soc.*, 1989, **111**, 6471.
214. D. H. R. Barton, J. M. Beaton, L. E. Geller, and M. M. Pechet, *J. Am. Chem. Soc.*, 1961, **83**, 4076.
215. P. Kabasakalian, E. R. Townley, and M. D. Yudis, *J. Am Chem. Soc.*, 1962, **84**, 2716.
216. M. J. Robins, Z. Guo, M. C. Samano, and S. F. Wnuk, *J. Am. Chem. Soc.*, 1996, **118**, 11 317.
217. G. Harris, G. W. Ashley, M. J. Robins, R. L. Tolman, and J. Stubbe, *Biochemistry*, 1987, **26**, 1895.
218. M. Tamba and M. Quintilliani, *Radiat. Phys. Chem.*, 1984, **23**, 259.
219. I. L. Cartwright, D. W. Hutchinson, and V. W. Armstrong, *Nucleic Acid Res.*, 1976, **3**, 2331.
220. J. V. Staros, H. Bayley, D. N. Standring, and J. R. Knowles, *Biochem. Biophys. Res. Commun.*, 1978, **80**, 568.
221. S. Steenken, *Landolt-Börnstein*, 1985, **13e**, 147.
222. H. A. Schwarz and R. W. Dodson, *J. Phys. Chem.*, 1989, **93**, 409.
223. M. S. Akhlaq, C. P. Murthy, S. Steenken, and C. von Sonntag, *J. Phys. Chem.*, 1989, **93**, 4331.
224. C. Chatgilialoglu and K. D. Asmus, "Sulfur-Centered Reactive Intermediates in Chemistry and Biology," Plenum Press, New York, 1990, vol. 197, p. 155.
225. C. von Sonntag, in "Sulfur-Centered Reactive Intermediates in Chemistry and Biology," eds. C. Chatgilialoglu and K. D. Asmus, Plenum Press, New York, 1990, vol. 197, p. 359.

5.09
Radical Reactions Featuring Lysine 2,3-Aminomutase

PERRY A. FREY

University of Wisconsin–Madison, WI, USA

5.09.1 INTRODUCTION

Most enzymatic reactions proceed by polar reaction mechanisms. However, it is increasingly recognized that certain important biochemical reactions cannot be understood on the basis of polar chemical mechanisms and must take place by way of organic radical intermediates. Some of these reactions are crucially important to the survival of living organisms, examples being the reduction of ribonucleotides to deoxyribonucleotides and the rearrangement reactions catalyzed by adenosyl-cobalamin-dependent enzymes. The deoxyribonucleotides are required for DNA biosynthesis; and adenosylcobalamin, a vitamin B_{12} coenzyme, is essential for certain rearrangement reactions that take place in microorganisms and higher animals. Certain steps in the biosynthesis of steroid hormones and in the detoxification of xenobiotics are also thought to take place by radical mechanisms.

In the cases of a few enzymes, convincing evidence of radical mechanisms is now available. One such enzyme is lysine 2,3-aminomutase, which is the principal topic of this chapter. Because of the nature of its cofactors, this enzyme displays properties that make it nearly ideal for probing an enzymatic radical mechanism. The results of spectroscopic studies have led to a remarkably detailed and interesting picture of an enzymatic radical rearrangement, in which several of the participating paramagnetic species have been observed by EPR, including two organic radical intermediates. Although the three-dimensional structure of lysine 2,3-aminomutase is not yet available, very detailed structural and kinetic information on organic radical intermediates at the active site are available from spectroscopic data. Lysine 2,3-aminomutase will be the focus of the first and major section of this chapter.

In Section 5.09.3, two activating enzymes from *Escherichia coli* will be described. The activating enzymes generate stable protein radicals at glycine sites in the enzymes pyruvate formate lyase and anaerobic ribonucleotide reductase. The chemistry of radical initiation in the reaction of lysine 2,3-aminomutase appears to be employed in the active sites of these two activating enzymes to generate the stable glycyl radicals in activated pyruvate formate lyase and anaerobic ribonucleotide reductase. The last section is a brief overview of issues in the mechanism of oxygenation by methane mono-oxygenase.

5.09.2   LYSINE 2,3-AMINOMUTASE

The interconversion of L-lysine (*S*-lysine) and L-*β*-lysine (*S*-*β*-lysine) is catalyzed by lysine 2,3-aminomutase, hereafter referred to as 2,3-aminomutase, in Clostridia according to Equation (1). The enzyme has been purified from *Clostridium subterminale* SB4 and characterized.[1,2] In Clostridia, 2,3-aminomutase catalyzes the first step in lysine metabolism to acetyl CoA and ammonia, which supply all of the energy, carbon skeletons, and nitrogen required for cell growth.[3] In Streptomyces, *β*-lysine is a secondary metabolite that is used as an aminoacyl substituent in the assembly of a number of antibiotics.[4–7]

$$\text{(structure of lysine)} \quad \rightleftharpoons \quad \text{(structure of β-lysine)} \tag{1}$$

The mechanism of the 2,3-aminomutase reaction is interesting because of the problems posed by the necessity to break the unactivated C-3—H bond to make way for the transfer of the amino group from C-2 to C-3. In general, polar reaction mechanisms do not allow the cleavage of unactivated C—H bonds in enzymatic processes. Biological reactions such as the rearrangement of lysine to *β*-lysine, in which a hydrogen atom and a group bonded to an adjacent carbon exchange places (Equation (2)), are characteristic of adenosylcobalamin-dependent enzymes; however, the activity of Clostridial 2,3-aminomutase does not require adenosylcobalamin or any other derivative of vitamin B_{12}.

$$-\overset{|}{\underset{|}{C}}_{\alpha}-\overset{|}{\underset{|}{C}}_{\beta}- \quad \rightleftharpoons \quad -\overset{|}{\underset{|}{C}}_{\alpha}-\overset{|}{\underset{|}{C}}_{\beta}- \tag{2}$$
$$\quad\; H \;\; X \qquad\qquad\qquad X \;\; H$$

5.09.2.1   Molecular and Catalytic Properties

Purified Clostridial 2,3-aminomutase was first described in 1970.[1,2] The enzyme was shown to be composed of subunits of molecular weight $\sim 47\,000$. The overall molecular weight of the purified enzyme was estimated by ultracentrifugation to be about $285\,000$, suggesting that the enzyme consists of six subunits in aggregation. Cross-linking and light-scattering experiments supported this formulation.[8]

The reaction was shown to take place according to Equation (1), in which the hydrogen transfer proceeded without exchange of substrate hydrogens with solvent hydrogens.[1] The equilibrium constant for the transformation of lysine into *β*-lysine was found to be 7 at pH 8.0. The favorable position of *β*-lysine in this equilibrium is likely to be due to the different values of pK_a for the carboxyl group and the *α*- and *β*-amino groups in lysine and *β*-lysine, respectively. In lysine, the pK_a for the carboxylic acid group is 2.18, whereas that for the carboxyl group in *β*-lysine will be similar to that of *β*-alanine, which is 3.60.[9] In lysine, the value of pK_a for the *α*-amino group is 8.95, whereas that for the *β*-amino group in *β*-lysine is likely to be similar to that of *β*-alanine, which is 10.19.[9]

Clostridial 2,3-aminomutase does not contain a corrinoid and is not activated by adenosyl-cobalamin.[1,10] The enzyme was shown to contain iron and pyridoxal-5′-phosphate (PLP) and to be activated by *S*-adenosylmethionine (SAM).[1,8] The roles of the cofactors were not known for many years. Their original identification did not provide an obvious rationale for the mechanism of the reaction. As the functions of these cofactors in the reaction of 2,3-aminomutase are now understood, they violate all of the usual rules for the actions of SAM, PLP, and iron in enzymatic reactions. SAM is known as the biological methyl group donor. The most obvious chemical property of SAM is its alkylating reactivity, and this allows it to serve as the methylating agent for DNA and the biosynthesis of hormones and neurotransmitters and for the methylation of proteins in bacterial chemotaxis. However, in the reaction of 2,3-aminomutase, the adenosyl moiety of SAM mediates hydrogen transfer. The usual biological function of PLP is to stabilize anionic intermediates in numerous enzymatic reactions of amino acids, including decarboxylations, transaminations, α,β-eliminations, β,γ-eliminations, and aldol reactions. In the 2,3-aminomutase reaction, PLP facilitates a radical rearrangement. The usual function of iron in iron–sulfur clusters is to mediate electron transfer. The aminomutase catalyzes an isomerization, yet it contains an iron–sulfur cluster. The cluster probably facilitates an internal, transient, and reversible electron transfer; however, there is reason to believe that the mechanism by which this takes place is novel for iron–sulfur clusters. The elucidation of the mechanism of the 2,3-aminomutase reaction and the roles played by the cofactors creates a new chapter in biological mechanisms.

5.09.2.2   Stereochemistry

The stereochemistry of hydrogen transfer in the Clostridial 2,3-aminomutase reaction was shown to follow the course of Equation (1).[11] The chemical synthesis of stereospecifically deuterium-labeled lysine and β-lysine, coupled with NMR analysis, allowed the complete stereochemical course of the reaction to be unmasked. The reaction takes place with transfer of hydrogen from the 3-pro-*R* position of *S*-lysine to the 2-pro-*R* position of *S*-β-lysine. Therefore, both hydrogen transfer and amino group transfer proceed with inversion of stereochemical configuration at C-2 and C-3, the termini of hydrogen and amino group migration. Amino group transfer takes place intramolecularly, whereas hydrogen transfer takes place by a mechanism that allows both intermolecular and intra-molecular transfer.[12]

A 2,3-aminomutase activity that has been detected in extracts of Streptomyces was reported to convert *S*-lysine into *R*-β-lysine, an acyl substituent of the antibiotic bellenamine, an anti-HIV agent.[13] The hydrogen transfer stereochemistry is not yet known for the Streptomyces aminomutase. It seems likely that the chemical mechanism will be similar to that for the Clostridial enzyme. If so, then the hydrogen from the 3-pro-*S* position in *S*-lysine should be found to migrate to the 2-pro-*R* position of *R*-β-lysine. This expectation is based on the assumption that the basic chemical mechanisms followed by the two 2,3-aminomutases are essentially the same, and that the reactions differ essentially in the stereospecificity with which the initial hydrogen abstraction takes place at C-3 of *S*-lysine. Should this expectation not be sustained by experiment, the mechanism of the Streptomyces 2,3-aminomutase would either have to differ from that of the Clostridial enzyme, or the mechanism of action for the latter enzyme would have to be revised.

5.09.2.3   *S*-Adenosylmethionine as a Mediator of Hydrogen Transfer

The most striking features of 2,3-aminomutase are the chemical difficulties presented by the reaction it catalyzes (Equation (1)) and the nature of its coenzyme requirements. These features are related, as we shall see, because the mechanisms by which the coenzymes function allow the chemical problems to be overcome. One difficulty with the chemistry is the matter of cleaving the C-3—H bond of lysine, which is strong and unactivated. The cleavage of such bonds in biochemistry generally requires the participation of highly reactive organometallic cofactors.[14] Enzymatic rearrangement reactions such as Equation (1) that follow the pattern of Equation (2) generally require the par-ticipation of adenosylcobalamin, coenzyme B_{12}, to cleave the unactivated carbon–hydrogen bond and mediate hydrogen transfer.[14] Several adenosylcobalamin-dependent lysine aminomutases cat-alyze reactions patterned on Equation (2), including β-lysine 5,6-aminomutase, *R*-lysine 5,6-amino-mutase, and ornithine 4,5-aminomutase.[3,15] While the reaction of 2,3-aminomutase follows the same pattern, 2,3-aminomutase does not require added adenosylcobalamin, and the enzyme does not contain a corrinoid as a tightly bound cofactor. Nevertheless, the reaction is related to the adenosylcobalamin-dependent reactions, and a brief consideration of the role of adenosylcobalamin in reactions of this type will be useful in understanding the 2,3-aminomutase. Detailed discussions of adenosylcobalamin-dependent enzymes will be found in Chapter 5.08.

The function of adenosylcobalamin in enzymatic rearrangement reactions is understood to be the initiation of radical formation in substrates.[14] The cobalt–carbon bond in adenosylcobalamin is weak, in relation to most covalent bonds, and displays a bond dissociation energy of ~ 30 kcal mol^{-1}.[15,16] This bond is thought to be reversibly cleaved at enzymatic active sites to generate cobalamin(II) and the deoxyadenosyl radical, 5′-deoxyadenosine-5′-yl, according to Equation (3). The energy required to cleave the cobalt–carbon bond is thought to be linked to substrate binding forces.[14] Evidence supporting this concept is based on structural analysis of methylmalonyl-CoA mutase.[17] The deoxyadenosyl radical is thought to initiate substrate rearrangements through the general mechanism in Scheme 1. Abstraction of the transferable hydrogen from the substrate ((a) in Scheme 1) by the deoxyadenosyl radical generates 5′-deoxyadenosine (5′-dAdo) and a substrate radical (b). Rearrangement of the substrate radical to the product-related radical (c) follows, and in this process the group-X migrates from C_β to the adjacent carbon C_α. The product-related radical abstracts a hydrogen atom from C-5′ of 5′-dAdo to regenerate the deoxyadenosyl radical and form the product (d). In this way, adenosylcobalamin can initiate radical formation, and the rearrangements can take place by the isomerization of substrate-related radicals to product-related radicals.

$$(3)$$

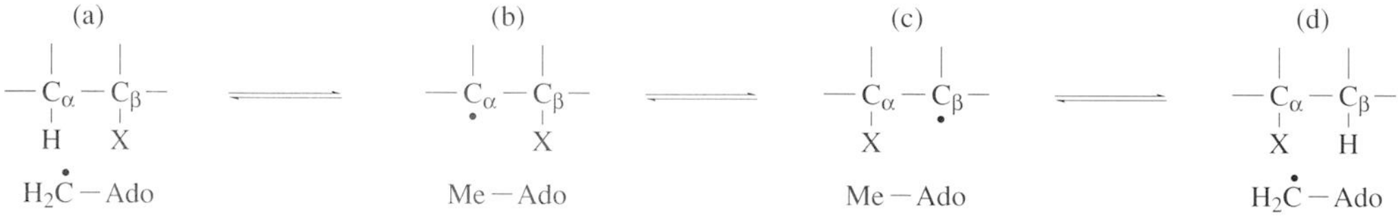

Scheme 1

In the case of the vitamin B$_{12}$-independent 2,3-aminomutase, SAM mediates hydrogen transfer in place of adenosylcobalamin. This is known from the activation of 2,3-aminomutase by *S*-[5′-³H] adenosylmethionine, which leads to the incorporation of tritium into lysine and β-lysine according to Equation (4),[18] just as the activation of B$_{12}$-dependent enzymes by [5′-³H]adenosylcobalamin leads to the incorporation of tritium into the reaction products.[19,20–23] (In Equation (4), E-SAM refers to the enzyme-SAM complex.) Reverse transfer of tritium from β-[3-³H]lysine into *S*-adenosylmethionine has been confirmed.[24] Furthermore, hydrogen transfer is both intermolecular and intramolecular in the course of a large number of turnovers,[12] as is the case with adenosylcobalamin-dependent rearrangements according to Scheme 1.[20] Reactions of *S*-[3-²H₂]lysine (Lys-d_2) mixed with *S*-lysine (Lys) in various ratios led to the production of β-Lys-d_2 (*S*-β-[2-²H₁,3-²H₁]lysine) as well as two species of β-Lys-d_1 (*S*-β-[2-²H₁]lysine and *S*-β-[3-²H₁]lysine) in corresponding ratios, showing that deuterium transfer takes place intermolecularly, as predicted on the basis of the reaction of [5′-³H]SAM. However, a plot of the ratio [β-Lys-d_2/β-Lys-d_1] against [Lys-d_2/Lys] gave a positive intercept when extrapolated to [Lys-d_2/Lys] = 0. Therefore, even when the concentration of Lys is increased to infinitely large values at a fixed concentration of Lys-d_2, a certain amount of β-Lys-d_2 is still produced (11%). Therefore, intramolecular deuterium transfer takes place by a mechanism that cannot be fully suppressed, even by an infinitely large amount of unlabeled Lys. This is in accord with Scheme 1, which requires that any one of the three 5′-methyl hydrogens can be transferred to the product radical in the second step of the reaction of a given substrate molecule.

$$\text{E-[5′-}^3\text{H]SAM} + \text{Lysine} \rightleftharpoons \text{E-SAM} + [^3\text{H}]\text{Lysine} + \beta\text{-}[^3\text{H}]\text{Lysine} \qquad (4)$$

While the hydrogen transfer data implicate the adenosyl moiety of SAM in mediating hydrogen transfer, presumably through the intermediate formation of 5'-dAdo and methionine, they do not provide information about the chemical mechanism by which this takes place. Independent experiments prove that SAM is cleaved to 5'-dAdo and methionine in the course of the reaction, although kinetic competence in the production of these species remains to be proven.[25] The question of the chemical mechanism by which SAM is cleaved to the putative deoxyadenosyl radical in a chemical reaction with the iron–sulfur cluster is deferred (Section 5.09.2.7).

5.09.2.4 Role of PLP: An Azocyclopropylcarbinyl Radical Rearrangement

PLP is a required coenzyme for 2,3-aminomutase, as well as for the adenosylcobalamin-dependent aminomutases.[1,3,19] A mechanistic role for PLP in the aminomutase reactions was first put forward in connection with the reaction of 2,3-aminomutase.[18] In the aminomutase reactions, the chemical mechanism for the participation of PLP differs from conventional PLP mechanisms, in which the coenzyme stabilizes amino acid carbanions. PLP is postulated to facilitate a radical rearrangement, as illustrated for 2,3-aminomutase in Figure 1. The reaction is postulated to proceed through lysine radical intermediates, the formation of which is initiated by the putative deoxyadenosyl radical derived from SAM (or from adenosylcobalamin in the cases of vitamin B_{12}-dependent aminomutases). Lysine is assumed to be bound to the enzyme in the form of an external aldimine with PLP, which is a very tightly bound coenzyme, as shown in the upper left of Figure 1. The putative deoxyadenosyl radical abstracts the 3-pro-*R* hydrogen from the lysyl side chain to form radical (**1**), the C-3-radical of lysyl-PLP external aldimine. Radical (**1**) undergoes isomerization to radical (**2**) through a pairing of the unpaired electron in (**1**) with a π-electron in the imine linkage, leaving the unpaired electron on pyridoxyl-C-4' of radical intermediate (**2**). The aziridylcarbinyl intermediate (**2**) is quasi-symmetric with respect to its potential to undergo cleavage of the aziridyl ring. It may open by reversal to intermediate (**1**) or by a forward opening to intermediate (**3**), the 2-radical of β-lysyl-PLP external aldimine. Intermediate (**3**) is the product-related radical, and it abstracts a hydrogen atom from the 5'-methyl group of 5'-dAdo to form the external imine of β-lysine and PLP. β-Lysine is then released from the enzyme, which in turn binds lysine to start another cycle of catalysis.

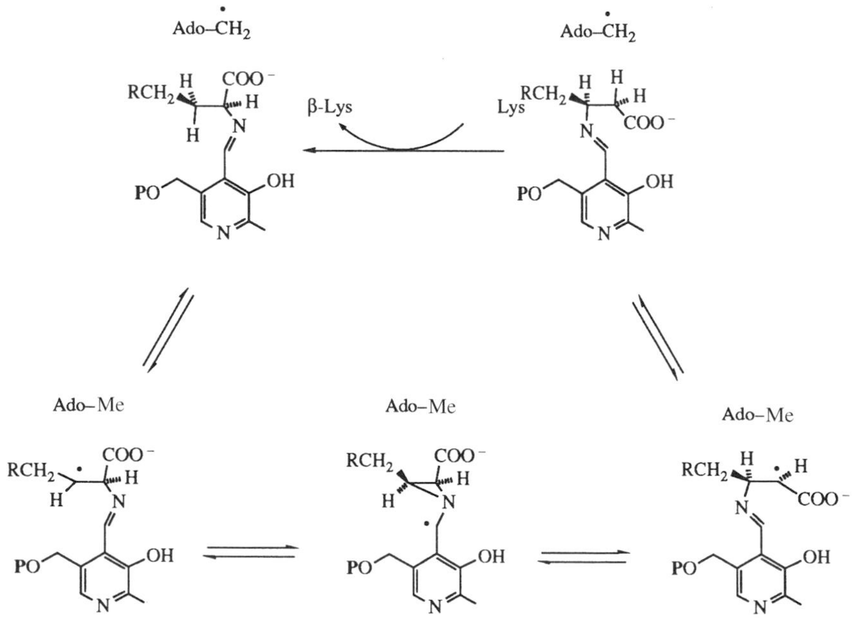

Figure 1 The mechanism of the rearrangement catalyzed by lysine 2,3-aminomutase. Ado-CH_2· refers to the 5'-deoxyadenosyl radical, which is postulated to arise from a chemical reaction between SAM and the (+)-form of the iron–sulfur cluster in lysine 2,3-aminomutase. The side chain substituent R is aminopropyl.

The mechanism in Figure 1 simultaneously accounts for the amino group migration and PLP requirement in the 2,3-aminomutase reaction and in the adenosylcobalamin-dependent amino-mutase reactions as well. Support for this mechanism has been obtained in studies of chemical models and through the direct EPR spectroscopic observation of intermediates (1) and (3) in Figure 1.

5.09.2.5 Chemical Models for the Aziridylcarbinyl Radical Rearrangement

The concept of the aziridylcarbinyl radical rearrangement was inspired by the well-known cyclo-propylcarbinyl radical rearrangements.[26,27] That the substitution of a nitrogen atom for carbon in the central position would allow the rearrangement was not obvious. A possible chemical model was described in the reaction of the aziridylcarbinyl radical in liquid cyclopropane according to Equation (5).[28] The aziridylcarbinyl radical was generated in liquid cyclopropane by the reaction of N-methylaziridine with the t-butoxy radical. Both radicals in Equation (5) were observed and characterized by EPR spectroscopy, the aziridylcarbinyl radical at temperatures lower than $-130\,^{\circ}$C and the ring-opened radical at temperatures higher than $-130\,^{\circ}$C. The experiments proved that an aziridylcarbinyl radical easily undergoes the ring opening shown for intermediate (2) in Figure 1, even at very low temperatures.

$$\triangleright N - \overset{\bullet}{C}H_2 \quad \rightleftharpoons \quad \overset{\bullet}{\diagdown} N = CH_2 \qquad (5)$$

$$< -130\,^{\circ}C \qquad\qquad > -130\,^{\circ}C$$

A chemical model that is more closely related to the 2,3-aminomutase reaction is the rearrange-ment of 2-methyl-3-bromo(N-benzaldimino)alanine methyl ester (4) in the presence of AIBN and tributyltin hydride, shown in Scheme 2.[29] Abstraction of the bromine atom from (4) generates the corresponding C-3-radical (5). At a low concentration of tributyltin hydride, free radical (5) is sufficiently long-lived to undergo rearrangement to the C-2-radical (6), in which the imino nitrogen has migrated to C-3. Eventual quenching of the radicals by tributyltin hydride gives a 1 : 13 mixture of the unrearranged (7) and isomerized compound (8) as the only detectable products. The formation of (7) and (8) in Scheme 2, and the absence of other products, indicates that the rearrangement does not take place by fragmentation and reassociation of the initial radical (5). Instead, an internal rearrangement, presumably through the aziridylcarbinyl radical analogous to (2) in Figure 1, accounts most straightforwardly for the results. The chemical models taken together prove that the aziridylcarbinyl radical rearrangement mechanism in Figure 1 is chemically allowed and reasonable.

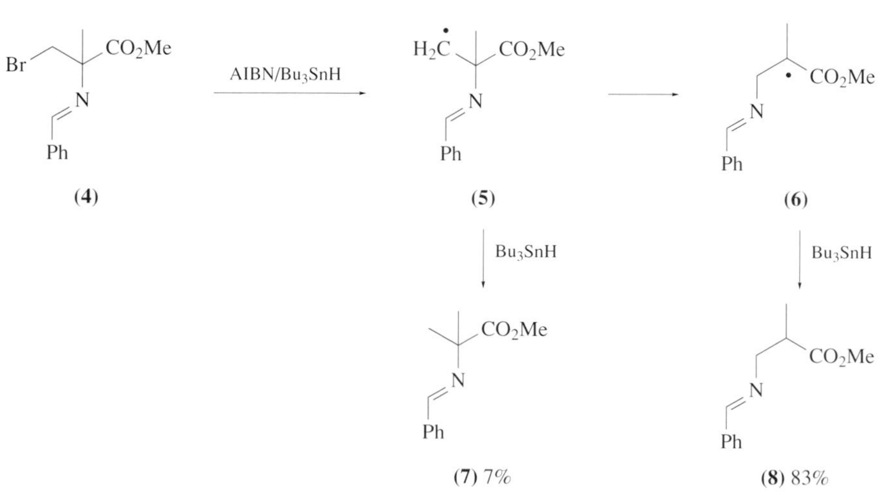

Scheme 2

5.09.2.6   Spectroscopic Characterization of Substrate Radical Intermediates

Considerable evidence has been accumulated in support of the mechanism in Figure 1, mainly by EPR techniques. The most direct and convincing proof of the participation of radicals in a reaction mechanism is the observation and assignment of EPR spectra corresponding to the structures of the proposed radicals. In the case of 2,3-aminomutase, lysine-based radicals corresponding to intermediates (**1**) and (**3**) in Figure 1 have been observed and characterized.

Upon mixing active aminomutase at 40–80 mM with lysine at 200 mM inside an EPR tube, followed by freezing at 77 K in liquid N_2, within 30–40 s in the steady state, a prominent EPR spectrum was observed at 77 K.[30] The spectrum was centered at $g = 2.001$, and it displayed a complex pattern of couplings. The g value corresponded to that of an organic radical. Because it appeared only in the presence of 2,3-aminomutase, SAM, and lysine, and was absent when any one of these components was omitted, a paramagnetic species appeared to represent a component in the enzymatic reaction mechanism. The concentration of the radical was highest within the first 30 s after the addition of SAM and lysine to the enzyme, as determined by integration of the EPR signal, and it decreased to a lower value within a few minutes at the same rate at which the equilibrium between free lysine and β-lysine was attained.[30] Therefore, the species eliciting the radical EPR spectrum behaved in the manner of an intermediate in the transformation of lysine into the equilibrium mixture of lysine and β-lysine.

Analysis of the effects of isotopic labeling in the lysine used to generate the EPR spectrum allowed it to be assigned to intermediate (**3**) in Figure 1. The representative EPR spectra shown in Figure 2 were obtained. The spectrum was dramatically broadened when [2-^{13}C]lysine was used as the substrate. The degree of broadening is characteristic of the effects of hyperfine splittings observed with ^{13}C at the central atom in analogous π-radicals.[31] The spectral envelope was simplified and narrowed when [2-^{2}H]lysine was used as the substrate, and this indicated that much of the proton coupling in the original spectrum could be attributed to the C-2-H of lysine. Substitution of this proton with deuterium simplified the spectrum because of the small value of the deuterium nuclear hyperfine coupling constant relative to that for protium.[31]

A detailed analysis of the spectrum for intermediate (**3**) in Figure 1 allowed its conformation to be deduced. Because of the intensity of the spectrum for (**3**) and its low ratio of noise/signal, it could be digitized conveniently and subjected to a resolution enhancement analysis. By this means, the spectral envelope could be resolved into all of its components.[32] Each of the components of the resolution-enhanced spectrum could be assigned from consideration of the resolution-enhanced spectra obtained with [2-^{2}H]lysine, perdeuterolysine, and [2-^{2}H,2-^{15}N]lysine. This allowed all of the relevant coupling constants to be determined. The coupling constants, when considered in the light of literature values for the angular dependence of coupling constants in related π-radicals, allowed the conformation of radical intermediate (**3**) in Figure 1 to be deduced. The conformation is as shown in structure (**9**), in which the dihedral angle between the π-orbital and β-H is about 70° and that between the π-orbital and β-N is about 10°.

(**9**)

While the EPR analysis allowed the basic structural linkage and conformation of the β-imino π-radical from C-1 to C-3 to be specified, it could not implicate PLP in the structure of intermediate (**3**) of Figure 1. [4'-^{2}H]PLP and electron spin echo envelope modulation (ESEEM) spectroscopy were exploited to address this issue. The PLP in a sample of 2,3-aminomutase was allowed to exchange with [4'-^{2}H]PLP to introduce deuterium into the carboxaldehyde carbon of radical intermediate (**3**). The [4'-^{2}H]PLP-enzyme was mixed with SAM and lysine to prepare radical (**3**) and then frozen at 77 K in the steady state. Analysis of the sample by deuterium ESEEM spectroscopy revealed a strong doublet at the NMR frequency of deuterium, and from the coupling constant the separation between the unpaired electron and the carboxaldehyde-deuterium nucleus was estimated to be 3.4 Å.[33] On this basis, PLP was concluded to be connected to the radical intermediate, most likely as the external aldimine shown as (**3**) in Figure 1.

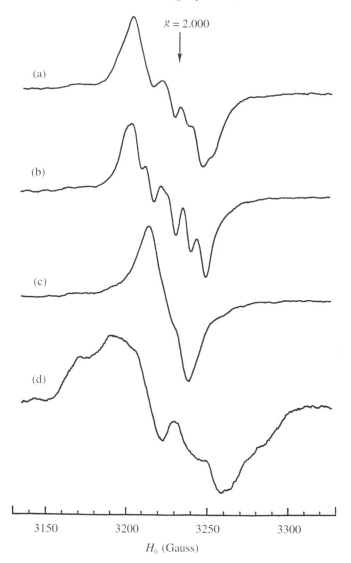

Figure 2   Characterization of radical intermediate (**3**) of Figure 1 by EPR spectroscopy. Shown are X-band EPR spectra (77 K) of samples containing 30–35 μM 2,3-aminomutase, 1.2 mM SAM, and 200 mM lysine (unlabeled or labeled) at pH 8.0 (50 mM Tris buffer) and 30 mM sodium dithionite: (a) *S*-lysine; (b) *S*-[3,3,4,4,5,5,6,6-^{2}H$_8$]lysine; (c) *RS*-[2-^{2}H]lysine; (d) *S*-[2-^{13}C]lysine. (Reproduced by permission of the American Chemical Society from *Biochemistry*, 1992, **31**, 10 782.)

The kinetic competence of radical (**3**) was established in rapid mix-freeze quench EPR experiments, in which the turnover rate for the radical was shown to be indistinguishable from the turnover rate for the enzyme.[34] The kinetic competence of radical (**3**) proves that it is either a compulsory catalytic intermediate or is in equilibrium with an intermediate.

The deoxyadenosyl radical and putative radical intermediates (**1**) and (**2**) in Figure 1 were not observed in EPR experiments in which the substrate was lysine. Radical intermediate (**3**) was observed, presumably because it was stable enough to exist at concentrations that could be detected by EPR. The stability of radical (**3**) can be attributed to the fact that the unpaired electron is significantly delocalized into the lysyl carboxyl group. The unpaired electrons in the deoxyadenosyl radical and intermediate (**1**) are not delocalized, and so are presumably too unstable to exist at concentrations that would be detectable by EPR spectroscopy, especially against the background of a fairly high concentration of the more stable radical (**3**). The unpaired electron in the putative intermediate (**2**) may be delocalized into the pyridine ring, depending on its conformation, and this would be a stabilizing factor; however, the three-membered aziridyl ring is highly strained and would be a destabilizing factor. In the balance, radical (**2**) is apparently less stable than radical (**3**) and is not observed.

An alternative rationale for the failure to observe one or another of the radicals in the mechanism of Figure 1 is that it might not have been a discrete intermediate. In the case of radical (1), its status as an intermediate has been supported by results obtained with an alternative substrate in place of lysine, in which a structural modification stabilized the unpaired electron on C-3 enough to make the corresponding analogue of (1) the dominant radical in the steady state.[35]

4-Thia-L-lysine reacts according to Scheme 3 as a substrate for 2,3-aminomutase.[35] The maximum rate is about 3% of that for the reaction of lysine, and the value of K_m for 4-thia-L-lysine is 5.5 mM, which is similar to the value of 7 mM for lysine at pH 8.5. A strong EPR spectrum appears in the steady state of the reaction of 4-thia-L-lysine. Representative spectra are shown in Figure 3. The spectrum is centered at $g = 2.003$, as in the reaction of lysine; however, the spectral envelope differs from that in the reaction of lysine, indicating that the coupling constants differ. Furthermore, the spectrum is more intense than that of radical intermediate (3). The spectrum elicited by 4-thia-L-[3-^{13}C]lysine is dramatically broadened, and that observed with 4-thia-L-[3-^{2}H$_2$]lysine is narrowed. The effects of 3-^{13}C and 3-^{2}H$_2$ prove that the unpaired electron is substantially localized on C-3 of 4-thialysine. This confirms that the dominant radical in the steady state of the reaction is the 4-thia analogue of radical (1) in Figure 1. The unpaired electron is stabilized at C-3 by orbital overlap with a doubly occupied *p*-orbital on the 4-thia group. This interpretation is strengthened by the fact that the EPR spectrum for the species derived from 4-thia-L-lysine is significantly narrowed and simplified by the substitution of 4-thia-L-[5,6-^{2}H$_4$]lysine, indicating the existence of significant nuclear hyperfine coupling of the unpaired electron with protons at C-5, and possibly also C-6. Inasmuch as the ^{13}C-broadening and C-3-^{2}H-narrowing are so prominent, the 4-thia analogue of radical (1) in Figure 1 must be the dominant organic radical in solution. Therefore, the participation of intermediate (1) in the 2,3-aminomutase mechanism is supported.

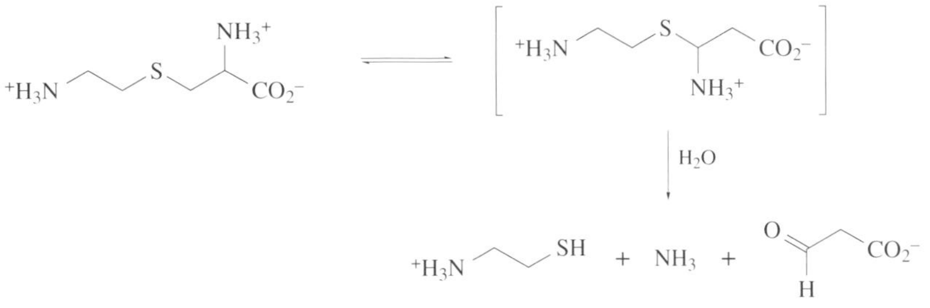

Scheme 3

5.09.2.7 SAM and the Iron–Sulfur Cluster in Radical Initiation

The UV/VIS absorption spectrum of 2,3-aminomutase is characteristic of proteins that contain iron–sulfur clusters.[8] An absorption maximum at 410 nm and the long-wavelength charge-transfer band first indicated the presence of iron–sulfur clusters, and this was confirmed by an analysis for inorganic sulfide which showed that sulfide was present in stoichiometric equivalence with the iron originally found associated with this enzyme.[10] The enzyme preparation that was originally described had been purified under a blanket of argon, that is, with limited exposure to dioxygen.[1] This preparation contained about 3 g atoms iron per mole of hexameric enzyme, and its specific enzymatic activity was reported to be 3–6 μmol min^{-1} mg^{-1} (protein). Purification by an updated method inside an anaerobic chamber increased the iron and sulfide content to 10–12 gram-atoms mol^{-1} and the specific activity to a maximum of 40 μmol min^{-1} mg^{-1} (protein), without significantly improving the homogeneity of the protein.[10,36] The enzyme preparations originally described contained less than the maximum amount of iron, presumably because of the effects of adventitious traces of air that could not be excluded outside of an anaerobic chamber. A loss of iron due to irreversible oxidation presumably led to low activity in the original preparations. On this basis, the catalytic activity of 2,3-aminomutase has been linked to its iron and sulfide content, in addition to its requirements for SAM and PLP.

2,3-Aminomutase also contains zinc and cobalt in a divalent cation binding site. The content of (Zn + Co) is about 5.5 mol^{-1} of hexameric enzyme.[10] Contrary to a report in 1991[10] later experiments indicate that the activity of 2,3-aminomutase does not depend on its cobalt content and that the Zn-rich enzyme is approximately as active as the cobalt-rich enzyme.[36]

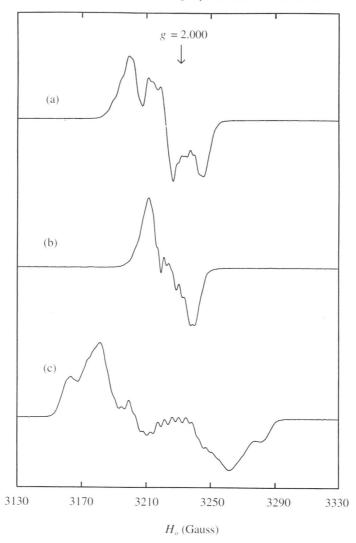

$g = 2.000$

(a)

(b)

(c)

3130   3170   3210   3250   3290   3330

H_o (Gauss)

Figure 3 Characterization of the 4-thia analogue of radical intermediate (**1**) in Figure 1 by EPR spectroscopy. X-band EPR spectra (77 K) are shown of samples prepared from lysine 2,3-aminomutase as in Figure 2, except that 10–20 mM 4-thialysine (unlabeled or labeled) was the substrate: (a) 4-thia-*RS*-lysine; (b) 4-thia-*RS*-[3-^{2}H$_2$]lysine; (c) 4-thia-*S*-[3-^{13}C]lysine. (Reproduced by permission of the American Chemical Society from *Biochemistry*, 1992, **31**, 10 782.)

Four states of the iron–sulfur cluster have been observed and partially characterized by EPR spectroscopy. Some of the properties of these states and their interrelationships are given in Figure 4. The three main oxidation states are designated (+), (2+), and (3+); the fourth state is a [3Fe–4S] cluster derived from the (3+)-state. As purified under anaerobic conditions, 2,3-aminomutase exists as a mixture of the (2+) and (3+)-forms.[37] The (3+)-form displays an EPR spectrum at 11 K that is centered at $g = 2.007$. The field position and line shape are not characteristic of [2Fe–2S] clusters. The spectrum for the (3+) state is unusual but compatible with a [4Fe–4S]-cluster; however, the best indication that it is a [4Fe–4S]-cluster is its transformation by controlled oxidation into a form that displays an EPR spectrum centered at $g = 2.015$ that is typical of [3Fe–4S]-clusters.[37] The latter form can be converted back into an active form by the addition of ferrous iron and a reducing agent.

Thiol reducing agents transform the (3+)-form into an EPR-silent form designated (2+). The reduction potential for this process is not yet known; however, it appears to be more positive than -370 mV based on the partitioning of the (3+) and (2+)-forms in the purification buffers.[37]

The catalytically functional form of the iron–sulfur cluster is that designated as (+) in Figure 4, and it can be obtained by reduction of the (2+)-form with a strong reducing agent in the presence of SAM. Both SAM and a low-potential reducing agent are required to transform the cluster into

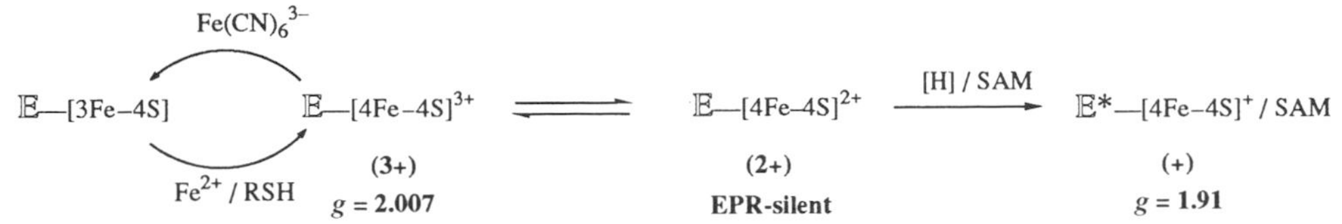

Figure 4 Four forms of the iron–sulfur cluster in lysine 2,3-aminomutase. The four species of the iron–sulfur cluster have been characterized by EPR spectroscopy. The forms designated $(3+)$ and $(2+)$ exist in the purified enzyme under anaerobic conditions. The catalytically functional form is the $(+)$-cluster $(g+1.91)$, which is produced by the combined actions of a strong reducing agent and SAM.

the active form, which displays a characteristic broad EPR spectrum centered at $g = 1.91$.[36] That the $(+)$-form of the cluster is catalytically functional is shown by gel permeation chromatography, removing excess reducing agents and SAM. The gel filtered enzyme displays the characteristic broad spectrum of the $(+)$-form centered at $g + 1.91$, and it is fully active without additional SAM or reducing agent.

Iron–sulfur clusters are typically observed in two oxidation states, whereas that in 2,3-aminomutase appears to exist in three oxidation states. It is significant that reduction to the $(+)$-state requires the presence of SAM. (S-Adenosylhomocysteine (SAH) will also potentiate reduction of $[4Fe–4S]^{2+}$ to a similar form of the cluster, as indicated by its EPR spectrum.[36]) The most obvious rationale for the SAM-requirement to reduce the $(2+)$-cluster is that the binding of SAM alters the reduction potential of the iron–sulfur cluster. In the altered state, the reduction potential of the $(2+)$ form is raised enough to allow conventional low-potential reducing agents such as dithionite to reduce the cluster to the $(+)$-form. The hypothetical mechanism is described by Equations (6) and (7).

$$\text{E}-[4Fe–4S]^{2+} \ + \ \text{SAM} \quad \Longleftrightarrow \quad \text{E}^*-[4Fe–4S]^{2+} \ / \ \text{SAM} \qquad (6)$$

$$\text{E}^* \quad [4Fe–4S]^{2+} \ / \ \text{SAM} \quad \xrightarrow{[H]} \quad \text{E}^* \quad [4Fe–4S]^{+} \ / \ \text{SAM} \qquad (7)$$

Results from diverse experiments indicate that Equation (7) is practically irreversible, so that once SAM is bound and the iron–sulfur cluster is reduced to the $(+)$-state, the enzyme is active and functions in the catalytic cycle many times without reverting to the $(2+)$-state or releasing SAM. The lines of evidence supporting this include the following:

(i) In the hydrogen transfer experiments, all of the tritium bound to C-5′ of $[5′-^3H]$SAM can be transferred to lysine and β-lysine only when the stoichiometric ratio of SAM to enzymatic active sites is equal to or less than 1.0. Excess $[5′-^3H]$SAM does not participate in the hydrogen transfer process.[12]

(ii) The activated enzyme designated as the $(+)$-state in Figure 4 can be separated from free SAM and reducing agents by gel permeation chromatography under anaerobic conditions and retains full activity.[36]

(iii) Substitution of SAH for SAM in the reduction of the $(2+)$-form leads to an inactive SAH-analogue of the $(+)$-form, based on the similarity of its low-temperature EPR spectrum to that of the $(+)$-form. In this complex, $[^{14}C]$SAH is noncovalently bound to the enzyme and dissociates only very slowly, as indicated by gel permeation chromatography.[36]

According to the working hypothesis, the form of 2,3-aminomutase shown in Figure 4 to contain SAM and $[4Fe–4S]^{+}$ is active. To be active, it must generate the substrate radicals (**1**) and (**3**) in Figure 1, which have been identified as catalytic intermediates. The simplest route to these species is by way of the reversible cleavage of SAM to the deoxyadenosyl radical. Such a process is illustrated in Figure 5, in which the complex designated as $(+)$ in Figure 4 is postulated to be in equilibrium with a form in which SAM is cleaved to methionine and the deoxyadenosyl radical, and the iron–sulfur cluster is oxidized by one electron to a $(2+)$-state. The deoxyadenosyl radical could abstract hydrogen from the bound substrate and initiate the rearrangement illustrated in Figure 1.

The mechanism by which the cleavage of SAM would take place is not known. The mechanism must be one that is readily reversible and can be controlled by interactions of the iron–sulfur cluster and SAM with the enzymatic active site. Four hypotheses for the mechanism of such a cleavage are illustrated in Figure 6, in all of which the $(+)$-cluster $([4Fe–4S]^{+})$ reacts reversibly with SAM to produce the deoxyadenosyl radical $(Ado-CH_2\cdot)$ and $[4Fe–4S]^{2+}$. In mechanism 1, the $(+)$-cluster

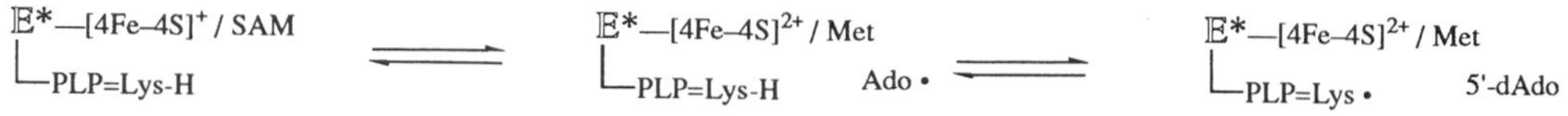

$$\mathbb{E}^*—[4Fe–4S]^+ / SAM \rightleftharpoons \mathbb{E}^*—[4Fe–4S]^{2+} / Met \rightleftharpoons \mathbb{E}^*—[4Fe–4S]^{2+} / Met$$

Figure 5 Reversible reaction of SAM with the reduced iron–sulfur cluster to generate substrate-based radicals. The deoxyadenosyl radical is postulated to arise from a reversible reaction between SAM and the (+)-form of the iron–sulfur cluster, accompanied by the one-electron oxidation of the cluster. The deoxyadenosyl radical is postulated to initiate the isomerization of lysine by abstracting a hydrogen atom.

initiates the cleavage by transferring an electron to the sulfur atom of SAM. Sulfur in SAM has a closed shell of electrons, so the addition of another electron will require a strong reducing agent, and the reduced SAM will be unstable to fragmentation, probably into radicals such as the deoxyadenosyl radical. This mechanism can be expected to lead to the fragmentation of SAM; however, it is not clear how enzymatic interactions could control the course of such a fragmentation and maintain reversibility. Mechanism 2 has been described.[38] A sulfur atom in the (+)-form of the iron–sulfur cluster acts as a nucleophile in undergoing alkylation by the deoxyadenosyl moiety of SAM. The resulting adenosyl-cluster undergoes fragmentation to the (2+)-cluster and the deoxyadenosyl radical. There are advantages to mechanism 2, in that it makes use of the natural alkylation reactivity of SAM and the intrinsic nucleophilic reactivity of sulfur. In addition, it offers obvious means for reversibility and control through enzymatic binding interactions of SAM and the cluster. However, numerous efforts in this laboratory to detect an adenosyl-cluster of the type shown in mechanism 2 have failed to provide any chemical evidence for bonding of the adenosyl moiety to sulfur. In mechanism 3, an iron in the (+)-form of the cluster is alkylated by the adenosyl group of SAM, and the adenosyl ligand is then reversibly separated as a radical, with concomitant formation of the (2+)-cluster. In mechanism 4, SAM is pictured as interacting directly, albeit weakly, with Fe^{2+} in the (+)-cluster through long-range coordination with the nonbonding electron pair of the sulfonium center. In a concerted process, electron transfer from the cluster to the sulfonium center cleaves SAM into methionine and the deoxyadenosyl radical, and the resulting thioether-sulfur of methionine becomes more strongly coordinated to iron in the (2+)-cluster.

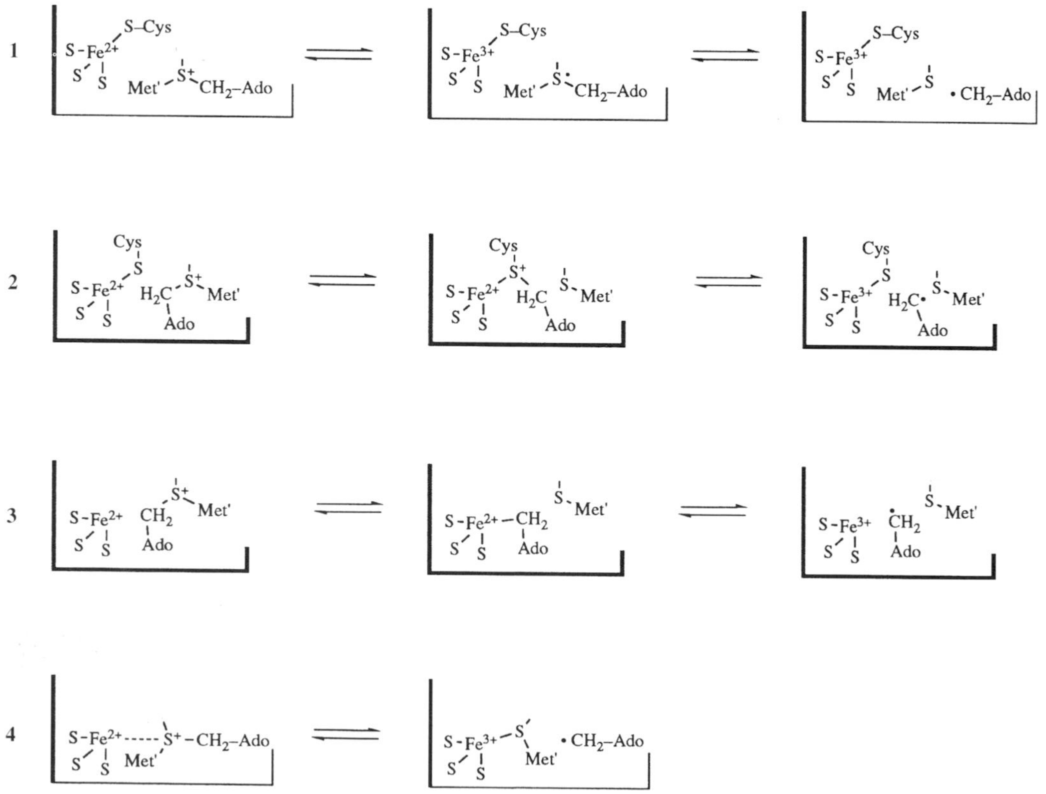

Figure 6 Hypothetical mechanisms for the cleavage of SAM by the reduced iron–sulfur cluster at the active site of lysine 2,3-aminomutase. In the representations of the structure of SAM, Met′ refers to carbons 1–4 and Me to the methyl group of the methionine moiety, and Ado-CH₂ refers to the 5′-deoxyadenosyl moiety.

Mechanism 4 is a controllable variation of mechanism 1, with some additional advantages. Control can be exerted through binding interactions of SAM with the enzyme that would be such as to allow the adenosyl moiety to occupy the leaving position dictated by stereoelectronic factors in the ligation of methionine to the interacting iron of the cluster. An additional advantage is that the increased strength of coordination to methionine relative to SAM as the reaction proceeds can provide the driving force to overcome some of the energy barrier to electron transfer. Additional experimentation will be required to determine whether any of the mechanisms in Figure 6 can be accepted as a true picture of the interactions leading to the cleavage of SAM at the active site of 2,3-aminomutase.

5.09.3 *S*-ADENOSYLMETHIONINE AND ENZYME-RADICAL FORMATION

SAM participates in the formation of radicals in two other enzymatic systems, the activation of pyruvate formate lyase (PFL) from *E. coli* and the activation of anaerobic ribonucleotide reductase (ARR) from *E. coli*. These enzymes are found in *E. coli* only when the cells are cultured under anaerobic conditions, and both of them harbor glycyl radicals in their amino acid sequences, Gly734 in the case of PFL and Gly681 in the case of ARR. The structure of the glycyl radical in proteins is illustrated below in structure (**10**).

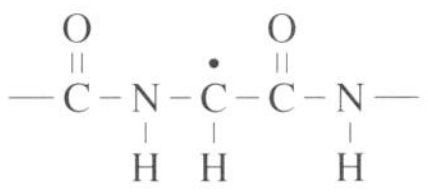

$$
\begin{array}{ccc}
\overset{\displaystyle O}{\overset{\displaystyle \|}{}} & & \overset{\displaystyle O}{\overset{\displaystyle \|}{}} \\
-C-N-\overset{\bullet}{C}-C-N- \\
\quad | \quad | \quad \quad | \\
\quad H \quad H \quad \quad H
\end{array}
$$

(**10**)

5.09.3.1 Pyruvate Formate Lyase Activating Enzyme (PFL-AE)

PFL catalyzes the reaction of pyruvate with coenzyme A to produce carbon dioxide and acetyl CoA according to Equation (8). The active enzyme contains a glycyl radical at position 734 in the amino acid sequence.[39–41]

$$
\overset{O}{\underset{CO_2^-}{\|}} \quad + \quad CoA\text{–}SH \quad \longrightarrow \quad \overset{O}{\underset{S\text{–}CoA}{\|}} \quad + \quad HCO_2^- \tag{8}
$$

The newly translated protein for PFL is inactive and contains a conventional glycine residue at position 734. The enzyme is activated by the action of an activating enzyme, PFL-AE, which catalyzes Equation (9), in which an α-hydrogen is abstracted from Gly734 and SAM is cleaved into methionine and 5′-dAdo.

$$
PFL\text{–}Gly734\text{–}H_\alpha \ + \ SAM \ + \ e^- \ \longrightarrow \ PFL\text{–}Gly734\bullet \ + \ Met \ + \ 5'\text{-}dAdo \tag{9}
$$

It should be noted that the balancing of Equation (9) requires the addition of an electron on the left side. The most obvious role for the electron is in the cleavage of SAM to methionine and the deoxyadenosyl radical, which in turn abstracts a hydrogen atom from C-2 of Gly734. Cleavage of SAM to methionine and 5′-dAdo in connection with radical formation has been postulated in Figure 5 as an intermediate step in the overall catalytic mechanism of 2,3-aminomutase. The electron required for this cleavage is derived from the iron–sulfur cluster in its (+)-state. The activation of PFL in Equation (9) is analogous to radical generation in Figure 5, and it seems likely that the two processes should take place by similar chemical mechanisms. Support for this expectation is provided by the results of experiments showing that PFL-AE is an iron–sulfur protein that contains a [4Fe–4S] cluster.[42] PFL-AE is a dimeric protein, and the cubane-like iron–sulfur center seems to be assembled at the interface of subunits from two [2Fe–2S]-clusters. The [4Fe–4S]-cluster in PFL-AE is also subject to SAM-dependent reduction to a (+)-form that displays an EPR spectrum similar to that of the (+)-form of 2,3-aminomutase.

5.09.3.2 Anaerobic Ribonucleotide Reductase Activating Enzyme (ARR-AE)

Ribonucleotide reductases produce 2′-deoxyribonucleotides by catalyzing the reduction of ribonucleotides. Three of the four classes of ribonucleotides contain radicals centered on amino acid side chains, and the fourth is adenosylcobalamin-dependent. The mechanism of enzymatic ribonucleotide reduction is in many respects similar for all four classes of enzymes. The ARR differs significantly from the others in two respects: the enzymatic radical is a glycyl residue at position 681, and the reducing substrate is formate.[43,44] The other reductases either harbor a tyrosyl radical or make use of adenosylcobalamin as the radical initiator, and the others use a vicinal dithiol compound as the reducing agent. The reaction catalyzed by ARR is shown in Equation (10).

$$PPPO\text{—}\!\!\!\begin{array}{c} O \\ \end{array}\!\!\!\text{B} \quad + \quad HCO_2^- \quad + \quad H^+ \quad \longrightarrow \quad PPPO\text{—}\!\!\!\begin{array}{c} O \\ \end{array}\!\!\!\text{B} \quad + \quad CO_2 \quad + \quad H_2O \qquad (10)$$

The radical-initiated mechanism of enzymatic ribonucleotide reduction is discussed in Chapter 5.09. The posttranslational modification of the inactive ARR-protein to introduce the glycyl radical at position 681 is in some respects analogous to the modification of PFL. The activating enzyme ARR-AE differs in that it is a component of the reductase enzyme itself and is bound as a dimer to ARR. The activating enzyme catalyzes Equation (11), which is strictly analogous to Equation (9) for the activation of PFL. The ARR activating enzyme also contains an iron–sulfur cluster that presumably participates in the cleavage of SAM and the abstraction of hydrogen from C-2 of Gly681.[45]

$$ARR\text{–}Gly681\text{–}H_\alpha \quad + \quad SAM \quad + \quad e^- \quad \longrightarrow \quad ARR\text{–}Gly681\bullet \quad + \quad Met \quad + \quad 5'\text{-dAdo} \qquad (11)$$

5.09.4 OXYGENATION BY METHANE MONOOXYGENASE

5.09.4.1 Molecular and Catalytic Properties

Accumulating evidence implicates radical mechanisms for a number of other enzymatic reactions. Ribonucleotide reductases, PFL, and prostaglandin H synthase are examples that are discussed in Chapters 5.08 and 5.09, and adenosylcobalamin-dependent reactions are discussed in Chapter 5.10. The mechanism of oxygenation catalyzed by methane monooxygenase (MMO) is controversial. The reaction has been suggested to involve substrate radical intermediates in a rebound mechanism similar to the mechanism generally accepted for cytochrome P-450 monooxygenases. The validity of the radical mechanism has been challenged and an alternative nonsynchronous concerted mechanism has been advanced. The issue of the radical or concerted oxygenation mechanism will be explored in this section.

Methanotrophic bacteria can grow on methane as their sole source of carbon and energy. This is made possible by the action of MMOs, which catalyze Equation (12). MMOs are known in soluble and membrane-bound forms, and the soluble enzymes have been extensively studied. They are iron metalloproteins that contain μ-oxo diiron complexes. The membrane-bound or particulate MMOs are copper metalloproteins. This discussion will be limited to the soluble MMOs, about which there is considerable structural and mechanistic information.

$$CH_4 \quad + \quad O_2 \quad + \quad NADH \quad + \quad H^+ \quad \longrightarrow \quad MeOH \quad + \quad H_2O \quad + \quad NAD^+ \qquad (12)$$

Soluble MMOs consist of three protein components: MMOH, the μ-oxo-Fe$_2$ containing hydroxylase (MW 245 kDa); MMOR, the flavin and iron–sulfur containing reductase (MW 38 kDa); and MMOB, a coupling factor (MW 15 kDa) that increases the rate of electron transfer from MMOR to MMOH.[46] The hydroxylase component consists of six subunits of three types $(\alpha\beta\gamma)_2$; its reduced form carries out the oxygenation of substrates by dioxygen, and its oxidized form accepts electrons for this process from the reductase component. The reductase component dehydrogenates NADH and transfers reducing equivalents to the hydroxylase component, thereby reducing the iron cofactor in the hydroxylase to the correct redox state for the oxygenation process. Little is known about the details of electron transfer. The hydroxylase component (MMOH) carries out the oxygenation of a wide variety of substrates in addition to methane. Essentially any hydrocarbon up to about eight carbons in length will serve as a substrate, with alkanes being oxygenated to alcohols, 1-alkenes to epoxides, and benzenes to phenols.[47]

The two soluble MMOHs that have been studied in detail are from *Methylococcus capsulatus* (Bath) and *Methylosinus trichosporium* (OB3b).[49–52] Their structures are very similar, and their catalytic properties are also similar, while differing significantly in quantitative details. The μ-oxo-Fe_2 complex is the oxygenating cofactor and is bound to the enzyme through ligation to histidine and glutamate side chain functional groups, as shown in (**11**). Spectroscopic and kinetic analysis of the oxygenation process has led to the formulation shown in Figure 7 for the catalytic cycle.[51] The oxidized cofactor H° [Fe_2^{III}] at the upper left of Figure 7 is first reduced to H^r [Fe_2^{II}] by the action of MMOR and MMOB, with NADH serving as the reducing agent. Reaction with dioxygen produces an oxidized peroxo-bridged complex P [Fe_2^{III}], by way of an intermediate noncovalent dioxygen-Fe_2^{II} complex O. The peroxo-bridged complex undergoes dehydration to complex Q, a μ-dioxo-Fe_2^{IV}. Complex Q is the oxygenating species and inserts an oxygen atom into a C—H bond of a substrate to produce complex T, consisting of the alcohol product and the μ-oxo-Fe_2^{III} cofactor. Release of the alcohol and binding of water regenerates the starting complex H°.

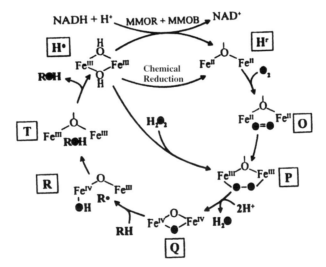

(**11**)

Figure 7 Steps in the oxygenation of alkanes by methane monooxygenase (reproduced by permission of the American Chemical Society from *Biochemistry*, 1997, **36**, 5223).

A point of contention is the mechanism by which complex Q carries out the oxygenation. Figure 7 shows this proceeding by a mechanism analogous to the rebound mechanism of cytochrome P-450. Abstraction of a hydrogen atom from the substrate produces an alkane radical and complex R, a hydroxy-mixed valent diiron complex, which in the rebound step immediately quenches the radical in a covalent capture by the hydroxy-ligand to form complex T. The competing mechanistic proposal holds that hydroxylation is concerted; that is, in the formulation of Figure 7, the transformation of Q to T is direct and no intermediate radical or mixed valent diiron complex exists.

5.09.4.2 Radical Trap Substrates

The evidence on the question of a stepwise or concerted oxygenation consists of the results of experiments designed to detect a radical intermediate or to observe a stepwise mechanism. Because the putative alkyl-radical intermediates would be too unstable to observe by EPR spectroscopy, experiments to detect a radical intermediate employed substrates that could serve as internal radical traps; that is, substrates that serve as "radical clocks" in nonenzymatic reactions.[27] Observation of a radical rearrangement product would indicate a radical intermediate in the mechanism. Experiments designed to observe a stepwise mechanism employed chiral-methyl substrates, principally *R*- and *S*-[1-^{2}H,^{3}H]ethane. Observation of racemized products would indicate a stepwise mechanism of oxygenation.

A clear resolution of the oxygenation mechanism has not emerged because of two complications. First, the results of radical trapping and stereochemical experiments differed significantly in studies of the *M. trichosporium* and *M. capsulatus* enzymes. Second, alternative interpretations of the results have been advanced.

The evidence from oxygenation of internal radical traps can be summarized as follows. Oxygenation of 1,1-dimethylcyclopropane by the MMO from *M. trichosporium* gave the results in Scheme 4.[53] The observation of 13% 1-methylcyclobutanol (**13**) and 6% 2-methyl-1-buten-4-ol (**14**), in addition to 81% conventional oxygenation product (**12**), clearly indicated partitioning of one or more intermediates to several products. Experiments with $H_2{}^{18}O$ and $^{18}O_2$ showed that oxygen in all of the products was derived from dioxygen, although 1-methylcyclobutanol contained significant oxygen from water. The production of the homoallylic alcohol (**14**) implicated a cyclopropylcarbinyl radical rearrangement (Equation (13)) and indicated a radical intermediate. Oxygenation of *trans*-2-phenylmethylcyclopropane (**15**) gave 3% of ring-opened alcohol in the case of *M. trichosporium* MMO and no detectable ring opening with *M. capsulatus* MMO.[54] Based on the isomerization rate constants for the putative radical species in free solution, rate constants for the rebound step were calculated to be 10^{12} to $>10^{13}$ s^{-1}, depending on the species of soluble MMO and the radical probe.[54] These values approached vibrational frequencies and were regarded as evidence for a nonsynchronous concerted mechanism, such as that in Equation (14), where an alternative formulation for intermediate Q is adopted. The formation of radical products was attributed to side reactions or radical-like reactivity in the transition state.

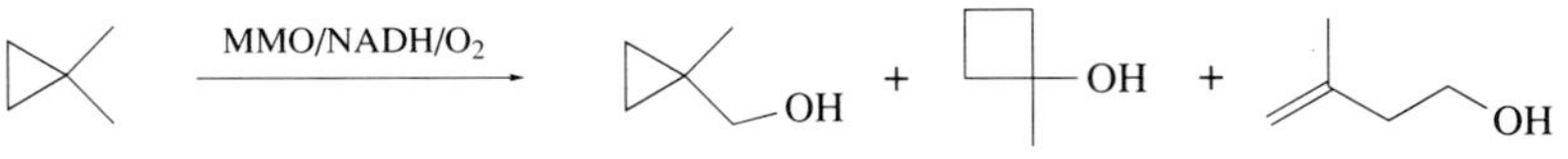

(12) 81%   **(13)** 13%   **(14)** 6%

Scheme 4

(13)

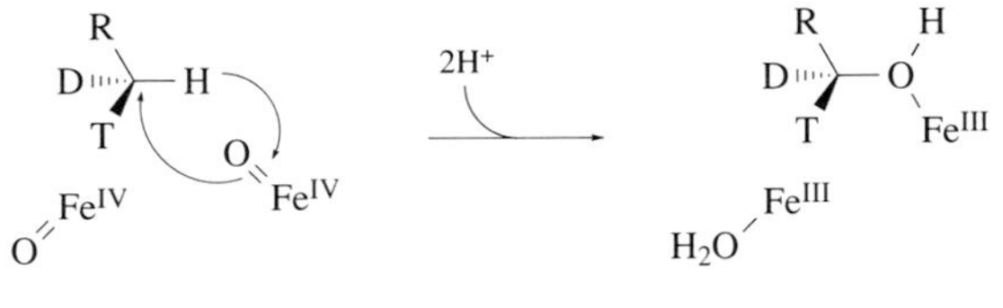

(14)

Implicit in the calculation of rebound rate constants by the foregoing method are two assumptions: (i) a putative radical species will undergo rearrangements within the confines of an active site at the same rate as in free solution, and (ii) the rearranged radical will be quenched in the rebound step at the same rate as the initial radical and cannot revert to the initial radical. The latter assumption is often valid for nonenzymatic reactions in homogeneous media; however, it is problematic for enzymatic reactions. The regiospecificity of enzymes is likely to undermine radical lifetime calculations. Consideration of Equation (13) will emphasize the translocation of the radical center from C-1 to C-4 in the rearrangement of the 1-methylcyclopropylcarbinyl radical. In the enzyme-bound state, hydrogen abstraction from C-1 created the radical and C-1 must be proximal to the locus of oxygenation. Upon ring opening to the homoallylic radical, the radical center at C-4 is likely to be remote from the oxygenating species. Therefore, it is unlikely to react as fast with the oxygenating species as the cyclopropylcarbinyl radical, and a lifetime calculation will give an artificially low value. That MMO acts regiospecifically on these substrates is proven by the report that the oxygenation of DL-*trans*-2-phenylmethylcyclopropane (**15**) proceeds with oxygenation of the methyl group in one enantiomer and the phenyl group in the other, and also by the fact that methylcubane is oxygenated exclusively on the methyl group.[55] Stereochemical evidence supports the interpretation that selective oxygenation of ring-closed forms can occur at enzymatic sites. The oxygenation of either *cis-* or *trans-*2-(2,3-dihydroxyphenyl)cyclopropane-1-carboxylate by extradiol catechol dioxygenase proceeds with *cis–trans-*isomerization of the cyclopropyl ring, with no observable ring-opened product being formed.[56]

(15)

5.09.4.3 Stereochemical Probes

The stereochemical experiments with *R-* and *S-*[1-^{2}H$_1$,1-^{3}H]ethane and both the *M. trichosporium* and *M. capsulatus* MMO gave the results in Equation (15).[57,58] The observation of 70% retention and 30% inversion of configuration indicated either the participation of a stereochemically labile intermediate, such as a radical, or the involvement of parallel reaction pathways differing in stereochemical specificity. Oxygenation of chiral [1-^{2}H$_1$,1-^{3}H]butane proceeded with 77% retention. (Oxygenation by the membrane-bound, particulate MMO proceeded with 100% retention of configuration of *R-* and *S-*[1-^{2}H,^{3}H]ethane.[59]) The composition of partitioning products in the stereochemical studies allowed the calculation of a value for the rebound step of oxygenation based on the known barrier to C—C bond rotation for the ethyl radical in its free state, assuming a barrier of 0.15 kcal mol^{-1} for rotation about a C—C bond in a π-radical. According to this calculation, the rebound step would have to be governed by a rate constant of $\sim 10^{13}$ s^{-1} at 45 °C. Again, this value is at the bond vibrational frequency and could not represent a rate constant for the reaction of a discrete intermediate. On this basis, a complex mechanism was presented in which the basic mechanism is postulated to be a nonsynchronous concerted process (Equation (14)), and partial racemization was attributed to parallel concerted reaction pathways having opposite stereochemical consequences.[59]

Implicit in the foregoing calculation is the assumption that C—C rotation in the active site would be insensitive to local polar, magnetic, and steric effects. Polar and steric effects may not be an issue; however, the reaction is taking place in the presence of a nearby paramagnet, the oxygenation cofactor. An answer to the question of the degree to which a local magnetic field might affect the rotational barrier would be worthwhile to pursue.

Advocates of the stepwise rebound mechanism in Figure 7 regard the formation of radical rearrangement products and partial racemization as positive experimental results supporting the rebound mechanism. They question whether reliable rebound rate constants within the confines of active sites can be calculated based on rate constants measured in free states. Advocates of the concerted mechanism in Equation (15) accept the assumptions on which the calculated rebound rate constants are based. It remains for the future to resolve the issue of whether radicals are discrete intermediates in the reactions of MMO.

(15)

ACKNOWLEDGMENTS

The author is pleased to thank the National Institutes of Health of the US Public Health Service for its continuing support of his research. In the areas covered by this chapter, support has been provided through Grant No. DK 28607 from the National Institute of Diabetes and Digestive and Kidney Diseases. Research on lysine 2,3-aminomutase in the author's laboratory was carried out with the following co-workers: Marcia Moss, Janina Baraniak, Robert Petrovich, Bin Song, Oksoo

Han, Marcus Ballinger, Kafryn Lieder, Christopher Chang, and Squire Booker. The author is also grateful to his collaborators George H. Reed, Helmut Beinert, Frank J. Ruzicka, and Russell LoBrutto.

5.09.5 REFERENCES

1. T. P. Chirpich, V. Zappia, R. N. Costilow, and H. A. Barker, *J. Biol. Chem.*, 1970, **245**, 1778.
2. V. Zappia and H. A. Barker, *Biochim. Biophys. Acta*, 1970, **207**, 505.
3. T. C. Stadtman, *Adv. Enzymol. Relat. Areas Mol. Biol.*, 1973, **38**, 413.
4. J. C. French, Q. R. Bartz, and H. W. Dion, *J. Antibiot. (Japan)*, 1973, **26**, 272.
5. T. H. Haskell, S. A. Fusari, R. P. Frohardt, and Q. R. Bartz, *J. Am. Chem. Soc.*, 1952, **74**, 599.
6. S. J. Gould, K. J. Martinkus, and C.-H. Tann, *J. Am. Chem. Soc.*, 1981, **103**, 2871.
7. S. J. Gould and T. K. Thiruvengadam, *J. Am. Chem. Soc.*, 1981, **103**, 6752.
8. K. B. Song and P. A. Frey, *J. Biol. Chem.*, 1991, **266**, 7651.
9. W. P. Jencks and J. Regenstein, in "Handbook of Biochemistry and Molecular Biology," ed. H. A. Sober, Chem. Rub. Co., Cleveland, OH, 1970, pp. J186–226.
10. R. M. Petrovich, F. J. Ruzicka, G. H. Reed, and P. A. Frey, *J. Biol. Chem.*, 1991, **226**, 7656.
11. D. J. Aberhart, S. J. Gould, H.-J. Lin, T. K. Thiruvengadam, and B. H. Weiller, *J. Am. Chem. Soc.*, 1983, **105**, 5461.
12. J. Baraniak, M. L. Moss, and P. A. Frey, *J. Biol. Chem.*, 1989, **264**, 1357.
13. Y. Ikeda, H. Naganawa, D. Ikeda, and S. Kondo, Tennen Yuki Kagobutsu Toronkai Koen Yoshishu, 1992, 34th, 204–211.
14. P. A. Frey, *Chem. Rev.*, 1990, **90**, 1343.
15. J. Halpern, S.-H. Kim, and T. W. Leung, *J. Am. Chem. Soc.*, 1984, **106**, 8317.
16. R. J. Finke and B. P. Hay, *Inorg. Chem.*, 1984, **23**, 3041.
17. F. Mancia, N. H. Keep, A. Nakagawa, P. F. Leadley, S. McSweeney, B. Rasmussen, P. Bösecke, O. Diat, and P. R. Evans, *Structure*, 1996, **4**, 339.
18. M. L. Moss and P. A. Frey, *J. Biol. Chem.*, 1987, **262**, 14 859.
19. J. J. Baker and T. C. Stadtman, in "Amino Mutases in B₁₂: Biochemistry and Medicine," ed. D. Dolphin, Wiley, New York, 1982, vol. 2, pp. 203–232.
20. P. A. Frey, M. K. Essenberg, and R. H. Abeles, *J. Biol. Chem.*, 1967, **242**, 5369.
21. H. A. Barker, *Biochem. J.*, 1967, **105**, 1.
22. J. Retey and D. Arigoni, *Experientia*, 1966, **22**, 783.
23. B. Babior, *Biochim. Biophys. Acta*, 1968, **167**, 456.
24. J. Kilgore and D. J. Aberhart, *J. Chem. Soc., Perkin Trans. 1*, 1991, 79.
25. M. L. Moss and P. A. Frey, *J. Biol. Chem.*, 1990, **265**, 18 112.
26. J. K. Kochi, P. J. Kusic, and D. R. Eaton, *J. Am. Chem. Soc.*, 1969, **91**, 1877.
27. D. Griller and K. U. Ingold, *Acc. Chem. Res.*, 1980, **13**, 317.
28. W. C. Danen and C. T. West, *J. Am. Chem. Soc.*, 1974, **96**, 2447.
29. O. Han and P. A. Frey, *J. Am. Chem. Soc.*, 1990, **112**, 8982.
30. M. D. Ballinger, G. H. Reed, and P. A. Frey, *Biochemistry*, 1992, **31**, 949.
31. J. A. Weil, J. R. Bolton, and J. E. Wertz, "Electron Paramagnetic Resonance," Wiley, New York, 1994, p. 534.
32. M. D. Ballinger, P. A. Frey, and G. H. Reed, *Biochemistry*, 1992, **1**, 10 782.
33. M. D. Ballinger, P. A. Frey, G. H. Reed, and R. LoBrutto, *Biochemistry*, 1995, **34**, 10 086.
34. C. H. Chang, M. D. Ballinger, G. H. Reed, and P. A. Frey, *Biochemistry*, 1996, **34**, 11 081.
35. W. Wu, K. W. Lieder, G. H. Reed, and P. A. Frey, *Biochemistry*, 1996, **34**, 10 532.
36. K. W. Lieder, S. Booker, H. Beinert, G. H. Reed, and P. A. Frey, *Biochemistry*, 1998, **37**, 2578.
37. R. M. Petrovich, F. J. Ruzicka, G. H. Reed, and P. A. Frey, *Biochemistry*, 1992, **31**, 10 774.
38. P. A. Frey and G. H. Reed, *Adv. Enzymol. Relat. Areas Mol. Biol.*, 1993, **66**, 1.
39. J. Knappe, F. A. Neugebauer, H. P. Blaschekowski, and M. Gänzler, *Proc. Natl. Acad. Sci. USA*, 1984, **81**, 1332.
40. A. F. Volker Wagner, M. Frey, F. A. Neugebauer, W. Schäfer, and J. Knappe, *Proc. Natl. Acad. Sci. USA*, 1992, **89**, 996.
41. J. Knappe, S. Elberet, M. Frey, and A. F. V. Wagner, *Biochem. Soc. Trans.*, 1993, **21**, 731.
42. J. B. Broderick, R. E. Duderstadt, D. C. Fernandez, K. Wojtuszewski, T. F. Henshaw, and M. K. Johnson, *J. Am. Chem. Soc.*, 1997, **119**, 7396.
43. R. Eliasson, M. Fontecave, H. Jörnvall, M. Krook, E. Pontis, and P. Reichard, *Proc. Natl. Acad. Sci. USA*, 1990, **87**, 3314.
44. X. Sun, S. Ollagnier, P. P. Schmidt, M. Atta, E. Mulliez, L. Lepape, R. Eliasson, A. Gräslund, M. Fontecave, P. Reichard, and B. M. Söberg, *J. Biol. Chem.*, 1996, **271**, 6827.
45. S. Ollangnier, E. Mulliez, J. Gaillard, R. Eliasson, M. Fontecave, and P. Reichard, *J. Biol. Chem.*, 1996, **271**, 9410.
46. Y. Liu, J. C. Nesheim, K. E. Paulsen, M. T. Stanovich, and J. D. Lipscomb, *Biochemistry*, 1997, **36**, 5223.
47. H. Dalton, *Adv. Appl. Microbiol.*, 1980, **26**, 71.
48. A. C. Rosenzweig, C. A. Frederick, S. J. Lippard, and P. Nordlund, *Nature*, 1993, **366**, 537.
49. A. C. Rosenzweig, P. Nordlund, P. M. Takahara, C. A. Frederick, and S. J. Lippard, *Chem. Biol.*, 1995, **2**, 409.
50. B. G. Fox, W. A. Froland, J. E. Dege, and J. D. Lipscomb, *J. Biol. Chem.*, 1989, **264**, 10 023.
51. B. J. Waller and J. D. Lipscomb, *Chem. Rev.*, 1996, **96**, 2625.
52. N. Elango, R. Radjakrishman, W. A. Froland, B. J. Wallar, C. A. Earhart, J. D. Lipscomb, and D. H. Ohlendorf, *Protein Sci.*, 1997, **6**, 556.
53. F. Ruzicka, D.-S. Huang, M. I. Donnelly, and P. A. Frey, *Biochemistry*, 1990, **29**, 1696.
54. K. E. Liu, C. C. Johnson, M. Newcomb, and S. J. Lippard, *J. Am. Chem. Soc.*, 1993, **115**, 939.

55. S.-Y. Choi, P. E. Eaton, P. F. Hollenberg, K. E. Liu, S. J. Lippard, M. Newcomb, D. A. Putt, S. P. Upadhyaya, and Y. Xiong, *J. Am. Chem. Soc.*, 1996, **118**, 6547.

56. E. L. Spence, G. J. Langley, and T. D. H. Bugg, *J. Am. Chem. Soc.*, 1996, **118**, 8336.

57. N. D. Priestly, H. G. Floss, W. A. Froland, J. D. Lipscomb, P. G. Williams, and H. Morimoto, *J. Am. Chem. Soc.*, 1992, **114**, 7561.

58. A. M. Valentine, B. Wilkinson, K. E. Liu, S. Komar-Panicucci, N. D. Priestly, P. G. Williams, H. Morimoto, H. G. Floss, and S. J. Lippard, *J. Am. Chem. Soc.*, 1997, **119**, 1818.

59. B. Wilkinson, M. Zhu, N. D. Priestly, H.-H. T. Nguyen, H. Morimoto, P. G. Williams, S. L. Chan, and H. G. Floss, *J. Am. Chem. Soc.*, 1996, **118**, 921.

5.10
Structure, Function, and Inhibition of Prostaglandin Endoperoxide Synthases

LAWRENCE J. MARNETT, DOUGLAS C. GOODWIN,
SCOTT W. ROWLINSON, AMIT S. KALGUTKAR, and
LISA M. LANDINO
Vanderbilt University School of Medicine, Nashville, TN, USA

5.10.1   METABOLIC AND BIOLOGICAL ROLES

Three separate classes of oxygenases participate in arachidonic acid metabolism to prostaglandins, thromboxane, prostacyclin, leukotrienes, etc. Cytochromes P450 introduce one atom of O_2 to form epoxy- or hydroxyarachidonic acid derivatives;[1] lipoxygenases incorporate one molecule of O_2 to form hydroperoxyarachidonic acid derivatives;[2] and cyclooxygenases carry out the double dioxygenation of arachidonic acid to form the hydroperoxy endoperoxide, prostaglandin G_2 (PGG_2) (Scheme 1).[3,4] PGG_2 is reduced to the hydroxy endoperoxide, prostaglandin H_2 (PGH_2), by a peroxidase that is on the same protein as the cyclooxygenase.[5,6] The ability of this protein to carry out two enzymatic reactions has led to its designation by multiple names such as prostaglandin endoperoxide synthase (PGH synthase) and PGG/H synthase. This chapter will focus on the structure and function of prostaglandin endoperoxide synthase and will use the abbreviation PGH synthase for the enzyme and the terms cyclooxygenase and peroxidase to refer to the reactions that it catalyzes.

5.10.1.1   PGH Synthase Isoforms

There are two isoforms of PGH synthase that carry out the same chemical reactions but are regulated differently (Scheme 2).[7,8] PGH synthase-1 is generally considered to be a constitutively expressed housekeeping gene, although there is some evidence that it may be differentially regulated during development.[9–11] Its presence is detectable in many different tissues and it is particularly abundant in platelets, gastric epithelial cells, and vascular endothelial cells *inter alia*.[8] In platelets, PGH synthase-1 leads to production of the potent platelet aggregation factor, thromboxane A_2 (TxA_2).[12] TxA_2 triggers platelet aggregation and thrombus formation, which can lead to ischemic heart disease. TxA_2 biosynthesis can be reduced through use of aspirin, a nonsteroidal anti-inflammatory drug (NSAID).[13,14] However, continued use of high doses of aspirin can lead to deleterious side affects due to decreased production of the gastric cytoprotective agent prostaglandin E_2 (PGE_2). Gastrointestinal side effects are exhibited by all currently available NSAIDs (see below).[15]

PGH synthase-2 is the product of an immediate early gene and is rapidly synthesized and degraded.[16] It is not normally detectable in tissues unless they have been exposed to some type of stimulus such as a growth factor, hormone, cytokine, etc.[7,8] This is the case in inflammatory states such as rheumatoid arthritis, asthma, and inflammatory bowel disease.[17–19] PGH synthase-2 appears to contribute to hyperalgesia (sensitivity to pain), as PGH synthase-2 mRNA levels (but not PGH synthase-1 levels) are elevated in the lumbar section of the spinal cord following adjuvant-induced inflammation at a peripheral site.[20] There is increasing evidence for a correlation between PGH synthase-2 expression and the risk of developing colorectal cancer. This was first revealed by epidemiological studies which showed that NSAID intake correlated with reduced mortality from colon cancer[21,22] and is supported by the observation that colorectal tumors have elevated levels of PGE_2 compared with normal colorectal tissue.[23] It now appears that this increased prostaglandin production is associated with PGH synthase-2 gene expression because its mRNA and protein are detected in greater than 80% of human colorectal cancers.[24,25] In contrast, PGH synthase-2 expression is undetectable in normal colonic mucosa. PGH synthase-1 is highly expressed in normal colon but its levels do not increase in benign or malignant colon tumors.[25]

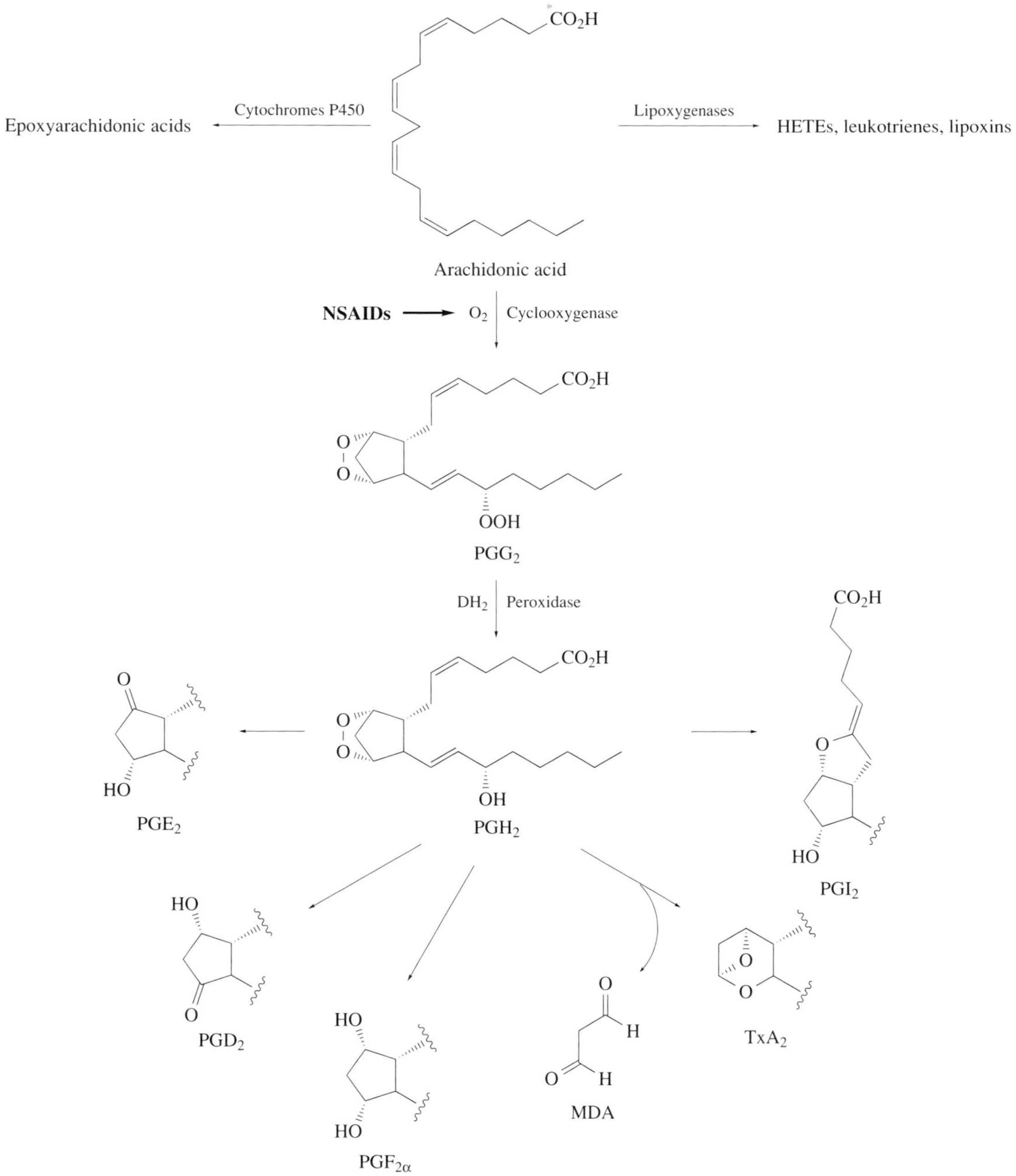

Scheme 1

5.10.1.2 Targeted Deletion of PGH Synthase Genes

The development of PGH synthase-2 knockout mice was reported simultaneously by two groups.[26,27] PGH synthase-2 gene deletion results in renal abnormalities that are fatal within a few weeks of birth. PGH synthase-2 null mice have a greater incidence of cardiac fibrosis than control animals and female mice are infertile. The inflammatory response to treatment with either arachidonic acid or tetradecanoylphorbol acetate is the same in knockout mice as in control mice, suggesting that an active PGH synthase-2 gene is not essential for inflammation in these two models. It is possible that constitutively expressed PGH synthase-1 may compensate for the loss of PGH synthase-2 in the induction of inflammation.

PGH synthase-1 knockout mice exhibit fewer renal abnormalities than PGH synthase-2 deficient mice which confers a substantially longer lifespan on these animals.[28] Perinatal pup survival in homozygous matings of PGH synthase-1-deficient mice (but not homozygous/heterozygous

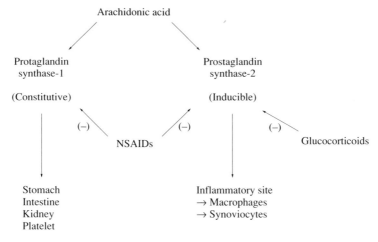

Scheme 2

matings) is reduced. This is presumably due to the adverse affects of prostaglandin deficiency on parturition and is consistent with similar reproductive deficiencies in mice bearing homozygous deletions in the prostaglandin $F_{2\alpha}$ ($PGF_{2\alpha}$) receptor.[29] Platelet aggregation is reduced in PGH synthase-1-deficient mice, which is consistent with the involvement of PGH synthase-1 in TxA_2 biosynthesis in platelets. PGH synthase-1 knockout mice are similar to control animals in their inflammatory response to tetradecanoylphorbol acetate but their response to arachidonic acid administration is attenuated. This is possibly due to the fact that tetradecanoylphorbol acetate is able to induce PGH synthase-2 expression in both PGH synthase-1 knockout mice and control animals, whereas arachidonic acid does not induce PGH synthase-2 in either genetic background. Thus, only the PGH synthase-1 present only in the control animals is able to metabolize the administered arachidonic acid.

One surprising observation made with the PGH synthase-1 knockout mice is the lack of gastro-intestinal lesions.[28] Gastrointestinal lesions are expected in these animals based on a significant body of *in vivo* data showing that NSAIDs, all of which inhibit both isoforms to some extent, cause stomach ulceration.[15] The absence of gastric lesions in PGH synthase-1 knockout mice suggests this pathophysiological response is multifactorial in nature and that mice may not be good models for human beings in this regard.

5.10.2   GENE/PROTEIN STRUCTURE

5.10.2.1   Gene/mRNA

The PGH synthase-1 gene maps to human chromosome 9[30] and is approximately 22 kb in length. It contains 11 exons and 10 introns.[31,32] Transcription of this gene produces a 2.8 kb mRNA that codes for a 599 amino acid protein with a 23 amino acid signal peptide (Figure 1). Its 5′-control sequence contains multiple transcription start sites, has a GC-rich region[32] but does not contain a TATA box, which is typical of developmentally regulated genes. Although analysis of the 5′ untranslated region has identified several potential promoter enhancer elements,[33] only one study appears to have identified control elements that regulate basal human PGH synthase-1 activity. These are two Sp1 elements at -610 to -604 and -111 to -105 that when removed or mutated result in a 70–75% loss of promoter activity.[34] The inability to completely abrogate activity by removal of these two sites suggests that other promoter enhancer elements are operative.

The human PGH synthase-2 gene is located on chromosome 1[35] and differs from PGH synthase-1 in that it is much smaller (~ 8 kb). It contains only 10 exons and 9 introns,[36–38] and encodes two different-sized transcripts (2.8 kb and 4.6 kb) owing to alternative polyadenylation in the 3′ untranslated region. Cloning of the 5′ flanking region allowed identification of putative transcriptional regulatory elements.[39] Since numerous growth factors and cytokines involved in inflammation, hyperalgesia, etc., exert their actions via signal transduction pathways that alter

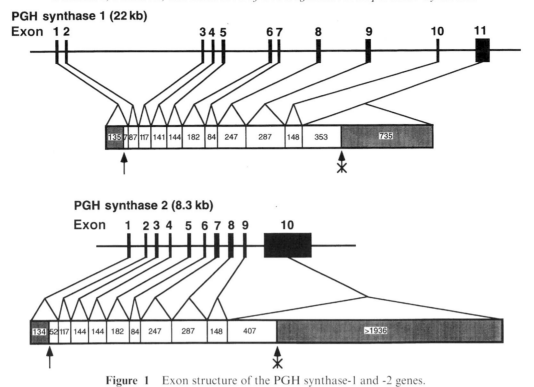

Figure 1 Exon structure of the PGH synthase-1 and -2 genes.

transcription,[7,8] transcriptional control is a major means by which the PGH synthase-2 message is regulated. Functional analysis of this 5′ untranslated region has been performed in a number of species, providing evidence that the following sites are indeed regulatory elements, the E box, CRE (cAMP response element), C/EBP (CAAT/enhancer bonding protein), NF-IL6 (C/EBPβ), NFκB (nuclear factor κB), and Sp1 (sequence specific transcription factor).[7,8,39,40]

In addition to transcriptional regulation, post-transcriptional regulation is believed to be a major means of controlling PGH synthase-2 expression. Unlike PGH synthase-1 mRNA, human PGH synthase-2 mRNA contains 22 copies of the RNA instability sequence, AUUUA, in its 3′ untranslated region, which is believed to be responsible for the short half-life (∼ 30 min) observed for PGH synthase-2 mRNA transcripts.[41,42] Interestingly, either interleukin-1β or interleukin-1α extends the half-life of PGH synthase-2 mRNA message in human ECV304 cells, providing evidence for post-transcriptional regulation of PGH synthase-2 levels. Evidence has also been provided for negative post-transcriptional regulation of the PGH synthase-2 gene. It was shown by Ristimaki *et al.* that elevated PGH synthase-2 levels resulting from pretreatment of cells with interleukin-1β could be downregulated by treatment with dexamethasone.[43] Thus, destabilization of PGH synthase-2 mRNA may be an important means through which anti-inflammatory glucocorticoids exert their actions.

5.10.2.2 Protein

Despite differences in the size of PGH synthase genes and mRNA, the mature proteins contain the same number of amino acids (Figure 2). Both proteins contain a short signal sequence at their N-termini, which is removed during maturation.[36,44-48] Deletion of exon-2 of PGH synthase-2 results in the deletion of 14 amino acids near the N-terminus of the mature protein relative to PGH synthase-1. This deletion is compensated for by a 14 amino acid insertion near the C-terminus of PGH synthase-2. Thus, most of the amino acid residue numbers of PGH synthase-1 are 14 units higher than the corresponding residue numbers of PGH synthase-2. By convention, the residues of both proteins are referred to by the residue numbers of PGH synthase-1.

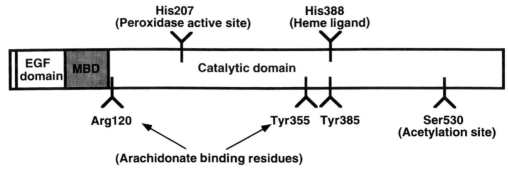

Figure 2   Domain structure and position of important amino acids in PGH synthase: MBD, membrane binding domain; EGF, epidermal growth factor.

Crystal structures have been determined for sheep PGH synthase-1[49] and mouse and human PGH synthase-2 (Figure 3).[50,51] The primary sequences of PGH synthases-1 and -2 are approximately 60% identical within species,[8] so it is not surprising that there is a high level of structural conservation between the two forms. Comparison of backbone atoms of sheep PGH synthase-1 and human PGH synthase-2 shows an r.m.s deviation of 0.9 Å.[51] PGH synthases exist as homodimers related by a noncrystallographic twofold symmetry axis.[49] Although the interfaces between PGH synthases-1 and -2 are well conserved, it is believed that there are enough differences to preclude heterodimerization of the isozymes.[51]

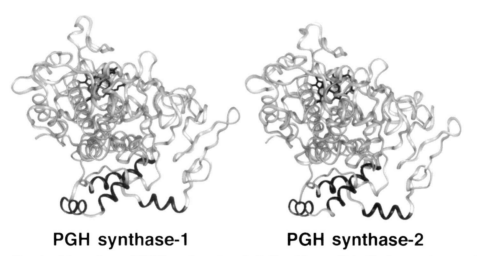

Figure 3   Structural homology of PGH synthase-1 and -2. Brookhaven Data Bank accession numbers are 1PGE and 3PGH, respectively.

A single monomer of PGH synthase consists of three domains, an N-terminal epidermal growth factor-like domain (residues 34–72), a membrane binding domain (residues 73–116), and a catalytic domain (residues 117–600). The function of the epidermal growth factor domain is not known but it contains the disulfide-linked β-sheet typical of the epidermal growth factor folding pattern. This domain is secured to the main body of the protein by a disulfide bond from Cys37 to Cys159.

5.10.2.2.1   *Membrane-binding domain*

The membrane-binding domain consists of four amphipathic α-helices that form a novel motif for insertion of the protein into one side of the membrane bilayer (Figure 4).[49] Three of these helices, B (residues 86–92), C (residues 97–105), and D (residues 108–122), combine to form a three-walled U-shaped structure that is believed to form the beginning of the arachidonate access channel. Helix A (residues 74–82), which immediately follows the epidermal growth factor domain, is slightly displaced from helices B, C, and D. This leaves the lowest part of the arachidonate access channel in an open-sided conformation. Since PGH synthases are homodimers in which the subunits are related to each other by a C-2 axis of symmetry, two sets of hydrophobic helices participate in stabilization of the protein in the membrane. Indeed, PGH synthases exhibit tight membrane association characteristic of integral membrane proteins. Dimerization appears to be important for membrane insertion and also for stabilizing the protein. Thermal denaturation studies indicate that dimer–monomer dissociation precedes irreversible denaturation.[52]

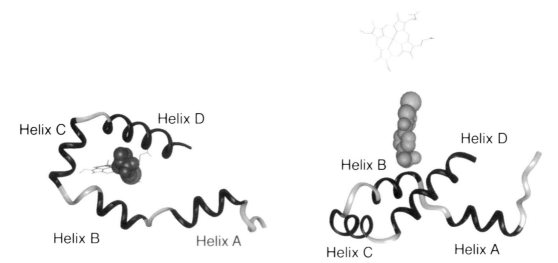

Figure 4   Membrane view (left) and side view (right) of the four membrane-bound amphipathic α-helices, iodosuprofen, and heme from PGHS-1/iodosuprofen complex (PBD accession 1PGE).

The amphipathic helices that represent the membrane association site for PGH synthases also circumscribe the substrate access channel to the cyclooxygenase active site.[49] Polyunsaturated fatty acids are the substrates for oxygenation by the cyclooxygenase but they do not exist free in cells.[53,54] Rather, they are covalently attached to the 2-hydroxyl group of phospholipid molecules. Release of arachidonic acid or other polyunsaturated fatty acids by the action of phospholipases constitutes a major control point in the biosynthesis of prostaglandins.[53–55] Because of the orientation of PGH synthases in the membrane, arachidonic acid and other polyunsaturated fatty acids have ready access to the cyclooxygenase binding site through the membrane-associated substrate access channel. This appears to occur in the endoplasmic reticulum and nuclear membranes where PGH synthases have their major concentration.[56–58]

5.10.2.2.2   Catalytic domain

The large catalytic domain contains both the peroxidase and cyclooxygenase active sites. Comparisons between the structures of PGH synthase-1 and -2 should be carefully interpreted because the sheep PGH synthase-1 complex has only been determined in an inhibitor-bound state,[49,59] thus inhibitor-induced conformational changes may mask the true unbound relaxed conformation of this isoform. Nevertheless, it appears that the positioning of residues such as Arg120, Tyr355, and Tyr385, that have been implicated as functionally important in binding and catalysis, are conserved between PGH synthases-1 and -2.[60] There are some major differences in the arachidonate access channel, which are evident despite the high degree of alignment between the backbone atoms of PGH synthase-1 and -2 and any possible differences that may be associated with the inhibitor-bound PGH synthase-1 isoform.

The volume of the arachidonate-binding channel in PGH synthase-1 is smaller than that in PGH synthase-2.[51] This is significant because it is on the basis of steric differences that certain classes of inhibitors (e.g., diarylheterocycles) exhibit PGH synthase-2 selectivity. The best characterized of these differences involves the Ile substitution of PGH synthase-1 to Val of PGH synthase-2 at position 523.[61–64] This substitution, along with other closely associated side chains that include positions 434 and 513 result in an increased volume (estimated as nearly 80 $\AA^3$) of the PGH synthase-2 active site.[50,51] Another amino acid substitution that appears responsible for isoform-specific inhibition occurs at position 503.[51] In PGH synthase-1, this is a Phe, whereas in PGH synthase-2 it is Leu. The Phe-to-Leu substitution results in an increased volume of the top part of the PGH synthase-2 substrate binding channel.

The active sites for the cyclooxygenase and peroxidase reations occur in different locations of the PGH synthase molecules (Figure 5). The cyclooxygenase active site occurs near the top of the fatty acid access channel underneath the heme prosthetic group, whereas the peroxidase active site is on the opposite side of the protein above the heme.[49] In Figure 5 a molecule of the cyclooxygenase inhibitior, iodosuprofen, is shown bound in the cyclooxygenase active site. The potential access of this site to the membrane through the four hydrophobic helices is apparent. A wide opening to the peroxidase active site allows a variety of primary and secondary hydroperoxides to approach the heme group. A loop that sits above the opening to the peroxidase active site is apparent in both PGH synthases but its role is unknown. The loop is protease-sensitive in apoPGH synthase-1 but not in holoPGH synthase-1 or in apo or holoPGH synthase-2.[65-68]

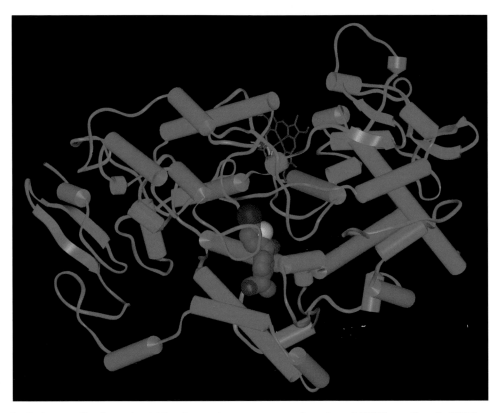

Figure 5   Iodosuprofen bound within the cyclooxygenase active site of PGH synthae-1 (PBD accession 1PGE).

5.10.2.2.3   Heme-binding site

Comparison of the heme-binding sites of PGH synthases-1 and -2 reveals no significant differences in structure (Figure 6). Both isoforms contain the proximal iron ligand His388 and the closely associated His386, whereas W387 and the putative active site residue Tyr385 are orientated away from the peroxidase site in both the PGH synthase-1 and -2 structures. The hydroperoxide-reducing sites of PGH synthases lie on the surface of the enzymes on the distal sides of the liganded heme prosthetic group. This site contains the conserved His207 and Gln203 side chains that have been shown to be important for peroxidase activity.[69,70] Differences in peroxidase activity of the two PGH synthase isoforms have been reported but they cannot be attributed to any obvious structural differences.[71,72]

The mechanism of hydroperoxide reduction by peroxidases requires that the terminal peroxide oxygen directly contact the heme iron and there appears to be ample room for this to happen in PGH synthases. PGH synthases reduce a broader range of organic hydroperoxides than most other mammalian peroxidases which may reflect the generous size of the opening to the peroxidase active site.[5,73,74] Tertiary hydroperoxides such as cumene hydroperoxide or *t*-butyl hydroperoxide are poorer substrates than primary or secondary hydroperoxides.[73]

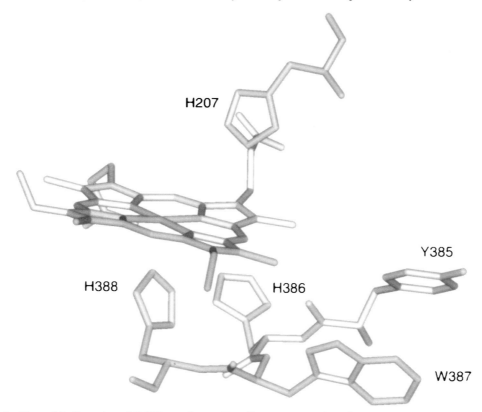

Figure 6 Heme-binding site of PGH synthase. Coordinates were taken from the structure of ovine PGH synthase-1 complexed with iodosuprofen deposited by Garavito and co-workers[59] (PDB accession 1PGE).

Examination of the crystal structure in Figure 5 indicates that there is no place in the cyclo-oxygenase active site (indicated by iodosuprofen) where the polyunsaturated fatty acid substrate can make direct contact with the heme prosthetic group. This implies that the oxidizing agent which oxidizes the fatty acid is protein derived rather than heme derived. Furthermore, there is no obvious pathway that physically connects the cyclooxygenase active site to the peroxidase active site so it is currently thought that PGG_2 must diffuse back down, then out of the fatty acid access channel to reach the peroxidase active site.

5.10.2.2.4 *Structural homology*

The tertiary structures of PGH synthases are very similar to the structure of the mammalian peroxidase, myeloperoxidase, despite the fact that the proteins are only 14% identical in primary structure (Figure 7).[49,75] Most of the amino acid identity between PGH synthases and myeloperoxidase occurs in a region of the proteins well removed from the heme-binding sites.[69] This region is near the top of two long α-helices that constitute the distinguishing feature of the "peroxidase" folding pattern. In addition to these gothic helices suspended over the heme group, α-helices run across the heme (one above and two below) to stabilize it and provide ligands to the iron. The major difference between PGH synthases and myeloperoxidase is the presence of the membrane-binding domain near the N-terminus of PGH synthase.

5.10.3 REACTIONS CATALYZED

5.10.3.1 Fatty Acid Oxygenation

In 1967, Hamberg and Samuelsson proposed the mechanism illustrated in Scheme 3 to explain the oxygenation of arachidonic acid by cyclooxygenase.[76] Their hypothesis was based on the findings

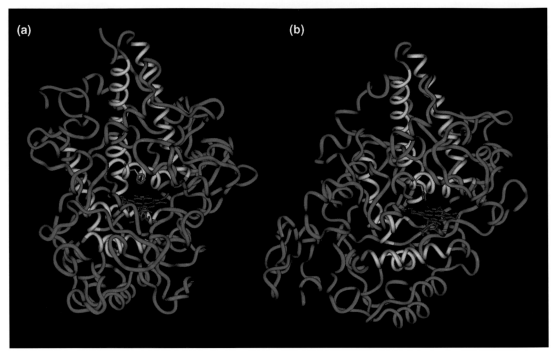

Figure 7 Crystal structures of myeloperoxidase (a) and PGH synthase (b). Coordinates for the structure of myeloperoxidase (a) were taken from Zeng and Fenna[75] (PDB accession 1MHL) and the structure of PGH synthase (b) was produced using coordinates determined by Garavito and co-workers[49] (PDB accession 1PRH). The heme prosthetic group (red) as well as proximal and distal histidines (purple) are indicated.

that O_2 or molecular oxygen is the source of the oxygen atoms at the 9, 11, and 15 positions in prostaglandins and that a significant isotope effect is observed for removal of the 13-*proS*-hydrogen atom.[76,77] The essential features of the mechanism depicted in Figure 2 have been confirmed by the isolation of PGG_2 as an intermediate in catalysis[3,4] by the trapping of carbon-centered radicals during the incubation of arachidonic acid with cyclooxygenase under anaerobic conditions,[78] and by the demonstration that peroxyl free radical derivatives of polyunsaturated fatty acids undergo serial cyclization to form the bicyclic peroxide moiety typical of prostaglandin endoperoxides.[79–81] The major roles of PGH synthase in catalyzing the reactions depicted in Scheme 3 are the binding of arachidonic acid in an extended conformation and stereospecific removal of the *proS*-hydrogen atom from carbon-13.

5.10.3.2 Hydroperoxide Reduction

Reduction of the hydroperoxide group of PGG_2 to the alcohol of PGH_2 is catalyzed by a typical heme peroxidase mechanism (Scheme 4).[74] The hydroperoxide substrate is reduced by the heme prosthetic group to an alcohol with concomitant generation of a heme higher oxidation state.[82,83] The two oxidizing equivalents of the hydroperoxide substrate are transferred to the enzyme, producing a metal–oxo derivative known as Compound I.[83] The ferric iron of the resting heme is oxidized to the ferryl state and the other oxidizing equivalent resides either in the porphyrin ring as a cation radical or in an amino acid residue as a protein radical. The formation of a peroxidase Compound I in which the second oxidizing equivalent is either a porphyrin radical cation or a protein radical depends upon the presence of an easily oxidizable amino acid residue in close proximity to the heme. For example, a tryptophan residue (Trp191) sits under the heme group of cytochrome c peroxidase and is oxidized to a radical as part of the formation of Compound I.[84] The corresponding residue in lignin peroxidase is a phenylalanine which is not oxidized, so the lignin peroxidase Compound I contains the second oxidizing equivalent as a porphyrin cation radical (Figure 8).[85] The spectral properties of PGH synthase Compound I indicate that it contains a porphyrin cation radical capable of intramolecular electron transfer under certain conditions (see below).

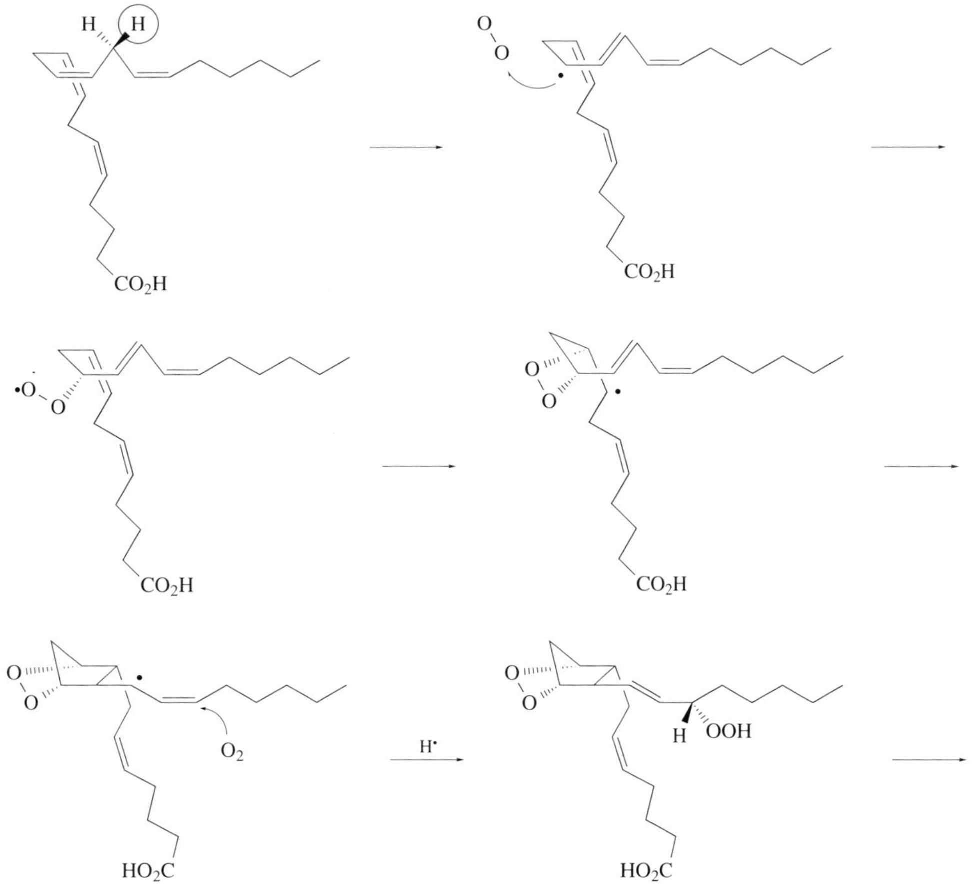

Scheme 3

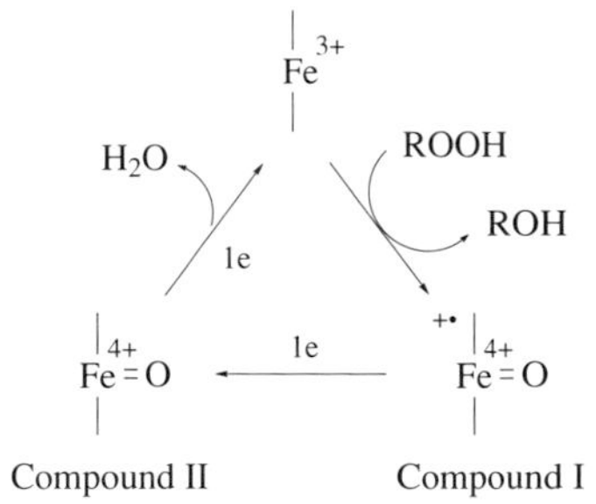

Scheme 4

The peroxidase of PGH synthase reacts with a range of hydroperoxides but at considerably different rates.[86] The rate coefficient for reaction with PGG_2 to form peroxidase Compound I and PGH_2 is $1.4 \times 10^7 \, M^{-1}s^{-1}$, whereas the rate coefficients for reaction with H_2O_2 and t-butyl-hydroperoxide are $4.6 \times 10^5 \, M^{-1}s^{-1}$ and $5.9 \times 10^4 \, M^{-1}s^{-1}$, respectively.[86,87] The value for PGG_2 is comparable to the rate coefficients for reaction of H_2O_2 with other peroxidases such as horseradish peroxidase.[88] The rate of reaction of PGH synthase with H_2O_2 is unusually low for a heme peroxidase. This may reflect the fact that PGH synthase evolved to reduce fatty acid hydroperoxides but not H_2O_2. Interestingly, the rate coefficient for reaction with peroxynitrous acid ($1.5 \times 10^7 \, M^{-1}s^{-1}$ at pH 7.0) is equal to that of PGG_2 and approximately 30-fold higher than the value for H_2O_2.[89] The high rate of reaction of PGH synthase with peroxynitrous acid may reflect a role for this inorganic

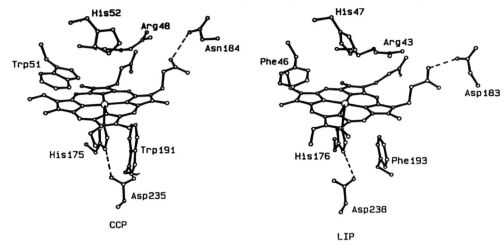

Figure 8 Comparison of the active site structures of cytochrome c peroxidase (CCP) and lignin peroxidase (LIP).

hydroperoxide in regulating prostaglandin biosynthesis. In fact, peroxynitrous acid has been implicated as a physiological activator of cyclooxygenase activity in inflammatory cells.[90]

The pH–rate profiles for hydroperoxide reduction reveal the requirement for a basic group with a pK_a of approximately 4.0.[86] A similar acid dissociation dependence has been detected in the formation of Compound I of other peroxidases in the pH range 2.5–5.3. The group responsible for this pH dependence is believed to be the imidazole group of the distal histidine which acts as a general base to facilitate hydroperoxide ligation to the heme iron. As with other peroxidases, the preferred form of the hydroperoxide substrate is the protonated form.[86]

Reconstitution of apoPGH synthase with Mn^{3+}–protoporphyrin IX produces an enzyme with significantly reduced peroxidase activity.[5] Mn^{3+}–PGH synthase reacts with polyunsaturated fatty acid hydroperoxides to form a higher oxidation state but the rate coefficients are several orders of magnitude slower than those of the iron enzyme.[91–93] A similar observation has been made for horseradish peroxidase and the difference in peroxidase activities has been attributed to the presence of a water molecule as the sixth ligand to the metal center in mangano-horseradish peroxidase but not in ferric-horseradish peroxidase.[94] The rate of water dissociation from metal–porphyrin complexes is sufficiently slow that it constitutes a kinetic barrier to reaction of the metal center of the peroxidase with hydroperoxides.[94] A water ligand to Fe^{3+}–PGH synthase has been detected by resonance Raman analysis of PGH synthase,[95] which seems inconsistent with the high rates of reaction of PGH synthase peroxidase with fatty acid hydroperoxides. In addition, the crystal structure of a PGH synthase-2–inhibitor complex determined at 2.6 Å resolution reveals a water molecule bound as the sixth ligand to iron.[96] Thus, despite the high rate of reaction of iron PGH synthase with PGG_2, there does appear to be a water molecule as the distal ligand to the heme iron.

5.10.3.3   Reduction of Peroxidase Higher Oxidation States

A major difference between the peroxidase of PGH synthase and the classic peroxidase horseradish peroxidase is the stability of their Compounds I. Horseradish peroxidase Compound I is quite stable and solutions of it can be readily prepared and its reactions studied independently.[88] In contrast, PGH synthase Compound I is extremely short-lived and begins to decay within 10 ms after its formation.[83] This may be due to differences in hydrogen bonding to the proximal histidine ligands of the two enzymes or to the rapid intramolecular electron transfer that occurs with PGH synthase Compound I. In addition, PGH synthase Compound I reacts with hydroperoxides to form Compound II and peroxyl radicals. This not only further shortens the half-life of Compound I but induces decomposition of the hydroperoxide substrate to the nonalcohol product. The induced decomposition is particularly evident with PGH synthase-2.[70] The inherently short half-life of PGH synthase Compound I, its ability to intramolecularly oxidize protein residues, and its ability to oxidize hydroperoxides have greatly complicated the kinetic analysis of its formation and reaction. No doubt, this has contributed to the controversy surrounding the mechanism of peroxidase activation of cyclooxygenase activity (see below).

Catalytic turnover of the peroxidase requires reduction of the higher oxidation states to resting enzyme, with an input of two electrons. One-electron reduction of Compound I produces Compound II, which comprises a ferryl–oxo complex and a fully covalent porphyrin.[83,97,98] Reduction of Compound II by a single electron regenerates the resting enzyme. Reduction of peroxidase higher oxidation states generally proceeds in a sequential manner (i.e., two one-electron transfers), but, in some cases, two-electron reduction from a single donor occurs.[74,88]

The instability of PGH synthase Compound I has prevented an extensive kinetic investigation of its reaction with reducing substrates. Measurements with diethyldithiocarbamate as reductant indicate a bimolecular rate coefficient for reaction of $0.58–1.8 \times 10^7 \, M^{-1}s^{-1}$.[99] The rate coefficient for reduction of PGH synthase Compound II with diethyldithiocarbamate is $1 \times 10^5 \, M^{-1}s^{-1}$.[99] The differential in rate coefficients for reduction of Compounds I and II is typical of that seen with other peroxidases.[88] Normally, reduction of Compound II constitutes the rate-limiting step in the peroxidase catalytic cycle. The sole determinant of the reduction of peroxidase higher oxidation states by reducing substrates is their redox potential.[100] The ρ value for reduction of PGH synthase Compound II by a series of substituted phenols is -2.0, which compares to values of -6.9 and -4.6 for reduction of horseradish peroxidase Compounds I and II, respectively.[100] The ρ value for the overall process of reduction of the organic hydroperoxide 5-phenyl-4-pentenylhydroperoxide by substituted thioanisoles is -0.8.[101]

Concomitant with the reduction of peroxidase higher oxidation states, the reducing substrate is oxidized.[82,102–104] One-electron oxidation is the predominant mode of oxidation. This produces a one-electron oxidized derivative of the reducing substrate and Compound II. The one-electron oxidized form of the substrate is a carbon- or oxygen-centered radical or a nitrogen-centered cation radical (Scheme 5).[103–105] Carbon- or oxygen-centered radicals appear to diffuse from the peroxidase active site and react with oxygen or other molecules in solution.[106,107] For carbon-centered radicals, coupling to O_2 to form peroxyl radicals is a common mode of reaction.[107,108] The peroxyl radicals can be reduced to form hydroperoxides, some of which are themselves substrates for the peroxidase.[106] Phenoxyl radicals also react with O_2 but more commonly they couple to other phenoxyl radicals to form dimers or higher order oligomers (e.g., guaiacol–tetraguiaicol).[109,110] Phenoxyl radical coupling products are often reducing substrates for further oxidation by peroxidases so the pattern of products generated from phenols by peroxidase oxidation is frequently complex.[111]

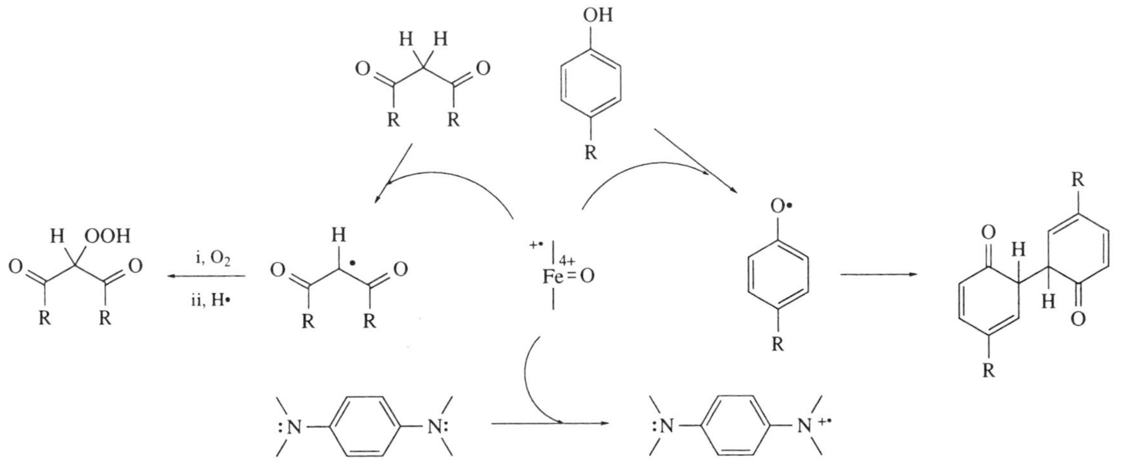

Scheme 5

Sulfur- or nitrogen-centered radical cations also can diffuse from the active site to participate in reactions in solution. In some cases, the radical cations are stable and constitute the final products of reaction (e.g., tetramethylphenylenediamine → Wurster's blue cation).[111] However, sulfur- and nitrogen-centered cation radicals can also couple to the oxo ligand of Compound II which results in oxygen transfer to the heteroatom (Equation (1)). Studies with isotopically labeled substrates indicate that the oxygen atom delivered to the heterotaom is derived from the hydroperoxide group as required by the mechanism outlined in Equation (1).[112]

$$R^{18}O\text{--}^{18}OH \quad + \quad \underset{\text{(Me--S--phenyl)}}{} \quad \xrightarrow{\text{Peroxidase}} \quad R^{18}OH \quad + \quad \underset{\text{(Me--S(=}^{18}O)\text{--phenyl)}}{} \qquad (1)$$

The identity of the physiologic reducing substrate for PGH synthase has been the subject of speculation but little hard evidence is available to judge if a single reductant is used. Among a series of naturally occurring molecules, epinephrine and uric acid appear to be the best PGH synthase reducing substrates.[82] Uric acid is only a fair reducing substrate but it is present in high concentrations in plasma ($\sim 300\ \mu M$).[113,114] Oxidation of cellular polyhydroxylated compounds such as epinephrine is also anticipated. In addition, glutathione, present in millimolar concentrations in cells, can react with electron-deficient intermediates to form the thiyl radical, GS·.[115] Thus, PGH synthase turnover can deplete cellular antioxidants. The identity of the physiologic reducing substrate is expected to vary with cell and tissue type.

Hydroperoxide-dependent oxidation of a range of foreign compounds including drugs, toxicants, and carcinogens is catalyzed by the PGH synthase peroxidase.[104,105] The oxidized derivatives of many of these molecules are reactive electrophiles capable of covalent attachment to intracellular nucleophiles such as proteins and nucleic acid. The structural alterations induced by covalent modifications can lead to loss of protein function or induction of mutations, resulting in cell death or transformation.[116,117] PGH synthase-dependent cooxidation of xenobiotics has been proposed to play a role in metabolic activation leading to toxicity, carcinogenicity, and teratogenicity in tissues that express high levels of PGH synthase.[118–121]

5.10.4 PEROXIDASE ACTIVATION OF CYCLOOXYGENASE

5.10.4.1 Hydroperoxide-dependent Activation

Considerable evidence implicates peroxidase turnover in the activation of cyclooxygenase catalysis:

(i) Blockade of peroxidase function by coordination of CN^- to the heme iron inhibits cyclooxygenase activity.[122,123]

(ii) Depletion of hydroperoxides by the addition of glutathione and glutathione peroxidase induces a lag phase for induction of maximal cyclooxygenase activity.[124] Lag phases are also induced by exposure to high concentrations of aspirin-inhibited PGH synthase in the presence of a reducing substrate.[125] Aspirin-inhibited PGH synthase has high peroxidase activity but no cyclooxygenase activity.

(iii) The sensitivity of different PGH synthases to inhibition by glutathione/glutathione peroxidase is inversely related to their level of peroxidase activity. For example, Fe-PGH synthase, which has a peroxidase activity that is 125-fold higher than Mn-PGH synthase, requires 250-fold more glutathione peroxidase to be inhibited.[91,92]

(iv) Mutations that abolish peroxidase activity induce a lag phase for cyclooxygenase activity that can be overcome by the addition of hydroperoxides (Figure 9). These mutations include the distal histidine of PGH synthase-2 (His207) and a histidine located on the proximal side of the heme group of PGH synthase-1 (His386).[69,70]

(v) Once the lag phases in cyclooxygenase activity exhibited by peroxidase-deficient PGH synthases are overcome, the rates of cyclooxygenase activity are comparable (within an order of magnitude) to that of wild type.[70] This suggests that the peroxidase functions only in the activation of cyclooxygenase activity but not in its continual turnover.

These observations suggest that activation of cyclooxygenase catalysis is somehow dependent on the formation of peroxidase higher oxidation states. It is unlikely that the heme–oxo intermediates directly oxidize arachidonic acid because the heme group is completely shielded by protein residues from the cyclooxygenase active site.[49–51] This implies that the peroxidase higher oxidation states

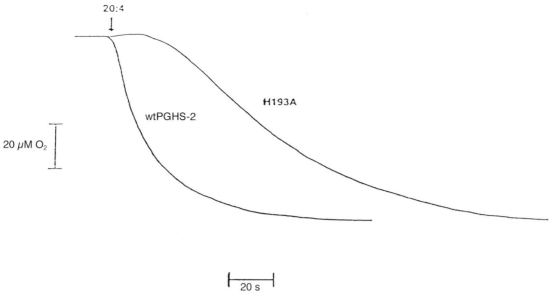

Figure 9 Oxygenation of arachidonic by wild-type (wtp9HS-2) and the His193Ala mutant (H193A) of PGH synthase-2.

oxidize a protein residue, which then oxidizes arachidonic acid. Because continual peroxidase turnover is not required for cyclooxygenase catalysis, the oxidized protein residue must be regenerated at a late stage in the cyclooxygenase catalytic cycle. As discussed above, intramolecular oxidation of protein residues by peroxidase higher oxidation states is well precedented and, in fact, peroxidases participate in long-range electron transfer as well. For example, cytochrome c peroxidase oxidizes the ferrous form of cytochrome c across a distance of 24 Å, which makes the direct transfer of electrons from cytochrome c to cytochrome c peroxidase extremely unlikely.[127] In this case, protein residues appear to act as participants in electron transfer. As mentioned previously, the cyclooxygenase and peroxidase active sites of PGH synthase are physically separated from one another by over 10 Å.

The identity of the physiological hydroperoxide activator of PGH synthase has been the subject of considerable investigation. Attention has focused primarily on fatty acid hydroperoxides such as 15-hydroperoxy-5,8,11,13-eicosatetraenoic acid (15-HPETE) because the K_m values for reaction of 15-HPETE with the PGH synthase peroxidases are very low (5–20 μM).[128,129] However, these hydroperoxides are excellent substrates for glutathione peroxidase (GSH-Px), the selenium-containing peroxidase that uses reduced glutathione as a cosubstrate. Bimolecular rate constants for the reaction of fatty acid hydroperoxides with GSH-Px are as high as 4×10^7 $M^{-1}s^{-1}$.[130] Incubation of PGH synthase with GSH-Px and GSH inhibits prostaglandin formation because hydroperoxide contaminants in commercial preparations of arachidonic acid are reduced to the corresponding alcohols, thereby preventing peroxidase-dependent cyclooxygenase activation as outlined above.[129]

Work in the authors' laboratory has implicated peroxynitrite as a possible physiologic hydroperoxide activator of PGH synthase.[90] Evidence linking nitric oxide (NO) and prostaglandin biosynthesis in a variety of cell types is considerable;[131–134] however, support for a direct interaction of NO with PGH synthase is inconclusive.[135–137] Peroxynitrite anion (ONOO⁻) is formed by the diffusion-controlled reaction of NO with superoxide anion (Scheme 6) and ONOO⁻ is a hydroperoxide substrate for the peroxidase activities of both PGH synthase isoforms.[90,138] The direct reaction of ONOO⁻/ONOOH with PGH synthase-1 to form Compound I occurs with a rate coefficient of 1.5×10^7 $M^{-1}s^{-1}$ at pH 7.[139] ONOO⁻ activates the cyclooxygenase activity of GSH-Px/GSH-inhibited PGH synthase-1 and -2 by reacting with the PGH synthase peroxidase active site. Under conditions where ONOO⁻ is generated *in situ* from NO and superoxide anion, Cu,Zn-superoxide dismutase completely inhibits prostaglandin formation; thus, NO alone cannot activate PGH synthase. In cultured mouse macrophages, membrane-permeant superoxide dismutase mimetic agents reduce prostaglandin biosynthesis by 80%.[90]

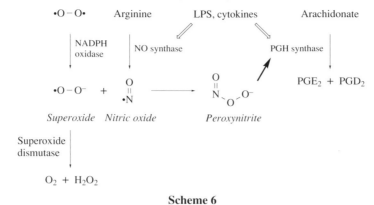

Scheme 6

5.10.4.2 Generation of Protein Radicals

5.10.4.2.1 Detection of tyrosyl radicals

The formation of a PGH synthase-derived protein radical was first demonstrated by Ruf and colleagues who reported detection by EPR of a tyrosyl radical when PGH synthase-1 was incubated with PGG_2.[140,141] Addition of organic hydroperoxides to either PGH synthase-1 or -2 causes the disappearance of the EPR signal of ferric high-spin heme ($g = 6.8$–5.5) and the appearance of a signal typical of organic free radicals ($g = 2.005$).[140] The spectrum of this new radical signal is very similar to that of protein tyrosyl radicals, in particular, the protein tyrosyl radical of ribonucleotide reductase. The changes in electronic absorption spectra that parallel generation of the EPR signal reveal that formation of the PGH synthase tyrosyl radical coincides with the appearance of PGH synthase Compound II (Figure 10).[141] The detection and assignment of this tyrosyl radical coupled to the related changes in the oxidation state of the heme led Ruf to propose the currently accepted model for arachidonic acid (AA) oxygenation by PGH synthase (Scheme 7). Although some controversy surrounds the identity of the spectrally detectable tyrosyl radicals,[126,142–148] the basic tenets of the Ruf mechanism appear consistent with the bulk of the experimental data. At the heart of the mechanism is the oxidation of a PGH synthase tyrosine by Compound I to yield a Compound II-like heme intermediate and a protein-bound tyrosyl radical. The tyrosyl radical then oxidizes arachidonic acid to initiate cyclooxygenase catalysis. In the last step of the cyclooxygenase catalytic cycle, a peroxyl radical precursor to PGG_2 reoxidizes the tyrosine to a tyrosyl radical concomitant with the generation of PGG_2. This fulfills the criterion of regeneration of the tyrosyl radical independent of peroxidase turnover.

The model suggested by Ruf provides an explanation for many observations about PGH synthase catalysis. First, this model explains the absolute requirement for the heme prosthetic group. Without it, there is no peroxidase activity so there can be no cyclooxygenase activation. Naturally, this also explains the absolute requirement for a hydroperoxide activator. Ruf's model also provides an explanation of how the two catalytic activities interact via the heme group and a mechanism for participation of the peroxidase in cyclooxygenase activation but not catalytic turnover.

5.10.4.2.2 Time-dependent changes in tyrosyl radicals

Although the Ruf model is attractive, it has been difficult to unequivocally identify the spectroscopically detectable tyrosyl radical as a catalytic intermediate. Tyrosyl radicals have been detected for both PGH synthase isoforms, but the radical signals undergo time-dependent changes in linewidth, hyperfine structure, g value, and amplitude.[143,147,149] Tyrosyl radical signals from PGH synthase-1 are the best characterized. The first radical signal observed upon addition of an organic hydroperoxide is a broad doublet ($g = 2.005$) (Figure 11).[140] The hyperfine structure of the initial doublet radical bears striking resemblance to the tyrosyl radical observed for ribonucleotide reductase from *Escherichia coli*.[150] Indeed, it was this similarity that enabled initial assignment of the PGH synthase radical as tyrosine centered. It has been shown that replacement of tyrosines in recombinant PGH synthase-1 with perdeuterated tyrosine results in collapse of the observed doublet to a narrow singlet.[126] This confirms that the radical observed during arachidonic acid oxygenation

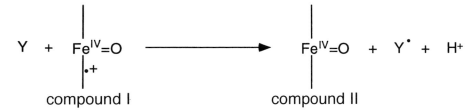

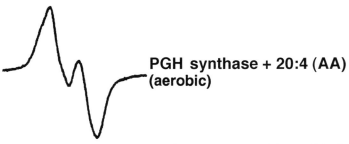

Figure 10 PGH synthase-1 tyrosyl radical observed upon reaction with arachidonic acid under aerobic conditions. The intramolecular electron transfer believed to generate the tyrosyl radical is also depicted.

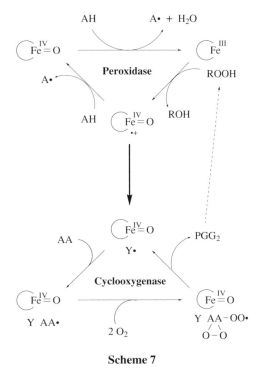

Scheme 7

by PGH synthase is a tyrosyl radical. With time, the initial doublet is replaced by a highly stable, broad singlet ($g = 2.004$), similar to that observed for the tyrosyl radical of photosystem II.[140,142]

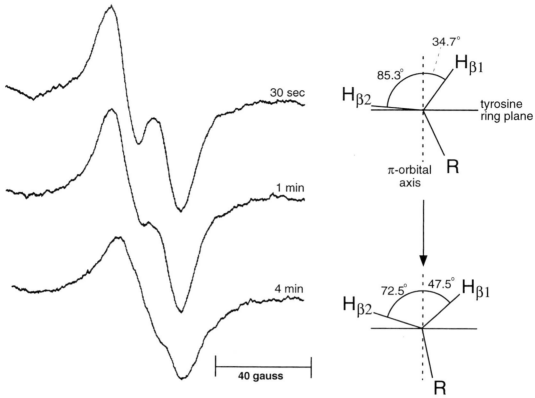

Figure 11   Changes in PGH synthase tyrosyl radical EPR signal with time. Spectra of ovine PGH synthase-1 recorded at the indicated times during reaction with hydroperoxide (PPHP).[142] The approximate dihedral angles of the methylene protons of tyrosyl radicals shown to the right of the spectra were determined by computer simulation for similar spectra in comparable experiments.

The time-dependent changes in the tyrosyl radical signal can be explained by one of two possibilities: either different tyrosines are oxidized at different times during PGH synthase catalysis, or a single tyrosyl radical is formed, and the change from doublet to singlet is the result of a change in the conformation of the tyrosine in question. The dihedral angle of the C-7 methylene protons relative to the axis of the π-electron system of the tyrosine ring is the major factor determining the hyperfine structure of a protein tyrosyl radical. A C-7 methylene proton will give the greatest hyperfine splitting when it is parallel with the axis of the π-electron system (perpendicular to the plane of the ring). The proton will contribute least to hyperfine splitting when it is perpendicular to the axis of the π-electron system (parallel to the plane of the ring). The initial appearance of a doublet tyrosyl radical signal suggests that one of the methylene protons is nearly coplanar with the phenolic ring (little hyperfine interaction), whereas the other is roughly at a 70° angle with the phenolic ring (substantial hyperfine splitting). The effective splitting by one methylene proton yields the doublet signal. When both protons are roughly at a 30° angle to the ring, there is relatively little hyperfine splitting from either proton. Computer simulation shows that this conformation produces a singlet tyrosyl radical signal nearly identical to the experimentally observed radical.[143]

Clearly, the time-dependent changes in PGH synthase tyrosyl radical hyperfine structure could be explained by simple rotation of the phenolic ring possibly induced by conformational changes in the protein. However, collapse of the doublet radical to the singlet is insufficient evidence for this conclusion. It is possible that two tyrosines are oxidized to radicals during the course of arachidonic acid oxygenation by PGH synthase. Indeed, there are six tyrosines in PGH synthase-1 in which the phenoxyl oxygen is within 9–13.5 Å of the heme, which is the heme–radical distance suggested by power saturation experiments.[140,142] Interestingly, both the singlet and doublet radicals show nearly identical power saturation characteristics, suggesting that both radicals are approximately the same distance from the iron atom of the heme prosthetic group.[142]

As mentioned previously, tyrosyl radicals also are observed during AA oxygenation by the inducible isozyme, PGH synthase-2.[149,151] However, there are a number of differences in the properties of the observed radical. Unlike PGH synthase-1, a broad singlet is first detected upon addition of arachidonic acid or ethylhydroperoxide.[149] With time, this signal gains slightly more doublet character, apparently opposite to the changes observed for PGH synthase-1.[149] However, the time-dependent changes in the PGH synthase-2 radical signal are not nearly as dramatic as those observed for PGH synthase-1. Also, the initial tyrosyl radical observed for PGH synthase-2 appears to be formed more efficiently than the corresponding radical from PGH synthase-1. Xiao *et al.* have reported that 0.94 spin/heme is obtained upon reaction of PGH synthase-2 with ethylhydroperoxide, whereas only 0.5–0.7 spin/heme is observed for PGH synthase-1.[152] The implications that the time-dependent changes in tyrosyl radicals for each isozyme, and the differences in tyrosyl radical yields have for the mechanisms of catalysis and possible suicide inactivation have yet to be determined.

5.10.4.2.3 *Identification of oxidized tyrosines*

Since Ruf's suggestion that a tyrosyl radical may play a critical role in PGH synthase catalysis, identification of the tyrosine oxidized to this radical intermediate has been a central question. Smith and co-workers reported that tetranitromethane inactivates cyclooxygenase and nitrates several protein tyrosine residues.[153] Nitration of three of these residues is prevented by the cyclooxygenase inhibitor indomethacin, suggesting that these tyrosines are in or near the cyclooxygenase active site. Site-directed mutagenesis of all three residues reveals that replacement of Tyr385 in PGH synthase-1 with Phe completely abolishes cyclooxygenase activity but not peroxidase activity.[153] These data suggested that Tyr385 is the tyrosine oxidized to a radical during PGH synthase turnover. Subsequent determination of the crystal structure of PGH synthase-1 supported this hypothesis because Tyr385 is located between the heme prosthetic group and the putative arachidonic acid binding site (Figure 6 and see below).[49] The analogous tyrosine in PGH synthase-2 occupies the same position in its three-dimensional structure and is positioned 3 Å from C-13 of arachidonic acid.[50,51]

Although the results of protein modification, site-directed mutagenesis, and crystallography are consistent with a role for Tyr385 in PGH synthase-1 catalysis, it has been difficult to establish that this residue forms a spectroscopically detectable tyrosyl radical. Mutagenesis of Tyr385 to Phe abolishes cyclooxygenase activity, but the mutant protein still forms EPR-detectable tyrosyl radicals.[126,154] The spectral properties of these radicals are similar to those of the radical formed with wild-type PGH synthase. Replacement of tyrosines in the PGH synthase-1 Tyr385Phe mutant with perdeuterated tyrosines results in a narrowing of the signal and loss of the doublet components of the spectrum, confirming that the radical of wild-type PGH synthase is centered on a tyrosine residue.[126] These observations suggest that either the tyrosyl radical does not participate in the cyclooxygenase reaction, or that in the absence of Tyr385 other tyrosines are oxidized to radicals that are catalytically inactive. The slight differences in EPR spectra of the wild-type and Tyr385Phe radicals are not sufficient to conclude that these radicals arise from the oxidation of different tyrosines.

Two explanations have been put forward to reconcile this apparent paradox: (i) the observed radical does not participate in cyclooxygenase catalysis and is therefore formed with wild-type enzyme or the Tyr385Phe mutant, or (ii) the radical observed with wild-type PGH synthase is derived from Tyr385 but another tyrosine is oxidized when Tyr385 is mutated to Phe. Hypothesis (i) is supported by the fact that another catalytically active mutant, His386Ala, does not produce a spectrally detectable tyrosyl radical.[8] His386Ala and the analogous mutant in PGH synthase-2 (H372A) maintain cyclooxygenase activity.[69,155] Hypothesis (ii) is supported by the demonstration that the tyrosyl radical formed with wild-type enzyme can oxidize arachidonic acid. Tsai *et al.* have reported that, in the absence of O_2, the PGH synthase tyrosyl radical is replaced by a carbon-centered radical upon addition of arachidonic acid.[148,156] Changes in the hyperfine structure of this radical when the tyrosyl radical is reacted with 5,6,8,9,11,12,14,15-octadeuteroarachidonic acid indicate that it is derived from the fatty acid substrate. Thermodynamically, a tyrosyl radical is able to oxidize a polyunsaturated fatty acid. However, it has been reported that tyrosyl radicals produced in the PGH synthase-2 Tyr385Phe mutant do not oxidize arachidonic acid to a radical intermediate.[156] Apparently, the tyrosyl radical(s) generated when this protein is oxidized are not accessible to arachidonic acid. This is supporting evidence that the tyrosyl radicals observed with the wild-type PGH synthases are derived from different tyrosines than those generated in site-directed mutants of Tyr385.

   Despite the attention that has been given to this aspect of PGH synthase catalysis, unequivocal evidence to show that Tyr385 is oxidized to radical intermediates during turnover of these enzymes has remained elusive. The apparent limitation of site-directed mutagenesis in combination with EPR for identification of PGH synthase tyrosyl radicals prompted us to use a different approach to solve this problem. The reaction of tyrosyl radicals with NO to form nitrosocyclohexadienone occurs with other enzymes such as ribonucleotide reductase[157,158] and photosystem II.[159,160] It has been demonstrated that NO also traps the tyrosyl radicals from both PGH synthase-1 and -2.[149,161] However, unlike the ribonucleotide reductase and photosystem II reactions, the reactions between PGH synthase tyrosyl radicals and NO appears irreversible due to the eventual formation of nitrotyrosine (Scheme 8).[149,161] By using the formation of nitrotyrosine as a marker for peptide mapping, the authors' laboratory has demonstrated that Tyr385 of PGH synthase-1 is nitrated during arachidonic acid oxidation in the presence of NO.[161] The same phenomenon has been observed when an organic hydroperoxide is used in place of arachidonic acid. This strongly suggests that Tyr385 is oxidized to a radical intermediate during PGH synthase turnover.

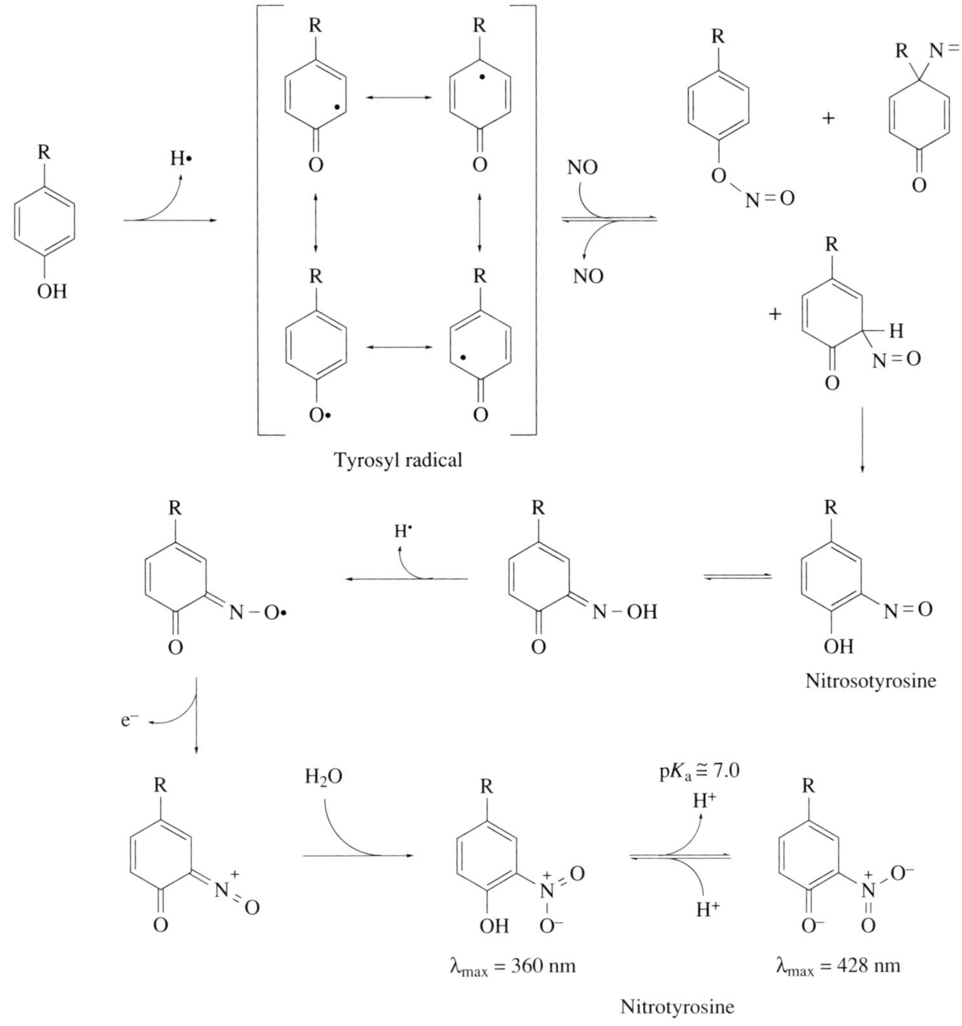

Scheme 8

5.10.5 KINETIC MODELS OF CYCLOOXYGENASE/PEROXIDASE INTEGRATION AND ARACHIDONIC ACID OXYGENATION

   Attempts have been made to integrate the peroxidase and cyclooxygenase activities of PGH synthase into a single mechanistic scheme.[87,162,163] The most common mechanism, first proposed by

Ruf and co-workers, is the branched-chain mechanism (Scheme 9(a)).[141] In general, this mechanism suggests that the peroxidase activity of PGH synthase is required for the activation of cyclo-oxygenase, but upon activation, the cyclooxygenase operates independently of the peroxidase. An alternative mechanism has been proposed by Bakovic and Dunford.[163] These investigators suggest that the peroxidase and cyclooxygenase are tightly coupled (Scheme 9(b)). That is, the peroxidase must reactivate the cyclooxygenase for each turnover. Thus, according to Dunford, PGH synthase is critically dependent on fatty acid hydroperoxide (in particular, PGG_2) not only to initiate arachidonic acid oxygenation, but for continued turnover of the enzyme.

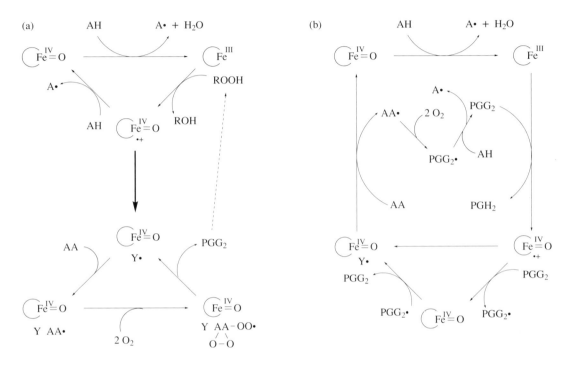

Scheme 9

According to both mechanisms, a tyrosyl radical (Tyr385) is required for hydrogen abstraction from arachidonic acid, and both mechanistic hypotheses postulate a role for the peroxidase in the generation of this tyrosyl radical. The critical difference between the two mechanisms is that in the branched-chain mechanism the tyrosyl radical is regenerated at the end of each catalytic cycle by a reaction of Tyr385 with a PGG_2 peroxyl radical. Thus, once the peroxidase forms the tyrosyl radical, its activity no longer is necessary to regenerate the radical. The tightly coupled mechanism suggests that the PGG_2 peroxyl radical does not oxidize Tyr385 at the end of each cyclooxygenase catalytic cycle, but reacts with a reducing agent in bulk solution. This leaves Try385 fully covalent, and requires that the peroxidase reoxidize it for the next cyclooxygenase catalytic cycle to occur.

Much work has been done to determine the validity of each model. In support of the branched-chain model, it has been noted that a significant amount of PGG_2 accumulates during PGH synthase turnover.[164] This indicates that the cyclooxygenase is able to synthesize PGG_2 in excess of that able to be processed by the peroxidase before its inactivation, and suggests a degree of independence between the peroxidase and cyclooxygenase activities of the enzyme.

As mentioned previously, peroxide scavengers such as glutathione/glutathione peroxidase inhibit cyclooxygenase activity, however, the inhibition is evident as a lag phase and arachidonic acid oxygenation is eventually observed.[124,129] As such, the peroxide scavenger is a much more effective inhibitor of the peroxidase than the cyclooxygenase. Similar effects have been observed with the site-directed mutants of PGH synthase H193A and H386A.[69,70] These substitutions almost completely block the peroxidase activity but only introduce a temporary lag in cyclooxygenase activity. Again, this suggests that once the cyclooxygenase is activated, continuous production of PGG_2 is not required to sustain prostaglandin synthesis.

Kinetic modeling has also been utilized as a means of resolving this mechanistic issue. Wei *et al.* modeled reactions of PGH synthase containing excess concentrations of arachidonic acid and peroxidase reducing substrate.[164] The consumption of arachidonic acid, O_2, and peroxidase reducing substrate as well as the production of PGG_2 and PGH_2 were all modeled. The results indicate that the rates of reaction predicted by the tightly coupled model are far too slow to account for the experimentally observed values. Conversely, the branched-chain mechanism more accurately predicts the experimentally observed rates of arachidonic acid oxygenation, especially when irreversible inactivation of PGH synthase is taken into account. Interestingly, Wei *et al.* found that inclusion of high concentrations (10^{-5} M) of hydroperoxide in the model reaction for the tightly coupled mechanism is sufficient to obtain reasonable rates of arachidonic acid oxygenation. This also disagrees with experimental observation. Addition of high concentrations of a fatty acid hydroperoxide such as 15-HPETE does not stimulate cyclooxygenase activity, but inhibits it by about 50%. Added hydroperoxide in the range 10^{-9}–10^{-6} M is predicted to have no effect on the rate of arachidonic acid oxygenation using the branched-chain model, and only a slight stimulation is predicted at 10^{-5} M hydroperoxide. Thus, the effect of added hydroperoxide is not accurately predicted by either model, but in modeling the effect of the hydroperoxide, a reaction for irreversible inactivation of the enzyme was not included in either model. Even so, it seems that the branched-chain model better predicts experimental observation than the tightly-coupled mechanism.

It is important to point out that the branched-chain mechanism does not accurately predict all aspects of cyclooxygenase catalysis. For example, based on the mechanism shown in Scheme 9(a), one would expect that the presence of a peroxidase reducing substrate would have only an inhibitory effect on arachidonic acid oxygenation. This is clearly not the case. Reducing substrates such as phenol have long been known to dramatically stimulate cyclooxygenase activity.[124,165,166] Only at very high concentrations of these reductants is any inhibition observed.[166] It is important to mention that little is known about the mechanism of PGH synthase irreversible inactivation, and likewise, the mechanism of its prevention by reductants. Delineation of these mechanisms is likely to assist in refinement of the branched-chain mechanism of prostaglandin synthesis by PGH synthase.

5.10.5.1 Alternate Mechanisms for Activation

Penning and co-workers have questioned the involvement of Compound I or Compound II either directly or indirectly in cyclooxygenase activation of PGH synthase.[167] They have shown by presteady-state kinetic analysis, that Compound I and Compound II of PGH synthase-2 appear at earlier time points in a reaction with arachidonic acid than with ethyl hydroperoxide or 5-phenyl-4-pentenylhydroperoxide (PPHP). The widely accepted model for PGH synthase catalysis suggests that the cyclooxygenase is activated through the formation of Compound I by reaction of the ferric enzyme with hydroperoxide. Penning and co-workers assert that if this mechanism is correct, Compound I and Compound II should be observed at an earlier time point with added hydroperoxide than with arachidonic acid which contains only a trace amount of fatty acid hydroperoxide. Interestingly, these investigators also detect what they assign as a transient ferrous enzyme intermediate upon reaction of aspirin acetylated PGH synthase-2 with arachidonic acid.[167] Aspirin-acetylated PGH synthase-2 retains full peroxidase activity and shows a 15-(R)-lipoxygenase activity in place of cyclooxygenase.[152,169] The appearance of the ferrous intermediate coincides with the initiation of oxygen consumption, and as observed with the unmodified enzyme, Compound I and Compound II are detected at much later times when reacted with either ethyl hydroperoxide or PPHP.[167] These studies would seem to indicate that the reaction of ferric PGH synthase with fatty acid hydroperoxides to form ferryl–oxo derivatives is not sufficient to explain the activation of cyclooxygenase turnover. Indeed, Penning and co-workers propose that, upon binding arachidonic acid, ferric PGH synthase reacts with a fatty acid hydroperoxide to form the ferrous intermediate and a peroxyl radical.[167] The peroxyl radical then oxidizes a PGH synthase tyrosine, presumably Tyr385, to initiate the cyclooxygenase catalytic cycle. However, it remains to be explained why the ferrous intermediate is detected with aspirin-acetylated PGH synthase-2 but not the unmodified enzyme. Furthermore, any peroxyl radicals generated by hydroperoxide reduction of the ferric heme would be on the wrong side of the protein to react with Tyr385 and would need to diffuse from the peroxidase active site to the cyclooxygenase active site to do so. It seems unlikely that a peroxyl radical generated in the reducing environment of a cell could survive long enough to do this.

5.10.6 ARACHIDONIC ACID ENTRY, OXYGENATION, AND EXIT FROM PGH SYNTHASE-2

Hypotheses have been advanced to describe the binding of arachidonic acid in the substrate access channel. Picot *et al.* determined from the crystal structure of a PGH synthase-1/(*S*)-flurbiprofen complex that the carboxyl group of the inhibitor is ion-paired to Arg120, the only positively charged residue in the substrate access channel.[49] The involvement of Arg120 in inhibitor binding was confirmed by the structure of the PGH synthase-1/bromoacetylsalicylic acid complex (Figure 12) in which the carboxylate group of salicylate forms a salt bridge with Arg120.[169] Since the carboxylate groups of both (*S*)-flurbiprofen and aspirin interact with Arg120, it was predicted that the carboxylate moiety of arachidonic acid also binds to this residue. Mutagenesis of Arg120 to Lys or Gln confirmed this hypothesis as these substitutions raised the K_m of the protein for arachidonic acid from 4 μM to 87 μM and 3300 μM, respectively.[60]

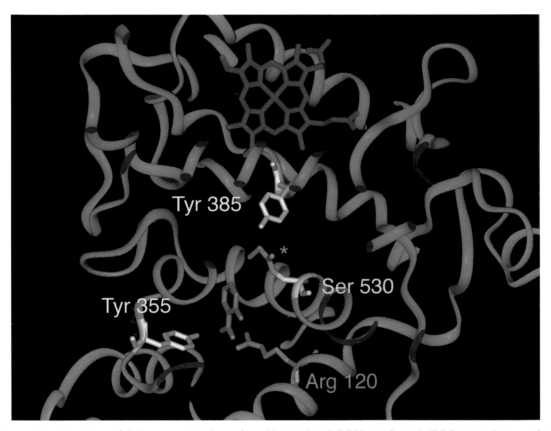

Figure 12 Structure of 2-bromoacetoxybenzoic acid acetylated PGH synthase-1 (PDB accession number 1PTH). Highlighted are the bromoacetyl adduct on Ser530 (in green and denoted with a green *) and the carboxylic acid moiety of salicylic acid (pink) interacting with Arg120.

Analysis of the crystal structure of the PGHS-1/bromoacetylsalicylic acid complex suggests that Tyr355 and Glu524 also play a role in arachidonic acid binding. Tyr355 is located within hydrogen-bonding distance of the carboxylate group of bromoacetylsalicylic acid, and Glu524 forms a salt bridge with Arg120. Interestingly, these three side chains are directed into the channel which reduces the channel volume and confers a constriction near its lower end.

5.10.7 INHIBITION OF CYCLOOXYGENASE ACTIVITY BY NSAIDS

5.10.7.1 Nonselective Inhibition

Vane and co-workers proposed in 1971 that the anti-inflammatory and analgesic properties of NSAIDs are due to inhibition of prostaglandin biosynthesis.[170–172] This was based on the discovery that aspirin or indomethacin inhibited PGE_2 biosynthesis by homogenates of guinea pig lung *in vitro*. An enormous amount of work has confirmed the basic tenets of this hypothesis and identified the cyclooxygenase activity of PGH synthase(s) as the target(s) for inhibition.[124] Among the 25 different NSAIDs currently in worldwide use, only aspirin covalently modifies the PGH synthase protein to inactivate cyclooxygenase activity; all the other NSAIDs bind tightly but noncovalently.[13,173] All NSAIDs inhibit the cyclooxygenase activity of the protein; the peroxidase activity remains unaffected.[174,175] The interaction of most NSAIDs with cyclooxygenase follows a two-step kinetic sequence as shown in Equation (2). The first step involves a rapid interaction between enzyme (E) and inhibitor (I), resulting in a reversible (E·I) complex. The second step is the conversion of the initial (E·I) complex to a form in which the inhibitor is bound more tightly to the enzyme or to one in which the inhibitor forms a covalent bond (for aspirin) with the enzyme as depicted by (E·I)*. Actual kinetic measurements of enzyme–inhibitor association kinetics indicate that the actual mechanism of inhibition may be more complex than the simple two-step mechanism.[176]

The kinetic model originally proposed by Lands classifies NSAIDs into three categories based on their mode of PGH synthase inhibition (Equation (2)).

$$ \mathrm{E} \ + \ \mathrm{I} \ \underset{k_{-1}}{\overset{k_1}{\rightleftharpoons}} \ (\mathrm{E\bullet I}) \ \overset{k_2}{\longrightarrow} \ (\mathrm{E\bullet I})^* \qquad (2) $$

$$ K_i = k_{-1}/k_1 \qquad\qquad K_{\mathrm{inact}} = k_{-1} + k_2/k_1 $$

5.10.7.1.1 *Competitive inhibitors*

This class comprises purely reversible cyclooxygenase inhibitors. Most compounds in this category compete reversibly with arachidonic acid for binding at the cyclooxygenase active site. Examples of this category include ibuprofen, mefenamic acid, flufenamic acid, piroxicam, sulindac sulfide, naproxen, and several phenolic derivatives (Figure 13).[165,177–179]

5.10.7.1.2 *Slow, tight-binding inhibitors*

Slow, tight-binding cyclooxygenase inhibitors exhibit more complex kinetics than simple, competitive inhibitors.[180] Examples of NSAIDs in this category include indomethacin, flurbiprofen, meclofenamate, and diclofenac.[174,175] Substrate analogues like SQ29,535 also display potent cyclooxygenase inhibitory properties.[181] Like competitive inhibitors, they too form the initial, reversible (E·I) complex but this complex is converted to a more stable (E·I)* complex in a time-dependent fashion at an enzyme–inhibitor ratio of one to one.[174,175,182] The (E·I)* complex presumably results from a conformational change in the protein leading to tighter but noncovalent binding of the inhibitor to the protein. *In vitro*, dissociation of the inhibitor from the (E·I)* complex occurs very slowly. *In vivo*, dissociation appears to be more rapid so the duration of inhibition is inversely related to the pharmacokinetics of inhibitor metabolism.[183,184] Although the dynamics and the nature of the interactions that lead to formation of the (E·I)* complex remain poorly understood, it is clear that an inhibitor-induced conformational change leads to formation of the (E·I)* complex.[175,182] Evidence for inhibitor-induced conformational changes is provided by the induction of resistance to limited proteolysis by inhibitor binding.[65,185,186] The conformational change responsible for trypsin resistance is different from the one responsible for tight binding of the inhibitor.

Garavito and co-workers have solved the crystal structures of PGH synthase-1 complexed with NSAIDs such as flurbiprofen,[49] bromoacetylsalicyclic acid,[169] iodoindomethacin,[59] and iodosuprofen.[59] Despite their structural diversity and mechanistic differences, all NSAIDs bind in the substrate access channel with their carboxylate ion-paired to Arg120 as outlined above. The bound drugs prevent access of arachidonate to the active site cavity. That the guanadino moiety of Arg120 ion pairs to the carboxylate of arachidonate and inhibitors has also been verified by site-directed mutagenesis experiments in PGH synthase-1. The Arg120Gln mutant is resistant to inhibition by carboxylic acid-containing NSAIDs[60,187] but remains sensitive to inhibition by noncarboxylic acid-containing inhibitors such as the diarylheterocycles (see below).

Another residue of interest in the PGH synthase-1 structure, particularly in NSAID binding, was identified as Tyr355. As described earlier, the phenolic side chain of Tyr355 resides at the opening

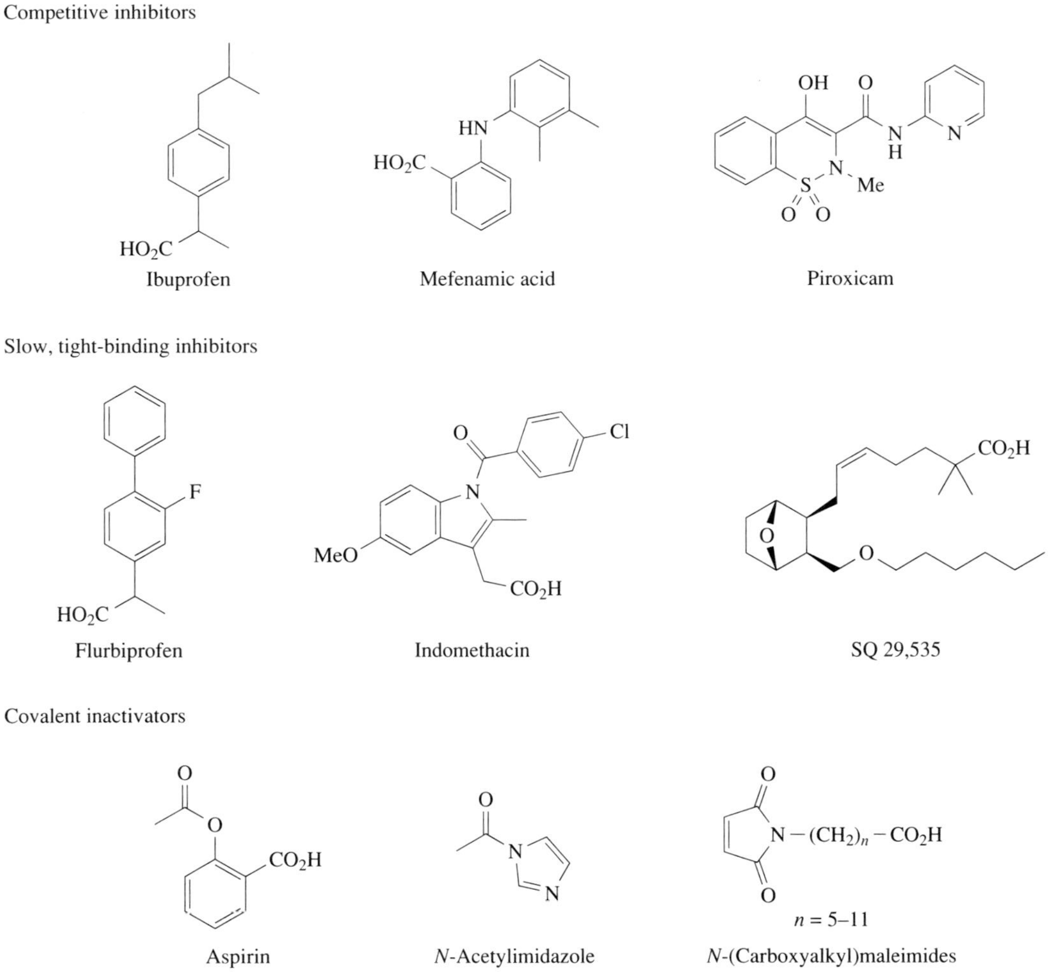

Competitive inhibitors

Slow, tight-binding inhibitors

Covalent inactivators

Figure 13 Inhibitors of PGH synthases.

of the substrate channel opposite Arg120 and together these residues form a narrow constriction at the mouth of the channel. Tyr355 is also responsible for the stereospecificity of inhibition of PGH synthase-1 by 2-(S)-arylpropionates. The Tyr355Phe mutant is inhibited equipotently by both S and R enantiomers, revealing a loss of stereoselective inhibition.[60] These results suggest the imposition of steric constraints by the phenolic OH group in Tyr355 on the R-α-methyl group but not the S-α-methyl enantiomers of 2-arylpropionates and provides an explanation for the stereochemical preference of PGH synthase for the S-enantiomers.

5.10.7.1.3 Covalent inactivators

Aspirin is the only NSAID that covalently modifies PGH synthase-1 and -2 (Equation (3)). The mechanism of cyclooxygenase inhibition involves initial, reversible binding at the active site ($K_i \sim 20$ mM),[174,175] followed by irreversible acetylation of Ser530.[188-190] Acetylation of PGH synthase-1 by aspirin completely inactivates the cyclooxygenase activity but does not inhibit the peroxidase activity.[124] In contrast, aspirin acetylation of PGH synthase-2 does not result in complete loss of oxygenase activity. Instead, acetylated PGH synthase-2 converts arachidonic acid into 15-(R)-hydroxyeicosatetraenoic acid (15-(R)-HETE) instead of PGG$_2$.[168,177] The biological role of 15-(R)-HETE, which differs from the lipoxygenase-derived 15-(S)-HETE in stereochemical configuration, remains undetermined.

$$\text{(structure of aspirin reacting with HO–Enz to form acetylated enzyme)}\qquad(3)$$

Ser530 is not important for cyclooxygenase activity; mutagenesis of this serine to alanine does not affect catalytic activity or arachidonate binding. This suggests that acetylation by aspirin inhibits cyclooxygenase activity by blocking the approach of arachidonate to the catalytic tyrosine in the cyclooxygenase active site channel. Indeed, substitution of bulkier side chains at positions 530 or 516 in the two PGH synthases inhibits arachidonate oxygenation.[168,191] Substitution of larger residues is required to inhibit PGH synthase-2 relative to PGH synthase-1, which suggests that the arachidonate access channel of PGH synthase-2 is a bit larger. This is confirmed by comparison of the crystal structures of the two proteins.[51]

The selective acetylation of a single serine hydroxyl group by aspirin is surprising given the low inherent nucleophilicity of the hydroxyl group relative to other protein nucleophiles. This suggests either that Ser530 is unusually nucleophilic or that acetylation is entropically driven as a result of binding of the salicylate moiety of aspirin in the vicinity of these hydroxyl groups. Both possibilites may be important. Although Ser530 is not part of a charge relay network such as that used by proteolytic enzymes to enhance serine nucleophilicity, it sits in the middle of a very hydrophobic channel, which might increase its reactivity. In addition, Ser530 is located above Arg120, which binds the carboxylate of the salicylate moiety and increases the effective concentration of aspirin near Ser530. Reaction of PGH synthase-1 with *N*-acetylimidazole acetylates several residues, but not Ser530, which underscores the importance of the salicylate binding site in the juxtaposition of aspirin's acetyl group to Ser530.[192] Indeed, crystallization of PGH synthase-1 acetylated with bromoacetylsalicylic acid not only confirms Ser530 acylation but also reveals a salicylate ion-paired to Arg120.[169]

Studies from the authors' laboratory have led to the identification of *N*-(carboxyalkyl)maleimides as covalent modifiers of PGH synthase-1 and -2.[193,194] Inactivation of the cyclooxygenase and peroxidase activities occurs in a biphasic manner with extremely rapid inhibition followed by slow, time-dependent inactivation. The presence of the carboxylic acid moiety is crucial for rapid inactivation, as the corresponding *N*-alkylmaleimides only display time-dependent inhibition. The most potent inhibitor in this class is *N*-(carboxyheptyl)maleimide (IC$_{50}$ ~ 0.1 μM (cyclooxygenase); IC$_{50}$ ~ 3 μM (peroxidase)). Inhibition is extremely sensitive to the chain length of *N*-(carboxyalkyl)maleimides and shortening or increasing the alkyl chain by one methylene unit drastically reduces inhibitory potency. Rapid inactivation of murine and human PGH synthase-2 by *N*-(carboxyheptyl)maleimide is also discernible, albeit at higher concentrations (IC$_{50}$ mouse PGH synthase-2 ~ 4.5 μM; IC$_{50}$ human PGH synthase-2 ~ 14 μM). The corresponding *N*-(carboxyheptyl)succinimide derivative does not display cyclooxygenase or peroxidase inhibition, suggesting that covalent modification is critical for rapid as well as time-dependent inhibition. Although labeling of *holo*- and *apo*-PGH synthase-1 with [^{14}C]-*N*-(carboxyheptyl)maleimide results in incorporation of radioactivity, the modified amino acid residue(s) is unstable to the conditions of peptide mapping and the site of covalent modification has not been conclusively identified.

5.10.7.2 Development of Selective Inhibitors of PGH Synthase-2

The discovery of the inducible isozyme PGH synthase-2 has led to initiation of substantial drug discovery programs committed to the identification of selective PGH synthase-2 inhibitors. The driving force for these efforts is the observation that PGH synthase-2 is expressed in macrophages and brain but not in the gastrointestinal tract. Thus, PGH synthase-2 inhibitors should be anti-inflammatory and analgesic without inducing the gastrointestinal side effects typical of currently available NSAIDs. Early efforts at drug discovery focused on the modification of two leads compounds, DuP 697 (5-bromo-2(4-fluorophenyl)-3-(4-methylsulfonyl)thiophenol) and NS-398 (N-(2-cyclohexyloxy-4-nitrophenyl)methane sulfonamide) (Figure 14).[195,196] Structurally, these compounds differ greatly from the current NSAIDs, most of which are arylacetic acid or anthranilic acid derivatives. These studies have resulted in an impressive array of potent and selective PGH synthase-2 inhibitors, most of which have IC$_{50}$ values for PGH synthase-2 inhibition in the sub-nanomolar range and selectivity ratios toward PGH synthase-2 ranging from 1000 to 10 000. Some of these compounds are well advanced in human clinical trials. All of these compounds inhibit PGH synthase-2 in a time-dependent manner, typical of slow, tight-binding inhibitors. The basis for selective PGH synthase-2 inhibition is, in fact, derived from the slow, time-dependent phase of binding to PGH synthase-2; all of these agents retain their ability to inhibit PGH synthase-1 in a competitive fashion. A brief description of some of the reported structural classes is included below.

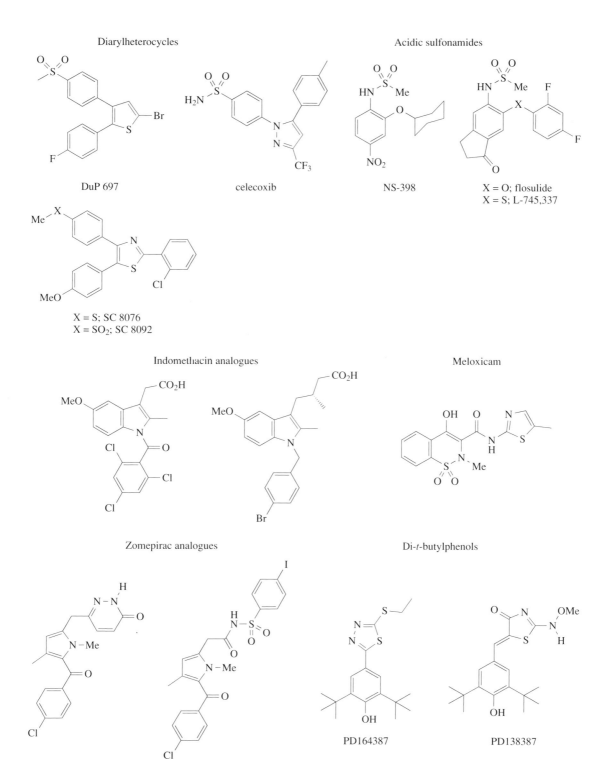

Figure 14 Selective inhibitors of PGH synthase-2.

5.10.7.2.1   Diarylheterocycles

The common structural motif in this class includes a *cis*-stilbene moiety and the presence of a 4-methylsulfonyl or sulfonamide substituent in one of the pendant phenyl rings. The oxidation state of the sulfur is a very important determinant for selectivity. For example, conversion of the methylsulfonyl moiety in SC58092, a PGH synthase-2 selective inhibitor, to a methylsulfide functionality affords SC58076, a time-dependent PGH synthase-1 selective inhibitor. The ring system that is fused to the stilbene framework has been extensively manipulated to include every imaginable heterocyclic and carbocyclic skeleton (reviewed by Prasit and Riendeau).[197] In this series, celecoxib, a 1,5-diarylpyrazole, is currently undergoing phase III clinical evaluation.[198] Celecoxib has been reported to exhibit a 325-fold selectivity *in vitro*, and exhibits efficacy in a postdental pain model of analgesia and in the treatment of the symptoms of osteoarthritis.

5.10.7.2.2   Acidic sulfonamide derivatives

Structural modification of flosulide and NS-398 led to the identification of L-745,337 as a potent and selective PGH synthase-2 inhibitor (IC_{50} (PGH synthase-1) $> 10\ \mu M$; IC_{50} (PGH synthase-2) ~ 50 nM).[199] A detailed structural analysis of this class is provided by Li *et al.*[200] Continued interest in this class, most notably by Japanese groups, has led to the identification of T-614 and FR115068 as selective PGH synthase-2 inhibitors with potent antiinflammatory properties and low ulcerogenicity profiles.[201,202]

5.10.7.2.3   Di-t-butylphenols

The Parke-Davis group has disclosed a new class of selective PGH synthase-2 inhibitors, comprising di-*t*-butylphenol derivatives, notable examples of which include PD164387 and PD138387.[203] Both analogues were identified to be potent and selective PGH synthase-2 inhibitors *in vitro* (PD 164387: IC_{50} PGH synthase-1 $> 100\ \mu M$, PGH synthase-2 $\sim 0.14\ \mu M$; PD138387: IC_{50} PGH synthase-1 $> 100\ \mu M$, PGH synthase-2 $\sim 0.98\ \mu M$) and orally active *in vivo*.

5.10.7.2.4   Modification of nonselective NSAIDs to selective PGH synthase-2 inhibitors

Structural modification of two nonselective NSAIDs, zomepirac and indomethacin, into selective PGH synthase-2 inhibitors has also been recently reported. Unlike diarylheterocycles and acidic sulfonamides, these compounds do not contain the methylsulfonyl or the methylsulfonamide moiety. Thus, replacement of the chlorobenzoyl moiety in indomethacin with a trichlorobenzoyl functionality or reduction of the chlorobenzoyl group to the corresponding chlorobenzyl substituent gave highly selective and potent PGH synthase-2 inhibitors.[204] Similarly, derivatization of the carboxylate in zomepirac by a pyridazinone moiety afforded a highly selective series of PGH synthase-2 inhibitors.[51] Most of the lead compounds were as potent as the parent NSAID, but were devoid of gastrointestinal toxicity even at dosages exceeding their therapeutic efficacy. Structure–activity studies on the enol-carboxamide class of NSAIDs which includes piroxicam led to the discovery of meloxicam as a selective PGH synthase-2 inhibitor with a selectivity ratio ranging from 3 to 50 depending on the type of inhibition assays employed.[205–207]

5.10.7.2.5   Selective covalent inactivators of PGH synthase-2

Aspirin is the only example of an NSAID that covalently modifies PGH synthase. Unfortunately, aspirin displays a greater potency against PGH synthase-1 than PGH synthase-2 which undoubtedly contributes to its ulcerogenic activity. Furthermore, aspirin-acetylated PGH synthase-2 retains some oxygenase activity although it does not make PGG_2, suggesting that its cyclooxygenase active site is larger than that of PGH synthase-1. This result is consistent with the broader fatty acid substrate specificity of PGH synthase-2.[208] The presumption of a larger active site in PGH synthase-2 led to the evaluation of a series of acyl salicylates with side chains larger than acetate as potential selective

Acyl salicylates

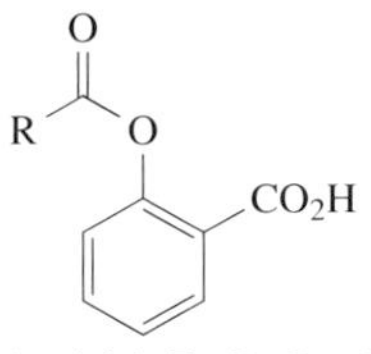

R = Me (aspirin), Et, Pr, Bu, Pen, Hex

2-(Acyloxyphenyl)alkyl sulfide analogues

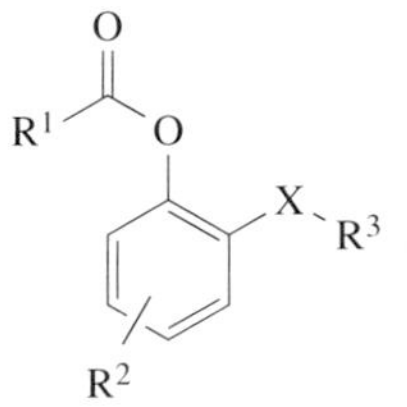

R^1 = Me, Et, CF_3, NH_2
R^2 = H or F
R^3 = alkyl, alkenyl, alkynyl, cycloalkyl, aryl
X = S, Se, CH_2, O, NMe

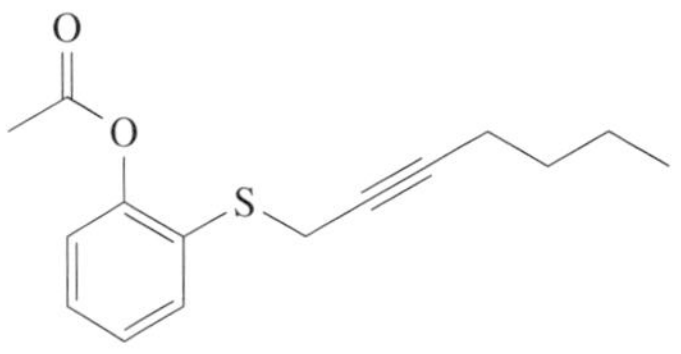

2-(Acetoxyphenyl)hept-2-ynyl sulfide

Figure 15 Selective covalent inactivators of PGH synthases.

inactivators (Figure 15). None of the compounds exhibited selective inhibition of PGH synthase-2.[209]

The authors' laboratory undertook a different approach to the discovery of PGH synthase-2 acetylators.[210] The carboxylate moiety in aspirin was replaced with the alkylsulfide or alkylsulfone moieties, which are required for isozyme specificity in the diarylheterocycle series. The most potent compound in the series that was prepared, 2-(acetoxyphenyl)hept-2-ynyl sulfide, was 60-fold more reactive against PGH synthase-2 than aspirin and 100-fold more selective (see Figure 15). Structure–activity studies indicate that the oxidation state of the sulfur is very crucial for inhibition; the sulfoxide or sulfone derivatives are inferior to the sulfide in potency. This effect is opposite to that observed in the diarylheterocyle series of noncovalent inhibitors. Furthermore, changes in the heteroatom from sulfur to oxygen, nitrogen, or carbon results in less potent compounds. As in earlier studies on N-substituted maleimides, inhibitory potency is dependent on the chain length of the alkyl group attached to sulfur and inclusion of a triple bond in the optimized chain length (C-7) enhances potency and selectivity.

Peptide mapping studies indicate that 2-(acetoxyphenyl)hept-2-ynyl sulfide acetylates the same serine residue (Ser530) as aspirin but site-directed mutagenesis studies reveal that the molecular basis for selective modification of Ser530 is different from the events that lead to the modification of Ser530 by aspirin. Furthermore, residues that are key determinants of the binding of diaryl-heterocycles are not important in the acetylation of Ser530 by the acetoxyphenylalkyl sulfides. This is the first description of a selective covalent modifier of PGH synthase-2 and site-directed mutagenesis studies strongly suggest that these compounds bind to previously uncharacterized domains of the protein, which can be exploited in the development of selective PGH synthase-2 inhibitors.

5.10.7.3 Structural Basis for Selective PGH Synthase-2 Inhibition

The solution of the crystal structures of human and murine PGH synthase-2 with bound inhibitors was reported by the Roche and the Monsanto groups in 1996.[50,51] The Roche group disclosed two human PGH synthase-2 structures complexed with novel PGH synthase-2-selective inhibitors that were derived from modification of the nonselective PGH synthase inhibitor, zomepirac (Figure 14).[51] Preliminary comparison of the crystal structures of sheep PGH synthase-1 and human PGH synthase-2 revealed that the structures are highly homologous, particularly in the cyclooxygenase active sites (Figure 16). The two zomepirac analogues bind in the same site in PGH synthase-2 as the classical NSAIDs do in PGH synthase-1. The nitrogen and oxygens of the acylsulfonamide moiety in the *p*-(iodophenyl)sulfonamide derivative ion pair to Arg120 and the chlorophenyl ring is positioned near Tyr385. The latter is very similar to the position of the chlorophenyl ring in indomethacin and the 4-phenyl ring in flurbiprofen complexed to PGH synthase-1. Interestingly, the *p*-iodophenyl ring extends past the constriction site between Arg120 and Tyr355 and resides just at the top of the membrane binding helices. The structure around the narrow constriction remains relatively undisturbed by the extension of the *p*-iodophenyl moiety.

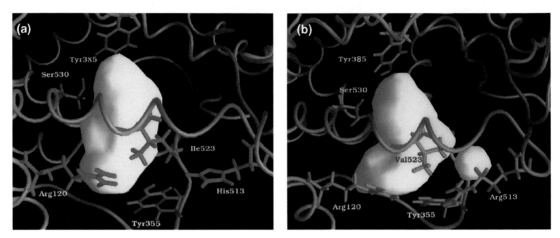

Figure 16 Comparison of the accessible volume of the NSAID binding sites of PGH synthase-1 (a) and PGH synthase-2 (b) (reproduced by permission of *Nature Struct. Biol.*, 1996, **3**, 927, © Nature Publishing Co.).

Although the structural basis for selective PGH synthase-2 inhibition by the acylsulfonamide is not evident from the cocrystal structure, the interactions of the smaller pyridazinone inhibitor with PGH synthase-2 reveal dramatic differences between the inhibitor-bound enzymes. Although the agent binds in the same general region as the acylsulfonamide in PGH synthase-2 and flurbiprofen in PGH synthase-1, the binding of the pyridazinone leads to major conformational perturbations in Glu524, Arg120, and most of the membrane binding domain.[51] As discussed above, structures of PGH synthase-2 complexed with the acylsulfonamide or PGH synthase-1 with NSAIDs reveals a hydrogen bonding network between Arg120, Glu524, Tyr355, and the inhibitors, which causes the constriction site to close down on the bound inhibitor. In contrast, the structure of PGH synthase-2 complexed to the pyridazinone reveals a widening of the constriction site. The pyridazinone moiety disrupts the hydrogen bonding network and causes the helix that forms part of the active site to unwind, as a result of which, Arg120 points away from the drug and salt bridges to two backbone carbonyl oxygens. These results unambiguously demonstrate that the membrane binding domain undergoes significant conformational changes upon inhibitor binding and these alterations may determine the structural and molecular basis for selective inhibition for some compounds.

The Monsanto group disclosed the crystal structures of native murine PGH synthase-2 and murine PGH synthase-2 bound with flurbiprofen, indomethacin, or SC-558, a selective PGH synthase-2 inhibitor belonging to the diarylheterocycle class.[50] The binding of flurbiprofen or indomethacin to PGH synthase-2 is very similar to their binding in PGH synthase-1 which prevents fatty acid binding to Arg120. SC-558, a 1,5-diarylpyrazole derivative, also binds in the cyclooxygenase active site. Several structural features of SC-558 binding to PGH synthase-2 are common to the structure of PGH synthase-2 bound to flurbiprofen or indomethacin. For example, the *p*-bromophenyl ring in SC-558 occupies a similar region as the *p*-phenyl moiety in flurbiprofen and the *p*-chlorophenyl group in indomethacin; the pyrazole portion of SC-558 a similar space as the fluorophenyl ring of flurbiprofen; and the carboxylate of flurbiprofen and indomethacin bind in the same region of the protein as the trifluoromethyl group in SC-558. The major difference between the binding of SC-558 and the arylacetic acids is the insertion of the phenylsulfonamide of SC-558 into a hydrophobic side pocket that extends off the main cyclooxygenase channel in PGH synthase-2 (Figure 17). The sulfonamide portion extends beyond the hydrophobic side pocket and one of the oxygen atoms on the sulfonamide forms a hydrogen bond to Arg513, whereas the other oxygen is linked by a hydrogen bond to His90. The sulfonamide nitrogen is hydrogen-bonded to the carbonyl group of Phe518. Arg513 in PGH synthase-2 is replaced by histidine in PGH synthase-1 and superposition of the two enzymes suggests that the imidazole ring of this histidine does not interact with the sulfonamide group in PGH synthase-1. These results explain the critical requirement of the sulfonyl moiety of SC-558 for selective PGH synthase-2 inhibition.

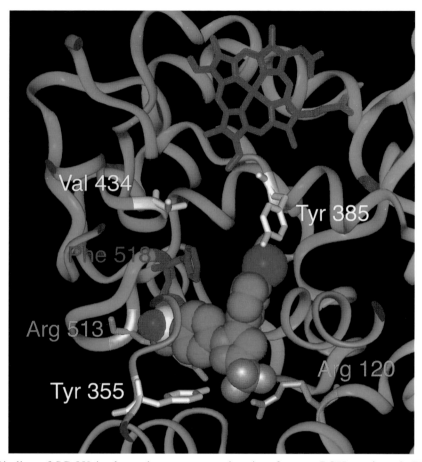

Figure 17   Binding of SC-558 in the cyclooxygenase active site of mouse PGH synthase-2 (PBD accession 1CX2).

The structural basis for the PGH synthase-2 selectivity of SC-558 arises from binding of the sulfonamide moiety in the hydrophobic side pocket. The side pocket is more accessible to inhibitors in PGH synthase-2 primarily because of two valine residues at positions 523 and 434, which are isoleucines in PGH synthase-1. The smaller size of these valine residues allows the phenylsulfonamide moiety to insert into the side pocket. Although a similar side pocket is discernible in PGH synthase-1, the bulkier isoleucines at positions 523 and 434 provide steric hindrance to entry by the diaryl-heterocycles. Site-directed mutagenesis studies have shown that the Val523Ile mutant as well as the Val523Ile/Arg513His double mutant and the Val434Ile/Arg513His/Val523Ile triple mutant of PGH synthase-2 are resistant to inhibition by diarylheterocycles and the acidic sulfonamides but not 2-arylpropionate-type NSAIDs.[61–63] The Monsanto group disclosed the structure of a cocrystal of PGH synthase-2 complexed with NS-398.[211] Interestingly, the sulfonamide moiety of NS-398 does not bind in the side pocket as do the diarylheterocycles; rather, it ion pairs with Arg120 and the *O*-cyclohexyl group extends into the hydrophobic side pocket. However, the *p*-nitrosulfonamido moiety of NS-398 is highly acidic relative to the sulfonamide group of SC-558 and is likely to be ionized at physiological pH.

Comparison of the crystal structures of native PGH synthase-2 and PGH synthase-2 complexed to inhibitors does not reveal any significant conformational changes that are expected by the biphasic kinetics of inhibition. Therefore, the structural features that account for the selective time-dependent inhibition must be subtle and involve dynamic events between the inhibitors and the protein occurring in the cyclooxygenase channel. Inhibitor binding does not greatly perturb the resting structure. The narrow constriction site between Arg120 and Tyr355 is slightly more restricted and Arg120 still retains the salt bridge with Glu524 in most of the structures. Overall, these results suggest that the constriction must widen in order to accommodate bulky inhibitors such as indomethacin and then close down on the inhibitor. Although models for these events have been proposed, the exact mechanism remains poorly understood.[176,212,213]

5.10.8   STEROID INHIBITION OF PGH SYNTHASE-2 EXPRESSION

Glucocorticoids are powerful anti-inflammatory agents *in vivo* but they do not inhibit cyclo-oxygenase activity *in vitro*. Multiple mechanisms have been proposed to relate glucocorticoid action to alterations in arachidonic acid metabolism. There is some evidence that glucocorticoids control arachidonic acid release by regulating the levels of both secretory PLA_2 and cytosolic PLA_2.[214–216] But there is even stronger evidence that glucocorticoids attenuate prostaglandin levels directly by regulating PGH synthase expression. The discovery that PGH synthase-2 is an immediate early gene suggested that steroids may affect the levels of the enzyme because they are powerful transcriptional regulators.[48] Evidence for this has been provided by DeWitt and Meade[217] who showed that glucocorticoids inhibit serum-stimulated PGH synthase-2 transcription by 70%, leading to a con-comitant 70% reduction in PGH synthase-2 levels. However, a steroid response element that could account for transcriptional inhibition has not been found in the 5′ regulatory region of the PGH synthase-2 gene. This suggests that glucocorticoids may act by an indirect mechanism involving interaction of glucocorticoid receptors with other transcriptional elements. Alternatively, steroids may act by stimulation of expression of another gene that is the actual negative regulator of PGH synthase-2 transcription or enhancer of mRNA turnover.[42]

5.10.9   CONCLUSIONS

The study of arachidonic acid oxygenation and its pharmacological manipulation has been going on for a very long time. The biological activities of prostaglandins were first reported in the 1920s, their structures were elucidated in the 1950s, and inhibition of their formation was identified as the mode of action of NSAIDs in the 1960s. The latter finding linked the study of arachidonic acid metabolism to the natural product chemistry and pharmacology of salicylates, which dates to ancient Egypt. Viewed from this perspective, it is remarkable to note the progress that has been made in the mid 1990s in our understanding of the structure, function, and inhibition of PGH synthases. A coherent picture is emerging of the biochemical and kinetic mechanism of oxygenation of arachidonic acid by this protein, the structural basis for selective inhibition, and the regulation of enzyme levels and activity. The authors look forward to using that knowledge to generate novel pharmaceutical agents to treat pain and inflammation coincident with the new millennium.

However, much remains to be learned about the physiological and pathophysiological functions of the two different PGH synthases and the molecular mechanisms by which these functions are acheived. For example, it is still not understood why there are multiple forms of this enzyme or why both PGH synthases exist in the same cell type. Many cells that express PGH synthase-2 following stimulation with growth factors, cytokines, etc., already express detectable levels of PGH synthase-1. Despite the fact that the levels of PGH synthase-2 may increase dramatically, the actual stimu-lation of prostaglandin biosynthesis may only be by a factor of two or three. Why does the cell go to the trouble of making a new enzyme to double or triple the level of eicosanoids?

Reddy and Herschman have proposed one solution to this riddle which relates to the utilization of different pools of arachidonic acid by the two enzymes. They find that in mast cells arachidonic acid released by signal transduction and cytosolic phospholipase A_2 activation is channeled to PGH synthase-2, whereas arachidonic acid released by secretory phospholipase A_2 is utilized by PGH synthase-1.[218] Does the cell channel arachidonic acid to one of two enzymes located in the same cellular compartment by using specific protein molecules or does this differential substrate utilization reflect differences in the kinetics of arachidonic acid binding or cyclooxygenase activation by the two different isoforms?

Do the two different PGH synthases channel their endoperoxide product PGH_2 to different metabolizing enzymes in the same cell so that each PGH synthase may produce a different metabolic product? Do PGH synthases participate in protein–protein associations with other molecules unrelated to the arachidonic acid cascade? Are there functions for PGH synthases unrelated to their ability to oxygenate arachidonic acid or reduce fatty acid hydroperoxides? With regard to inhibition of PGH synthases, how much selectivity for PGH synthase-2 must a compound exhibit in an *in vitro* or cellular assay to be an effective anti-inflammatory and analgesic agent without inducing gastric side effects? Is the slow reversibility of inhibition of PGH synthases by slow, tight-binding inhibitors *in vitro* reflected in the kinetics of inhibitor dissociation *in vivo* or is loss of inhibition more closely related to pharmacokinetics of inhibitor metabolism and excretion? Are there PGH synthases-3, 4, 5, etc.? Given the speed with which questions are asked and answered in contemporary biomedical science, the field of arachidonic acid oxygenation promises to be a dynamic and exciting one for years to come.

5.10.10 REFERENCES

1. J. H. Capdevila, J. R. Falck, and R. W. Estabrook, *FASEB J.*, 1992, **6**, 731.
2. S. Yamamoto, *Biochim. Biophys. Acta*, 1992, **1128**, 117.
3. M. Hamberg, J. Svensson, T. Wakabayashi, and B. Samuelsson, *Proc. Natl. Acad. Sci. USA*, 1974, **71**, 345.
4. D. H. Nugteren and E. Hazelhof, *Biochim. Biophys. Acta*, 1973, **326**, 448.
5. S. Ohki, N. Ogino, S. Yamamoto, and O. Hayaishi, *J. Biol. Chem.*, 1979, **254**, 829.
6. W. R. Pagels, R. J. Sachs, L. J. Marnett, D. L. DeWitt, J. S. Day, and W. L. Smith, *J. Biol. Chem.*, 1983, **258**, 6517.
7. H. R. Herschman, *Biochim. Biophys. Acta Lipids Lipid Metab.*, 1996, **1299**, 125.
8. W. L. Smith, R. M. Garavito, and D. L. DeWitt, *J. Biol. Chem.*, 1996, **271**, 33 157.
9. T. S. Brannon, A. J. North, L. B. Wells, and P. W. Shaul, *J. Clin. Invest.*, 1994, **93**, 2230.
10. I. Chakraborty, S. K. Das, J. Wang, and S. K. Dey, *J. Mol. Endocrinol.*, 1996, **16**, 107.
11. P. Toth, X. Li, Z. M. Lei, and Ch. V. Rao, *J. Clin. Endocrinol. Metab.*, 1996, **81**, 1283.
12. M. Hamberg, J. Svensson, and B. Samuelsson, *Proc. Natl. Acad. Sci. USA*, 1975, **72**, 2994.
13. G. J. Roth and P. W. Majerus, *J. Clin. Invest.*, 1975, **56**, 624.
14. A. L. Willis, *Prostaglandins*, 1974, **5**, 1.
15. M. J. S. Langman, J. Weil, P. Wainwright, D. H. Lawson, M. D. Rawlins, R. F. A. Logan, M. Murphy, M. P. Vessey, and D. G. Colin-Jones, *Lancet*, 1994, **343**, 1075.
16. D. A. Kujubu, B. S. Fletcher, B. C. Varnum, R. W. Lim, and H. R. Herschman, *J. Biol. Chem.*, 1991, **266**, 12 866.
17. L. J. Crofford, R. L. Wilder, A. P. Ristimäki, H. Sano, E. F. Remmers, H. R. Epps, and T. Hla, *J. Clin. Invest.*, 1994, **93**, 1095.
18. L. Pang and A. J. Knox, *Br. J. Pharmacol.*, 1997, **121**, 579.
19. J. Hendel and O. H. Nielsen, *Am. J. Gastroenterol.*, 1997, **92**, 1170.
20. F. Beiche, S. Scheuerer, K. Brune, G. Geisslinger, and M. Goppelt-Struebe, *FEBS Lett.*, 1996, **390**, 165.
21. G. A. Kune, S. Kune, and L. F. Watson, *Cancer Res.*, 1988, **48**, 4399.
22. M. J. Thun, M. M. Namboodiri, and C. W. Heath, Jr., *N. Engl. J. Med.*, 1991, **325**, 1593.
23. T. Narisawa, H. Kusaka, Y. Yamazaki, M. Takahashi, H. Koyama, K. Koyama, Y. Fukaura, and A. Wakizaka, *Dis. Colon Rectum*, 1990, **33**, 840.
24. C. E. Eberhart, R. J. Coffey, A. Radhika, F. M. Giardiello, S. Ferrenbach, and R. N. DuBois, *Gastroenterology*, 1994, **107**, 1183.
25. S. L. Kargman, G. P. O'Neill, P. J. Vickers, J. F. Evans, J. A. Mancini, and S. Jothy, *Cancer Res.*, 1995, **55**, 2556.
26. J. E. Dinchuk, B. D. Car, R. J. Focht, J. J. Johnston, B. D. Jaffee, M. B. Covington, N. R. Contel, V. M. Eng, R. J. Collins, P. M. Czerniak, S. A. Gorry, and J. M. Trzaskos, *Nature*, 1995, **378**, 406.
27. S. G. Morham, R. Langenbach, C. D. Loftin, H. F. Tiano, N. Vouloumanos, J. C. Jennette, J. F. Mahler, K. D. Kluckman, A. Ledford, C. A. Lee, and O. Smithies, *Cell*, 1995, **83**, 473.
28. R. Langenbach, S. G. Morham, H. F. Tiano, C. D. Loftin, B. I. Ghanayem, P. C. Chulada, J. F. Mahler, C. A. Lee, E. H. Goulding, K. D. Kluckman, H. S. Kim, and O. Smithies, *Cell*, 1995, **83**, 483.
29. Y. Sugimoto, A. Yamasaki, E. Segi, K. Tsuboi, Y. Aze, T. Nishimura, H. Oida, N. Yoshida, T. Tanaka, M. Katsuyama, K. Hasumoto, T. Murata, M. Hirata, F. Ushikubi, M. Negishi, A. Ichikawa, and S. Narumiya, *Science*, 1997, **277**, 681.
30. C. D. Funk, L. B. Funk, M. E. Kennedy, A. S. Pong, and G. A. FitzGerald, *FASEB J.*, 1991, **5**, 2304.
31. C. Yokoyama and T. Tanabe, *Biochem. Biophys. Res. Commun.*, 1989, **165**, 888.
32. L.-H. Wang, A. Hajibeigi, X.-M. Xu, D. Loose-Mitchell, and K. K. Wu, *Biochem. Biophys. Res. Commun.*, 1993, **190**, 406.
33. H. Inoue, C. Yokoyama, S. Hara, Y. Tone, and T. Tanabe, *J. Biol. Chem.*, 1995, **270**, 24 965.
34. X. M. Xu, J. L. Tang, X. M. Chen, L. H. Wang, and K. K. Wu, *J. Biol. Chem.*, 1997, **272**, 6943.
35. D. A. Jones, D. P. Carlton, T. M. McIntyre, G. A. Zimmerman, and S. M. Prescott, *J. Biol. Chem.*, 1993, **268**, 9049.
36. T. Hla and K. Neilson, *Proc. Natl. Acad. Sci. USA*, 1992, **89**, 7384.
37. S. B. Appleby, A. Ristmäki, K. Neilson, K. Narko, and T. Hla, *Biochem. J.*, 1994, **302**, 723.
38. T. Kosaka, A. Miyata, H. Ihara, S. Hara, T. Sugimoto, O. Takeda, E. Takahashi, and T. Tanabe, *Eur. J. Biochem.*, 1994, **221**, 889.
39. X. H. Yang, F. X. Hou, L. Taylor, and P. Polgar, *Biochim. Biophys. Acta Gene Struct. Expression*, 1997, **1350**, 287.

40. J. K. Morris and J. S. Richards, *J. Biol. Chem.*, 1996, **271**, 16 633.
41. S. K. Srivastava, T. Tetsuka, D. Daphna-Iken, and A. R. Morrison, *Am. J. Physiol.*, 1994, **36**, F504.
42. A. Ristimaki, S. Garfinkel, J. Wessendorf, T. Maciag, and T. Hla, *J. Biol. Chem.*, 1994, **269**, 11 769.
43. A. Ristimaki, K. Narko, and T. Hla, *Biochem. J.*, 1996, **318**, 325.
44. J. P. Merlie, D. Fagan, J. Mudd, and P. Needleman, *J. Biol. Chem.*, 1988, **263**, 3550.
45. C. Yokoyama, T. Takai and T. Tanabe, *FEBS Lett.*, 1988, **231**, 347.
46. W. Xie, J. Chipman, D. L. Robertson, and D. L. Simmons, *FASEB J.*, 1991, **5**, A475 (Abstract).
47. D. A. Kujubu, S. T. Reddy, B. S. Fletcher, and H. R. Herschman, *J. Biol. Chem.*, 1993, **268**, 5425.
48. M. K. O'Banion, H. B. Sadowski, V. Winn, and D. A. Young, *J. Biol. Chem.*, 1991, **266**, 23 261.
49. D. Picot, P. J. Loll, and R. M. Garavito, *Nature*, 1994, **367**, 243.
50. R. G. Kurumbail, A. M. Stevens, J. K. Gierse, J. J. McDonald, R. A. Stegeman, J. Y. Pak, D. Gildehaus, J. M. Miyashiro, T. D. Penning, K. Seibert, P. C. Isakson, and W. C. Stallings, *Nature*, 1996, **384**, 644.
51. C. Luong, A. Miller, J. Barnett, J. Chow, C. Ramesha, and M. F. Browner, *Nature Struct. Biol.*, 1996, **3**, 927.
52. G. Xiao, W. Chen, and R. J. Kulmacz, *J. Biol. Chem.*, 1998, **273**, 6801.
53. W. E. M. Lands and B. Samuelsson, *Biochim. Biophys. Acta*, 1968, **164**, 426.
54. H. Vonkeman and D. A. Van Dorp, *Biochim. Biophys. Acta*, 1968, **164**, 430.
55. C. Galli, G. Galli, and G. Porcellati, in "Phospholipases and Prostaglandins. Advances in Prostaglandin and Thromboxane Research," Raven Press, New York, 1978, Vol. 3.
56. I. Song and W. L. Smith, *Arch. Biochem. Biophys.*, 1996, **334**, 67.
57. I. Morita, M. Schindler, M. K. Regier, J. C. Otto, T. Hori, D. L. DeWitt, and W. L. Smith, *J. Biol. Chem.*, 1995, **270**, 10 902.
58. M. K. Regier, D. L. DeWitt, M. S. Schindler, and W. L. Smith, *Arch. Biochem. Biophys.*, 1993, **301**, 439.
59. P. J. Loll, D. Picot, O. Ekabo, and R. M. Garavito, *Biochemistry*, 1996, **35**, 7330.
60. D. K. Bhattacharyya, M. Lecomte, C. J. Rieke, R. M. Garavito, and W. L. Smith, *J. Biol. Chem.*, 1996, **271**, 2179.
61. J. K. Gierse, J. J. McDonald, S. D. Hauser, S. H. Rangwala, C. M. Koboldt, and K. Seibert, *J. Biol. Chem.*, 1996, **271**, 15 810.
62. Q. P. Guo, L. H. Wang, K. H. Ruan, and R. J. Kulmacz, *J. Biol. Chem.*, 1996, **271**, 19 134.
63. E. Wong, C. Bayly, H. L. Waterman, D. Riendeau, and J. A. Mancini, *J. Biol. Chem.*, 1997, **272**, 9280.
64. Q. Guo, L.-H. Wang, K.-H. Ruan, and R. J. Kulmacz, *J. Biol. Chem.*, 1996, **271**, 19 134.
65. Y.-N. P. Chen, M. J. Bienkowski, and L. J. Marnett, *J. Biol. Chem.*, 1987, **252**, 16 892.
66. B. G. Titus and W. E. M. Lands, *Methods Enzymol.*, 1982, **86**, 69.
67. J. Sirois and J. S. Richards, *J. Biol. Chem.*, 1992, **267**, 6382.
68. Q. P. Guo, S. L. Chang, L. Diekman, G. H. Xiao, and R. J. Kulmacz, *Arch. Biochem. Biophys.*, 1997, **344**, 150.
69. T. Shimokawa and W. L. Smith, *J. Biol. Chem.*, 1991, **266**, 6168.
70. L. M. Landino, B. C. Crews, J. K. Gierse, S. D. Hauser, and L. J. Marnett, *J. Biol. Chem.*, 1997, **272**, 21 565.
71. J. H. Capdevila, J. D. Morrow, Y. Y. Belosludtev, D. R. Beauchamp, R. N. DuBois, and J. R. Falck, *Biochemistry*, 1995, **34**, 3325.
72. R. J. Kulmacz and L. H. Wang, *J. Biol. Chem.*, 1995, **270**, 24 019.
73. L. J. Marnett and G. A. Reed, *Biochemistry*, 1979, **18**, 2923.
74. L. J. Marnett and K. Maddipati, in "Peroxidases in Chemistry and Biology," eds. J. Everse, K. E. Everse, and M. B. Grisham, CRC Press, Boca Raton, FL, 1991, Vol. 1, p. 293.
75. J. Zeng and R. E. Fenna, *J. Mol. Biol.*, 1992, **226**, 185.
76. M. Hamberg and B. Samuelsson, *J. Biol. Chem.*, 1967, **242**, 5336.
77. B. Samuelsson, *J. Am. Chem. Soc.*, 1965, **87**, 3011.
78. J. Schreiber, T. E. Eling, and R. P. Mason, *Arch. Biochem. Biophys.*, 1986, **249**, 126.
79. N. A. Porter and M. O. Funk, *J. Org. Chem.*, 1975, **40**, 3614.
80. W. A. Pryor and J. P. Stanley, *J. Org. Chem.*, 1975, **40**, 3615.
81. N. A. Porter, *Acc. Chem. Res.*, 1986, **19**, 262.
82. C. M. Markey, A. Alward, P. E. Weller, and L. J. Marnett, *J. Biol. Chem.*, 1987, **262**, 6266.
83. A. M. Lambeir, C. M. Markey, H. B. Dunford, and L. J. Marnett, *J. Biol. Chem.*, 1985, **260**, 14 894.
84. T. L. Poulos, S. T. Freer, R. A. Alden, S. L. Edwards, U. Skogland, K. Takio, B. Eriksson, N.-H. Xuong, T. Yonetani, and J. Kraut, *J. Biol. Chem.*, 1980, **255**, 575.
85. T. L. Poulos, S. L. Edwards, H. Wariishi, and M. H. Gold, *J. Biol. Chem.*, 1993, **268**, 4429.
86. M. Bakovic and H. B. Dunford, *Biochem. Cell Biol.*, 1996, **74**, 117.
87. M. Bakovic and H. B. Dunford, *J. Biol. Chem.*, 1996, **271**, 2048.
88. H. B. Dunford and J. S. Stillman, *Coord. Chem. Rev.*, 1976, **19**, 187.
89. L. Landino, unpublished results.
90. L. M. Landino, B. C. Crews, M. D. Timmons, J. D. Morrow, and L. J. Marnett, *Proc. Natl. Acad. Sci. USA*, 1996, **93**, 15 069.
91. S. Strieder, K. Schaible, H.-J. Scherer, R. Dietz and H. H. Ruf, *J. Biol. Chem.*, 1992, **267**, 13 870.
92. R. Odenwaller, K. R. Maddipati, and L. J. Marnett, *J. Biol. Chem.*, 1992, **267**, 13 863.
93. R. J. Kulmacz, G. Palmer, C. Wei, and A. L. Tsai, *Biochemistry*, 1994, **33**, 5428.
94. R. K. Gupta, A. S. Mildvan, and G. R. Schonbaum, *Arch. Biochem. Biophys.*, 1980, **202**, 1.
95. S. Gaspard, G. Chottard, J. P. Mahy and D. Mansuy, *Eur. J. Biochem.*, 1996, **238**, 529.
96. R. Kurumbail, unpublished results.
97. P. J. O'Brien and A. D. Rahimtula, in "Advances in Prostaglandin and Thromboxane Research," eds. B. Samuelsson, P. W. Ramwell, and R. Paoletti, Raven Press, New York, 1980, vol. 6, p. 145.
98. R. J. Kulmacz, Y. Ren, A.-L. Tsai, and G. Palmer, *Biochemistry*, 1990, **29**, 8760.
99. Y. Hsuanyu and H. B. Dunford, *Biochem. Cell Biol.*, 1990, **68**, 965.
100. Y. Hsuanyu and H. B. Dunford, *J. Biol. Chem.*, 1992, **267**, 17 649.
101. P. Ple and L. J. Marnett, *J. Biol. Chem.*, 1989, **264**, 13 983.

102. L. J. Marnett, P. Wlodawer, and B. Samuelsson, *J. Biol. Chem.*, 1975, **250**, 8510.
103. L. J. Marnett, in "Free Radicals in Biology," ed. W. A. Pryor, Academic Press, New York, 1984, Vol. 6, p. 63.
104. L. J. Marnett and T. E. Eling, in "Reviews in Biochemical Toxicology," eds. E. Hodgson, J. R. Bend, and R. M. Philpot, Elsevier/North Holland, New York, 1983, p. 135.
105. T. E. Eling, D. C. Thompson, G. L. Foureman, J. F. Curtis, and M. F. Hughes, *Annu. Rev. Pharmacol. Toxicol.*, 1990, **30**, 1.
106. L. J. Marnett, M. J. Bienkowski, W. R. Pagels, and G. A. Reed, in "Advances in Prostaglandin and Thromboxane Research," eds. B. Samuelsson, P. W. Ramwell, and R. Paoletti, Raven Press, New York, 1980, Vol. 6, p. 149.
107. G. A. Reed, E. A. Brooks, and T. E. Eling, *J. Biol. Chem.*, 1984, **259**, 5591.
108. V. M. Samokyszyn and L. J. Marnett, *J. Biol. Chem.*, 1987, **262**, 14 119.
109. B. C. Saunders, in "Inorganic Biochemistry," ed. G. L. Eichhorn, Elsevier, Amsterdam, 1975, p. 988.
110. D. R. Doerge, R. L. Divi, and M. I. Churchwell, *Anal. Biochem.*, 1997, **250**, 10.
111. B. C. Saunders, A. G. Holmes-Siedel, and B. P. Stark, "Peroxidases," Butterworths, London, 1964.
112. R. W. Egan, P. H. Gale, W. J. Vandenheuvel, E. M. Baptista, and F. A. Kuehl, *J. Biol. Chem.*, 1980, **255**, 323.
113. N. Ogino, S. Yamamoto, O. Hayaishi, and T. Tokuyama, *Biochem. Biophys. Res. Commun.*, 1979, **87**, 184.
114. B. Frei, R. Stocker and B. N. Ames, *Proc. Natl. Acad. Sci. USA*, 1988, **85**, 9748.
115. T. E. Eling, J. F. Curtis, L. S. Harman, and R. P. Mason, *J. Biol. Chem.*, 1986, **261**, 5023.
116. L. J. Marnett, G. A. Reed, and D. J. Dennison, *Biochem. Biophys. Res. Commun.*, 1978, **82**, 210.
117. I. G. C. Robertson, K. Sivarajah, T. E. Eling, and E. Zeiger, *Cancer Res.*, 1983, **43**, 476.
118. L. J. Marnett, *Life Sci.*, 1981, **29**, 531.
119. L. J. Marnett (ed.), "Arachidonic Acid Metabolism and Tumor Initiation," Martinus Nijhoff, Boston, MA, 1985.
120. R. R. Arlen and P. G. Wells, *J. Pharmacol. Exp. Ther.*, 1996, **277**, 1649.
121. L. M. Winn and P. G. Wells, *Free Radic. Biol. Med.*, 1997, **22**, 607.
122. M. E. Hemler and W. E. M. Lands, *J. Biol. Chem.*, 1980, **255**, 6253.
123. R. J. Kulmacz and W. E. M. Lands, *Prostaglandins*, 1985, **29**, 175.
124. W. L. Smith and W. E. M. Lands, *J. Biol. Chem.*, 1971, **246**, 6700.
125. R. J. Kulmacz, Jr., J. F. Miller, and W. E. M. Lands, *Biochem. Biophys. Res. Commun*, 1985, **130**, 918.
126. A.-L. Tsai, L. C. Hsi, R. J. Kulmacz, G. Palmer, and W. L. Smith, *J. Biol. Chem.*, 1994, **269**, 5085.
127. T. L. Poulos and J. Kraut, *J. Biol. Chem.*, 1980, **255**, 10 322.
128. R. J. Kulmacz, *Arch. Biochem. Biophys.*, 1986, **249**, 273.
129. R. J. Kulmacz and W. E. M. Lands, *Prostaglandins*, 1983, **25**, 531.
130. F. Ursini and A. Bindoli, *Chem. Phys. Lipids*, 1987, **44**, 255.
131. D. Salvemini, T. P. Misko, J. L. Masferrer, K. Seibert, M. G. Currie, and P. Needleman, *Proc. Natl. Acad. Sci. USA*, 1993, **90**, 7240.
132. D. Salvemini, S. L. Settle, J. L. Masferrer, K. Seibert, M. G. Currie, and P. Needleman, *Br. J. Pharmacol.*, 1995, **114**, 1171.
133. D. Salvemini, P. T. Manning, B. S. Zweifel, K. Seibert, J. Connor, M. G. Currie, and P. Needleman, *J. Clin. Invest.*, 1995, **96**, 301.
134. D. Salvemini and J. L. Masferrer, *Methods Enzymol.*, 1996, **269**, 12.
135. D. P. Hajjar, H. M. Lander, S. F. A. Pearce, R. K. Upmacis, and K. B. Pomerantz, *J. Am. Chem. Soc.*, 1995, **117**, 3340.
136. A.-L. Tsai, C. Wei and R. J. Kulmacz, *Arch. Biochem. Biophys.*, 1994, **313**, 367.
137. J. F. Curtis, N. G. Reddy, R. P. Mason, B. Kalyanaraman, and T. E. Eling, *Arch. Biochem. Biophys.*, 1996, **335**, 369.
138. W. H. Koppenol and R. Kissner, *Chem. Res. Toxicol.*, 1998, **11**, 87.
139. L. Landino, unpublished results.
140. R. Karthein, R. Dietz, W. Nastainczyk, and H. H. Ruf, *Eur. J. Biochem.*, 1988, **171**, 313.
141. R. Dietz, W. Nastainczyk, and H. H. Ruf, *Eur. J. Biochem.*, 1988, **171**, 321.
142. G. Lassmann, R. Odenwaller, J. F. Curtis, J. A. DeGray, R. P. Mason, L. J. Marnett, and T. E. Eling, *J. Biol. Chem.*, 1991, **266**, 20 045.
143. J. A. DeGray, G. Lassmann, J. F. Curtis, T. A. Kennedy, L. J. Marnett, T. E. Eling, and R. P. Mason, *J. Biol. Chem.*, 1992, **267**, 23 583.
144. G. Lassmann, R. Odenwaller, J. F. Curtis, J. A. DeGray, R. P. Mason, L. J. Marnett, and T. E. Eling, in "Eicosanoids and Other Bioactive Lipids in Cancer and Radiation Injury," eds. S. Nigam, K. V. Honn, L. J. Marnett, and T. L. Walden, Jr., Kluwer, Norwell, 1993, p. 51.
145. R. J. Kulmacz, G. Palmer, and A.-L. Tsai, *Mol. Pharmacol.*, 1991, **40**, 833.
146. W. L. Smith, T. E. Eling, R. J. Kulmacz, L. J. Marnett, and A. Tsai, *Biochemistry*, 1992, **31**, 3.
147. A.-L. Tsai, G. Palmer, and R. J. Kulmacz, *J. Biol. Chem.*, 1992, **267**, 17 753.
148. A.-L. Tsai, R. J. Kulmacz, and G. Palmer, *J. Biol. Chem.*, 1995, **270**, 10 503.
149. M. R. Gunther, L. C. Hsi, J. F. Curtis, J. K. Gierse, L. J. Marnett, T. E. Eling, and R. P. Mason, *J. Biol. Chem.*, 1997, **272**, 17 086.
150. A. Graslund, M. Sahlin, and B.-M. Sjoberg, *Environ. Health Perspect.*, 1985, **64**, 139.
151. L. C. Hsi, C. W. Hoganson, G. T. Babcock, and W. L. Smith, *Biochem. Biophys. Res. Commun.*, 1994, **202**, 1592.
152. G. Xiao, A.-L. Tsai, G. Palmer, W. C. Boyar, P. J. Marshall, and R. J. Kulmacz, *Biochemistry*, 1997, **36**, 1836.
153. T. Shimokawa, R. J. Kulmacz, D. L. DeWitt, and W. L. Smith, *J. Biol. Chem.*, 1990, **265**, 20 073.
154. L. C. Hsi, C. W. Hoganson, G. T. Babcock, R. M. Garavito, and W. L. Smith, *Biochem. Biophys. Res. Commun.*, 1995, **207**, 652.
155. D. Goodwin, unpublished results.
156. A.-L. Tsai, G. Palmer, G. Xiao, D. C. Swinney, and R. J. Kulmacz, *J. Biol. Chem.*, 1998, **273**, 3888.
157. M. Lepoivre, J. M. Flaman, P. Bobe, G. Lamaire, and Y. Henry, *J. Biol. Chem.*, 1994, **269**, 21 891.
158. B. Roy, M. Lepoivre, Y. Henry, and M. Fontecave, *Biochemistry*, 1995, **34**, 5411.
159. V. A. Szalai and G. W. Brudvig, *Biochemistry*, 1996, **35**, 15 080.

160. Y. Sanakis, C. Goussias, R. P. Mason, and V. Petrouleas, *Biochemistry*, 1997, **36**, 1411.
161. D. C. Goodwin, M. H. Gunther, L. H. Hsi, B. C. Crews, T. E. Eling, R. P. Mason, and L. J. Marnett, *J. Biol. Chem.*, 1998, **273**, 8903.
162. A.-L. Tsai, G. Wu, and R. J. Kulmacz, *Biochemistry*, 1997, **36**, 13 085.
163. M. Bakovic and H. B. Dunford, *Biochemistry*, 1994, **33**, 6475.
164. C. Wei, R. J. Kulmacz, and A.-L. Tsai, *Biochemistry*, 1995, **34**, 8499.
165. F. A. Dewhirst, *Prostaglandins*, 1980, **20**, 209.
166. Y. Hsuanyu and H. B. Dunford, *J. Biol. Chem.*, 1992, **267**, 17 649.
167. M. S. Tang, R. A. Copeland, and T. M. Penning, *Biochemistry*, 1997, **36**, 7527.
168. M. Lecomte, O. Laneuville, C. Ji, D. L. DeWitt, and W. L. Smith, *J. Biol. Chem.*, 1994, **269**, 13 207.
169. P. J. Loll, D. Picot, and R. M. Garavito, *Nature Struct. Biol.*, 1995, **2**, 637.
170. J. R. Vane, *Nature New Biol.*, 1971, **231**, 232.
171. J. B. Smith and A. L. Willis, *Nature*, 1971, **231**, 235.
172. S. H. Ferreira, S. Moncada, and J. R. Vane, *Nature*, 1971, **231**, 237.
173. G. J. Roth, N. Stanford, and P. W. Majerus, *Proc. Natl. Acad. Sci. USA*, 1975, **72**, 3073.
174. L. H. Rome and W. E. M. Lands, *Proc. Natl. Acad. Sci. USA*, 1975, **72**, 4863.
175. R. J. Kulmacz and W. E. M. Lands, *J. Biol. Chem.*, 1985, **260**, 12 572.
176. C. A. Lanzo, J. Beechem, J. Talley, and L. J. Marnett, *Biochemistry*, 1998, **37**, 217.
177. E. A. Meade, W. L. Smith, and D. L. DeWitt, *J. Biol. Chem.*, 1993, **268**, 6610.
178. O. Laneuville, D. K. Breuer, D. L. DeWitt, T. Hla, C. D. Funk, and W. L. Smith, *J. Pharmacol. Exp. Ther.*, 1994, **271**, 927.
179. M. S. Tang, L. J. Askonas, and T. M. Penning, *Biochemistry*, 1995, **34**, 808.
180. O. H. Callan, O. Y. So, and D. C. Swinney, *J. Biol. Chem.*, 1996, **271**, 3548.
181. I. Pal, R. Odenwaller, and L. J. Marnett, *J. Med. Chem.*, 1992, **35**, 2340.
182. R. A. Copeland, J. M. Williams, J. Giannaras, S. Nurnberg, M. Covington, D. Pinto, S. Pick, and J. M. Trzaskos, *Proc. Natl. Acad. Sci. USA*, 1994, **91**, 11 202.
183. A. Rane, O. Olez, J. C. Frolich, H. W. Seyberth, B. J. Sweetman, J. T. Watson, G. R. Wilkinson, and J. A. Oates, *Clin. Pharmacol. Ther.*, 1978, **23**, 658.
184. A. R. Brash, D. E. Hickey, T. P. Graham, M. T. Stahlman, J. A. Oates, and R. B. Cotton, *N. Engl. J. Med.*, 1981, **305**, 67.
185. R. J. Kulmacz and K. K. Wu, *Arch. Biochem. Biophys.*, 1989, **268**, 502.
186. R. J. Kulmacz and W. E. M. Lands, *Biochem. Biophys. Res. Commun.*, 1982, **104**, 758.
187. J. A. Mancini, D. Riendeau, J.-P. Falgueyret, P. J. Vickers, and G. P. O'Neill, *J. Biol. Chem.*, 1995, **270**, 29 372.
188. F. J. Van Der Ouderaa, M. Buytenhek, D. H. Nugteren, and D. A. Van Dorp, *Eur. J. Biochem.*, 1980, **109**, 1.
189. G. J. Roth, E. T. Machuga, and J. Ozols, *Biochemistry*, 1983, **22**, 4672.
190. D. L. DeWitt and W. L. Smith, *Proc. Natl. Acad. Sci. USA*, 1988, **85**, 1412.
191. T. Shimokawa and W. L. Smith, *J. Biol. Chem.*, 1992, **267**, 12 387.
192. I. Wells and L. J. Marnett, *Biochemistry*, 1992, **31**, 9520.
193. A. S. Kalgutkar and L. J. Marnett, *Biochemistry*, 1994, **33**, 8625.
194. A. S. Kalgutkar, B. C. Crews, and L. J. Marnett, *J. Med. Chem.*, 1996, **39**, 1692.
195. K. R. Gans, W. Galbraith, R. J. Roman, S. B. Haber, J. S. Kerr, W. K. Schmidt, C. Smith, W. E. Hewes, and N. R. Ackerman, *J. Pharmacol. Exp. Ther.*, 1990, **254**, 180.
196. N. Futaki, S. Takahashi, M. Yokoyama, I. Arai, S. Higuchi, and S. Otomo, *Prostaglandins*, 1994, **47**, 55.
197. P. Prasit and D. Riendeau, *Ann. Rep. Med. Chem.*, 1997, **32**, 211.
198. T. D. Penning, J. J. Talley, S. R. Bertenshaw, J. S. Carter, P. W. Collins, S. Docter, M. J. Graneto, L. F. Lee, J. W. Malecha, J. M. Miyashiro, R. S. Rogers, D. J. Rogier, S. S. Yu, G. D. Anderson, E. G. Burton, J. N. Cogburn, S. A. Gregory, C. M. Koboldt, W. E. Perkins, K. Seibert, A. W. Veenhuizen, Y. Y. Zhang, and P. C. Isakson, *J. Med. Chem.*, 1997, **40**, 1347.
199. C.-C. Chan, S. Boyce, C. Brideau, A. W. Ford-Hutchinson, R. Gordon, R. Guay, R. G. Hill, C.-S. Li, J. Mancini, M. Penneton, P. Prasit, R. Rasori, D. Riendeau, P. Roy, P. Tagari, P. Vickers, E. Wong, and I. W. Rodger, *J. Pharmacol. Exp. Ther.*, 1995, **274**, 1531.
200. C. S. Li, W. C. Black, C. C. Chan, A. W. Ford-Hutchinson, J. Y. Gauthier, R. Gordon, D. Guay, S. Kargman, C. K. Lau, J. Mancini, N. Ouimet, P. Roy, P. Vickers, E. Wong, R. N. Young, R. Zamboni, and P. Prasit, *J. Med. Chem.*, 1995, **38**, 4897.
201. K. Tanaka, H. Kawasaki, K. Kurata, Y. Aikawa, Y. Tsukamoto, and T. Inaba, *Jpn. J. Pharmacol.*, 1995, **67**, 305.
202. K. Nakamura, K. Tsuji, N. Konishi, H. Okumura, and M. Matsuo, *Chem. Pharm. Bull.*, 1993, **41**, 894.
203. Y. Song, D. T. Connor, R. J. Sorenson, R. Doubleday, A. D. Sercel, P. C. Unangst, R. B. Gilbertsen, K. Chan, D. A. Bornmeier, and R. D. Dyer, *Inflamm. Res.*, 1997, **46**, S141.
204. W. C. Black, C. Bayly, M. Belley, C.-C. Chan, S. Charleson, D. Denis, J. Y. Gauthier, R. Gordon, D. Guay, S. Kargman, C. K. Lau, Y. Leblanc, J. Mancini, M. Ouellet, D. Percival, P. Roy, K. Skorey, P. Tagari, P. Vickers, E. Wong, L. Xu, and P. Prasit, *Bioorg. Med. Chem. Lett.*, 1996, **6**, 725.
205. G. Engelhardt, R. Bogel, Chr. Schnitzler, and R. Utzmann, *Biochem. Pharmacol.*, 1996, **51**, 21.
206. G. Engelhardt, R. Bogel, Chr. Schnitzler, and R. Ultzmann, *Biochem. Pharmacol.*, 1996, **51**, 29.
207. E. S. Lazer, C. K. Miao, C. L. Cywin, R. Sorcek, H.-C. Wong, Z. Meng, I. Potocki, M. A. Hoermann, R. J. Snow, M. A. Tschantz, T. A. Kelly, D. W. McNeil, S. J. Coutts, L. Churchill, A. G. Graham, E. David, P. M. Grob, W. Engel, H. Meier, and G. Trummlitz, *J. Med. Chem.*, 1997, **40**, 980.
208. O. Laneuville, D. K. Breuer, N. Xu, Z. H. Huang, D. A. Gage, J. T. Watson, M. Lagarde, D. L. DeWitt, and W. L. Smith, *J. Biol. Chem.*, 1995, **270**, 19 330.
209. D. K. Bhattacharyya, M. Lecomte, J. Dunn, D. J. Morgans, and W. L. Smith, *Arch. Biochem. Biophys.*, 1995, **317**, 19.
210. A. S. Kalgutkar, B. C. Crews, S. W. Rowlinson, C. Garner, K. Seibert, and L. J. Marnett, *Science*, 1998, **280**, 1268.
211. R. Kurumbail, Structural studies of COX-2 with substrate, 1998, unpublished results.

212. A. S. Kalgutkar, B. C. Crews, and L. J. Marnett, *Biochemistry*, 1996, **35**, 9076.
213. V. Houtzager, M. Ouellet, J. P. Falgueyret, L. A. Passmore, C. Bayly, and M. D. Percival, *Biochemistry*, 1996, **35**, 10 974.
214. T. Nakano, O. Ohara, H. Teraoka, and H. Arita, *J. Biol. Chem.*, 1990, **265**, 12 745.
215. K. Gewert and R. Sundler, *Biochem. J.*, 1995, **307**, 499.
216. M. Goppelt-Struebe, *Biochem. Pharmacol.*, 1997, **53**, 1389.
217. D. L. DeWitt and E. A. Meade, *Arch. Biochem. Biophys.*, 1993, **306**, 94.
218. S. T. Reddy and H.R. Herschman, *J. Biol. Chem.*, 1997, **272**, 3231.

5.11

The Chemistry and Enzymology of Cobalamin-dependent Enzymes

YAPING XU and CHARLES B. GRISSOM
University of Utah, Salt Lake City, UT, USA

5.11.1 INTRODUCTION

Cobalamin (B_{12}) has intrigued chemists since its recognition as an essential compound for the maintenance of human health and its identification as the "extrinsic factor" in liver that cures pernicious anemia. Cobalamin is manufactured exclusively by prokaryotes, including *Klebsiella*, *Salmonella*, and related bacteria. These organisms produce B_{12} as a necessary cofactor for enzymatic reactions that produce branched-chain acids and provide oxidizing equivalents for growth under anaerobic conditions. In humans, cobalamin is a necessary cofactor for two enzymatic reactions. As methylcobalamin (MeCbl), it is a covalently bound cofactor in the cytosolic methionine synthase reaction, wherein it serves as the carrier of a methyl cation equivalent in the conversion of homocysteine to methionine. As adenosylcobalamin (AdoCbl), it is required as a dissociable cofactor in the mitochondrial methylmalonyl-CoA mutase reaction, wherein homolysis of the Co—C bond initiates a 1,2-rearrangement that involves cob(II)alamin and a substrate-derived intermediate with radical character. Reviews of cobalamin chemistry, enzymology, and physiology offer additional perspectives on the reactivity of B_{12} cofactors and proteins.[1-6]

At least 10 cobalamin-dependent enzymes catalyze the vicinal rearrangement of a hydrogen atom and another substituent on adjacent carbon atoms of the substrate, as depicted in Equation (1).

$$\underset{|}{\overset{X}{-C_1}}\!\!-\!\!\underset{|}{\overset{H}{C_2}}\!-\ \xrightarrow[\text{Adenosylcobalamin}]{\text{Enzyme}}\ \underset{|}{\overset{H}{-C_1}}\!\!-\!\!\underset{|}{\overset{X}{C_2}}\!- \qquad (1)$$

The group X can be a heteroatom functional group such as -OH or -NH$_2$, or it can be a bulky carbon-containing fragment such as -CH(NH$_2$), -COOH, -COSCoA, or -C(=CH$_2$)COOH. The cobalamin-dependent enzymatic reactions that progress through a vicinal rearrangement, as well as the cobalamin-dependent methyltransferases, are summarized in Table 1.

Table 1 Coenzyme B$_{12}$-dependent enzymatic reactions.

Enzyme	Reaction
Methylmalonyl-CoA mutase	HO$_2$C–CH$_2$–CH$_2$–CO–SCoA ⇌ HO$_2$C–CH(CH$_3$)–CO–SCoA
Glutamate mutase	HO$_2$C–CH$_2$–CH$_2$–C(=NH$_2$)–CO$_2$H ⇌ HO$_2$C–CH(CH$_3$)–C(=NH$_2$)–CO$_2$H
2-Methyleneglutarate mutase	HO$_2$C–CH$_2$–CH$_2$–C(=CH$_2$)–CO$_2$H ⇌ HO$_2$C–CH(CH$_3$)–C(=CH$_2$)–CO$_2$H
D- and L-Lysine mutase	H$_2$N–(CH$_2$)$_3$–CH(NH$_2$)–CO$_2$H ⇌ CH$_3$–CH(NH$_2$)–CH$_2$–CH(NH$_2$)–CO$_2$H
Ornithine mutase	H$_2$N–CH$_2$–CH$_2$–CH(NH$_2$)–CO$_2$H ⇌ CH$_3$–CH(NH$_2$)–CH(NH$_2$)–CO$_2$H
Leucineamino mutase	(CH$_3$)$_2$CH–CH$_2$–CH(NH$_2$)–CO$_2$H ⇌ (CH$_3$)$_2$CH–CH(NH$_2$)–CH$_2$–CO$_2$H
Ethanolamine ammonia lyase	H$_2$N–CH$_2$–CH$_2$–OH → CH$_3$CHO
Diol dehydrase	CH$_3$–CH(OH)–CH$_2$–OH → CH$_3$CH$_2$CHO
Glycerol dehydrase	HO–CH$_2$–CH(OH)–CH$_2$–OH → HO–CH$_2$–CH$_2$–CHO

Early difficulty in understanding the mechanism of these unusual reactions arose partly because of a lack of close model reactions in simple organic systems and partly because of an incomplete understanding of the unusual chemistry afforded by the cobalamin cofactor. Detailed physical studies of cobalamins, as well as studies of model reactions involving B$_{12}$ analogues, have greatly enhanced the understanding of cobalamin-dependent reaction mechanisms. In addition, the structure of several B_{12}-dependent enzymes has been determined by X-ray crystallography.[7,8] These structures revealed the unexpected coordination of a histidine imidazole side chain in at least two cobalamin-dependent enzymes, methionine synthase and methylmalonyl-CoA mutase.[7,8]

In this chapter, we will focus on the mechanism of cobalamin-dependent enzymes and the role of the B_{12} cofactor, particularly as revealed by new techniques and approaches that have been applied to methylmalonyl-CoA mutase, ethanolamine ammonia lyase, diol dehydrase, glutamate mutase, and methionine synthase. The reader is directed to Chapter 5.08 on ribonucleotide reductase for a detailed discussion of the mechanism of this B_{12}-dependent enzyme.

5.11.2 COBALAMIN CHEMISTRY AND REACTIVITY

The structure of cobalamin is shown in Figure 1. It is convenient to think of this molecule as having four moieties that dominate control of reactivity.

(i) The corrin ring is a nitrogen-containing macrocycle that resembles the porphyrin ring of heme, but is more reduced and is more selective for cobalt as the metal atom.

(ii) The seven amide side chains that adorn the corrin ring are important recognition elements for B_{12} transport proteins. Furthermore, they partially control the dynamics and reactivity of the corrin ring and provide sites for multiple hydrogen-bonding interactions that facilitate catalysis through the utilization of binding energy.

(iii) The α-axial (lower) ligand to cobalt is one of the nitrogens of 5,6-dimethylbenzimidazole (DMB) that is attached to ribose, forming a nucleotide loop from the corrin ring.

(iv) The β-axial (upper) ligand to cobalt can be one of four commonly occurring groups: a cyanide ligand, in the vitamin form of B_{12} that is available as a dietary supplement; a hydroxyl group, as found in hydroxocobalamin; a methyl group, as found in methylcobalamin; and a 5'-deoxyadenosyl

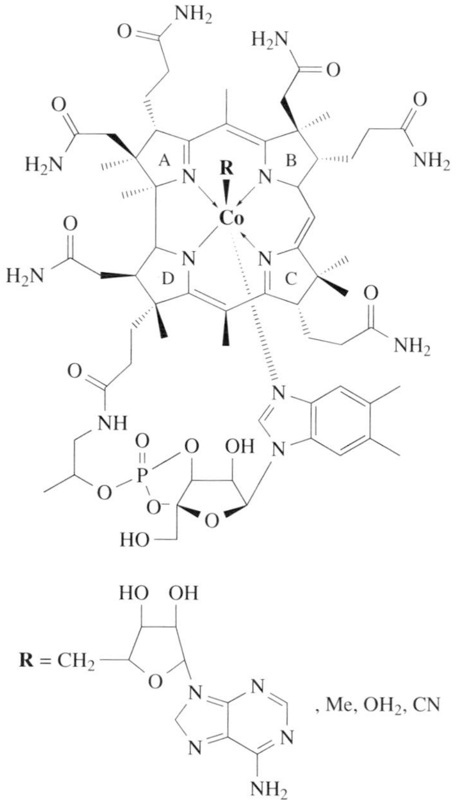

Figure 1 Structure of adenosylcobalamin, methylcobalamin, aquocobalamin, and cyanocobalamin.

group, as found in coenzyme B_{12} (aptly named for its role as a dissociable cofactor for all of the cobalamin-dependent enzymes except methionine synthase).

Although the corrin ring itself is nonpolar, corrinoids are generally very hydrophilic because of the amide side chains that adorn the corrin macrocycle. The presence of the α and β axial ligands introduces steric requirements that lead to distortion (butterfly-like puckering) of the corrin ring.[9–11]

Cobalt occurs as Co^I, Co^{II}, and Co^{III} in biologically relevant cobalamins and corrinoids. Co^{III} is the most stable oxidation state with six coordinating ligands, as found in nearly all stable alkylcobalamins. Co^{II} is paramagnetic with significant radical character and five coordinating ligands. Co^I has four coordinating ligands and is an extraordinarily potent nucleophile. The Co—C bond in adenosylcob(III)alamin ($AdoCbl^{III}$) is one of the weakest covalent bonds known, with a bond dissociation energy (BDE) of about 31 kcal mol^{-1}.[12–26] The Co—C BDE of methylcobalamin is somewhat higher, at 37 kcal mol^{-1}.[27–29] The Co—C bond is unusual because of the combination of relatively low $3d$–$4s/4p$ promotion energy and the accessibility of a homolytic cleavage pathway that results in cob(II)alamin and two heterolytic cleavage pathways that allow the formation of cob(I)alamin or cob(III)alamin.

The unusual lability of the Co—C bond is the product of several interrelated factors, including the stereoelectronic properties of the β-ligand and the pucker of the corrin ring. X-ray crystallographic studies of various B_{12} analogues show this pucker to be the result of steric demands imposed by the 5,6-dimethylbenzimidazole group and this factor leads to an unusually long Co—N bond.[30] The replacement of DMB by a smaller substituent decreases puckering of the corrin ring and alters the thermodynamic properties of the Co—C bond.

Thermodynamic parameters were determined for a homologous series of 5′-deoxyadenosyl-cobinamide analogues with α-axial bases of varying steric demands and pK_a values. These data suggest that the α-axial ligand provides significant control of reactivity in the free cofactor, with the importance of the α-axial base being less clear when the cofactor is bound in the enzyme active site.[31,32] An extended critical analysis of the role of the axial base can be found in a report by Garr et al.[31]

Several groups examined the steric influence of the α- and β-ligands on the stability of B_{12} model compounds.[33,34] Their work showed the length of the Co—R bond, where R is an alkyl group as the β-ligand, is determined mainly by the bulk of the β-alkyl group, and to a lesser extent by the bulk of the α-ligand. The Co—L bond distance, where L is the α-ligand, is influenced by the steric bulk of L, as well as by the electron donating ability of the β-ligand. Similarly, this study showed that the steric bulk of L inversely correlates with Co—C bond stability in the cobaloxime B_{12} analogues trans-(NH_2Ph) $(Me)Co(dmgH)_2$ and trans-$(NH_2Ph)(CH_2C(CO_2Et)_2Me)$-$Co(dmgH)_2$, where dmgH is the monoanion of dimethylglyoxime. The X-ray crystal structure of these compounds clearly shows the equatorial moiety deforming upwards in a "butterfly" bend. Similar butterfly puckering is seen in the X-ray crystallographic structure of AdoCbl. Interaction of the corrin ring with DMB base causes one side of the strained corrin ring to distort upward, resulting in unfavorable interactions with the adenosine moiety. The opposing repulsive effect decreases the upward distortion of the ring, thereby suggesting that strain in AdoCbl helps to lower the Co—C bond dissociation energy and destabilize the organometallic bond.[35,36]

A key question concerning alkylcobalamin reactivity is how various enzymes direct bond cleavage towards homolysis or heterolysis. Formation of Co^I from B_{12} probably involves breaking or weakening the Co—N (DMB) bond.[37] As Co—C bond homolysis occurs, the Co—N bond from DMB becomes stronger, and removal of the DMB ligand favors heterolytic cleavage, since this leads to the more stable four-coordinate Co^I species. EXAFS and X-ray crystallographic studies show the distance between the DMB ligand and the corrin ring is 2.20 Å. This unusually long Co—N bond is consistent with DMB being a poor electron donor in alkylcobalamins because of steric repulsion. X-ray edge spectroscopy was used to examine the edge shift in AdoCbl and MeCbl.[38] It was concluded that the adenosyl moiety in AdoCbl is a stronger electron-donating group than the methyl group in MeCbl, thereby reducing the effective charge on cobalt from the expected value for Co^{III}. This observation may be relevant to the differences observed in homolytic vs. heterolytic cleavage among the two cofactors.

The X-ray crystal data also show that cobalt moves towards the DMB ligand and out of plane with the corrin ring by 0.13 Å in cob(II)alamin when compared to cob(III)alamin. Other studies have shown that base-off forms of cobalamin are resistant to homolytic thermolysis relative to the base-on species.[39]

Daikh and Finke examined reversible Co—C bond rearrangement in the B_{12} model compound $C_6H_5CH_2Co[C_2(DO)(DOH)_{Pn}]I$ (1) $\rightleftharpoons$ $Co[C_2(DO)(DOH)_{Pn}CH_2C_6H_5]I$ (2) (Equation (2)).[40]

$$\Delta \text{ or } h\nu \;\; \rightleftharpoons$$

(2)

(1)   (2)

The presence of freely diffusing radicals in this alkyl migration reaction was demonstrated with TEMPO as a radical trap. The ΔG value of -0.28 kcal mol^{-1} is small, presumably due to the conformational flexibility in (2), the favorable cage effect imposed by the solvent, and the benzyl–carbon bond which has nearly the same dissociation energy as the Co—C bond.[40] The question remains as to whether B$_{12}$ itself is capable of participating in a similar intramolecular migration reaction.

In the crystal structure of *trans*-bis(dimethylglyoximato)isopropyl(2-aminopyridine)Co(III), the bond from cobalt to the endocyclic nitrogen of 2-aminopyridine is 2.19 Å.[41] This unusually long bond is consistent with steric repulsion between the 2-amino group and the equatorial ligand.[41] This steric repulsion prevents the metal ion from forming an ideal bond with the *trans* ligand. This places a weaker electron donor in the *trans* position, thereby weakening the corresponding Co—C bond in an example of the *trans steric effect* across the metal ion. It has been suggested that cobalamin enzymes may distort the corrin ring and thereby increase the C—N (DMB) bond length and thus weaken the Co—C bond.

Perhaps the most controversial aspect of alkylcobalamin reactivity has been the accurate determination of Co—C bond dissociation energies.[12–29,42] Finke and co-workers determined a value of 31 kcal mol^{-1} for the Co—C BDE in adenosylcobalamin, whereas Halpern determined a slightly lower value approaching 26 kcal mol^{-1}.[12–29] Both methods rely on thermolysis of the Co—C bond to yield cob(II)alamin and the corresponding alkyl radical, with trapping of the alkyl radical by TEMPO or excess cob(II)alamin, respectively. The 5 kcal mol^{-1} difference derives, in part, from different assumptions regarding the efficiency of radical pair recombination in the solvent cage. With fractional cage recombination efficiencies of greater than 90% under some circumstances, correction factors for ΔH can be several kcal mol^{-1}. The value of 31 kcal mol^{-1}, as determined by Finke and reproduced by other laboratories,[42] appears to have the most precise correction for solvent cage effects and is generally considered closer to the true value of the Co—C BDE in adenosylcobalamin.

The low BDE of alkylcobalamins makes these compounds ideal free-radical precursors, with their apparent thermal stability derived in part from the reversibility of Co—C bond homolysis. If the alkyl radical–cob(II)alamin radical pair is not intercepted by another reactant or a radical scavenger, efficient cage recombination regenerates the starting alkylcob(III)alamin. To probe the radical initiation and hydrogen abstraction steps of cobalamin-dependent mutases, a model reaction for the methylmalonyl-CoA mutase rearrangement was studied (Figure 2).[43,44] In this model reaction, the abstraction of a hydrogen atom is coupled to migration of the bulky alkyl group that leads to rearrangement of the carbon skeleton. This experiment suggests that a radical rearrangement pathway is a competent mechanism by which product can form. An electron transfer step that leads to carbanion formation in the penultimate intermediate is also viable.[43]

A detailed analysis of cobalamin structure and reactivity provides important insight into the unique chemistry that is possible with this organometallic cofactor, but a more complete understanding of cobalamin reactivity emerges when the cobalamin-dependent enzymatic reactions are examined in detail.

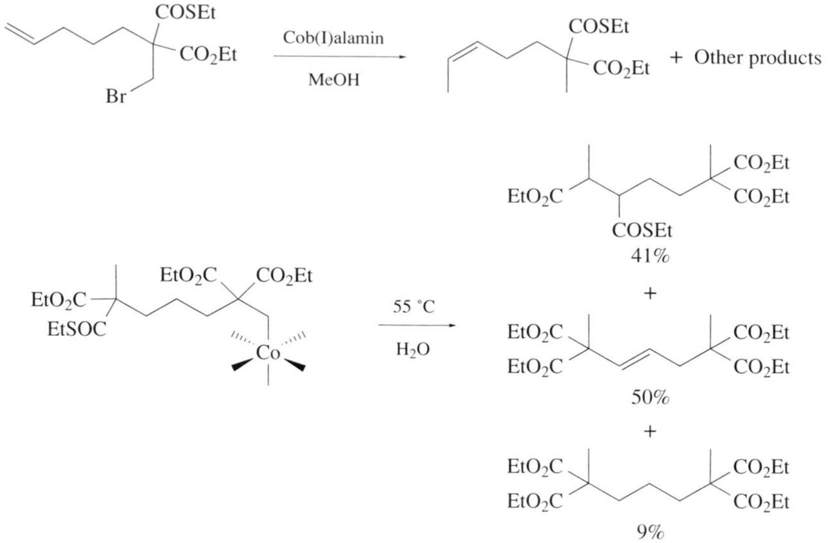

Figure 2 Model reactions for methylmalonyl-CoA mutase (after Dowd *et al.*[43]). Cob(I)alamin in methanol (top) promotes abstraction of an allylic hydrogen to facilitate carbon skeleton rearrangement to the *cis*-alkene. Radical-promoted rearrangement resulted in only the *trans*-alkene. This result supports the possibility of a carbanionic intermediate in methylmalonly-CoA mutase. Thermolysis of an alkylcob(III)alamin (bottom) also promotes carbon skeleton rearrangement.

5.11.3   ROLE OF THE PROTEIN—INTERACTIONS BETWEEN ENZYME, COENZYME, AND SUBSTRATE

In all cases, the coenzyme is bound to apoenzyme without a profound change in the absorption spectrum. This observation suggests that the enzyme does not distort the corrin ring from its ground-state solution conformation. However, no information on the identity of the α-axial ligand is available from the UV–visible absorption spectrum of the coenzyme–protein complex. Studies have shown that the α-axial 5,6-dimethylbenzimidazole ligand is replaced with a histidine side chain from the respective protein in at least three cobalamin-dependent enzymes: methionine synthase,[4,7] methylmalonyl-CoA mutase,[8,45,46] and glutamate mutase.[47] In contrast, the 5,6-benzimidazole ligand from cobalamin still occupies the α-axial ligand position in AdoCbl bound to diol dehydrase[48] and ribonucleotide reductase (see Chapter 5.08).

To accommodate coordination by the histidine side chain, the "nucleotide loop" of the 5,6-dimethylbenzimidazole base, the ribofuranosylphosphodiester, and the propionamide side chain must swing away from the corrin ring and insert into a binding pocket within the protein. The consensus sequence **DxHxxG** has been identified in cobalamin-dependent enzymes that substitute a histidine side chain for the α-axial ligand.[7,49,50] The histidine that coordinates to the cobalt atom most certainly interacts with the aspartate residue through a hydrogen bond. Beyond the role of a coordinating ligand, the exact function of the His/Asp dyad as a modulator of enzymatic activity is not yet clear.[49] Other residues in the active site may also participate in an extended hydrogen-bonding network to control reactivity at the metal center.[7]

Adenosylcobalamin-dependent enzymes must increase the rate of Co—C bond homolysis by at least a factor of 10^{12} to achieve the observed rate of catalysis.[14] Labilization of adenosylcobalamin demands weakening of the Co—C bond to allow homolytic fission. Although not a proven concept in cobalamin-dependent enzymes, sufficient energy to distort (weaken) the Co—C bond could easily come from the utilization of excess binding energy that is derived from multiple hydrogen-bonding interactions to the cobalamin cofactor, including the amide side chains that decorate the corrin ring.[51] The amide side chains are also important recognition elements for binding to the non-enzymatic cobalamin transport proteins transcobalamin, haptocorrin, and intrinsic factor.[52] Removal or chemical derivatization of specific amide side chains alters the binding to transcobalamin by 3–40 fold[52] and complete removal of the c-amide side chain allows the introduction of an additional double bond in the macrocyclic skeleton, thereby flattening the corrin ring with an accompanying red shift of the absorption maximum.[53]

In addition to promoting the desired reaction, adenosylcobalamin-dependent enzymes perform a second important function by preventing the unwanted side reactions that often accompany radical reactions.[54] This function of negative catalysis[54] demands a highly constrained free energy surface that might be envisioned as a deep canyon through which the reaction progresses along the free energy surface. This canyon must feature steep walls to prevent side reactions and constrain the highly reactive radical intermediate. At the end of the reaction sequence, the radical intermediate is quenched by recombination of the 5'-deoxyadenosyl radical and cob(II)alamin, thereby regenerating the ground state (resting state) of the adenosylcob(III)alamin cofactor. If the enzyme invests 25 kcal mol^{-1} to promote homolysis of the C—Co bond to begin the catalytic cycle, this energy must be recovered at the end of the reaction sequence. Recovery of the energy would not be possible if an undesirable sequence of rearrangements occurred to produce a radical that is thermodynamically more stable than the 5'-deoxyadenosyl radical. Therefore, the enzyme must prevent the alkyl radical from rearranging to a low-energy intermediate, or migrating through the protein scaffold to form a stable radical that cannot participate in catalysis.[55]

A paramagnetic species does not form until all reaction components are present in the enzyme–substrate complex, as formation of a coenzyme-derived radical in the absence of substrate could lead to unwanted side reactions. EPR experiments with methylmalonyl-CoA mutase[46,55] confirm that homolysis of the Co—C bond occurs only after the addition of substrate, as indicated by the appearance of a product-like EPR signal.[46] Similarly, stopped-flow spectrophotometric experiments with ethanolamine ammonia lyase show that the signature of cob(II)alamin appears in the visible spectrum only after enzyme and coenzyme are combined with saturating levels of substrate.[56–59]

Photolysis, thermolysis, and enzyme-promoted homolysis of the Co—C bond of adenosylcobalamin results in the singlet radical pair consisting of cob(II)alamin and the 5'-deoxyadenosyl radical.[60,61] Since all of these processes lead to formation of the same radical pair, information from the reaction dynamics of photolysis studies can be related to proposed enzymatic reaction mechanisms. Photohomolysis of adenosylcobalamin begins with an electronic $\pi \to \pi^*$ promotion in the corrin ring and must involve an intermediate charge-transfer state on the way to Co—C bond homolysis.[60,61] Picosecond photolysis studies of adenosylcobalamin reveal geminate radical pair recombination rates of 10^9 s^{-1}, with fractional cage recombination efficiencies, Fc, of about 94%.[42,62–64] Nanosecond and continuous-wave photolysis studies of cobalamins confirm efficient radical pair recombination in the geminate cage, with an overall photochemical quantum yield of about 20% for adenosylcobalamin and 35% for methylcobalamin.[65,66] In the absence of enzyme, a large fraction of the geminate radical pairs recombines before significant diffusional separation occurs. To stabilize the 5'-deoxyadenosyl radical and promote catalysis, the enzyme must increase the radical pair separation distance, probably through a conformational change. Whatever the mechanism, the high rate of geminate recombination in the radical pair demands that one of the enzyme's functions is to separate the radicals or temporarily trap one of the radicals to prevent premature recombination.

Although the resulting singlet {5'-deoxyadenosyl radical:cob(II)alamin} radical pairs have identical electronic states,[59–61,67] enzymatic homolysis of the Co—C bond cannot access an excited electronic state and must begin with another process to weaken the Co—C bond and displace the equilibrium towards dissociation (see Section 5.11.2). Strain-induced cleavage will not only result in homolysis of the Co—C bond, but this process will also increase the radical pair separation distance if a component of the distortion occurs along the apical cobalt axis. This brute-force approach of stretching the Co—C bond may be the most efficient method to achieve both homolysis and a net decrease in the rate of radical pair recombination.

It is also possible for the enzyme to strengthen the Co—C bond and disfavor homolysis by compressing the apical Co–α–N interaction. The thermodynamic properties of adenosylcobinamide analogues [AdoCbi · N-MeIm]$^+$ and [AdoCbi-pyridine]$^+$ were studied.[68] A stronger Co—C bond was observed in [AdoCbi · N-MeIm]$^+$ with a shorter Co—N bond distance when compared to pyridine as the α-axial base. The stronger Co—C bond leads to a significant shift towards heterolysis as the preferred pathway. In this model system, the enzyme under study, methylmalonyl-CoA mutase, must either prevent Co—C heterolysis or follow a heterolytic pathway.[68]

In methionine synthase, the corrin portion of cobalamin is sandwiched between two domains of a 27 kDa fragment, with the nucleotide tail penetrating into a deep pocket formed by residues of the C-terminal domain.[7,69,70] This sequence contains a region of moderate hydrophobicity that is flanked by extended hydrophilic segments.[71] Structural similarities are expected among the domains that interface with the corrin ring. The α/β domain that binds the lower half of the corrin ring and accommodates the extended nucleotide tail (phosphodiester moiety and 5,6-dimethylbenzimidazole group) in methionine synthase and methylmalonyl-CoA mutase is also expected in glutamate mutase (*vide infra*).

5.11.4 REACTION MECHANISM OF COBALAMIN-DEPENDENT ENZYMES

Classification of cobalamin-dependent enzymes requires that an arbitrary prejudice or mechanistic bias be applied. Classification by cofactor requirement is straightforward, but this approach is of limited utility, as there are only two categories: adenosylcobalamin vs. methylcobalamin. Classification by base-on 5,6-dimethylbenzimidazole α-coordination vs. **DxHxxG** histidine coordination is similarly straightforward, but the mechanistic significance of this structural feature is unclear. Perhaps the most enlightening classification has been put forth by Buckel and Golding, and is reproduced in Table 2.[6]

Table 2 Cobalamin enzyme classification scheme of Buckel and Golding.

$$\begin{array}{ccc} \overset{\displaystyle H}{\underset{\displaystyle Y}{a-C-C-b}} & \rightleftharpoons & \overset{\displaystyle X\;H}{\underset{\displaystyle Y\;H}{a-C-C-b}} \end{array}$$

Substituent	a	b	X	Y
Class I: carbon skeleton mutases				
(1) Glutamate mutase	CO_2^-	H	2-Glycinyl	H
(2) 2-Methyleneglutarate mutase	CO_2	H	2-Acrylate	H
(3) Methylmalonyl-CoA mutase	CO_2H	H	Formyl-CoA	H
(4) Isobutyryl-CoA mutase	Me	H	Formyl-CoA	H
Class II: eliminases				
(5) Propanediol dehydrase	H	H, Me, CF_3	OH	OH
(6) Glycerol dehydrase	H	CH_2OH	OH	OH
(7) Ethanolamine ammonia-lyase	H	H, Me	NH_3^+	OH
(8) Ribonucleotide-triphosphate reductase	C-4′ of ribonucleotide	C-1′ of ribonucleotide	OH	OH
Class III: aminomutases				
(9) β-Lysine-5,6-aminomutase	4-(3-Aminobutyrate)	H	NH_3^+	H
(10) D-Ornithine-4,5-aminomutase	3-D-Alanine 4-(D-2-Aminobutyrate)	H	NH_3^+	H

a, b, and Y are variable substitutes; X is the migrating group -OH, NH_3^+ or a carbon-centered group.

In the Buckel and Golding scheme for classifying adenosylcobalamin-dependent enzymes, Class I enzymes are carbon skeleton mutases, Class II enzymes are eliminases, and Class III enzymes are aminomutases.[6] An examination of this classification scheme reveals that methylmalonyl-CoA mutase and glutamate mutase, two of the enzymes with the **DxHxxG** motif and histidine coordination, are Class I enzymes. In contrast, the **DxHxxG** motif has NOT been identified in any of the Class III enzymes, and histidine coordination to cobalt has been ruled out in diol dehydrase and ribonucleotide reductase.

Doubt occasionally surfaces as to whether radical signals observed by EPR represent catalytically significant intermediates along the reaction coordinate of all adenosylcobalamin-dependent enzymes. The catalytic cycle of all Class I, II, and III enzymes appears to begin with homolysis of the Co—C bond, with subsequent hydrogen atom abstraction from substrate (Figure 3). Much less is known about migration of the substrate X-group.

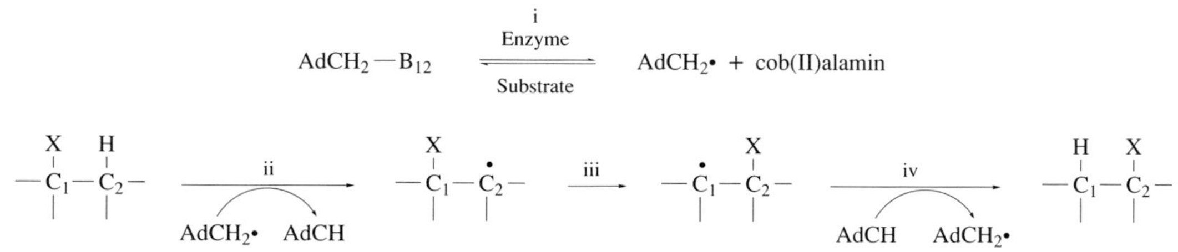

Figure 3 General mechanism for adenosylcobalamin-dependent enzymes. (i) Homolysis of the Co—C bond produces the 5′-deoxyadenosyl radical and cob(II)alamin as the enzyme-bound radical pair; (ii) abstraction of a hydrogen atom from substrate generates the substrate radical; (iii) rearrangement of the substrate-derived radical leads to formation of the product-like radical; (iv) reverse hydrogen-atom abstraction forms the closed-shell product and regenerates the 5′-deoxyadenosyl radical.

Evidence for the above general mechanism in many adenosylcobalamin-dependent enzymes includes:

(i) EPR spectrum indicates the presence of cob(II)alamin and an organic radical after the addition of substrate to holoenzyme.

(ii) UV–Visible absorption spectrum indicates that a catalytically-competent amount of cob(II) alamin is formed during the reaction.[72]

(iii) Reversible formation of 5'-deoxymethyladenosine during the course of the enzymatic reaction.[73,74]

(iv) Isotopically labeled hydrogen is transferred from the substrate to the 5'-carbon of the 5'-deoxyadenosyl radical to form 5'-deoxymethyladenosine. Isotopic scrambling occurs among the 5'-hydrogen atoms, such that any one of the three hydrogens can be returned to the final product.[75–81] Further kinetic discrimination against return of the isotopic label to the substrate or product radical produces an unusually large tritium kinetic isotope effect in excess of 120 when a protein-based radical is an intermediate radical carrier (as in ethanolamine ammonia lyase and diol dehydrase), but a more normal deuterium kinetic isotope effect of about 7 or smaller when direct hydrogen atom abstraction occurs (as in methylmalonyl-CoA mutase and glutamate mutase).[75–81]

Details of the radical rearrangement (X-group migration) that leads to product radical formation are not known. The substrate radical may rearrange directly to the product radical, or it may be converted to an intermediate carbocation, carbanion, or even an organocobalt adduct of cob(II) alamin. The possibility of direct substrate radical rearrangement to the product radical in the methylmalonyl-CoA mutase reaction was evaluated in the model compound $EtS(C=O)C(Me)$ $(CH_2Br)COOEt$ (Figure 4).[20]

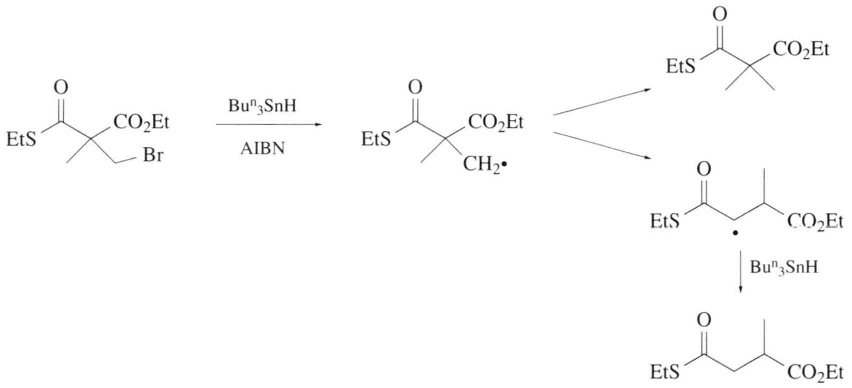

Figure 4 Methylmalonyl-CoA mutase model compound reaction mechanism.

In this reaction, only products derived from direct trapping of the substrate radical and 1,2-migration of the thioester group were observed. This result suggests that the rearrangement process is compatible with the radical lifetime in this model system. Carbanion formation by reduction of the same starting material with sodium naphthenide at $-78\,^{\circ}C$ yields a greater range of products. Rearrangement through a carbocation process is possible, with 1,2-migration of NH_2 or OH, but there is no evidence for the formation of cob(I)alamin—a species that has a distinct spectral signature above 600 nm and can be trapped by N_2O.[23,82] Rearrangement via an organocobalt intermediate is supported by studies with some model compounds, but a substrate–B_{12} or product–B_{12} adduct has never been isolated.[83] In the following sections, the catalytic mechanism of methylmalonyl-CoA mutase, glutamate mutase, ethanolamine ammonia lyase, diol dehydrase, and methionine synthase will be considered in detail to highlight important experimental results that have contributed to the current understanding of cobalamin-dependent catalysis.

5.11.4.1 Methylmalonyl-CoA Mutase

Methylmalonyl-CoA mutase is present in both mammals and bacteria. In *Propionibaterium shermanii*, the enzyme is required for the fermentation of pyruvate to propionate. In mammalian mitochondria, the enzyme is required for the degradation of odd-chain fatty acids and branched-chain amino acids through the conversion of methylmalonyl-CoA to succinyl-CoA.

The bacterial enzyme is a heterodimer with the α-subunit being 728 amino acid residues ($M_r = 80\,147$) and the β-subunit being 638 amino acid residues ($M_r = 69\,465$).[45] The native enzyme, when normally isolated, is a mixture of apoenzyme and cofactor-bound holoenzyme. The two polypeptides have a high degree of homology, particularly in the middle of the sequences, where about 80% of the amino acids are identical or substituted by similar residues. The binding site for cobalamin may be located at the C-terminal region of the α-subunit.[8] The heterodimer dissociates with increasing ionic strength and concomitant loss of enzyme activity.[84] Methylmalonyl-CoA mutase exhibits a high degree of specificity for L-methylmalonyl-CoA as substrate, and hydrogen atom migration is stereospecific, with retention of configuration.[85,86]

The catalytic mechanism of methylmalonyl-CoA mutase is still controversial. Abstraction of a hydrogen atom from the methyl group of the substrate has not been detected in any model compound. The holoenzyme (active dimer plus stoichiometric adenosylcobalamin) exhibits no EPR signal. Upon addition of substrate, a broad signal appears at about $g = 2.11$ and 2.00, with 50 G hyperfine features from cobalt.[55,87–89] The organic radical component of this signal has not been identified, but probably corresponds to a strongly coupled substrate or product-derived radical.[89,90]

The crystal structure of methylmalonyl-CoA mutase has been determined at 2.0 Å resolution.[8] In addition to showing replacement of the 5,6-dimethylbenzimidazole ligand by histidine (see Section 5.11.3), the structure clearly shows the corrin ring and the nucleotide loop of B_{12} buried in the pocket of an αβ (TIM) barrel. This structure provides a narrow access channel to a deeply buried active site, thereby protecting reactive intermediates from side reactions. The active-site cavity is formed by loops at the C-termini of the β-strands. The corrin ring of cobalamin and its amide side chains restrict access to the interior of the barrel. The active site is completely inaccessible to solvent, except through the CoA channel along the barrel axis. Such a deeply buried active site may be necessary to protect reactive radical intermediates. Binding interactions involving the adenosine ligand are not resolved.[8]

5.11.4.2   Glutamate Mutase

Glutamate mutase catalyzes the first step in the fermentation of glutamate by *Clostridium tetanomorphum* through the conversion of L-glutamate to threo-β-methyl-L-aspartate.[91] Glutamate mutase consists of two readily separable protein components, E and S, with component S being a small monomeric protein of MW = 15 kDa and component E being a homodimer with a subunit MW = 50 kDa. Component E was initially thought to bind both substrate and cofactor,[49,92] but later studies showed that the S component actually contains the **DxHxxG** cobalamin binding motif for B_{12}. The enzyme is highly specific for L-glutamate as substrate.

A deuterium kinetic isotope effect on $V_{\max}$ and $V_{\max}/K_m$ for L-Glu have nearly equal values of about 7.[6,93] Since the exchange of hydrogen from coenzyme to substrate is calculated to have a similar deuterium kinetic isotope effect of 6–7.4 (determined by the rate of tritium washout from 5′-^{3}H-adenosylcobalamin), the intermediacy of a protein-based radical carrier can be excluded.[6,93]

The EPR spectrum of glutamate mutase exhibits a low-spin Co^{II} signal at $g_{xy} = 2.1$ and $g_z = 2.0$ in the presence of adenosylcobalamin and *L*-Glu.[94] The broad g_{xy} value suggests either a distorted six-coordinate Co^{II} intermediate, with the sixth ligand being the substrate or an amino acid from the enzyme, or the more likely case of a strongly coupled radical pair, as seen for methylmalonyl-CoA mutase. More recent EPR studies feature better resolution of the broad Co^{II} signal and show distinct lines at $g_{xy} = 2.1$ and $g_z = 1.985$, with 50 G hyperfine coupling from cobalt.[95,96] Spectral simulation suggests that this may be reminiscent of methylmalonyl-CoA mutase, with a strongly coupled radical pair that is separated by less than 6 Å.[6] The EPR spectrum of glutamate mutase incorporating ^{15}N-histidine in the protein clearly shows the superhyperfine doublets expected from coordination by a β-axial histidine side chain.[47]

The stereochemical outcome of glutamate mutase can be explained by fragmentation of a substrate-derived radical to yield acrylate and the 2-glycinyl radical.[6] Fragmentation of the substrate-derived radical requires a specific conformation in which the C-2–C-3 bond is nearly parallel to the *p*-orbital at the radical center. Addition of the 2-glycinyl radical to the *Re* face at C-2 of acrylate leads to a product-related radical. This process leads to inversion of configuration at the carbon atom from which hydrogen is abstracted, and the center to which the X-group migrates.[6] The proposed reaction pathway is shown in Figure 5.[6] Analogous reaction schemes can be drawn for methylmalonyl-CoA mutase and 2-methyleneglutarate mutase–all class I carbon skeleton mutases in the Buckel and Golding classification.[6,96]

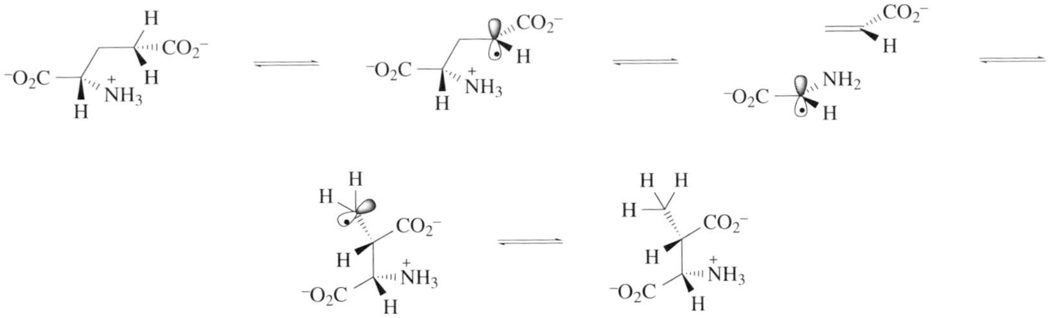

Figure 5 Proposed reaction mechanism for glutamate mutase. Fragmentation of a substrate-derived radical gives acrylate and the 2-glycinyl radical. The C-2–C-3 bond must be nearly parallel to the *p*-orbital at the radical center to allow addition of the 2-glycinyl radical to the *Re* face at C-2 of acrylate (after Buckel and Golding[6]).

5.11.4.3 Ethanolamine Ammonia Lyase

Ethanolamine ammonia lyase (EAL) is a bacterial enzyme that catalyzes the conversion of ethanolamine to acetaldehyde and ammonia. It is an $\alpha_6\beta_6$ multimeric protein with an α subunit MW of 35 kDa and a β-subunit MW of 55 kDa.[97,98] A number of bacterial species express the enzyme when fermented in the presence of ethanolamine as nitrogen source.[97,99] A cluster of six genes is located at 50 min in the *Salmonella typhimurium* genome, although only two structural genes are needed for ethanolamine ammonia lyase production. Sequence comparison shows that the large β-subunit is distantly related to both subunits of methylmalonyl-CoA mutase from *P. shermanii*. Kinetic analysis of EAL and titration of the apoenzyme with CNCbl showed that the protein contains two active sites per $\alpha_6\beta_6$ multimer.[98] The proposed catalytic mechanism for EAL is shown in Figure 6.[59,67]

No doubt remains that an essential radical pair is formed in the catalytic cycle. In the presence of ethanolamine, adenosylcobalamin, and EAL, a signal from Co^{II} is seen in the freeze-quench EPR spectrum, at $g = 2.34$ and 2.08.[56] Experiments with isotopically labeled substrate show that the high-field doublet at $g = 2.007$ corresponds to a radical at C-1 of the substrate.[100]

Stopped-flow kinetic studies in which enzyme is rapidly mixed with substrate and cofactor show cob(II)alamin growing in at a rate that is kinetically competent to support the formation of product.[56–59] When the reaction reaches equilibrium with near-total consumption of substrate, the regeneration of Co^{III} from Co^{II} is observed.[56] Premixing of enzyme and coenzyme before rapid mixing with the substrate increases the first-order rate of cob(II)alamin formation from 5 s^{-1} to 90 s^{-1}, but no cob(II)alamin is formed in the absence of substrate. Rapid-scanning stopped-flow techniques, coupled with single-value decomposition analysis of the resulting absorption spectrum, allow extraction of cob(III)alamin and cob(II)alamin absorption spectra. Careful analysis of these data show no evidence for cob(I)alamin, or a spectroscopically prominent tryptophan or tyrosine radical that might be observable if it existed.[59]

Unambiguous evidence for the homolysis of 5-deoxyadenosylcobalamin comes from the magnetic field dependence of the steady-state and stopped-flow kinetic parameters of EAL.[59,67,101] In chemical and enzymatic reactions with a weakly coupled radical pair that is separated by 8–12 Å, a d.c. magnetic field in the range 10–2000 G will alter the rate of intersystem crossing between the singlet and triplet spin states of the radical pair.[102,103] Since only the singlet radical pair can undergo nonproductive recombination to regenerate adenosylcob(III)alamin, the net forward flux through the catalytic cycle will decrease, with a concomitant decrease in V_{max}/K_m.[67] The rate of cob(II)alamin formation, as monitored by stopped-flow kinetic techniques, also depends on the applied magnetic field, thereby identifying the magnetic field sensitive step as recombination of the cob(II)alamin: 5'-deoxyadenosyl radical pair.[59]

No magnetic field dependence is observed in the methylmalonyl-CoA mutase reaction because the radical pair is tightly coupled with a separation distance of only 5–6 Å and a large electron exchange interaction.[101] The energy gap between singlet and triplet electron spin states of a strongly coupled radical pair is too large for intersystem crossing to reversibly populate both states.[102] Therefore, no d.c. magnetic field effect on the kinetic parameters of methylmalonyl-CoA mutase can be observed.[102]

In EAL, there is a large tritium isotope effect of about 160 on the exchange of hydrogen label from the 5'-methylene group of adenosylcobalamin and substrate.[104] This large kinetic isotope effect

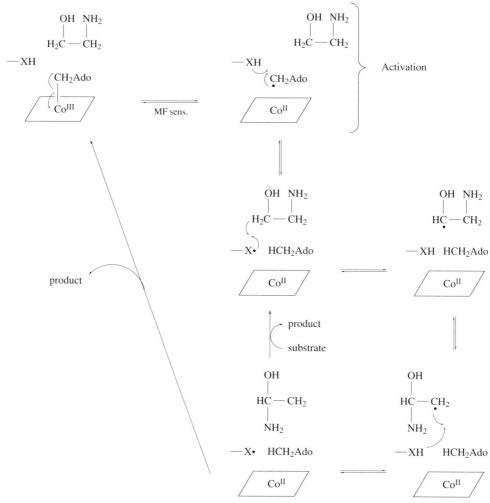

Figure 6 Mechanism for ethanolamine ammonia lyase (after Harkins and Grissom[59]). Homolytic cleavage of the Co—C bond produces the 5′-deoxyadenosyl radical and cob(II)alamin. Tritium kinetic isotope effects on the washout of label from the cofactor suggests the 5′-deoxyadenosyl radical may abstract a hydrogen atom from an amino acid side chain to generate a protein radical, X·. In close analogy to a free-radical chain process, the protein radical X· abstracts H· from ethanolamine to generate a substrate radical. 1,2-Migration of the amine group produces the carbinolamine radical. Reverse H· abstraction from XH yields the closed-shell carbinolamine. Subsequent catalytic cycles occur in a radical chain process via X· without the need to reform or break the C—C bond of adenosylcob(III)alamin.

is the product of the primary kinetic isotope effect on hydrogen atom transfer and the statistical discrimination against a given hydrogen migrating from a "pool" of hydrogens in the active site.[105] This result most certainly demands that a protein residue be the intermediate radical carrier that propagates subsequent catalytic cycles under conditions of saturating substrate.[105]

5.11.4.4 Diol Dehydrase

Diol dehydrase converts vicinal diols such as 1,2-propanediol and 1,2-ethanediol to a geminal diol that dehydrates to the corresponding aldehyde. The enzyme provides *Clostridium, Klebsiella,* and *Salmonella* with oxidizing equivalents under conditions of anaerobic growth.[106] The enzyme was first isolated from *Klebsiella oxytoca,* and characterized by Abeles and co-workers, but only small quantities are available from natural sources.[75,76,107,108]

Recombinant diol dehydrase derived from both *Klebsiella oxytoca* and *Salmonella typhimurium* is now available.[109,110] The minimum composition for active diol dehydrase is three types of subunits of MW 60 kDa, 24 kDa, and 19 kDa, in an $(\alpha\beta\gamma)_4$ arrangement.[110] One cofactor molecule binds to each $(\alpha\beta\gamma)_2$ oligomer.[110] EPR studies show that Co—C bond homolysis forms Co^{II} with $g = 2.22$ and an organic radical at $g = 2.003$.[111]

Hydrogen migration and *gem*-diol dehydration are stereospecific. Aldehyde formation occurs by oxygen migration, followed by the elimination of water from the *gem*-diol, rather than possible β-elimination of water with subsequent tautomerization of the resulting enol. Hydrogen atom transfer between product and enzyme-bound cofactor is reversible, with a large tritium kinetic isotope effect of 125 which suggests the intermediacy of a protein-based radical, as postulated for ethanolamine ammonia lyase.[105,112]

The activity of many adenosylcobalamin analogues has been investigated in the diol dehydrase reaction.[113–115] Adenosylcobalamin analogues with modifications to the ribose moiety still function as cofactors, but with decreased catalytic activity. Analogues in which the 5′-methylene linker to cobalt has been extended inhibit the reaction and do not support catalysis. Modification of the amide side chains that adorn the corrin ring appear to affect coenzyme binding more than catalysis. An adenosylcobalamin analogue in which the e-propionamide is hydrolyzed to a carboxylic acid still supports catalysis, but at a lower overall rate.

Several important amino acid residues have been identified in diol dehydrase by chemical modification.[116–120] The arginine-specific reagents 2,3-butanedione and phenylglyoxal inactivate apo-enzyme but not holoenzyme. Protection by phosphate suggests that an essential arginine residue is located in the active site and may be involved in binding coenzyme. Thiol modification blocks coenzyme binding and leads to loss of catalytic activity, with the rate of inactivation being higher above pH 8.5.[116] Chemical modification studies in which histidine is selectively modified shows that those residues are also important for cobalamin binding.[120]

5.11.4.5 Methionine Synthase

In contrast to the other enzymes discussed in this chapter, methionine synthase utilizes methylcobalamin as cofactor. This enzyme has a critical role in one-carbon metabolism because it converts homocysteine to methionine.[121,122] It is this requirement for methylcobalamin to support methionine synthase that often identifies a physiological B_{12} deficiency through incompetent hematopoiesis and neurological defects that are thought to be caused by insufficient nerve myelination.[123]

The tightly bound cofactor serves as an intermediary to accept a methyl group from methyl tetrahydrofolate and transfer this methyl group to homocysteine, as shown in Figure 7. Methyl transfer is concomitant with cob(I)alamin generation, and this reactive nucleophile is oxidized to cob(II)alamin in about one out of every 2000 turnovers. Thus, a mechanism for reductive methylation must also exist to restore cob(II)alamin to methylcob(III)alamin prior to the next catalytic cycle.[124–126]

Methionine synthase has a modular structure with distinct domains that are responsible for substrate binding, cobalamin binding, and reductive reactivation with adenosylmethionine.[125] Separation of these fragments by proteolysis facilitated crystallization of the cobalamin-binding domain and subsequent structure analysis, including the first direct observation of His759 coordination as a substitute for the 5,6-dimethylbenzimidazole ligand.[7]

In methionine synthase, His759 is linked by hydrogen bonds to Asp757 and Ser810, with protonation of the His-Asp-Ser triad increasing the strength of imidazole nitrogen coordination to cob(I)alamin following methyl transfer.[125,126] The site-directed mutant H759G is completely inactive towards methyl transfer, whereas D757N and S810A can still support methyl transfer, but have lower rates of reactivation following cob(II)alamin formation.

5.11.5 CONCLUSIONS

The availability of large quantities of recombinant cobalamin-dependent enzymes, along with knowledge of the methionine synthase structure, rekindled interest in cobalamin biochemistry, with additional insight coming from EPR data, chemical analogue data, and visible spectroscopic data. It is clear that a kinetically competent substrate radical is involved in several adenosylcobalamin-dependent reactions, but the role of the protein in facilitating the radical rearrangement is unknown. The intermediacy of a protein radical in some adenosylcobalamin-dependent reactions seems certain, but it is unclear why a protein side chain is not a universal feature of adenosylcobalamin-dependent enzymes.

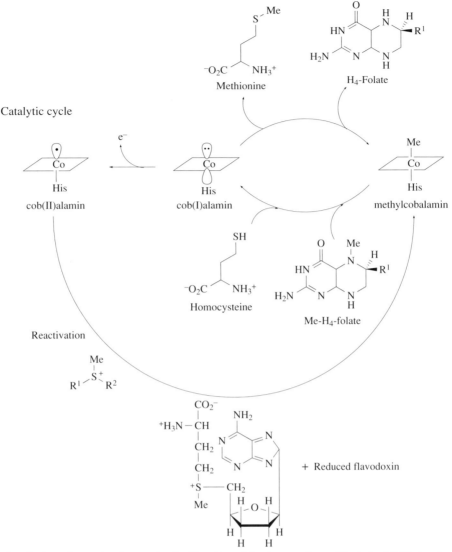

Figure 7 Catalysis and reactivation cycles in *E. coli* methionine synthase. Methylcob(III)alamin transfers a methyl group to homocysteine to yield methionine and cob(I)alamin. Methyltetrahydrofolate, MeH$_4$-folate, remethylates cob(I)alamin to regenerate methylcob(III)alamin. Cob(II)alamin that is formed in a minor side reaction is reduced to cob(I)alamin at the expense of (*S*)-adenosylmethionine and reduced flavin in a reactivation pathway (after Drennan *et al.*[124]).

Structural studies of cobalamin-dependent enzymes will continue to reveal important features of plumbing and architecture, but a complete understanding of the catalytic mechanism will only emerge when all the techniques of bioorganic chemistry, enzymology, and molecular biology are brought to bear on each system. Although cobalamin-dependent enzymes have been known for more than 30 years, classification according to reaction mechanism (i.e., homolytic vs. heterolytic chemistry) is subject to revision as new information emerges. Cobalamin is considered to be a cofactor of primordial origin, but our knowledge of enzyme reaction mechanisms involving B$_{12}$ is still incomplete.

5.11.6   REFERENCES

1. D. Dolphin (ed.), "B$_{12}$," Wiley, New York, 1982, vols. 1, 2.
2. Z. Schneider and A. Stroinski (eds.), "Comprehensive B$_{12}$," de Gruyter, Berlin, 1987.
3. B. Kräutler, D. Arigoni, and B. T. Golding (eds.), "Vitamin B$_{12}$ and B$_{12}$ Proteins," VCH Weinheim, 1998.
4. M. L. Ludwig and R. G. Matthews, *Annu. Rev. Biochem.*, 1997, **66**, 269.

5. B. T. Golding and D. N. R. Rao, in "Enzyme Mechanisms," eds. M. I. Page and A. Williams, Royal Society of Chemistry, London, 1987, p. 405.
6. W. Buckel and B. T. Golding, *Chem. Soc. Rev.*, 1996, **26**, 329.
7. C. L. Drennan, S. Huang, J. T. Drummond, R. G. Matthews, and M. L. Ludwig, *Science*, 1994, **266**, 1669.
8. F. Mancia, N. H. Keep, A. Nakagawa, P. F. Leadlay, S. McSweeney, B. Rasmussen, P. Bösecke, O. Diat, and P. R. Evans, *Structure*, 1996, **4**, 339.
9. J. Halpern, *Science*, 1985, **227**, 869.
10. M. S. A. Hamza and J. M. Pratt, *J. Chem. Soc., Dalton Trans.*, 1994, 1377.
11. J. M. Pratt, "Metal Ions in Biological Systems," eds. H. Sigel and A. Sigel, Marcel Dekker, New York, 1993, **29**, 229.
12. M. D. Waddington and R. G. Finke, *J. Am. Chem. Soc.*, 1993, **115**, 4629.
13. B. P. Hay and R. G. Finke, *J. Am. Chem. Soc.*, 1987, **109**, 8012.
14. B. P. Hay and R. G. Finke, *J. Am. Chem. Soc.*, 1986, **108**, 4820.
15. B. P. Hay and R. G. Finke, *Polyhedron*, 1988, **7**, 1469.
16. T. W. Koenig, B. P. Hay, and R. G. Finke, *Polyhedron* 1988, **7**, 1499.
17. R. G. Finke and B. P. Hay, *Inorg. Chem.* 1984, **23**, 3041.
18. T.-T. Tsou, M. Loots, and J. Halpern, *J. Am. Chem. Soc.*, 1982, **104**, 623.
19. J. Halpern, S. -K. Kim, and T. W. Leung, *J. Am. Chem. Soc.*, 1984, **106**, 8317.
20. S. Wollowitz and J. Halpern, *J. Am. Chem. Soc.*, 1984, **106**, 8319.
21. J. Halpern, *Polyhedron*, 1988, **7**, 1483.
22. J. Halpern, *Bull. Chem. Soc. J.*, 1988, **61**, 13.
23. J. Halpern, in "B₁₂," ed. D. Dolphin, Wiley, New York, 1982, vol. 1, p. 502.
24. J. M. Wood, in "B₁₂," ed. D. Dolphin, Wiley, New York, 1982, vol. 2, Chap. 6.
25. J. F. Endicott, K. P. Balakrishnan, and C.-L. Wong, *J. Am. Chem. Soc.*, 1980, **102**, 5519.
26. L. E. H. Gerards, H. Bulthuis, M. W. G. de Bolster, and S. Balt, *Inorg. Chim. Acta*, 1991, **190**, 47.
27. B. D. Martin and R. G. Finke, *J. Am. Chem. Soc.*, 1990, **112**, 2419.
28. B. D. Martin and R. G. Finke, *J. Am. Chem. Soc.*, 1992, **114**, 585.
29. P. J. Toscano, A. L. Seligson, M. T. Curran, A. T. Skrobutt, and D. C. Sonnenberger, *Inorg. Chem.*, 1989, **28**, 166.
30. S. Geremia, M. Mari, L. Randaccio, and E. Zangrando, *J. Organomet. Chem.*, 1991, **408**, 95.
31. C. D. Garr, J. M. Sirovatka, and R. G. Finke, *Inorg. Chem.*, 1996, **35**, 5912.
32. C. D. Garr, Ph.D. Thesis, University of Oregon, Eugene, OR, 1992.
33. S. Geremia, L. Randaccio, E. Zangrando, and L. Antolini, *J. Organomet. Chem.*, 1992, **425**, 131.
34. B. Krautler, R. Konrat, E. Stupperich, G. Farber, K. Gruber, and C. Kratky, *Inorg. Chem.*, 1994, **33**, 4128.
35. V. Pett, M. Liebman, P. Murray-Rust, K. Prasad, and J. P. Glusker, *J. Am. Chem. Soc.*, 1987, **109**, 3207.
36. P. G. Lenbert, *Proc. R. Soc. London, Ser. A*, 1968, **303**, 45.
37. M. D. Wirt, I. Sagi, and M. R. Chance, *Biophys. J.*, 1992, **63**, 412.
38. M. D. Wirt, I. Sagi, E. Chen, S. M. Frisbie, R. Lee, and M. R. Chance, *J. Am. Chem. Soc.*, 1991, **113**, 5299.
39. S. M. Chemaly and J. M. Pratt, *J. Chem. Soc., Dalton Trans.*, 1980, 2274.
40. B. E. Daikh and R. G. Finke, *J. Am. Chem. Soc.*, 1991, **113**, 4160.
41. M. T. Summers, P. J. Toscano, N. Bresciani-Pahor, G. Nardin, L. Randaccio, and L. G. Marzilli, *J. Am. Chem. Soc.*, 1983, **105**, 6259.
42. W. B. Lott, A. M. Chagovetz, and C. B. Grissom, *J. Am. Chem. Soc.*, 1995, **117**, 12 194.
43. P. Dowd, B. Wilk, and B. K. Wilk, *J. Am. Chem. Soc.*, 1992, **114**, 7949.
44. M. He and P. Dowd, *J. Am. Chem. Soc.*, 1996, **118**, 711.
45. F. Francalanci, N. K. Davis, J. Q. Fuller, D. Murfitt, and P. F. Leadlay, *Biochem. J.*, 1986, **236**, 489.
46. R. Padmakumar, S. Taoka, R. Padmakumar, and R. Banerjee, *J. Am. Chem. Soc.*, 1995, **117**, 7033.
47. O. Zelder, B. Beatrix, F. Kroll, and W. Buckel, *FEBS Lett.*, 1995, **369**, 252.
48. M. Yamanishi, S. Yamada, H. Muguruma, Y. Murakami, T. Tobimatsu, A. Ishida, J. Yamauchi, T. Toraya, *Biochemistry*, 1998, **37**, 4799.
49. E. N. G. Marsh and D. E. Holloway, *FEBS Lett.*, 1992, **310**, 167.
50. H. -P. Chen and E. N. G. Marsh, *Biochemistry*, 1997, **36**, 7884.
51. E.-I. Ochiai, in "Metal Ions in Biological Systems," eds. H. Sigel and A. Sigel, Marcel Dekker, New York, 1994, p. 255.
52. P. M. Prathare, D. S. Wilbur, S. Heusser, E. V. Quadros, P. McLoughlin, and A. C. Morgan, *Bioconjugate Chem.*, 1996, **7**, 217.
53. K. L. Brown, S. Cheng, J. D. Zubkowski, and E. J. Valente, *Inorg. Chem.*, 1996, **35**, 3442.
54. J. Retey, *Angew. Chem., Int. Ed. Engl.*, 1990, **29**, 355.
55. Y. Zhao, P. Such, and J. Retey, *Angew. Chem., Int. Ed. Engl.*, 1992, **31**, 215.
56. K. N. Joblin, A. W. Johnson, and M. F. Lappert, *FEBS Lett.*, 1975, **53**, 193.
57. M. R. Hollaway, H. A. White, K. N. Joblin, A. W. Johnson, M. F. Lappert, and O. C. Wallis, *Eur. J. Biochem.*, 1978, **82**, 143.
58. J. S. Krouwer, B. Holmquist, R. S. Kipnes, and B. M. Babior, *Biochim. Biophys. Acta.* 1980, **612**, 153.
59. T. T. Harkins and C. B. Grissom, *J. Am. Chem. Soc.*, 1995, **117**, 566.
60. E. Natarajan and C. B. Grissom, *Photochem. Photobiol.*, 1996, **64**, 286.
61. M. B. Taraban, T. V. Leshina, M. A. Anderson, and C. B. Grissom, *J. Am. Chem. Soc.*, 1997, **119**, 5769.
62. J. F. Endicott and T. L. Netzel, *J. Am. Chem. Soc.*, 1979, **101**, 4000.
63. L. A. Walker, II, J. T. Jarrett, N. A. Anderson, S. H. Pullen, R. G. Matthews, and R. J. Sension, *J. Am. Chem. Soc.*, 1998, **120**, 3597.
64. C. D. Garr and R. G. Finke, *J. Am. Chem. Soc.*, 1992, **114**, 10 440.
65. E. Chen and M. R. Chance, *J. Biol. Chem.*, 1990, **265**, 12 987.
66. E. Chen and M. R. Chance, *Biochemistry*, 1993, **32**, 1480.
67. T. T. Harkins and C. B. Grissom, *Science*, 1994, **263**, 958.
68. J. M. Sirovatka and R. G. Finke, *J. Am. Chem. Soc.*, 1997, **119**, 3057.
69. J. T. Jarrett, M. Amaratunga, C. L. Drennan, J. D. Scholten, R. H. Sands, M. L. Ludwig, and R. G. Mathews, *Biochemistry*, 1996, **35**, 2464.

70. J. T. Drummond, R. R. Ogorzalek Loo, and R. G. Matthews, *Biochemistry*, 1993, **32**, 9282.
71. R. V. Banerjee, N. L. Johnston, J. K. Sobeski, P. Datta, and R. G. Matthews, *J. Biol. Chem.*, 1989, **264**, 13 888.
72. B. M. Babior and J. S. Krouwer, *CRC Crit. Rev. Biochem.*, 1979, 35.
73. B. Zagalak and W. Friedrich (eds.), "Vitamin B$_{12}$," de Gruyter, Berlin, 1979.
74. B. M. Babior, *Acc. Chem. Res.*, 1975, **8**, 376.
75. P. A. Frey, M. K. Essenberg, and R. H. Abeles, *J. Biol. Chem.*, 1967, **242**, 5369.
76. P. A. Frey and R. H. Abeles, *J. Biol. Chem.*, 1966, **241**, 2732.
77. J. Retey and D. Arigoni, *Experientia*, 1966, **22**, 783.
78. R. L. Switzer, B. G. Baltimore, and H. A. Barker, *J. Biol. Chem.*, 1969, **244**, 5236.
79. B. M. Babior, *Biochim. Biophys. Acta*, 1968, **167**, 456.
80. H. P. C. Hogenkamp, R. K. Ghambeer, C. Brownson, R. L. Blakley, and E. Vitols, *J. Biol. Chem.*, 1968, **243**, 799.
81. H. P. C. Hogenkamp, R. K. Ghambeer, C. Brownson, and R. L. Blakley, *Biochem. J.*, 1967, **103**, 5c.
82. J. T. Drummond and R. G. Matthews, *Biochemistry*, 1994, **33**, 3742.
83. R. B. Silverman and D. Dolphin, *J. Am. Chem. Soc.*, 1973, **95**, 1686.
84. E. N. Marsh, S. E. Harding, and P. F. Leadlay, *Biochem. J.*, 1989, **260**, 353.
85. M. Sprecher, M. J. Clark, and B. D. Sprinson, *J. Biol. Chem.*, 1966, **241**, 872.
86. M. Sprecher, M. J. Clark, and B. D. Sprinson, *Biochem. Biophys. Res. Commun.*, 1964, **15**, 581.
87. N. H. Keep, G. A. Smith, M. C. W. Evans, G. P. Diakun, and P. F. Leadlay, *Biochem. J.*, 1993, **295**, 387.
88. Y. Zhao, A. Abend, M. Kunz, P. Such, and J. Retey, *Eur. J. Biochem.*, 1994, **225**, 891.
89. R. Padmakumar and R. Banerjee, *J. Biol. Chem.*, 1995, **270**, 9295.
90. R. Banerjee, in "Vitamin B$_{12}$ and B$_{12}$ Proteins," eds. B. Kräutler, D. Arigoni, and B. T. Golding, VCH Weinheim, 1998, p. 189.
91. F. Suzuki and H. A. Barker, *J. Biol. Chem.*, 1966, **241**, 878.
92. E. N. Marsh, N. McKie, N. K. Davis, and P. F. Leadlay, *Biochem. J.*, 1989, **260**, 345.
93. E. N. G. Marsh, *Biochemistry*, 1995, **34**, 7542.
94. U. Leutbecher, S. P. J. Albracht, and W. Buckel, *FEBS Lett.*, 1992, **307**, 144.
95. O. Zelder, B. Beatrix, U. Leutbecher, and W. Buckel, *Eur. J. Biochem.*, 1994, **226**, 577.
96. B. Beatrix, O. Zelder, K. Kroll, G. Orlygsson, B. T. Golding, and W. Buckel, *Angew. Chem., Int. Ed. Engl.*, 1995, **34**, 2398.
97. L. P. Faust, J. A. Connor, D. M. Roof, J. A. Hoch, and B. M. Babior, *J. Biol. Chem.*, 1990, **265**, 12 462.
98. L. P. Faust and B. M. Babior, *Arch. Biochem. Biophys.*, 1992, **294**, 50.
99. D. M. Roof and J. R. Roth, *J. Bacteriol.*, 1988, **170**, 3855.
100. B. M. Babior, T. H. Moss, W. H. Orme-Johnson, and H. Beinert, *J. Biol. Chem.*, 1974, **249**, 4537.
101. S. Taoka, R. Padmakumar, C. B. Grissom, and R. Banerjee, *Bioelectromagnetics*, 1997, **18**, 506.
102. C. B. Grissom, *Chem. Rev.*, 1995, **95**, 3.
103. C. B. Grissom and E. Natarajan, *Methods Enzymol.*, 1997, **281**, 235.
104. D. A. Weisblat and B. M. Babior, *J. Biol. Chem.*, 1971, **246**, 6064.
105. W. W. Cleland, *CRC Crit. Rev. Biochem.*, 1982, **13**, 385.
106. T. Toraya, in "Metal Ions in Biological Systems," eds. H. Sigel and A. Sigel, Marcel Dekker, New York, 1995, **30**, 217.
107. B. Zagalak, P. A. Frey, G. L. Karabatsos, and R. H. Abeles, *J. Biol. Chem.*, 1966, **241**, 3028.
108. R. H. Abeles and B. Zagalak, *J. Biol. Chem.*, 1966, **241**, 1245.
109. T. Tobimatsu, T. Hara, M. Sakaguchi, Y. Kishimoto, Y. Wada, M. Isoda, T. Sakai and T. Toraya, *J. Biol. Chem.*, 1995, **270**, 7142.
110. Y. Xu, T. Bobik, C. B. Grissom, and J. Roth, manuscript in preparation.
111. T. H. Finlay, J. Valinsky, A. S. Mildvan, and R. H. Abeles, *J. Biol. Chem.*, 1973, **248**, 1285.
112. M. K. Essenberg, P. A. Frey, and R. H. Abeles, *J. Am. Chem. Soc.*, 1971, **93**, 1242.
113. T. Toraya, K. Ushio, S. Fukui, and H. P. C. Hogenkamp, *J. Biol. Chem.*, 1977, **252**, 963.
114. T. Toraya and S. Fukui, *J. Biol. Chem.*, 1980, **255**, 3520.
115. T. Toraya and R. H. Abeles, *Arch. Biochem. Biophys.*, 1980, **203**, 174.
116. S. Kuno, T. Toraya, and S. Fukui, *Arch. Biochem. Biophys.*, 1980, **205**, 240.
117. S. Kuno, T. Toraya, and S. Fukui, *Arch. Biochem. Biophys.*, 1981, **211**, 722.
118. S. Kuno, T. Toraya, and S. Fukui, *Arch. Biochem. Biophys.*, 1981, **210**, 474.
119. T. Toraya, N. Watanabe, K. Ushio, T. Matsumoto, and S. Fukui, *J. Biol. Chem.*, 1983, **258**, 9296.
120. S. Kuno, S. Fukui, and T. Toraya, *Arch. Biochem. Biophys.*, 1990, **277**, 211.
121. M. A. Foster, M. J. Dilworth, and D. D. Woods, *Nature*, 1964, **201**, 39.
122. R. Banerjee and R. G. Matthews, *FASEB J.*, 1990, **4**, 1450.
123. S. S. Pant, A. K. Asbury, and E. P. Richardson, *Acta Neurol. Scand.*, 1968, **44**, 7.
124. C. L. Drennan, M. M. Dixon, D. M. Hoover, J. T. Jarrett, C. W. Goulding, R. G. Matthews, and M. L. Ludwig, in "Vitamin B$_{12}$ and B$_{12}$ Proteins," eds. B. Kräutler, D. Arigoni, and B. T. Golding, VCH Weinheim, 1998, p. 133.
125. J. J. Jarrett, M. Amaratunga, C. L. Drennan, J. D. Scholten, R. H. Sands, M. L. Ludwig, and R. G. Matthews, *Biochemistry*, 1996, **35**, 2464.
126. M. Amaratunga, K. Fluhr, J. T. Jarrett, C. L. Drennan, M. L. Ludwig, R. G. Matthews, and J. D. Scholten, *Biochemistry*, 1996, **35**, 2453.

5.12
Glycosyl Transferase Mechanisms

DAVID L. ZECHEL and STEPHEN G. WITHERS
University of British Columbia, Vancouver, BC, Canada

5.12.1 INTRODUCTION

5.12.1.1 Definitions and Categorization of Glycosyl Transferases

Glycosyl transferases and hydrolases catalyze the transfer of a glycosyl residue from their specific donor to an acceptor. They therefore play crucial roles in the biosynthesis and degradation of a large range of biological structures. These include polysaccharides, oligosaccharides, saponins, antibiotics, glycolipids, glycoproteins, proteoglycans and peptidoglycans. They are also important in a number of other processes including the remodeling of cell walls, the damage and repair of DNA by base excision or transfer, and other toxin mechanisms. Given this large range of roles, a wide variety of enzyme structures can be anticipated and, as is shown later, this is indeed the case. However the number of different mechanisms is very limited. A number of useful reviews on these enzymes, or aspects thereof, have been published of which the following are a sample: more general reviews;[1-11] amylases;[12] glycogen phosphorylase;[13-21] thermophilic glycosidases.[22]

The term glycosyl transferase can be used broadly to describe any enzyme catalyzing a glycosyl transfer reaction. In the case of hydrolases transfer occurs, of course, to water, whereas for other transferases the acceptor is most commonly an alcohol functionality, typically one from another sugar. However, transfer can occur to nitrogen or sulfur nucleophiles in some cases. Most commonly, the term glycosyl transferase is reserved for enzymes catalyzing glycosyl transfer to an acceptor other than water, while the term glycosidase or glycoside hydrolase is used for enzymes catalyzing the hydrolytic process.

Within the class of glycosyl transferases, the enzymes can be categorized broadly on the basis of the identity of the donor. One class of transferases is that of the nucleotide phosphate-using transferases. These are enzymes involved strictly in biosynthetic processes that use an activated glycosyl species as the glycosyl donor in order to ensure that the reaction proceeds in the direction of glycosyl transfer. In most cases, the donor is a nucleotide diphosphate sugar such as uridine diphospho-α-D-glucose (**1**), thus the leaving group will be a nucleotide diphosphate (Nature's tosylate). In one case, however, CMP-*N*-acetyl neuraminic acid (**2**), the donor is a nucleotide monophosphate. Different nucleotides are used for the transfer of different sugars, and Table 1 shows those most commonly employed in mammalian systems. However, the identity of the specific nucleotide employed can vary in different processes and is often different in the plant versus animal kingdoms. Another type of donor employed for syntheses catalyzed by many membrane bound glycosyl transferases is that of the dolichol phosphate sugars (**3**), themselves synthesized from a nucleotide phosphate sugar.

Table 1 Sugar nucleotides commonly used by mammalian glycosyl transferases.

UDP	GDP	CMP
N-Acetylgalactosamine	Fucose	Sialic acid
N-Acetylglucosamine	Mannose	
N-Acetylmuramic acid		
Galactose		
Glucose		
Glucuronic acid		
Xylose		

(**1**)   (**2**)

(**3**)

A second, broad class of glycosyl transferases exists in which transfer occurs from a species other than a nucleotide diphosphate sugar. The most commonly occurring of these are those in which the donor is an oligosaccharide or polysaccharide and transfer occurs to another sugar or to a phosphate moiety. The best-studied examples, which are discussed within this chapter, are those of enzymes involved in α-glucan conversion. These are cyclodextrin glucanotransferase, which carries out an intramolecular glycosyl transfer, thereby making cyclic oligosaccharides, and glycogen phosphorylase, which carries out transfer of glucosyl units from glycogen to phosphate as an acceptor. Also discussed in some detail are nucleoside phosphorylases, which transfer a ribosyl moiety from a nucleoside to phosphate with release of the heterocyclic base.

The reaction catalyzed in all cases is a substitution reaction at a chiral acetal or ketal center. This displacement can occur with either inversion or retention of configuration at the sugar anomeric center and, as noted below, this demands different mechanisms. The determination of whether the glycosidase is an inverting or retaining enzyme is therefore of paramount importance in any mechanistic characterization. Here, we describe glycosidases in terms of the anomeric configuration of substrate and initial product, using e and a to indicate equatorial and axial. Thus enzymes will be either $e \rightarrow e$ or $a \rightarrow a$ retaining, or $e \rightarrow a$ or $a \rightarrow e$ inverting glycosidases.

5.12.1.2 Classification of Glycosyl Hydrolases and Transferases

The traditional classification of glycosidases has been on the basis of the actual reaction catalyzed. Thus enzymes were named and classified on the basis of the anomeric stereochemistry of the substrate cleaved (α or β), and the glycone type cleaved the most rapidly (D-glucoside, L-arabinoside, etc.), resulting in the commonly used names such as β-D-glucosidase or α-L-arabinosidase. This very pragmatic naming system has served well, but is not without its problems since many glycosidases have a relatively broad specificity, yet the name is often derived from the substrate first found to be rapidly cleaved; this is not necessarily the best substrate. Classification on the basis of substrate specificity has provided the foundation for the IUBMB designation according to EC numbers. Thus glycosidases cleaving O-glycosides have the designation (3.2.1.X), in which the final digit "X" generally provides information on the substrate specificity. While this classification is very useful in practical terms for defining the specificity, it takes no account of the structural or mechanistic relationships between enzymes, thus is of limited value in any structure/function analyses or in assessing evolutionary relationships. This problem is well-discussed in a recent review.[3]

A solution to this problem has come from classification on the basis of sequence similarities. Thus, the glycosidases for which sequence information is available have been arranged into over 60 different families based upon sequence similarities.[23-25] Approximately 1000 enzymes have been categorized in this way at this stage, with the number increasing daily. The full listing of the members of these families, along with links to the sequence data base, information on whether enzymes within the family follow a retaining or an inverting mechanism, and information on the identities of key amino acid residues, where known, is available on the Internet (URL: http://www.expasy.ch/cgi-bin/lists?glycosid.txt).

The implication of the existence of these families is that all enzymes within a family likely have a similar structure (i.e. a similar protein "fold"). Further, if the structures are the same it is probable that, within a family, enzymes would follow a similar mechanism. These expectations have turned out to be correct in all cases investigated to date. The conservation of mechanism (inverting versus retaining) was first investigated within the cellulase members of these families by use of ¹H NMR to determine the stereochemical outcomes of hydrolysis catalyzed by a range of such enzymes.[26] In all cases investigated, enzymes within a family were shown to follow the same stereochemical outcome, and this has held true in all other cases investigated since then. Structural similarity for different enzymes within a family has been demonstrated in a number of cases; indeed molecular replacement procedures have often been used to solve phase problems in solving structures of other members of the same family. This is well-illustrated in a recent review of glycosidase structures and their family relationships.[2]

Not surprisingly, enzymes within a family generally catalyze the same reaction, and the presence of a range of different, yet related enzymes, simply reflects their different sources. However, there can be a significant range of different specificities within a family. These can be relatively small variations in specificity such as those in Family 1 where enzymes typically have both β-glucosidase and β-galactosidase activity, but some are very specific for the hydrolysis of 6-phospho-β-glucosides and galactosides. An apparently wider range of activities is seen within Family 13, which includes both hydrolases and transferases. The majority of the family members are α-glycosidases such as α-amylases and α-glucosidases. However, the family also includes pure glycosyl transferases such as cyclodextrin glucanotransferase and glycogen debranching enzyme. Three-dimensional structures are available for several hydrolases and transferases within this family and these confirm the structural similarities, consistent with the expected mechanistic similarities of such hydrolases and transferases. Apparently greater diversity is present within members of Family 27 which includes both α-galactosidases and α-N-acetylgalactosaminidases; similarly, and perhaps most surprisingly, Family 39 contains both β-D-xylosidases and α-L-iduronidases. While the apparent α/β difference is actually simply a function of the vagaries of carbohydrate nomenclature, it is surprising that a pentopyranoside hydrolase and a glycuronidase should show such sequence similarity. In addition to this diversity within some families, there is, in fact, considerable structural similarity between several families, which is not readily apparent at the sequence level, but is clearly present at the three-dimensional level. As a consequence, a number of these families have been assigned as members of so-called "clans" with similar three-dimensional structures.[27] The biggest of these clans identified to date is that of clan GH-A, the members of which all hydrolyze β-glycosides with net retention of configuration. This clan is alternatively known as the 4/7 clan because the enzymes all have an $(\alpha/\beta)_8$ fold with the key carboxylic acid residues located on the carboxyl-terminal ends of the 4- and 7-β-strands.[28]

As noted above, this family classification also includes glycosyl transferases in which transfer occurs between sugar residues. Such enzymes clearly use essentially identical mechanisms to the analogous retaining glycosidases. The mechanistic differences between these two classes are minimal, being determined by whether the glycosyl enzyme intermediate can react simply with water, or whether binding interactions with a second sugar are required. This does not appear to be the case for the nucleotide sugar-using glycosyl transferases, since no homology has yet been noted between catalytic domains of this class of enzymes and any glycosidases. Mechanistic insight into this class of enzymes is limited, there having been very few detailed studies, and essentially all of these on the inverting class of glycosyl transferases. Structural studies are even sparser, the only structure yet reported being that of a DNA β-glucosyl transferase,[29] which appears unrelated to other glycosyl transferases studied. However, this situation is likely to change rapidly as the pace of research in this area has picked up substantially of late, particularly in the cloning and expression of these important, and potentially commercially valuable enzymes. Indeed a sequence-based classification has now been proposed in which some 556 proteins have been placed into 26 families on the basis of sequence similarities.[28] The assignment of glycosidases into families has proved extraordinarily useful in bringing order into structural and mechanistic studies of this enormous class of enzymes. Similar benefits can be anticipated for the glycosyl transferases.

5.12.1.3 Structures of Glycosidases

Until 1990, three-dimensional structures for only a very few glycosidases were available, most notably hen egg white lysozyme, bacteriophage T4 lysozyme, and Taka α-amylase. However, there has been an incredible increase in the number of available structures with representatives of 27 of the 60 sequence related families being structurally defined. Representations of these structures are shown in Figure 1. In a number of cases, the structures shown are those of the catalytic domains of the enzymes, since the intact modular enzymes containing, for example, binding and recognition domains, frequently do not crystallize. Structures for most of these glycosidases, along with those of a host of other enzymes, are conveniently available at the following web site maintained by the Biomolecular structure and modelling group at University College, London (URL: http://www.biochem.ucl.ac.uk/bsm/enzymes/index.html).

Perhaps the most striking feature of this collection of structures is their incredible structural diversity with some being almost entirely composed of β-sheet, others being primarily α-helical, and then a range of structures of mixed α-helix and β-sheet in between. This may seem surprising initially given that all carry out the same chemical transformation, hydrolysis of an acetal. However the diversity observed is presumably in part a reflection of the diverse structures of the glycoside substrate being cleaved, and in part a reflection of the different solutions arrived at evolutionarily to the problem of construction of a catalytic site. A review[2] summarizing the structure/function relationships points out, amongst other things, that the active site topologies of these enzymes fall into three classes which reflect the cleavage patterns. For those glycosidases which have a purely exo-mode of action (they cleave only from the ends of chains, typically releasing either a mono-

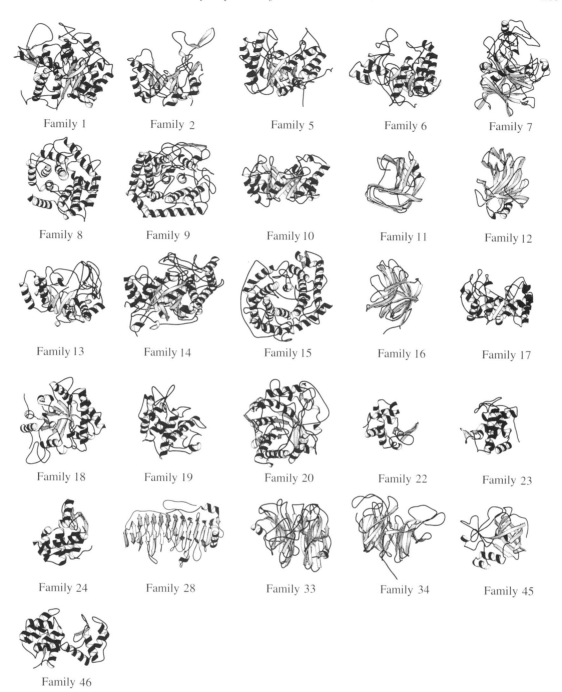

Figure 1 Representative structures of glycosidases, generously provided by G. Davies (Dept of Chemistry, University of York, Heslington, York): Family 1, *Trifolium repens* β-glucosidase;[30] Family 2, *Escherichia coli* β-galactosidase;[31] Family 5, *Clostridium cellulolyticum* endoglucanase A;[32] Family 6, *Trichoderma reesei* cellobiohydrolase II[33]; Family 7, *Trichoderma reesei* cellobiohydrolase I;[34] Family 8, *Clostridium thermocellum* endoglucanase celA;[35] Family 9, *Clostridium thermocellum* endoglucanase D;[36] Family 10, *Streptomyces lividans* xylanase A;[37] Family 11, *Bacillus circulans* xylanase;[38] Family 12, *Streptomyces lividans* endocellulase celB;[39] Family 13, *Aspergillus oryzae* α-amylase;[40] Family 14, *Glycine max* β-amylase;[41] Family 15, *Aspergillus awamori* glucoamylase;[42] Family 16, *Bacillus* sp. β-1,3-1,4 glucanase;[43] Family 17, *Hordeum vulgare* β-1,3-1,4 glucanase;[44] Family 18, *Serratia marcesans* chitinase A;[45] Family 19, *Hordeum vulgare* chitinase;[46] Family 20, *Serratia marcesans* chitiobiase;[47] Family 22, hen egg-white lysozyme;[48] Family 23, goose egg-white lysozyme;[49] Family 24, bacteriophage T4 lysozyme;[50] Family 28, *Aspergillus aculeatus* rhamnogalacturonase;[51] Family 33, *Salmonella typhimurium* sialidase;[52] Family 34, Influenza virus B neuraminidase;[53] Family 45, *Humicola insolens* endoglucanase V;[54] Family 46, *Streptomyces* N174 chitosanase.[55]

saccharide or in some cases a disaccharide) the active sites are a pocket into which the end of the chain fits. Cleavage of the glycoside would require release of the monosaccharide product before a second cleavage event on the oligosaccharide could occur. Examples of this class of enzyme are the Family 1 exo-glucosidases, the Family 14 β-amylases, or the Family 15 glucoamylases. For glycosidases with an endo-mode of action (cleavage can occur within a polysaccharide chain) the active site is in a cleft on the enzyme surface. Examples of this class of enzyme are the Family 22–24 lysozymes, Family 18 and 19 chitinases, the Family 13 α-amylases and several families of endo-cellulases. A third type of active site is that of the tunnel through which it appears the polymeric substrate must thread. It is assumed that this allows the enzyme to remain bound to the substrate after cleavage, enhancing the processional nature of the hydrolytic process. A good example of this is seen in the Family 6 cellobiohydrolases. Threading of the substrate may, however, not be explicitly required as it appears likely that the loops forming the roof of the tunnel can unfold allowing access and egress of the polymeric chain, but possibly on a less frequent basis than threading.

5.12.1.4 Reaction Mechanisms

As noted earlier, there are two basic categories of glycosidases, those that hydrolyze the glycosidic bond with net retention of anomeric configuration and those that do so with net inversion. Accordingly, for the most part, there appear to be two basic mechanisms, as first suggested by Koshland well over 40 years ago,[56] one (inversion) involving a single chemical step and the other (retention) involving two. In terms of basic mechanism, it seems to make no significant difference whether the glycosidase is cleaving an axial or an equatorial bond, the mechanisms are essentially the same. Similarly, nonnucleotide sugar-using glycosyl transferases appear to follow essentially identical mechanisms to the corresponding glycosidases. In this section, therefore, we shall therefore consider all glycosidases as belonging to two mechanistic groups, the only substantial variation being an alternative "substrate-assisted" mechanism seen for certain retaining glycosidases.

5.12.1.4.1 *Inverting glycosidases*

Catalysis by this class of enzymes occurs within an enzyme active site containing, in all cases structurally examined to date, a pair of carboxylic acids, typically at a separation (oxygen atoms to oxygen atoms) of approximately 9.5 Å.[5,57] Such an arrangement allows one of the carboxylic acids to function as a general acid catalyst, assisting the departure of the aglycone, while the other functions as a general base catalyst, deprotonating the nucleophilic water molecule in a concerted manner as it attacks at the sugar anomeric center. This mechanism is shown in Scheme 1. A key feature of all displacement reactions at the anomeric centers of sugars is the dissociative nature of the mechanism. Thus reactions proceed via transition states with substantial oxocarbenium ion character since the oxygen adjacent to the reaction center stabilizes the developing positive charge by release of electrons from its lone pairs. The formation of such an oxocarbenium ion-like species necessarily requires the flattening of the sugar ring to accomodate the partial double-bond character developed between the ring oxygen and the anomeric center. This could be best accomplished in a pyranose within either a half-chair or a boat conformation and in a furanose within an envelope. Contrary to what was generally thought to be the case some years ago, and to what still appears in many biochemistry text books, such oxocarbenium ions are unlikely to be present as discrete intermediates. Lifetimes of oxocarbenium ions in aqueous solution have been measured and found to be typically in the range of 10^{-10}–10^{-14} s, a value far too low for realistic consideration as a discrete intermediate,[58,59] as discussed in detail elsewhere.[8] The function of the enzyme is therefore to provide a binding site incorporating the correct arrangement of this pair of carboxylic acids while also providing an optimal environment for the binding of the sugar in its transition state conformation and charge distribution. This must be provided by an exquisite array of hydrogen bonding, hydrophobic, and electrostatic interactions.

5.12.1.4.2 *Retaining glycosidases*

The required double-displacement mechanism occurs within an enzyme active site containing, in all cases except those involving substrate-assisted catalysis, a pair of carboxylic acids separated by

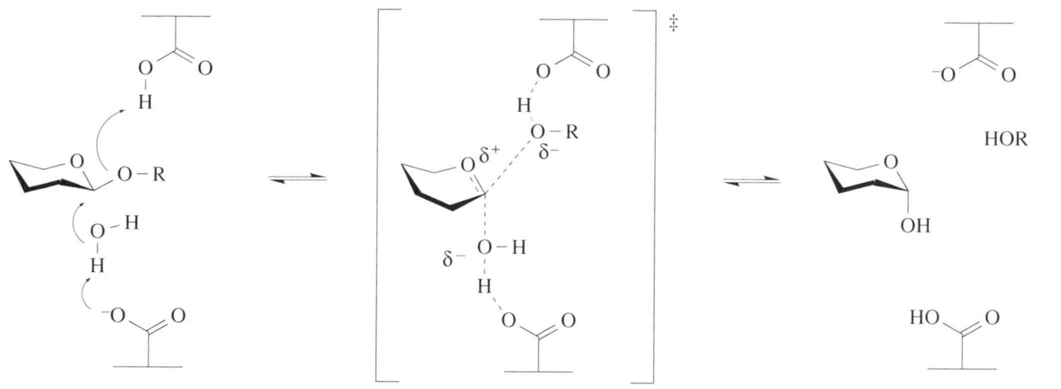

Scheme 1

approximately 5.5 Å. The shorter distance between the carboxyl groups than that seen in inverting glycosidases reflects the different roles played. In this case, one carboxylic acid residue functions as a general acid catalyst, protonating the departing glycosidic oxygen, while the other functions as a nucleophile, attacking directly at the sugar anomeric center with displacement of the aglycone leaving group. This results in the formation of a covalent glycosyl-enzyme intermediate as shown in Scheme 2. The reaction is completed by the general base-catalyzed attack of water at the anomeric center of the glycosyl-enzyme intermediate with displacement of the enzymic carboxylate and release of the sugar with retained anomeric configuration. As for the inverting glycosidases, each of these steps is dissociative in character, thus oxocarbenium ion-like transition states are once more involved. Evidence for this transition state structure derives from secondary deuterium kinetic isotope effects, from the substantially reduced rates of hydrolysis of glycosides with fluorine substituted close to the reaction center, and from the linear free energy relationship observed between rates of spontaneous hydrolysis of a series of 2,4-dinitrophenyl deoxy- and deoxyfluoroglycosides and k_{cat}/K_m values for their enzymatic cleavage, as described in Section 5.12.2.1.1.

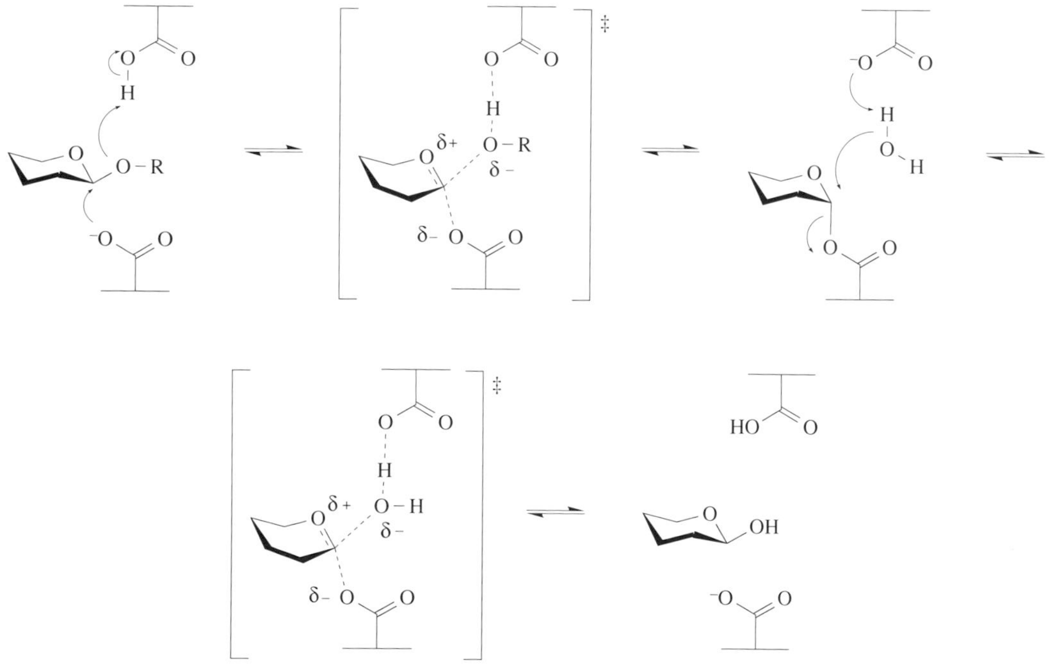

Scheme 2

Evidence for the covalent nature of the glycosyl-enzyme intermediate has come from a number of studies. First, the measurement of secondary deuterium kinetic isotope effects of $k_H/k_D > 1.0$, rather than inverse isotope effects, as discussed in Section 5.12.2.1.1, is consistent only with an sp^3 hybridized, covalent intermediate. Second, intermediates have been trapped by the use of 2-deoxy-2-fluoroglycosides, and their structures probed by ^{19}F NMR, mass spectrometry, amino acid sequencing and even, more recently, three-dimensional structures of such trapped glycosyl-enzyme intermediates have been determined, as discussed in Sections 5.12.2.1.1, 5.12.2.1.2, and 5.12.3.2.

Insights into the role of the general acid/base catalyst and the ways in which its function is optimized for catalysis have come out of several kinetic studies on mutants as well as measurements of active-site pK_a values. Thus the importance of general acid catalysis has been probed through kinetic studies on mutants in which that residue has been replaced by a nonionizable amino acid. This has been carried out on a range of glycosidases, and, as described for a Family 1 β-glucosidase in Section 5.12.2.1.1 and for Family 10 and 11 xylanases in Section 5.12.2.1.2, this residue is critical, with rate reductions of at least 10^5-fold obtained for the natural substrates when it is missing. General base catalysis appears to be less important, its absence dropping rates typically 10^3–10^4-fold and indeed, in the case of the Family 1 thioglucosidases, the residue seems to be completely absent.[60–62] The precise positioning of the general acid/base catalyst and the nucleophile have been probed in a Family 11 xylanase by shortening and lengthening the side-chain through a combination of mutagenesis and chemical modification, as reported in Section 5.12.2.1.2,[63,64] with the location of the nucleophile proving more critical than of the acid/base. Furthermore, ^{13}C NMR measurements of the pK_a of the general acid/base catalyst of this same xylanase have shown it to cycle during catalysis, dropping by 2.5 pH units between the free enzyme and the glycosyl-enzyme intermediate, consistent with its role in each step (see Section 5.12.2.1.2).[65]

An alternative mechanism is proposed for some retaining glycosidases in which a neighboring group on the substrate itself participates in the catalytic mechanism through so-called substrate-assisted catalysis. The best example of this mechanistic class is that of the retaining hexosaminidases. As is shown in Equation (1) for a β-hexosaminidase, the enzyme contains at least one carboxyl group that functions as a general acid/base catalyst. As with normal retaining glycosidases, the first step involves a general acid-catalyzed cleavage of the glycosidic bond with release of the aglycone. However, in this case, instead of an enzymic nucleophile attacking at the sugar anomeric center, the neighboring sugar acetamide group itself attacks with the formation of an enzyme-bound oxazoline or oxazolinium ion intermediate. In a second step, general base-catalyzed attack of water occurs at the anomeric center of this oxazoline or oxazolinium ion intermediate, thereby opening the ring and releasing the sugar with retained anomeric configuration. This mechanism is well-precedented in the nonenzymatic case,[66,67] and although proposed[68,69] and discarded for lysozyme 30 years ago, considerable knowledge has been developed from structure/activity relationships, and kinetic/structural studies of inhibitor binding, as described in Section 5.12.2.1.4.

(1)

(5)

5.12.1.5 Strategies for Identification of Active Site Residues

5.12.1.5.1 *The nucleophile in retaining glycosidases*

Several approaches have been developed in the past for specifically tagging and thereby identifying the active-site nucleophile, but many of these have proved unreliable, often hitting other residues in the active site.[10] The only reliable method for such identification has involved the trapping of the glycosyl-enzyme intermediate itself, and this has been achieved in several ways. Denaturation trapping of the intermediate was first achieved on the Family 2 β-glucosidase from *Aspergillus wentii* by Legler using substrates for which deglycosylation is slowed, allowing identification of an aspartic acid as shown in Table 2.[74,75] A similar approach, but using the natural substrates, was used to identify an aspartic acid on a sucrose α-glucosyl transferase[88,91] and on sucrose phosphorylase,[92] while the intermediates on glycogen debranching enzyme[82] and cyclodextrin glucanotransferase[84] were trapped and characterized without the need for acid denaturation by use of a nonnucleophilic acceptor sugar. Fluorinated glycosides with excellent aglycone leaving groups have proved particularly valuable reagents for trapping covalent intermediates. The basic principle involved is that introduction of an electronegative fluorine close to the reaction center will inductively destabilize the electron-deficient oxocarbenium ion transition states for formation and hydrolysis of the intermediate, thereby slowing both steps. Inclusion of a good leaving group will accelerate the glycosylation step, rendering the intermediate kinetically accessible and thereby permitting its trapping. The first such reagents introduced were the 2-deoxy-2-fluoroglycosides with dinitrophenolate or fluoride as leaving groups. These inactivators have been universally successful in trapping intermediates on β-glycosidases, thereby labeling the nucleophile. The relevance and catalytic competence of the intermediate so trapped have been proven in a number of cases, as described in Section 5.12.2.1.1 for a Family 1 β-glucosidase.[93] Table 2 provides a listing of enzymatic nucleophiles so identified, as reviewed previously.[10] The 2-deoxy-2-fluoroglycosides were not, however, successful reagents for labeling α-glycosidases, and to tackle this problem two new classes of fluoroglycosides were developed based upon the same mechanistic principles: the trinitrophenyl 2-deoxy-2,2-difluoroglycosides[94] and the 5-fluoroglycosyl fluorides.[95] Both were successful in trapping intermediates, but the 5-fluoroglycosyl fluorides have proved the more valuable in actual peptide identification studies.[83] Kinetic analysis of candidate mutants has also proved useful in the identification of the catalytic nucleophile, with the nucleophile mutants (when carefully prepared to be devoid of wild-type enzyme) typically exhibiting activities some 10^6–10^7-fold lower than wild-type. Furthermore, almost wild-type levels of activity can be restored to the Ala mutant at that position by the addition of azide (or formate) as an alternate nucleophile, with formation of the glycosyl azide product of the opposite configuration to the normal substrate, as described in Section 5.12.2.1.1.[57,96]

Table 2 Active site nucleophiles of retaining glycosidases and glycosyl transferases identified from trapped intermediates.

Family	Enzyme	Residue	Sequence	Ref.
1	*Agrobacterium* sp. β-glucosidase	Glu358	YITENGA	70
	Staphylococcus aureus 6-phospho β-glucosidase	Glu375	TYTENGL	71
	Sweet almond β-glucosidase	Glu	ITENGVD	72
2	*Aspergillus niger* β-glucosidase	Asp	VMSDW	73
	Aspergillus wentii β-glucosidase	Asp	VMSDW	74,75
3	*Escherichia coli* (*lac z*) β-galactosidase	Glu 537	ILCEYAH	76
5	*Clostridium thermocellum* CelC endoglucanase	Glu280	YCGEF	77
7	*Fusarium oxysporum* endoglucanase I	Glu197	VCCNEL	78
10	*Cellulomonas fimi* exo-glycanase Cex	Glu230	YITELDA	79
11	*Bacillus circulans* endo-xylanase	Glu78	SPLIEY	80
12	*Streptomyces lividans* endocellulase Celβ	Glu120	EIMIW	81
13	Rabbit glycogen debranching enzyme	Asp549	VRLDNCHS	82
	Saccharomyces cerevisiae α-glucosidase	Asp214	FRIDTAGL	83
	Bacillus circulans CGTase	Asp229	DAVKHM	84
	Streptococcus sobrinus α-glucosyl transferase	Asp	IRVDAVD	85
30	Human β-glucocerebrosidase	Glu340	FASEA	86
35	Human β-galactosidase	Glu268	LINSEF	87
38	Jack bean α-mannosidase	Asp	WQIDPF	88
39	*Thermoanaerobacterium saccharolyticum* β-xylosidase	Glu277	ITEYNTS	89

5.12.1.5.2 The general acid and base catalysts

No reagent for the reliable labeling of this key residue has yet been identified, although, on occasion, the glycosyl epoxides, conduritol epoxides, and *N*-bromoacetylglycosylamines have caused selective derivatization, as reviewed previously.[10] Beyond identification by X-ray crystallography, the most reliable approach yet devised involves the systematic mutation of all conserved glutamic and aspartic acids to alanine followed by detailed kinetic analysis of the mutants so formed (as described in Section 5.12.2.1.1 and 5.12.2.1.2 for retaining glycosidases and Section 5.12.2.2.1 for inverting glycosidases). Of particular value are the azide discrimination protocols described therein.

5.12.1.6 Probes of Mechanism

5.12.1.6.1 *Stereochemistry*

The single most important piece of information in the mechanistic characterization of a glycosidase is the stereochemical outcome of the reaction catalyzed. This has been determined in a number of ways. Initially, the use of optical rotation measurements found favor, but especially after several incorrect diagnoses were made[97,98] other approaches have dominated. One approach involves the use of HPLC systems that allow the resolution of anomeric forms[99] while another employs ^{1}H NMR spectroscopy to directly follow the reaction stereochemistry, particularly by monitoring the anomeric proton. This latter method is probably the most reliable, and has been used widely.[26,100]

5.12.1.6.2 *Hammett relationships*

The relatively broad aglycone specificity of most glycosidases, allowing the use of aryl glycoside substrates, has made the study of their mechanisms through Hammett/Brønsted relationships feasible. The observation of a strong dependence of the kinetic parameter on the leaving group ability of the phenol provides evidence that the cleavage of the glycosidic bond is rate-determining for those substrates. Further, the slope of the plot (ρ or β_{lg}) provides insight into the extent of charge development on the leaving group (phenolate) oxygen. This in turn provides some measure of the extent of bond cleavage and/or the degree of proton donation at the transition state. A further measure of the degree of bond cleavage is provided by secondary deuterium kinetic isotope effects on these same substrates (*vide infra*), thereby allowing better interpretation of the β_{lg} value. The absence of any dependence suggests either that some step other than bond cleavage is rate-determining, or if isotope effects decree that bond cleavage is rate-limiting, then there must be no significant charge accumulation on the oxygen, hence there must be extensive proton donation. Plots of $\log(k_{cat}/K_m)$ provide insight into the first irreversible step, commonly assumed to be the glycosidic bond cleavage, whereas plots of $\log k_{cat}$ report on the rate-limiting step. It is not uncommon to see bent Hammett plots, with a concave downward appearance indicating a change in rate-determining step as the leaving group ability changes. Examples of this are described in Section 5.12.2.1.1.

5.12.1.6.3 *Kinetic isotope effects*

Measurements of primary kinetic isotope effects on glycosidases have been quite limited, largely owing to the difficulty in accurate measurement of such heavy-atom isotope effects on enzymatic systems. They have, however, been applied to the hydrolysis of aryl glycosides by *E. coli* β-galactosidase, yeast α-glucosidases, and hen egg white lysozyme for example.[101–103] Care must be taken in the interpretation of such isotope effects to account for the significant equilibrium isotope effect on the ionization of the phenol. Apart from providing insight into whether or not bond cleavage is rate-limiting for the substrate studied, they can also provide detailed insight into transition-state structure, especially when combined with secondary deuterium kinetic isotope effect data. This approach has been refined considerably by the group of Schramm in probing mechanisms of nucleoside hydrolases and phosphorylases, as described in Section 5.12.3.4.

Secondary α-deuterium kinetic isotope effects have generally proved more valuable. These provide insight into changes in geometry at the reaction center between the reaction ground state and transition state. If the anomeric carbon atom undergoes a change from sp^3 to full sp^2 geometry then an isotope effect of $k_H/k_D = 1.1$–1.3 will be observed, with lower values indicating partial development of sp^2 hybridization. Thus the isotope effect measured reports on the extent of oxocarbenium ion character developed at the transition state under investigation. These have been used extensively on a variety of enzymes as described in Sections 5.12.2.1.1, 5.12.2.1.2, 5.12.2.1.3, 5.12.2.2.2, and 5.12.3.4.

5.12.2 TRANSFERS TO WATER

5.12.2.1 Retaining Glycosidases

5.12.2.1.1 Agrobacterium *sp.* β-glucosidase *and* E. coli, lac Z β-galactosidase

Many of the features of Koshland's double-displacement mechanism[56] have been testable in *Agrobacterium* sp. β-glucosidase (Abg), a Family 1 glycosidase derived from an organism originally typed as *Alcaligenes faecalis*. This is primarily the result of the enzyme's tolerance of a wide range of aryl-β-D-glucopyranoside substrates, which has allowed the construction of an extensive Brønsted relationship. The downward break observed in the Brønsted plot of $\log k_{cat}$ versus pK_a of the aglycone phenol identified substrates for which glycosylation and deglycosylation were rate-limiting (Figure 2).[104] Accordingly, the α-secondary deuterium kinetic isotope effects (KIEs) for the aryl glucopyranosides were divided into two groups. Highly reactive substrates (aglycone $pK_a < 8$), for which the deglycosylation step was rate-limiting, exhibited large α-secondary KIEs ($k_H/k_D = 1.05$–1.07), reflecting a transition state with substantial oxocarbenium ion character. However, less reactive substrates (aglycone $pK_a > 8$), for which glycosylation was rate limiting, appear to have later transition states with significantly more S_N2 character ($k_H/k_D = 1.10$–1.12), as expected for greater preassociation of the catalytic nucleophile. Indeed, the large negative slope of the leaving group-dependent portion of the Brønsted plot ($\beta_{lg} = -0.7$) implies a product-like transition state with substantial glycosidic bond cleavage. More importantly, the normal α-secondary deuterium KIE observed for the deglycosylation step indicates a $sp^3 \rightarrow sp^2$ rehybridization of the anomeric carbon in the transition state, consistent with a glucosyl-enzyme intermediate that is covalent. If the intermediate had been a glucosyl oxocarbenium ion, an inverse isotope effect ($k_H/k_D < 1$) would be expected, corresponding to $sp^2 \rightarrow sp^3$ rehybridization. Unfortunately, substrate specificity often precludes modulation of the rate-determining step in other retaining glycosidases, and at present the deglycosylation steps of only seven others have been characterized by normal α-secondary deuterium KIEs, these being *E. coli, lacZ* β-galactosidase ($k_H/k_D = 1.2$–1.25),[105] *E. coli, ebg*[a], *ebg*[b], and *ebg*[ab] β-galactosidases ($k_H/k_D = 1.10$),[106] *Botryodiplodia theobromae* β-glucosidase ($k_H/k_D = 1.09$),[107] *Stachybotrys atra* β-glucosidase ($k_H/k_D = 1.11$)[108] and *Cellulomonas fimi* β-(1,4) exo-glycanase ($k_H/k_D = 1.10$–1.12).[109]

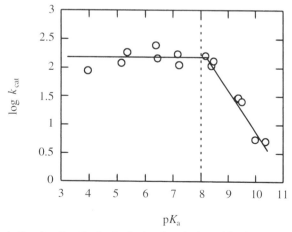

Figure 2 Brønsted plot relating k_{cat} for the hydrolysis of aryl glucosides by *Agrobacterium* sp. β-glucosidase with the leaving group ability of the phenol.

The most satisfying evidence for a covalent intermediate is derived from direct observation of the structure. In the case of Abg this was achieved by trapping the intermediate with the mechanism-based inactivator, 2′,4′-dinitrophenyl-2-deoxy-2-fluoro-β-D-glucopyranoside.[93] The inactivation of Abg followed a first-order dependence, which was accompanied by a burst of aglycone that was stoichiometric with enzyme concentration. Also indicative of active-site titration was the protection afforded by reversible inhibitors. However, the most definitive feature of the trapped intermediate was its ability to turn over slowly hydrolytically, or more rapidly in the presence of acceptor glucosides via transglycosylation.[70] The catalytic competence of this intermediate confirms that it is part of the normal reaction pathway, and that the labeled residue is catalytically relevant. The α-stereochemistry of the acylal linkage was determined from the ^{19}F NMR chemical shift for the corresponding 2-deoxy-2-fluoromannosyl-enzyme intermediate, which was more chemical shift sensitive to anomeric configuration (by virtue of the 1,2-*trans* stereochemistry) than the corresponding glucosyl analogue.[110] Finally, the nucleophile was determined to be Glu358 by use of a radiolabeled version of the inactivator,[70] a form of detection that is no longer necessary due to the success of detecting the loss of the acylal-linked label from a peptide with HPLC-MS/MS.[111]

Unfortunately, a glycosidase inactivator does not yet exist which will specifically and reliably label the general acid/base catalyst. Nevertheless, this residue can be identified with confidence by deleting the conserved candidate and assessing the consequence of the change in three key kinetic experiments: leaving group dependence, pH-rate dependence, and anion competition. In the case of the E170G mutant of Abg, the glycosylation rate (determined from the pre-steady state) with 2′,4′-dinitrophenyl-β-D-glucopyranoside (2,4-DNPG) was reduced only 2-fold relative to the wild-type rate, whereas with 4-nitrophenyl-β-D-glucopyranoside the reduction was $\geqslant 10^4$-fold.[61] The deglycosylation rate, reflected by k_{cat} for the steady-state hydrolysis of 2,4-DNPG (for which the deglycosylation step is rate limiting), was reduced 1760-fold. These results are consistent with Glu170 functioning as a general acid catalyst in the glycosylation step, because substrates with an excellent leaving group (e.g. 2,4-dinitrophenol, pK_a = 3.96), which do not require protonic assistance were hydrolyzed by the mutant at nearly wild-type rates, whereas substrates with poor leaving groups (e.g., 4-nitrophenol, pK_a = 7.18) requiring acid catalysis for departure were hydrolyzed at vastly reduced rates. Likewise, the deglycosylation rate, for which base-catalyzed attack of water on the glucosyl enzyme intermediate is critical, was also greatly affected by the mutation. Also consistent with these results was the disappearance of a basic ionization (pK_a = 8.1) in the pH versus k_{cat} plot for the hydrolysis of 2,4-DNPG by the E170G mutant. Moreover, the addition of azide as a competitive nucleophile increased the mutant k_{cat} values 100–300-fold for substrates whose rate-limiting step was deglycosylation, yielding β-D-glucosyl azide. Azide had no effect on the wild-type enzyme. Unlike water, azide does not require base catalysis for nucleophilic attack and therefore can successfully compete with water for the covalent glucosyl-enzyme intermediate. The fact that only β-products were formed with the mutant, but not with the wild-type enzyme, suggests that the replacement of Glu170 by glycine creates a cavity sufficient to accommodate small anions directly adjacent to the β-face of the glucosyl-enzyme intermediate. Presumably, charge screening from the glutamate side-chain in the wild-type enzyme is sufficient to deny access to anions. These three experiments have proven to be of general utility in retaining glycosidases, having also been successfully applied to *Bacillus circulans* xylanase (Glu172),[64] *Cellulomonas fimi* exo-β-1,4-glycanase (Glu127),[96,112] and *E. coli*, *lacZ* β-galactosidase (Glu461).[113]

The catalytic activity in the Abg E170G mutant could also be restored with assistance from suitable substrates.[60] Intramolecular proton donation to the glycosidic oxygen is possible in the substrate 2′-carboxyphenyl β-D-glucopyranoside, and enhanced the glycosylation rate in the mutant to match wild-type levels (Scheme 3(i)). This corresponds to a 10^7-fold increase in the glycosylation rate over 4′-carboxyphenyl β-glucopyranoside, which cannot provide intramolecular proton donation. This artificial system may reflect the mechanism used by Family 1 thioglucosidases (myrosinases) which substitute Gln for the normal Glu general acid/base catalyst. General acid catalysis may be available in the substrates targeted by these enzymes, glucosinolates (4), which contain a sulfate group in their aglycone.[114]

The anion competition experiment can also be used to identify the catalytic nucleophile in a retaining glycosidase, thereby obviating the need for active-site labeling and sequencing. For example, the nonexistent activity of the Abg nucleophile mutant, E358A, towards 2,4-DNPG was restored to near wild-type levels upon the addition of azide.[57] The formation of strictly α-glucosyl azide as a product indicates that azide can successfully replace the catalytic nucleophile and displace the substrate aglycone, as shown in Scheme 3(ii). Effectively, this mutant acted as an inverting glycosidase, with azide replacing water as the nucleophile. Furthermore, the E358A mutant, despite retaining specificity for β substrates, also cleaved α-glucosyl fluoride (Scheme 4(i)). This catalytic flexibility is a typical trait of inverting glycosidases and occurs through a transglycosylation mechanism (Scheme 4(ii)).[115] It is noteworthy that the catalytic rescue of the Abg nucleophile mutant with small anions was not an isolated case; *Bacillus circulans* xylanase[116] and *Cellulomonas fimi* exoglycanase[96] have been converted to inverting glycosidases in the same manner.

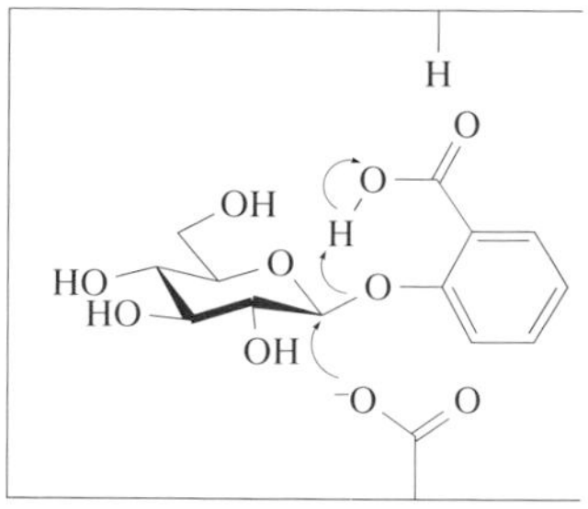

Scheme 3

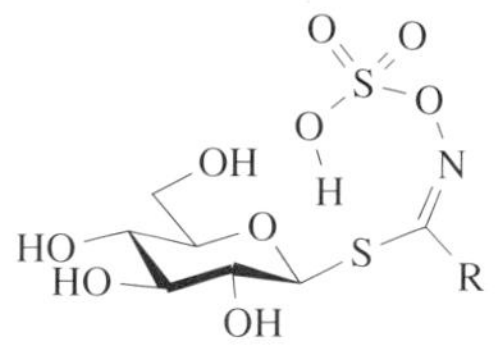

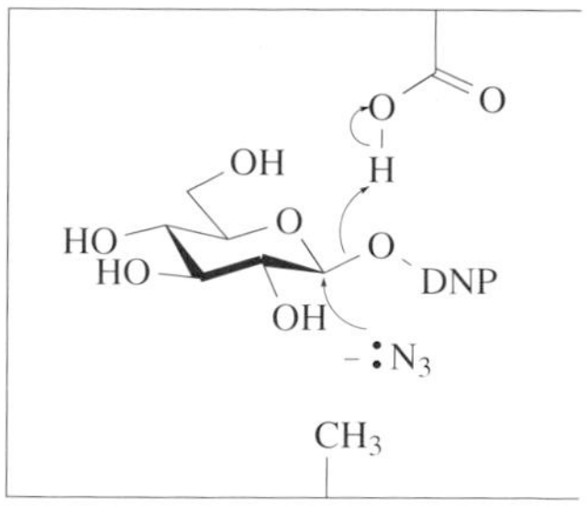

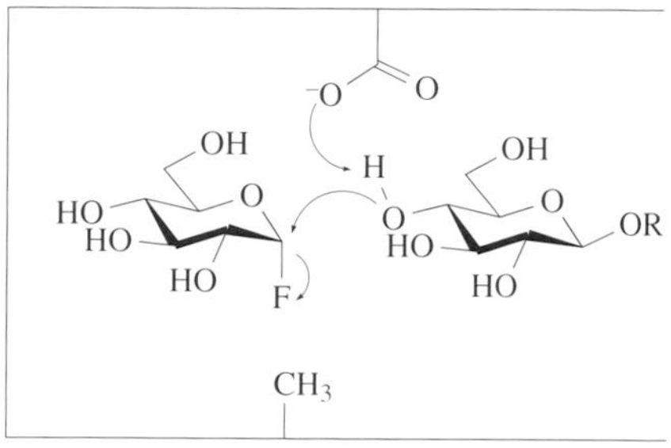

Scheme 4

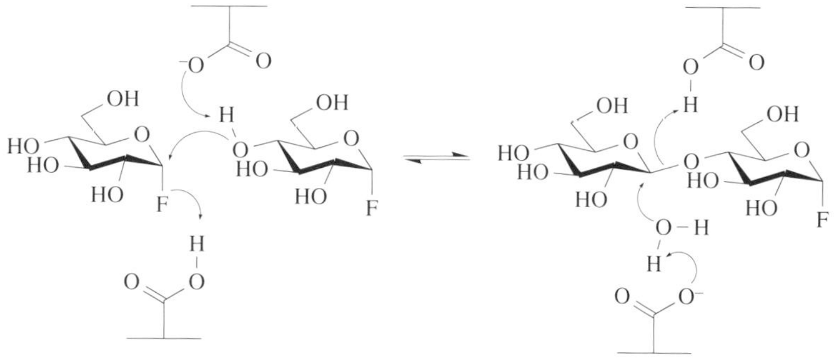

These results suggest that the anion competition experiment can serve as a general, rapid method for identifying the two key catalytic carboxylates in retaining glycosidases. Such an approach would involve identifying conserved carboxylates through sequence alignments then generating Ala mutants at these positions. Screening of these mutants with an activated chromogenic substrate in the presence and absence of azide would allow rapid identification of candidates: if the glycosyl azide product has the same stereochemistry as the starting substrate, then the residue mutated is assigned as the general acid/base catalyst; if it has the opposite stereochemistry, then the residue in question must be the catalytic nucleophile. This test is remarkably precise; when a third conserved carboxylate in *Cellulomonas fimi* exoglycanase, Asp123, was mutated to Ala, a modest rate enhancement for the hydrolysis of *p*-nitrophenyl-β-D-cellobioside in the presence of azide was observed.[96] However, the reaction product was determined to be cellobiose, not cellobiosyl azide, thus the behavior of Asp123 was clearly different from the acid/base and nucleophile residues. Indeed, a crystal structure of a cellobiosyl-Cex intermediate indicated that Asp123 was part of a His-Asp dyad that appeared to influence the ionization of the nucleophile (see Section 5.12.2.1.2).[117]

The transition states for the double-displacement mechanism are generally assumed to involve concerted general acid or base catalysis. The assumption may not be a rigorous one, however, as inferred from anion competition experiments with the general acid/base mutant (E461G) of *E. coli*, *lacZ* β-galactosidase.[118] It was determined that deglycosylation of the covalent mutant-galactosyl intermediate exhibited azide selectivities ($k_{az}/k_{H_2O} = 14\,000\,M^{-1}$) approaching that expected for the partitioning of stable carbocations in water ($k_{az}/k_{H_2O} \approx 10^7\,M^{-1}$). In contrast, a drastic reverse selectivity was observed for the wild-type enzyme, with anionic nucleophiles presumably being excluded from the active site by the ionized general acid/base catalyst. It was argued that this selectivity was evidence for the formation of an oxocarbenium ion intermediate during the deglycosylation step, the result of general acid-catalyzed departure of the aglycone preceding attack of the nucleophile, Glu537 (Figure 3). In light of this proposal, it is interesting to note that the α-secondary isotope effect measured on the deglycosylation step for β-galactosidase ($k_H/k_D = 1.2$–1.25)[105] is significantly larger than for Abg ($k_H/k_D = 1.10$–1.12), perhaps reflecting a more fully expressed oxocarbenium ion verging on a real existence.

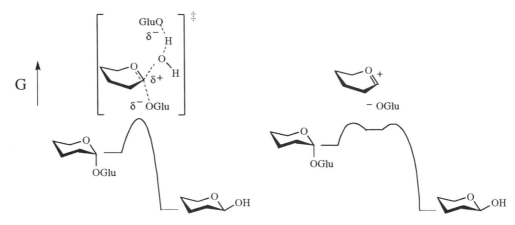

Figure 3 Concerted versus stepwise deglycosylation in the retaining glycosidase mechanism.

Noncovalent interactions between the enzyme and substrate, which are optimally expressed at the reaction transition state, provide enormous contributions to the catalytic efficiencies of glycosidases. These have been probed in detail for Abg by steady-state and presteady-state kinetic analysis of the hydrolysis of a series of mono-substituted 2,4-dinitrophenyl glucopyranosides in which the individual hydroxyl groups on the glycone were substituted by hydrogen or fluorine.[119] Of note is the linear free energy relationship observed between $\log k_{cat}$ and $\log k_{H_2O}$ for this series. Spontaneous glycoside hydrolysis is known to proceed through an oxocarbenium transition state,[120,121] therefore this correlation reflects a similar transition state for the enzymic reaction. This correlation was also observed for *E. coli*, *lacZ* β-galactosidase with the analogous series of 2,4-dinitrophenyl galactopyranosides.[122] The contributions of each hydroxyl group to ground state binding are relatively weak ($\leqslant 0.8$ kcal mol^{-1}) but are much greater at both the glycosylation and deglycosylation transition states. Interactions at the 3 and 6-positions each contribute at least 2.2 kcal mol^{-1} to transition state stabilization while those at the 2-position contribute at least 5 kcal mol^{-1}. The likely source of this strong interaction is commented upon in the next section on a related β-glycosidase, Cex.

5.12.2.1.2 Cellulomonas fimi *exoglycanase and* Bacillus circulans *xylanase*

Substantial research has focused on the mechanism of *C. fimi* exoglycanase/cellulase (Cex) (Family 10) and *B. circulans* xylanase (BCX) (Family 11) with the hope of developing potential applications for industrial biomass degradation and fuel conversion. The catalytic nucleophile and acid/base catalysts have been identified in both cases (Glu233 and Glu127, respectively, in Cex; Glu78 and Glu172, respectively, in BCX) through labeling and mutagenesis studies analogous to those described for Abg.[112,79,80,123] The 3-dimensional structure of the trapped 2-deoxy-2-fluorocellobiosyl Cex intermediate has been determined recently by crystallography,[117] providing a snapshot of a critical moment along the reaction coordinate (Figure 4). The anomeric carbon formed an α-acyl linkage to Glu233 which was *syn* to the ester group, a preferential location for nucleophilic attack by a carboxylate.[124] Furthermore, the nucleophile formed a hydrogen bond to His205, which in turn formed a hydrogen bond to Asp235. This His-Asp pair may be responsible for regulating the nucleophilicity of Glu233, which is an interesting contrast to the Tyr residue used by Abg for this purpose.[125] Likewise, the His-Asp pair may improve the leaving-group ability of Glu233 in the covalent intermediate by stabilizing the negative charge on the departing carboxylate. The structure also suggested that the O^{ϵ^2} of Glu233 would form a hydrogen bond to the 2-OH of the attached saccharide, which may account for the significant contribution of the 2-OH to transition state stabilization observed in Cex ($\sim$ 9 kcal mol^{-1}),[126] as well as in other retaining glycosidases such as *E. coli*, *lacZ* β-galactosidase ($\sim$ 8 kcal mol^{-1})[122] and Abg ($\sim$ 5 kcal mol^{-1}).[119] The occurrence of unfavorable steric interactions between the C-5 hydroxymethyl group, Gln87, and Trp281 corresponds well with the 85-fold greater activity of Cex against xylobiosides over cellobiosides.[109] Also, significant hydrogen bonding to the distal saccharide may explain the specificity and greater activity of Cex for the hydrolytic release of disaccharides rather than monosaccharides.[79] This interaction also has a remarkable effect on the transition state for glycosylation, with cellobiosides ($\beta_{lg} = -0.3$, $k_H/k_D = 1.06$) proceeding through an earlier, less-charged transition state than glucopyranosides ($\beta_{lg} = -1.0$, $k_H/k_D = 1.12$).[109] However, the deglycosylation transition state is similar ($k_H/k_D = 1.10–1.12$) for the hydrolysis of either intermediate. Therefore, the interactions with the distal saccharide appear to increase the rate of formation of the glycosyl–enzyme intermediate through improved acid catalysis and greater nucleophile preassociation, without affecting its rate of decomposition. Interestingly, removal of the acid/base catalyst (E127A Cex) results in a much greater aglycone sensitivity ($\beta_{lg} \approx -1$) for aryl cellobiosides, directly consistent with the loss of significant proton donation.

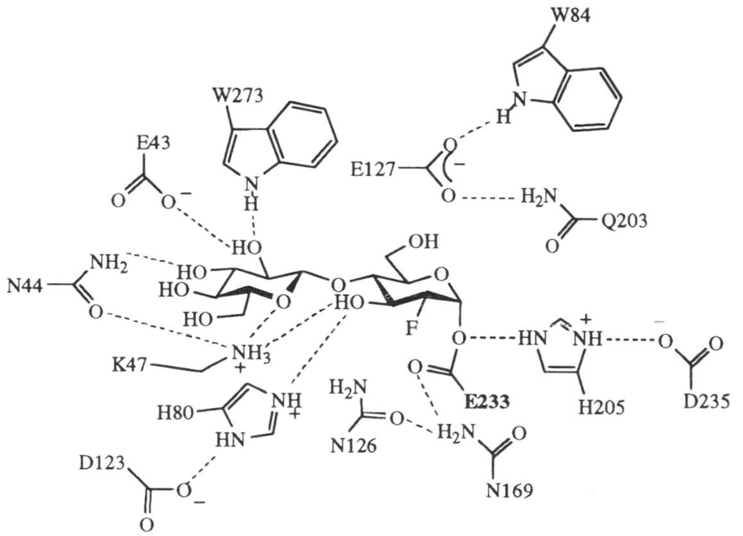

Figure 4 Active site structure of trapped *Cellulomonas fimi* β-1,4-endoglycanase.

The dual role of the acid/base catalyst in the double-displacement mechanism demands that the enzyme control its ionization state in a dynamic manner. Insight into this control is obtained from a study performed with BCX.[65] The acid/base catalyst (Glu172) and nucleophile (Glu78) comprise the only glutamic acid residues in the enzyme, allowing specific ^{13}C-enrichment of the δ-carboxyl group in these residues. Titration of both residues in the free enzyme, and the acid/base catalyst in the trapped fluoroxylobiosyl-enzyme complex, was monitored by ^{13}C NMR. In the unbound state, Glu78 and Glu172 have pK_a values of 4.6 and 6.7 respectively, consistent with the ionization states required for catalysis of the glycosylation step at the pH optimum of the enzyme (pH = 5.7). However, titration of the fluoroxylobiosyl-enzyme covalent intermediate revealed that the pK_a of the acid-base catalyst dropped to 4.2, thereby producing the ionization state required for general base catalysis in the subsequent deglycosylation step. The high pK_a of Glu172 in the free enzyme has been ascribed to charge repulsion from the nucleophile, Glu78; the drop in pK_a is therefore likely a consequence of removal of the charge on Glu78 in the intermediate.[127] This "pK_a cycling" of the acid/base catalyst is therefore an inherent component of the mechanism, fine-tuning the acidity of this key residue, and could be expected to be a general feature of retaining glycosidases.

As described in the previous section, the mechanism followed by a glycosidase, retaining or inverting, is determined in part by the distance spanning the two catalytic carboxylates. BCX has served as an excellent model of the positional requirements of the two catalytic residues, which can be shortened $\sim$1 Å by mutation to Asp, or lengthened $\sim$1.6 Å by mutation to Cys followed by alkylation with iodoacetate. Shortening the nucleophile reduced the glycosylation rate for aryl-xylobiosides 1600–5000-fold ($\beta_{lg}(k_{cat}/K_m) = -0.7$), which is in the same range observed for the corresponding nucleophile mutants in Abg (2500-fold, $\beta_{lg}(k_{cat}/K_m) = -0.6$)[128] and Cex (3000–4000-fold, $\beta_{lg}(k_{cat}/K_m) = -0.3$),[96] corresponding to a ~ 5 kcal mol^{-1} increase in the glycosylation transition state. Interestingly, the Brønsted constants derived for these mutants were identical to the corresponding wild-type values, suggesting that the shortening of the nucleophile does not affect the degree of negative charge development on the departing phenolate at the glycosylation transition state, and hence the degree of C—O bond cleavage or proton donation. Likewise the α-secondary KIE for deglycosylation was virtually unchanged in the Cex E233D mutant ($k_H/k_D = 1.09$), indicating that the extent of oxocarbenium ion development at the transition state for the hydrolysis of the cellobiosyl-enzyme intermediate is also unaffected by the mutation.[96] These results are consistent with the critical role of the nucleophile in stabilizing the oxocarbenium ion transition state, which is destabilized greatly when the negative charge of the nucleophile is distanced from the development of positive charge. Accordingly, the binding of transition-state analogue inhibitors (e.g., gluconolactone) by Abg was impaired to a much greater extent by the E358D mutation than was the binding of ground-state inhibitors (e.g., β-glucosyl benzene). In contrast, lengthening the nucleophile in BCX had a much smaller effect on the glycosylation rate (16–100-fold reduction).[63] Presumably the lengthened nucleophile can fold into an appropriate position that can still effectively stabilize the developing charge of the transition state.

These results contrast sharply with the relaxed positional requirements for proton transfer by the acid/base catalyst.[64] Altering the length of Glu172 in BCX reduced the glycosylation rate only 2–23-fold for aryl-xylobioside hydrolysis, and the Brønsted constant was unchanged ($\beta_{lg}(k_{cat}/K_m) = -0.7$), indicating that the position of the acid catalyst does not seriously affect the degree of proton donation at the transition state. This confirms the expectation that the positional requirements for carbon–carbon bond formation are much greater than that for proton donation.

5.12.2.1.3 Sialidases

Sialidases (also known as neuraminidases) are responsible for the cleavage of sialic acid residues from various glycoconjugates, and therefore play a significant part in creating the sialic acid topology exposed on cell surfaces. In animals particularly, such sialylation provides a critical form of cellular regulation, and is also the target of viruses and bacteria, which express their own sialidases to assist in their pathogenesis.[129] Any discussion of sialidase mechanism must first consider the unusual nature of the substrate. Glycosides of sialic acid present a highly reactive anomeric center for hydrolytic attack, with formation of a sialosyl cation facilitated by the absence of a hydroxyl group at C-3 and the presence of an α-carboxylate. This carboxylate can assist spontaneous solvolysis either anchimerically for poor leaving groups (Scheme 5(i)),[130] or in an electrostatic sense by stabilizing the oxocarbenium ion transition state and subsequent sialosyl cation intermediate for more labile aglycones (Scheme 5(ii)).[131] In fact, it has been estimated that the α-carboxylate stabilizes the oxocarbenium ion in water by 17 kcal mol^{-1} relative to the H-substituted analogue.[132] As a result, the sialosyl cation has a lifetime that allows equilibration with the bulk solution,[131] which is not the case with other less stable glycosyl oxocarbenium ions.[59] Therefore, the catalytic demands facing a sialidase are significantly different than for other glycosidases.

Three-dimensional structures for a number of bacterial and viral sialidases have now been solved, including those of influenza A and B viral sialidases complexed with numerous ligands,[133–141] *Vibrio cholerae* sialidase,[142] *Salmonella typhimurium* LT2 sialidase in native and inhibitor complexed forms,[52,143] and *Micromonospora viridifaciens* sialidase.[144] Despite a lack of significant sequence homology between bacterial (Family 33) and viral (Family 34) enzymes ($\sim 15\%$), the active-site structures are remarkably uniform, consisting of a fold often described as a β-propeller or superbarrel.[24,52,129] Likewise, after substantial debate, all sialidases examined thus far hydrolyze α-sialosides with retention of anomeric configuration.[98,145,146] The difficulty in determining the stereo-chemistry of sialidase-catalyzed hydrolysis stems from the instability of the α-product, which rapidly mutarotates to the equilibrium value of $\sim 90\%$ β and $\sim 10\%$ α; for this reason ^{1}H NMR spectroscopy has been the method of choice. Interestingly, the structural similarity of sialidase active sites does not appear to translate into identical kinetics. Bacterial sialidases typically turn over 1000-times faster than the influenza enzymes, and do not bind sialic acid, whereas the influenza enzymes do so with a $K_i \approx 1$ mM.[129] Negative Brønsted constants have been obtained for leech ($\beta_{lg}(V_{max}/K_m) = -0.60$, $\beta_{lg}(V_{max}) = -0.5$),[147] influenza A ($\beta_{lg}(V_{max}/K_m) = -0.46$, $\beta_{lg}(V_{max}) = -0.11$),[98] *V. cholerea* ($\beta_{lg}(V_{max}/K_m) = -0.73$, $\beta_{lg}(V_{max}) = -0.25$),[148] and *S. typhimurium* ($\beta_{lg}(V_{max}/K_m) = -0.80$,

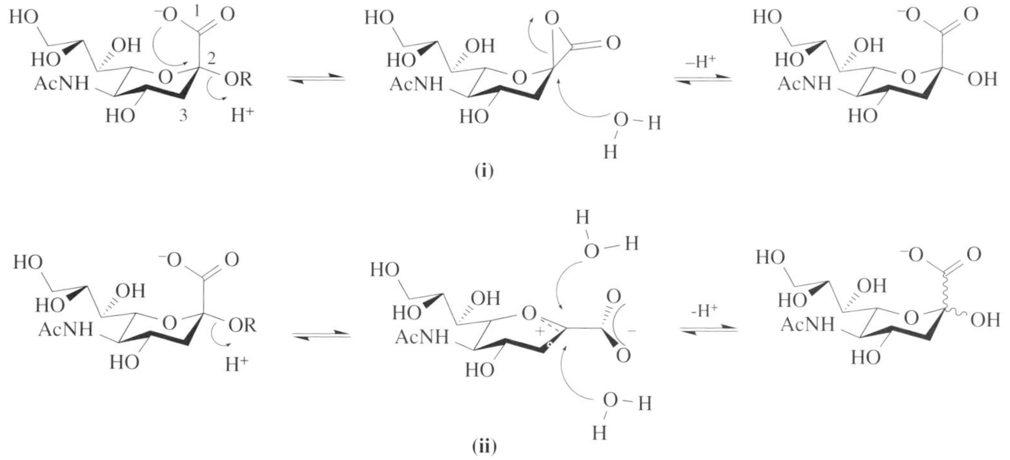

(i)

(ii)

Scheme 5

$\beta_{lg}(V_{max}) = -0.53)^{98}$ sialidases, indicating that the glycosylation step is rate determining for aryl-α-D-sialosides, with varying degrees of glycosidic bond cleavage in the transition state. As well, the substantial β-dideuterium secondary KIEs (V_{max}/K_m) determined for leech ($k_H/k_D = 1.07$),[147] *V. cholerae* ($k_H/k_D = 1.05$),[148] influenza ($k_H/k_D = 1.13–1.15$),[98,149] and *S. typhimurium* ($k_H/k_D = 1.09$)[98] sialidases correspond to a glycosylation transition state with substantial, although variable, oxocarbenium ion character. Likewise, dissection of the contributions of the *pro-R* and *pro-S* deuterons has yielded a spectrum of sialidase transition state geometries:[98] a boat conformation ($B_{2,5}$) for influenza A and possibly for *S. typhimurium*, a ground state $^2C_5$ chair for *V. cholerae*, and a completely inverted $^5C_2$ chair for leech. Interestingly, this latter enzyme, which converts glycosides of *N*-acetylneuraminic acid to 2,7-anhydro-*N*-acetylneuraminic acid, appears to bind the small population of substrate that adopts this disfavored conformation.[147]

The reaction geometry calculated for influenza A sialidase corresponds well with the crystal structures of influenza enzymes complexed with sialic acid, which is bound exclusively as the unfavorable α-anomer.[137,138,141] The product ligand is forced into a boat ($B_{2,5}$) conformation by a strong salt linkage formed between the anomeric carboxylate and a triad of Arg residues (Arg292, Arg371, Arg118; A/Tokyo/3/67 strain numbering) that is conserved in all influenza strains, suggesting that distortion may contribute to catalysis. Asp151, located on the α-face of the ligand, forms a hydrogen bond to the axial 2-OH, implicating this residue as the acid/base catalyst. This contribution to catalysis does not appear to be as significant as the binding energy of the Arg cluster, since an ionizable residue with a pK_a of 9, most likely arising from one of the three Arg residues, dictated inhibition and substrate catalysis in the influenza A sialidase.[150] A corresponding carboxyl residue acting as a nucleophile is not observed on the opposite side of the ligand. Instead, the phenolic oxygen of Tyr406, its phenolate character presumably enhanced by a hydrogen bond to Glu277, is observed within ~3 Å of the β-face of the anomeric carbon atom. This distance is reduced to ~2.8 Å in complexes with the transition state analogue 2-deoxy-2,3-dihydro-*N*-acetylneuramic acid (DANA), a general sialidase inhibitor ($K_i \approx 10^{-6}$ M) that binds the active site of the influenza enzyme 1000-fold tighter than sialic acid.[137,138] Interestingly, influenza sialidase can synthesize this inhibitor from sialic acid by elimination of water, resulting in time dependent inhibition.[137] Whether Tyr406 forms a covalent intermediate, or stabilizes a sialosyl cation intermediate along with the anomeric carboxylate, is not certain (Scheme 6). A small inverse β-secondary dideuterium KIE for V_{max} (assumed to represent deglycosylation) observed for the hydrolysis of 4-methylumbelliferyl-α-D-neuraminide ($k_H/k_D = 0.979 \pm 0.007$) was offered as evidence for a sialosyl cation intermediate.[149] However, this was questioned with a normal (although ambiguous) KIE obtained for V_{max} with *p*-nitrophenyl-α-D-sialoside ($k_H/k_D = 1.0095 \pm 0.011$).[98] A nucleophilic role for a Tyr residue in a retaining glycosidase is not completely without precedent; a nucleophile mutant of Abg (E358D) forms a glucosyl-enzyme intermediate through the occasional attack of Tyr298, a residue postulated to position the nucleophile, although the intermediate so formed does not turn over.[125] In light of the fact that the sialosyl cation can have a real, solvent-equilibrated lifetime in solution,[131] the existence of such an intermediate in the sialidase mechanism cannot be discounted. For this reason, a classical nucleophile (a carboxylate) may not be required by sialidases for catalysis. It has been suggested that the anomeric carboxylate could act as the nucleophile in a form of substrate-assisted catalysis,[98,130] but the movement and energy required to form a strained α-lactone intermediate is difficult to reconcile with how rigidly the anomeric carboxylate appears to be held by the Arg cluster.

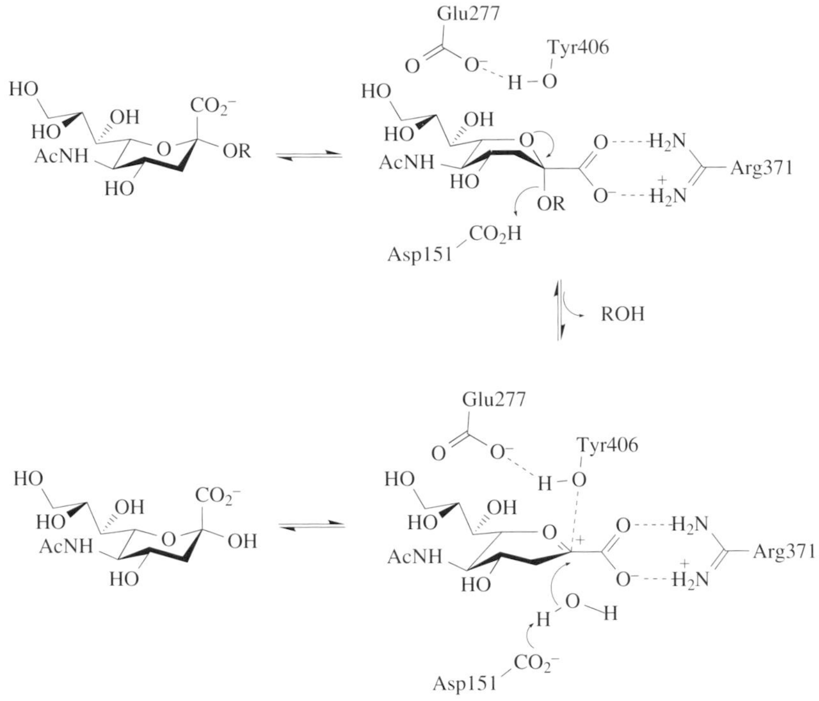

Scheme 6

Mutagenesis studies with the influenza sialidase N2 A/Tokyo/3/67 support the perceived importance of the these catalytic residues.[151] The conservative mutation of Arg371 to Lys, which forms the strongest salt-linkage to the anomeric carboxylate, reduced activity to 5–10% of wild-type levels, while the Y406F and E277Q(or D) mutants were completely inactive. Additionally, the binding of the C-6 glycerol side-chain in a bidentate mode by Glu276, and the C-5 acetamido group by Trp178 and Arg152 must be substantial, as mutation of these residues also destroyed all activity. The 4-OH is accomodated by a polar pocket formed in part by Glu119. This pocket has been a fruitful target for influenza sialidase drug design: substitution of the 4-OH of DANA by an amino or a guanidino group improved inhibition potency to $K_i = 50$ nM and 0.2 nM, respectively.[152] Crystallographic studies verified that this increased potency arose from a salt-bridge to Glu119, and also Glu227. Surprisingly, the subtle hydrophobic potential of the glycerol side-chain binding pocket has been exploited successfully to improve inhibitor potency. A carbocyclic analogue of DANA, incorporating an amino group at C-4 and an isopentyloxy group in place of the glycerol side-chain, inhibited influenza A/PR/8/34 sialidase with an IC_{50} of 1 nM.[133] The structure of the enzyme with this inhibitor revealed that Glu276 was forced to bend away from the isopentyloxy group to form a large hydrophobic surface with its hydrocarbon chain, along with Ala246, Arg224, and Ile222.

5.12.2.1.4 *Hen egg white and bacteriophage T4 lysozymes, N-Acetyl-hexosaminidases and chitinases*

Despite continued debate,[7,8] the original mechanism proposed by Phillips[153] for the hydrolytic action of hen egg white lysozyme (HEWL) has persevered, and is historically cited (perhaps erroneously) as a paradigm for retaining glycosidases. To initiate the hydrolysis of NAG-NAM bonds, the Phillips mechanism holds that HEWL binds a NAG residue in a distorted half-chair conformation,[48,154,155] with the glycosidic oxygen in a pseudoaxial position near the general acid/base catalyst, Glu35. Expulsion of the glycosidic oxygen occurs with general-acid catalysis and nearly complete bond cleavage in the rate determining transition state.[102] Traditionally it has been thought that this leads to a fully formed oxocarbenium ion intermediate, stabilized electrostatically by an ionized Asp52. General-base-catalyzed attack of water at the β-face of the intermediate leads to a hydrolyzed product with retained stereochemistry (Equation (2)). The distressing part of this mechanism is the formidable challenge of stabilizing the highly reactive oxocarbenium ion for a length of time sufficient to allow diffusion of the aglycone from the active site, with the return of a water molecule to complete hydrolysis with retention. Unfortunately, kinetic evidence for or against the existence of an ion pair has been impossible to obtain because there are no substrates known for which hydrolysis of the glycosyl-enzyme intermediate is rate determining.[8,101,102,156]

$$(2)$$

However, recent mechanistic and structural studies with other retaining glycosidases which cleave 2-acetamido-2-deoxy-β-D-glucopyranosides have convincingly established the function of the acetamido group in substrate assisted catalysis in those cases. In an abortive complex formed by *Serratia marcesens* chitobiase with chitobiose, the nonreducing end NAG residue was observed to be distorted into a sofa conformation with the acetamido oxygen oriented towards the anomeric carbon.[47] The acetamido oxygen could provide anchimeric assistance for the expulsion of the aglycone, forming an oxazolinium ion (**5**), as shown in Equation (1). Supporting this role is the absence of a classical enzyme nucleophile near the α-face of the substrate. Glu540 forms a hydrogen bond to the glycosidic oxygen and is poised to act as an acid/base catalyst. A similar mechanism has been proposed for hevamine, a plant glycosidase with chitinase activity, based on the crystal structure of a complex formed with the tight binding inhibitor, allosamidin (**6**).[157] It is believed that the high affinity for allosamidin stems from its resemblance to the oxazolinium-like transition state and intermediate. Acetamido participation has also been suggested for the soluble lytic transglycosylase, based on a structure of the enzyme complexed with bulgecin, a naturally occurring inhibitor.[158] This form of catalysis has also been inferred from kinetic studies with modified substrates. For example, *Aspergillus niger* β-*N*-acetylglucosaminidase was subjected to linear free energy relationship using *p*-nitrophenyl-2-deoxy-*N*-trifluoroacetyl, *N*-difluoroacetyl, and *N*-acetyl-β-D-glucosaminide as substrates.[159] A strongly negative reaction constant ($\rho^* = -1.41$) was obtained from the plot of $\log V_{max}/K_m$ versus σ^*, as expected for anchimeric assistance when the basicity of the carbonyl oxygen is varied. A reduction in basicity can also be obtained by replacing the carbonyl oxygen with sulfur, which resulted in a 253-fold decrease in k_{cat}/K_m, although this rate reduction could also be a consequence of the greater size of sulfur. Strong evidence for acetamido participation was also inferred from the potent inhibition of jack bean hexosaminidase by NAG-thiazoline (**8**), a stable sulfur analogue of the cyclic intermediate ($K_i = 280$ nM).[160] The catalytic competence of this intermediate was demonstrated by the observation that the jack bean enzyme could synthesize this inhibitor from the corresponding thioacetamido β-glucoside (**7**), resulting in time-dependent inactivation (Equation (3)). Acetamido participation also appears to operate in *Bacillus* sp. chitinase, which can apparently synthesize chitin quantitatively from an activated chitobiose monomer (**9**) incorporating the oxazoline ring (Equation (4)).[161]

(**6**)

(3)

(7)   (8)

(9)

The evidence for acetamido participation spurs a similar interpretation of the mechanism for HEWL, as originally suggested by Lowe.[69] That this mechanism should be conserved among glycosidases which cleave *N*-acetylhexosaminides with retention should not be a surprise, since the acetamido function contributes significantly to the spontaneous solvolysis of glucosaminides,[66,67,162] and one might expect a glycosidase to evolve to enhance this inherent catalysis. However, anchimeric assistance in HEWL has not gained widespread favor, possibly because the acetamido group has not been sufficiently resolved in HEWL crystal structures to ascertain its position precisely,[48,155] and in particular because substrates lacking an acetamido side-chain are also hydrolyzed by HEWL (although 2-deoxy derivatives should hydrolyze rapidly due to reduced electron withdrawal from the anomeric center).[163,164] Although a nucleophilic role for Asp52 cannot be dismissed, the carboxylate appears to be at a distance from the anomeric center that exceeds that of a covalent bond,[48,155] although a conformational change in the enzyme could breach this gap. However, mutations of Asp52 have modest effects on activity,[165–167] which is inconsistent with effects on other β-glycosidases for which a nucleophilic function is clear. It is quite possible that Asp52 stabilizes the positive charge of the oxazolinium intermediate.

Despite the anchimeric propensity of the acetamido group, not all hexosaminidases cleave glycosaminides with retention. The bacteriophage T4 lysozyme (T4L) was once thought to be a retainer, based on the structural homology of the active site with HEWL.[168] This appeared to draw support from the significant activity and wild-type pH-rate dependence observed for the D20C mutant, leading to the conclusion that Asp20 was acting as a nucleophile rather than a counterion,[169] though it was subsequently shown that the D20A mutant was similarly active.[170] Interestingly, when a nearby residue, Thr26, was replaced with glutamic acid, the resulting mutant was rapidly inactivated by *E. coli* cell walls, resulting from the formation of a glycosyl-enzyme covalent adduct via an α-acyal linkage to the newly installed residue.[50] This mutation had effectively created an incompetent retaining glycosidase. A comparison of the mutant-adduct structure with the wild-type enzyme revealed that the mutated residue had displaced a water molecule normally held in place by Asp20 and Thr26; an optimal position for displacement of a substrate aglycone with inversion. Indeed, a careful HPLC analysis of the wild-type (and the E20C mutant) hydrolysis products indicated that only the α-anomer was released, thus establishing T4 lysozyme as an inverter.[171] A competent retaining glycosidase (as shown by HPLC) was obtained by substituting a histidine residue at position 26. Again, this residue was thought to displace the nucleophilic water, in this case forming a covalent intermediate which turned over due to the greater reactivity of the imidazolium leaving group.[171] This mutant also had substantial transglycosylase activity, indicating that the intermediate has sufficient stability to favour transfer to the nonreducing end of another substrate bound in the aglycone subsites.[172]

5.12.2.1.5 The α-amylase superfamily

The glucano hydrolases and transferases comprising the large α-amylase family (Family 13) are responsible for the processing of α-(1,4) linked glucan polymers, such as starch and glycogen. Glucosyl transfer to water or to the 4′-hydroxyl of another glucan chain (see Section 5.12.3.2) occurs with retention of anomeric configuration, consistent with a double-displacement mechanism analogous to that described for β-glycosidases. The members of this family, which include cyclodextrin glucanotransferases (CGTase's), glycogen debranching enzyme (Glix), porcine pancreatic α-amylase, *Aspergillus oryzae* α-amylase (TAKA-amylase), and yeast α-glucosidase, share four short consensus sequences that are localized to a common TIM barrel catalytic fold.[173,174] Located within these sequences are three strictly conserved carboxylic acid residues that have been shown to be catalytically essential.[173] A number of three-dimensional structures of complexes with acarbose (**10**), a naturally occurring inhibitor whose acarviosine ring resembles an oxocarbenium transition state, have helped to define the roles of these residues. A 2.2 Å porcine amylase structure[175] suggested that the acarbose chain had been transformed, via transglycosylation, into a pentasaccharide occupying subsites -3 to $+2$. This placed the acarviosine ring in the -1 subsite within the proximity of Asp197, Glu233, and Asp300. The former pair of residues occupy opposite sides of the binding cleft, with Glu233 positioned to act as a general acid catalyst by forming a hydrogen bond to the acarviosine "glycosidic" nitrogen. The carboxyl oxygen of the Asp197 was located within 3.3 Å of the β-face of the acarviosine C-1, as anticipated for nucleophilic attack to form a β-glycosyl-enzyme intermediate. While maintaining a hydrogen bond through a water molecule to Glu233, Asp300 rotated upon inhibitor binding to form a strong interaction with the acarviosine C-2 and C-3 hydroxyl groups. Thus this residue appears to fulfill two roles: maintaining the ionization state of Glu233 and substrate/transition state binding. The ionization of Glu233 appears to be further perturbed by a nearby chloride ion, which may explain the observed alkaline shift in the pH optimum for this enzyme in the presence of aqueous chloride. A similar arrangement of the homologous residues was observed in the acarbose complexes of TAKA amylase (Asp206, Glu230, Asp297)[176] and *B. circulans* CGTase (Asp229, Glu257, Asp328).[177-179] Interestingly, in all of the substrate/inhibitor complexes formed by porcine amylase, CGTase, or TAKA amylase, the ligand is observed to bind in a curved conformation, resulting in a kink at the scissile glycosidic bond. This disrupts the internal hydrogen bond between the C-2 and C3- hydroxyl groups of the $+1$ and -1 sugar units, which normally stabilizes the helical structure of α-glucan polymers. Such distortion may therefore be designed to expose the scissile glycosidic bond to the catalytic machinery.

(**10**)

Strong evidence now suggests that the nucleophile indeed forms a stable covalent bond upon glycosidic bond cleavage, as opposed to stabilizing an oxocarbenium ion intermediate as an ion pair. The nucleophile in the transferase active site of Glix from rabbit muscle, Asp549, was trapped as a covalent intermediate using 4-deoxymaltotriosyl-α-fluoride as a substrate.[82] The reactive fluoride leaving group ensured formation of an intermediate, which could not subsequently turn over via transglycosylation to another substrate molecule due to the absence of a 4′-hydroxyl group. This same substrate also served to trap the intermediate in the acid-base mutant of *B. circulans* CGTase (Glu257Gln) and identified Asp229 as the nucleophile.[84] Yeast α-glucosidase was inactivated with 5-fluoro-α-D-glucosyl fluoride, which acts analogously to the 2-deoxy-2-fluoroglycosides, due to labelling of the nucleophile, Asp214.[95] Interestingly, the C-5 epimer of this compound, 5-fluoro-β-L-idosyl fluoride, also rapidly inactivated the enzyme. While the intermediate in the porcine α-amylase has yet to be trapped, low temperature ^{13}C NMR experiments have suggested the formation of a covalent acylal linkage, presumably to Asp197.[180]

The members of the α-amylase family also share two histidine residues located in the active site, with a third histidine being conserved among α-amylases and CGTases.[173] Replacement of each of the three histidine residues (His140, His233, His327) in *Bacillus* sp. 1011 CGTase with asparagine significantly reduced the k_{cat} for the formation, coupling, and hydrolysis of β-cyclodextrin, whereas the K_m for these activities was essentially unchanged, leading to the conclusion that these residues are critically involved in transition state stabilization.[181] Furthermore, His327 was found to be essential for catalysis over an alkaline pH range. Indeed, the crystal structure of CGTase complexed with the transition-state analogue, acarbose, supports these results, revealing a number of hydrogen bonds formed by these residues to the ligand at the $+1$ and -1 subsites, which may assist the stabilization of a planar oxocarbenium ion transition state. Also, an ionized His327 appears to elevate the pK_a of the acid catalyst, Glu257, through a hydrogen bond.[179] Similar interactions are observed for the homologous histidine residues in the structure of porcine amylase complexed with acarbose.[175]

5.12.2.2 Inverting Glycosidases

5.12.2.2.1   (e → a) Inverters: **Cellulomonas fimi** *endoglucanase A,* **Trichoderma reesei**
cellobiohydrolase II, and **Thermomonospora fusca** *endocellulase E2*

A number of bacterial and fungal organisms secrete sets of cellulases that act in a synergistic manner to degrade cellulose into nutritive glucose.[182] Retaining and inverting cellulases are often found in the same set, which is an interesting issue in itself, since it is not immediately apparent how an organism would benefit from the enzymatic hydrolysis of a β-(1,4) glucan linkage with two opposite stereochemical outcomes. As suggested by Wood, this may stem from the structure of a cellulose chain in which each glucose residue is flipped $180°$ relative to its neighbor, thereby presenting two orientations of the β-(1,4) linkage for attack.[183]

Four inverting cellulases have been described in atomic detail, these being *Humicola insolens* endoglucanase V,[54,184] *Clostridium thermocellum* endocellulase D,[36,185] *Thermomonospora fusca* endo-cellulase E2,[186] and *Trichoderma reesei* cellobiohydrolase II (CBHII).[33] The cellulases *Th. fusca* E2 and *T. ressei* CBHII belong to Henrissat's Family 6, and appear to offer a structural basis for endo and exo-hydrolytic behavior.[187] Although E2 and CBHII share a TIM barrel catalytic domain, extra loop sequences are inserted into the CBHII domain which fold over the active-site cleft to form a tunnel.[33] It has been suggested that the nonreducing end of the glucan polymer is inserted into this tunnel and cleaved in an exo-fashion to release cellobiose as the predominant product.[33,186,188] A tube, orthogonal to this tunnel and leading to the enzyme surface, may shuttle water molecules to the enclosed active site. In contrast, the more open binding cleft of E2 may allow this enzyme to bind freely along the glucan polymer and cleave in an endo-fashion. The importance of active site accessibility is illustrated by the increase in endoglucanase activity by *Th. fusca* exoglucanase CbhA upon deletion of one of its active site loops. However, CBHII is not a strict endocellulase and displays some exo-hydrolytic behavior towards cellooligosaccharides.[188] This may reflect a degree of flexibility of the active site loops in CBHII, which may unfold to allow full access to the binding cleft, or may be a function of the tunnel length, which is shorter than that found in the strict exocellulase *T. reesei* CBHI. Additionally, a penchant for endo or exo-behavior may arise from differences in substrate affinity at subsites on either side of the scissile bond. In any case, an endo or exo-designation is not absolute. In reality, a continuum of overlapping specificities exists, with some enzymes tending to one extreme.[187]

Koshland's original hypothesis for an inverting glycosidase mechanism calls for a pair of catalytic carboxyl groups, one acting as a general base and one as a general acid, to effect the direct displacement of the aglycone by attack of water at the anomeric carbon atom.[56] In all inverting cellulases examined thus far, candidates for both roles have been found. Four aspartates are conserved in the active-site domains of Family 6 cellulases. In the active-site structure determined for E2, Asp265 (Asp401 in CBHII) and Asp117 (Asp221 in CBHII) were positioned on opposite sides of the binding cleft, with their carboxyl oxygens within ~ 5 Å of the reducing end anomeric carbon of a bound cellobiose molecule.[186] The total distance between the carboxyl oxygens (~ 10 Å) is significantly greater than that observed for retaining glycosidases (~ 5 Å) and is consistent with Koshland's mechanism in which a water molecule and the substrate simultaneously occupy the active site. The function of each residue was initially deduced from their local environments. Asp117 was assigned as the general acid due to the proximity of a third conserved carboxyl residue, Asp156 (Asp 263 in CBHII), presumably in an ionized state, which would maintain Asp117 in a protonated state as required. Asp265 appeared to be stabilized in an ionized state by a salt-linkage to Arg221, consistent with a general base role for this residue. Despite some doubt as to the identity of the general base in CBHII,[33,189] a comparison of the active site with that of E2 reveals a remarkable homology in the arrangement of the four conserved carboxyl groups, with Asp401 residing in a position that is obviously that of the general base. As with E2, it is likely that this residue is ionized due to a salt-linkage to Arg353 and Lys395. The fourth conserved residue in E2, Asp79 (Asp175 in CBHII), is positioned on a loop at the cusp of the active-site cleft and somewhat removed from the other catalytic residues. However, the homologue of this residue in the active site of CBHII is also positioned on a loop that is arranged to place this residue near the acid catalyst, Asp221. Therefore, in both enzymes, this fourth conserved carboxyl may fulfill a role in maintaining the protonated state of the acid catalyst. The different positions of the loops in each enzyme may reflect a dynamic active-site behavior that is common to this family.

The functions of these conserved carboxyl residues in Family 6 have been corroborated through mutagenesis of the homologous residues in another member of this family, *C. fimi* endoglucanase CenA. Replacement of the putative acid catalyst, Asp252, with alanine reduced k_{cat} for the hydrolysis of carboxymethyl cellulose (CM-cellulose) 2×10^5-fold, but only 1.2-fold for 2,4-dinitrophenyl β-cellobioside (2,4-DNPC),[190] a substrate that does not require general acid catalysis. Furthermore, the pH dependence of k_{cat} for the hydrolysis of 2,4-DNPC by the D252A mutant indicated the loss of a more basic ionization, as expected for an acid catalyst role.[191] In contrast, mutation of the homologous general base catalyst, Asp392, caused a significant decrease in k_{cat} for both 2,4-DNPC and CM-cellulose (2.2×10^4-fold and 3.4×10^4-fold, respectively). This is consistent with the fact that deprotonation of water is critical in the inverting mechanism, regardless of leaving group ability. Finally, mutation of the other two conserved carboxyl groups, Asp287 and Asp216 (Asp156 and Asp79 in E2, respectively), resulted in a similar dependence of k_{cat} on leaving group ability to that observed for the acid catalyst, but to a lesser degree, as expected for residues that are not directly involved in catalysis, but only maintain an ionization state. Analogously, deletion of the general acid catalyst, Asp221, in CBHII resulted in complete loss of activity against cello-oligosaccharides, whereas the D175A mutant retained significant activity.[33,192] Surprisingly, no kinetic studies on mutants of CBHII modified at position 401, the likely base catalyst, have been reported.

Inspection of the crystal structures of E2 and CBHII, in addition to studies on the cleavage patterns of cello-oligosaccharides,[188,192] reveals that the substrate binding cleft of Family 6 cellulases comprises four subsites (-2 to $+2$). Some disparity is observed in the contribution of the $+2$ subsite to catalysis when CenA and CBHII are compared. Substrate binding in all four subsites is important for CenA, as illustrated by the 800-fold greater value of k_{cat} for cellotetraose ($k_{cat} = 220$ s^{-1}) relative to cellotriose ($k_{cat} = 0.27$ s^{-1}).[191] The latter substrate binds to subsites -2 to $+1$, as indicated by its cleavage to cellobiose and glucose.[193] In contrast, the k_{cat} observed for the hydrolysis of cellotetraose by CBHII ($k_{cat} = 3.3$ s^{-1}) was only 100-fold greater than for cellotriose ($k_{cat} = 0.03$ s^{-1}),[189] reflecting a smaller requirement for binding in subsite $+2$. As noted above, this difference in aglycone binding strength may dictate the degree of endo or exo-behavior exhibited by each enzyme. The contribution to catalysis by the $+1$ subsite in CenA is equally impressive, as illustrated by the comparable hydrolysis rates of 2,4-DNPC ($k_{cat}/K_m = 2.2$ mM^{-1} s^{-1}), β-cellobiosyl fluoride ($k_{cat}/K_m = 1.5$ mM^{-1} s^{-1}), and cellotriose ($k_{cat}/K_m = 0.47$ mM^{-1} s^{-1}).[191] If it is assumed that CenA does not greatly assist the hydrolysis of the artificial substrates through specific aglycone interactions, it is evident that CenA improves the leaving group ability of a glucose aglycone (leaving group $pK_a \approx 16$) to equal that of 2,4-dinitrophenol (leaving group $pK_a = 4$). When the $+1$ and $+2$ subsites are filled, as is the case for cellotetraose ($k_{cat}/K_m = 1800$ mM^{-1} s^{-1}), the leaving group ability of cellobiose ($pK_a \approx 16$) approaches that of a phenol with a pK_a of about zero (as estimated from an extrapolation of a Brønsted plot for a series of aryl cellobiosides).

The significantly greater substrate affinity in the $+1$ and $+2$ subsites of CenA relative to CBHII may also explain some apparent mechanistic differences between CenA and CBHII. CenA can hydrolyze α-cellobiosyl fluoride, a substrate with the "wrong" anomeric configuration, through a transglycosylation-hydrolysis mechanism,[191] as originally proposed by Hehre.[194] In contrast, CBHII appears to hydrolyze this same substrate with normal Michaelis-Menten kinetics and 20-fold faster than the "natural" anomer, β-cellobiosyl fluoride.[189] Presumably the greater binding energy obtained in the aglycone subsites of CenA favors the binding of a second substrate as required by the Hehre mechanism. This may not be possible for CBHII at similar substrate concentrations. To account for this difference, an S_N1 mechanism was proposed for CBHII in which a solvent-separated ion pair is formed and then rapidly hydrolyzed with retention of anomeric configuration, analogous to the trifluoroethanolysis of α-glucosyl fluoride.[195] This hypothesis was further used to account for the absence of a base catalyst in the active site of CBHII, although comparison of the active site with E2 clearly indicates this not to be the case.

Very few kinetic data are available describing the transition state structure for inverting cellulases. By analogy to the better characterized glucoamylase and β-amylase inverting enzymes, the transition state is most likely oxocarbenium in nature. This is reflected somewhat in a Brønsted relationship for CenA, correlating k_{cat}/K_m for a series of aryl-cellobiosides.[191] The large slope ($\beta_{lg} = -0.9$) is consistent with substantial charge development on the leaving group oxygen, signifying nearly complete bond cleavage with little proton donation.

5.12.2.2.2 *(a → e) Inverters:* **Aspergillus** *glucoamylase and soybean* β-*amylase*

The nomenclature reserved for amylolytic glycosidases can be confusing to those who are new to the literature because the prefixes α and β are applied inconsistently. Therefore, a brief clarification is worthwhile. As described previously, α-amylase refers to any enzyme that will cleave α-(1,4) glucosidic linkages within amylose to release products with retained anomeric configuration. In contrast, glucoamylases (sometimes called amyloglucosidases or glucodextrinases) hydrolyze amylose in an exo-fashion to release β-D-glucose, and therefore operate by an inverting mechanism. The β-amylases also hydrolyze amylose in an exo-fashion with inversion, the prefix β referring to the anomeric configuration of the maltose product that is released, as opposed to the stereochemistry of the substrate glycosidic linkage.

Analogous to the inverting cellulases, *Aspergillus* glucoamylase (Family 15) uses two carboxyl groups for general acid/base catalysis. These have been unambiguously defined in the crystal structures solved for the *A. awamori* enzyme complexed with 1-deoxynojirimycin[196] and acarbose.[197] With 1-deoxynojirimycin bound, a water molecule was positioned within 3 Å of the β-face of the C-1 carbon atom via hydrogen bonds to the C-6 hydroxyl and Glu400, identifying this residue as the general base. An analogous arrangement was seen about the acarviosine ring of acarbose. The general acid was convincingly shown to be Glu179, which was located on the α-face of the ring and formed a strong hydrogen bond (2.66 Å) to the acarviosine amino group.

Kinetic isotope effects and inhibitor studies with the *Aspergillus* enzyme describe an oxocarbenium-like transition state in the inverting mechanism. Consistent with $sp^3 \rightarrow sp^2$ hybridization of the anomeric carbon during the enzymatic hydrolysis of α-glucosyl fluoride were significant ^{14}C and α-^3H effects ($V_{max}/K_m(6\text{-}^3H, 1\text{-}^{14}C) = 1.030$ and $V_{max}/K_m(6\text{-}^{14}C, 1\text{-}^3H) = 1.192$).[198,199] Also suggestive of this transition state character was the 2700-fold reduction in k_{cat} observed for the hydrolysis of 1,1-difluoro-D-glucopyranoside relative to α-glucosyl fluoride, a consequence of the nondeparting fluorine disfavouring the development of positive charge at the anomeric carbon.[200] Moreover, the tight binding of acarbose ($K_i = 10^{-12}$ M), can be attributed to the planar acarviosine ring which bears some resemblance to an oxocarbenium-ion transition state.[197,201] Of note is the observation of a remote isotope effect at the C-6 hydroxyl group ($V_{max}/K_m(6\text{-}^3H, 6\text{-}^{14}C) = 1.041$),[198] which is consistent with the hydrogen bond that is formed to the attacking water molecule, and may also arise in part from distortion of the pyranoside ring towards a half-chair conformation, as suggested by the structure of the enzyme with the ground-state analogue, D-*gluco*-dihydroacarbose ($K_i = 10^{-8}$ M).[197] The torque applied to the C-6 hydroxyl may be derived from a steric interaction with Trp52 and a hydrogen bond to Asp55. Also consistent with a role for Asp55 in transition state stabilization is the large reduction in k_{cat} (230 to 365-fold), without a change in K_m, observed for the hydrolysis of malto-oligosaccharides when this residue was mutated to glycine.[202]

The general base in glucoamylase, Glu400, appears to be less crucial for catalysis than the analogous residue in inverting cellulases. Mutation of this residue to glutamine resulted in a mere 35 to 60-fold decrease in k_{cat} for the hydrolysis of maltose and maltoheptaose, respectively.[203] However, this is not generally the case for $a \rightarrow e$ inverters since reductions of $>10^4$-fold were seen on mutating the general base catalyst in β-amylase.[204,205] In contrast, the enzyme is very dependent upon the general acid as indicated by the 2000-fold decrease in k_{cat} for the hydrolysis of maltoheptaose when Glu179 was mutated to glutamine.[206] This has been offered as evidence for a nonconcerted hydrolysis mechanism, in which an oxocarbenium ion intermediate is formed following general acid-catalyzed aglycone expulsion.[196,203] Water would react rapidly with this intermediate without requiring significant base catalysis, whereas concerted base catalysis would be required in the absence of a significant lifetime for the oxocarbenium ion.[207,208] This may explain why a proton inventory for the hydrolysis of *p*-nitrophenyl-α-D-glucopyranoside indicated that the observed solvent isotope effect ($k_{cat}(H_2O)/k_{cat}(D_2O) = 1.8$) arose from a single proton transfer, presumably from the general acid, Glu179.[209] Because this is a poor substrate for glucoamylase, bond breaking was assumed to be rate determining and the isotope effect would presumably reflect a proton transfer in the hydrolytic step. As suggested by others, this may leave Glu400 to play a lesser role in maintaining the active-site structure through a hydrogen bond network composed of the attacking water, Tyr48, Ser411, and Gln401.[203] Additionally, Glu400 may stabilize the transition state electrostatically along with other local carboxylate residues such as Asp55, Asp176, Glu179 and Glu180.[197,210] Indeed, mutations of Tyr48,[201,203] Asp55,[202] and Asp176[211] reduce k_{cat} to a degree that matches or surpasses that caused by mutating Glu400. Furthermore, the stabilization of positive charge development at the anomeric center by a cluster of carboxylates is suggested by the tighter binding of isofagomine ($K_i = 3.7$ μM) than 1-deoxynojirimycin ($K_i \approx 100$ μM).[212]

Pre-steady-state kinetic analyses with malto-oligosaccharide substrates have suggested a three-step hydrolytic mechanism for glucoamylase.[213,214] According to one interpretation[215] the first step is believed to involve the fast formation of an enzyme–substrate complex. The second step is thought to involve bond hydrolysis as evidenced by the two-fold decrease in the forward and reverse rate constants in D_2O. However, only a small solvent isotope effect was observed for the hydrolysis of maltose ($k_{cat}(H_2O)/k_{cat}(D_2O) = 1.1$), suggesting that the rate-determining step is not bond hydrolysis. This is believed to be the third step, product release, mediated predominantly by Trp120,[216] which was observed to stack hydrophobically with the third sugar ring of acarbose in the $+2$ subsite.[197,217] The authors interpret the observation that the mutation W120F had no effect on the hydrolysis of α-glucosyl fluoride, whereas hydrolysis of maltose and longer substrates was strongly affected, as being supportive of this mechanism, since only with the longer substrates would product release be rate determining.[210,214–216] However, no account was taken of the inherently greater leaving group ability of fluoride than of a sugar alcohol. Others, although in agreement with a three-step model, believe the first two steps involve the formation of the Michaelis complex, with a slow catalytic step comprising the third.[211] In this model products are released rapidly and enzyme–product complexes do not accumulate.

An analogous mechanistic picture has been developed for the β-amylases. Two carboxyl residues in soybean β-amylase, Asp101 and Glu186, which are highly conserved in the β-amylase family, were demonstrated through mutagenesis to be absolutely essential for activity.[205] The inactivator 2,3-epoxypropyl α-D-glucopyranoside had previously been used to label Glu186,[218] and more recently to label the homologous residue in sweet potato β-amylase, Glu187.[219] Four crystal structures of soybean β-amylase have recently been solved: a 2.0 Å structure of the enzyme complexed with the weak competitive inhibitor, α-cyclodextrin,[41] and 1.9–2.2 Å structures of the enzyme in its free form, or bound with β-maltose or maltal.[220] The catalytic domain consists of an $(α/β)_8$ fold, which contains a cleft that runs into a pocket containing the two catalytically essential carboxyl residues, a structure that is consistent with the exo-glucanase activity of this enzyme. Interestingly, α-cyclodextrin formed an inclusion complex with Leu383 (also a strictly conserved residue in β-amylases) and was some distance (~ 7 Å) from the catalytic residue Glu186, but effectively blocked access to the active-site pocket. The mode of binding of this inhibitor may reflect the binding of the natural substrate, amylose, which adopts a helical structure in free solution.[41] Two adjacent residues of β-maltose or maltal bound in the active-site pocket, thereby identifying at least four subsites. That maltose is the chief product released by this enzyme is neatly explained by the location of Glu187 and Glu380 (also conserved) at the reducing end of the more deeply bound maltose molecule (-1 and -2 subsites). These residues were positioned on opposite sides of the pocket, with Glu187 presumably acting as the general acid catalyst from its position on the β-face of the maltose molecule, and Glu380 acting as the general base, positioned on the α-face.[220] Despite its essential function in catalysis, Asp101 did not appear to be directly involved in the hydrolysis of a glycosidic bond, but instead resided on a hinged loop that closed over the binding cleft when a ligand was bound.[220] With the loop in the closed position, Asp101 interacted with the maltose molecule occupying the $+1$ and $+2$ subsites, and therefore may contribute to catalysis by interacting with the aglycone.

Glu380 also formed a strong hydrogen bond to a water molecule in the free enzyme. This hydrogen bond and the water molecule were absent in the β-2-deoxymaltose complex of the enzyme, the product derived from the hydration of maltal from the β-face. In D_2O, this prochiral substrate was hydrated with a large solvent isotope effect ($V_H/V_D = 6.5$) involving a single proton transfer.[221] Deuterium was exclusively transferred to the axial position at C-2, indicating that protonation also occurred from the β-face of maltal.[222] Presumably Glu380 is the source of this proton, initially forming an oxocarbenium ion intermediate that rapidly collapses to the *cis*-hydrated product in the presence of the nearby water molecule. That Glu380 is the proton source, rather than Glu186, is another example of the adaptability of the catalytic carboxyl residues employed by inverting glycosidases when faced with an unnatural substrate.[222] In this case, it has been suggested that the pK_a of Glu380 is elevated by the proximity of the hydrophobic double bond of bound maltal.[220]

5.12.3 OTHER GLYCOSYL TRANSFERS

5.12.3.1 Glycogen Phosphorylase (Rabbit Muscle)

Mammalian glycogen phosphorylase (GP) works in concert with glycogen debranching enzyme to liberate glucosyl units from glycogen for consumption by active muscle cells. The enzyme consists of two identical α/β domains that can exist in two forms, differing only in the phosphorylation state of Ser14 (designated GPa or GPb, the former being the phosphorylated form).[223] Each form can, in turn, exist in inactive or active conformational states (T and R, respectively) as determined by the presence of various allosteric effectors,[18,224] with the notable difference being that the GPa form is not dependent on the presence of AMP for activation. The structural details of allosteric activation have been intensively studied.[225-229] The most significant changes noted were the movement and ordering of the N-terminal tails upon phosphorylation,[18,228] rotation of the amino and carboxy-terminal sub-domains upon binding AMP,[227] and the displacement of a loop from the active site (residues 282 to 286).[13,18,229]

The activated forms of GP attack the nonreducing ends of glycogen, catalyzing the phosphorolysis of α-(1,4) glucoside bonds to release glucose-1-phosphate with retention of anomeric configuration. This product specificity, as well as the inability of GP to cleave limit dextrin, is dictated by the five subsites comprising the binding cleft (-1 to $+4$).[230] In the absence of orthophosphate, GP can also operate in the synthetic direction using glucose-1-phosphate as a substrate. The specifics of the catalytic mechanism, however, remain unclear, and lengthy debate has been given to the role of the active site pyridoxal phosphate (PLP) cofactor (**11**), and the form of the glycosyl-enzyme inter-mediate that is required by a retaining transfer.[16,231] It has been argued that PLP acts as a general acid/base catalyst based upon: (i) the proximity of the cofactor phosphate to the phosphate group of glucose-1-phosphate analogues bound in the active site (held in place by a salt-bridge mediated by Arg569)[232-234] and (ii) the observed A-S_E2 addition of orthophosphate to highly reactive enol ethers such as D-glucal and heptenitol.[16,234] However, substantial evidence argues against this meta-bolically expensive cofactor fulfilling a simple general acid/base role which could easily be performed by an appropriate amino acid side-chain. Parrish demonstrated that the activity of GP reconstituted with pyridoxal could be restored with equal catalytic effectiveness by adding fluorophosphate (pK_{a2} = 4.8) or phosphite (pK_{a2} = 6.6) to assay solutions.[235] Presumably these activator anions replaced the cofactor 5′-phosphate moiety. Likewise, the pH rate profiles determined for each activator anion were essentially identical.[236] Neither behavior is expected if the anions were participating in a proton shuttle. However, crystallographic studies indicated that these activator anions bound at a site 1.2 Å removed from the position of the cofactor 5′-phosphate, and may merely fulfill a structural role.[237] Such mobility was not possible in GP reconstituted with phosphonate (**12**) (pK_{a2} = 7.2) and α-difluorophosphonate (**13**) (pK_{a2} = 4.2) analogues of PLP, which retained significant activity (25–30%) with glucose-1-phosphate as a substrate.[238] Despite having widely different solution pK_a values for the second ionization of the 5′-phosphonic acid moiety, the pH dependencies of k_{cat}, K_m, and k_{cat}/K_m for each reconstituted enzyme were virtually indistinguishable, a behavior that is, as noted by Parrish, inconsistent with an acid/base function for PLP. In fact, the low pK_{a2} of the α-difluoro-phosphonate analogue ensured that this analogue bound to GP as a dianion under neutral assay conditions. A dianionic state for PLP in the GP active site had also been inferred from solid-state ^{31}P NMR studies.[239,240]

(**11**)   (**12**)   (**13**)

Two mechanistic interpretations of a dianionic PLP have been forwarded. The first treats the phosphate of PLP as an electrophilic catalyst, accepting a negatively charged oxygen from orthophosphate to form a five-membered bipyramidal species.[241-243] This could assist departure or attack of the substrate phosphate by quenching excess negative charge. However, the negative charge repulsion between the phosphate oxygen atoms would make the approach of the nucleophilic phosphate highly unfavorable. Alternatively, it is possible that this charge repulsion provides ground state destabilization of the enzyme–substrate complex[244] that could be relieved by movement of the substrate phosphate away from the dianionic cofactor 5′-phosphate in the transition state.[238] This could complement the destabilization applied to glucose-1-phosphate in the form of a weakened *exo*-anomeric effect, arising from an unfavorable glycosidic bond dihedral angle stabilized by a hydrogen bond between the 2′-OH and a phosphate oxygen (Scheme 7).[16,234,245] Both effects could assist the expulsion of the anomeric phosphate in the synthetic direction of GP.

The form of the glucosyl-enzyme intermediate is unclear due to the absence of a classical nucleo-phile near the β-face of the substrate. This has led to a mechanistic interpretation drawn from the classical Phillips mechanism for lysozyme, in which aglycone departure is purely S_N1 with internal return of orthophosphate, or a 4′-OH, to the same face of the stabilized oxocarbenium ion inter-mediate.[16] It is certain that the transition state for phosphorolysis is oxocarbenium-like, as illustrated

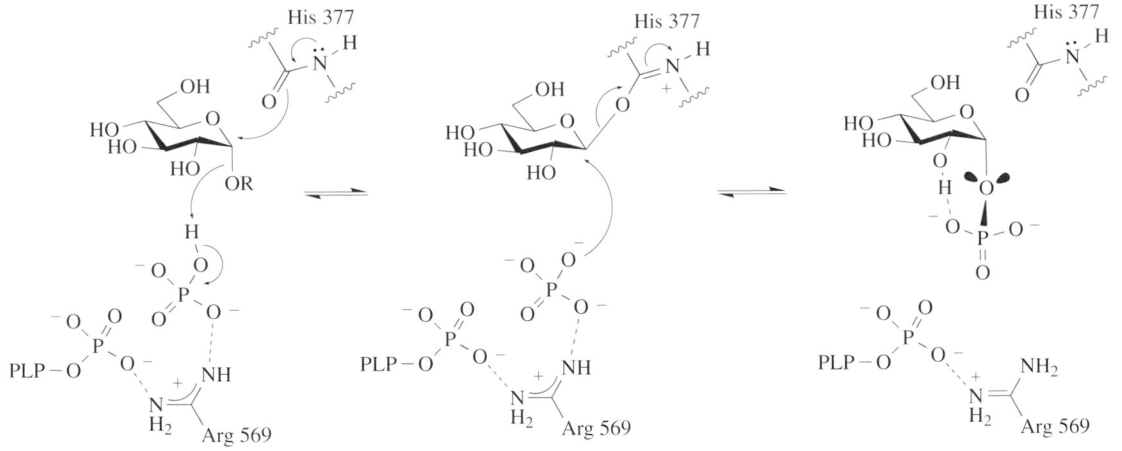

Scheme 7

by the strong synergistic inhibition of GP by nojirimycin tetrazole (an oxocarbenium-ion analogue) in the presence of phosphate ($K_i = 53$ μM).[232] Furthermore, a strong correlation is observed between the log k_{cat}/K_m values for a series of deoxy- and deoxyfluoroglucose-1-phosphate substrates and logk_{H_2O} for their acid-catalyzed hydrolysis, a reaction known to involve an oxocarbenium ion-like transition state.[246] However, it is unlikely that an oxocarbenium ion could exist as a real intermediate in the presence of a negatively charged phosphate.[207] Consequently, it is tempting to assign a nucleophilic role to the backbone carbonyl group of His377 (Scheme 7). Crystallographic data indicate that this carbonyl oxygen atom is directed towards the β-face of the anomeric carbon in nojirimycin tetrazole,[232] and contributes to the potency of inhibitors containing an aglycone amide functionality by accepting a hydrogen bond.[247–249] While this may seem unusual at first glance, this is not an unprecedented role for an amide carbonyl, recalling acetamido participation in the mechanism of jack bean hexosaminidase and various chitinases. Of course, other nucleophile candidates cannot be excluded.

The other amino acid residues involved in substrate binding and transition state stabilization have been defined from crystallographic studies of complexes with substrates and transition-state analogues.[232,250] Additionally, the contributions of individual sugar hydroxyl groups to binding in the ground and transition states have been evaluated from kinetic studies with modified substrates.[246,252] Interactions with the 3 and 6-hydroxyl groups were identified as the most important, consistent with the presence of charged hydrogen-bonding partners on the enzyme.

Aglycone binding also contributes significantly to catalysis by GP. This has been revealed in a recent 3 Å crystal structure of *E. coli* maltodextrin phosphorylase complexed with maltohexaose, the first time an oligosaccharide has been observed bound to the active site of an α-glucan phosphorylase.[253] An equivalent structure could not be obtained for rabbit muscle GP, possibly due to the significantly lower affinity for linear oligosaccharides ($K_m \sim 30$ mM) relative to the bacterial enzyme ($K_m \sim 0.39$ mM). Despite this difference, and the fact that the bacterial enzyme does not have any allosteric properties, the two enzymes have remarkably conserved active sites (98% sequence identity). Tyr280 (rabbit muscle GP numbering) formed a stacking interaction with the sugar residue in the +2 subsite. This contributes significantly to catalysis, as illustrated by the 10^4-fold decrease in k_{cat}/K_m when this residue is replaced by alanine. Although Tyr280 is at the start of the dynamic active-site loop in rabbit muscle GP, no change in the position of this residue was observed in the native and ligand-bound forms of the bacterial enzyme. In the native rabbit muscle GP (T or R states), Tyr280 is turned away from subsite +2. It is thought that a change in the position of the active-site loop upon substrate binding reorients Tyr280 over the +2 subsite. The contact that Tyr280 forms with the substrate appears to break the preferred helical conformation of successive α-(1,4) linked sugars. By modeling a glucose residue into the −1 subsite, it was determined that this binding interaction would distort the glycosidic bond across the −1 and +1 subsites so that it would be more exposed to phosphorolytic attack, analogous to the kink induced in the scissile bond by α-amylases.

5.12.3.2 Cyclodextrin Glucanotransferase (*Bacillus circulans*) and Glycogen Debranching Enzyme (Rabbit Muscle)

Despite sharing a common set of catalytic machinery, the α-amylase superfamily is remarkable in the divergent strategies that have evolved for the capture of the covalent glycosyl-enzyme intermediate (see also Section 5.12.2.1.5). This has been the focus of substantial study with CGTases, which normally form mixtures of α, β, and γ-cyclodextrins (cyclic malto-oligosaccharides 6, 7, and 8 residues in size, respectively), with the hope of improving their specificity for industrial applications. A structure of *B. circulans* CGTase complexed with an acarbose-based maltononaose inhibitor has revealed how this enzyme brings about an intramolecular transglycosylation.[254] The inhibitor occupied seven subsites beyond the scissile glycosidic bond on the nonreducing side, leaving a maltose unit to be cleaved from the reducing end. This agrees with the observation that β-cyclo-dextrin is the main product produced by this CGTase. Moreover, the inhibitor bound in a circular conformation, with a distance of 23 Å separating the nonreducing end 4′-hydroxyl in subsite +7 from the anomeric carbon of the scissile bond in subsite +1. That CGTase is capable of transferring this hydroxyl over such a vast distance following cleavage of the glycosidic bond is satisfactorily explained by the stability of the covalent intermediate that is formed. Numerous aromatic residues line the binding site and stack with the monosaccharide rings to assist induction of a cyclic conformation.[255] Of note is Tyr195, which appears to act as a hydrophobic post around which the cyclodextrin ring forms, and indeed mutation of this residue exerts dramatic effects on the cyclization characteristics of CGTase.[256] Additionally, this residue may protect the glycosyl-enzyme inter-mediate from hydrolytic attack.[256]

Another notable transfer strategy is found in glycogen debranching enzyme, which utilizes two separate active sites in its task of linearizing the α-(1,6) branch points of limit dextrin. The transferase site employs the homologous catalytic machinery described above to transfer a maltotriosyl unit from an α-(1,6) branch to the nonreducing end of the glucan polymer with retention of configur-ation.[257] The use of a separate active site to cleave the remaining α-(1,6) linked glucosyl unit is substantiated by the observation that hydrolysis occurs with inversion, and by the finding that the transferase activity can be inactivated with carbodiimide without affecting the glucosidase activity.[257] How the switch from one activity to the other comes about has been inferred from inhibition of the α-(1,6) glucosidase activity with oligosaccharides of different lengths, using α-glucosyl fluoride as a substrate.[257] Glucose, maltose, and maltotriose are competitive inhibitors, whereas oligosaccharides with four or more glucose units activate the glucosidase activity with uncompetitive kinetics. This would imply that the activators do not bind at the same site as does α-glucosyl fluoride, but bind elsewhere and induce a conformational change in the enzyme. However, all oligosaccharides tested were competitive inhibitors of the transferase activity, implying that the binding site is common to both activities. These results can be explained if it is assumed that the activator and α-glucosyl fluoride bind to the enzyme in a fashion that resembles an α-(1,6) branch point following the transfer reaction, and promote a conformational change in the enzyme that reorients the substrate for the glucosidase reaction.

5.12.3.3 *N*-Ribosyl Hydrolases and Phosphorylases (*Crithidia fasciculata* and Calf Spleen)

Nucleoside hydrolases are ubiquitous in protozoan parasites, utilized for the salvage of purines and pyrimidines from mammalian hosts. One such enzyme, the inosine-uridine nucleoside hydrolase from *Crithidia fasciculata* (IU-NH), has become a model system for the application of transition-state analysis and inhibitor design.[259] This enzyme accepts all natural nucleosides as substrates, cleaving the *N*-glycosidic bonds with inversion followed by random release of products.[260] A remarkably detailed description of the transition state was obtained from KIEs (V_{max}/K_m) with labeled inosine.[261] The highly developed oxocarbenium ion ($[1'-^3H] = 1.150$, $[2'-^3H] = 1.161$) forms an elongated bond (~ 2 Å) to a protonated hypoxanthine leaving group, simultaneously stabilized by preassociation with the incoming water nucleophile (as calculated from $[1'-^{14}C] = 1.044$, $[9-^{15}N] = 1.026$). This water molecule must be bound tightly in position, as reflected by the lack of competition by stronger nucleophiles such as methanol. A remote isotope effect observed at the hydroxymethyl group ($[5'-^3H] = 1.051$, $[4'-^3H] = 0.992$) suggests that the enzyme uses this group as a lever to distort the ribose ring. Model building implied that this is the 3′-*exo*-ribose ring conformer, which aligns the C-2—H bond with the C-1 *p*-orbital of the developing oxocarbenium ion. This would result in a stabilizing hyperconjugation effect, and may explain the large β-tritium KIE that is observed.[262] This is an interesting contrast to the transition states for the hydrolysis of pyranosides, because the oxocarbenium ion charge seems to be distributed over the ribosyl ring, rather than localized to a $C{=}O^+ \leftrightarrow C^+{-}O$ resonance pair (Equation (5)).[262]

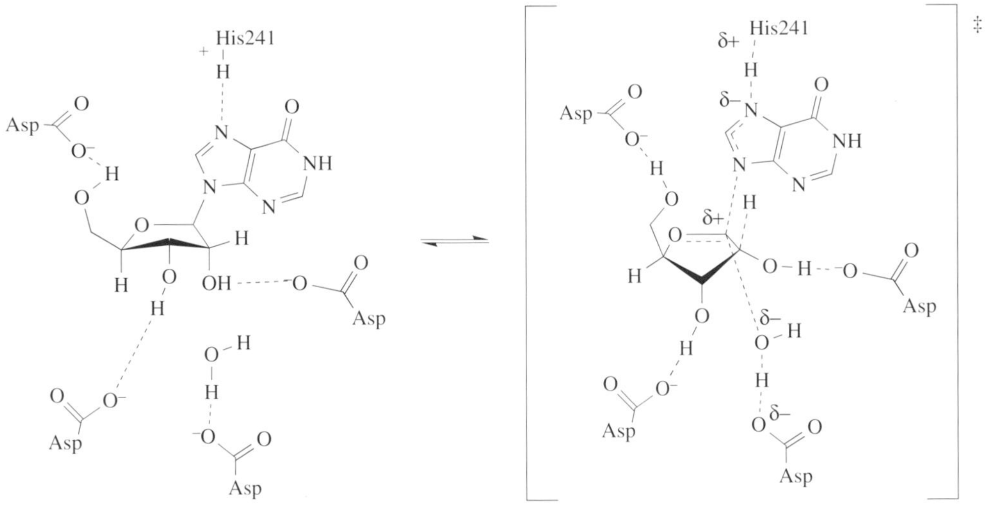

(5)

The transition-state geometry calculated from KIEs has assisted the design of nucleoside hydrolase inhibitors. The most potent inhibitor, *p*-nitrophenyl-riboamidrazone (**14**) ($K_i = 2\,\text{nM}$), incorporates the two major features of the transition state: positive charge localized at the anomeric center, and an aglycone functionality which can interact with the general acid catalyst.[263] Slow-onset inhibition was observed, which was attributed to a conformational change in the enzyme to form a more tightly bound complex. This could include a proton transfer to the imino nitrogen to form the charged transition state analogue.[264] The resonance form of the bound inhibitor was probed with Raman spectroscopy, and determined to be the neutral zwitterion (**14**).[265] The topology of this electron delocalization matches that of the natural transition state predicted by isotope effects.[261]

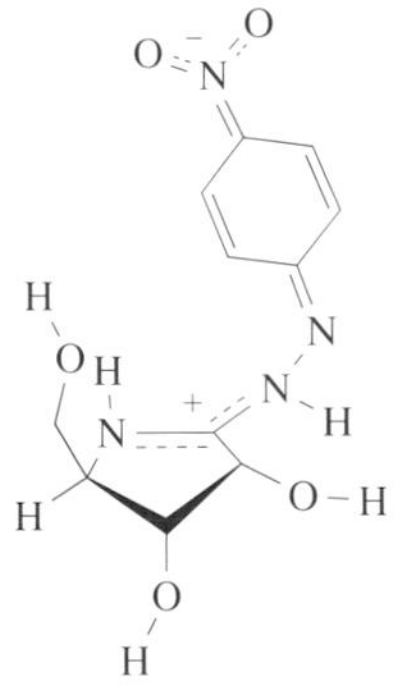

(**14**)

Solution of the crystal structure for IU-NH revealed a novel active-site architecture.[266] The bottom of the binding site is lined with four aspartate residues, one of which may act as the base catalyst inferred from pH studies.[260] The other aspartate residues have the potential to form strong hydrogen bonds to the ribosyl hydroxyl groups and electrostatically stabilize the transition state. His241, demonstrated by mutagenesis studies to be the general acid catalyst,[267] is also located within the active site. Other than a hydrogen bond to the acid catalyst, the binding of the purine ring is proposed to consist primarily of hydrophobic and stacking interactions, thereby accounting for the relaxed aglycone specificity displayed by this enzyme. Such a lack of specificity is also consistent with the relatively small loss of transition state stabilization ($4.6\,\text{kcal mol}^{-1}$) observed for the hydrolysis of inosine by the His241Ala mutant, and the return of wild-type activity with the more reactive substrate, *p*-nitrophenyl riboside. With enzymatic stabilization of the transition state estimated to be $17.7\,\text{kcal mol}^{-1}$, the mutagenesis study implies that the bulk of this stabilization ($\sim 13\,\text{kcal mol}^{-1}$) is derived from interactions with the ribosyl ring.[267]

The purine nucleoside phosphorylase from calf spleen stabilizes similar transition-state geometry, but with phosphate replacing water as the nucleophile. Moreover, stabilizing interactions are applied primarily to the aglycone rather than the glycone.[268,269] The high affinity for the aglycone is evident in the phosphorolysis of inosine, for which the release of hypoxanthine is rate determining (K_d = 1.3×10^{-12} M, k_{off} = 1.3×10^{-4} s^{-1} at 30 °C).[268] This enzyme can also catalyze a slow hydrolytic reaction in the absence of phosphate, but only through one catalytic cycle because the hypoxanthine remains tightly bound, perhaps due to an enzyme conformation with residual transition state affinity.[268] Occupation of the nucleophile binding site by phosphate or arsenate seems to facilitate relaxation of this tight state allowing aglycone release (k_{off} = 3×10^{-4} s^{-1}, 5 mM arsenate, 30 °C) and catalytic turnover. The substrate *p*-nitrophenyl riboside also illustrates the dependence on aglycone activation.[270] Because the labile aglycone does not require acid catalysis for departure, this substrate should function best with those enzymes which interact primarily with the ribosyl ring to stabilize the oxocarbenium ion transition state. Accordingly, this was an ideal substrate for *C. fasciculatata* IU-NH (k_{cat}/K_m = 4.1×10^6 M^{-1} s^{-1}), but not so with calf spleen nucleoside phosphorylase (k_{cat}/K_m = 0.89 M^{-1} s^{-1}).

5.12.3.4   Nucleotide-sugar Utilizing Transferases: Fucosyltransferases

While the mechanistic and structural detail of glycosidases steadily accumulates, the current knowledge surrounding glycosyltransferases remains in its infancy. This is despite the fact that over 500 sequences for glycosyltransferases are now available, and arranged into 26 families on the basis of sequence similarities.[271] The only crystal structure available for a glycosyl transferase is that of a DNA glycosyl transferase.[29] Furthermore, the nontrivial purification of these enzymes and the requirement of expensive nucleotide substrates has discouraged *in vitro* mechanistic study. Almost all eukaryotic glycosyltransferases are membrane bound, with an independent, lumenally oriented, C-terminal catalytic domain anchored by a single transmembrane α-helix. Catalysis involves the transfer of a monosaccharide unit to an oligosaccharide, lipid, or protein acceptor. Such transfers occur with either inversion or retention of anomeric configuration, providing one level of classification. The anomeric center of the monosaccharide in the glycosyl donor is activated towards nucleophilic attack by a high energy phosphoryl linkage to either a nucleoside or lipid carrier.

The central importance of fucosylation of cell surface oligosaccharides to cell–cell interactions has driven the mechanistic study of α-(1,3)-fucosyltransferase V (FucT V), an inverting transferase of transferase Family 10, beyond most others. The five homologous forms of fucosyltransferase have been cloned, all differing in acceptor specificity through a hypervariable region in the catalytic domain that immediately precedes the transmembrane segment.[272] FucT V transfers fucose from GDP-fucose to LacNAc of sialyl LacNAc to form Lewis X and sialyl Lewis X, respectively. The latter ligand has been implicated in tumor metastasis, and also causes leucocyte rolling by binding E-selectin as part of the inflammatory response. The transfer occurs with inversion at the fucose anomeric center and follows an ordered Bi-Bi kinetic scheme with GDP-fucose binding first.[273] A pH-rate profile identified an ionizable catalytic group with a pK_a of 4.1. This is most likely a carboxylate that deprotonates the acceptor hydroxyl group for attack on the activated fucose.[273] A proton inventory also suggested a single proton "in flight" during the catalytic step.[273] Unlike inverting glycosidases which activate the glycosidic bond with acid catalysis, FucT V employs a Mn^{2+} cofactor to electrophilically stabilize the departing nucleotide. A bidentate chelation mode by the two phosphate residues was inferred from the relative inhibition of the guanosine series: GTP ~ GDP > GMP ≫ guanosine.[273] A large α-secondary KIE was observed with GDP-[1-^{2}H]-fucose as a substrate (D_V = 1.32, $D_{V/K}$ = 1.27), consistent with an advanced oxocarbenium ion transition state (Equation (6)).[274]

The development of inhibitors for FucT V has benefited from a transition state that is also common to glycosidases. Aza sugars inhibit a variety of glycosidases, including fucosidases, by mimicking the charge of the transition state when the endocyclic nitrogen is protonated. Although an aza analogue of fucose, homofuconojirimycin, inhibits FucT V poorly on its own (IC_{50} = 71.5 mM), the presence of GDP improves binding substantially (IC_{50} = 1.54 mM, 30 μM GDP).[275] This stronger affinity for the active site is attributed to a synergistic interaction between the positively charged aza sugar and the negatively charged GDP, which mimics the charge distribution in the normal transition state. Furthermore, this synergism is not observed with GMP, illustrating the

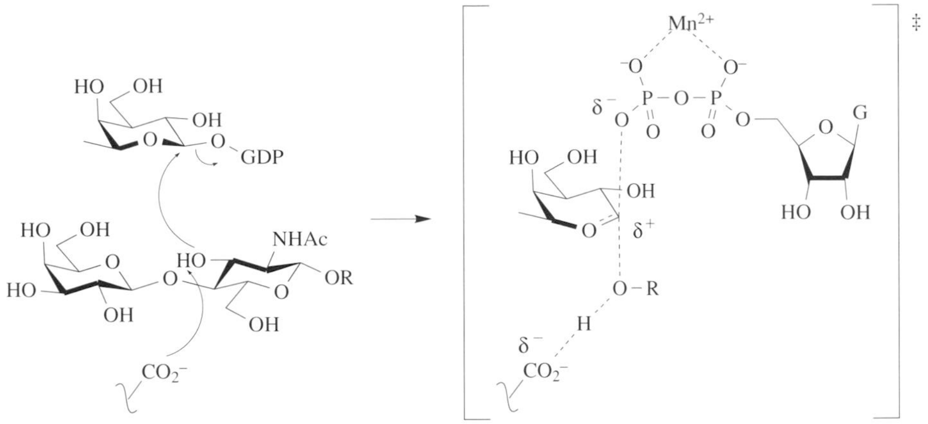

$$ (6) $$

sensitivity of the interaction to the distance between the two charges.[273] Even greater potency and enzyme specificity was achieved by coupling an acceptor analogue of LacNAc to homo-fuconojirimycin through an ethylene bridge (15) (IC_{50} = 31 μM).[275] However, the most potent reversible inhibitor developed to date is GDP-2-deoxy-2-fluorofucose (16) (K_i = 4.2 μM),[274] which is remarkable for an inhibitor that is a ground-state analogue. This could be the result of the relatively low transition-state affinity estimated for this enzyme (8.6×10^{-1} M).[273,276]

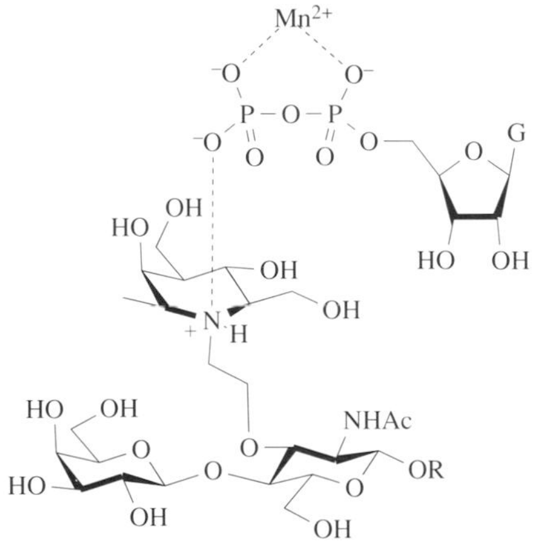

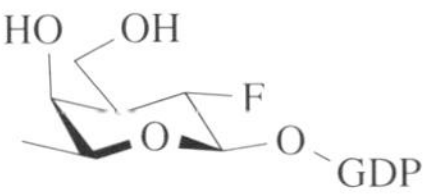

(16)

(15)

Efforts have been made to map the importance of interactions with individual hydroxyl groups by measuring reaction rates with a series of acceptor analogues in which each hydroxyl group is individually modified.[277–279] Lower resolution insights have been obtained from molecular biological approaches. For example, 75 amino acid residues could be removed from the N terminus of human Fuc T V without loss of significant activity, but none from the C terminus.[280] Interestingly, human Fuc T III (an α-(1,4) transferase) is very similar in sequence, most of the differences lying near the N terminus. Indeed, an enzyme with both activities was generated by swapping part of the N terminus.[280]

5.12.4 SUMMARY

There are few enzyme families that rival the glycosidases in terms of structural diversity and mechanistic understanding. This can be attributed, at least in part, to the relative ease of purification of many glycosidases, and the fact that chromogenic substrates for a number of these enzymes are readily available. In these circumstances, the tools of physical organic chemistry have been applied with great success to define the oxocarbenium ion structure of the transition states for glycosyl transfer to water, and to allow study of the various Michaelis complexes and reaction intermediates. When coupled with the power of site directed mutagenesis, X-ray crystallography, NMR spectroscopy, as well as advances in inhibitor and inactivator design, a remarkably detailed picture of

310   *Glycosyl Transferase Mechanisms*

substrate binding and transition-state stabilization has been obtained in a number of cases. Foremost among these, perhaps, is the proof for the covalent structure of the glycosyl-enzyme intermediate formed by the retaining glycosidases. Moreover, with the organization of glycosidases into families based on sequence alignment, the mechanistic detail obtained for a specific glycosidase may be extended to others by analogy. Remarkably, this wealth of mechanistic data has done little to change Koshland's original theory for the inverting and retaining mechanisms for glycosidases, which he postulated 45 years ago. It is unfortunate, then, that Koshland's theory has been overshadowed by a great irony in the history of glycosidase study, this being the presentation of hen egg white lysozyme in many biochemistry textbooks as a paradigm for the retaining mechanism, despite being, arguably, one of the most inaccessible glycosidases to mechanistic study. Indeed, by analogy to hexosaminidases and chitinases, the lysozyme mechanism appears to be a variation of the retaining mechanism rather than a general case.

Although the mechanisms of enzymic glycosyl transfer to water are, for the most part, well understood, much remains to be defined for the process of glycosyl transfer to other acceptor molecules. A prime example here is glycogen phosphorylase. Despite the wealth of structural information available for this enzyme, the exact details of glucosidic bond phosphorolysis with retention of anomeric configuration remain elusive. However, the next mechanistic frontier belongs to the nucleotide-dependent glycosyl transferases. Given the large number of transferase sequences now available, their organization into families, and the commercial interest in this class of enzyme, it is certain that considerable advances will be made in the study of structures and mechanism within the next few years.

5.12.5   REFERENCES

1. G. Semenza, in "Mammalian Ectoenzymes", eds. A.J. Kenny and A.J. Turner, Elsevier, Amsterdam, 1987, p. 265.
2. G. Davies and B. Henrissat, *Structure*, 1995, **3**, 853.
3. B. Henrissat and G. J. Davies, *Curr. Opin. Struct. Biol.*, 1997, **7**, 637.
4. G. Legler, *Adv. Carbohydr. Chem. Biochem.*, 1990, **48**, 319.
5. J. D. McCarter and S. G. Withers, *Curr. Opin. Struct. Biol.*, 1994, **4**, 885.
6. G. Mooser, in "The Enzymes", ed. D. S. Sigman, Academic Press, New York, 1992, vol. 20, p. 187.
7. M. L. Sinnott, in "Enzyme Mechanisms," eds. M. I. Page and A. Williams, Royal Society of Chemistry, University Press Ltd., London, 1987, p. 259.
8. M. L. Sinnott, *Chem. Rev.*, 1990, **90**, 1171.
9. B. Svensson and M. Sogaard, *J. Biotechnol.*, 1993, **29**, 1.
10. S. G. Withers, in "ACS Symposium Series: Biochemical and Biotechnological Applications of Electrospray Ionization Mass Spectrometry," American Chemical Society, Washington, DC, 1995, p. 365.
11. G. Davies, M. L. Sinnott, and S. G. Withers, in "Comprehensive Biological Catalysis," ed. M. L. Sinnott, Academic Press, New York, 1997, vol. 1, p. 119.
12. B. Svensson, *Plant Mol. Biol.*, 1994, **25**, 141.
13. M. F. Browner and R. J. Fletterick, *Trends Biochem. Sci.*, 1992, **17**, 66.
14. R. J. Fletterick and S. R. Sprang, *Acc. Chem. Res.*, 1982, **15**, 361.
15. D. J. Graves and J. H. Wang, in "The Enzymes," ed. P. D. Boyer, Academic Press, New York, 1972, vol. 7, p. 435.
16. E. J. Helmreich, *Biofactors*, 1992, **3**, 159.
17. L. N. Johnson, J. Hajdu, K. R. Acharya, D. I. Stuart, P. J. McLaughlin, N. G. Oikonomakous, and D. Barford, in "Allosteric Proteins," ed. G. Herve, CRC Press, Boca Raton, FL, 1987.
18. L. N. Johnson, *FASEB*, 1992, **6**, 2274.
19. C. B. Newgard, P. K. Hwang, and R. J. Fletterick, *CRC Crit. Rev. Biochem. Mol. Biol.*, 1989, **24**, 69.
20. N. B. Madsen and S. G. Withers, *Prog. Clin. Biol. Res.*, 1984, **144a**, 117.
21. N. B. Madsen and S. G. Withers, in "Coenzymes and Cofactors," eds. D. Dolphin, R. Poulsen, and O. Avramovic, Wiley, New York, 1986, vol. 1, part B, p. 355.
22. M. W. Bauer, S. B. Halio, and R. M. Kelly, *Adv. Protein Chem.*, 1996, **48**, 271.
23. B. Henrissat, *Biochem. J.*, 1991, **280**, 309.
24. B. Henrissat and A. Bairoch, *Biochem. J.*, 1993, **293**, 781.
25. B. Henrissat and A. Bairoch, *Biochem. Lett.*, 1996, **316**, 695.
26. J. Gebler, N. R. Gilkes, M. Claeyssens, D. B. Wilson, P. Beguin, W. Wakarchuk, D. G. Kilburn, R. C. J. Miller, R. A. J. Warren, and S. G. Withers, *J.Biol. Chem.*, 1992, **267**, 12 259.
27. B. Henrissat, I. Callebaut, S. Fabrega, P. Lehn, J.-P. Mornon, and G. Davies, *Proc. Natl. Acad. Sci.*, 1995, **92**, 7090.
28. J. Jenkins, L. L. Leggio, G. Harris, and R. Pickersgill, *FEBS Letters*, 1995, **362**, 281.
29. A. Vrielink, W. Ruger, H. P. Driessen, and P. S. Freemont, *EMBO J.*, 1994, **13**, 3413.
30. T. Barrett, C. G. Suresh, S. P. Tolley, E. J. Dodson, and M. A. Hughes, *Structure*, 1995, **3**, 951.
31. R. H. Jacobson, X.-J. Zhang, R. F. DuBose, and B. W. Matthews, *Nature*, 1994, **369**, 761.
32. V. Ducros, M. Czjzek, A. Belaich, C. Gaudin, H. P. Fierobe, L. P. Belaich, G. J. Davies, and R. Haser, *Structure*, 1995, **3**, 939.
33. J. Rouvinen, T. Bergfors, T. Teeri, J. K. Knowles, and T. A. Jones, *Science*, 1990, **249**, 380.
34. C. Divne, J. Ståhlberg, T. Reinikainen, L. Ruohonen, G. Pettersson, J. K. Knowles, T. T. Teeri, and T. A. Jones, *Science*, 1994, **265**, 524.

35. P. M. Alzari, H. Souchon, and R. Dominguez, *Structure*, 1996, **4**, 265.
36. M. Juy, A. G. Amit, P. M. Alzari, R. J. Poljak, M. Claeyssens, P. Béguin, and J.-P. Aubert, *Nature*, 1992, **357**, 89.
37. U. Derewenda, L. Swenson, R. Green, R. Wei, F. Morosoli, F. Shareck, D. Kluepfel, and Z. S. Derewenda, *J. Biol. Chem.*, 1994, **269**, 20 811.
38. W. W. Wakarchuk, R. L. Campbell, W. L. Sung, J. Davoodi, and M. Yaguchi, *Protein Sci.*, 1994, **3**, 467.
39. G. Sulzenbacher, F. Shareck, R. Morosoli, C. Dupont, and G. J. Davies, *Biochemistry*, 1997, **36**, 16 032.
40. E. Boel, L. Brady, A. M. Brzozowski, Z. Derewenda, G. G. Dodson, V. J. Jensen, S. B. Petersen, H. Swift, L. Thim, and H. F. Woldike, *Biochemistry*, 1990, **29**, 6244.
41. B. Mikami, E. J. Hehre, M. Sato, Y. Katsube, M. Hirose, Y. Morita, and J. C. Sacchettini, *Biochemistry*, 1993, **32**, 6836.
42. A. Aleshin, A. Golubev, L. M. Firsov, and R. B. Honzatko, *J. Biol. Chem.*, 1992, **267**, 19 291.
43. T. Keitel, O. Simon, R. Borriss, and U. Heinemann, *Proc. Natl. Acad. Sci. U.S.A.*, 1993, **90**, 5287.
44. J. N. Varghese, T. P. Garrett, P. M. Colman, L. Chen, P. B. Hoj, and G. B. Fincher, *Proc. Natl. Acad. Sci. U.S.A.*, 1994, **91**, 2785.
45. A. Perrakis, I. Tews, Z. Dauter, A. B. Oppenheim, I. Chet, K. S. Wilson, and C. E. Vorgias, *Structure*, 1994, **2**, 1169.
46. P. J. Hart, A. F. Monzingo, M. P. Ready, S. R. Ernst, and J. D. Robertus, *J. Mol. Biol.*, 1993, **229**, 189.
47. I. Tews, A. Perrakis, A. Oppenheim, Z. Dauter, K. S. Wilson, and C. E. Vorgias, *Nature Struct. Biol.*, 1996, **3**, 638.
48. N. C. J. Strynadka and M. N. G. James, *J. Mol. Biol.*, 1991, **220**, 401.
49. L. H. Weaver, M. G. Grutter, and B. W. Matthews, *J. Mol. Biol.*, 1995, **245**, 54.
50. R. Kuroki, L. H. Weaver, and B. W. Matthews, *Science*, 1993, **262**, 2030.
51. T. N. Petersen, S. Kauppinen, and S. Larsen, *Structure*, 1997, **5**, 533.
52. S. J. Crennell, E. F. Garman, W. G. Laver, E. R. Vimr, and G. L. Taylor, *Proc. Natl. Acad. Sci. U.S.A.*, 1993, **90**, 9852.
53. J. N. Varghese, W. G. Laver, and P. M. Colman, *Nature*, 1983, **303**, 35.
54. G. J. Davies, G. G. Dodson, R. E. Hubbard, S. P. Tolley, Z. Dauter, K. S. Wilson, C. Hjort, J. M. Mikkelsen, G. Rasmussen, and M. Schülein, *Nature*, 1993, **365**, 362.
55. E. M. Marcotte, A. F. Monzingo, S. R. Ernst, R. Brzezinski, and J. D. Robertus, *Nat. Struct. Biol.*, 1996, **3**, 155.
56. D. E. Koshland, *Biol. Rev.*, 1953, **28**, 416.
57. Q. Wang, R. W. Graham, D. Trimbur, R. A. J. Warren, and S. G. Withers, *J. Am. Chem. Soc.*, 1994, **116**, 11 594.
58. P. R. Young and W. P. Jencks, *J. Am. Chem. Soc.*, 1977, **99**, 8238.
59. T. L. Amyes and W. P. Jencks, *J. Am. Chem. Soc.*, 1989, **111**, 7888.
60. Q. Wang and S. G. Withers, *J. Am. Chem. Soc.*, 1995, **117**, 10 137.
61. Q. Wang, D. Trimbur, R. Graham, R. A. J. Warren, and S. G. Withers, *Biochemistry*, 1995, **34**, 14 554.
62. W. Burmeister, S. Cottaz, H. Driguez, R. Iori, S. Palmieri, and B. Henrissat, *Structure*, 1997, **5**, 663.
63. S. L. Lawson, W. W. Wakarchuk, and S. G. Withers, *Biochemistry*, 1996, **35**, 10 110.
64. S. L. Lawson, W. W. Wakarchuk, and S. G. Withers, *Biochemistry*, 1997, **36**, 2257.
65. L. P. McIntosh, G. Hand, P. E. Johnson, M. D. Joshi, M. Korner, L. A. Plesniak, L. Ziser, W. W. Wakarchuk, and S. G. Withers, *Biochemistry*, 1996, **35**, 9958.
66. D. Piszkiewicz and T. C. Bruice, *J. Am. Chem. Soc.*, 1967, **89**, 6237.
67. D. Piszkiewicz and T. C. Bruice, *J. Am. Chem. Soc.*, 1968, **90**, 2156.
68. G. Lowe, G. Sheppard, M. L. Sinnott, and A. Williams, *Biochem. J.*, 1967, **104**, 893.
69. G. Lowe and G. Sheppard, *J. Chem. Soc., Chem. Commun.*, 1968, 529.
70. S. G. Withers, R. A. J. Warren, I. P. Street, K. Rupitz, J. B. Kempton, and R. Aebersold, *J. Am. Chem. Soc.*, 1990, **112**, 5887.
71. P. Stadtler, S. Honig, R. Frank, S. G. Withers, and W. Hengstenberg, *Eur. J. Biochem.*, 1995, **232**, 658.
72. S. He and S. G. Withers, *J. Biol. Chem.*, 1997, **272**, 24 864.
73. S. G. Withers, unpublished.
74. E. Bause and G. Legler, *Hoppe-Seyler's Z. Physiol. Chem.*, 1974, **355**, 438.
75. G. Legler, K. R. Roeser, and H. K. Illig, *Eur. J. Biochem.*, 1979, **101**, 85.
76. J. C. Gebler, R. Aebersold, and S. G. Withers, *J. Biol. Chem.*, 1992, **267**, 11 126.
77. Q. Wang, D. Tull, A. Meinke, N. R. Gilkes, R. A. J. Warren, R. Aebersold, and S. G. Withers, *J. Biol. Chem.*, 1993, **268**, 14 096.
78. L. F. Mackenzie, G. J. Davies, M. Schülein, and S. G. Withers, *Biochemistry*, 1997, **36**, 5893.
79. D. Tull, S. G. Withers, N. R. Gilkes, D. G. Kilburn, R. A. J. Warren, and R. Aebersold, *J. Biol. Chem.*, 1991, **266**, 15 621.
80. S. Miao, L. Ziser, R. Aebersold, and S. G. Withers, *Biochemistry*, 1994, **33**, 7027.
81. D. L. Zechel, S. He, C. Dupont, S. G. Withers, *Biochem. J.*, 1998, in press.
82. C. Braun, T. Lindhorst, N. B. Madsen, and S. G. Withers, *Biochemistry*, 1996, **35**, 5458.
83. J. D. McCarter and S. G. Withers, *J. Biol. Chem.*, 1996, **271**, 6889.
84. R. Mosi, S. He, J. Vitdehaag, B. W. Dijkstra, and S. G. Withers, *Biochemistry*, 1997, **36**, 9927.
85. G. Mooser, S. A. Hefta, R. J. Paxton, J. E. Shively, and T. D. Lee, *J. Biol. Chem.*, 1991, **266**, 8916.
86. S. Miao, J. D. McCarter, M. Grace, G. Grabowski, R. Aebersold, and S. G. Withers, *J. Biol. Chem.*, 1994, **269**, 10 975.
87. J. D. McCarter, D. L. Burgoyne, S. Miao, S. Zhang, J. W. Callahan, and S. G. Withers, *J. Biol. Chem.*, 1997, **272**, 396.
88. S. Howard, S. He, and S. G. Withers, *J. Biol. Chem.*, 1998, **273**, 2067.
89. D. J. Vocadlo, L. F. Mackenzie, S. He, and S. G. Withers, *Biochem. J.*, in press.
90. G. Mooser and K. R. Iwaoka, *Biochemistry*, 1989, **28**, 443.
91. G. Mooser, S. A. Hefta, R. J. Paxton, S. J. E., and T. D. Lee, *J. Biol. Chem.*, 1991, **266**, 8916.
92. J. G. Voet and R. H. Abeles, *J. Biol. Chem.*, 1970, **245**, 1020.
93. I. P. Street, J. B. Kempton, and S. G. Withers, *Biochemistry*, 1992, **31**, 9970.
94. C. Braun, G. Brayer, and S. G. Withers, *J. Biol. Chem.*, 1995, **270**, 26 778.
95. J. D. McCarter and S. G. Withers, *J. Am. Chem. Soc.*, 1996, **118**, 241.
96. A. M. MacLeod, D. Tull, K. Rupitz, R. A. J. Warren, and S. G. Withers, *Biochemistry*, 1996, **35**, 13165.

97. T. E. Nelson and J. Larner, *Biochim. Biophys. Acta*, 1970, **198**, 538.
98. X. Guo, W. Laver, E. Vimr, and M. Sinnott, *J. Am. Chem. Soc.*, 1994, **116**, 5572.
99. C. Braun, A. Meinke, L. Ziser, and S. G. Withers, *Anal. Biochem.*, 1993, **212**, 259.
100. S. G. Withers, D. Dombroski, L. A. Berven, D. G. Kilburn, R. C. Miller, Jr., R. A. J. Warren, and N. R. Gilkes, *Biochem. Biophys. Res. Commun.*, 1986, **139**, 487.
101. S. Rosenberg and J. F. Kirsch, *Biochemistry*, 1981, **20**, 3189.
102. S. Rosenberg and J. F. Kirsch, *Biochemistry*, 1981, **20**, 3196.
103. L. Hosie and M. L. Sinnott, *Biochem. J.*, 1985, **226**, 437.
104. J. B. Kempton and S. G. Withers, *Biochemistry*, 1992, **31**, 9961.
105. M. L. Sinnott, *FEBS Lett.*, 1978, **94**, 1.
106. B. F. Li, C. A. Holdup, C. A. Morton, and M. L. Sinnott, *Biochem. J.*, 1989, **260**, 109.
107. G. M. Umezurike, *Biochem. J.*, 1988, **254**, 73.
108. E. van Doorslaer, O. van Opstal, H. Kersters-Hilderson, and C. K. De Bruyne, *Bioorg. Chem.*, 1984, **12**, 158.
109. D. Tull and S. G. Withers, *Biochemistry*, 1994, **33**, 6363.
110. S. G. Withers and I. P. Street, in "ACS Symposium Series: Biogenesis and Biodegradation of Plant Cell Wall Polymers," American Chemical Society, Washington, DC, 1989, vol. 399, p. 597.
111. D. Tull, D. L. Burgoyne, D. T. Chow, S. G. Withers, and R. Aebersold, *Anal. Biochem.*, 1996, **234**, 119.
112. A. M. MacLeod, T. Lindhorst, S. G. Withers, and R. A. J. Warren, *Biochemistry*, 1994, **33**, 6371.
113. J. P. Richard, R. E. Huber, S. Lin, C. Heo, and T. L. Amyes, *Biochemistry*, 1996, **35**, 12 377.
114. S. Cottaz, B. Henrissat, and H. Driguez, *Biochemistry*, 1996, **35**, 15 256.
115. E. J. Hehre, S. Kitahata, and C. F. Brewer, *J. Biol. Chem.*, 1986, **261**, 2147.
116. S. Lawson, Ph.D. Thesis, University of British Columbia, 1997.
117. A. White, D. Tull, K. Johns, S. G. Withers, and D. R. Rose, *Nature Struct. Biol.*, 1996, **3**, 149.
118. J. P. Richard, R. E. Huber, C. Heo, T. L. Amyes, and S. Lin, *Biochemistry*, 1996, **35**, 12 387.
119. M. N. Namchuk and S. G. Withers, *Biochemistry*, 1995, **34**, 16 194.
120. B. Capon, *Chem. Rev.*, 1969, **69**, 407.
121. A. J. Bennet and M. L. Sinnott, *J. Am. Chem. Soc.*, 1986, **108**, 7287.
122. J. D. McCarter, M. J. Adam, and S. G. Withers, *Biochem. J.*, 1992, **286**, 721.
123. M. R. Bray and A. J. Clarke, *Biochem. J.*, 1990, **270**, 91.
124. R. D. Gandour, *Bioorg. Chem.*, 1981, **10**, 169.
125. J. C. Gebler, D. E. Trimbur, R. A. J. Warren, R. Aebersold, M. Namchuk, and S. G. Withers, *Biochemistry*, 1995, **34**, 14 547.
126. D. Tull, Ph.D. Thesis, Dept. of Chemistry, University of British Columbia, Vancouver, 1995.
127. J. Davoodi, W. W. Wakarchuk, R. L. Campbell, P. R. Carey, and W. K. Surewicz, *Eur. J. Biochem.*, 1995, **232**, 839.
128. S. G. Withers, K. Rupitz, D. Trimbur, and R. A. J. Warren, *Biochemistry*, 1992, **31**, 9979.
129. G. Taylor, *Curr. Opin. Struct. Biol.*, 1996, **6**, 830.
130. M. Ashwell, X. Guo, and M. L. Sinnott, *J. Am. Chem. Soc.*, 1992, **114**, 10 158.
131. B. A. Horenstein and M. Bruner, *J. Am. Chem. Soc.*, 1996, **118**, 10 371.
132. B. A. Horenstein, *J. Am. Chem. Soc.*, 1997, **119**, 1101.
133. C. U. Kim, W. Lew, M. A. Williams, H. Liu, L. Zhang, S. Swaminathan, N. Bischofberger, M. S. Chen, D. B. Mendel, C. Y. Tai, G. Laver, and R. C. Stevens, *J. Am. Chem. Soc.*, 1997, **119**, 681.
134. A. T. Baker, J. N. Varghese, W. G. Laver, G. M. Air, and P. M. Colman, *Proteins: Struct. Funct. Genet.*, 1987, **2**, 111.
135. P. Bossart-Whitaker, M. Carson, Y. S. Babu, C. D. Smith, W. G. Laver, and G. M. Air, *J. Mol. Biol.*, 1993, **232**, 1069.
136. W. P. Burmeister, R. W. Ruigrok, and S. Cusack, *EMBO J.*, 1992, **11**, 49.
137. W. P. Burmeister, B. Henrissat, C. Bosso, S. Cusak, and R. Ruigrok, *Structure*, 1993, **1**, 19.
138. M. N. Janakiraman, C. L. White, W. G. Laver, G. M. Air, and M. Luo, *Biochemistry*, 1994, **33**, 8172.
139. M. J. Jedrzejas, S. Singh, W. J. Brouillette, W. G. Laver, G. M. Air, and M. Luo, *Biochemistry*, 1995, **34**, 3144.
140. J. N. Varghese and P. M. Colman, *J. Mol. Biol.*, 1991, **221**, 473.
141. J. N. Varghese, J. L. McKimm-Breschkin, J. B. Caldwell, A. A. Kortt, and P. M. Colman, *Proteins: Struct. Funct. Genet.*, 1992, **14**, 327.
142. S. J. Crennell, E. F. Garman, W. G. Laver, E. R. Vimr, and G. L. Taylor, *Structure*, 1994, **2**, 535.
143. S. J. Crennell, E. F. Garman, W. G. Laver, E. R. Vimr, G. Phillipon, A. Vassella, and G. L. Taylor, *J. Mol. Biol.*, 1996, **259**, 264.
144. A. Gaskell, S. J. Crennell, and G. L. Taylor, *Structure*, 1995, **3**, 1197.
145. T. Terada, K. Kitajima, S. Inoue, J. C. Wilson, A. K. Norton, D. C. M. Kong, R. J. Thomson, M. von Itzstein, and Y. Inoue, *J. Biol. Chem.*, 1997, **272**, 5452.
146. J. C. Wilson, D. I. Angus, and M. von Itzstein, *J. Am. Chem. Soc.*, 1995, **117**, 4214.
147. M. L. Sinnott, X. Guo, S.-C. Li, and Y.-T. Li, *J. Am. Chem. Soc.*, 1993, **115**, 3334.
148. X. Guo and M. L. Sinnott, *Biochem. J.*, 1993, **294**, 653.
149. A. K. Chong, M. S. Pegg, N. R. Taylor, and M. von Itzstein, *Eur. J. Biochem.*, 1992, **207**, 335.
150. A. K. Chong, M. S. Pegg, and M. von Itzstein, *Biochem. Int.*, 1991, **24**, 165.
151. M. R. Lentz, R. G. Webster, and G. M. Air, *Biochemistry*, 1987, **26**, 5351.
152. M. von Itzstein, W.-Y. Wu, G. Kok, M. Pegg, J. Dyason, B. Jin, T. Phan, M. Smythe, H. White, S. Oliver, P. M. Colman, J. Varghese, M. Ryan, J. Woods, R. Bethell, V. Hotham, J. Cameron, and C. Penn, *Nature*, 1993, **363**, 418.
153. D. C. Phillips, *Proc. Natl. Acad. Sci. U.S.A.*, 1967, **57**, 484.
154. I. Matsumura and J. F. Kirsch, *Biochemistry*, 1996, **35**, 1890.
155. A. T. Hadfield, D. J. Harvey, D. B. Archer, D. A. MacKenzie, D. J. Jeenes, S. E. Radford, G. Lowe, C. M. Dobson, and L. N. Johnson, *J. Mol. Biol.*, 1994, **243**, 856.
156. F. W. Dahlquist, T. Rand-Meir, and M. A. Raftery, *Biochemistry*, 1969, **8**, 4214.
157. A. C. Terwissccha van Scheltinga, S. Armand, K. H. Kalk, A. Isogai, B. Henrissat, and B. W. Dijkstra, *Biochemistry*, 1995, **34**, 15 619.
158. A.-M. W. H. Thunnissen, H. J. Rozeboom, K. H. Kalk, and B. W. Diijkstra, *Biochemistry*, 1995, **34**, 12 729.
159. C. Jones and D. Kosman, *J. Biol. Chem.*, 1980, **255**, 11861.

160. S. Knapp, D. Vocadlo, Z. Gao, B. Kirk, J. Lou, and S. G. Withers, *J. Am. Chem. Soc.*, 1996, **118**, 6804.
161. S. Kobayashi, T. Kiyosada, and S. Shoda, *J. Am. Chem. Soc.*, 1996, **118**, 13 113.
162. R. M. Dunn and T. C. Bruice, *Adv. Enzymol.*, 1973, **37**, 1.
163. M. A. Raftery and T. Rand-Meir, *Biochemistry*, 1968, **7**, 3281.
164. T. Rand-Meir, F. W. Dahlquist, and M. A. Raftery, *Biochemistry*, 1969, **8**, 4206.
165. Y. Hashimoto, K. Yamada, H. Motoshima, T. Omura, H. Yamada, T. Yasukochi, T. Miki, T. Ueda, and T. Imoto, *J. Biochem.*, 1996, **119**, 145.
166. I. Matsumura and J. F. Kirsch, *Biochemistry*, 1996, **35**, 1881.
167. B. A. Malcolm, S. Rosenberg, M. J. Corey, J. S. Allen, A. de Baetselier, and J. F. Kirsch, *Proc. Natl. Acad. Sci., USA*, 1989, **86**, 133.
168. W. F. Anderson, M. G. Grutter, S. J. Remington, L. H. Weaver, and B. W. Matthews, *J. Mol. Biol.*, 1981, **147**, 523.
169. L. W. Hardy and A. R. Poteete, *Biochemistry*, 1991, **30**, 9457.
170. D. Rennell, S. E. Bouvier, L. W. Hardy, and A. R. Poteete, *J. Mol. Biol.*, 1991, **222**, 67.
171. R. Kuroki, L. H. Weaver, and B. W. Matthews, *Nature Struct. Biol.*, 1995, **2**, 1007.
172. R. Kuroki, L. H. Weaver, and B. W. Matthews, in "Second Carbohydrate Bioengineering Meeting, La Rochelle, France," eds. Pr. M. D. Legoy, E. Bossuet-Girard, C. Brunet, and O. Whitechurch, University of La Rochelle, France, 1997.
173. B. Svensson, *Plant Mol. Biol.*, 1994, **25**, 141.
174. H. M. Jespersen, E. A. MacGregor, B. Henrissat, M. R. Sierks, and B. Svensson, *J. Protein Chem.*, 1993, **12**, 791.
175. M. Qian, R. Haser, G. Buisson, E. Duee, and F. Payan, *Biochemistry*, 1994, **33**, 6284.
176. A. M. Brzozowski and G. J. Davies, *Biochemistry*, 1997, **36**, 10 837.
177. B. Strokopytov, D. Penninga, H. J. Rozeboom, K. H. Kalk, L. Dijkhuizen, and B. W. Dijkstra, *Biochemistry*, 1995, **34**, 2234.
178. R. M. Knegtel, B. Strokopytov, D. Penninga, O. G. Faber, H. J. Rozeboom, K. H. Kalk, L. Dijkhuizen, and B. W. Dijkstra, *J. Biol. Chem.*, 1995, **270**, 29 256.
179. C. Klein, J. Hollender, H. Bender, and G. E. Schulz, *Biochemistry*, 1992, **31**, 8740.
180. B. Y. Tao, P. J. Reilly, and J. F. Robyt, *Biochim. Biophys. Acta*, 1989, **995**, 214.
181. A. Nakamura, K. Haga, and K. Yamane, *Biochemistry*, 1993, **32**, 6624.
182. A. J. Clarke, in "Biodegradation of Cellulose: Enzymology and Biotechnology", Technomic Publishing Company, Inc., Lancaster, PA, 1997, p. 69.
183. T. M. Wood, *Biochem. Soc. Trans.*, 1985, **13**, 407.
184. G. J. Davies, S. Tolley, B. Henrissat, C. Hjort, and M. Schülein, *Biochemistry*, 1995, **34**, 16 210.
185. P. M. Alzari, M. Juy, and H. Souchon, in "Proceedings of the 2nd TRICEL Symposium on *Trichoderma reesei* Cellulases and other Hydrolases, Espoo, 1993", eds. P. Suominen and T. Reinikainen, Foundation for Biotechnical and Industrial Fermentation Research, Helsinki, 1993, p. 73.
186. M. Spezio, D. B. Wilson, and P. A. Karplus, *Biochemistry*, 1993, **32**, 9906.
187. T. T. Teeri, *Trends Biotechnol.*, 1997, **15**, 160.
188. P. Biely, M. Vrsanska, and M. Claeyssens, in "Proceedings of the 2nd TRICEL Symposium on *Trichoderma reesei* Cellulases and other Hydrolases, Espoo, 1993," eds. P. Suominen and T. Reinikainen, Foundation for Biotechnical and Industrial Fermentation Research, Helsinki, 1993, p. 99.
189. A. K. Konstantinidis, I. Marsden, and M. L. Sinnott, *Biochem. J.*, 1993, **291**, 883.
190. H. G. Damude, S. G. Withers, D. G. Kilburn, R. C. Miller, Jr., and R. A. J. Warren, *Biochemistry*, 1995, **34**, 2220.
191. H. G. Damude, V. Ferro, S. G. Withers, and R. A. J. Warren, *Biochem. J.*, 1996, **315**, 467.
192. L. Ruohonen, A. Koivula, T. Reinikainen, R. Valkeajärvi, A. Teleman, M. Claeyssens, M. Szardenings, T. A. Jones, and T. T. Teeri, in "Proceedings of the 2nd TRICEL Symposium on *Trichoderma reesei* Cellulases and other Hydrolases, Espoo, 1993," eds. P. Suominen and T. Reinikainen, Foundation for Biotechnical and Industrial Fermentation Research, Helsinki, 1993, p. 87.
193. M. Claeyssens and B. Henrissat, *Protein Sci.*, 1992, **1**, 1293.
194. E. J. Hehre, H. Matsui, and C. F. Brewer, *Carbohydr. Res.*, 1990, **198**, 123.
195. M. L. Sinnott and W. P. Jencks, *J. Am. Chem. Soc.*, 1980, **102**, 2026.
196. E. M. Harris, A. E. Aleshin, L. M. Firsov, and R. B. Honzatko, *Biochemistry*, 1993, **32**, 1618.
197. A. E. Aleshin, B. Stoffer, L. M. Firsov, B. Svensson, and R. B. Honzatko, *Biochemistry*, 1996, **35**, 8319.
198. Y. Tanaka, W. Tao, J. S. Blanchard, and E. J. Hehre, *J. Biol. Chem.*, 1994, **269**, 32 306.
199. H. Matsui, J.S. Blanchard, C. F. Brewer, and E. J. Hehre, *J. Biol. Chem.*, 1989, **264**, 8714.
200. A. Konstantinidis and M. L. Sinnott, *Biochem. J.*, 1991, **279**, 587.
201. C. R. Berland, B. W. Sigurskjold, B. Stoffer, T. P. Frandsen, and B. Svensson, *Biochemistry*, 1995, **34**, 10 153.
202. M. R. Sierks and B. Svensson, *Biochemistry*, 1993, **32**, 1113.
203. T. P. Frandsen, C. Dupont, J. Lehmbeck, B. Stoffer, M. R. Sierks, R. B. Honzatko, and B. Svensson, *Biochemistry*, 1994, **33**, 13 808.
204. N. Uozumi, T. Matsuda, N. Tsukagoshi, and S. Udaka, *Biochemistry*, 1991, **30**, 4594.
205. A. Totsuka, V. H. Nong, H. Kadokawa, C. S. Kim, Y. Itoh, and C. Fukazawa, *Eur. J. Biochem.*, 1994, **221**, 649.
206. M. R. Sierks, C. Ford, P. J. Reilly, and B. Svensson, *Protein Eng.*, 1990, **3**, 193.
207. N. S. Banait and W. P. Jencks, *J. Am. Chem. Soc.*, 1991, **113**, 7951.
208. N. S. Banait and W. P. Jencks, *J. Am. Chem. Soc.*, 1991, **113**, 7958.
209. T. Selwood and M. L. Sinnott, in "Molecular Mechanisms in Bioorganic Processes," eds. B. T. Golding and C. Bleasdale, Royal Society of Chemistry, Cambridge, 1990.
210. M. R. Sierks and B. Svensson, *Biochemistry*, 1996, **35**, 1865.
211. U. Christensen, K. Olsen, B. B. Stoffer, and B. Svensson, *Biochemistry*, 1996, **35**, 15 009.
212. W. L. Dong, T. Jespersen, M. Bols, T. Skrydstrup, and M. R. Sierks, *Biochemistry*, 1996, **35**, 2788.
213. K. Olsen, B. Svensson, and U. Christensen, *Eur. J. Biochem.*, 1992, **209**, 777.
214. K. Olsen, U. Christensen, M. R. Sierks, and B. Svensson, *Biochemistry*, 1993, **32**, 9686.
215. S. K. Natarajan and M. R. Sierks, *Biochemistry*, 1996, **35**, 15 269.
216. S. Natarajan and M. R. Sierks, *Biochemistry*, 1996, **35**, 3050.

217. A. E. Aleshin, L. M. Firsov, and R. B. Honzatko, *J. Biol. Chem.*, 1994, **269**, 15 631.
218. Y. Nitta, Y. Isoda, H. Toda, and F. Sakiyama, *J. Biochem. (Tokyo)*, 1989, **105**, 573.
219. H. Toda, Y. Nitta, S. Asanami, J. P. Kim, and F. Sakiyama, *Eur. J. Biochem.*, 1993, **216**, 25.
220. B. Mikami, M. Degano, E. J. Hehre, and J. C. Sacchettini, *Biochemistry*, 1994, **33**, 7779.
221. S. Kitahata, S. Chiba, C. F. Brewer, and E. J. Hehre, *Biochemistry*, 1991, **30**, 6769.
222. E. J. Hehre, S. Kitahata, and C. F. Brewer, *J. Biol. Chem.*, 1986, **261**, 2147.
223. L. N. Johnson and D. Barford, *Protein Sci.*, 1994, **3**, 1726.
224. L. N. Johnson, D. Barford, R. Acharya, N. G. Oikonomakos, and J. L. Martin, in "Proceedings of The Robert A. Welch Foundation Conference on Chemical Research: Regulation of Proteins by Ligands, Houston, 1992", Robert A. Welch Foundation, Houston, TX, 1992, vol. 36, p. 17.
225. M. F. Browner, D. Hackos, and R. J. Fletterick, *Nature Struct. Biol.*, 1994, **1**, 327.
226. M. F. Browner, P. K. Hwang, and R. J. Fletterick, *Biochemistry*, 1992, **31**, 11 291.
227. S. R. Sprang, S. G. Withers, E. J. Goldsmith, R. J. Fletterick, and N. B. Madsen, *Science*, 1991, **254**, 1367.
228. D. Barford, S. H. Hu, and L. N. Johnson, *J. Mol. Biol.*, 1991, **218**, 233.
229. J. Buchbinder and R. J. Fletterick, *J. Biol. Chem.*, 1996, **271**, 22 305.
230. P. J. Kavinsky, N. B. Madsen, J. Sygusch, and R. J. Fletterick, *J. Biol. Chem.*, 1978, **253**, 3343.
231. D. Palm, H. W. Klein, R. Schinzel, M. Buehner, and E. J. M. Helmreich, *Biochemistry*, 1990, **29**, 1099.
232. E. P. Mitchell, S. G. Withers, P. Ermert, A. T. Vasella, E. F. Garman, N. G. Oikonomakos, and L. N. Johnson, *Biochemistry*, 1996, **35**, 7341.
233. E. M. Duke, S. Wakatsuki, A. Hadfield, and L. N. Johnson, *Protein Sci.*, 1994, **3**, 1178.
234. L. N. Johnson, K. R. Acharya, M. D. Jordan, and P. J. McLaughlin, *J. Mol. Biol.*, 1990, **211**, 645.
235. R. F. Parrish, R. J. Uhing, and D. J. Graves, *Biochemistry*, 1977, **16**, 4824.
236. S. G. Withers, S. Shechosky, and N. B. Madsen, *Biochem. Biophys. Res. Commun.*, 1982, **108**, 322.
237. N. G. Oikonomakos, S. E. Zographos, K. E. Tsitsanou, L. N. Johnson, and K. R. Acharya, *Protein Sci.*, 1996, **5**, 2416.
238. W. G. Stirtan and S. G. Withers, *Biochemistry*, 1996, **35**, 15 057.
239. R. Challoner, C. A. McDowell, W. Stirtan, and S. G. Withers, *Biophys. J.*, 1993, **64**, 484.
240. J. E. Taguchi, S. J. Heyes, D. Barford, L. N. Johnson, and C. M. Dobson, *Biophys. J.*, 1993, **64**, 492.
241. S. G. Withers, N. B. Madsen, B. D. Sykes, M. Takagi, S. Shimomura, and T. Fukui, *J. Biol. Chem.*, 1981, **256**, 10 759.
242. S. G. Withers, N. B. Madsen, and B. D. Sykes, *Biochemistry*, 1981, **20**, 1748.
243. S. R. Sprang, N. B. Madsen, and S. G. Withers, *Protein Sci.*, 1992, **1**, 1100.
244. W. P. Jencks, *Adv. Enzymol.*, 1975, **43**, 219.
245. J. L. Martin, L. N. Johnson, and S. G. Withers, *Biochemistry*, 1990, **29**, 10 745.
246. I. P. Street, K. Rupitz, and S. G. Withers, *Biochemistry*, 1989, **28**, 1581.
247. C. J. F. Bichard, E. P. Mitchell, M. R. Wormald, K. A. Watson, L. N. Johnson, S. E. Zographos, D. D. Koutra, N. G. Oikonomakos, and G. W. J. Fleet, *Tetrahedron Lett.*, 1995, **36**, 2145.
248. T. M. Krulle, K. A. Watson, M. Gregoriou, L. N. Johnson, S. Crook, D. J. Watkin, R. C. Griffiths, R. J. Nash, K. E. Tsitsanou, S. E. Zographos, N. G. Oikonomakos, and G. W. J. Fleet, *Tetrahedron Lett.*, 1995, **36**, 8291.
249. K. A. Watson, E. P. Mitchell, L. N. Johnson, J. C. Son, C. J. Bichard, M. G. Orchard, G. W. Fleet, N. G. Oikonomakos, D. D. Leonidas, and M. Kontou, *Biochemistry*, 1994, **33**, 5745.
250. S. G. Withers, N. B. Madsen, S. R. Sprang, and R. J. Fletterick, *Biochemistry*, 1982, **21**, 5372.
251. S. R. Sprang, E. J. Goldsmith, R. J. Fletterick, S. G. Withers, and N. B. Madsen, *Biochemistry*, 1982, **21**, 5364.
252. I. P. Street, C. R. Armstrong, and S. G. Withers, *Biochemistry*, 1986, **25**, 6021.
253. M. O'Reilly, K. A. Watson, R. Schinzel, D. Palm, and L. N. Johnson, *Nature Struct. Biol.*, 1997, **4**, 405.
254. B. Strokopytov, R. M. A. Knegtel, D. Penninga, H. J. Rozeboom, K. H. Kalk, L. Dijkhuizen, and B. W. Dijkstra, *Biochemistry*, 1996, **35**, 4241.
255. A. Nakamura, K. Haga, and K. Yamane, *Biochemistry*, 1994, **33**, 9929.
256. D. Penninga, B. Strokopytov, H. J. Rozeboom, C. L. Lawson, B. W. Dijkstra, J. Bergsma, and L. Dijkhuizen, *Biochemistry*, 1995, **34**, 3368.
257. W. Liu, N. B. Madsen, C. Braun, and S. G. Withers, *Biochemistry*, 1991, **30**, 1419.
258. W. Liu, N. B. Madsen, B. Fan, K. A. Zucker, R. H. Glew, and D. E. Fry, *Biochemistry*, 1995, **34**, 7056.
259. V. L. Schramm, B. A. Horenstein, and P. C. Kline, *J. Biol. Chem.*, 1994, **269**, 18 259.
260. D. W. Parkin, B. A. Horenstein, D. R. Abdulah, B. Estupinan, and V. L. Schramm, *J. Biol. Chem.*, 1991, **266**, 20 658.
261. B. A. Horenstein, D. W. Parkin, B. Estupinan, and V. L. Schramm, *Biochemistry*, 1991, **30**, 10 788.
262. B. A. Horenstein and V. L. Schramm, *Biochemistry*, 1993, **32**, 7089.
263. M. Boutellier, B. A. Horenstein, A. Semenyaka, V. L. Schramm, and B. Ganem, *Biochemistry*, 1994, **33**, 3994.
264. D. W. Parkin and V. L. Schramm, *Biochemistry*, 1995, **34**, 13 961.
265. H. Deng, W.-Y. Chan, C. K. Bagdassarian, B. Estupinan, B. Ganem, R. H. Callender, and V. L. Schramm, *Biochemistry*, 1996, **35**, 6037.
266. M. Degano, D. N. Gopaul, G. Scapin, V. L. Schramm, and J. C. Sacchettini, *Biochemistry*, 1996, **35**, 5971.
267. D. N. Gopaul, S. L. Meyer, M. Degano, J. C. Sacchettini, and V. L. Schramm, *Biochemistry*, 1996, **35**, 5963.
268. P. C. Kline and V. L. Schramm, *Biochemistry*, 1992, **31**, 5964.
269. P. C. Kline and V. L. Schramm, *Biochemistry*, 1995, **34**, 1153.
270. L. J. Mazzella, D. W. Parkin, P. C. Tyler, R. H. Furneaux, and V. L. Schramm, *J. Am. Chem. Soc.*, 1996, **118**, 2111.
271. J. A. Campbell, G. J. Davies, V. Bulone, and B. Henrissat, *Biochem. J.*, 1997, **326**, 929.
272. D. J. Legault, R. J. Kelly, Y. Natsuka, and J. B. Lowe, *J. Biol. Chem.*, 1995, **270**, 20 987.
273. B. W. Murray, T. Shuichi, J. Schultz, and C.-H. Wong, *Biochemistry*, 1996, **35**, 11 183.
274. B. W. Murray, V. Wittmann, M. D. Burkart, S.-C. Hung, and C.-H. Wong, *Biochemistry*, 1997, **36**, 823.
275. L. Qiao, B. W. Murray, M. Shimazaki, J. Schultz, and C. -H. Wong, *J. Am. Chem. Soc.*, 1996, **118**, 7653.
276. A. Radzicka and R. Wolfenden, *Science*, 1995, **267**, 90.
277. M. Du and O. Hindsgaul, *Carbohydr. Res.*, 1996, **286**, 87.
278. S. Gosselin and M. M. Palcic, *Bioorg. Med. Chem.*, 1996, **4**, 2023.
279. T. de Vries, C. A. Srnka, M. M. Palcic, S. J. Swiedler, D. H. van den Eijnden, and B. A. Macher, *J. Biol. Chem.*, 1995, **270**, 8712.
280. Z. Xu, L. Vo, and B. A. Macher, *J. Biol. Chem.*, 1996, **271**, 8818.

5.13
Electrophilic Alkylations, Isomerizations, and Rearrangements

JULIA M. DOLENCE and C. DALE POULTER
University of Utah, Salt Lake City, UT, USA

5.13.1 INTRODUCTION

5.13.1.1 General

Of the many biological reactions that have been studied at the enzyme level, relatively few proceed by electrophilic mechanisms. A prominent example is the glycosyl transfer and hydrolysis reactions in carbohydrate metabolism. As will be discussed in Chapter 5.11, the formation and hydrolysis of complex carbohydrates proceed through transition states with considerable oxocarbonium ion character at the anomeric carbon in the sugar (Scheme 1).[1,2] Similar mechanisms are seen for reactions that proceed with either inversion or retention of configuration at the anomeric center, although net retention of configuration is the result of two inversion steps with transient formation of a glycosyl–enzyme intermediate. Another example is the electrophilic alkylation of a wide variety of nucleophiles catalyzed by *S*-adenosylmethionine methyl transferases. When the nucleophile is a carbon–carbon double bond, as shown in Scheme 2 for the side chain alkylation of higher plant steroids such as *β*-sitosterol and cycloartenol,[3,4] the C-24–C-25 double bond reacts with the activated methyl group of *S*-adenosylmethionine to generate a carbocationic intermediate. A third example is the electrophilic alkylation of nucleophiles catalyzed by prenyltransferases. As the name implies, these enzymes transfer the hydrocarbon moiety of an allylic isoprenoid diphosphate to an electron-rich acceptor. The most prominent example of a prenyltransferase is farnesyl diphosphate synthase. The enzyme catalyzes the basic chain elongation reaction in the isoprenoid biosynthetic pathway

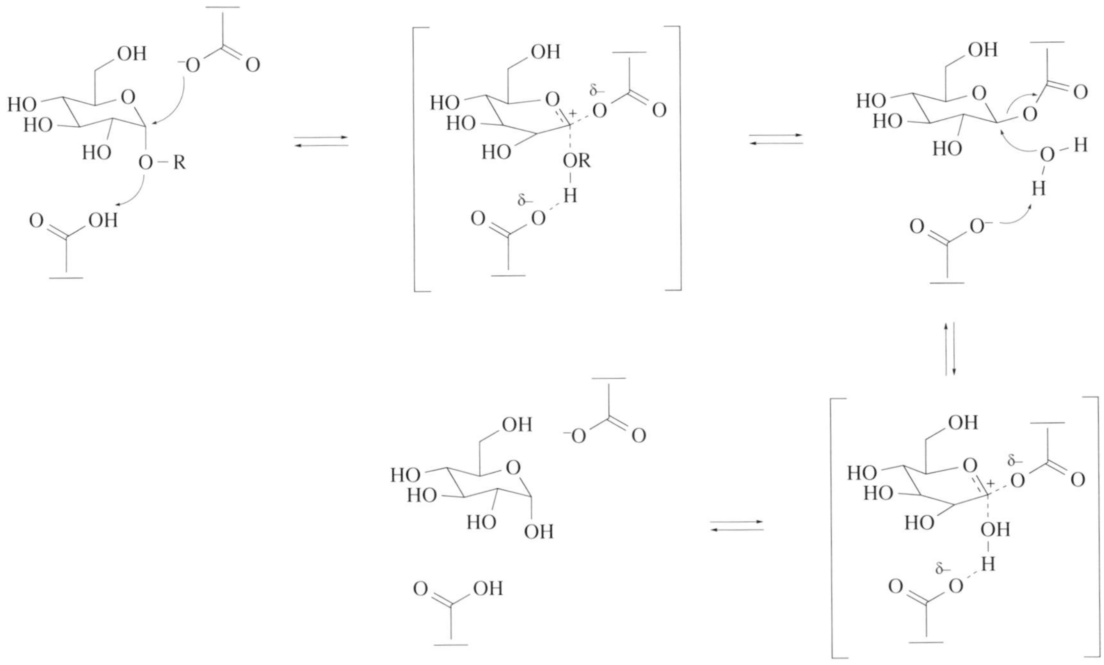

Scheme 1

by an electrophilic alkylation of the double bond in isopentenyl diphosphate to form the next higher homologue of the allylic substrate (Scheme 3). Electrophilic reactions in isoprenoid metabolism are the subject of this chapter.

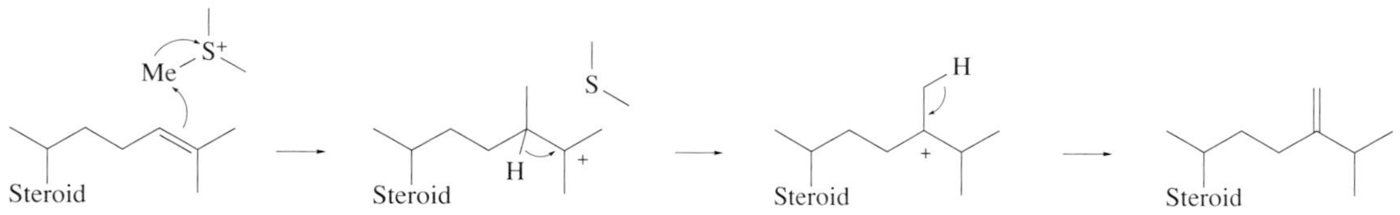

Scheme 2

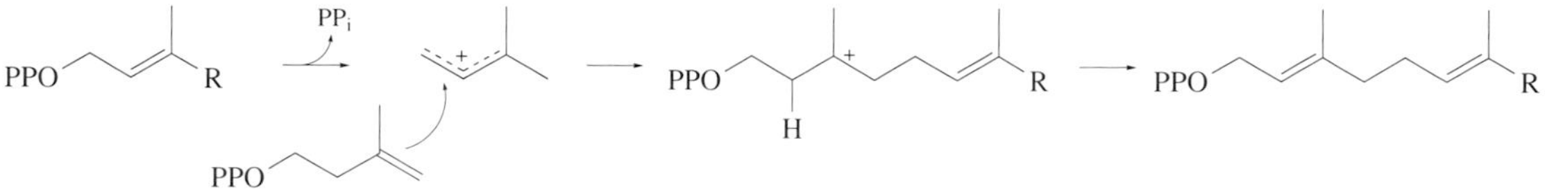

Scheme 3

5.13.1.2 Methods for Studying Electrophilic Reactions

Several methods have been used to study the mechanisms of electrophilic enzymatic reactions. Some of the more common techniques will be briefly discussed here and then presented in greater detail for individual prenyltransferases in subsequent sections of this chapter. Mechanistic studies of enzyme-catalyzed reactions face several limitations not encountered in studies of uncatalyzed reactions. Typically only modest changes can be made in the conditions for enzymatic reactions because most proteins are denatured, or at best minimally reactive, outside of the rather narrow range of temperatures, pHs, ionic strengths, or "solvents" that they experience in their host organisms. Care must be exercised when interpreting kinetic parameters because the catalytic cycle includes steps for binding of substrates, chemistry, and release of products. Often the chemical step is not, or only partially, rate limiting! Finally, enzyme-catalyzed reactions occur within the confines of an active site. Except for temperature, the microenvironment experienced by a bound substrate molecule cannot be predicted from the bulk properties of the buffer used for the reaction. Furthermore, the topology of the active site in an enzyme imposes steric and chiral constraints not encountered in solution.

Linear free energy correlations between enzyme-catalyzed reactions and nonenzymatic models whose mechanisms are well established are important tools in studying electrophilic transformations. These studies typically involve the use of Hammett correlations to compare substituent effects for the substrates in solution under conditions more amenable to a detailed mechanistic evaluation with those for the catalyzed reactions. When designing these experiments, it is important to minimize steric perturbations introduced by the substituents, to maximize the spread of reactivities consistent with being able to conduct kinetic measurements at the extremes with the set of substrates being studied, to verify that the rates for the chemical steps are being compared in both cases, and to establish the structures of the products.

Fluorine has proved to be a particularly valuable probe for determining the extent to which positive charge develops at the transition state for enzyme-catalyzed reactions. Fluorine is the most electrophilic of the elements. The powerful electron-withdrawing effects of fluorine can lead to pronounced effects on the rates of reactions that develop positive charge at the transition state through destabilizing inductive effects when a hydrogen atom near a developing cationic center is replaced by fluorine. The magnitude of the effect typically parallels the extent to which positive charge develops at the transition state. Fluorine is a particularly valuable substituent for enzymatic studies because the atom has a van der Waals radius (~ 1.2 Å) similar to that of hydrogen (~ 1.35 Å). Although substitution of hydrogen by fluorine often does not substantially alter substrate binding, one must exercise caution because of changes in bond polarity that may alter conformational preferences and effects that arise from the lone pair electrons on fluorine.

Much of the catalytic power of enzyme-catalyzed transformations comes from the selective stabilization of the transition state relative to the ground state as bound substrates undergo reactions.[5] Stable molecules whose structures closely approximate transition state structures are often potent inhibitors with dissociation constants which, in the most favorable cases, are several orders of magnitude lower than those of the normal substrates. Compounds containing electron deficient heteroatoms in place of the carbon atoms that develop positive charge at the transition state are often good inhibitors for enzymes that catalyze electrophilic reactions. Potent inhibition by transition state analogues is often used as support for an electrophilic mechanism.[6] Ammonium ion analogues such as (1)–(3) are commonly used for studies of prenyl transfer reactions where the transition states have substantial carbocationic character in the allylic substrate, although sulfonium and arsonium have been employed as well.

(1)   (2)   (3)

5.13.1.3   Isoprenoid Enzymes

With over 23 000 different structures now identified, isoprenoid compounds comprise the most chemically diverse collection of small molecules found in nature. They serve as reproductive hormones, as attractants, as defensive agents, as regulators in signal transduction, as carbohydrate carriers during glycoprotein biosynthesis, as photoprotective agents, and in other roles too numerous to list. Isopentenyl diphosphate (IPP), the fundamental five-carbon building block for more complex isoprenoids, is synthesized in cells by two independent routes. In archaebacteria and eukaryotes, IPP is assembled from three molecules of acetyl CoA by the well-characterized two-carbon mevalonate pathway.[7] In other bacteria and in plant chloroplasts, the carbon atoms in IPP are derived from a three-carbon precursor, perhaps dihydroxyacetone phosphate, by a biosynthetic route that has yet to be firmly established.[8]

Beyond IPP all organisms use the same general set of reactions to incorporate the basic isoprenoid unit into more complex metabolites. Typically, the catalytic properties of isoprenoid enzymes that catalyze the same reaction are similar regardless of the host organism that produced the protein. For at least one enzyme in the isoprenoid pathway, farnesyl diphosphate (FPP) synthase, a phylogenetic correlation of primary amino acid sequences for the enzyme from organisms in each of the three major kingdoms (Archae, Bacteria, and Eukaryae) suggests a common ancestor.[9] However, the amount of genetic information about the pathway is rather limited. There are no organisms where all of the genes for the pathway have been identified, even those whose genomes have been completely sequenced, and relatively little is known about how the pathway is regulated beyond the pioneering work of Brown and Goldstein on cholesterol metabolism in human fibroblasts.[10] At this point, perhaps most is known at the genetic level in *Saccharomyces cerevisiae* (baker's yeast), where all but a few of the genes for known steps in the pathway have been identified and sequenced. The branches of the pathway expressed in yeast are typical of those found in most eukaryotic organisms, and each step, along with the genes which code for the associated enzymes that catalyze the step, are given in Scheme 4.

5.13.2   FPP SYNTHASE

5.13.2.1   Mechanistic Studies

5.13.2.1.1   Early mechanistic studies

FPP synthase (EC2.5.1.1) catalyzes the irreversible sequential addition of IPP to dimethylallyl diphosphate (DMAPP) and geranyl diphosphate (GPP) to form FPP. The substrates for these reactions were first characterized in the laboratories of Bloch[11] and Lynen.[12] They recognized the electrophilic nature of allylic isoprenoid diphosphates and suggested that the chain elongation involved an electrophilic alkylation of the double bond in IPP. As part of their pioneering work

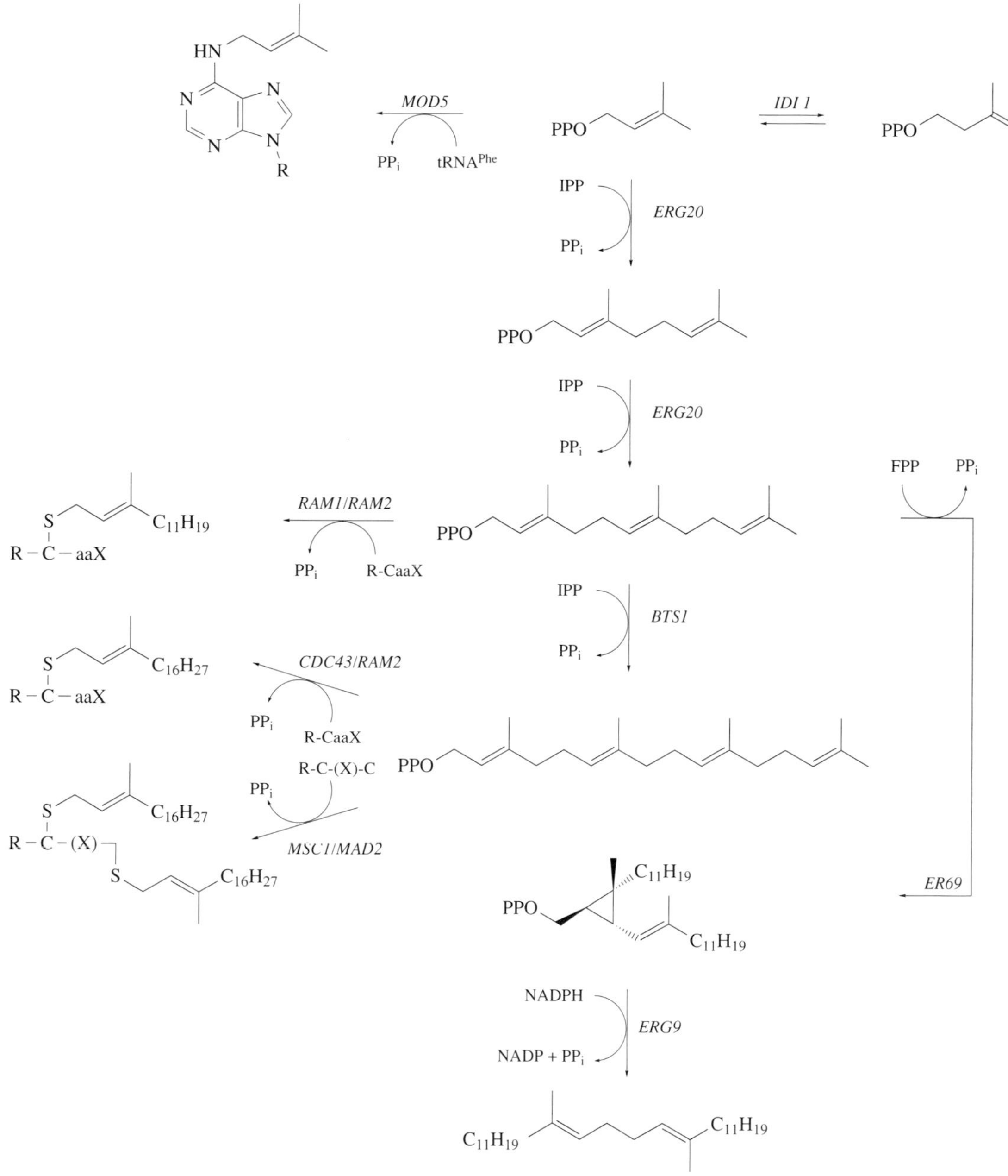

Scheme 4

on squalene biosynthesis from mevalonic acid, Popjak and Cornforth established the absolute stereochemistry for the two chain elongation steps catalyzed by FPP synthase.[13,14] They determined that the allylic substrate adds to the si face of the double bond of IPP with inversion of configuration at C-1 and that the pro-R proton is removed from C-2 of IPP to generate the new (*E*) C-2–C-3 double bonds in FPP. The overall stereochemistry for the addition–elimination to IPP is suprafacial. They rationalized the stereochemistry by a *trans* intermolecular S$_N$2′ displacement where an unknown nucleophile, or "X" group, adds to the re face of the double bond in IPP with concerted formation of the new single bond between the isoprenoid units and elimination of the diphosphate group from the allylic substrate. Subsequently, a *trans* elimination of "X" and the proton from C-2 generates the new (*E*) double bond. The X-group mechanism rationalizes the stereochemistry not the nature of the transition state for the prenyl transfer reaction. Although the stereochemistry of the prenyl transfer reaction is nicely accommodated, there is, in fact, no stereochemical imperative for the X-group mechanism, and the involvement of an X group introduces unnecessary complications. More simply, one could assume that the stereochemistry for chain elongation is dictated by the chiral nature of the active site through noncovalent interactions between the enzyme and the reactants. This ambiguity always arises for enzyme-catalyzed reactions and places rather stringent restrictions on the conclusions one can draw about the mechanism from stereochemical studies. It is important to note that the observations which led to the formulation of the X-group mechanism only require that the new bonds between isoprenoid residues form before loss of stereochemistry at C-1 of the allylic substrate and do not provide information about the extent of formation or rupture of bonds at the transition state.

5.13.2.1.2 *Linear free energy studies*

As mentioned earlier, Bloch and Lynen first suggested that allylic diphosphates are excellent substrates for electrophilic reactions. The diphosphate unit becomes an increasingly powerful leaving group as negative charge is neutralized,[15] and the double bond stabilizes developing position charge in the allylic moiety. The first direct evidence for an electrophilic reaction was obtained by using 3-trifluoro-2-butenyl diphosphate and 2-fluorogeranyl diphosphate as alternate allylic substrates.[13,16] FPP synthase catalyzed condensation of the analogues. However, the rate of the reaction for the fluoro analogues was depressed more than three orders of magnitude for the 2-fluoro analogue and by more than six orders of magnitude for the trifluoro derivative than when GPP was the allylic substrate. These dramatic decreases in the reactivity of the allylic substrate upon replacement of hydrogen atoms by fluorine were similar to those seen for solvolysis of the corresponding methanesulfonates. In contrast, when geranyl chloride and 2-fluorogeranyl chloride were treated with cyanide in acetone under conditions more suitable for a direct nucleophilic displacement, the fluoro analogue was twofold more reactive. These results suggested that there is a substantial development of positive charge at the transition state for the chain elongation reaction (Scheme 3).

Additional evidence for an electrophilic mechanism was obtained from a Hammett analysis of the prenyl transfer reaction by comparing the rates of prenyl transfer to IPP for a series of fluorinated GPP analogues ((4)–(6)) with those for solvolysis of the corresponding methanesulfonates.

(4) R = CH_2F
(5) R = CHF_2
(6) R = CF_3

Substitutions in the geranyl moiety were designed to minimize steric clashes between the analogues and active site residues, to minimize steric interactions between the two substrates, and to maximize the kinetic response. This was accomplished by replacing the methyl group at C-3 in GPP by a hydrogen atom and mono-, di-, and trifluoromethyl groups.[17] The first-order rate constants, k_s, for solvolysis of the methanesulfonate derivatives of (4)–(6) in acetone–water spanned over six orders of magnitude. The maximal velocities (V_{max}) for condensation of the GPP analogues with IPP also covered a wide range. However, release of products, not chemistry, is rate limiting for the normal prenyl transfer reaction, and it was necessary to calculate the rate constants for the chemical step in the catalytic cycle for the Hammett correlation. The values for the condensation of (4)–(6) with IPP using avian liver FPP synthase covered almost eight orders of magnitude. A Hammett plot of k_{chem} for 1′–4 condensation vs. k_s is shown in Figure 1. The slope of the plot is slightly less than unity and suggests that the 1′–4 condensation is a bit more sensitive to the electron-withdrawing groups at C-3 in the geranyl analogues than is the model solvolysis reaction. The large spread in reactivities between methyl and trifluoromethyl indicates a highly developed allylic carbocation at the transition state for both reactions.

The linear free energy studies with FPP synthase do not directly address the question of whether breaking the carbon–oxygen bond in the allylic substrate is concerted with formation of the carbon–carbon bond to IPP. If the reaction is not concerted, the carbocationic intermediate must have a finite lifetime under the reaction conditions. How long a carbocation exists depends on the degree of stabilization afforded by the substituents attached to the electron-deficient carbon and the reactivity of nearby nucleophiles. The lifetime of the dimethylallyl cation in water is approximately

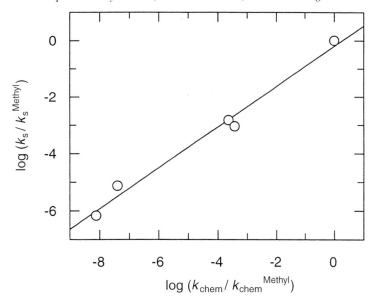

Figure 1 Hammett plot of k_{chem} vs. k_s for FPP synthase.

10^{-8} s,[18] and the predicted lifetime when water is replaced by a trisubstituted alkene whose substitution pattern is similar to that of the double bond in IPP is 10^{-7} s.[19] While the conditions that exist within the active site of FPP synthase are clearly different from those in bulk solvent, the predicted lifetime is six orders of magnitude larger than estimated for the borderline between concerted and nonconcerted reactions for the dimethylallyl cation.

5.13.2.1.3 *Bisubstrate analogues*

Interest in determining the conformations of IPP and DMAPP and the relative orientations of the hydrocarbon moieties of the two substrates when bound in the catalytic site led to the design and synthesis of bisubstrate analogues (**7**) and (**8**), where the hydrocarbon chains are joined by a one-carbon bridge between the Z-methyl at C-3 of DMAPP and C-4 of IPP.[20] Both compounds were alternate substrates for avian FPP synthase. The catalytic efficiencies for processing the two analogues, as determined by the ratio of the catalytic constant (k_{cat}) and the Michaelis constant (K_m) were similar to those measured for the normal substrates. However, an analysis of the products revealed that the reactions of the analogues were more complex. Two products (**9**) and (**10**) were obtained when (**7**) was incubated with FPP synthase (Scheme 5), and three products (**9**), (**10**), and (**11**) were formed from (**8**). Formation of cyclohexenyl diphosphates (**9**) and (**11**) parallels the normal reaction, where C-1 of the allylic substrate (from cation (**12**)) adds to C-4 of IPP with concomitant elimination of a proton from C-2 of IPP. The tether between the hydrocarbon moieties in the two substrates results in the formation of a cyclohexene ring as a consequence of the normal prenyl transfer reaction. The most likely route to cyclohexadiene (**10**) also involves cyclization to a cyclohexene ring. However, instead of elimination of a hydrogen from C-2 in the IPP chain, a 1,2-hydrogen migration from C-4 to C-3 (generating cations (**13**) and (**14**), respectively) is followed by an elimination. This sequence was established with deuterium-labeled substrate, where a deuteron at C-4 in (**7**) migrated into the side chain in cyclohexadiene (**10**). The 1,2-hydrogen shift is strong evidence for the presence of cation (**13**) as an *intermediate* in the reaction that partitions between elimination and rearrangement (Scheme 6).

(7)

(8)

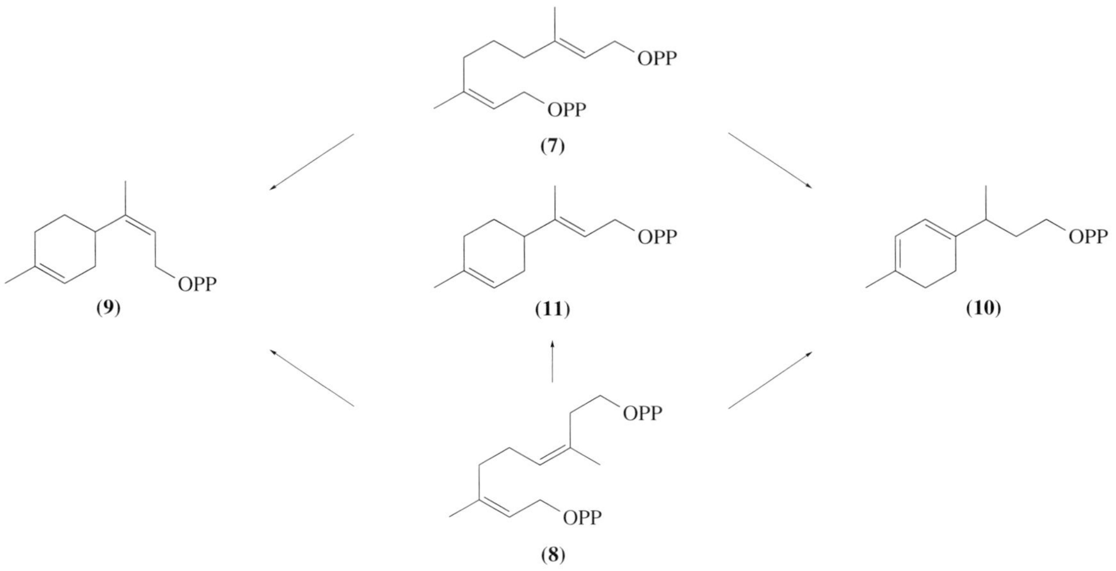

(7)

(9)

(11)

(10)

(8)

Scheme 5

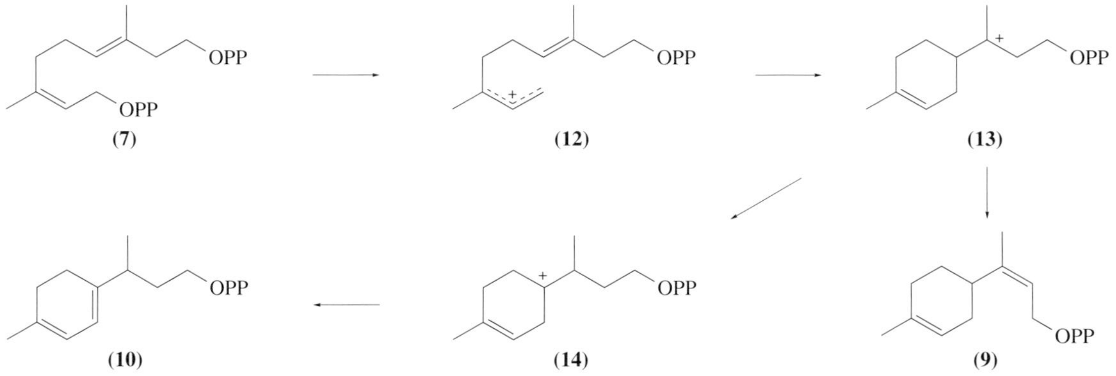

(7)

(12)

(13)

(10)

(14)

(9)

Scheme 6

These results demonstrate that formation of the carbon–carbon bond between the two hydrocarbon chains is not concerted with the final elimination of a hydrogen and suggest a similar scenario during chain elongation for the normal substrates. Since the 1,2-hydrogen migration requires an unfavorable endothermic rearrangement of a tertiary carbocation to its less stable secondary isomer for the intermolecular condensation between IPP and DMAPP, the rearrangement should not occur for the normal enzyme-catalyzed reaction. Furthermore, stereochemical studies with the bisubstrate analogues revealed that each of the products are generated from discrete enzyme–substrate complexes formed from discretely bound conformers of the substrate analogues.[21]

The enzymatic conversion of (7) and (8) to cyclic products is analogous to the cyclizations that produce the *p*-menthane and sabinane monoterpenes. In effect, the inherent prenyltransferase activity of FPP synthase has been transformed into cyclase activity by using the binding interactions of the diphosphate moieties in the bisubstrate analogue to position the hydrocarbon chain in a conformation to facilitate cyclization by an electrophilic alkylation. There is genetic evidence for a connection between the prenyltransferases that catalyze chain elongation and cyclases. FPP synthase and its relatives contain two highly conserved aspartate-rich regions that bind the magnesium salts of IPP and the allylic substrates (see Section 5.13.2.2.2). Similar sequences are found in several sesquiterpene cyclases.[22] While additional phylogenetic and structural information will be required to establish an evolutionary link between the two classes of enzymes, it is straightforward to envision how sesquiterpene cyclase activity could evolve from a prenyltransferase by mutations that deleted the IPP binding site and altered the conformation of the hydrocarbon chain of FPP to facilitate cyclization of the allylic carbocation with a distal double bond. In fact, Saito and Rilling found that FPP synthase catalyzed the cyclization of FPP when IPP was not present during the incubation.[23]

5.13.2.2 FPP Synthase—Enzymatic Studies

5.13.2.2.1 *Production of the enzyme*

Several purifications of FPP synthase have been reported from fungal and liver sources where the activity of the enzyme is high. These include the enzymes from *Saccharomyces cerevisiae*,[24] avian liver,[25] porcine liver,[26] and human liver.[27] The procedure described for the avian liver enzyme is typical.[28] The cell-free homogenate is fractionated by precipitation with ammonium sulfate, followed by ion exchange chromatography on DE52, adsorption chromatography on hydroxyapatite, and chromatofocusing followed by gel filtration to remove the ampholites used to generate the pH gradient on the focusing column. These procedures yielded approximately 2–4 mg of homogeneous protein from 4 kg of avian liver.

All of the FPP synthases characterized thus far are homodimers with molecular masses typically ranging from 70 to 90 kDa. Each subunit has a single catalytic site that catalyzes the condensation of IPP with both DMAPP and GPP. The enzyme has an absolute requirement for a divalent metal ion cofactor, typically Mg^{2+}, although Mn^{2+} and Zn^{2+} will substitute.[29] The metal ion does not bind to the apo-enzyme. However, up to 2 mol of metal are bound per catalytic site in the presence of substrate.[30]

More convenient sources of FPP synthase are now available from specially engineered strains of *Escherichia coli*. Recombinant proteins for yeast,[31] avian,[32] rat,[33] human,[34] plant,[35] and bacterial forms[36,37] of the enzyme have been produced and purified to homogeneity. In favorable cases, the enzyme is synthesized as a soluble, fully active homodimer that constitutes up to 50% of the total soluble protein in the bacterium. Purification typically requires only two steps, chromatography on DE52 and hydroxyapatite. The kinetic properties of the recombinant enzyme are virtually identical to their nonrecombinant counterparts.

5.13.2.2.2 *Phylogenetic relationships*

Beginning in the late 1980s the nucleotide sequences of open reading frames coding for the prenyltransferases of chain elongation have appeared in the literature with increasing frequency. As a result, the deduced amino acid sequences are now available for FPP synthases, geranylgeranyl diphosphate synthases (GGPP synthases), and longer chain synthases that synthesize isoprenoids with *E* double bonds. Examples include representatives from Archaea, Bacteria, and Eukarya. A phylogenetic tree constructed from multiple sequence alignments suggests that the enzymes evolved from a common ancestor which was present at the very beginning of cellular life. All members of the family contain five regions that have been preserved throughout evolution, with a substantial percentage of completely conserved residues in each region (Figure 2). Two of the regions contain aspartate-rich Asp–Asp–Xxx–Xxx–(Xxx–Xxx)–Asp motifs, where Xxx denotes a general amino acid.[38] The aspartate-rich motifs are sites for binding the magnesium salts of the diphosphate substrates (see Section 5.13.2.2.4) and are also found in several sesquiterpene cyclases.

5.13.2.2.3 *Site-directed mutagenesis*

The highly conserved aspartate residues were obvious initial targets for site-directed mutagenesis experiments. A series of mutants for rat FPP synthase were constructed where the negatively charged carboxylate groups for some of the conserved aspartates were preserved but the side chain was lengthened by a carbon atom by changing aspartate to glutamate.[39,40] Two additional charge-neutral mutations converted two nearby conserved arginines in the first aspartate region to lysine. Each mutant enzyme was overproduced in *E. coli* and purified. The D104E (aspartate at position 104 mutated to glutamate), D107E, R112K (arginine at position 112 mutated to lysine), R113K, and D243E mutants all gave approximately 1000-fold lower values for V_{max} compared with the wild-type (wt) enzyme with no significant changes in the K_m-values for either IPP or GPP.

Highly conserved charged amino acids were also mutated in yeast FPP synthase.[41] In this case, negatively charged aspartates were replaced with neutral alanine residues, and positively charged arginines and lysines were replaced by glutamines or alanines. The recombinant enzyme contained a C-terminal–Glu–Glu–Phe α-tubulin epitope to facilitate rapid purification of the protein using immunoaffinity chromatography and to remove any contaminating activity from native *E. coli* FPP

```
Region 1                 Region 2

FPP   GK.nRgl            FPP   E.lqayfLv.DD...MD.s.tRRG
GGPP  gkr.Rp.            GGPP  e..h..sL.hDD.p..D...lRRG
LCPP  GKr.Rp.            LCPP  E.IHtAsL.HDDv...De...RRG



Region 3                 Region 4

FPP   GQ..D              FPP   KT
GGPP  GQ..d              GGPP  kt
LCPP  GE..q              LCPP  KT



Region 5

FPP   G..fQ.qDD.LD.g.....GK..G.D....K
GGPP  G..fQ..DD.........gk....d....k
LCPP  G..fQ..DD.LD......GKp...D.....
```

Figure 2 Alignment of isoprene diphosphate synthases. The five conserved regions are shown. Capital letters denote residues that are completely conserved. Lower case letters are residues which are mostly conserved and dots are residues which are not conserved.

synthase. The six conserved asparate residues in regions 2 and 5 were changed to alanines, and the three conserved arginines were changed to glutamate. In addition, Arg^{350}, located near the C-terminus of the protein, and Lys^{254}, located in the second asparate-rich region, were converted to alanine. Mutations in any of the aspartates and arginines in the first asparate-rich region or Asp^{240} and Asp^{241} in the second asparate-rich region drastically lowered k_{cat} (10^{-4}–10^{-7}) relative to wt FPP synthase. The k_{cat} values for the D244A and K254A mutants were substantially lower (10–1000-fold), while the activity of the R350A mutant was similar to wild type. It is clear from these results that the conserved aspartate and arginine residues in the first region as well as the first two aspartates in the second region are important components of the catalytic machinery of FPP synthase.

5.13.2.2.4   *Crystallographic studies*

The availability of large amounts of recombinant FPP synthase provided an opportunity to obtain an X-ray structure for the enzyme. Subsequently, a crystal structure was determined at 2.6 Å resolution for the avian protein.[42] Avian FPP synthase crystallized as a homodimer with an approximate two-fold axis of symmetry for the two subunits. The protein has a previously unreported all α-helix (and loop) structure with 10 core helices arranged around a large central cavity whose dimensions suggested that it is the binding pocket for the allylic substrate. The two conserved aspartate-rich sequences are located on opposite walls near the entrance to the cavity. Repeated attempts to co-crystallize wt FPP synthase with either IPP or an allylic substrate were unsuccessful.

An analysis of the X-ray structure of the avian enzyme revealed that the side chain of Phe^{113} formed the "floor" of the putative allylic binding pocket and was "buttressed" by the aromatic ring of Phe^{112}. An analysis of this region in multiple sequence alignments for the published amino acid sequences of the chain elongation prenyltransferases revealed an interesting pattern. Eukaryotic FPP synthases had either phenylalanine or tyrosine residues at positions corresponding to 112 and 113 in the avian enzyme. GGPP and longer chain length synthases had residues with smaller side chains, typically alanine–serine or alanine–methionine, at these two positions. Bacterial and archaebacterial FPP synthases had a phenylalanine at position 112 and serine or threonine at position 113. Interestingly, the archaebacterial synthase from *Methanobacterium thermoautotrophicum* is a bifunctional enzyme that produces both FPP required for the biosynthesis of squalene and GGPP needed for the biosynthesis of archaebacterial lipids.

Involvement of the amino acids at positions 112 and 113 in regulating the chain length of the products was examined by site-directed mutagenesis.[43] The three possible combinations where phenylalanine at position 112 in the avian enzyme was changed to alanine, and the phenylalanine at position 113 was changed to serine were produced. In each case, the chain length of the product was altered. The F112A mutant catalyzed the synthesis of GGPP from DMAPP with an efficiency (V_{max}/K_m) similar to that for the formation of FPP by the wt avian enzyme. The F113S mutant converted DMAPP to the C_{25} product geranylfarnesyl diphosphate (GFPP), again with a catalytic efficiency similar to that for the synthesis of FPP by wt FPP synthase. Presumably, removing the "buttressing" phenylalanine partially collapsed the floor of the allylic binding pocket in order to accommodate a farnesyl residue in order to generate GGPP as the final product. In a similar manner, substitution of the phenyl ring of F113 in the "floor" of the pocket by a hydrogen atom (A113) lengthened the pocket in order to bind a geranylgeranyl residue for the final elongation step to GFPP.

Analysis of the products from an incubation of IPP and DMAPP with the F112A/F113S double mutant gave an unanticipated result. Instead of a single product, the enzyme synthesized a family of longer chain length allylic diphosphates ranging from four to 14 isoprene units. In contrast to the F112A and F113S mutants, the kinetic constants for the double mutant indicated a substantial loss of catalytic efficiency. In addition, the enzyme was inhibited by product, and the rate of the reaction decreased substantially after several turnovers.

An X-ray structure of the F112A/F113S protein explained the unanticipated loss of chain length control.[43] The substitutions of the phenyl rings in phenylalanine by hydrogen (position 112) and hydroxy (position 113) substituents deepened the hydrophobic isoprenoid chain binding pocket. Soaking experiments with DMAPP, GPP, FPP, and GGPP showed that the allylic substrates bound with the diphosphate residues located near the aspartate residues in the first aspartate-rich region and the hydrocarbon tails inserted into the hydrophobic pocket. By rotating the side chains of just two residues that form the new "floor" of the pocket, Tyr[154] and Glu[150], a passageway is opened through the monomer subunit to the dimer interface where the same passageway from the other subunit terminated. Small rotations of eight more side chains open a channel that leads to the outer solvent accessible surface of the homodimer. All of these side chain rotations can be accomplished without interference with the side chains of neighboring residues.

The structures obtained with DMAPP, GPP, and FPP also showed two magnesium atoms that served as cationic bridges between the diphosphate moiety and the aspartates (Figure 3). There were additional interactions between two nonbridging oxygens attached to the P-2 phosphorus atom and the conserved arginine at position 126 and between a nonbridging oxygen of the P-1 phosphorus atom and the side chain ammonium of Lys[214]. These interactions create a positively charged field around the diphosphate of the allylic substrate that should help stabilize the developing negative charge in the leaving group of the transition state for the alkylation reaction. Since we were able to obtain a structure of the enzyme–allylic substrate complex, it is likely that the binding of IPP induces a conformational change that provides the final trigger to initiate cleavage of the bond between the diphosphate moiety and the allylic isoprenoid unit.

5.13.3 ISOPRENE CYCLASES

5.13.3.1 Mechanistic Studies

5.13.3.1.1 *Monoterpene cyclases*

Although monoterpenoids only contain 10 carbon atoms, they form a large family of considerable structural diversity whose functions are still mostly unknown. The majority of the several hundred known naturally occurring monoterpenes are cyclic. The biochemistry and mechanism of action of monoterpene cyclases have been reviewed periodically.[44,45] Studies of the stereochemistries of the reactions catalyzed by cyclases for the biosynthesis of *p*-menthane, bornane, camphane, pinane, fenchane, and thujane classes of monoterpenes have provided important insights into how these enzymes function.[45] A considerable body of evidence now exists which clearly demonstrates that GPP is the natural substrate for these enzymes. However, the geranyl skeleton itself cannot be directly involved in the cyclization reaction because the *Z* C-2–C-3 double bond would generate a prohibitively strained *trans*-cyclohexene structure. The monoterpene cyclases finesse the problem by first catalyzing the isomerization of GPP to its allylic isomer linalyl diphosphate (LPP). At this point, a rotation about the C-2–C-3 single bond places the hydrocarbon chain of LPP in a conformation that can close to a *p*-menthane structure. The mechanism for the allylic transposition of the diphosphate is thought to involve ruptures of the carbon–oxygen bond in GPP to form an allylic

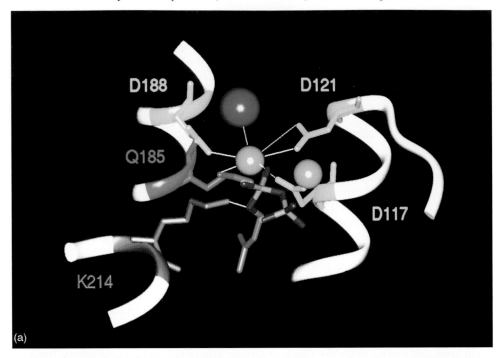

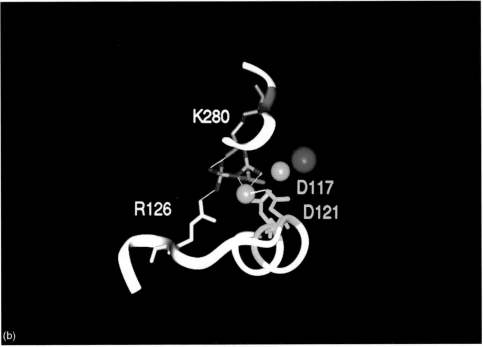

Figure 3 Three-dimensional structure of F112A/F113S FPP synthase mutant. Yellow spheres are magnesium ions and the blue spheres are water molecules. (a) Residues D188, Q185, D121, D117, and K214 bound to DMAPP. (b) View showing residues R126, K280, D117, and D121 bound to DMAPP.

cation–pyrophosphate ion pair which then collapses to LPP (Scheme 7). Following rotation about the C-2–C-3 bond, a similar heterolytic cleavage in LPP generates an isomeric allylic cation–pyrophosphate ion pair capable of alkylating the C-6–C-7 double bond. These steps, isomerization, bond rotation, and cyclization, appear to be employed by all isoprenoid cyclases that form "*p*-menthane-like" structures from allylic substrates. Subsequent steps, involving further electrophilic cyclizations, rearrangements, and hydride shifts, may occur before the reaction is terminated by deprotonation to form an alkene or capture of the cation by a nucleophile.

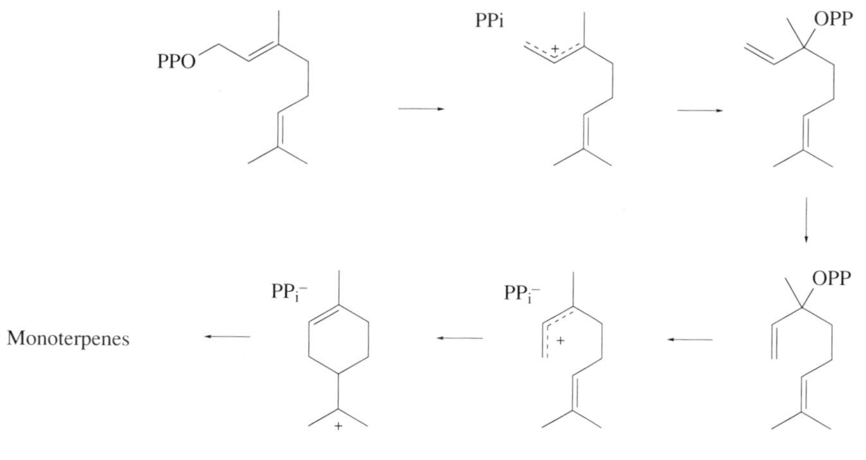

Scheme 7

The enzyme-catalyzed allylic isomerizations of GPP are stereospecific. The enantiomer of LPP formed is dependent on the initial folding of the geranyl chain in the active site, as are the subsequent cyclizations of the LPP enantiomer. The stereoselectivity of the isomerization sets the stage for the stereochemistries of the cyclization and subsequent steps. The steps following the initial isomerization/cyclization lead to the immense structural diversity found in the monoterpenes.

The electrophilic mechanisms for isomerization and cyclization are believed to be common to all monoterpene cyclases. Fluorinated alternate substrates similar to those developed for studying the mechanism for chain elongation by FPP synthase have provided important insights about monterpene cyclases.[45,46] 2-FluoroGPP and 2-fluoroLPP have been used as alternative substrates for several of the enzymes. Decreases in the rates of catalysis of two orders of magnitude relative to those for GPP were observed with the fluorinated analogues. Although detailed linear free energy correlations have not been reported, the magnitudes of the effects are similar to those of the related prenyltransferases and the model solvolysis reactions.

Several different replacements for GPP have been used to probe the electrophilic isomerizations and cyclizations.[47] These include DMAPP and FPP and analogues lacking a double bond (6,7-dihydroGPP (**15**) and 2,3-dihydroGPP (**16**)). Substrate and inhibition studies were performed using partially purified (+)-α-pinene cyclase and (+)-bornyl diphosphate cyclase.

The shorter (DMAPP) and longer (FPP) isoprenes were inhibitors. In addition, DMAPP was hydrolyzed to the corresponding alcohol at 95% of the rate of cyclization. FPP was converted to a mixture of farnesol and *trans*-nerolidol at 26% the rate of cyclization. 6,7-DihydroGPP (**15**) produced a 1:3 mixture of alcohols (dihydrogeranool, dihydrolinalool, and dihydronerol) and olefinic materials. 2,3-DihydroGPP (**16**) inhibited the cyclization reactions but was not hydrolyzed. The cyclase-catalyzed hydrolysis of the substrate analogues support the electrophilic nature of the cyclization process.

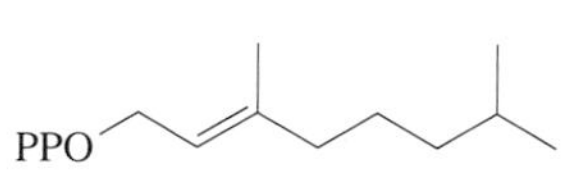

(**15**) 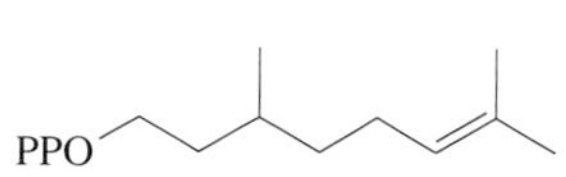   (**16**)

Although there is strong circumstantial evidence for the rearrangement of GPP to LPP during the enzyme-catalyzed cyclization, an enzyme-bound LPP intermediate has not been detected nor is LPP released during the conversion of GPP to cyclic products. 2,3-CyclopropylGPP (**17**) was designed to provide additional support for an electrophilic process.[48] The cyclopropane ring stabilizes positive charge at adjacent electron-deficient centers by resonance, and the reactivity of diphosphate (**17**) is comparable to that of GPP. However, the homoallylic isomer generated by a 1,3 rearrangement of the diphosphate moiety is not sufficiently reactive to serve as a substrate for the subsequent electrophilic cyclization, thereby decoupling the isomerization and cyclization steps.

Upon incubation of cyclopropylcarbinyl analogue (17) with a partially purified preparation of (+)-α-pinene cyclase and (+)-bornyl diphosphate cyclase from sage (Scheme 8), the products one would anticipate from an enzyme-catalyzed isomerization (derived from cation (18)) were isolated as a mixture of trienes (21)–(23) (31%), alcohols (19), (20) (58%), and the homoallylic diphosphate (24) produced by internal return of the leaving group (10%)

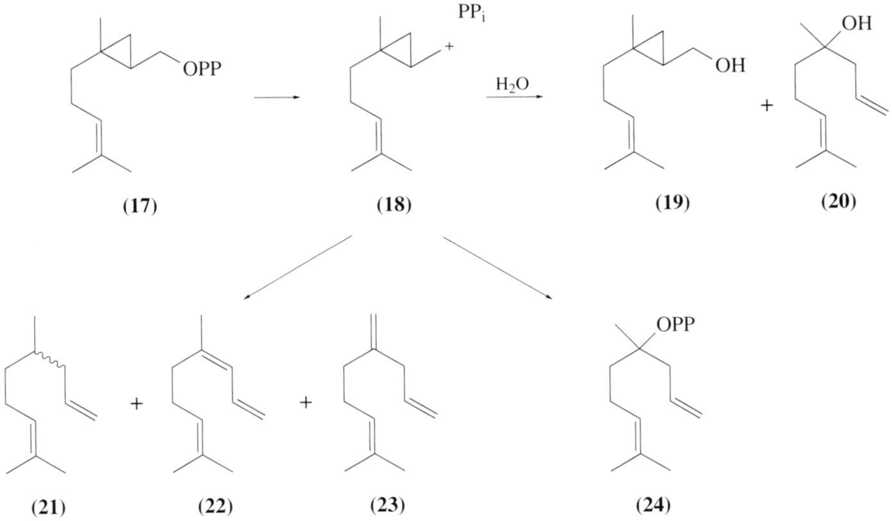

Scheme 8

Sulfonium ion analogues (25) and (26) were designed to mimic the structures of linalyl and α-terpinyl carbocations proposed as intermediates in the isomerization and cyclization reactions.[49] Both compounds were potent competitive inhibitors of cylase preparations from sage with K_i values in the micromolar range. However, when 50 μM inorganic pyrophosphate (PP$_i$) was added to the assay buffer, the potency of (25) and (26) increased to submicromolar levels. In addition, the K_i value for PP$_i$, which was only a modest inhibitor, decreased several fold in the presence of 5 μM (25) or (26). The synergy between the sulfonium ion analogues and PP$_i$ suggests that the cyclases bind the cation–PP$_i$ pair more tightly than either component alone and strongly implicates the presence of a structurally related set of ion pairs in the cyclization of GPP.

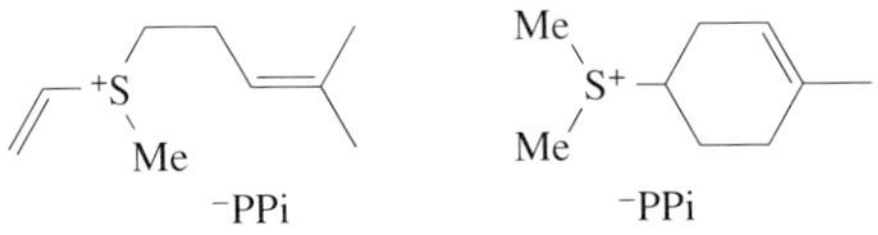

An analogue of GPP, in which the terminal dimethyl groups were replaced by a cyclopropyl moiety (27), was a potent irreversible inhibitor of several monoterpene cyclases.[50] Inhibition was time- and concentration-dependent, and the natural substrate, GPP, provided protection against inactivation by the cyclopropyl analogue. When a crude enzyme extract was incubated with (27), selective and covalent labeling of limonene synthase was demonstrated by sodium dodecyl sulfate (SDS) gel electrophoresis, radiofluorography, and immunoblotting. The inhibitor–enzyme complex was stable to extended dialysis and boiling in the presence of 2% SDS. The analogue is cleverly designed to undergo the normal allylic isomerization reaction, followed by cyclization of the linyl intermediate. However, electrophilic addition of the allylic carbocation to the remote double bond now generates a reactive cyclopropyl cation, which upon electrocyclic rearrangement produces an allylic cation whose substitution pattern precludes any reaction channel except combination with a nucleophile. In the case of limonene synthase, the most likely nucleophile is the putative base in the active site normally involved in deprotonation of the methyl group to generate limonene. In addition to providing mechanistic insight about the electrophilic cyclization, the inhibitor could be used to identify the residue in limonene synthase that was covalently modified.

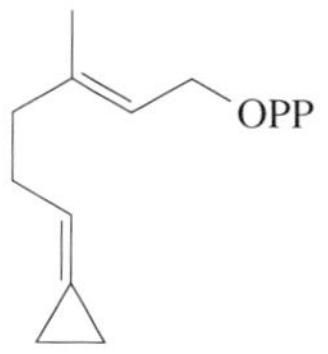

(27)

5.13.3.1.2 Sesquiterpene cyclase

Sesquiterpene cyclases catalyze the conversion of FPP to a variety of C_{15} terpenoids. To date, over 200 different structures have been isolated from a variety of sources. FPP is the substrate for sesquiterpene cyclases, and cyclization can occur at the proximal C-6–C-7 or the distal C-10–C-11 double bond. Cyclization at the distal double bond generates a macrocyclic structure that will accommodate the stereochemistry of the C-2–C-3 double bond in FPP without the need for a coupled isomerase activity. However, like monoterpene cyclases, cyclization to the proximal double bond can only occur after the sesquiterpene cyclase catalyzes the isomerization of FPP to its allylic isomer nerolidyl diphosphate (NPP) (Scheme 9).[51] Electrophilic attack of the resulting allylic cations from FPP at the distal (paths a and b) or NPP at the central (paths c and d) double bond followed by further cyclizations and rearrangements, and ultimately quenching with loss of a proton or capture by a nucleophile, gives rise to the wide variety of sesquiterpenes.

A variety of studies, many of which are similar to those described for the monoterpene cyclases in Section 5.13.3.1.1, were used to provide evidence for the electrophilic cyclizations in Scheme 9. Although reports include work with a variety of sesquiterpene cyclases, the majority describe experiments with trichodiene synthase and serve to illustrate the approaches that have been used to determine the mechanisms of action for sesquiterpene cyclases.

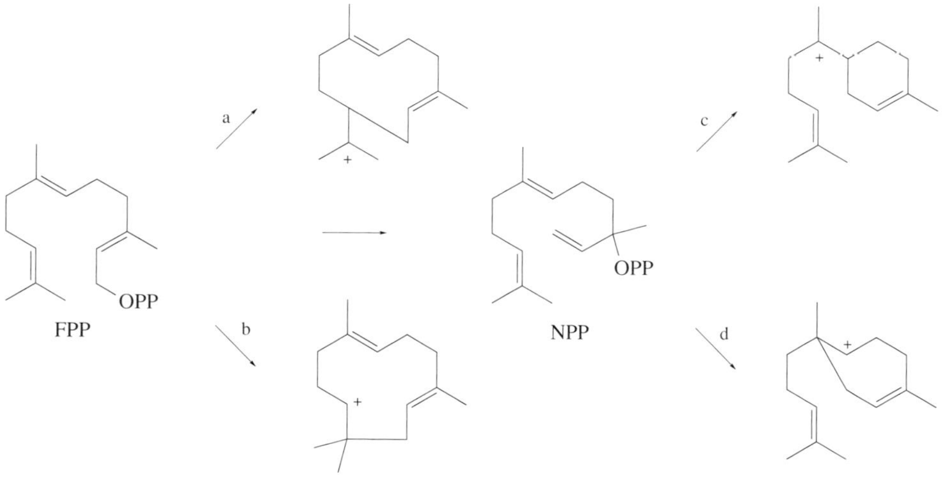

Scheme 9

Trichodiene is the parent hydrocarbon for the trichothecane family of mycotoxins and antibiotics.[52] Formation of trichodiene from FPP involves an electrophilic cyclization with attachment of C-1 to C-6 in the proximal double bond. This reaction occurs with retention of both protons at C-1 and with net retention of configuration. Like the reactions discussed for monoterpene cyclases, FPP cannot be the immediate precursor of the cyclic products because of steric constraints. Thus, trichodiene synthase, like its monoterpene cyclase counterparts, must possess an isomerase activity. A series of studies established that the mechanisms for the isomerization reactions are similar.[51–53] As observed for the monoterpene cyclases, NPP is an alternate substrate for trichodiene synthase, is not released from the active site, and the isomerization of FPP to NPP cannot be directly observed. Strong evidence for the isomerization of FPP to NPP was obtained with 6,7-dihydroFPP,[53] where the isomerization and cyclization steps were uncoupled by removing the proximal double bond. Both (7S)-E-6,7-dihydroFPP and (7R)-E-6,7-dihydroFPP were competitive inhibitors against

FPP and were alternate substrates for trichodiene synthase. Incubation of either enantiomer resulted in a mixture of acyclic sesquiterpene trienes (80–85%), acyclic allylic alcohols (15–20%), and the *Z* isomer of *cis*-6,7-dihydroFPP (24%) by analogy to the results shown in Scheme 8 for monoterpene cyclases. These results can be explained by initial isomerization of 6,7-dihydroFPP via an allylic cation to give an allylic intermediate, followed by rotation about the C-2–C-3 bond, and formation of the isomeric allylic cation. The products are typical of those expected when the normal cyclization reaction is blocked by removing the 6,7-double bond.

Additional support for an electrophilic mechanism was obtained by studying a series of aza analogues ((**28**), (**29**)).[54] In contrast to the potent inhibition seen for the sulfonium inhibitors of monoterpene cyclases, the aza sesquiterpene analogues were remarkably poor inhibitors. However, like their monoterpene counterparts, all four compounds were more potent when inorganic diphosphate was added. These results suggest that trichodiene synthase has a special affinity for the analogue–PP$_i$ ion pair. Since sesquiterpene derivatives (*S*)-(**29**) and (*R*)-(**29**) were more potent than their monoterpene counterparts, hydrophobic interactions as well as electrostatic interactions appear to be important for optimal binding. Interestingly, there was little preference for one enantiomer over another. This is consistent with a model active site which is able to accommodate a variety of intermediates of varying geometry and charge during the conversion of FPP to trichodiene.

(*R,S*)-(**28**)   (*R,S*)-(**29**)

5.13.3.2 Isoprene Cyclases—Enzymology

5.13.3.2.1 *Production of protein*

Studies of monoterpene cyclases are hampered by the availability of protein. The enzymes are not abundant in most plant tissues, and most of the studies reported for these enzymes involve partially purified preparations. However, over 20 monoterpene cyclases from *Salvia*, *Mentha*, *Foeniculum*, *Tanacetum*, *Pinus*, and *Citrus* species have been studied.[44] A major problem encountered with impure preparations is the presence of high levels of phosphatase activity, which hydrolyzes substrate and complicates kinetic and mechanistic studies. More recently, oil glands in the leaves of mint were identified as the primary site of monoterpene biosynthesis and procedures were developed for removing the glands from leaf material by gentle abrasion.[55] These preparations are substantially enriched in monoterpene cyclase activity.

The cyclases studied to date are soluble enzymes with molecular masses ranging from 50 to 100 kDa. Like most enzymes which process diphosphate esters, a divalent metal, typically Mg^{2+}, is required for activity. pH optima typically range from 6 to 7, and turnover rates are rather slow (0.01–1.0 s^{-1}). Although most monoterpene cyclases can use NPP or LPP as alternate substrates, the normal substrate is GPP with K_ms in the low micromolar range.

More progress has been made in the area of sesquiterpene synthases. Trichodiene synthase, pentalene synthase, aristolochene synthase, 5-epi-aristolochene synthase, and patchoulol synthase have been purified to apparent homogeneity.[51] Trichodiene synthase and patchoulol synthase are homodimers, while the other enzymes are apparently monomers as deduced by gel filtration and SDS–PAGE experiments. All of these enzymes have a divalent metal requirement, again typically Mg^{2+}.

Trichodiene synthase has been cloned from several sources.[51] The nucleotide sequence of the protein from *Fusarium sporotrichioides* was encoded by an 1182 nucleotide open reading frame containing 60 nucleotide in-frame intron. Following removal of the intron, the open reading frame was subcloned into an *E. coli* expression vector, and recombinant trichodiene synthase was produced as a soluble protein.[56] The gene for trichodiene synthase has also been cloned from *Gibberella pulcaus*.[51] More recently, an improved strain for production of recombinant trichodiene synthase was reported[57] using a vector which carried the promoter and translational leader sequence from phage T7 gene 10. With this construct, transformants produced trichodiene synthase as 20–30% of the total soluble protein. The genes for other sesquiterpene synthases, including pentalene synthase[58] and vetispiradiene synthase,[59] have also been cloned and the encoded proteins overproduced in *E. coli.*

Alignments of amino acid sequences for trichodiene synthase from different organisms show 96% identity. However, a comparison of trichodiene synthase sequences with those for other isoprenoid enzymes showed little similarity except for two basic-rich sequences DRRYR and DHRYR[22] which may be involved in diphosphate binding but differ in composition and charge from the DDXXD motifs found in the chain elongation prenyltransferases.

5.13.4 DIMETHYLALLYLTRYPTOPHAN SYNTHASE

5.13.4.1 Mechanistic Studies

Dimethylallyltryptophan (DMAT) synthase catalyzes the alkylation of the C-4 in L-tryptophan by DMAPP. This is the first pathway-specific step in the biosynthesis of ergot alkaloids.[60] An electrophilic mechanism was proposed for DMAT synthase. Although C-4 is not the preferred site of electrophilic attack on the indole ring, the position is activated toward electrophilic aromatic substitution relative to benzene by a resonance interaction with the indole nitrogen through the pyrrole double bond. Evidence for the electrophilic mechanism was obtained in linear free energy correlations that involved substitutions in the allylic substrate, like those described for FPP synthase, and in the aromatic acceptor using analogues of DMAPP and tryptophan.[61] The allylic analogues were the *E* and *Z* isomers of fluoromethyl and difluoromethyl DMAPP analogues (30)–(32) and five 7-substituted analogues of L-tryptophan ((33), X = OCH₃, Me, F, CF₃, and NO₂).

(30) X = Me
(31) X = CH₂F
(32) X = CHF₂

(33) Y = H, OMe, Me, F, CF₃, NO₂

Kinetic constants for the enzyme-catalyzed reactions were measured for *Claviceps purpurea* DMAT synthase for incubations of the DMAPP analogues with L-tryptophan and the L-tryptophan analogues with DMAPP. First-order rate constants for the model solvolysis reactions were measured for methanesulfonate derivatives of the fluorinated dimethylallyl analogues. Hammett σ values were used for the tryptophan derivatives. A Hammett plot for the solvolysis and enzyme-catalyzed reactions for the dimethyl analogues gave a linear correlation similar to that seen for FPP synthase. A ρ value of -2.0, deduced from a σ/ρ plot of the rates for prenyltransfer of DMAPP to the tryptophan analogues, is at the low end of the range normally seen for benzene derivatives.[62] However, the negative slope is clear evidence for an electrophilic alkylation, and the lower sensitivity toward electron-withdrawing substituents in the indole ring may be the result of an enhanced ability of the indole nitrogen to stabilize developing positive charge relative to benzene.

5.13.4.2 Purification of DMAT Synthase

DMAT synthase has been purified from mycelia of *Claviceps purpurea* ATCC 26245.[63] Gel filtration and electrophoresis experiments indicated that the enzyme is a homodimer with 52 kDa subunits. Unlike other prenyl transferases, DMAT synthase is active in metal-free buffers, although the activity is enhanced upon addition of Ca^{2+} or Mg^{2+}.[64] In 1995, the *dmaW* gene encoding DMAT synthase was isolated using a degenerate oligonucleotide probe that covered all possible codons encoding a peptide fragment obtained from cleavage of DMAT synthase with cyanogen bromide.[64] The *dmaW* gene was cloned into a yeast expression vector, and extracts from the yeast transformants catalyzed the alkylation of L-tryptophan by DMAPP.

5.13.5   PROTEIN FARNESYLTRANSFERASE

5.13.5.1   Mechanistic Studies

5.13.5.1.1   *General*

Protein farnesyltransferase (PFTase) catalyzes the post-translational modification of a variety of eukaryotic proteins by the attachment of a farnesyl group to a cysteine residue to form a thioether bond.[65] Farnesylated proteins perform a variety of functions in cells, including roles as enzyme catalysts, structural proteins, and components in signal transduction networks. The discovery that farnesylation is required for oncogenic forms of Ras proteins to promote unregulated cell division has prompted a widespread interest in protein prenylation.[65,66] The substrates for PFTase are FPP and proteins or peptides with a Cys–Aaa–Aaa–Xxx tetrapeptide sequence at their C-terminus, where Cys is cysteine, Aaa denotes an aliphatic amino acid, and Xxx is either methionine, alanine, serine, or glutamine. When Xxx is leucine, the tetrapeptide becomes the substrate for a related enzyme, protein geranylgeranyltranferase-type I (PGGTase-I), which transfers a geranylgeranyl group to the cystein residue. A second protein geranylgeranyltransferase (PGGTase-II) adds two geranylgeranyl residues to cysteines in C-terminal–CCXX, –CXCX, –XXCC, and –XCXC motifs.[65] The stereo-chemistry of PFTase reaction was determined using the enantiomers of [1-^{2}H]FPP as substrates in separate experiments.[67] Analysis of the NMR spectra of the products from incubation of the deuteriated derivatives and tetrapeptide dansyl–NH–Cys–Val–Ile–Met–CO$_2$H with recombinant human PFTase revealed that alkylation of the sulfhydryl moiety occurred with inversion of configuration.

5.13.5.1.2   *Linear free energy studies*

In contrast to the reaction catalyzed by FPP synthase where a carbon–carbon double bond is the nucleophilic acceptor, the thiol or thiolate moiety that is alkylated by the PFTase reaction is a potent nucleophile. Based on the lifetime of the dimethylallyl cation in water and the nucleophilicity–rate correlations of Mayr and Patz[19] discussed in Section 5.13.2.1.2, the predicted lifetime of an allylic farnesyl cation–thiolate ion pair is less than that required for vibration of a carbon–sulfur single bond. In this situation, the existence of the allylic cation as a discrete intermediate in the reaction becomes problematic. Thus, it is unclear precisely how the change in nucleophilicity has changed the degree of bonding between C-1 in FPP, the sulfur nucleophile, and the diphosphate leaving group at the transition state.

These questions were addressed by a linear free energy study of the reaction catalyzed by yeast PFTase between dansyl–NH–Cys–Val–Ile–Ala and a series of FPP analogues with hydrogen, monofluoromethyl, difluoromethyl, and trifluoromethyl groups in place of the C-3 methyl. FPP analogues (**34**)–(**37**) were alternate substrates for yeast PFTase.[68,69] The rate of the reaction decreased as the electron-withdrawing effects of the C-3 substituent increased, with the CF$_3$ analogue (**36**) being 770-fold less reactive than FPP. A Hammett plot of model solvolysis rate constants k_s vs. k_{cat} for farnesylation of the cysteine in the peptide substrate was linear, and the slope was positive. Thus, like the chain elongation reaction, alkylation of cysteine by FPP is electrophilic with a substantial degree of positive charge developed in the allylic moiety at the transition state. However, the magnitude of the slope was only 0.3, indicating that substantially less positive charge develops in the allylic moiety in the presence of the sulfur nucleophile.

(**34**) R = CH$_2$F
(**35**) R = CHF$_2$
(**36**) R = CF$_3$

(**37**)

The effects of nucleophilicity were also compared with a model system where azide was the nucleo-phile. Nucleophilic rate constants, k_{nuc}, were measured for the *p*-methoxysulfonate derivatives in the presence of sodium azide. Inclusion of azide increased the rates of the solvolysis reaction. A plot of k_{nuc} vs. k_{cat} for the enzyme-catalyzed reaction was linear with a slope of 0.6 (Figure 4). Thus, the compression of substituent effects by azide in the model solvolysis studies resulted from a change in mechanism from a stepwise reaction to a concerted process through an enforced transition state.[70] A similar trend was seen for alkylation of cysteine by PFTase, suggesting that the enzymatic reaction is an enforced process.

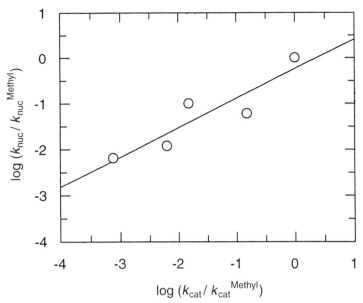

Figure 4   Hammett plot of k_{nuc} vs. k_{cat} for PFTase.

5.13.5.1.3   *Transition state analogues*

A class of bisubstrate analogues, based on L-β-farnesyl aminoalanine (faA), was designed to mimic the electrophilic nature of the transition state during prenyl transfer.[71] Tetrapeptide faA-Val-Ile-Ala (**38**) was an inhibitor of yeast PFTase with an IC_{50} of 14 μM. In contrast, faA and farnesylamine were only modest inhibitors, while *S*-farnesyl-Cys-Val-Ile-Ala was not an inhibitor. Inclusion of inorganic pyrophosphate ($IC_{50} = 2.4$ mM) in the assay buffer at a concentration of 1 mM decreased the IC_{50} of faAVIA 35-fold. The overall patterns of inhibition were similar to those seen for related transition state inhibitors of isoprenoid enzymes. However, the degree of synergism seen here is less than described previously for ammonium analogue inhibitors of trichodiene synthase[54] and squalene synthase[72] and may reflect a lower degree of charge separation between PP_i and the developing allylic carbocation in a less polar transition state.

(**38**)

5.13.5.2   PFTase—Enzymatic Studies

5.13.5.2.1   *Production of protein*

PFTase was initially purified from rat brain and from bovine brain $\sim 60\,000$-fold using peptide affinity columns based on the Cys–Aaa–Aaa–Xxx recognition motif.[73] The enzyme contains a tightly bound atom of zinc which may bind the sulfur in the peptide substrate as a thiolate. Mg^{2+}, presumably as the Mg^{2+} salt of FPP, is also necessary for activity. Upon separation of the subunits, the binding and catalytic functions of the enzyme are destroyed. The α subunit of PFTase is identical to the α subunit of PGGTase-I.[65] Thus, the different preferences shown by the two enzymes toward their respective substrates must reside in their distinctive β subunits. While the α and β subunits of PGGTase-II are different from those for PFTase and PGGTase-I, all these enzymes share a substantial amount of similarity at the amino acid sequence level.[74]

Initial attempts to engineer a system for production of recombinant PFTase met with limited success, presumably because of problems associated with balancing the levels of synthesis of the two subunits. The recombinant yeast[75] and human[76] PFTases were successfully overproduced in *E. coli* using a plasmid with a synthetic operon in which the *RAM*1 and *RAM*2 genes encoding the two subunits were translationally coupled by overlapping the TAATG stop–start codons. A Glu–Glu–Phe tripeptide was attached to the C-terminus of the Ram1 protein. The fusion facilitated purification of the enzyme by immunoaffinity chromatography on an anti-α-tubulin antibody column. In addition the nucleotides coding for the glutamates were selected to embed a consensus *E. coli* ribosomal binding site in the 3′ end of the open reading frame for Ram1 protein located in the proper position upstream of the translocation initiation codon for the Ram2 protein to maximize ribosomal binding and facilitate a balanced synthesis of the two proteins.

5.13.5.2.2 Site-directed mutagenesis

Several groups have prepared and studied mutant forms of PFTase to identify catalytic site amino acids important for binding and catalysis.[77] Several deletion and point mutations were made in the α subunit of the enzyme. Deletion of 106 residues from the amino terminus was needed to abolish activity, while removal of only five amino acids at the carboxy terminus produced inactive enzyme. Mutation of several internal residues (N164K, Y166F, R172E, N199D, W203H, and N199K) affected PFTase activity to varying degrees, but no clear trends were discerned. An alignment of amino acid residues for the β subunits of PFTase, PGGTase-I, and PGGTase-II showed five regions with a high degree of similarity. Thirteen of the conserved charged or polar residues in yeast PFTase were changed to nonpolar residues by site-directed mutagenesis.[78] Five of these mutant enzymes, R211Q, D307A, C309A, Y310F, and H363A, had substantially lower values for k_{cat} (10^{-3} relative to wt-PFTase). A metal analysis of the wild type protein and the five mutants showed that the substitutions had compromised Zn^{2+} binding in the D307A, C309A, and H363A enzymes. An X-ray structure of rat PFTase published in 1997 shows a bound zinc ligated to the carboxyl of Asp[307], the sulfhydryl of Cys[309], an imidazole nitrogen of His[363], and a molecule of water, which is presumably displaced by the sulfur in the peptide substrate upon binding.[79]

5.13.6 ISOPENTENYL ISOMERASE

5.13.6.1 Mechanistic Studies

5.13.6.1.1 General

IPP:DMAPP isomerase (EC 5.3.3.2) catalyzes the interconversion of IPP and DMAPP. This reaction creates the electrophilic dimethylallyl unit needed for the synthesis of more complex molecules in the isoprenoid pathway. Labeling experiments and stereochemical studies revealed that the pro-R hydrogen at C-2 of IPP is removed and a hydrogen ultimately derived from water is added to the re face of C-4 in the C-3–C-4 double bond.[80] The net 1,3 transposition of hydrogen from C-2 to C-4 is antarafacial, and the new methyl group in DMAPP is located in the E position.

Of the several mechanisms that have been proposed for the isomerization, only a protonation–deprotonation mechanism has received experimental support.[81] As shown in Scheme 10, a conjugate acid of an active site residue protonates the double bond in IPP to produce a tertiary carbocation, which then loses a proton from C-2 to complete the reaction. Although the manner in which the carbocation is generated by isomerase differs from that for prenyl transfer reactions, the techniques used to study prenyltransferases are generally applicable to isomerase as well. In direct analogy to the linear free energy experiments discussed for prenyltransferases in Sections 5.13.2.1.2, 5.13.4.1, and 5.13.5.1.2, substitution of hydrogen atoms on carbons adjacent to C-3 in IPP by fluorine will destabilize the tertiary carbocation and thereby retard the rate of isomerization, decreasing the rate of the protonation step. Several fluorinated analogues of IPP were synthesized and studied. (Z)-3-(trifluoromethyl)-2-butenyl diphosphate (**39**) was not a substrate for the enzyme and was estimated to be at least 2×10^{-6} times less reactive than IPP. While an effect by the trifluoromethyl group of this magnitude is consistent with a carbocationic mechanism for the isomerization, the experiment does not provide positive evidence about the mechanism of the reaction. An attempt to study monofluoro derivatives of IPP was unsuccessful for reasons outlined in the following section.

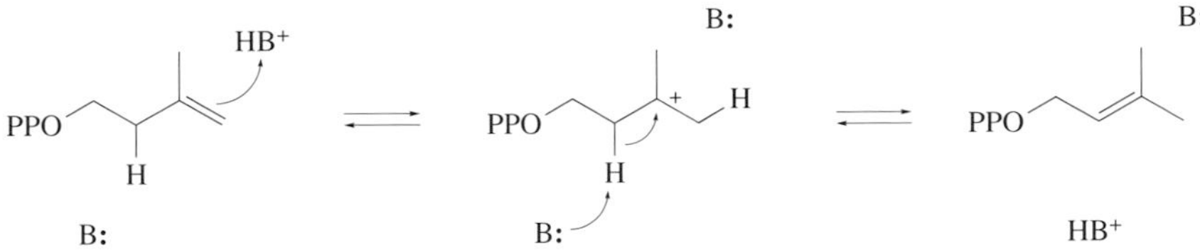

Scheme 10

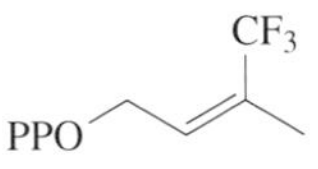

(**39**)

5.13.6.1.2 Irreversible inhibitors

Allylic fluorides (**40**) and (**41**) were originally synthesized for a linear free energy study of isomerase from *Claviceps* sp. However, both compounds inhibited the enzyme in a pseudo first-order time-dependent manner typical of irreversible inhibitors.[81,82] The 1:1 inhibitor–enzyme complex was stable to prolonged dialysis, treatment with 6 M urea, and SDS gel electrophoresis. Stoichiometric release of fluoride ion during the reaction was verified by a colorimetric assay[82] and by ^{19}F NMR spectroscopy.[28] These experiments indicated that (**40**) and (**41**) inhibited IPP isomerase by forming a covalent bond to a residue in the active site with release concomitant with release of fluoride ion.

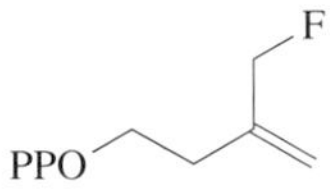

(**40**) 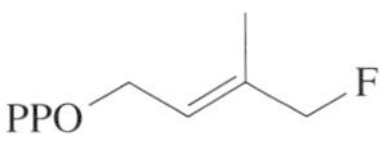   (**41**)

The observations with the monofluoro analogues suggested that epoxide derivatives of IPP and DMAPP might be mechanism-based inhibitors if isomerase. The mechanism for inhibition would involve protonation of the epoxide oxygen by the active site acid that normally protonates the double bond of IPP. Nucleophilic attack on the protonated epoxide by an active site base would lead to covalent attachment. Epoxides (**42**) and (**43**) were potent active site directed irreversible inhibitors.

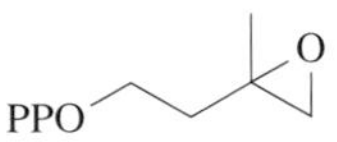

(**42**) 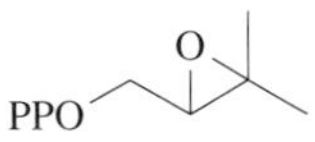   (**43**)

Finally, a "functionally" irreversible noncovalent inhibitor was discovered during a search for a transition state/reactive intermediate analogue that mimicked the shape and charge of the putative tertiary carbocationic intermediate in the normal isomerization. The compound, 2-(dimethyl-amino)ethyl diphosphate (**44**), was a potent inhibitor of *Claviceps* IPP isomerase at pH 7.0,[80–82] inactivated the enzyme in a time-dependent manner, and formed a stable 1:1 complex with the enzyme ($k_{off} < 1.1 \times 10^{-7}$ s^{-1} and $k_D < 10^{-10}$ M for *Claviceps* isomerase).[82] The dissociation constant for (**44**) with the yeast enzyme was estimated to be even lower.[80] The enzyme inhibitor complex was stable in 6 M urea, but the inhibitor was slowly released intact upon treatment with 2-thioethanol and SDS and was readily released when the pH was raised above 13. Presumably the ammonium moiety is deprotonated at higher pH whereupon the neutral amine is readily released (Scheme 10).

$$PPO\frown \overset{\overset{\displaystyle Me}{|}}{\underset{\displaystyle Me}{\overset{+}{N}}}\diagdown H$$

(44)

5.13.6.2   Isopentenyl Isomerase—Enzymology

5.13.6.2.1   *Protein purification*

Isomerase was first purified from various eukaryotic sources that had substantial levels of activity.[28,83] Oligonucleotide probes based on the N-terminal sequence of the *Saccharomyces cerevisiae* enzyme were used to isolate the fungal gene.[73] The gene, *IDI1*, was cloned into an *E. coli* expression vector, and transformants produced active enzyme as 30–35% of the total soluble protein.[84] Disruption experiments in yeast established that *IDI1* is an essential single copy gene.[85] Open reading frames from human[86] and *Schizosaccharomyces pombe*[87] isomerase have also been cloned, and catalytically active protein overproduced in *E. coli*. A database search with peptide sequences based on highly conserved regions in the eukaryotic enzymes uncovered an open reading frame encoding a bacterial IPP isomerase in the photosynthetic gene cluster of *Rhodobacter capsulatus*.[88] The open reading frame was inserted into an *E. coli* expression vector, and the recombinant protein had isomerase activity.

5.13.6.2.2   *Identification of active site residues*

The antarafacial stereochemistry for the isomerization is most simply explained by an active site that contains two residues to facilitate the protonation/deprotonation steps, one located on each face of the substrate. Incubation of yeast IPP isomerase with radiolabeled FIPP (40), digestion of the inactivated protein with trypsin, and a determination of the amino acid sequence of the labeled peptide revealed that two residues were alkylated, Cys[138] and Cys[139].[84] In similar experiments with epoxide (42) only Cys[139] was labeled.[89] The sulfhydryl moieties in both amino acids were replaced by site-directed mutagenesis. The C138V mutant was only slightly less active than wild type yeast isomerase, while the C139V mutant was inactive. The C139S mutant retained $< 10^{-4}$ of wild type activity.[90] Inactivation of the C139S protein with (40) resulted in covalent modification of E207. The E207Q and E207V mutants were both inactive, suggesting a catalytic role for the glutamate residue as well. Additional work will be needed to determine the specific roles of C139 and E207 in catalysis.

5.13.7   SQUALENE SYNTHASE AND PHYTOENE SYNTHASE

5.13.7.1   Mechanistic Studies

5.13.7.1.1   *General*

Squalene synthase (EC 2.5.1.21) catalyzes the synthesis of squalene from FPP in two distinct steps (Scheme 11). In the first reaction, two molecules of FPP are condensed to form presqualene diphosphate (PSPP). In the second, the diphosphate moiety of PSPP is expelled, the two newly formed cyclopropane bonds are cleaved, and the farnesyl residues are rejoined to form squalene with addition of hydride from NADPH. This is the first pathway-specific transformation in sterol metabolism. A similar series of reactions produces phytoene in the first pathway-specific transformations of carotenoid biosynthesis, where two molecules of the C_{20} isoprenoid GGPP condense to form prephytoene diphosphate followed by a similar rearrangement to give phytoene. Phytoene synthase does not require NADPH, and instead a proton is eliminated to produce the double bond between the geranylgeranyl residues.

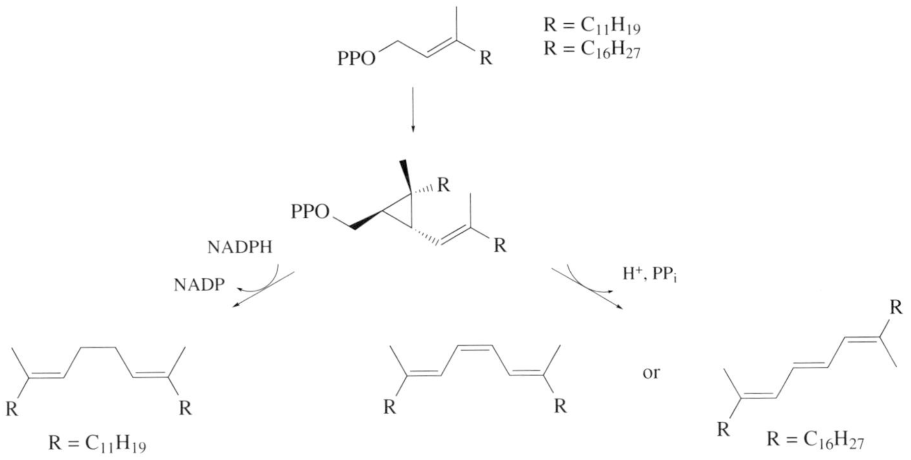

R = C_{11}H_{19}
R = C_{16}H_{27}

R = C_{11}H_{19}

R = C_{16}H_{27}

Scheme 11

The mechanisms of action for squalene and phytoene synthase have been intensely studied; however, until recently, this work was hampered by difficulties in obtaining pure enzyme preparations. Several mechanisms have been proposed for the cyclopropanation reaction, although there is little direct evidence to distinguish among the possibilities. Several mechanisms have also been proposed for rearrangement of the cyclopropane ring to the final products. Although there are significant differences in the proposals, each is based on the well-established propensity of cyclopropylcarbinyl cations to rearrange.[91] Model studies with the chrysanthemyl system, a simple C_{10} model for PSPP, to determine the feasibility of the proposed rearrangements under nonenzymatic conditions, gave mixed results (Scheme 12). While primary cyclopropylcarbinyl cation (**45**) rearranged to (**48**), only a very small amount (~0.04%) of the product corresponding to squalene or phytoene was detected.[92] Most of the products were derivatives of cation (**46**) (Scheme 12). These results illustrate the problems often encountered with models for enzyme-catalyzed reactions. It is rare indeed when the models duplicate the selectivity of their enzymatic counterparts, and the investigator is then left to rationalize how an enzyme might overcome the "chemical deficiencies" of the model.

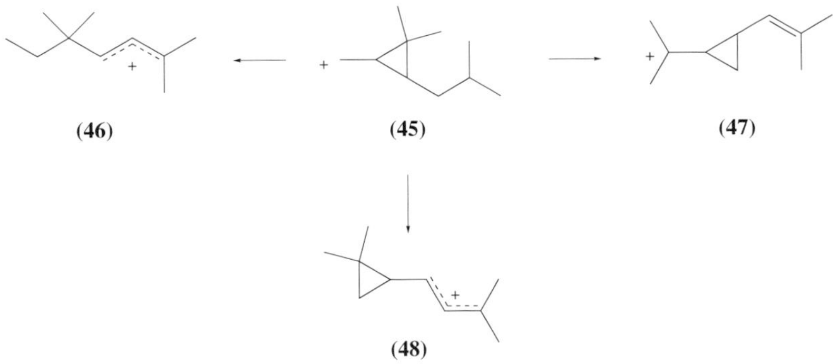

(**46**)   (**45**)   (**47**)

(**48**)

Scheme 12

5.13.7.1.2 *Ammonium ion analogues*

The first experimental evidence with squalene synthase to lend support to the mechanism involving cyclopropylcarbinyl cations was obtained with ammonium derivatives designed to mimic carbocationic intermediates like those discussed previously for FPP synthase, trichodiene synthase, and IPP isomerase.[72] Although the squalene synthase analogues contained a positively charged tetrahedral nitrogen in place of the trigonal carbon in the putative carbocations, space filling models suggested that the overall differences in shape were small. Surprisingly, when ammonium ion analogues (**49**) and (**50**) were tested, neither compound inhibited squalene synthase at concentrations up to 170 μM. However, when 1 mM PP$_i$ was added to the assay buffer, the potency of compound (**50**) increased dramatically. At 3 μM, ammonium analogue (**50**) inhibited the reaction by 50% and by more than 90% at 20 μM. Compound (**49**), an ammonium ion analogue of the primary cation, gave similar results. These results are consistent with a synergistic binding of PP$_i$ and the ammonium analogue by squalene synthase in place of the corresponding carbocation–PP$_i$ ion pairs in the normal enzyme-catalyzed reaction. Compound (**51**), in which a phosphonophosphate moiety was tethered to the amino group, was synthesized to test this hypothesis. The tethered analogue was a potent inhibitor in pyrophosphate-free buffer. When the synthesis of PSPP and squalene were measured simultaneously in the presence of (**51**), it was found that the inhibitor depressed both reactions to the same extent. These results strongly suggest that formation of PSPP and its subsequent rearrangement of squalene are catalyzed in the same active site.

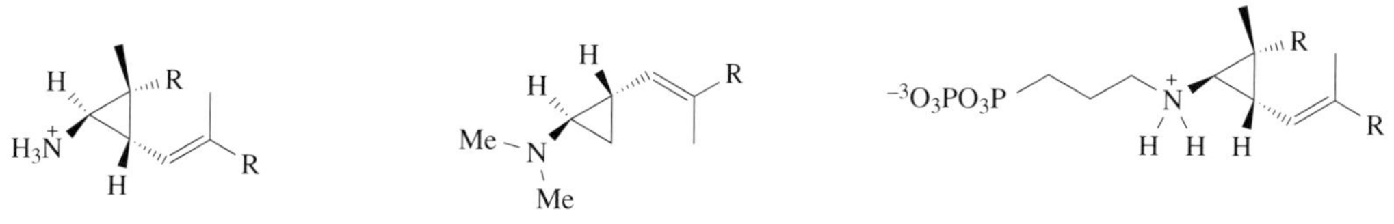

(49)   (50)   (51)

5.13.7.1.3 *Products studies*

When FPP was incubated with squalene synthase in the absence of NADPH, PSPP formed rapidly.[93] However, upon prolonged incubation, PSPP was converted to a mixture of three compounds (Scheme 13): *cis*-dehydrosqualene (**56**, $\sim$13%), botryococcenol (**54**, $\sim$26%), and squalenol (**55**, $\sim$61%). These three compounds constituted $\sim$97% of the total product mixture. The farnesyl units in (**55**) and (**56**) are joined in a 1′–1 fashion like the farnesyl and geranylgeranyl moieties in squalene and phytoene. The 1′–3 linkage in (**54**) is identical to that seen for the major products in the model reactions with the chrysanthemyl system. The formation of these products from PSPP by squalene synthase provides an unambiguous link between the model studies and the enzyme-catalyzed reaction. Furthermore, it demonstrates that the catalytic machinery of squalene synthase and phytoene synthase is capable of altering the "normal" course of the cyclopropylcarbinyl cationic rearrangements to produce squalene and phytoene in preference to the 1′–3 products. These products provide the strongest link between the enzymatic and model studies and powerful evidence for the mechanism shown in Scheme 11.

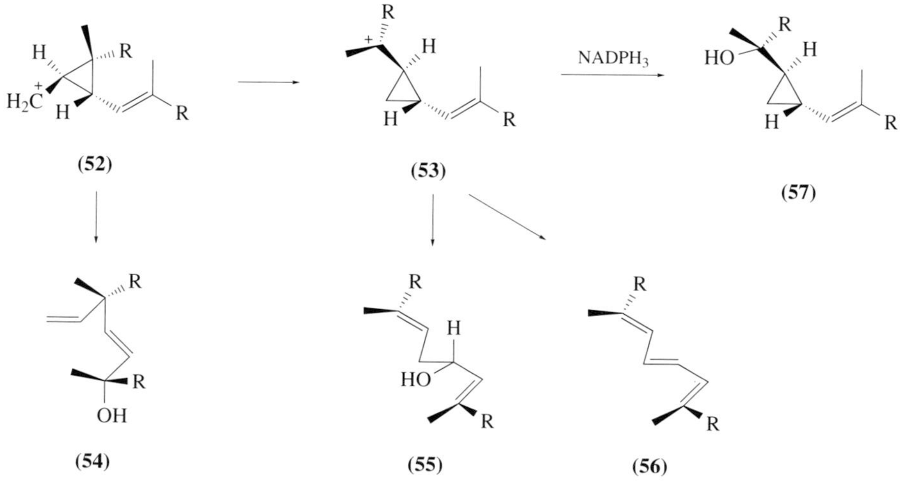

(52)   (53)   (57)

(54)   (55)   (56)

Scheme 13

Additional support came from experiments with an unreactive dihydroanalogue of NADPH (NADPH$_3$). Incubation of FPP with squalene synthase in the presence of the analogue gave a substantially different distribution of products: **56** (39%), **55** (31%), and a previously unseen compound, rillingol (**57**, 29%), all derived from tertiary cyclopropylcarbinyl cation (**52**) as shown in Scheme 13.[94] Formation of (**57**) as a major product in the presence of an unreactive NADPH analogue provides strong evidence for the existence of cyclopropylcarbinyl cation (**53**) during the rearrangement.

5.13.7.2 Squalene Synthase and Phytoene Synthase—Enzymology

Having a highly purified, soluble form of squalene synthase was a crucial part of the studies described in the preceding section. The enzyme is an intrinsic microsomal protein, and there have been many attempts to solubilize and purify the wild type protein.[91] Agnew and Popjak[95] achieved the first solubilization of squalene synthase with deoxycholate. However, the soluble enzyme was unstable and resisted purification. Later, Sasiak and Rilling succeeded in purifying small quantities of the *S. cerevisiae* enzyme after solubilization with *N*-octyl-β-D-glucopyranoside and Lubrol PX.[96] The enzyme was a monomer with a molecular mass of 47 kDa and required Mg^{2+} and NADPH or NADH for activity, but the protein lost activity upon attempts to remove the detergent. More recently, Shechter *et al.* purified a soluble form of rat squalene synthase followed release of the enzyme from microsomes with trypsin.[97]

The genes which encode squalene synthase have been cloned from bakers yeast, humans, rats, and *Schizosaccharomyces pombe*.[98] An analysis of the hydropathy plot for yeast squalene synthase revealed four regions of high hydrophobicity. One hydrophobic region of 25 amino acids was at the C-terminus of the protein and had polar flanking regions typical of those for a membrane spanning helix, suggesting that squalene synthase was anchored at the surface of the membrane by the helix. When the hydrophobic C-terminal domain was deleted, a soluble form of the protein was produced in *E. coli* at a level of 2–5% of the total soluble protein.[98] The recombinant enzyme was purified to >90% homogeneity in two steps by chromatography on hydroxyapatite and phenyl Superose. This protein had catalytic properties similar to those of wild type squalene synthase. A soluble form of phytoene synthase from *Capsicum* chloroplast stroma was purified by Dogbo *et al.*[99] Like squalene synthase, this enzyme was a monomer with a molecular mass of 47 000, but in contrast to squalene synthase, preferred Mn^{2+} rather than Mg^{2+} as a cofactor.

5.13.8 CONCLUSIONS

A wide variety of natural products are synthesized by enzymes whose reactions proceed via electrophilic mechanisms. Many of those studied to date are in the isoprenoid pathway and form their products by capture of an allylic cation by a wide variety of nucleophiles. Although the reactions catalyzed by the electrophilic enzymes discussed in this chapter are varied, and the properties of the enzymes themselves are very different, several techniques which have been discussed, including linear free energy studies, substrate analogues for cationic intermediates, and mechanism-based inhibitors, allowed for mechanistic similarities between the enzymes to be elucidated. In addition to these studies, one of the most important advances in the study of these enzymes has been the ability to obtain recombinant sources of protein. The availability of recombinant protein has facilitated large quantities of recombinant enzyme, in turn, facilitating a wide variety of studies.

5.13.9 REFERENCES

1. M. L. Sinnott, *Chem. Rev.*, 1990, **90**, 1171.
2. J. B. Kempton and S. G. Withers, *Biochemistry*, 1992, **31**, 9961.
3. M. Castle, G. Blondin, and W. R. Nes, *J. Am. Chem. Soc.*, 1963, **85**, 3306.
4. A. Rahier, J.-C. Genot, F. Schuber, P. Benveniste, and A. S. Narula, *J. Biol. Chem.*, 1984, **259**, 15 215.
5. R. Wolfenden and W. M. Kati, *Acc. Chem. Res.*, 1991, **24**, 209.
6. B. Ganem, *Acc. Chem. Res.*, 1996, **29**, 340.
7. G. Popjak, in "Natural Substances Formed Biologically from Mevalonic Acid," ed. T. W. Goodwin, Academic Press, New York, 1970, p. 17.
8. M. Rohmer, M. Seemann, S. Horbach, S. Bringer-Meyer, and H. Sahm, *J. Am. Chem. Soc.*, 1996, **118**, 2564.
9. A. Chen, P. A. Kroon, and C. D. Poulter, *Protein Sci.*, 1994, **3**, 600.

10. M. S. Brown and J. L. Goldstein, *Science*, 1986, **232**, 34.
11. H. C. Rilling and K. Bloch, *J. Biol. Chem.*, 1959, **234**, 1424.
12. F. Lynen, H. Eggerer, U. Henning, and I. Kessel, *Angew. Chem.*, 1958, **70**, 738.
13. C. D. Poulter and H. C. Rilling, *Acc. Chem. Res.*, 1978, **11**, 307.
14. C. D. Poulter and H. C. Rilling, in "Biosynthesis of Isoprenoid Compounds," eds. J. W. Porter and S. L. Spurgeon, Wiley, New York, 1981, vol. 1, p. 162.
15. F. Westheimer, *Science*, 1987, **235**, 1173.
16. C. D. Poulter and D. M. Satterwhite, *Biochemistry*, 1977, **16**, 5470.
17. C. D. Poulter, P. L. Wiggins, and A. T. Le, *J. Am. Chem. Soc.*, 1981, **103**, 3926.
18. C. L. Rodreiguez, Ph.D. Thesis, University of Utah, Salt Lake City, 1990.
19. H. Mayr and M. Patz, *Angew. Chem.*, 1994, **33**, 938.
20. V. J. Davisson, T. R. Neal, and C. D. Poulter, *J. Am. Chem. Soc.*, 1993, **115**, 1235.
21. V. J. Davisson and C. D. Poulter, *J. Am. Chem. Soc.*, 1993, **115**, 1245.
22. D. E. Cane, J. H. Shim, Q. Xue, B. C. Fitzsimons, and T. M. Hohn, *Biochemistry*, 1995, **34**, 2480.
23. A. Saito and H. C. Rilling, *Arch. Biochem. Biophys.*, 1981, **208**, 508.
24. N. L. Eberhardt and H. C. Rilling, *J. Biol. Chem.*, 1975, **250**, 863.
25. L. S. Yeh and H. C. Rilling, *Arch. Biochem. Biophys.*, 1977, **183**, 718.
26. G. F. Barnard, B. Langton, and G. Popjak, *Biochem. Biophys. Res. Commun.*, 1978, **85**, 1097.
27. G. F. Barnard, in "Methods in Enzymology," eds. J. H. Law and H. C. Rilling, Academic Press, London, 1985, vol. 110, p. 155.
28. C. D. Poulter in "Biochemistry of Cell Walls and Membranes in Fungi," eds. P. J. Kuhn, A. P. J. Trinci, M. J. Jung, M. W. Goosey, and L. G. Copping, Springer, Berlin, 1990, p. 169.
29. H. Sagami, K. Ogura, and C. D. Poulter, *Biochem. Inc.*, 1984, **8**, 661.
30. B. C. Reed and H. C. Rilling, *Biochemistry*, 1976, **15**, 3739.
31. M. S. Anderson, J. G. Yarger, C. L. Burck, and C. D. Poulter, *J. Biol. Chem.*, 1989, **264**, 19176.
32. D. N. Brems, E. Bruenger, and H. C. Rilling, *Biochemistry*, 1981, **20**, 3711.
33. J. H. Teruya, S. Y. Kutsunai, D. H. Spear, P. A. Edwards, and C. F. Clarke, *Mol. Cell. Biol.*, 1990, **10**, 2315.
34. B. T. Sheares, S. S. White, D. T. Molowa, K. Chan, V. D. H. Ding, P. A. Kroon, R. G. Bostedor, and J. D. Karkas, *Biochemistry*, 1989, **28**, 8129.
35. D. Delourme, F. Lacroute, and F. Karst, *Plant Mol. Biol.*, 1994, **26**, 1867.
36. S. Fujisaki, H. Hara, Y. Nishimura, K. Horiuchi, and T. Nishino, *J. Biochem.*, 1990, **108**, 995.
37. T. Koyama, S. Obata, M. Osabe, A. Takeshita, K. Yokoyama, M. Uchida, T. Nishino, and K. Ogura, *J. Biochem.*, 1993, **113**, 355.
38. M. W. Ashby and P. A. Edwards., *J. Biol. Chem.*, 1990, **265**, 13157.
39. A. Joly and R. A. Edwards, *J. Biol. Chem.*, 1993, **268**, 26983.
40. P. F. Marrero, C. D. Poulter, and P. A. Edwards, *J. Biol. Chem.*, 1992, **267**, 21873.
41. L. Song and C. D. Poulter, *Proc. Natl. Acad. Sci. USA*, 1994, **91**, 3044.
42. L. C. Tarshis, M. Yan, C. D. Poulter, and J. C. Sacchettini, *Biochemistry*, 1994, **33**, 10871.
43. L. C. Tarshis, P. J. Proteau, B. A. Kellogg, J. C. Sacchettini, and C. D. Poulter, *Proc. Natl. Acad. Sci. USA*, 1996, **93**, 15018.
44. R. Croteau, *Chem. Rev.*, 1987, **87**, 929.
45. R. Croteau, in "Biosynthesis of Isoprenoid Compounds," eds. J. W. Porter and S. L. Spurgeon, Wiley, New York, 1981, vol. 1, p. 225.
46. R. Croteau, W. R. Alonso, A. E. Koepp, and M. A. Johnson, *Arch. Biochem. Biophys.*, 1994, **309**, 184.
47. C. J. Wheeler and R. Croteau, *J. Biol. Chem.*, 1987, **262**, 8213.
48. C. J. Wheeler and R. Croteau, *Proc. Natl. Acad. Sci. USA*, 1987, **84**, 4856.
49. R. Croteau, C. J. Wheeler, R. Aksela, and A. C. Oehlschlager, *J. Biol. Chem.*, 1986, **261**, 7257.
50. R. Croteau, W. R. Alonso, A. E. Koepp, J. H. Shim, and D. E. Cane, *Arch. Biochem. Biophys.*, 1993, **307**, 397.
51. D. E. Cane, *Chem. Rev.*, 1990, **90**, 1089.
52. D. E. Cane, *Acc. Chem. Res.*, 1985, **18**, 220.
53. D. E. Cane, J. L. Pawlak, R. M. Horak, and T. M. Hohn, *Biochemistry*, 1990, **29**, 5476.
54. D. E. Cane, G. Yang, R. M. Coates, H.-J. Pyun, and T. M. Hohn, *J. Org. Chem.*, 1992, **57**, 3454.
55. J. Gershenzon, M. A. Duffy, F. Karp, and R. Croteau, *Anal. Biochem.*, 1987, **163**, 159.
56. T. M. Hohn, and R. D. Plattner, *Arch. Biochem. Biophys.*, 1989, **275**, 92.
57. D. E. Cane, Z. Wu, J. S. Oliver, and T. M. Hohn, *Arch. Biochem. Biophys.*, 1993, **300**, 416.
58. D. E. Cane, J.-K. Sohng, C. R. Lamberson, S. M. Rudnicki, Z. Wu, M. D. Lloyd, J. S. Oliver, and B. R. Hubbard, *Biochemistry*, 1994, **33**, 5846.
59. K. Back and J. Chappell, *J. Biol. Chem.*, 1995, **270**, 7375.
60. W. A. Cress, L. T. Chayet, and H. C. Rilling, *J. Biol. Chem.*, 1981, **256**, 10917.
61. J. C. Gebler, A. B. Woodside, and C. D. Poulter, *J. Am. Chem. Soc.*, 1992, **114**, 7354.
62. L. M. Stock, *Prog. Phys. Org. Chem.*, 1976, **12**, 21.
63. J. C. Gebler and C. D. Poulter, *Arch. Biochem. Biophys.*, 1992, **296**, 308.
64. H.-F. Tsai, H. Wang, J. C. Gebler, C. D. Poulter, and C. L. Schardl, *Biochem. Biophys. Res. Commun.*, 1995, **216**, 119.
65. S. Clarke, *Annu. Rev. Biochem.*, 1992, **61**, 355.
66. W. R. Schafer, R. Kim, R. Stern, J. Thorner, S.-H. Kim, and J. Rine, *Science*, 1989, **245**, 379.
67. Y. Mu, C. A. Omer, and R. A Gibbs, *J. Am. Chem. Soc.*, 1996, **118**, 1817.
68. J. M. Dolence and C. D. Poulter, *Proc. Natl. Acad. Sci. USA*, 1995, **92**, 5008.
68. C. D. Poulter, in "Biomedical Frontiers of Fluorine Chemistry," eds. I. Ojima, J. T. Welch, American Chemical Society, 1996, ACS Symposium series 639, p. 158.
70. W. P. Jencks, *Chem. Soc. Rev.*, 1981, **10**, 345.
71. P. B. Cassidy and C. D. Poulter, *J. Am. Chem. Soc.*, 1996, **118**, 8761.
72. C. D. Poulter, T. L. Capson, M. D. Thompson, and R. S. Bard, *J. Am. Chem. Soc.*, 1989, **111**, 3734.
73. D. L. Pompliano, E. Rands, M. D. Schaber, S. D. Mosser, N. J. Anthony, and J. B. Gibbs, *Biochemistry*, 1992, **31**, 3800.

74. M. S. Boguski, A. W. Murray, and S. Powers, *New Biologist*, 1992, **4**, 408.
75. M. P. Mayer, G. D. Prestwich, J. M. Dolence, P. D. Bond, H.-Y. Wu, and C. D. Poulter, *Gene*, 1993, **132**, 41.
76. C. A. Omer, A. M. Krai, R. E. Diehl, G. C. Prendergast, S. Powers, C. M. Allen, J. B. Gibbs, and N. E. Kohl, *Biochemistry*, 1993, **32**, 5167.
77. I. Sattler and F. Tamanoi, in "Regulation of the Ras Signaling Network," eds. H. Maruta and A. W. Burgess, Molecular Biology Intelligence Unit, R. G. Landes Co., Austin, TX, 1996, p. 95.
78. J. M. Dolence, D. B. Rozema, and C. D. Poulter, *Biochemistry*, 1997, **36**, 9246.
79. H. W. Park, S. R. Boduluri, J. F. Moomaw, P. J. Casey, and L. S. Beese, *Science*, 1997, **275**, 1800.
80. J. E. Reardon and R. H. Abeles, *Biochemistry*, 1986, **25**, 5609.
81. M. Muehlbacher and C. D. Poulter, *J. Am. Chem. Soc.*, 1985, **107**, 8307.
82. M. Muehlbacher and C. D. Poulter, *Biochemistry*, 1988, **27**, 7315.
83. M. S. Anderson, M. Muehlbacher, I. P. Street, J. Proffitt, and C. D. Poulter, *J. Biol. Chem.*, 1989, **264**, 19 169.
84. I. P. Street and C. D Poulter, *Biochemistry*, 1990, **29**, 7531.
85. M. P. Mayer, F. M. Hahn, D. J. Stillman, and C. D. Poulter, *Yeast*, 1992, **8**, 743.
86. F. M. Hahn, J. W. Xuan, A. F. Chambers, and C. D. Poulter, *Arch. Biochem. Biophys.*, 1996, **332**, 30.
87. F. M. Hahn and C. D. Poulter, *J. Biol. Chem.*, 1995, **270**, 11 298.
88. F. M. Hahn, J. A. Baker, and C. D. Poulter, *J. Bacteriol.*, 1996, **178**, 619.
89. X. J. Lu, D. J. Christensen, and C. D. Poulter, *Biochemistry*, 1992, **31**, 9955.
90. I. P. Street, H. R. Coffman, J. A. Baker, and C. D. Poulter, *Biochemistry*, 1994, **33**, 4212.
91. C. D. Poulter, *Acc. Chem. Res.*, 1990, **23**, 70.
92. C. D. Poulter, L. L. Marsh, J. M. Hughes, J. C. Argyle, D. M. Satterwhite, R. J. Goodfellow, and S. G. Moesinger, *J. Am. Chem. Soc.*, 1977, **99**, 3816.
93. D. Zhang and C. D. Poulter, *J. Am. Chem. Soc.*, 1995, **117**, 1641.
94. M. B. Jarstfer, B. S. J. Blagg, D. H. Rogers, and C. D. Poulter, *J. Am. Chem. Soc.*, 1996, **118**, 13 089.
95. W. S. Angew and G. Popjak, *J. Biol. Chem.*, 1978, **253**, 4574.
96. K. Sasiak and H. C. Rilling, *Arch. Biochem. Biophys.*, 1988, **260**, 622.
97. I. Shechter, E. Klinger, M. L. Rucker, R. G. Engstrom, J. A. Spirito, M. A. Islam, B. R. Boettcher, and D. B. Weinstein, *J. Biol. Chem.*, 1992, **267**, 8628.
98. D. Zhang, S. M. Jennings, G. W. Robinson, and C. D. Poulter, *Arch. Biochem. Biophys.*, 1993, **304**, 133.
99. O. Dogbo, A. Laferriere, A. d'Harlingue, and B. Camara, *Proc. Natl. Acad. Sci. USA*, 1988, **85**, 7054.

5.14
Catalysis by Chorismate Mutases

BRUCE GANEM
Cornell University, Ithaca, NY, USA

5.14.1 INTRODUCTION

The enzyme chorismate mutase (CM; EC 5.4.99.5) and its substrate, chorismic acid, play central roles in the shikimate pathway, which is nature's principal biosynthetic route to aromatic compounds in bacteria, fungi, and higher plants.[1] Chorismate stands at a key branchpoint of the shikimate pathway, and is involved in the biosynthesis of numerous mono- and polycyclic aromatic metabolites. These include the aromatic amino acids, folate coenzymes, iron-chelating agents such as enterobactin, and the ubiquinones, menaquinones, and plastoquinones involved in mitochondrial electron transport and oxidative phosphorylation (Scheme 1). At least five different enzymes (chorismate mutase, anthranilate synthase, *p*-aminobenzoate synthase, isochorismate synthase, arylamine synthase) must be carefully regulated in order to maintain proper partitioning of chorismate to various downstream intermediates. Both prokaryotes and eukaryotes have developed complex mechanisms for regulating the metabolic flow from chorismate through the multibranched pathway.

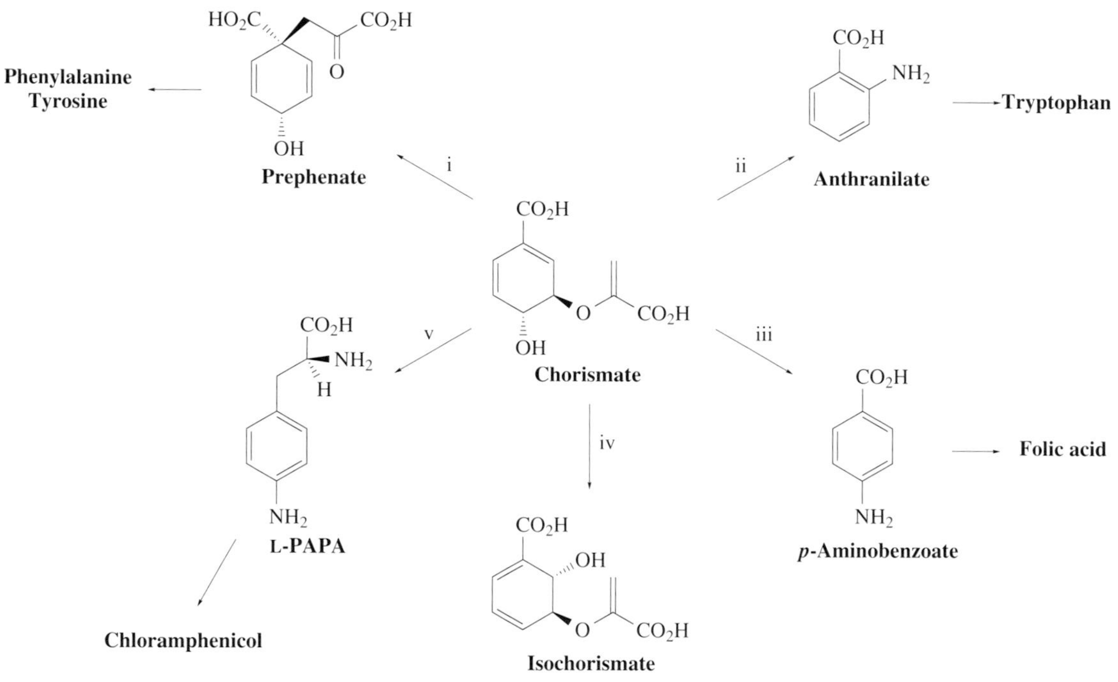

i, chorismate mutase; ii, anthranilate synthase; iii, *p*-aminobenzoate synthase; iv, isochorismate synthase; v, arylamine synthase

Scheme 1

Chorismate mutase catalyzes the rearrangement of chorismic acid to prephenic acid and represents the first committed step in the biosynthesis of phenylalanine (Phe) and tyrosine (Tyr, Scheme 2). The catalyzed reaction, which is formally a Claisen rearrangement, achieves a 10^6 rate acceleration over the uncatalyzed thermal process.[2] Both Phe and Tyr arise in subsequent reactions from prephenate, either by aromatization and transamination, or vice versa. Aromatization of prephenate to phenylpyruvate is catalyzed by prephenate dehydratase, while oxidative aromatization to *p*-hydroxyphenylpyruvate is catalyzed by prephenate dehydrogenase. Transamination of each α-keto acid leads to the respective α-amino acid.[1] Conversely, prephenate may first undergo transamination to arogenate, then conversion by arogenate dehydratase or arogenate dehydrogenase to Phe and

Tyr, respectively. One characteristic of higher plants is their general reliance on the L-arogenate pathway to Phe and Tyr.[3]

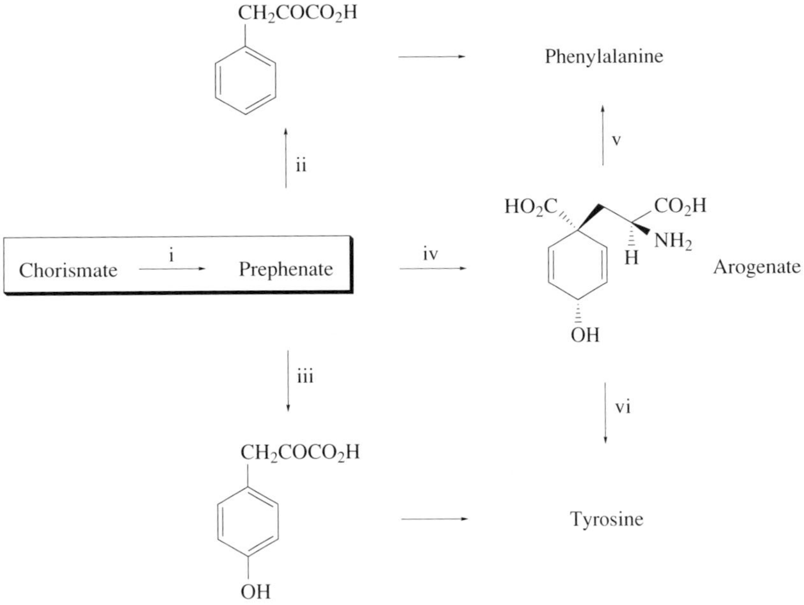

CH$_2$COCO$_2$H

Phenylalanine

ii

v

HO$_2$C   CO$_2$H

NH$_2$

H   Arogenate

Chorismate — i → Prephenate   iv

OH

iii

vi

CH$_2$COCO$_2$H

Tyrosine

OH

i, chorismate mutase; ii, prephenate dehydratase; iii, prephenate dehydrogenase; iv, prephenate aminotransferase; v, arogenate dehydratase; vi, arogenate dehydrogenase

Scheme 2

Prokaryotes, eukaryotes, and higher plants may use either route to phenylalanine and tyrosine. To achieve desired levels of regulation, many organisms have evolved isozymes of CM. Naturally occurring mutases fall into three categories: bifunctional mutases, monofunctional mutases subject to allosteric control, and monofunctional mutases lacking allosteric control.[4] These enzymes vary widely in their sizes and kinetic profiles, and are spread across many phylogenetic boundaries. Such structural diversity notwithstanding, the CM-promoted [3,3]-sigmatropic rearrangement of chorismate to prephenate has attracted much attention from chemists and biochemists because it represents the only example of an enzyme-mediated pericyclic process in primary metabolism.

A comprehensive monograph on the shikimate pathway was published in 1993,[1] and reviews describing recent progress in both the chemistry and enzymology of the pathway have appeared periodically.[4-7] The major objective of this chapter is to review what is known about plant and microbial CMs, with a principal focus on the three-dimensional structure and molecular mechanism of action of these enzymes.

5.14.1.1 Occurrence in Microorganisms and Plants

The first CMs to be characterized were of microbial origin. Cotton and Gibson identified the enzyme in *Escherichia coli* and *Aerobacter aerogenes* (now *Klebsiella pneumoniae*) as part of two bifunctional proteins.[8] In the P-protein, CM is coupled with prephenate dehydratase, which furnishes phenylpyruvate for the pathway to phenylalanine. The P-protein is subject to feedback inhibition by phenylalanine. In the T-protein, CM is combined with prephenate dehydrogenase to generate *p*-hydroxyphenylpyruvate, which in turn undergoes transamination to tyrosine. Tyrosine regulates the T-protein by feedback inhibition. A third bifunctional protein has since been discovered in mutants of the Marburg strain of *Bacillus subtilis* in which CM activity is fused with 3-deoxy-D-arabinoheptulosonate-7-phosphate synthase (DAHP synthase), which catalyzes an early step in the shikimate pathway.[9] As yet, no bifunctional CMs incorporating arogenate dehydratase or arogenate dehydrogenase activity have been reported.

Besides the known bifunctional CMs, which are designated CM-P, CM-T, and CM-DAHP (Table 1), microorganisms also produce two monofunctional mutases, designated CM-F and CM-R. The former is insensitive to Phe and Tyr, whereas the latter is sensitive to feedback inhibition by both amino acids.[10]

Table 1 Distribution and designation of various chorismate mutases.

Abbreviation	Microorganisms	Plants	Features
CM-P	✓		CM-prephenate dehydratase, Pre-inhibited
CM-T	✓		Bifunctional: CM-prephenate dehydrogenase, Tyr-inhibited
CM-DAHP	✓		CM-DAHP synthase
CM-F	✓		Monofunctional CM, unregulated
CM-R	✓		Monofunctional CM, Phe-, Tyr-inhibited
CM-1		✓	Trp-activated, Phe-, Tyr-inhibited
CM-2		✓	Unregulated
CM-3		✓	Trp, dimethoxycinnamate-activated, Phe-, Tyr-, ferulate-inhibited

Most plants possess two CM isozymes, which are typically monomeric proteins with isoelectric points (pI) between 4 and 6.[11] Members of the CM-1 family are usually activated by tryptophan (Trp) and feedback-regulated by Tyr and Phe, whereas the CM-2 family of enzymes is unregulated. A third isozyme (CM-3) has been reported in alfalfa, potato, and several other plant species, although in at least one case, only two CM isozymes were detected upon reinvestigation, where three had previously been reported.[12]

Table 2 summarizes the CMs that have been isolated and characterized in a variety of microorganisms and plants. The distribution of CM-1 and CM-2 in several plants has been examined, and seems to follow a general trend, with regulated (i.e., CM-1) mutase generally located in the chloroplast and unregulated (CM-2) mutase found in the cytosol. This distribution has been interpreted as evidence that two separate systems for aromatic amino acid biosynthesis have evolved in higher plants.[11] The biosynthesis of Phe, Tyr, and Trp may proceed without feedback inhibition in the cytosol, where additional precursors for secondary metabolites may also be in demand. The production of Phe, Tyr, and Trp in the chloroplast would principally meet protein synthesis needs.

5.14.1.2 Regulatory, Genetic, and Evolutionary Relationships

Aromatic amino acid biosynthesis is not regulated by repression of CM at the genetic level.[36] Instead, feedback inhibition of CM exerted by the products of multiple branchpoint pathways plays a critical role in the regulation of aromatic amino acid biosynthesis. In plants, for example, a boost in the production of CM-1 has been observed upon wounding of potato tubers, perhaps in response to new protein synthesis required for tissue repair.[31,37] Wounding of intact green leaves, where CM-2 represents >75% of mutase activity, has not been studied, but would provide further insight into tissue-specific expression of CM genes. It has been suggested that if promoters for CM-1 and CM-2 were chloroplast- and cytosol-specific, respectively, such promoters might also be useful in the tissue-specific expression of other proteins in plants.[11] The cloning of additional CM genes may ultimately lend further support to such a hypothesis.

At least one enteric bacterium, *Seratia rubidaea*, also possesses both regulated (CM-P and CM-T) and unregulated (CM-F) chorismate mutases.[10] Using this organism as a paradigm for the evolution of mutase structure, it has been proposed that CM-F, with its small size and lack of allosteric control, might most likely represent the primitive ancestral mutase. A combination of gene replication and gene fusion events could have led to additional catalytic and regulatory domains, as are present in CM-P and CM-T.

To date, only a few CM genes have been cloned and sequenced. The mutase domains of CM-P and CM-T in *E. coli* reside near the N-terminal region and are partially related, with 22 of the first 56 residues identical.[38] Such similarity likely reflects a common evolutionary origin. However, unlike other enzymes of aromatic acid biosynthesis, relatively little sequence similarity has been observed between mutases of different families. Yeast CM-R, which is the product of the *ARO7* gene in

Table 2 Chorismate mutase isotypes in microorganisms and plants.

Source	Type	M_r (subunit, kDa)	Structure	Ref.
Microorganisms				
Acinobacter calcoaceticus	CM-F	23		13
	CM-P	45	Dimer	
Aerobacter aerogenes	CM-T	76	Dimer	14
Azotobacter paspali	CM-P			
Bacillus subtilis	CM-F	14.5	Trimer	15
Candida maltosa	CM-F	63		
Escherichia coli	CM-P	42	Dimer	16,17
Erwinia herbicola	CM-F	34	Dimer	19
Neisseria gonorrhoeae	CM-P			20
Pichia guilliermondii	CM-F	63		21
Salmonella typhimurium	CM-P			22
	CM-T			
Serpens flexibilis	CM-P			23
Serratia marescens	CM-P			22
	CM-T			
Serratia rubidaea	CM-F	20	Dimer	10
	CM-P			24
	CM-T			24
Streptomyces aureofaciens	CM-F	63		25
Plants				
Alfalfa	CM-1	46		26
	CM-2	58		
	CM-3	69		
Mung beans	CM-1	50		27
	CM-2	36		
Oak	CM-1	45		28
	CM-2	n.d.		
Poppy	CM-1	42	Dimer	29
	CM-2	40	Dimer	
Potato	CM-1	55		30
	CM-2	52		31
Rue	CM-1	56		32
	CM-2	45		
Sorghum	CM-1	56		33
	CM-2	48		
Spinach	CM-1	59		34
	CM-2	48		
Tobacco	CM-1	52		12,35
	CM-2	65		

Saccharomyces cerevisiae, displays no significant homology with the N-terminal domains of the *E. coli* CM-P and CM-T proteins.[39,40] The *aroH* gene encoding CM-F in *B. subtilis* has been cloned and sequenced.[15] While minor similarities to the N-termini of *E. coli* CM-P and CM-T are observed, no significant resemblance is noted with yeast CM-R. Likewise, AroH displays no obvious similarity to the *AroQ* gene encoding CM-F in *Erwinia herbicola*.[19] The cDNA for *Arabidopsis thaliana* CM has been expressed in yeast, and the experimentally determined amino acid sequence reveals a 41% homology with yeast CM, but little similarity in the N-terminal region. No resemblance is found to any known bacterial mutases.[41]

It should be noted that the biosynthetic pathway to chloramphenicol may involve another distinct, mutase-like activity in the conversion of chorismate to *p*-amino-L-phenylalanine (L-PAPA, Scheme 1).[42] A plausible mechanism for this transformation was first proposed by Dardenne *et al.* (Scheme 3), involving amination of chorismate to 4-amino-4-deoxychorismate, followed by Claisen rearrangement to 4-amino-4-deoxyprephenate and subsequent oxidative aromatization.[43] The over-all process is catalyzed by arylamine synthase, and the activity of the crude enzyme preparation has been separated into three fractions.[44,45] While purification of the various fractions to homogeneity remains elusive, synthetic ($\pm$)-4-amino-deoxychorismate and ($\pm$)-4-amino-4-deoxyprephenate were incorporated by arylamine synthase into L-PAPA, thus lending credence to the possibility of an aminochorismate mutase.[46]

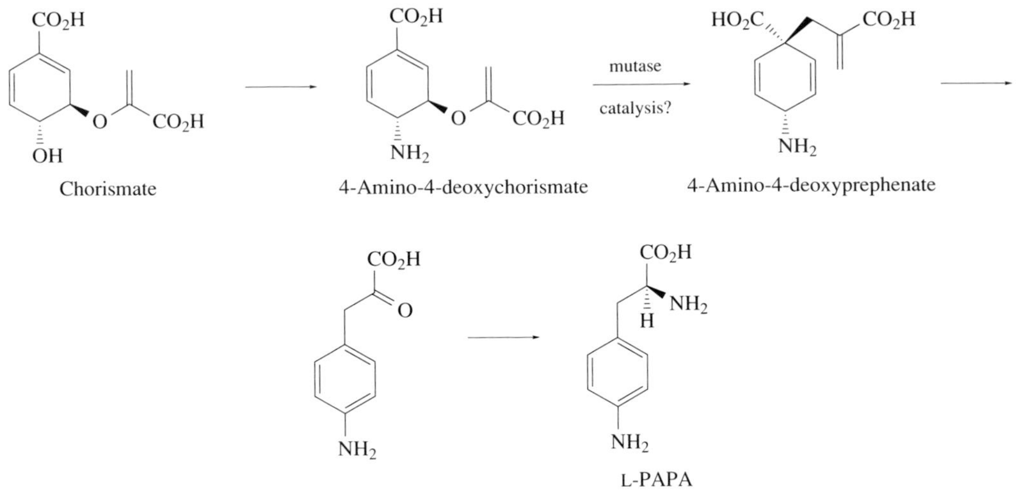

Chorismate   4-Amino-4-deoxychorismate   4-Amino-4-deoxyprephenate

L-PAPA

Scheme 3

5.14.1.3 The Uncatalyzed Rearrangement of Chorismate to Prephenate

In the absence of enzymes, the chemistry of chorismic acid is dominated by two simple thermal processes: aromatization to *p*-hydroxybenzoic acid by elimination of pyruvic acid and Claisen rearrangement. The rearrangement of chorismate to prephenate is a one-substrate, one-product process. It is one of very few chemical transformations where the enzymatic process may be compared directly with its unimolecular solution counterpart. Most enzyme-catalyzed reactions proceed at immeasurably slow rates in the absence of catalyst, so that direct mechanistic comparisons are difficult, if not impossible. However, the uncatalyzed rearrangement of chorismate to prephenate in water is a relatively facile process ($t_{1/2} = 90$ min at $50\,^{\circ}$C).[47] By way of comparison, aqueous chorismate rearranges approximately 4000 times faster than allyl vinyl ether (AVE) in the gas phase, and for this reason the inherently fast nonenzymic process has itself attracted a significant level of attention.[48,49] Besides being of intrinsic interest, the factors responsible for the rapid nonenzymic rearrangement of chorismate might well provide chemical clues to the action of chorismate mutases. It seems reasonable that understanding the intrinsically fast rearrangement of chorismate might shed light on, or provide clues about, the nature and role of substrate and transition-state binding interactions in the mutase process.

The nonenzymatic reaction proceeds by a chair-like transition structure in which the rearranging conformation adopts the pseudodiaxial arrangement of the hydroxy and enolpyruvyl groups (Equation (1)).[50] That conformation is readily accessible since 10–20% of chorismate exists as the pseudo-diaxial form in dynamic equilibrium with the pseudodiequatorial form in aqueous solution at room temperature.[47] Secondary tritium isotope effects on the uncatalyzed rearrangement are evident at C-5 ($k_H/k_T = 1.15$), but not at C-9, indicating an asymmetric transition state in which the breaking C—O bond has undergone substantial rupturing in advance of any new C—C bond formation.[51]

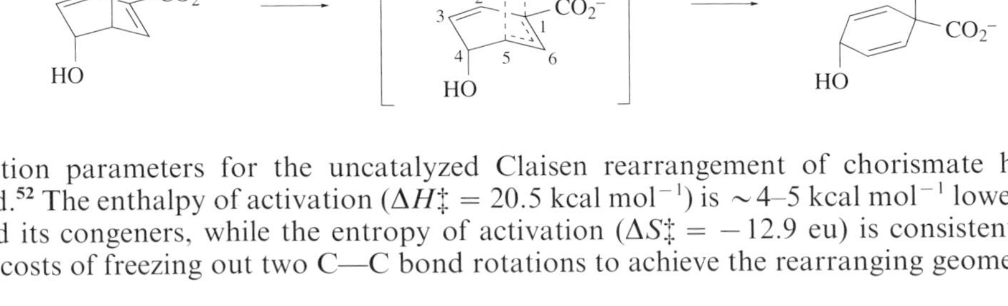

$$\tag{1}$$

Activation parameters for the uncatalyzed Claisen rearrangement of chorismate have been measured.[52] The enthalpy of activation ($\Delta H^{\ddagger} = 20.5$ kcal mol^{-1}) is $\sim$4–5 kcal mol^{-1} lower than for AVE and its congeners, while the entropy of activation ($\Delta S^{\ddagger} = -12.9$ eu) is consistent with the entropic costs of freezing out two C—C bond rotations to achieve the rearranging geometry.

5.14.1.4 Relationship with Other Claisen Rearrangements

The Claisen rearrangement of allyl vinyl ethers, which was first described in 1912, is the earliest recorded example of a [3,3]-sigmatropic reaction.[53] In its simplest form, the rearrangement is exemplified by the transformation of AVE to 4-pentenal.[54-56] The Claisen rearrangement is a classic example of a pericyclic process in which, by definition, bond breaking and bond making occur through a cyclic array of interacting orbitals. The transition state for such a process may exhibit either diradical or diyl character (Scheme 3).

Mechanistic and sterochemical information is available for a number of Claisen rearrangements. In the absence of overriding steric factors, chair or chair-like transition states are usually preferred, by up to 6 kcal mol^{-1}.[57] However, some geometrically constrained Claisen rearrangements may proceed either partially or completely via boat transition structures. Both the Claisen and Cope rearrangements are accelerated under high pressure.[58,59] The activation volumes for AVE (-18 cm^3 mol^{-1}) and for allyl phenyl ether (-7.7 cm^3 mol^{-1}) are in accordance with processes that display tightly bonded cyclic transition states. The reaction displays first-order kinetics,[60-62] and no crossover products are observed.[63]

Pericyclic processes are thought to proceed by concerted reaction mechanisms, with simultaneous, but not necessarily synchronous, bond making and bond breaking.[64] While the bonds undergoing chemical change might not necessarily be broken or formed to the same extent at every stage along the reaction coordinate, the changes must occur together in a single kinetic step, and there can be no discrete intermediates. In the case of AVE, secondary deuterium kinetic isotope effects reported by Gajewski and Conrad indicate that C—O bond breaking proceeds somewhat ahead of C—C bond making.[65,66]

Pericyclic reactions present the greatest challenge to mechanistic organic chemists, whose interest in chorismate mutases has been fueled by the suggestion that a better understanding of the nature of mutase catalysis may provide important new insights into the nature of pericyclic processes. In addition, a better understanding of catalysis gained from chorismate mutase could be of widespread importance in synthetic organic chemistry. With their highly ordered transition states, the Claisen and Cope rearrangements are especially useful in creating and controlling new stereocenters. Because the Claisen rearrangement also generates a new carbonyl group, it has become one of the most widely used methods of stereoselective C—C bond formation in synthetic chemistry. Several elegant transformations such as the Johnson orthoester,[67] Eschenmoser amide acetal,[68,69] and Ireland ester enolate[70,71] rearrangements rely on the [3,3]-sigmatropic rearrangement of substituted allyl vinyl ethers and variants thereof.

5.14.1.5 Potential Commercial and Therapeutic Interest

With a detailed knowledge of the structure and function of CM, it should be possible to develop new inhibitors that could provide important leads for new antibiotics. Indeed, as a general strategy, the design of inhibitors of key enzyme-catalyzed reactions in the shikimate pathway in microorganisms and plants has already proven successful in developing powerful new herbicidal and antibacterial agents. Glyphosate (*N*-phosphonomethylglycine) is the herbicidal component of Roundup, a broad-spectrum, postemergence weed killer.[72] With its good translocation in plants, rapid inactivation by soil microorganisms, and low toxicity to other organisms, Roundup became the first broad-spectrum herbicide to achieve $1 billion sales. The sulfa drugs sulfanilamide and sulfamethoxazole (which inhibit dihydropteroate synthase) can be used in combination with trimethoprim (which blocks the reduction of dihydro to tetrahydrofolate) to create a dual metabolic blockade of folic acid biosynthesis that is widely used in the treatment of bacterial infections (Septrin).[73] Given that chorismate occupies a strategic branchpoint position in the shikimate pathway, the development of CM inhibitors might well lead to potent therapeutic agents. Moreover, mutase inhibitors represent a form of treatment against which no bacterial resistance mechanisms have yet emerged.

5.14.2 FACTORS AFFECTING RATES OF CLAISEN REARRANGEMENTS

In contrast to most polar and radical reactions, pericyclic processes are relatively insensitive to solvent,[74,75] and are usually not catalyzed by acids, bases, or other agents.[76] However, the aliphatic Claisen rearrangement has long been known to exhibit sensitivity to solvents and catalysts. While the effect of catalysts on reaction mechanism remains poorly understood,[77] it should be noted that no evidence has been presented which suggests that accelerated Claisen rearrangements represent anything other than true pericyclic reactions.[78]

5.14.2.1 Catalysis

The first catalyzed Claisen rearrangement was discovered in 1912 by Ludwig Claisen, who observed that ammonium chloride accelerated the rearrangement of the allyl ether of ethyl aceto-acetate (Equation (2)).[79] Several additional aliphatic Claisen rearrangements have since been shown to undergo catalysis by solid NH_4Cl.[80]

(2)

A number of Brønsted and Lewis acids, and even weak acids such as silica gel or Celite,[81] have also been reported to accelerate Claisen rearrangements. Developments in this area prior to 1984 have been reviewed.[77] The most promising synthetic catalysts for regio- and stereospecific rearrangements now appear to be trivalent organoaluminum reagents.[82] Yamamoto and co-workers[83] developed a chiral aluminum naphthoxide (Scheme 4) displaying a cup-shaped, C_3 symmetric binding cleft whose three *p*-fluorophenyl groups effectively shield one face of the aluminum center. The chiral catalyst is capable of rearranging substituted allyl vinyl ethers to chiral aldehydes at $-78\,^{\circ}C$ with enantiomeric excesses up to 92%. Like other aluminum bis- and trisphenoxides, the catalyst in Scheme 4 is thought to function by coordination at the vinyl ether oxygen as shown. As a result, "charge-accelerated rearrangement" occurs as the allylically positioned charge become more delocalized in the rearrangement transition structure.[84-86]

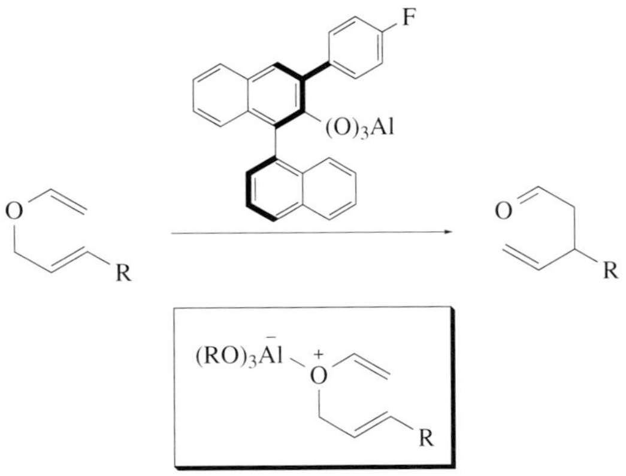

Scheme 4

A soluble urea has also been reported to catalyze the Claisen rearrangement of 6-methoxy-AVE (Scheme 5).[87] At $80\,^{\circ}C$, a 22-fold increase in the rate of rearrangement of 6-methoxy-AVE was reported, and the rate enhancement at $50\,^{\circ}C$ was estimated to be 34-fold. The rate acceleration was attributed to a bis-hydrogen-bonded transition state as shown (Scheme 5). Interestingly, the more acidic thiourea (not shown) had a weaker accelerating effect, further suggesting that hydrogen bonding, not acidity, was crucial for acceleration.

5.14.2.2 Solvent Effects

In an early study of medium effects on Claisen rearrangements, Kincaid and Tarbell found that the rate of conversion of allyl *p*-tolyl ether (ATE) to 2-allyl-4-methylphenol (Equation (3)) increased with time as concentrations of product increased.[88] Such autocatalysis is not observed in aliphatic

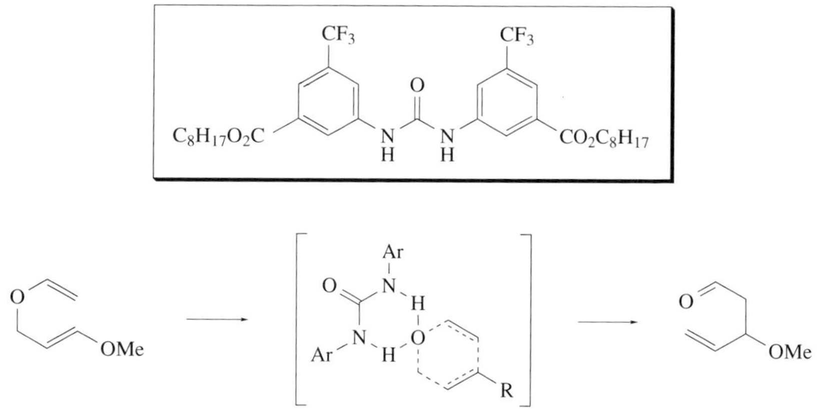

Scheme 5

Claisen systems because the enol tautomer of the initially formed carbonyl compound is a minor component at equilibrium. To test for simple acid catalysis, both acetic and chloroacetic acids were added; however, no rate enhancements were detected.

$$(3)$$

ATE   2-Allyl-4-methylphenol

Subsequent studies by White *et al.*[89] and by Goering and Jacobson[90] confirmed a significant solvent effect on the rate of rearrangement of ATE in various media, with the rearrangement of ATE ~30 times faster in phenol than in decalin. The more rapid rates in hydroxylic solvents led Goering and Jacobson to suggest, for the first time, a correlation with the solvent's hydrogen-bonding ability. A subsequent study of the rearrangement of ATE at 170 °C both in the gas phase and in 17 solvents of different polarities by White and Wolfarth revealed a substantial solvent effect, with rates ~300-fold faster in hydrogen-bonding solvents.[91]

Two groups have compared the effect of solvent on the aliphatic Claisen rearrangement. In one study, a twofold rate enhancement was noted for the rearrangement of AVE in methanol compared with benzene, whereas rate enhancements of 18–68-fold were measured for a series of 4-, 5-, and 6-alkoxy-AVE derivatives in the same two solvents.[92] The authors suggested that hydrogen bonding specifically to the enol ether oxygen of 4- and 6-alkoxy substrates was responsible for a major part of the increases observed in alcohol solvents. In an accompanying report, a nearly 60-fold rate enhancement for the AVE rearrangement was detected on going from dibutyl ether to 2:1 methanol–water.[48] Both studies provided experimental evidence for a dipolar or dissociative mechanism in which C—O heterolysis resulted in significant resonance-stabilized ion pair character in the rearrangement transition state.

Gajewski attempted to determine the extent to which polar solvent effects were composed of such factors as dielectric effects, solvent cohesive energy densities, hydrogen bond-donating effects, and hydrogen bond-accepting effects.[93] An equation was derived which, for the aliphatic Claisen rearrangement, revealed a specific correlation between rate and hydrogen bond-donating ability of the solvent. As further evidence against an ionic transition state, secondary deuterium KIEs measured at C-4 and C-6 of AVE both in *m*-xylene and in CD_3OD–D_2O mixtures indicated that bond breaking was no more extensive in aqueous media than in hydrocarbon solvents.[94]

A Monte Carlo simulation of the effect of hydration on Claisen rearrangements shed further light on the importance of hydrogen bonding. Severance and Jorgensen calculated that the transition state for rearrangement of AVE in water was 3.85 ± 0.16 kcal mol^{-1} better hydrated than the reactant.[95] Consistent with this model, the carbonyl oxygen of the product would be expected to form two hydrogen bonds. The authors suggested that in the transition state, two hydrogen bonds were donated by water molecules to the vinyl ether oxygen, with average strengths of -4.6 kcal mol^{-1} (Equation (4)). By contrast, AVE displayed only a single hydrogen bond with a strength of -3.4 kcal mol^{-1}. Such enhanced hydration could account for a 600–700-fold rate enhancement over the Claisen rearrangement of AVE in the gas phase. Moreover, the computed acceleration was not attributable to zwitterionic intermediates, tight ion pairs, or other C—O bond dissociative processes. A stimulation modeled at the *ab initio* level with two explicit water molecules reached essentially similar conclusions.[96]

$$\tag{4}$$

Besides stabilizing the transition state, hydrogen bonding to the reactant disrupts the well-known n–π conjugation in vinyl ethers,[97] raising the ground-state energy and further reducing the activation enthalpy for rearrangement.

5.14.2.3   Substituent Effects

In addition to solvent effects, aromatic Claisen rearrangements exhibit *para*-substituent effects. Electron-donating groups (OMe, NH_2) are moderately accelerating, whereas electron-withdrawing groups (NO_2, CN) are mildly retarding. These rate effects are complicated because *para* substituents influence the acidity of the product phenols, and acid-catalyzed enolization of the intermediate cyclohexadienone has not been ruled out in the overall rate-determining step of the rearrangement. These same substituent effects on acidity may also complicate the interpretation of solvent effects.[76] White and Wolfarth[91] found a reasonably good correlation between rearrangement rate and solvent polarity, as measured by Kosower's Z factors.[98]

Substituent effects on the aliphatic Claisen rearrangement have also been reported.[48,92] In the case of alkoxy-substituted allyl vinyl ethers, both 4- and 6-alkoxy derivatives rearranged between 9.5 and 159 times faster than the parent AVE.[92] The rearrangement of 5-methoxy-AVE occurred 40 times slower than that of AVE itself, and no appreciable solvent effect on its rate could be measured. A synergistic interaction of donor and acceptor substituents on the AVE framework was evident in the combined influence of a 4-alkoxy group with a 1-cyano or 1-carboethoxy group, which led to a further rate enhancement.

The thermal rearrangement of chorismic acid, dimethyl chorismate, and five additional chorismate analogues was monitored in 2:1 CD_3OD–D_2O at 75 °C to obtain the relative rates of rearrangement.[48] The observed substituent effects were consistent with a dissociative process, although the magnitude of the rate enhancements was not consistent with zwitterion or tight ion-pair formation. For example, the rate constant for push–pull ring opening of a substituted cyclopropane increased 32 000-fold upon changing the solvent from benzene to DMF.[99] By contrast, the largest solvent-induced rate enhancement in aliphatic Claisen rearrangements was a 200-fold effect on going from cyclohexane to water.[100]

Whatever its structure, the Claisen rearrangement transition state affected by polar hydroxylic solvents accumulated substantially less polar character than well-known zwitterions or tight ion pairs.[100] Of special interest was the fact that when measured in 2:1 methanol–water at 75 °C, the rearrangement of dimethyl chorismate was only 18 times faster than that of AVE.

5.14.3   STRUCTURAL REQUIREMENTS FOR CATALYSIS

5.14.3.1   Substrate Structural Parameters

Understanding what structural features of chorismate are required by the mutase may shed light on the mechanism of catalysis. In 1976, Haslam and co-workers reported studies with two chorismate derivatives shown in Scheme 6.[101] The chorismate monoester was not a substrate for chorismate mutase–prephenate dehydrogenase from *Klebsiella pneumoniae*, and it did not inhibit the enzymatic rearrangement of chorismate. The rearrangement of 5,6-dihydrochorismic acid was not accelerated by the mutase, but the compound was found to be a modest inhibitor. In a subsequent reinvestigation by Berchtold and co-workers, 5,6-dihydrochorismate proved to be a viable substrate for catalysis by the *E. coli* bifunctional enzyme chorismate mutase–prephenate dehydrogenase.[102] Since its un-catalyzed rearrangement is an exceptionally slow process, special experimental conditions were employed using a 20-fold increase in enzyme. Berchtold and co-workers studied both monoesters of chorismate, in addition to chorismate methyl ether and analogues lacking the ring carboxyl and ring hydroxyl groups. They concluded that the only functional groups on the allyl vinyl ether framework of chorismate that are required for active site binding and catalysis are the two carboxylic acid groups (see Scheme 6).[103] Whether cyclic substrates are required for catalysis, or whether other mutases besides the bifunctional T-protein have similar structural requirements, are questions that remain to be investigated.

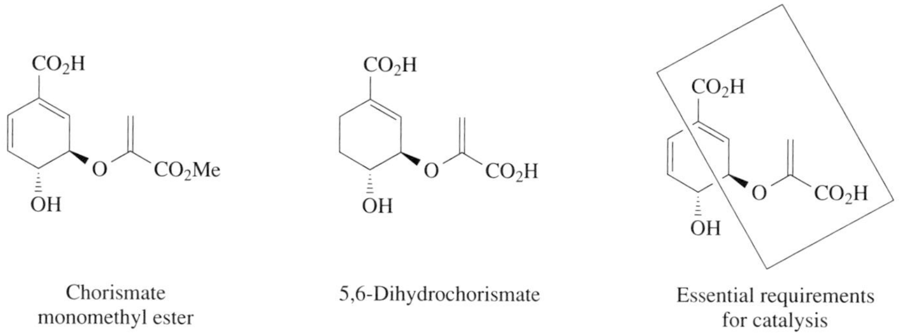

| Chorismate monomethyl ester | 5,6-Dihydrochorismate | Essential requirements for catalysis |

Scheme 6

5.14.3.2 Enzyme Structural Parameters

To identify key active site residues in chorismate mutase, a series of chemical modifications were carried out on both the *E. coli* P-protein[104–107] and T-protein.[108–112] In the case of the P-protein, which exists as a symmetric dimer, modification of a single lysine residue per subunit with 2,4,6-trinitrobenzene sulfonate completely inactivated the mutase with only a 20% loss of dehydratase activity. Further loss of the dehydratase activity occurred with modification of a second lysine residue.

Reaction of one cysteine per subunit with *N*-ethylmaleimide or with 5,5′-dithiobis(2-nitrobenzoic acid) (Nbs$_2$) resulted in complete inactivation of the dehydratase with only a 5% loss of mutase activity. Circular dichroism studies indicated that the Nbs$_2$ reaction caused significant changes in the secondary structure of the enzyme, resulting in longer and less distorted helices than in the native enzyme. Further modification of a second and third cysteine residue abolished both mutase activity and feedback inhibition by phenylalanine. However, when all four sulfhydryl groups of the P-protein subunit were modified with potassium tetrathionate, only 15% of the mutase activity was lost, suggesting that the cysteines were sensitive to sterically bulky modifications, but played no part in catalysis.

Exposure of the P-protein to tetranitromethane modified two tyrosine residues per subunit, and resulted in complete loss of dehydratase activity with only a 30% loss of mutase activity. Modification of the P-protein's tryptophan residues with dimethyl(2-hydroxy-5-nitrobenzyl)sulfonium bromide partially diminished both the mutase and dehydratase activities and desensitized both activities to phenylalanine.

Neither D-phenylalanine nor L-tryptophan inhibited the mutase and dehydratase activities, whereas varying degrees of inhibition were observed with *o*-, *m*-, and *p*-monosubstituted chloro-, fluoro-, and hydroxyphenylalanines.[107] Studies with other phenylalanine analogues revealed an absolute requirement for an unmodified α-carboxylic acid group.[107] Overall, results with the P-protein suggested that the mutase and dehydratase reactions were catalyzed at separate sites, and that feedback inhibition by phenylalanine occurred at a third, topologically distinct region of the polypeptide backbone.

Several studies on the dimeric T-protein investigated the catalytic role of sulfhydryl groups. Of the four cysteines identified in the T-protein subunit, one was found to be particularly reactive and was essential for both catalytic activities. Alkylation of that residue with iodoacetamide or Nbs$_2$ resulted in parallel loss of the mutase and dehydrogenase activities,[111] suggesting closely aligned, if not physically overlapping, catalytic sites.

Thus, the major difference between the P- and T-proteins occurred in the spatial relationship of the mutase site to its partner enzyme. The fact that P- and T-protein amino acid sequences showed similarity only in their N-termini indicated the likely location of the mutase active site. That observation led to the recombinant expression of a 113-residue N-terminal fragment of the *E. coli* P-protein as a fully active, monofunctional chorismate mutase.[113]

5.14.4 KINETIC PROPERTIES OF MUTASES

Like its nonenzymatic counterpart, the mutase-catalyzed conversion of chorismic acid to prephenic acid also proceeds by a chair conformation in the rearranging transition state.[114] Given that the rearrangement is an intramolecular process, stabilization of the transition state by the enzyme is likely to be the principal mechanism of catalysis. Kinetic measurements on various enzymes have been used to suggest how CM might achieve the experimentally observed rate enhancement.

5.14.4.1 Activation Parameters

Comparison of the activation for the uncatalyzed process with those for mutase-catalyzed rearrangements might be expected to shed light on the nature of enzyme catalysis. Accordingly, activation parameters for a variety of natural and synthetic mutase catalysts have been measured (Table 3).[115,116] The negative activation entropy for the uncatalyzed process (-12.7 eu) is consistent with the cost of restricting the rotation of two C—O bonds in the chair transition state. The near-zero entropy of activation for the process catalyzed by *K. pneumoniae*, *S. aureofaciens*, and *E. coli* enzymes suggests that the mutases fix chorismate in the chair conformation within the enzyme–substrate complex, and thus lower the entropy barrier. While this notion has generally been thought to play an important role in mutase catalysis, the activation entropy measured for the monofunctional mutase from *B. subtilis* is unfavorable, and comparable to that for the uncatalyzed reaction.[116] On the basis of that finding, it has been argued that while the enzyme may still exert much conformational control over the flexible substrate, the classical model of an entropy trap[117] may not apply to all CMs.

Table 3 Activation parameters for the rearrangement of chorismate to prephenate.

Catalyst	$\Delta H^{\ddagger}_{\ddagger}$ (kcal mol^{-1})	$\Delta S^{\ddagger}_{\ddagger}$ (eu)
Uncatalyzed	20.5	-12.9
S. aureofaciens	14.5	-1.6
K. pneumoniae	15.9	-1.1
B. subtilis	12.7	-9.1
E. coli mutase	16.3	1.5
11F1-2E11	18.3	-1.2
IF7	15.0	-22

Source: Sogo *et al.*[114] and Kast *et al.*[115]

The uncatalyzed activation enthalpy for the rearrangement of chorismate (20.5 kcal mol^{-1}) is ~ 5 kcal mol^{-1} lower than for the uncatalyzed rearrangement of allyl vinyl ether in the gas phase. That difference has been attributed largely to the effect of solvent hydrogen bonding in water, and in lesser part to substituent effects in the chorismate molecule, as was discussed in Sections 5.18.2.2 and 5.18.2.3. Activation enthalpies for the *K. pneumoniae*, *S. aureofaciens*, and *E. coli* CMs drop by another 5 kcal mol^{-1}, and an even greater reduction is evident in the case of the *B. subtilis* enzyme. Few studies have addressed the significance of enthalpic contributions to mutase catalysis, although further stabilization of the transition state through a variety of protein–substrate interactions is possible. Alternatively, the substrate-binding energy of the protein might be used to induce strain in the substrate.

Two catalytic antibodies have been reported with mutase-like activity. Antibody 11F1-2E11 accelerates the rearrangement 10^4-fold stereoselectively over the uncatalyzed process.[118] The activation entropy for this antibody is near zero, supporting the notion that conformational restriction is important in the catalytic mechanism. The modest reduction in activation enthalpy for 11F1-

2E11 (18.3 kcal mol^{-1}) compared with the uncatalyzed process has been attributed to antibody-induced strain or electrostatic stabilization of the transition state.[119] A second stereoselective antibody, 1F7, achieves a rate acceleration of 250-fold over the uncatalyzed reaction.[120,121] In the case of 1F7, however, the activation parameters indicate that the antibody functions by reducing the activation enthalpy for the reaction with a counterproductive drop in activation entropy.

5.14.4.2 Isotope Effects

Whereas secondary tritium isotope effects are evident in the uncatalyzed rearrangement of chorismate, the mutase-catalyzed rearrangement displays no comparable isotope effects.[51] The absence of an isotope effect at either bond-breaking or bond-making centers does not preclude similar pathways for the catalyzed and uncatalyzed rearrangements, since mechanistically relevant isotope effects may be masked by a prior, rate-limiting transition state involving the protein. In addition, a small ($k_H/k_T = 0.96$) inverse secondary isotope effect is observed on tritiation at C-4.[122] That effect has been rationalized in terms of the effect of tritium on substrate conformational equilibrium (tritium prefers to be equatorial, thus $k_H k_T = 0.95$).

The nonenzymic reaction is unaffected by D$_2$O; however, a D$_2$O solvent isotope effect is observed on the enzymic reaction, giving $^{D_2O}(K_m) = 1.06 \pm 0.31$, $^{D_2O}(V_{max}) = 2.23 \pm 0.19$, and $^{D_2O}(V_{max}/K_m) = 2.11 \pm 0.26$.[122] While the authors urge caution about interpreting such effects, the substantial change in k_{cat} but not in K_M suggests some protonic motion (i.e., general acid or general base catalysis) in the rate-limiting transition state.

5.14.4.3 Inhibitor Studies

The functional groups in chorismic acid offer several possibilities for noncovalent binding in the CM E–S complex. For example, the two charged carboxylate groups are capable of strong electrostatic interactions with protonated active site residues, and are known to be required for catalytic turnover by the *E. coli* T-protein.[103] Suitable partners may also form hydrogen bonds with chorismate's carboxyl and hydroxyl groups.

A family of 1-substituted adamantane derivatives were investigated as potential mimics of the chair transition state in the *E. coli* T-protein mutase-catalyzed rearrangement.[123] The potency of inhibition declined as a function of the substituent as follows: $AdPO_3^{2-} \gg AdP(OCH_3)O_2^- > AdCH_2CO_2^- > AdSO_2^- > AdSO_3^-$. Interestingly, 1-adamantanylacetic acid exhibited cooperative effects: at low concentrations it elevated both the mutase and dehydrogenase activities in the T-protein, whereas at higher concentration it functioned as an inhibitor.[124] Kinetic studies with several additional mono- and bicyclic aliphatic and aromatic diacids (Figure 1) indicated the importance of the vinyl ether oxygen,[125,126] and further suggested that hydrophobic forces may contribute to binding.[2] The possibility of π-electron interactions with chorismate's diene system has also been noted.[2] The most potent inhibitor currently known for any chorismate mutase is the *endo*-oxabicyclic diacid synthesized by Bartlett *et al.*[125] In assays using the *E. coli* T-protein, this compound binds competitively with substrate, and exhibits an inhibition constant (K_i) of 0.12 µmol L^{-1}, under conditions where chorismate shows a K_M value of 34 µmol L^{-1}.

5.14.4.4 NMR and IR Studies

Attempts using ^{13}C-NMR spectroscopy to observe the E–S complex and other ligand interactions between the monofunctional CM from *B. subtilis* have been reported with specifically ^{13}C-labelled chorismate, specifically labeled prephenate, and [U-^{13}C]-labeled prephenate.[127,128] All the enzyme-bound ^{13}C resonances for [U-^{13}C]prephenate were assigned and, where possible, vicinal C–C coupling constants were quantified. Neither free nor bound chorismate could be detected. Enzyme-bound prephenate exhibited different chemical shifts relative to free prephenate, and difference spectra revealed significant perturbations between C-5 and C-6 in bound prephenate, perhaps caused by changes in electron density and/or molecular geometry at those carbons. However, studies of pH and solvent dependence using various model compounds suggested that the observed chemical shift changes in enzyme-bound prephenate could not be rationalized solely on the basis of changing pK_a values of the carboxylic acid groups or of hydrophobic solvation at the active site. Based on these

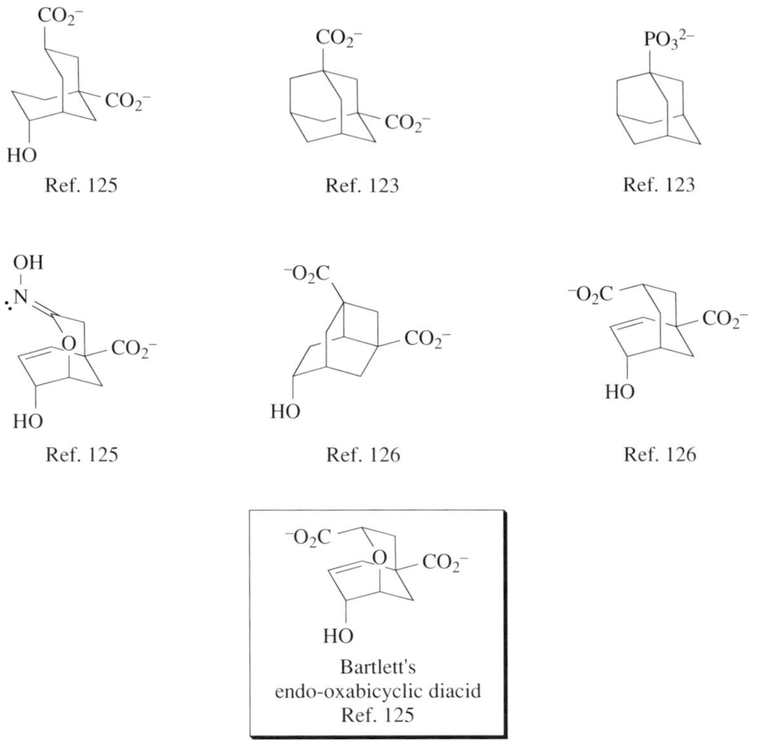

Figure 1   Representative inhibitors of chorismate mutase.

observations, NMR studies did not provide any evidence for a dissociative mechanism involving discrete intermediates. Moreover, no spectroscopic evidence indicated that the enzyme-catalyzed rearrangement of chorismate to prephenate involved anything other than a pericyclic process.

The ketonic carbonyl stretching frequency of isotopically labeled prephenate bound to *B. subtilis* CM has been observed at 1714 cm^{-1} by FTIR spectroscopy.[129] The carbonyl stretching frequency did not change significantly upon binding. However, marked differences were detected in the amide I$'$ vibration band in resolution-enhanced FTIR spectra of the unliganded mutase when compared with the CM–prephenate complex. Taken together, these data suggest that electrophilic catalysis is not involved in the CM-promoted rearrangement of chorismate, but that structural alterations may occur in the protein upon binding prephenate. This finding does not imply that no hydrogen bonds are formed by the protein to the carbonyl oxygen, but suggests that interactions on the enzyme are roughly equivalent to those in water. Analysis of the FTIR data further indicates that *B. subtilis* CM, although not catalyzing a dissociative reaction, likely promotes an asynchronous pericyclic rearrangement.

5.14.4.5   pH Dependence

Kinetic parameters of both the mutase and dehydrogenase reactions in the *E. coli* T-protein have been studied as a function of pH, with the aim of elucidating the role played by ionic amino acid residues in binding and catalysis.[130] For the mutase reaction, the variation with pH of the apparent first-order rate constant for the interaction of enzyme and chorismate (V/K) indicates the presence of three ionizing residues at the active site. Two of these residues, both with p$K \sim 7$, must be protonated, while the third, p$K \sim 6.3$, must be unprotonated. Since the maximum velocity of the mutase reaction is essentially independent of pH, all three ionic groups apparently participate in binding of chorismate to the enzyme. The variation of K_i with pH for Bartlett *et al.*'s mutase inhibitor (Figure 1) shows the same three ionizing groups as observed in the $V/K_{\text{chorismate}}$ profile, which confirms this conclusion.

5.14.4.6   Mechanistic Conclusions from Kinetic Properties

On the basis of the solvent deuterium isotope effect, tritium effects at C-4, C-5, and C-9 results from the nonenzymic reaction, and studies with substrate analogues, Guilford *et al.* presented four ways in which an enzyme might conceivably achieve the experimentally observed 10^6-fold rate enhancements of a typical mutase.[122] It was possible to exclude those pathways (i) that involve conformational isomerization of bound pseudodiequatorial substrate, (ii) that exploit a C-4 hydroxy group to generate a carbocation at that position, and (iii) that involve anchimeric assistance by the C-4 hydroxy group via an oxirinium ion intermediate. The authors favor a pathway in which attack of an enzymic nucleophile with rate-limiting heterolytic cleavage of the ether bond in chorismate gives an intermediate that collapses to product in an S_N2' process (Scheme 7).

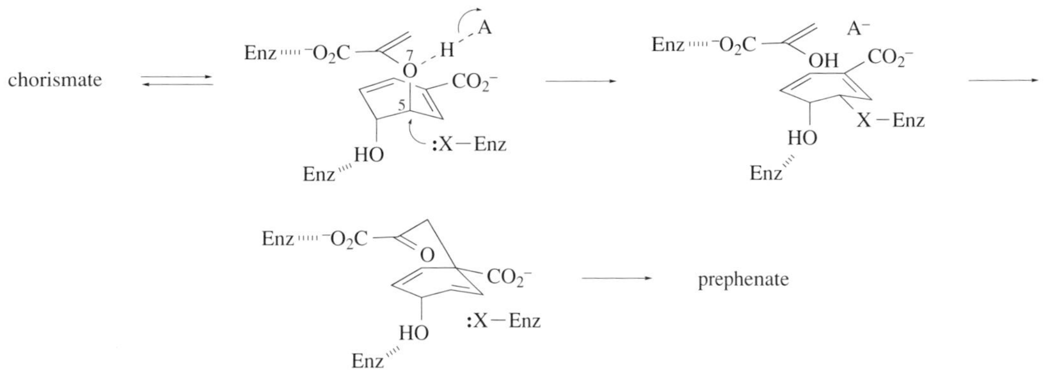

Scheme 7

The pH dependence studies of Turnbull *et al.*[130] led to a different mechanism for chorismate mutase (Scheme 8). Concluding that no ionizable group is required for the conversion of chorismate to prephenate, they proposed that one of the two protonated active site groups forms an ion pair with chorismate's ring carboxylate, while the other protonated group forms a hydrogen bond to chorismate's ether oxygen, ultimately promoting protonation at that site. The third active site residue, which must be ionized for binding, forms a hydrogen bond to chorismate's allylic hydroxy group.

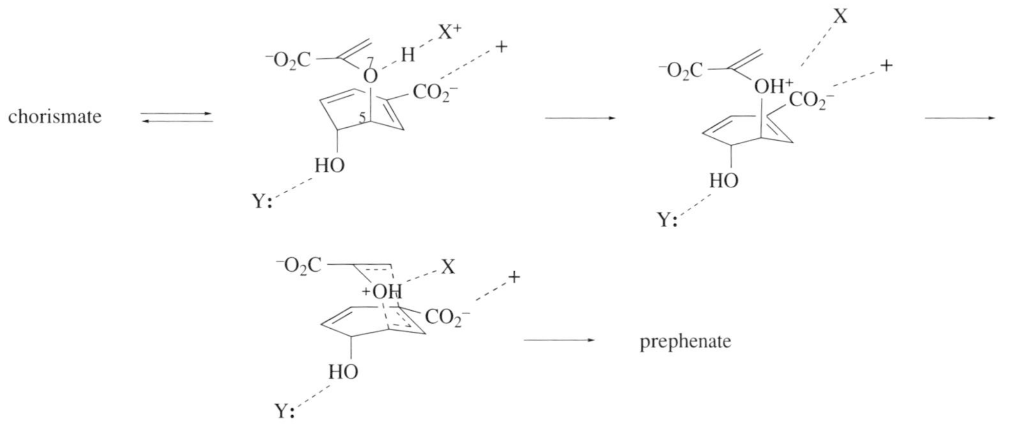

Scheme 8

5.14.5 CATALYSTS WITH MUTASE-LIKE ACTIVITY

5.14.5.1 Cryptands

Early efforts to mimic mutase catalysis using smaller molecules having appropriate three-dimensional structures were reported by Richards *et al.*[131] These were based on the catalytic model proposed in 1978 by Gorisch[2] that emphasized the complementarity of the mutase active site to the chair-like geometry of the pseudodiaxial conformation of chorismate. Syntheses of several crown ethers embodying two aryl rings, and also related cryptands,[131,132] were achieved. Association constants for 1:1 complexes with 2,2,2-trifluoroethanol and for various *para*-substituted phenols were measured,[131] but no mutase activity was reported.

5.14.5.2 Catalytic Antibodies

The remarkable ability of the immune system to mount a response to chemically designed mimics of reaction transition states has also led to the production of antibodies with catalytic properties. Because [3,3]-sigmatropic rearrangements represent unimolecular processes, antibody-catalyzed Claisen rearrangements became a focal point of early work in this field. Several catalytic antibodies were developed specifically for the CM reaction by two research groups. Antibody 11F1-2E11 displayed a 10^4-fold rate enhancement,[118,119] while antibody 1F7 catalyzed the rearrangement 250 times faster than the uncatalyzed process.[120,121] For a detailed discussion of these catalysts, the reader is referred to Chapter 5.17.

5.14.6 STRUCTURAL STUDIES

X-ray crystallography has been used to obtain structural information on several chorismate mutases as complexes with Bartlett *et al.*'s *endo*-oxabicyclic diacid inhibitor (Figure 1). Structures have been reported for (i) EcCM, the N-terminal chorismate mutase domain (residues 1–113), engineered from the bifunctional *E. coli* enzyme chorismate mutase–prephenate dehydratase (P-protein), (ii) BsCM, the 127-residue, monofunctional chorismate mutase from *B. subtilis*, (iii) ScCM, the allosterically regulated chorismate mutase from *S. cerevisiae* (no bound inhibitor), and (iv) 1F7, a catalytic antibody generated using the *endo*-oxabicyclic diacid inhibitor as a hapten. Comparison of these three-dimensional structures has been useful in identifying common active-site motifs and in evaluating mechanistic hypotheses for CMs.

5.14.6.1 *Bacillus subtilis* CM

The first detailed structural information on CM came from single-crystal X-ray diffraction analyses[133,134] of uncomplexed BsCM, the smallest naturally occurring monofunctional CM known to date, and of its prephenic acid and inhibitor complexes. The structure of the free enzyme was solved using the multiple isomorphous replacement method together with partial structure phase combination and noncrystallographic averaging. The final model was refined to 1.9 Å resolution (Figure 2). The free enzyme crystallizes with 12 monomers of protein in the asymmetric unit. Interpretation of the electron density map suggests that the enzyme is a solid-state trimer with the three monomers, related by pseudothreefold symmetry, packed to form a pseudo-α,β-barrel. Previous studies had suggested that BsCM was a homodimer[15] or homotrimer.[128]

The monomer, consisting of a single domain, embodies a five-stranded region of mixed β-sheet packed alongside an 18-residue α-helix (H1) and a two-turn 3_{10} helix (H2). An additional 3_{10} helix is found at the C-terminal (residues 112–115) protruding away from the β-sheet into the solvent. Strands I, II, and IV composed of residues 2–11, 44–50, and 88–94, respectively, occupy the interior of the β-sheet and are antiparallel. The two outer strands, III and V, composed of residues 73–76 and 107–109, respectively, are parallel. The H1 helix (residues 18–34) is joined to strands I and II by eight- and nine-residue loops, respectively. The H2 helix (residues 58–64) is connected to strands II and III by nine- and eight-residue loops, respectively.

The core structure of the BsCM trimer is mainly composed of the β-sheet regions of each monomer, which form a closed barrel of β-strands. The interfaces between pairs of adjacent subunits form three shallow clefts near the outer surface. The clefts, which are equivalent, share structural features from two neighboring polypeptide chains. The H2 helix, a loop (residues 66–71), and β-strand III from one subunit fit together with β-strands I, IV, and V from a neighboring subunit to create the wall region of the cleft. Each cleft is capped with a loop (residues 78–89) from the adjacent subunit.

The identity of this cleft as the BsCM active site was confirmed by X-ray analysis of the enzyme–inhibitor complex ($K_i = 3\ \mu\text{mol L}^{-1}$) shown in Figure 3, which was refined to 2.2 Å. The difference

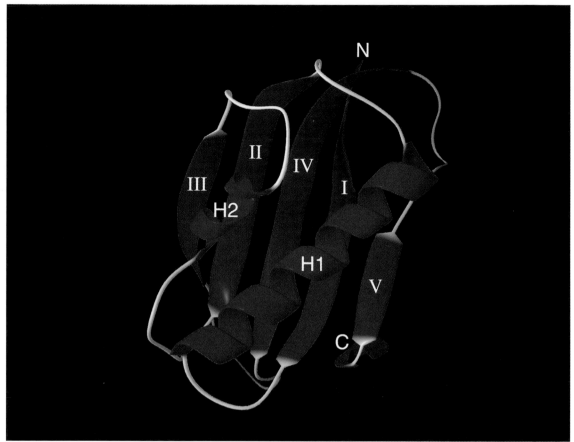

Figure 2 Structure of *B. subtilis* chorismate mutase.

Fourier map of free BsCM and its complex revealed little change in the enzyme's conformation upon inhibitor binding.

A number of residues in the active site interact with the bound inhibitor molecule, including Arg7, Glu78, Arg90, Tyr108, and Leu115 from one subunit together with Phe57′, Ala59′, Lys60′, Val73′, Thr74′, and Cys75′ from the adjacent subunit (Figure 4). A number of these contacts (not shown) involve hydrophobic interactions. Also notable are electrostatic interactions evident between the two arginine residues (Arg7, Arg90) and O-3 and O-4 corresponding to chorismate's enol pyruvyl side chain. Additional polar contacts (<3.5 Å distance) are formed involving (i) Arg90 with the ether oxygen O-7, (ii) Glu78 and Cys75′ with O-5, and (iii) Tyr108 with O-3 of the side chain carboxylate group, all of which are consistent with hydrogen-bonding interactions.

The structure of the enzyme–product complex of BsCM with prephenic acid has also been solved (Figure 5). Twelve enzyme monomers are found in the asymmetric unit, with bound prephenate occupying the same active site cleft. The overall structure of the BsCM–product complex is little changed from that of the free protein, and contacts between prephenate and the enzyme are similar to the contacts in the transition-state analogue complex. The authors concluded from the three crystallographic studies that the chorismate-binding active site is unlikely to be significantly different in structure and organization from that of the transition-state and product complexes. They also suggested that the rate-limiting transition state for the overall BsCM reaction is likely the encounter of the enzyme with the pseudodiaxial conformation of chorismate leading to the ES complex.

Noting that the absence of active site proton donors and nucleophiles leaves few possibilities for electrophilic and nucleophilic catalysis as earlier proposed by Guilford *et al.*,[122] the authors suggested that, in keeping with the effects of polar solvents on chorismate's nonenzymic rearrangement, the electrostatic features of the BsCM active site might aid in stabilizing a polar transition state for the pericyclic process on the enzyme. Specifically, the Arg90 guanidinium and the Glu78 carboxylate groups could stabilize incipient negative and positive charges on O-7 and C-5, respectively, in a polar C—O bond-breaking process (see Scheme 7).

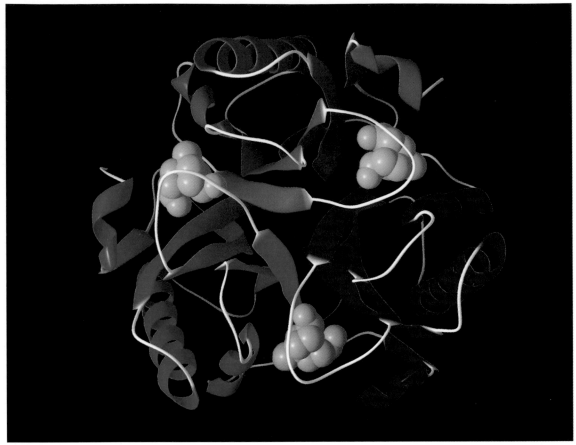

Figure 3 Active site of the *B. subtilis* chorismate mutase complexed with the oxabicyclic diacid inhibitor as defined by a 1.9 Å resolution X-ray diffraction analysis.

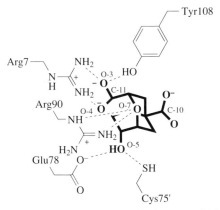

Figure 4 Schematic diagram of the hydrogen-bonding and electrostatic interactions of the transition-state inhibitor with the relevant side chains of BsCM.

5.14.6.2 *Escherichia coli* CM

The X-ray structure of the N-terminal chorismate mutase domain engineered from the bifunctional *E. coli* P-protein has also been reported as a complex with Bartlett *et al.*'s oxabicyclic diacid inhibitor (see Scheme 8).[135] The crystal structure was refined to 2.2 Å resolution and showed the enzyme to be a homodimer. The monomer unit (Figure 6) consists of a single polypeptide chain folded into three α-helical segments consisting of H1 (residues 6–42), H2 (residues 49–65), and H3 (residues 70–100) connected by two loops. The three α-helices, two long and one short, are arranged so as to cause the peptide backbone to adopt a figure-4 shape. The two monomers are related by a

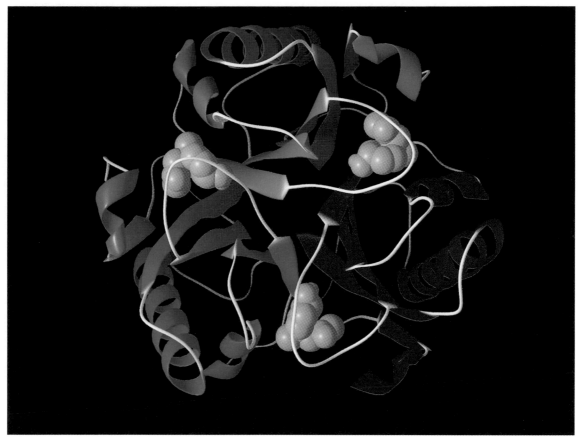

Figure 5 X-ray crystallographic structure of *B. subtilis* chorismate mutase complexed with prephenic acid.

noncrystallographic twofold axis where the helices of one monomer cross over those of the other. Almost one-third of the monomer's solvent-accessible surface area is buried in the dimer. The twofold molecular symmetry of the elongated dimer generates two antiparallel helix pair interactions between H1–H1′ and H3–H3′. Coiled-coil and helix–helix interactions between the two longest segments create a catalytically functional structure whose two equivalent, elbow-shaped active sites are highly charged and completely enclosed. The inhibitor binds near the middle of this relatively open helical bundle (Figure 7). Although access to the active site is possible from different directions, as might be expected in the parent bifunctional enzyme, charged Arg, Asp, and Glu side chain residues shield the catalytic region from solvent on different faces of the protein.

The bound inhibitor in EcCM identifies the enzyme's catalytic pocket (Figure 8), which can be compared and contrasted with the BsCM active site. Several similarities are evident. Hydrophobic interactions in the two structures are roughly equivalent, with Val85, Val35, Leu55, and Ile81 of EcCM (not shown) corresponding to Phe57, Leu115, Val73, and Ala 59 of BsCM. Both Arg11′ in EcCM and Arg7 in BsCM interact with O-3 and O-4 of the inhibitor, and both Glu52 in EcCM and Glu78 in BsCM interact with the C-4 hydroxy group. The Lys39 residue in EcCM adopts an equivalent position and identical charge to the Arg90 residue in BsCM, both interacting with O-4 and O-7. Both proteins concentrate positively charged residues along the left wall of their active sites, with Arg7 and Arg90 in BsCM, and Arg51, Arg11′, and Lys39 in EcCM directed towards the enol pyruvyl carboxylate group. Another notable feature of both mutases is the formation of hydrogen bonds to both lone electron pairs of the inhibitor's ether oxygen, using Lys39 and Gln88 in EcCM and Arg90 in BsCM.

By way of contrast, the tightly bound water molecule in EcCM bridging both inhibitor carboxyl groups and Arg51 finds no counterpart in the interactions of Tyr108 in BsCM. Another major structural difference is that the EcCM active site is completely enclosed, whereas the BsCM active site leaves approximately half of the bound inhibitor exposed to the medium. Consequently, a total of 14 residues are within 4 Å of bound inhibitor in EcCM, while 11 BsCM residues make similar contacts. Absent in BsCM are contacts between the protein and the C-11 carboxylate which in EcCM interacts strongly with Arg28 and Ser84.

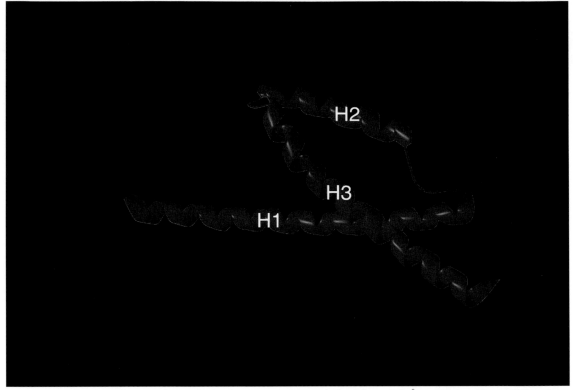

Figure 6   Subunit structure of *E. coli* chorismate mutase as defined by a 2.2 Å resolution X-ray diffraction
analysis.

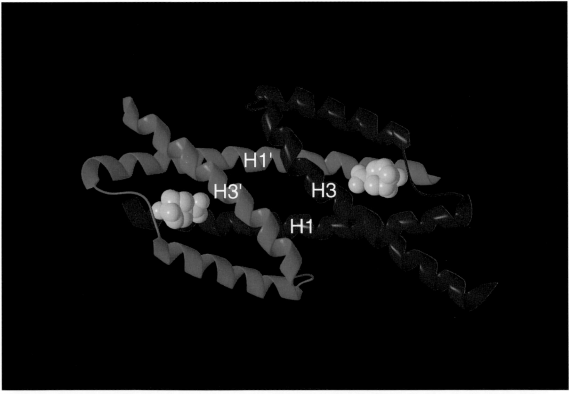

Figure 7   Structure of the *E. coli* chorismate mutase dimer complexed with the oxabicyclic diacid inhibitor as
defined by a 2.2 Å resolution X-ray diffraction analysis.

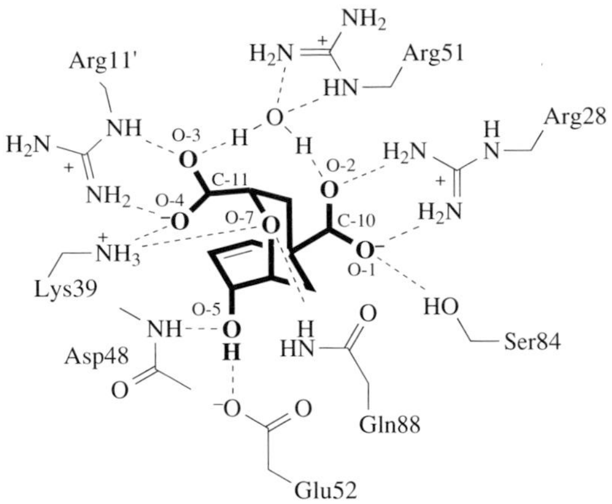

Figure 8 Schematic diagram of the hydrogen-bonding and electrostatic interactions of the transition-state inhibitor with the relevant side chains of EcCM.

Further analysis of the similarities and differences between crystal structures of BsCM and EcCM led to a mechanistic hypothesis for the chorismate mutase reaction.[136] The authors suggested that the positively charged residues in BsCM and EcCM promote E–S complex formation through electrostatic effects, and reduce the activation entropy by orienting and locking chorismate in the requisite chair conformer for rearrangement. On the basis of Monte Carlo simulations of the effect of water on Claisen rearrangements,[95] the authors also suggested that two hydrogen bonds formed between active site residues and O-7 of bound chorismate reduce the activation enthalpy and further accelerate the rearrangement.

5.14.6.3 *Saccharomyces cerevisiae* CM

The three-dimensional structure of an allosterically locked chorismate mutase from *S. cerevisiae* (ScCM) has been determined by X-ray crystallographic analysis.[137] A member of the CM-1 family encoded by the *ARO7* gene, wild-type ScCM is a dimer of two identical subunits (256 residues each, M_r 30 000 Da), each capable of binding one substrate, one activator, and one inhibitor molecule. The enzyme is activated by tryptophan and subject to feedback inhibition by tyrosine.[138] Sequencing studies on three mutant alleles of the *ARO7* gene that encode a constitutively activated CM have shown that a single substitution of isoleucine for threonine at residue 226 (Thr226Ile) is responsible for the loss of activation and feedback inhibition.[40]

The crystal structure of the mutant ScCM was determined at 2.2 Å resolution by the multiple isomorphous replacement method. The protein crystallized as a dimer in the shape of a bipyramid having overall dimensions of $90 \times 55 \times 55$ Å (Figure 9). Unlike BsCM, ScCM contains essentially no β-sheet structure, and helical regions account for $>70\%$ of the entire sequence. Four α-helical domains consisting of H2 (residues 14–33), H4 (residues 59–73), H8 (residues 140–171), and H11 (residues 195–211) are involved in the dimerization. Several polar and charged side chains from H2, H4, H5, and H8 combine to form a hydrophilic channel through the center of the dimer that is reinforced with solvent interactions. From the perspective of the hydrophobic channel, the dimer may be viewed as a four-helix bundle with antiparallel H2–H2′ and H8–H8′ interactions wrapped in a layer of remaining α-helical structures.

The four-helix bundle domain in ScCM closely resembles that found in EcCM. Superposition of both helical bundles places the EcCM inhibitor molecule squarely in the middle of the cavity formed by H2, H8, H11, and H12 (residues 227–251) within one monomer unit of ScCM, which is the likely catalytic site.[139] Least-squares fitting of the two domains results in an r.m.s. deviation of only 1.06 Å over a region spanning 94 amino acids, 21 of which are identical in the two enzymes.

The two active sites share several similarities. Four of the seven EcCM residues making contact with the inhibitor (Arg11′, Arg157, Lys39, Glu152) are conserved in ScCM (Arg16, Arg28, Lys168, Glu198), and a close correspondence may be seen in the interaction of the positively charged residues with the inhibitor's carboxylates (Figure 10). Notably different is the switch in residues making contact with the ether oxygen (O-7), from Gln88 in EcCM to Glu246 in ScCM. For a hydrogen bond to form between O-7 and ScCM, the authors observed that the carboxylate anion of Glu246

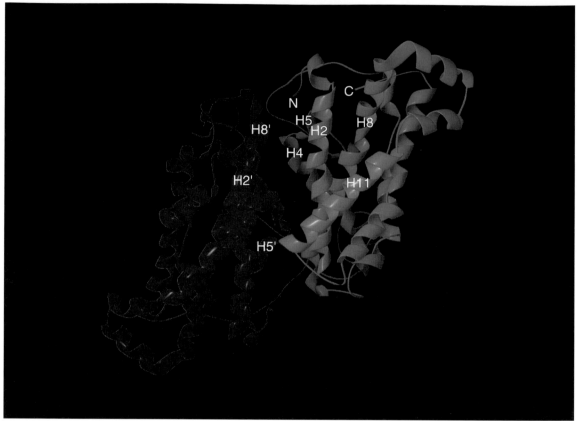

Figure 9 Structure of *S. cerevisiae* chorismate mutase as defined by a 2.2 Å resolution X-ray diffraction analysis.

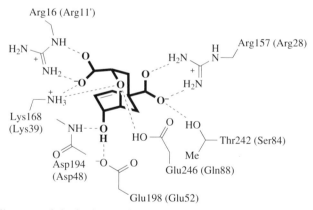

Figure 10 Schematic diagram of the hydrogen-bonding and electrostatic interactions of the transition-state inhibitor with the relevant side chains of ScCM.

would have to be protonated. Although CMs generally show little sequence similarity, alignments indicate that (i) Gln88 in the P-protein mutase is replaced by Glu86 in the T-protein mutase, and (ii) Glu246 in ScCM is replaced by Gln325 in the mutase from the higher plant *Arabidopsis thaliana* (AtCM). These variations raise the question of whether mutase catalysis involves any significant contribution from interaction of O-7 with a glutamine or glutamate residue, and whether different mechanisms operate in different mutases.

5.14.6.4 Antibody 1F7

The three-dimensional structure of the monoclonal catalytic antibody 1F7 has been determined to 3.0 Å resolution as the Fab′–hapten complex with the oxabicyclic diacid inhibitor.[140] The overall structure of the 1F7 antibody is similar to those of other immunoglobulin Fab′ fragments. Ligand binding occurs at the confluence of six loops made up of heavy (H1–H3) and light (L1–L3) chain variable domains. Approximately 90% of the solvent-accessible surface of the inhibitor is buried in the complex, indicating good shape complementarity with the Fab′ binding pocket. The active site displays a bowl- or cleft-like shape, and interacts with the inhibitor by a combination of electrostatic, hydrogen bonding, and hydrophobic effects. The dominant role of the heavy chain in the active site of 1F7 thus resembles EcCM and BsCM, whose dimeric active sites are also composed mainly of contributions from one monomer chain. As with most antibodies to small molecules, ligand complementarity resides mainly on the heavy chain; only a single residue of the light chain interacts with inhibitor

A total of 37 van der Waals and three hydrogen bonds are evident in the 1F3–inhibitor complex, whose relatively hydrophobic floor is composed of AsnH35, TrpH47, AlaH93, PheH100b, and LeuL96 side chains. The walls of the binding pocket include polar side chains from AsnH33, AsnH50, AsnH58, ArgH95, TyrH100, and TyrL94. It would appear that specific interactions with the inhibitor's two carboxylate groups largely dictate its orientation in the 1F7 pocket (Figure 11). The most buried part of the inhibitor is the C-11 carboxylate group, which forms part of a hydrogen-bonded network to an active site water molecule. No hydrogen bonding is evident at O-7. Although ArgH95 is the only formal positive charge in the active site, other nearby basic residues (HisH32, ArgH94, ArgH96) may enhance charge complementarity in the 1F7 active site.

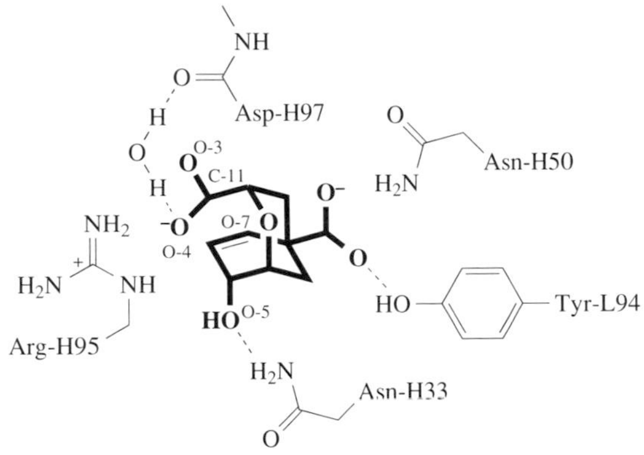

Figure 11 Schematic diagram of the hydrogen-bonding and electrostatic interactions of the oxabicyclic diacid inhibitor with antibody 1F7, which exhibits chorismate mutase activity.

5.14.7 OTHER MECHANISTIC INSIGHTS

In the first published effort to describe the mechanism of action of CM, Gorisch noted that chorismic acid's functional groups offered several possibilities for noncovalent binding in the E–S complex.[2] For example, the two charged carboxylate groups were considered capable of strong electrostatic interactions with protonated active site residues. Suitable residues might also form hydrogen bonds with both carboxyl and hydroxyl groups in chorismate. Hydrophobic interactions might also potentially contribute to binding. Finally, the possibility of π-electron interactions with chorismate's diene system was also noted.

5.14.7.1 From Kinetic Parameters

Activation parameters listed in Table 3 for a number of different mutases indicate that both entropic and enthalpic effects contribute to the 10^6-fold rate enhancement observed with most natural enzymes. Judging from the kinetic properties of the *E. coli*, *S. aureofaciens*, and *K. pneumoniae* enzymes, this 10^6-fold rate enhancement results from a near-zero activation entropy and a ~ 5 kcal mol^{-1} reduction in activation enthalpy. It remains unclear why activation parameters for the *B. subtilis* enzyme are significantly different from these values.

The reduction of $\Delta S\ddagger$ by 13–14 eu in the CM reaction would correspond to a $\sim 10^4$-fold rate enhancement. Interestingly, that is the best rate enhancement thus far achieved by antibody 11F1-2E11, the most efficient catalytic antibody yet engineered to promote the CM reaction. Since an ideal transition-state analogue may be expected to elicit maximum conformational control by the antibody's hypervariable region, rate enhancements of 10^4-fold may be the natural limit for catalyst development based purely on shape-complementarity selection experiments alone.

A 5 kcal mol^{-1} reduction in $\Delta H\ddagger$ would result in an additional 100-fold rate enhancement. The origin of that enthalpic effect has been attributed variously to a dissociative process,[122] hydrogen-bonding/acid catalysis,[130] or water-like hydrogen bonding in the enzymic transition state.[136] The smaller reduction in $\Delta H\ddagger$ observed with antibody 11F1-2E11 might reflect the use of antibody binding energy to induce strain in the substrate.[119]

5.14.7.2   From Molecular Mechanics Studies and Simulations

Theoretical treatments of the aliphatic Claisen rearrangement in solution have been published, concentrating primarily on the reaction of AVE.[95,141] Several research groups have used both semiempirical and *ab initio* theoretical methods to investigate the mutase-promoted rearrangement of chorismate to prephenate. In an early molecular orbital study of the gas-phase CM reaction, Andrews and Haddon used MINDO/2 and MINDO/3 methods to calculate detailed molecular geometries for the neutral and dianionic forms of chorismate and prephenate and for several possible rearrangement transition states.[142] Developing charge delocalization in the transition state, particularly involving the two carboxyl groups, was invoked as a potential source of catalysis. Mutase activity was attributed to two positively charged groups in the active site, with one possibility being the lysine implicated in earlier chemical modification studies.[106] More recently, full geometry optimizations of both chorismate and the oxabicyclic diacid inhibitor were carried out at the *ab initio* level, and an analysis of the binding of both the substrate and the transition-state analogue in the BsCM active site (calculated using the program AMBER) have been reported.[143]

Insights into CM catalysis have also been reported by Lyne *et al.* using a combination of quantum mechanical and molecular mechanical (QM/MM) methods, together with data from the BsCM X-ray crystal structure determination.[144] The hybrid computational approach, which was developed to study solution-phase and enzyme-mediated processes,[145] treats a subset of atoms quantum mechanically, and the remaining atoms using the molecular mechanics force field of CHARMM into which both semiempirical (AM1) and *ab initio* (GAMESS) MO methods have been incorporated. Computer simulations of the BsCM reaction indicate that the conformation of chorismate in the ES complex is considerably different compared with the calculated gas-phase reaction coordinate for rearrangement. Most notably, the two carbons forming the new C—C bond are 0.449 Å closer (with a corresponding change in the C-5—O-7 torsion angle) in the enzyme-assisted process. Calculations further indicate that the enzyme preferentially binds the transition state through substantial interactions with Arg90 and Glu78. Catalysis is rationalized on the basis of a combination of substrate strain and transition-state stabilization.

Wiest and Houk have reported *ab initio* calculations on the transition state for the gas-phase rearrangement of chorismate, with results at the RHF/6–31G level of theory successfully reproducing the experimental tritium secondary isotope effects.[146] In further work with other basis sets and using density functional theory, the effect of substituents on the Claisen rearrangement has been investigated using simple model systems.[147] To explore which amino acid residues play a role in catalysis by BsCM and by antibody 1F7, interactions of various functional groups found in the catalytic pockets with substituted AVEs were studied.

The most important factor in catalysis was found to be selective transition-state binding by appropriately positioned hydrogen bond donors, which lower the activation free energy by 6 kcal mol^{-1} in the case of AVE-2,6-dicarboxylate. Other important factors included charge complementarity between active site cation and substrate carboxylates, and also increased hydrogen bonding to the ether oxygen (O-7), which reduce the activation free energy by 1.7 and 2.0 kcal mol^{-1}, respectively. Stabilization of incipient partial (allylic) positive charge in the transition state had no significant catalytic effect, whereas strong hydrogen bonding to O-7 resulted in electrophilic catalysis of a dissociative mechanism. Antibody 1F7 was found to be lacking the amino acid residues that could provide the charge complementarity and/or increased hydrogen bonding deemed important for catalysis.

5.14.7.3   From Mutagenesis Studies

Cload *et al.* have reported the results of site-specific mutagenesis of active site residues in BsCM.[148] Mutant proteins containing six additional C-terminal histidines (for purification purposes) were expressed in *E. coli*, purified to homogeneity, and their structural integrity assayed by circular dichroism (CD). The CD spectra of all mutants were superimposable with wild-type BsCM, suggesting that the mutations had caused no significant structural changes.

The BsCM X-ray crystallographic analysis indicated that Tyr108, Arg116, Phe57, and Cys75 were all within hydrogen-bonding distance of the bound transition-state analogue (Figure 4). However, substitutions at each of these positions (Tyr108Phe, Arg116Lys, Phe57Trp, Cys75Ser, Cys75Ala, Cys75Asp) had little effect on catalysis, although the K_M values showed modest (4–17-fold) increases. The observed activity of mutant proteins did not support the importance of nucleophilic analysis by CM involving Cys75, and ruled out any role for Arg116 or Tyr108 in spatial orientation effects. The notion that developing positive charge was stabilized by Phe57 also gained no support from these data.

Mutagenesis of Arg7 resulted in a substantial loss of activity. In the Arg7Ala mutant, where the bidentate hydrogen-bonding interaction between the guanidine functional group and the inhibitor's C-11 carboxylate group has been removed, a 10^6-fold reduction in k_{cat}/K_M was observed. The Arg7Lys mutant displayed a 10^3-fold reduction in k_{cat}/K_M, perhaps indicating that some catalysis was restored through an interaction of the C-11 carboxylate with lysine's ε-amino group.

Mutagenesis of Arg90 also confirmed the importance of this residue in hydrogen bonding to O-7 and to the C-11 carboxylate. The Arg90Ala mutant showed no measurable activity, whereas the Arg90Lys displayed a 2.6×10^4-fold reduction in k_{cat}/K_M compared with wild-type BsCM.

The potential involvement of chorismate's allylic hydroxy group (O-5 in the bound inhibitor) in any general acid or general base catalysis of the CM reaction was also probed using mutagenesis. Glu78 is positioned 2.90 Å from the OH group and 3.2 Å from Arg90, with which it might also interact. Both the Glu78Gln mutant and the Glu78Ala mutant showed a $>10^4$-fold drop in activity. The Glu78Ala mutant gained a factor of 50 in rate by making a second change (Cys75Asp). Little variation in k_{cat} was observed, despite significant changes in K_M and K_i, suggesting that Glu78 contributes equally to the binding energy of both substrate and transition state. Cload *et al.*[148] concluded that a carboxy group in the vicinity of chorismate's secondary alcohol most likely oriented the substrate for a critical bidentate hydrogen-bonding interaction with Arg90.

Two groups have reported results of site-specific mutagenesis studies on EcCM.[149,150] Liu *et al.*[149] created a series of mutants to probe the importance of various active site residues. Substitutions at Arg11′ and Arg28 probed the extent to which hydrogen-bonding and electrostatic interactions with chorismate's two carboxyl groups contributed to catalysis. Values of k_{cat}/K_M for Arg11Lys and Arg11Ala mutants were 10^3- and 10^4-fold lower, respectively, than that for wild-type EcCM, indicating the importance of Arg11 hydrogen bonding to the C-11 carboxylate. A similar effect was noted for Arg7 in BsCM. In addition, Arg28Ala and Arg28Lys mutants both displayed 10^3-fold lower values of k_{cat}/K_M than wild-type EcCM, indicating that hydrogen bonds to the C-10 carboxylate also played a significant role.

Mutagenesis of Glu52, which is proximal to the inhibitor's hydroxy group (O-5), was used to study its involvement in general acid or general base catalysis. Liu *et al.* found that k_{cat} values for the Glu52Gln, Glu52Asp, and Glu52Ala mutants were 3.0-, 23-, and 150-fold lower, respectively, than that for wild-type EcCM. In view of the activity of the Glu52Gln mutant, catalysis involving the carboxylate group of Glu52 seemed unlikely, namely electrostatic stabilization of a dissociative transition state, nucleophilic participation, or general acid/general base catalysis.

The importance of Lys39 and Gln88 was also probed by both Liu *et al.*[149] and Zhang *et al.*[150] The critical role of Lys39, whose ε-amino group lies within hydrogen-bonding distance of O-7, was evident in the properties of Lys39Ala, Lys39Gln, and Lys39Arg mutants, which were 10^4–10^5-fold less active than wild-type EcCM. Zhang *et al.* noted a 335-fold drop in k_{cat} for the Lys39Arg mutant, suggesting that arginine marginally substituted for lysine as a positively charged and/or hydrogen-bonding residue in the catalytic mechanism. The longer side chain and branched guanidine functional group of Arg39 may distort its interaction with O-7 of the inhibitor and with other active residues.

Liu *et al.* found no significant differences in K_M or K_i between Lys39Arg, Lys39Asn, and wild-type EcCM, suggesting that Lys39 was more directly involved in catalysis than in substrate binding. Comparable mutations of Arg90 in BsCM that abolished the bridging hydrogen bonds to O-7 and C-11 also resulted in substantial decreases in k_{cat}/K_M.

Mutations to Gln88 shed further light on the importance of interactions between the mutase and the substrates ether oxygen (O-7). Liu *et al.* found that k_{cat}/K_M dropped by 10^4-fold for the Gln88Ala and Gln88Lys mutants. Both Liu *et al.* and Zhang *et al.* prepared the Gln88Glu mutant, and its activity was found to be highly pH dependent, rising dramatically under acidic conditions. At its optimum pH (4.5), the Gln88Glu mutant displayed 140% of the wild-type EcCM activity, largely due to enhancements in turnover ($k_{cat} = 9700$ min^{-1} for the mutant and 3700 min^{-1} for the wild type, both at pH 4.5).[150] Both Liu *et al.* and Zhang *et al.* attributed the dramatic increase to protonation of Glu88, which re-established the hydrogen bond to O-7.

Based on molecular modeling of the solid-state structure of uncomplexed ScCM with the structure of the EcCM–inhibitor complex, Zhang *et al.* further observed that a Glu246 in ScCM replaced Gln88 of EcCM. That change also significantly lowered the pH optimum of the enzyme, as would be expected in order for Glu to function as a hydrogen bond donor. The structure of the Gln88Glu mutant mutase was further probed in an energy-minimization study using the crystallographic coordinates of the wild-type EcCM complex as a reference.[150] Overall, mutagenesis studies on Gln88 provided convincing evidence that hydrogen bonding to chorismate's enolpyruvyl oxygen was important to both k_{cat} and K_M of EcCM.

5.14.8 CONCLUSIONS

Biological reactions, while varying greatly in their rates, also vary in the burden that they place on an efficient catalyst. With their ability to achieve 10^6-fold rate accelerations, CMs are not especially proficient enzymes in the sense defined by Radzicka and Wolfenden,[151] where catalytic proficiency measures the lower limit of an enzyme's affinity for substrate as it reaches the transition state. By that definition, proficient enzymes should be extremely powerfully inhibited by transition-state analogues. The best CM inhibitor to date closely resembles the putative chair transition state in the concerted, asynchronous [3,3]-sigmatropic rearrangement, and yet achieves only about 300-fold tighter binding to the enzyme.

Kinetic, spectroscopic, and isotope-labeling studies indicate that the mutase-catalyzed rearrangement cannot be distinguished from a classical pericyclic process. Sequence alignments of mutases reveal a diversity of active protein primary structures. Crystallographic studies further indicate that significant variations exist in overall protein folds and in the active-site organization of EcCM, BsCM, and ScCM. Mutagenesis studies on BsCM and EcCM argue against nucleophilic catalysis or general acid/general base catalysis in either enzyme. Taken as a whole, however, mutagenesis does point to the common importance of hydrogen bonding between mutases and chorismic acid's hydroxy, carboxylate, and enolpyruvyl groups. Hydrogen bonding to the allylic hydroxy group may be important in orienting the substrate within the pocket, whereas hydrogen bonding to the carboxylates likely reduces entropy and promotes catalysis by positioning the side chain atoms in the requisite chair conformation for rearrangement. Hydrogen bonding to the enolpyruvyl oxygen (O-7) further catalyzes the rearrangement. Several other residues appear to contribute equally both to substrate binding and to transition-state stabilization.

5.14.9 REFERENCES

1. E. Haslam, "Shikimic Acid Metabolism and Metabolites," Wiley, New York, 1993.
2. H. Gorisch, *Biochemistry*, 1973, **17**, 3700
3. N. Patel, D. L. Pierson, and R. A. Jensen, *J. Biol. Chem.*, 1977, **252**, 5839.
4. R. M. Romero, M. F. Roberts, and J. D. Phillipson, *Phytochemistry*, 1995, **40**, 1015.
5. B. Ganem, *Tetrahedron*, 1978, **34**, 3353.
6. U. Weiss and J. M. Edwards, "The Biosynthesis of Aromatic Compounds," Wiley-Interscience, New York, 1980.
7. P. M. Dewick, *Nat. Prod. Rep.*, 1992, **9**, 153.
8. R. G. H. Cotton and F. Gibson, *Biochim. Biophys. Acta*, 1965, **100**, 76.
9. W. M. Nakatsukasa and E. W. Nester, *J. Biol. Chem.*, 1978, **247**, 5972.
10. T. Xia and R. A. Jensen, *Arch. Biochem. Biophys.*, 1992, **294**, 147.
11. C. Poulsen and R. Verpoorte, *Phytochemistry*, 1991, **30**, 377.
12. S. K. Goers and R. A. Jensen, *Planta*, 1984, **162**, 109.
13. S. Amad, A.-T. Wilson, and R. A. Jensen, *Eur. J. Biochem.*, 1988, **176**, 69.
14. E. Hyde and J. F. Morrison, *Biochemistry*, 1978, **17**, 1573.
15. J. V. Gray, B. Golinelli-Pimpaneau, and J. R. Knowles, *Biochemistry*, 1990, **29**, 376.
16. K. H. C. Ma and B. E. Davidson, *Biochim. Biophys. Acta*, 1985, **827**, 1.
17. M.-J. H. Gething, B. E. Davidson, and T. A. A. Dolpheide, *Eur. J. Biochem.*, 1976, **71**, 317.
18. P. Sampathkumar and J. F. Morrison, *Biochim. Biophys. Acta*, 1982, **702**, 204.
19. T. Xia, J. Song, G. Zhao, H. Aldrich, and R. A. Jensen, *J. Bacteriol.*, 1993, **175**, 4729.
20. A. Berry, R. A. Jensen, and A. T. Hendry, *Arch. Microbiol.*, 1987, **149**, 87.
21. R. Bode, P. Koll, N. Prahl, and D. Birnbaum, *Arch. Microbiol.*, 1989, **1151**, 123.
22. S. Ahmad and R. A. Jensen, *Curr. Microbiol.*, 1988, **16**, 295.
23. S. Ahmad and R. A. Jensen, *Arch. Microbiol.*, 1987, **147**, 8.
24. S. Ahmad, W. G. Weisburg, and R. A. Jensen, *J. Bacteriol.*, 1990, **172**, 1051.
25. H. Gorisch and F. Lingens, *Biochemistry*, 1974, **13**, 3790.

26. T. S. Woodin and L. Nishioka, *Biochim. Biophys. Acta*, 1973, **309**, 211.
27. D. G. Gilchrist and J. A. Connelly, *Methods Enzymol.*, 1987, **142**, 450.
28. P. Gadal and H. Bouyssou, *Physiol. Plant.*, 1973, **28**, 7.
29. M. Benesova and R. Bode, *Phytochemistry*, 1992, **31**, 2983.
30. G. W. Kuroki and E. E. Conn, *Arch. Biochem. Biophys.*, 1988, **260**, 616.
31. G. W. Kuroki and E. E. Conn, *Plant Physiol.*, 1989, **89**, 472.
32. S. C. Hertel, M. Hieke, and D. Groger, *Acta Biotechnol.*, 1991, **11**, 39.
33. B. K. Singh, J. A. Connelly, and E. E. Conn, *Arch. Biochem. Biophys.*, 1985, **243**, 374.
34. C. L. Schmidt, D. Grundemann, G. Groth, B. Miller, H. Henning, and G. Schultz, *J. Plant Physiol.*, 1991, **138**, 51.
35. S. K. Goers and R. A. Jensen, *Planta*, 1984, **162**, 117.
36. M. Chu and J. M. Widholm, *Plant Physiol.*, 1972, **26**, 24.
37. G. W. Kuroki and E. E. Conn, *Plant Physiol.*, 1988, **86**, 895.
38. A. Maruya, M. J. O'Connor, and K. Backman, *J. Bacteriol.*, 1987, **169**, 4852.
39. S. G. Ball, R. B. Wickner, G. Cottarel, M. Schaus, and C. Tirtiaux, *Mol. Gen. Genet.*, 1986, **205**, 326.
40. T. Schmidheini, P. Sperisen, G. Paravicini, R. Hutter, and G. Braus, *J. Bacteriol.*, 1989, **171**, 1245.
41. J. Eberhard, H.-R. Raesecke, J. Schmid, and N. Amrhein, *FEBS Lett.*, 1993, **334**, 233.
42. L. C. Vining, V. S. Malik, and D. W. S. Westlake, *Lloydia*, 1968, **31**, 355.
43. G. A. Dardenne, P. O. Larsen, and E. Wieczorkowska, *Biochim. Biophys. Acta*, 1975, **381**, 416.
44. A. Jones and L. C. Vining, *Can. J. Microbiol.*, 1976, **22**, 237.
45. M. Francis and L. C. Vining, *Can. J. Microbiol.*, 1979, **25**, 1408.
46. C.-Y. P. Teng, B. Ganem, S. Z. Doktor, B. P. Nichols, R. K. Bhatnagar, and L. C. Vining, *J. Am. Chem. Soc.*, 1985, **107**, 5008.
47. S. D. Copley and J. R. Knowles, *J. Am. Chem. Soc.*, 1987, **109**, 5008.
48. J. J. Gajewski, J. Jurayj, D. R. Kimbrough, M. E. Gande, B. Ganem, and B. K. Carpenter, *J. Am. Chem. Soc.*, 1987, **109**, 1170.
49. C. J. Burrows and B. K. Carpenter, *J. Am. Chem. Soc.*, 1981, **103**, 6983.
50. S. D. Copley and J. R. Knowles, *J. Am. Chem. Soc.*, 1985, **107**, 5306.
51. L. Addadi, E. K. Jaffe, and J. R. Knowles, *Biochemistry*, 1983, **22**, 4494.
52. P. R. Andrews, G. D. Smith, and I. G. Young, *Biochemistry*, 1973, **12**, 3492.
53. L. Claisen, *Ber. Dtsch. Chem. Ges.*, 1912, **45**, 3157.
54. P. Wipf, in "Comprehensive Organic Synthesis," eds. B. M. Trost, I. Fleming, and L. A. Paquette, Pergamon, New York, 1991, Vol. 5, p. 827.
55. F. E. Ziegler, *Chem. Rev.*, 1988, **88**, 1423.
56. S. J. Rhoads, and N. R. Raulins, *Org. React.*, 1975, **22**, 1.
57. W. von E. Doering and W. R. Roth, *J. Am. Chem. Soc.*, 1962, **18**, 67.
58. C. Walling and M. Naiman, *J. Am. Chem. Soc.*, 1962, **84**, 2628.
59. M. K. Diedrich, D. Hochstrate, F.-G. Klarner, and B. Zimny, *Angew. Chem., Int. Ed. Engl.*, 1994, **33**, 1079.
60. F. W. Schuler and G. W. Murphys, *J. Am. Chem. Soc.*, 1950, **72**, 3155.
61. P. Vittorelli, T. Winkler, H.-J. Hansen, and H. Schmid, *Helv. Chim. Acta*, 1968, **51**, 1457.
62. H. J. Hansen and H. Schmid, *Tetrahedron*, 1974, **30**, 1959.
63. J. W. Ralls, R. E. Lundin, and G. F. Bailey, *J. Org. Chem.*, 1963, **28**, 3521.
64. T. H. Lowry and K. S. Richardson, "Mechanism and Theory in Organic Chemistry," 3rd edn., Harper and Row, New York, 1987, chap. 10.
65. J. J. Gajewski and N. D. Conrad, *J. Am. Chem. Soc.*, 1979, **101**, 2747.
66. J. J. Gajewski, *Acc. Chem. Res.*, 1980, **13**, 142.
67. W. S. Johnson, L. Wertheman, W. R. Bartlett, T. J. Brocksam, T.-T. Li, D. J. Faulkner, and M. R. Petersen, *J. Am. Chem. Soc.*, 1970, **92**, 741.
68. A. E. Wick, D. Felix, K. Steen, and A. Eschenmoser, *Helv. Chim. Acta*, 1964, **47**, 2425.
69. A. E. Wick, D. Felix, K. Gschwend-Steen, and A. Eschenmoser, *Helv. Chim. Acta*, 1969, **52**, 1030.
70. R. E. Ireland and R. H. Mueller, *J. Am. Chem. Soc.*, 1972, **94**, 5897.
71. R. E. Ireland, R. H. Mueller, and A. K. Willard, *J. Am. Chem. Soc.*, 1976, **98**, 2868.
72. J. E. Franz, in "The Herbicide Glyphosate," eds. E. Grossbard and D. Atkinson, Butterworth, Boston, 1985, p. 3.
73. T. J. Franklin and G. A. Snow, "Biochemistry of Antimicrobial Action," 2nd edn., Chapman & Hall, London, 1975, p. 146.
74. R. W. Alder, R. Baker, and J. M. Brown, "Mechanisms in Organic Chemistry," Wiley-Interscience, London, 1971, p. 239.
75. I. Fleming, "Frontier Orbitals and Organic Chemical Reactions," Wiley, London, 1976, p. 32.
76. B. Ganem, *Agnew. Chem., Int. Ed. Engl.*, 1996, **35**, 936.
77. R. P. Lutz, *Chem. Rev.*, 1984, **84**, 205.
78. J. V. Gray, D. Eren, and J. R. Knowles, *Biochemistry*, 1990, **29**, 8872.
79. L. Claisen, *Ber. Dtsch. Chem. Ges.*, 1912, **45**, 3157.
80. J. W. Ralls, R. E. Lundin, and G. F. Bailey, *J. Org. Chem.*, 1963, **28**, 3521.
81. R. D. H. Murray, M. Sutcliffe, and M. Hasegawa, *Tetrahedron*, 1975, **31**, 2966.
82. K. Nonoshita, H. Banno, K. Maruoka, and H. Yamamoto, *J. Am. Chem. Soc.*, 1990, **112**, 316.
83. K. Maruoka, S. Saito, and H. Yamamoto, *J. Am. Chem. Soc.*, 1995, **117**, 1165.
84. B. K. Carpenter, *Tetrahedron*, 1978, **34**, 1877.
85. R. Breslow and J. M. Hoffman, Jr., *J. Am. Chem. Soc.*, 1972, **94**, 2111.
86. H.-J. Hansen, B. Sutter, and H. Schmid, *Helv. Chim. Acta*, 1968, **51**, 828.
87. D. P. Curran and L. H. Kuo, *Tetrahedron Lett.*, 1995, **37**, 6647.
88. J. F. Kincaid and D. S. Tarbell, *J. Am. Chem. Soc.*, 1939, **61**, 3085.
89. W. N. White, D. Gwynn, R. Schlitt, C. Girard, and W. Fife, *J. Am. Chem. Soc.*, 1958, **80**, 3271.
90. H. L. Goering and R. R. Jacobson, *J. Am. Chem. Soc.*, 1958, **80**, 3277.
91. W. N. White and E. F. Wolfarth, *J. Org. Chem.*, 1970, **35**, 2196.

92. R. M. Coates, B. D. Rogers, S. J. Hobbs, D. R. Peck, and D. P. Curran, *J. Am. Chem. Soc.*, 1987, **109**, 1160.
93. J. J. Gajewski, *J. Org. Chem.*, 1992, **57**, 5500.
94. J. J. Gajewski and N. L. Brichford, *J. Am. Chem. Soc.*, 1994, **116**, 3165.
95. D. L. Severance and W. L. Jorgensen, *J. Am. Chem. Soc.*, 1992, **114**, 10966.
96. M. M. Davidson and I. H. Hillier, *J. Phys. Chem.*, 1995, **99**, 6784.
97. H. Dodziuk, H. Von Voithenberg, and N. L. Allinger, *Tetrahedron*, 1982, **38**, 2811.
98. E. M. Kosower, *J. Am. Chem. Soc.*, 1958, **80**, 3253.
99. A. B. Chmurny and D. J. Cram, *J. Am. Chem. Soc.*, 1973, **95**, 4237.
100. E. Brandes, P. A. Grieco, and J. J. Gajewski, *J. Org. Chem.*, 1989, **54**, 515.
101. R. J. Ife, L. F. Ball, P. Lowe, and E. Haslam, *J. Chem. Soc., Perkin Trans. 1*, 1976, 1776.
102. J. J. Delaney, III, R. E. Padykula, and G. A. Berchtold, *J. Am. Chem. Soc.*, 1992, **114**, 1394.
103. L. Pawlak, R. E. Padykula, J. D. Kronis, R. A. Aleksejczyk, and G. A. Berchtold, *J. Am. Chem. Soc.*, 1989, **111**, 3374.
104. M.-J. Gething and B. E. Davidson, *Eur. J. Biochem.*, 1976, **71**, 327.
105. M.-J. Gething and B. E. Davidson, *Eur. J. Biochem.*, 1977, **78**, 103.
106. M.-J. Gething and B. E. Davidson, *Eur. J. Biochem.*, 1977, **78**, 111.
107. T. A. A. Dopheide, P. Crewther, and B. E. Davidson, *J. Biol. Chem.*, 1972, **247**, 4447.
108. G. L. E. Koch, D. C. Shaw, and F. Gibson, *Biochim. Biophys. Acta*, 1972, **258**, 719.
109. E. Heyde, *Biochemistry*, 1979, **18**, 2766.
110. R. I. Christopherson, E. Heyde, and J. F. Morrison, *Biochemistry*, 1983, **22**, 1650.
111. G. S. Hudson, V. Wong, and B. E. Davidson, *Biochemistry*, 1984, **23**, 6240.
112. D. Christendat and J. Turnbull, *Biochemistry*, 1996, **35**, 4468.
113. J. D. Stewart, D. B. Wilson, and B. Ganem, *J. Am. Chem. Soc.*, 1990, **112**, 4582.
114. S. G. Sogo, T. S. Widlanski, J. H. Hoare, C. E. Grimshaw, G. A. Berchtold, and J. R. Knowles, *J. Am. Chem. Soc.*, 1984, **106**, 2701.
115. P. Kast, M. Asif-Ullah, and D. Hilvert, *Tetrahedron Lett.*, 1996, **37**, 2691.
116. C. C. Galopin, S. Zhang, D. B. Wilson, and B. Ganem, *Tetrahedron Lett.*, 1996, **37**, 8675; corrigendum: *Tetrahedron Lett.*, **38**, 1467.
117. F. H. Westheimer, *Adv. Enzymol. Related Areas Mol. Biol.*, 1962, **24**, 441.
118. D. Y. Jackson, J. W. Jacobs, R. Sugasawara, S. H. Reich, P. A. Bartlett, and P. G. Schultz, *J. Am. Chem. Soc.*, 1988, **110**, 4841.
119. D. Y. Jackson, M. N. Liang, P. A. Bartlett, and P. G. Schultz, *Angew. Chem., Int. Ed. Engl.*, 1992, **31**, 182.
120. D. Hilvert and K. N. Nared, *J. Am. Chem. Soc.*, 1988, **110**, 5593.
121. D. Hilvert, S. H. Carpenter, K. N. Nared, and M. M. Auditor, *Proc. Natl. Acad. Sci. USA*, 1988, **85**, 4953.
122. W. J. Guilford, S. D. Copley, and J. R. Knowles, *J. Am. Chem. Soc.*, 1987, **109**, 5013.
123. H. S.-I. Chao and G. A. Berchtold, *Biochemistry*, 1982, **21**, 2778.
124. R. I. Christopherson and J. F. Morrison, *Biochemistry*, 1985, **24**, 1116.
125. P. A. Bartlett, Y. Nakagawa, C. R. Johnson, S. H. Reich, and A. Luis, *J. Org. Chem.*, 1988, **53**, 3195.
126. T. Clarke, J. D. Stewart, and B. Ganem, *Tetrahedron*, 1990, **46**, 731.
127. J. V. Gray, D. Eren, and J. R. Knowles, *Biochemistry*, 1990, **29**, 8872.
128. J. S. Rajagopalan, K. M. Taylor, and E. K. Jaffe, *Biochemistry*, 1993, **32**, 3965.
129. J. V. Gray and J. R. Knowles, *Biochemistry*, 1994, **33**, 9953.
130. J. Turnbull, W. W. Cleland, and J. F. Morrison, *Biochemistry*, 1991, **30**, 7777.
131. T. I. Richards, K. Layden, E. E. Werminski, P. J. Milburn, and E. Haslam, *J. Chem. Soc., Perkin Trans. 1*, 1987, 2765.
132. M. Frederickson, N. A. Bailey, H. Adams, and E. Haslam, *J. Chem. Soc., Perkin Trans. 1*, 1990, 2353.
133. Y. M. Chook, H. Ke, and W. N. Lipscomb, *Proc. Natl. Acad. Sci. USA*, 1993, **90**, 8600.
134. Y. M. Chook, J. V. Gray, H. Ke, and W. N. Lipscomb, *J. Mol. Biol.*, 1994, **240**, 476.
135. A. Y. Lee, P. A. Karplus, B. Ganem, and J. Clardy, *J. Am. Chem. Soc.*, 1995, **117**, 3627.
136. A. Y. Lee, J. Clardy, J. D. Stewart, and B. Ganem, *Chem. Biol.*, 1995, **2**, 195.
137. Y. Xue, W. N. Lipscomb, R. Graf, G. Schnappauf, and G. Braus, *Proc. Natl. Acad. Sci. USA*, 1994, **91**, 10814.
138. S. Schmidheini, H.-U. Mosch, J. N. S. Evans, and G. Braus, *Biochemistry*, 1990, **29**, 3660.
139. Y. Xue and W. N. Lipscomb, *Proc. Natl. Acad. Sci. USA*, 1995, **92**, 10595.
140. M. R. Haynes, E. A. Strau, D. Hilvert, and I. A. Wison, *Science*, 1994, **263**, 646.
141. J. Gao, *J. Am. Chem. Soc.*, 1994, **116**, 1563.
142. P. R. Andrews and R. C. Haddon, *Aust. J. Chem.*, 1979, **32**, 1921.
143. M. M. Davidson, I. R. Gould, and I. H. Hillier, *J. Chem. Soc., Chem. Commun.*, 1995, 63.
144. P. D. Lyne, A. J. Mulholland, and W. G. Richards, *J. Am. Chem. Soc.*, 1995, **117**, 11345.
145. M. J. Field, P. A. Bash, and M. Karplus, *J. Comp. Chem.*, 1990, **11**, 700.
146. O. Wiest and K. N. Houk, *J. Org. Chem.*, 1994, **59**, 5782.
147. O. Wiest and K. N. Houk, *J. Am. Chem. Soc.*, 1995, **117**, 11628.
148. S. T. Cload, D. R. Liu, R. M. Pastor, and P. G. Schultz, *J. Am. Chem. Soc.*, 1996, **118**, 1787.
149. D. R. Liu, S. T. Cload, R. M. Pastor, and P. G. Schultz, *J. Am. Chem. Soc.*, 1996, **118**, 1789.
150. S. Zhang, P. Kongsaeree, J. Clardy, D. B. Wilson, and B. Ganem, *Biorg. Med. Chem.*, 1996, **4**, 1015.
151. A. Radzicka and R. Wolfenden, *Science*, 1995, **267**, 90.

5.15
Thymine Dimer Photochemistry: A Mechanistic Perspective

TADHG P. BEGLEY
Cornell University, Ithaca, NY, USA

5.15.1   INTRODUCTION

Ultraviolet radiation is the most abundant genotoxic agent to which living organisms are exposed; therefore, the evolution of effective repair strategies for UV-induced lesions in DNA has been essential for the survival of living systems. Exposure to the UV-B (290–320 nm) component of sunlight is a key event in the development of skin cancer in humans.[1,2] More than 800 000 people annually in the United States alone are expected to develop nonmelanoma skin cancer[3] and approximately one in four will get skin cancer at some point during their lifetimes.[4] Human health concerns about radiation damage to the genome have motivated research on the photochemistry of DNA and there is now a substantial body of information on the premutagenic lesions introduced into DNA by UV radiation.[5,6]

The nucleic acids constitute the dominant UV-absorbing chromophore in the cell. When DNA is irradiated, rapid energy transfer from the adenine, cytosine, and guanine triplet states to the lower energy thymine triplet state occurs. Consequentially, thymine (1) is the most photochemically reactive base in DNA.[7,8]

This chapter will focus on the mechanism of formation and repair of the premutagenic photo-lesions that form at adjacent thymines. These are the cyclobutane pyrimidine photodimer (2), the (6-4) photoproduct (3), the Dewar pyrimidinone (4), and the spore photoproduct (5) (Scheme 1). All four lesions can be efficiently repaired by an excision repair pathway in which the lesion is excised from the DNA and the resulting gap is filled in by DNA polymerase.

In addition, single enzymes exist for the conversion of photolesions (2), (3), and (5) directly back to two thymines.

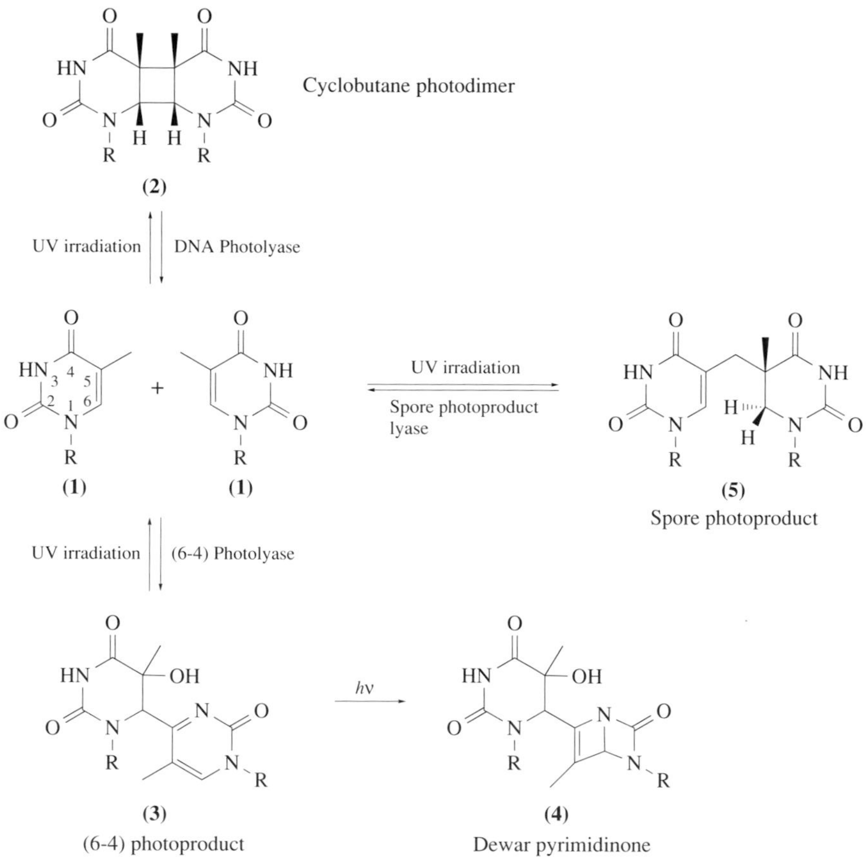

Scheme 1

5.15.2   THE CYCLOBUTANE PYRIMIDINE PHOTODIMER

5.15.2.1   Mechanism of Formation

The 5,6 double bonds of adjacent pyrimidines in DNA are well positioned to undergo a photochemical [2 + 2] cycloaddition reaction to give the cyclobutane pyrimidine photodimers (2) and (6)

(Table 1).[9-11] The thymine cyclobutane pyrimidine photodimer (2) is the highest quantum yield photolesion formed in DNA ($\Phi = 0.019$).[12] This dimerization reaction occurs from the thymine triplet state and is sensitized by triplet sensitizers such as acetone and acetophenone.[13]

A mechanism for the formation of the cyclobutane photodimer is outlined in Scheme 2. Addition of the C-5 or C-6 radical of the thymine triplet to the corresponding carbon of the adjacent thymine would give biradicals (11) or (12). Intersystem crossing to the singlet biradical, followed by radical recombination, would give the photodimer. It is not known whether the 5,5′ bond or the 6,6′ bond is formed first. Electronic structure calculations predict that both intermediates are close in energy for thymine dimerization.[14] The regiochemistry and stereochemistry are controlled by the geometry of the DNA double helix and the major isomer formed in DNA has the *cis–syn* stereochemistry shown for (2).

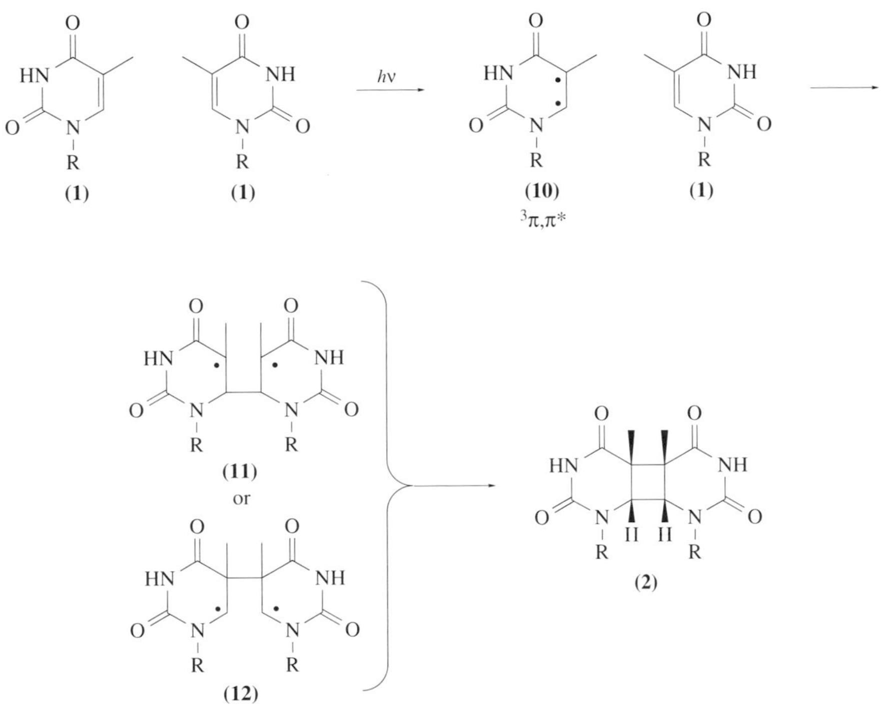

Scheme 2

UV-irradiation of thymine (13) in solution results in a mixture of all four possible thymine cyclobutane photodimer stereoisomers (14)–(17) (Equation (1)).[15]

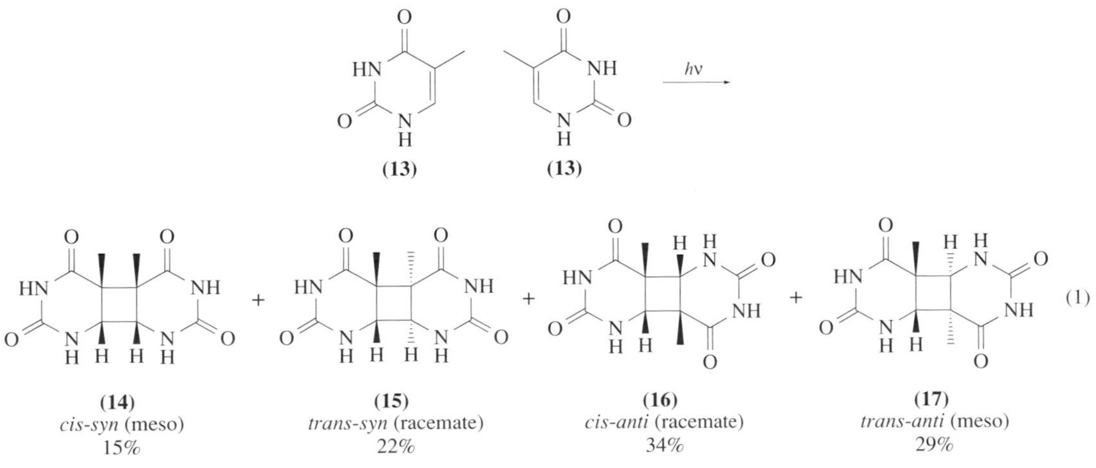

(1)

(14)
cis-syn (meso)
15%

(15)
trans-syn (racemate)
22%

(16)
cis-anti (racemate)
34%

(17)
trans-anti (meso)
29%

Table 1 Thymine derived photoproducts in *Escherichia coli* DNA and in dinucleotides.

Photoproduct	Relative yield in E. coli *DNA* (300 *nm*)	Quantum yield in E. coli *DNA* (280 *nm*)	Relative yield in dTpdN (300 *nm*)	Quantum yield in dTpdN (297 *nm*)
(2)	1	0.019	28	0.016
(6)	1.15	0.0027	6	0.0072
(3)	Not determined	0.0015 (3)+(7)	1	0.0005
(7)	Not determined	0.0015 (3)+(7)	1	0.00074
(4)	Not determined	Not determined	Not determined	0.02 (from the (6-4) photoproduct)
(8)	Not determined	Not determined	Not determined	0.018 (from the (6-4) photoproduct)
(9)	Not determined	Not determined	Not determined	0.00074

Table 1 (continued)

Photoproduct	Relative yield in E. coli DNA (300 nm)	Quantum yield in E. coli DNA (280 nm)	Relative yield in dTpdN (300 nm)	Quantum yield in dTpdN (297 nm)
(structure) (5)	0	0	0	0

The stereochemistry of the reaction can be controlled by constraining thymine in ice[16] or by linking the two thymines together with a short linker (Equation (2)).[17-19] Analogous CT and CC photodimers have also been synthesized.[20]

$$ (18) \xrightarrow[\text{Acetone/water}]{h\nu} (19) \quad (2) $$

UV-irradiation of the protected dTpdT dinucleotide gives the corresponding protected cyclobutane photodimer which has been used for the synthesis of photodimer-containing oligonucleotides.[21]

The crystal structures of the *cis–syn* 1,3-dimethylthymine dimer and the dTpdT dinucleotide photodimer have been determined.[22,23]

5.15.2.2 Photoreactivation

In the late 1940s, Albert Kelner observed that the lethal effects of UV irradiation on the fungus *Streptomyces griseus* could be reduced $1–4 \times 10^5$ fold by exposing the UV-irradiated cultures to visible light.[24] This remarkable phenomenon, later called photoreactivation, is the result of the repair of the cyclobutane photodimer by a light-dependent enzyme called DNA photolyase.

DNA photolyases have now been isolated from a large number of organisms. The *Escherichia coli* photolyase has been cloned and overexpressed at a high level[25,26] and is the best characterized system.[27-30] This enzyme is a 54 kD monomer, and binds to the photodimer in both single- and double-stranded DNA with comparable affinities ($K_d = 10^{-9}$ M^{-1} and 10^{-8} M^{-1}, respectively).[31,32] The binding to photodimer-free DNA is much weaker ($K_d = 10^{-4}$ M^{-1}).[31] The enzyme requires a photon in the 300–500 nm range as an essential component of the catalytic mechanism. The quantum yield of the monomerization reaction is high (0.7).[33]

DNA photolyase will also catalyze the cleavage of the dinucleotide cyclobutane photodimers of dTpdT, dUpdT, dUpdU, and dCpdC with quantum yields of 0.9, 0.8, 0.6, and 0.05 respectively.[34] For the dTpdT dinucleotide photodimer, the binding is greatly reduced compared to the binding of the enzyme to a thymine dimer in double-stranded DNA ($K_d = 10^{-4}$ M^{-1} and 10^{-9} M^{-1}, respectively).

The enzyme contains two cofactors: a 5,10-methenyltetrahydrofolylpolyglutamate (**20**) and a deprotonated reduced flavin (FADH$^-$, **21**).[35-38] The folate cofactor functions as an antenna chromophore by increasing the extinction coefficient of the enzyme. About 80% of the photons are absorbed by this chromophore[39] and this excitation is rapidly ($k = 4.6 \times 10^9$ s^{-1}) and efficiently (62%) transferred from the folate to FADH$^-$.[40] The photoexcited FADH$^-$ sensitizes the photodimer cleavage reaction.

(Glu)$_n$

(20)   **(21)** FADH$^-$

The crystal structure of the *E. coli* DNA photolyase has been determined and a model for the binding of the photodimer to the enzyme has been proposed.[41] In this model, the photodimer flips out from the double helix to bind to the flavin-containing active site where one of the thymines of the dimer is in van der Waals contact with the flavin chromophore. Asparagine 273, Glutamate 274, Asparagine 341, and the adenine of FADH$^-$ may form hydrogen bonding interactions with the dimer. The center of the folate chromophore is located 16.8 Å from the center of the flavin. Verification of this model and the identification of the interactions between the enzyme and the photodimer, which are important for catalysis, will require a structure of the enzyme substrate complex.

The mechanism by which the photoexcited flavin sensitizes the cleavage of the thymine photodimer is an intriguing mechanistic problem. The action spectrum for the enzyme (300–500 nm) rules out the possibility of photodimer cleavage by direct energy transfer from the photoexcited reduced flavin to the photodimer, because the photodimer does not have appreciable absorbance at $\lambda > 220$ nm. This suggests that the cleavage reaction occurs via radical ion intermediates. Since the oxidation and reduction potentials of the photodimer (2.0 V and -2.4 V vs. NHE, respectively)[42,43] lie well outside the physiologically accessible range, DNA photolyase requires light energy to enhance the redox potential of the enzyme even though the photodimer fragmentation reaction is exothermic by about 100 kJ mol^{-1}.[44,45]

In contrast to the high stability of the photodimer, its radical cation and anion are both highly reactive species and undergo facile monomerization reactions. These have therefore been proposed as possible intermediates during photoreactivation.[46]

5.15.2.3   The Cyclobutane Pyrimidine Photodimer Radical Cation

The thymine cyclobutane photodimer radical cation (23) can be generated by electron transfer to several photooxidizing agents (flavin,[47–51] quinones,[52–54] metal ions[55]). This radical is very unstable (lifetime < 1 ns)[56] and rapidly fragments. The CT and the CC cyclobutane photodimer radical cations also undergo facile fragmentation.[57] The mechanism of the fragmentation reaction is relatively well understood and is outlined in Scheme 3.

Cleavage of the 6,6 bond of (23) gives (24). Cleavage of the 5,5 bond of (24) followed by reduction of the thymine radical cation (25) by the sensitizer radical anion completes the reaction. This mechanism is consistent with the experiments described in the following sections.

5.15.2.3.1   *Trapping of the one-bond-cleaved intermediate*

A carbon–iodine bond beta to a radical center undergoes very fast ($k > 5 \times 10^9$ s^{-1}) homolysis.[58] This reaction has been used to trap the one-bond-cleaved intermediate (24) during the fragmentation of the photodimer radical cation (Scheme 4). When photodimer (27) was irradiated in the presence of dichlorodicyanobenzoquinone (DDQ) as the sensitizer, two reaction products (28) and (29) were isolated.

The formation of these products can be explained as follows. Cleavage of the 6,6 bond of (30) followed by cleavage of the 5,5 bond of (31) and back electron transfer from the sensitizer radical anion would give (28). In competition with this pathway, cleavage of the C—I bond of (31) would give acyl imminium ion (33) which, after cyclization and addition of water, would give (29).[59]

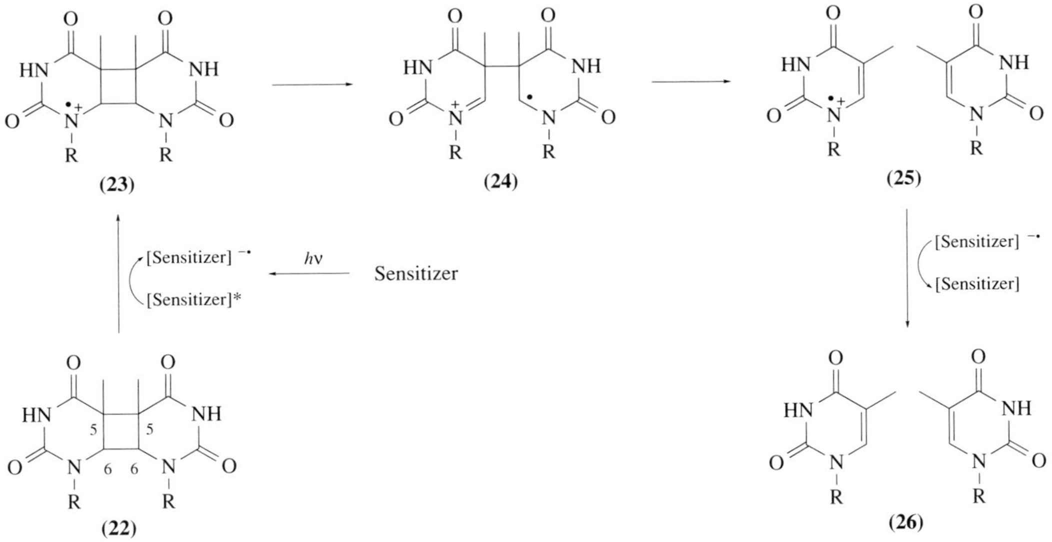

Scheme 3

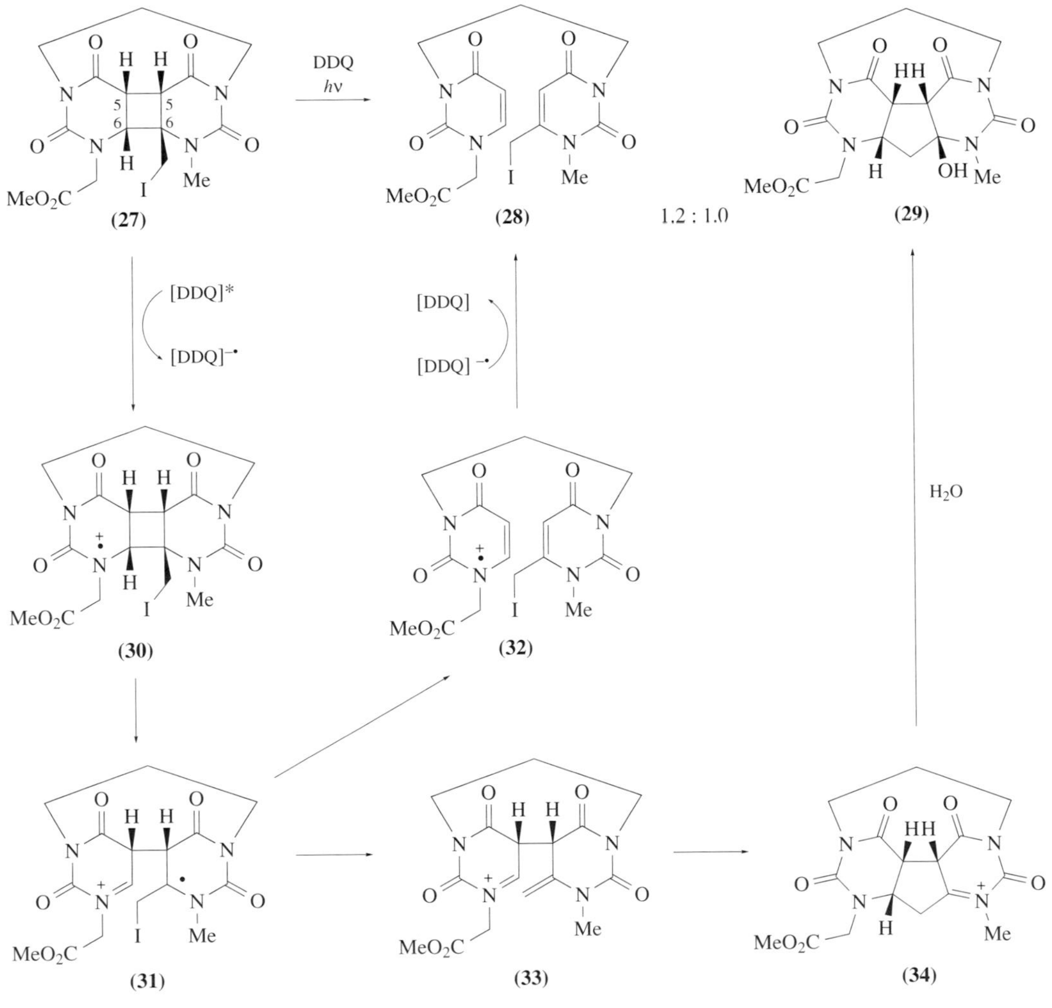

1.2 : 1.0

Scheme 4

Alkene extrusion from the radical cation of (35) also occurred by a stepwise mechanism as indicated by the formation of a mixture of (*E*)- and (*Z*)-alkene isomers (37)–(38) when (35) was irradiated in the presence of DDQ (Equation (3)).[60]

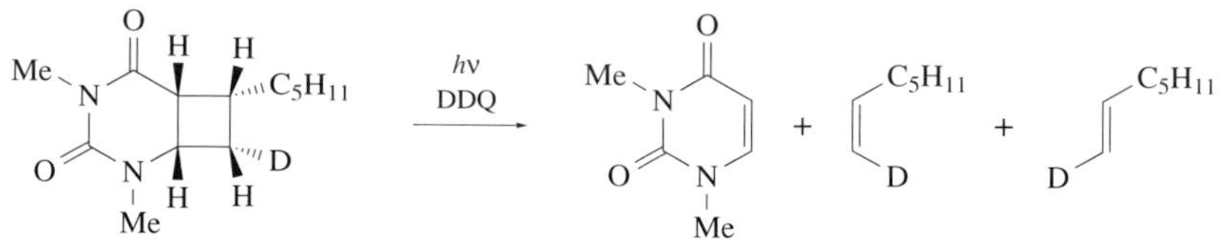

$$ (3) $$

5.15.2.3.2 *Isotope effects on the fragmentation*

The rate of a reaction involving a hybridization change can be retarded or accelerated, often in a predictable manner, by replacing hydrogen with deuterium. These deuterium isotope effects have been extensively used in mechanistic studies to determine the rate limiting step, or the first irreversible step, in a reaction sequence and to probe transition state structure.[61–64] Isotope effects measured under competitive conditions (i.e., using a mixture of protio and deuterio substrates) register hybridization changes occurring up to and including the first irreversible step in the reaction pathway. For these competitive isotope effects, hybridization changes after the first irreversible step, even if originating from the rate determining step in the reaction, are not registered.

The photodimer fragmentation reaction involves the conversion of four sp^3 centers to four sp^2 centers. This reaction should therefore show a normal (i.e., $k_H > k_D$) secondary deuterium isotope effect. These isotope effects, measured under competitive conditions, on the fragmentation of the radical cation of the model photodimer (39) are shown with the structure.[65] The observation of a substantial secondary deuterium isotope effect at the 6,6 positions and a negligible effect at the 5,5 positions is consistent with a fragmentation mechanism involving an irreversible cleavage of the 6,6 bond followed by cleavage of the 5,5 bond.

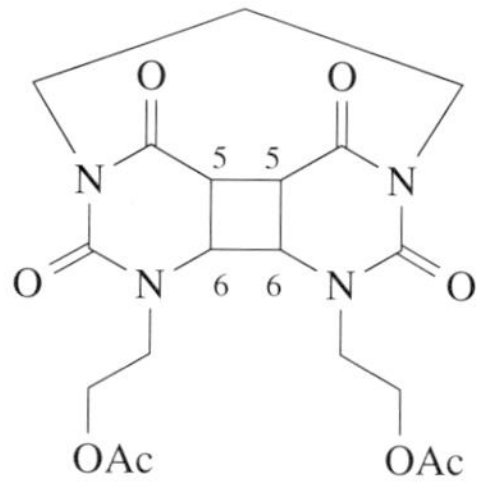

Isotope effects

6,6-D$_2$	1.19±0.02
5,5-D$_2$	1.03±0.02

Sensitizer : Anthraquinone

(39)

5.15.2.3.3 *Photo-CIDNP*

When the NMR spectrum of the *syn,cis* thymine photodimer (14) and anthraquinone-2-sulfonate is measured while the sample is being irradiated, a weakly enhanced absorption signal is observed for the C-5-Me protons of the thymine product. This modified NMR spectrum, called a photo-CIDNP spectrum, is due to nuclear spin sorting in the magnetic field of the instrument and demonstrates that the thymine radical cation is an intermediate in the anthraquinone-2-sulfonate sensitized photodimer cleavage reaction. The absence of enhanced NMR signals for the dimer demonstrates that the dimer radical cation has a lifetime that is too short for spin sorting to occur (i.e., <1 ns).[66]

5.15.2.3.4 *Electronic structure calculations*

Electronic structure calculations using the AM1 UHF method have been carried out on the fragmentation of the thymine photodimer radical cation.[67] The calculations support a mechanism involving cleavage of the 6,6 bond followed by cleavage of the 5,5 bond. There is no barrier for the cleavage of the 6,6 bond and the activation energy for the cleavage of the 5,5 bond is 59 kJ mol^{-1}. In contrast, the calculated barriers for the corresponding cleavage of the neutral photodimer are 96 kJ mol^{-1} for the initial cleavage of the 6,6 bond and 109 kJ mol^{-1} for the subsequent cleavage of the 5,5 bond. The barrier for the cleavage of the 5,5 bond of the biradical intermediate is too large suggesting that the calculation does not reliably predict the behavior of biradicals.

5.15.2.4 The Cyclobutane Pyrimidine Photodimer Radical Anion

The thymine cyclobutane photodimer radical anion can be generated by electron transfer from photoreducing agents such as indole,[68] dimethoxybenzene,[69] catalytic antibodies,[70] reduced flavin,[71–73] and dimethylaniline[74] and undergoes a facile fragmentation reaction ($k_{cleavage} = 1.8 \times 10^6$ s^{-1}).[74] The CT and the CC cyclobutane photodimer radical anions also undergo facile fragmentation.[75] Three possible mechanisms for this fragmentation are outlined in Scheme 5. Mechanism A is analogous to the fragmentation mechanism of the photodimer radical cation except that the 5,5 bond is cleaved before the 6,6 bond. Mechanism B is an asynchronous concerted fragmentation of the 5,5 and the 6,6 bonds[76] and mechanism C involves electron transfer from the one-bond-cleaved intermediate (41) back to the sensitizer radical cation before cleavage of the 6,6 bond.[77] Experiments to differentiate between these mechanisms are described in the following sections.

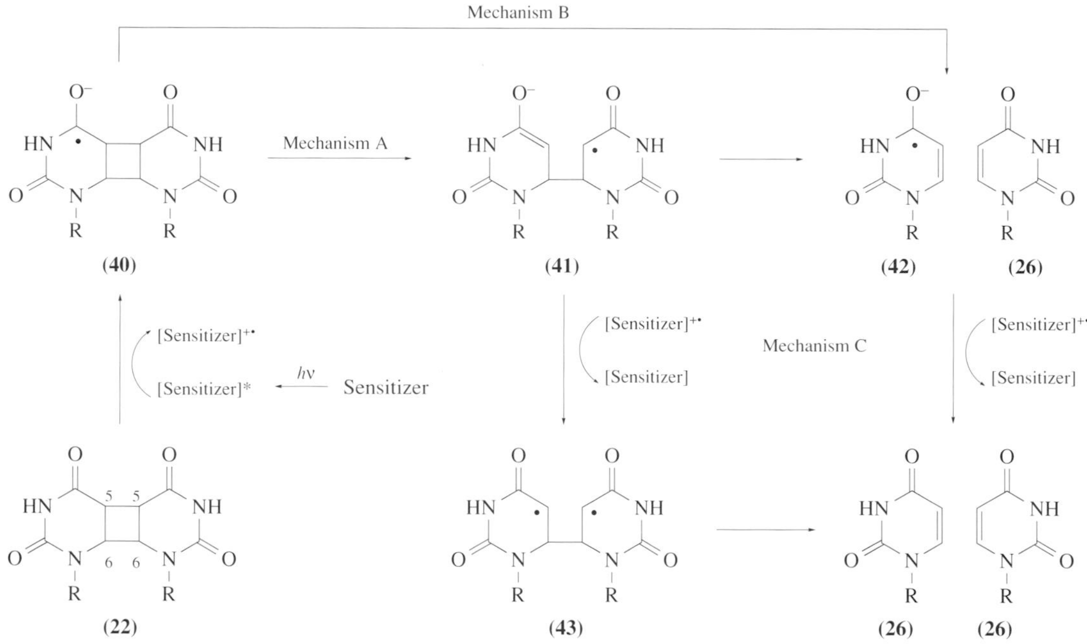

Scheme 5

5.15.2.4.1 Photo-CIDNP

In contrast to the photo-CIDNP spectrum for the anthraquinone-2-sulfonate sensitized cleavage of the *cis,syn* thymine photodimer (14), the photo-CIDNP spectrum of the *N*-acetyltryptophan sensitized cleavage reaction shows an intense emission signal for the C-6 proton of the thymine product.[78] This demonstrates that the thymine radical anion is an intermediate in the indole sensitized photodimer cleavage reaction.

The different photo-CIDNP spectra for the thymine radical anion and cation are a consequence of the different distribution of the unpaired electron in the two radicals. For the thymine radical cation, the single electron is localized primarily at C-5 and N-1; for the thymine radical anion, the single electron is localized primarily at C-4 and C-6.

5.15.2.4.2   *Detection of the thymine radical by EPR*

The radical anion of the *cys-syn* thymine dimer (14), generated by γ-radiolysis, underwent rapid fragmentation at 77 K to give the thymine radical anion which was detected by EPR.[79]

The EPR and photo-CIDNP experiments demonstrate that oxidation of (41) to the biradical (43) is not essential for the fragmentation to occur.

5.15.2.4.3   *Trapping of the one-bond-cleaved intermediate (41)*

Two experiments have been carried out to trap (41) and neither has been successful.

The first attempt involved the use of the well precedented methylcyclopropyl rearrangement to trap radical intermediates (Scheme 6).[80] When the cyclopropyl substituted photodimer (44) was irradiated in the presence of dimethylaniline as the reducing sensitizer, (45) was the only product formed. No products resulting from the methylcyclopropyl rearrangement (46)→(47) were detected. However, since the rate of rearrangement of (46) is $< 2.5 \times 10^4$ s^{-1}, this experiment does not rule out the possibility of a 5,5 bond-cleaved intermediate on the reaction pathway.[80]

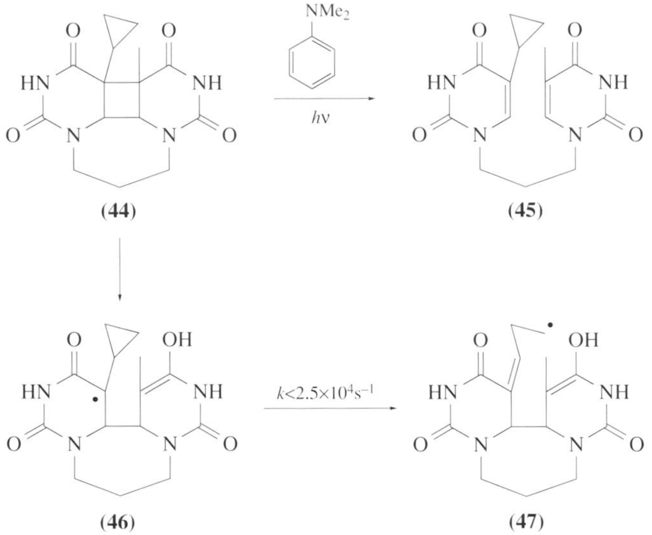

Scheme 6

A trapping experiment, analogous to that outlined in Scheme 4 for the radical cation, was not possible for the radical anion fragmentation reaction because of the reductive lability of the carbon–iodine bond.[81] Therefore an alternative trapping strategy, based on the rapid loss of a leaving group beta to a carbanion,[82] was developed. The rate constant for expulsion of thiophenolate from (48) is approximately 10^{10} s^{-1} (Equation (4)). This suggests that mesylate loss from (53) (Scheme 7) may occur at a faster rate because mesylate is a much better leaving group than thiophenylate.

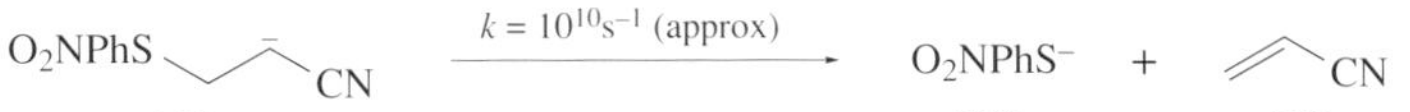

$$O_2NPhS \diagup^{\bar{}}\diagdown CN \xrightarrow{k = 10^{10}s^{-1} \text{ (approx)}} O_2NPhS^- + \diagup\diagdown CN \qquad (4)$$

$$(48) \qquad\qquad\qquad\qquad\qquad (49) \qquad (50)$$

When (51) was irradiated in the presence of *N*-phenylpiperazine as sensitizer, (59) was formed as the major reaction product (Scheme 7). Bispyrimidine (56), the expected product resulting from loss of mesylate from the one-bond-cleaved intermediate (53) followed by β-scission and hydrogen atom abstraction was not detected in the reaction mixture (detection limit $< 0.01\%$).[83] In a control experiment, (54) was independently generated and shown to convert cleanly to (56). This suggests, if the fragmentation proceeds via mechanism A, that the cleavage of the 6,6 bond of (53) is $> 10^4$ faster than mesylate loss and therefore that the rate of cleavage of the 6,6 bond is approximately 10^{14} s^{-1}. Since this rate constant is on the same time scale as a bond vibration, this experiment suggests the possibility that the photodimer radical anion fragmentation reaction may proceed by an asynchronous concerted mechanism. This proposal, however, is tentative until the rate of mesylate loss from (53) has been measured because (48) may not be a valid model for this reaction.

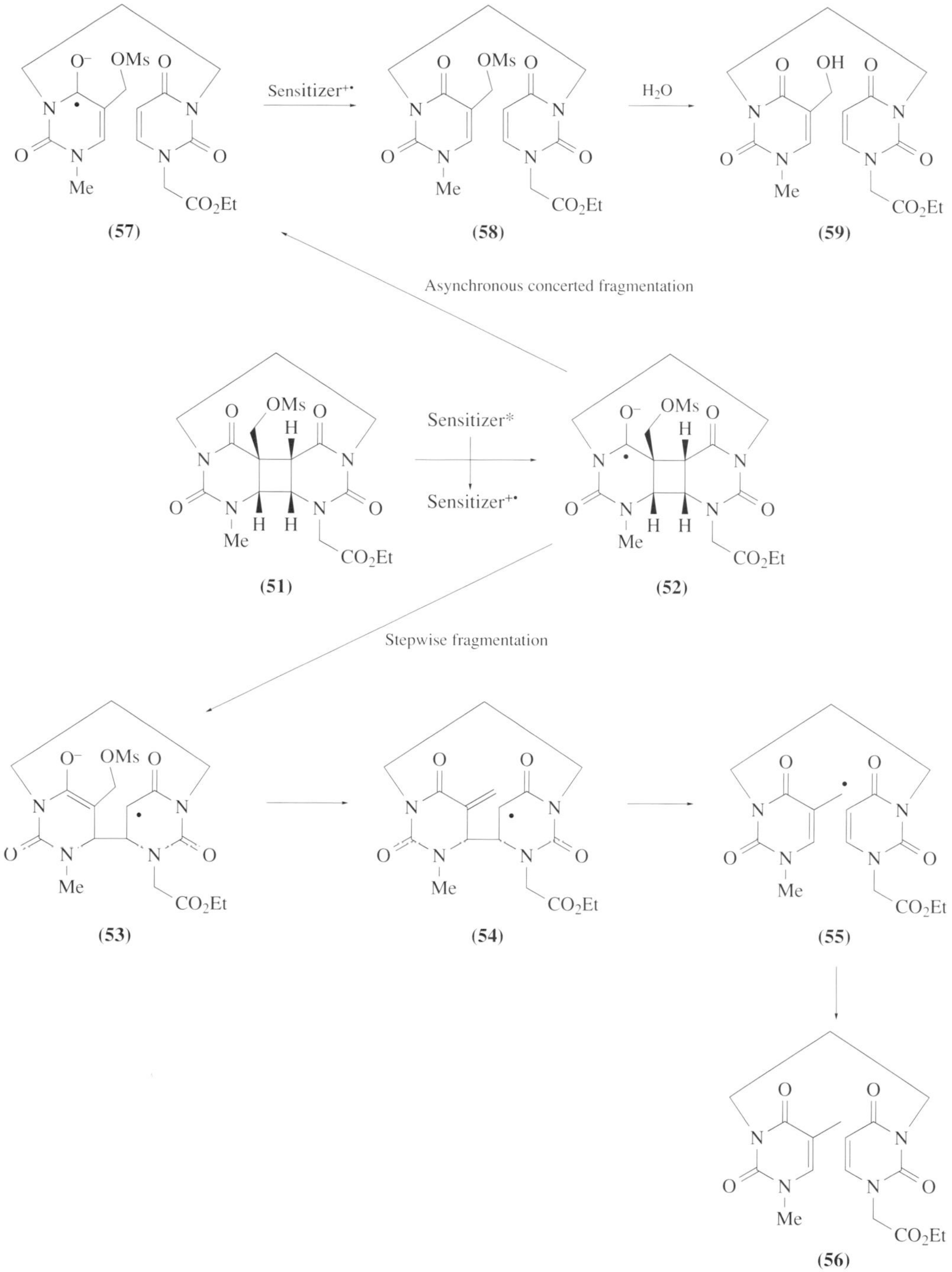

Scheme 7

5.15.2.4.4 Isotope effects on the fragmentation

The secondary deuterium isotope effects, measured under competitive conditions, for the fragmentation of the photodimer radical anion are shown with Structure (**60**).[84] Based on the isotope effects observed for the fragmentation of the photodimer radical cation, the relatively large isotope effect at the 6,6 positions was unexpected. This isotope effect is consistent with an asynchronous concerted fragmentation mechanism[85] or with a large β-isotope effect on the cleavage of the 5,5′ bond. The magnitude of a β-isotope effect on radical and carbocation formation depends on the overlap between the empty or partially filled p orbital at the α-carbon and the β C—H bond.[86] If this overlap is larger in the transition state for the cleavage of the 5,5 bond of the photodimer radical anion than in the transition state for the cleavage of the 6,6 bond of the photodimer radical cation, the β-isotope effect on the photodimer radical anion fragmentation will be larger than the β-isotope effect on the radical cation fragmentation.

A large isotope effect at the 6,6 positions for the antibody catalyzed cleavage of the photodimer has also been reported (V/K at the 5,5 and the 6,6 positions of 1.11 and 1.14, respectively). In this system, tryptophan is the sensitizer.[87] The difference in the magnitude of the isotope effects for the two systems (and for the enzymatic reaction which is described in Section 5.15.2.5.2) may be due to two factors: the transition state structure for the two fragmentation reactions may not be identical due to the short linker between the two pyrimidines of (60) and the relative kinetic significance of the fragmentation, compared to the electron transfer, is likely to be different for the two systems.

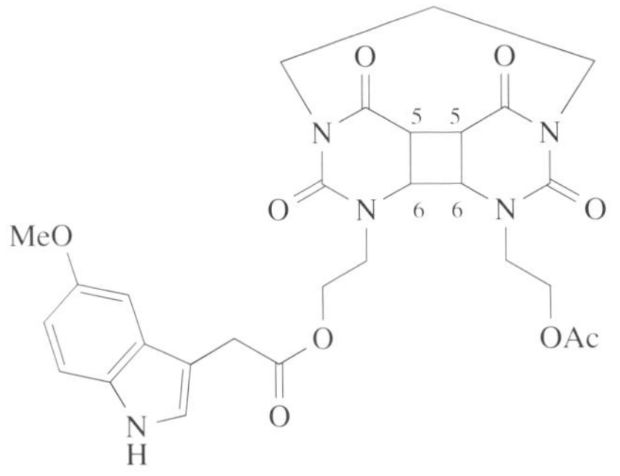

Isotope effects

5,5-D_2	1.17±0.01
6,6-D_2	1.08±0.01

(60)

5.15.2.4.5   *Electronic structure calculations*

Simple Hückel molecular orbital theory was used to predict that the energy barrier for the fragmentation of the photodimer via an asynchronous concerted mechanism or via a stepwise mechanism is reduced in the photodimer radical anion compared to the neutral species.[88] Electronic structure calculations using the AM1 UHF method have also been carried out on the fragmentation of the thymine cyclobutane photodimer radical anion.[89] The calculations support a mechanism involving the sequential cleavage of the 5,5 bond followed by cleavage of the 6,6 bond. However, this computational method is biased towards radical intermediates and therefore does not exclude the possibility of an asynchronous concerted fragmentation mechanism. Based on the calculated high electron affinity of the one-bond-cleaved intermediate, it was proposed that the cleavage of the 6,6 bond occurs from the radical anion and not from a biradical intermediate. The activation energies for the cleavage of the 5,5 and the 6,6 bonds were calculated to be 20 kJ mol^{-1} and 22 kJ mol^{-1}, respectively. The calculated barriers for the corresponding bond cleavage steps in the neutral photodimer were 118 kJ mol^{-1} and 31 kJ mol^{-1}.

5.15.2.5   The Enzymatic Reaction Mechanism

While the pyrimidine photodimer radical cation and anion can both undergo facile fragmentation reactions, there is now a substantial body of experimental evidence, which is summarized below, supporting the intermediacy of the photodimer radical anion for the enzymatic reaction.

5.15.2.5.1   *Thermodynamics*

The thermodynamic cycle for the FADH$^-$ sensitized cleavage of the dimethylthymine cyclobutane photodimer is shown in Figure 1.[90] Light absorption generates the high energy FADH$^-$ excited state which then undergoes exergonic electron transfer to the dimer. This is followed by exergonic fragmentation and back electron transfer from the monomer radical anion to the flavin. Electron transfer from FADH$^-$* to the photodimer is estimated to be exergonic by 63 kJ mol^{-1}. In contrast, electron transfer from the photodimer to $FADH^-*$ is estimated to be endergonic by 180 kJ mol^{-1}.[91] While there is still significant uncertainty about the exact reduction potential of dihydroflavin and the oxidation potential of the photodimer, it is unlikely that the errors in these numbers are sufficiently large to make formation of the radical cation thermodynamically favorable for the enzymatic reaction.

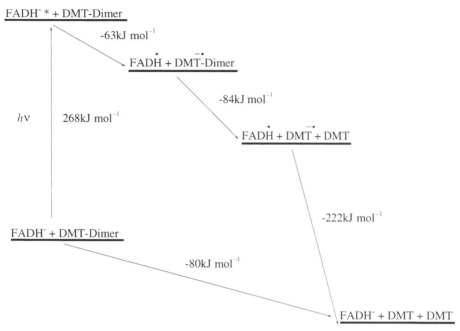

Figure 1 The estimated thermodynamic cycle for the $FADH^-$ sensitized cleavage of the dimethylthymine cyclobutane photodimer (DMT-dimer) via the radical anion pathway.

5.15.2.5.2 *Isotope effects on the fragmentation*

As discussed in Section 5.15.2.4.4, the fragmentation of the photodimer radical anion showed a relatively large isotope effect for the cleavage of the 6,6 bond which was not evident in the radical cation fragmentation reaction. Although it is still unclear whether this isotope effect is due to an asynchronous concerted fragmentation mechanism or to a large β-isotope effect, the presence of this effect is a useful diagnostic tool for differentiating between the radical cation and anion pathways for the enzyme. Thus, if the enzymatic reaction proceeds via the radical cation mechanism, the V/K isotope effect at the 5,5 positions should be small relative to the isotope effect at the 6,6 positions. However, if the reaction proceeds via the radical anion mechanism, the V/K isotope effect at the 6,6 positions should be substantial relative to the isotope effect at the 5,5 positions.

The results of this experiment using the uracil dinucleotide photodimer (**61**) are shown with the structure.[92] The relatively large isotope effect for the 6,6 bond cleavage is consistent with the photodimer radical anion fragmentation pathway.

Isotope effects

6,6-D$_2$	1.07±0.01
5,5-D$_2$	1.08±0.01

(61)

5.15.2.5.3 Trapping experiments

Trapping experiments, analogous to those carried out with the model systems using photodimers (62), (63), (64), and (65) have been attempted. The mesylate (62) is not a substrate for the enzyme. Iodomethyl substituted photodimers (64) and (65) were substrates; however, they were ambiguous mechanistic probes due to the possibility of competing direct reductive cleavage of the C—I bond by FADH⁻*. Attempts to synthesize the 6-iodomethyl substituted photodimer (63) were unsuccessful because irradiation of dinucleotides in which one of the pyrimidines has a substituent at C-6 gives the *trans–syn* photodimer (15).[93]

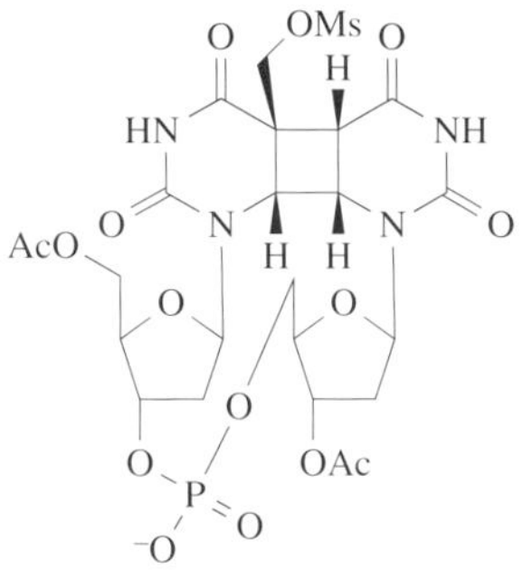

(62) 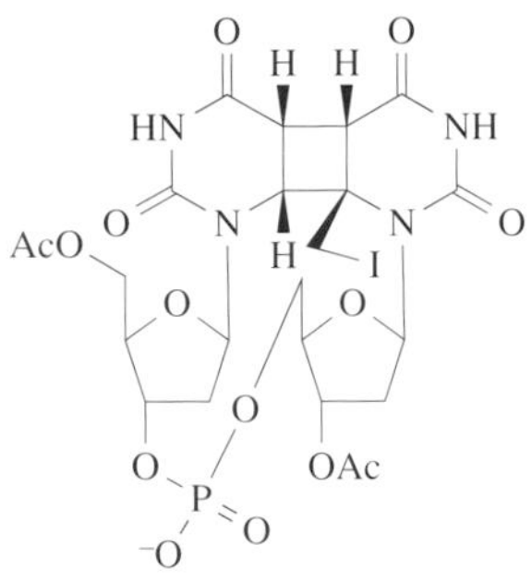   (63)

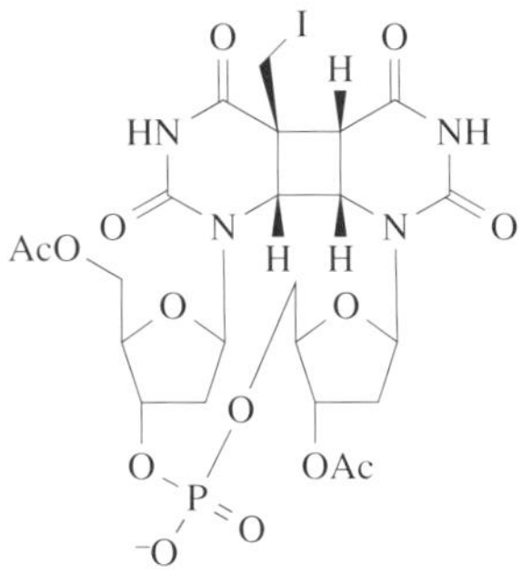

(64) 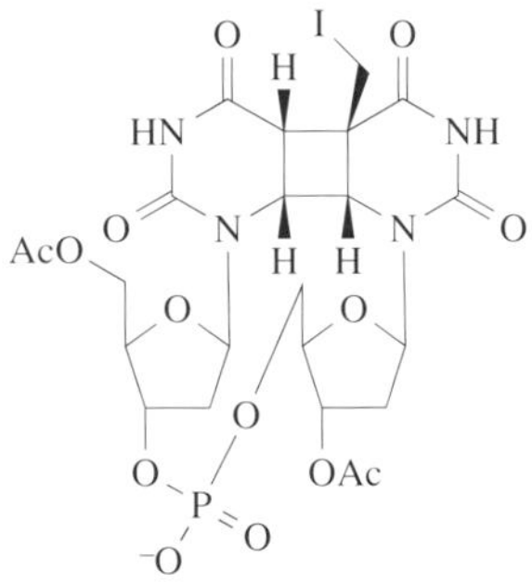   (65)

5.15.2.5.4 Photo-CIDNP

The radical cation and the radical anion fragmentation reactions give rise to completely different CIDNP spectra because the electron spin density distribution in the pyrimidine radical anion is different from the electron spin density distribution in the pyrimidine radical cation. Using this technique, it has been demonstrated that oxidized flavin sensitizes the cleavage of the photodimer via the radical cation pathway[94] and that reduced flavin sensitizes the cleavage reaction via the radical anion mechanism.[95]

5.15.2.5.5 Transient absorption spectroscopy

DNA photolyase is an ideal system for rapid kinetic studies using transient absorption spectroscopy because the enzyme substrate complex remains unreactive until irradiated. With this technique, the absorption spectra and the rates of formation and disappearance of the intermediates involved in the cleavage reaction can be probed.

When DNA photolyase, complexed to the dUpdU cyclobutane photodimer, is irradiated, the first intermediate to appear has a broad absorbance spanning 450–900 nm.[96] This absorbance has been assigned to the singlet excited state of the deprotonated reduced flavin (FADH⁻).[97–99] This signal decays with a rate constant of 5.5×10^9 s⁻¹, due to electron transfer from the reduced flavin to the photodimer and a new intermediate absorbing at 400 nm is observed to form over two nanoseconds.[100,101] Using the dUpdT cyclobutane photodimer as substrate, the lifetime of this intermediate has been estimated to be 0.5–2 ns[102] and has a shorter lifetime than the intermediate previously detected by ESR.[103] This intermediate was not observed when the dTpdT photodimer was used as a substrate, suggesting that the extra methyl group shifted the λ_{max} of the intermediate outside of the region probed (390–430 nm) or that the lifetime of the methyl substituted intermediate was too short for detection. It has been proposed that the 400 nm absorbing intermediate is due to the putative 5,5' bond cleaved intermediate (**41**), but in the absence of a more complete set of reference spectra, this assignment is tentative.

Electron transfer from FADH⁻ should give the flavin semiquinone, a species with a well characterized absorbance spectrum. This has not yet been detected by transient absorption spectroscopy. The transient spectroscopy studies also suggest that the photodimer fragmentation reaction is completed in the nanosecond time regime. This represents approximately a 1000-fold increase in rate over that observed for the nonenzymatic fragmentation reaction.[104] The specific interactions between the photodimer and the enzyme, required for this rate acceleration, have not yet been identified.

5.15.2.5.6 *Mechanistic proposal for the enzymatic reaction*

The current mechanistic proposal for the cleavage of the pyrimidine photodimer by DNA photolyase is outlined in Scheme 8. Reduction of the photodimer by the photoexcited FADH⁻ gives the photodimer radical anion (**66**). Cleavage of the 5,5' bond and the 6,6' bond, by an asynchronous concerted mechanism or by a stepwise mechanism, and back electron transfer to the flavin would complete the reaction.

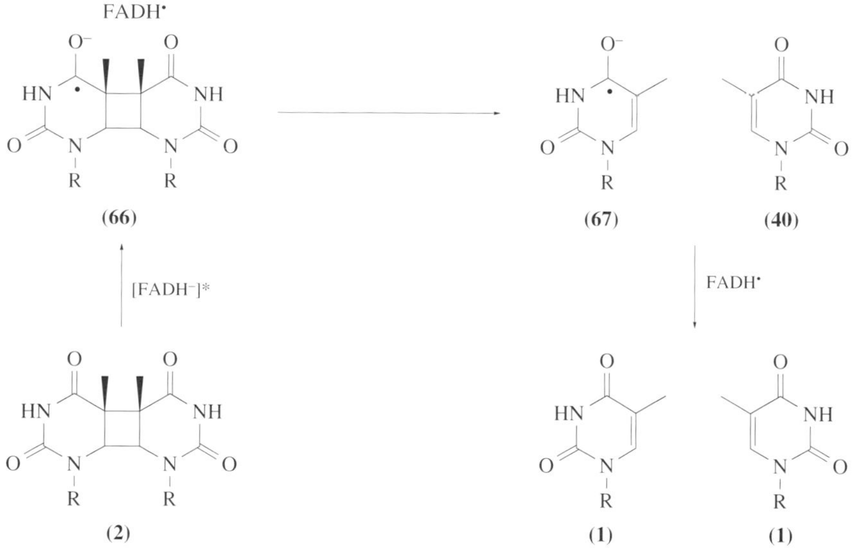

Scheme 8

Since oxidized flavin can cleave the photodimer via the radical cation pathway and reduced flavin can cleave the photodimer via the radical anion pathway, it is interesting to speculate as to why evolution has chosen the reductive pathway over the oxidative pathway. One possible explanation is based on the fact that proteins are very difficult to reduce but rather easy to oxidize. The photodimer radical cation is a very strong oxidizing agent (oxidation potential = 2.0 V vs. NHE)[105] and could oxidize tyrosine, tryptophan, histidine, cysteine, and methionine residues on the enzyme. Since electron transfer in proteins can occur over long distances, such oxidations would lower the quantum yield for the cleavage reaction and reduce the catalytic lifetime of the enzyme. Thus, it is likely that evolution has selected a photodimer repair pathway that is compatible with the redox properties of proteins.

5.15.3 THE (6-4) PHOTOPRODUCT

5.15.3.1 Formation of the (6-4) Photoproduct

In addition to the formation of the cyclobutane photodimer, UV irradiation of DNA results in the formation of a photolesion that makes the DNA sensitive to strand cleavage under basic conditions.[106,107]

This photolesion has been identified as the (6-4) photoproduct (**3**) and is the second most abundant photolesion formed at adjacent pyrimidines in UV-irradiated DNA (Table 1). The thymine (6-4) photoproduct has been synthesized and its structure determined.[108,109] UV-irradiation of the protected dTpdT dinucleotide gives the corresponding protected (6-4) photoproduct which has been incorporated into oligonucleotides.[110,111]

The mechanism for the formation of the thymine (6-4) photoproduct is outlined in Scheme 9. A [2 + 2] cycloaddition of the C-4 carbonyl group of one thymine across the 5,6 double bond of the adjacent thymine would give oxetane (**68**) which undergoes ring opening to give the (6-4) photoproduct (**3**). This can undergo acid catalyzed dehydration to give (**69**) which is the product isolated by the acid catalyzed hydrolysis of UV-irradiated DNA. In support of this proposal, irradiation of thymine at −196 °C results in the formation of an intermediate which is assumed to be the oxetane. This is converted to the (6-4) photoproduct, in the absence of light, when the temperature of the reaction is raised above −80 °C.[112]

Triplet sensitizers such as acetone and acetophenone inhibit the formation of the (6-4) photoproduct in DNA.[113,114] Therefore the cycloaddition reaction cannot be occurring from the thymine π,π^* triplet state.

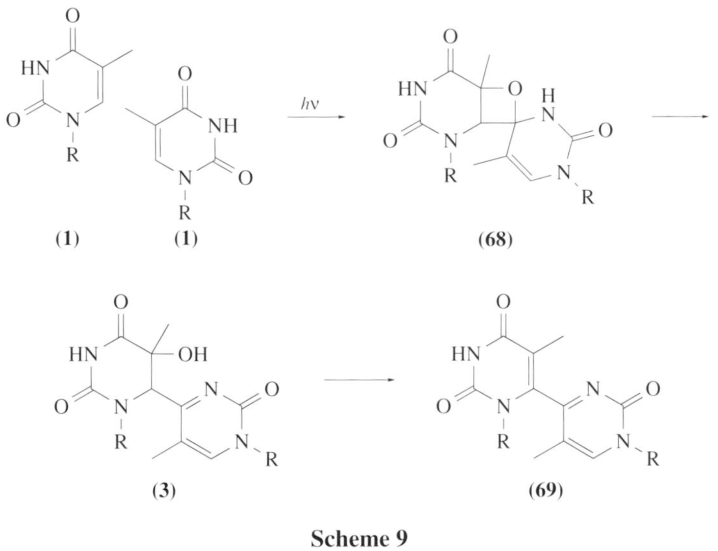

Scheme 9

5.15.3.2 The (6-4) Photoproduct Photolyase

A light-dependent (6-4) photoproduct lyase has been identified in *Drosophila melanogaster* and *Xenopus laevis*. This enzyme catalyzes the cleavage of the thymine (6-4) photoproduct. It has not been determined if the CT or the TC (6-4) photoproducts are substrates. The (6-4) photoproduct photolyase gene, from both systems, has been cloned, sequenced, and overexpressed, and the enzyme has been purified and characterized.[115,116] The enzyme from *X. laevis* has a predicted molecular mass of 60.6 kDa and contains an oxidized flavin (FAD) chromophore. The enzyme is activated by reduction[116] and the action spectrum for the repair reaction corresponds to the absorption spectrum of reduced flavin.[117] This suggests that reduced flavin is the catalytically relevant chromophore. The efficiency of the repair is 100-fold less than the efficiency of the repair of the cyclobutane photodimer by DNA photolyase.[118]

The mechanism of the (6-4) photoproduct photolyase has not yet been determined. A mechanism involving light absorption by the (6-4) photoproduct can be excluded because the action spectrum for the reaction does not correspond to the absorption spectrum of the photolesion. Four plausible mechanisms for the cleavage reaction can be considered. Mechanism A involves the reformation of the oxetane intermediate (**68**) followed by reduction with the photoexcited reduced flavin. Sequential cleavage of the C-5—O bond and the C-6—C-4′ bond of (**70**) and back electron transfer to the flavin would complete the reaction (Scheme 10).[118]

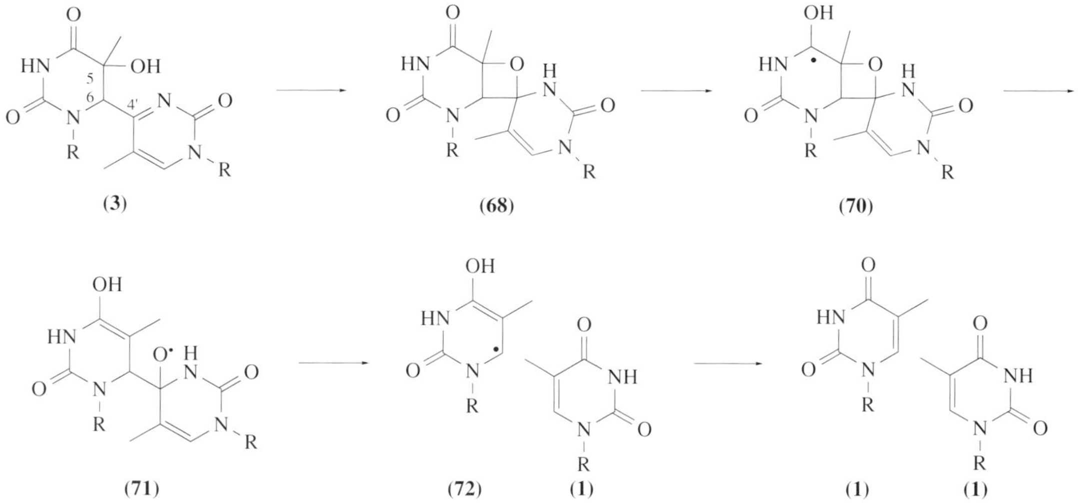

Scheme 10

In mechanism B, oxidation of oxetane (**68**), followed by sequential cleavage of the C-6—C-4′ bond and the C-5—O bond and back electron transfer, would complete the cleavage reaction (Scheme 11).[118]

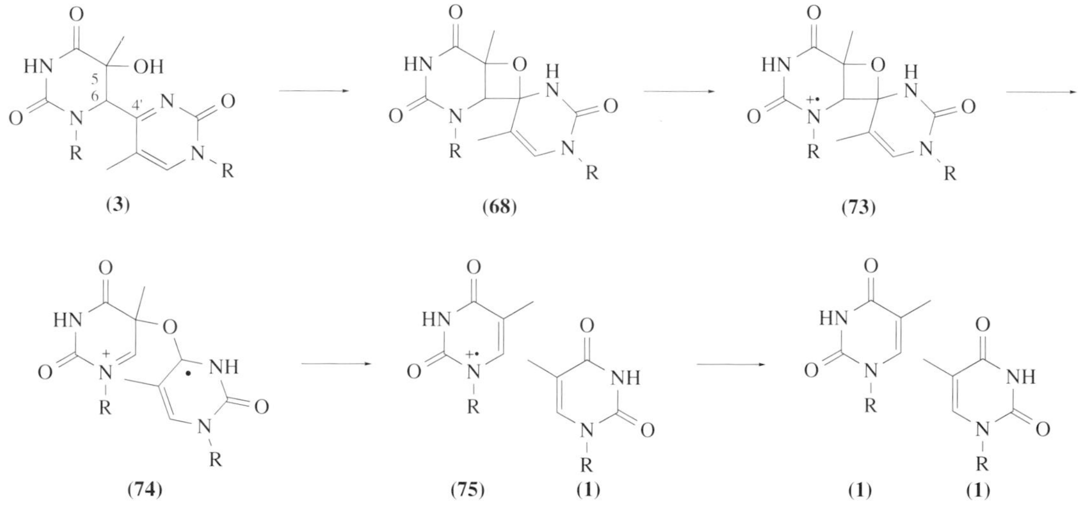

Scheme 11

In support of mechanisms A and B, it has been demonstrated that oxetane (**76**) can be cleaved using both oxidizing and reducing sensitizers (Equation (5)).[119]

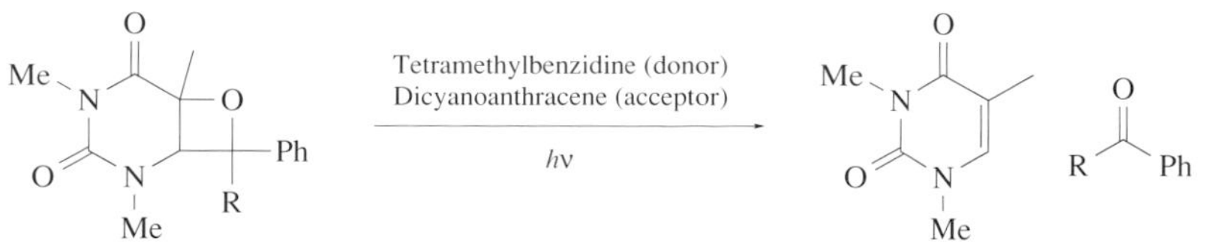

A major problem with these two proposals is that the oxetane is a high energy intermediate which is unstable above $-80\,°C$,[120] and has been estimated to be 61–69 kJ mol^{-1} higher in energy[121–123] than the (6-4) photoproduct. It is therefore unlikely that sufficient concentrations of oxetane can exist at the active site of the enzyme for efficient electron transfer as this requires that the enzyme-bound (6-4) photoproduct is primarily in the oxetane form.

An alternative radical cation fragmentation is outlined in mechanism C (Scheme 12).[124] Electron transfer from the (6-4) photoproduct to the photoexcited flavin would give (79) which should undergo cyclization more rapidly than the (6-4) photoproduct because of the positive charge on C-4′. Sequential cleavage of the C-6—C-4′ bond and the C-5—O bond of (80) and electron transfer would complete the fragmentation reaction.

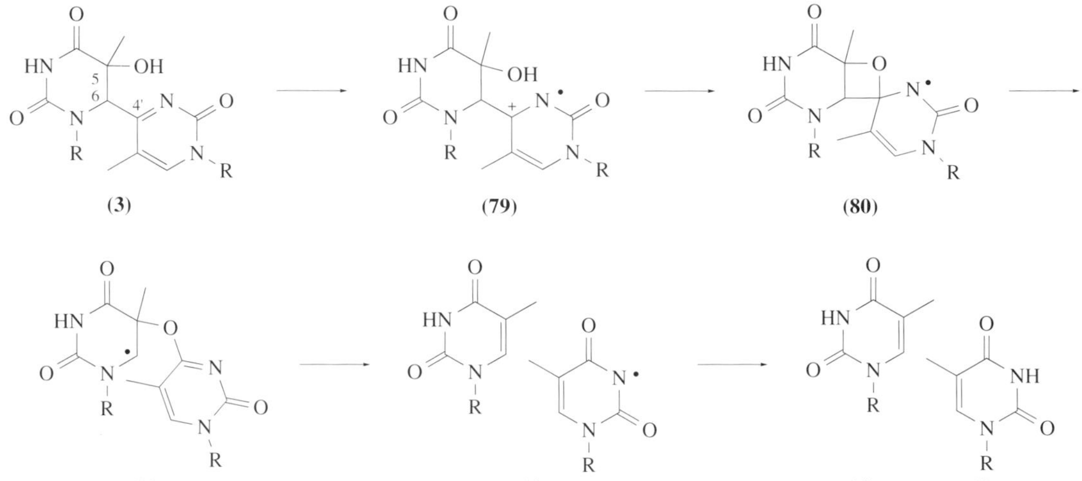

Scheme 12

An alternative radical anion fragmentation is outlined in mechanism D (Scheme 13). Addition of water to C-4′ of the (6-4) photoproduct would give (83). This reaction will be much more facile than oxetane formation because the reaction product is less strained. Reduction followed by loss of the C-5 hydroxyl from (84) would give (85). This reaction is analogous to one of the steps in the ribonucleotide reductase catalyzed reduction of nucleotides to deoxynucleotides.[125] Cleavage of the C-6—C-4′ bond of (85) and back electron transfer would complete the reaction.

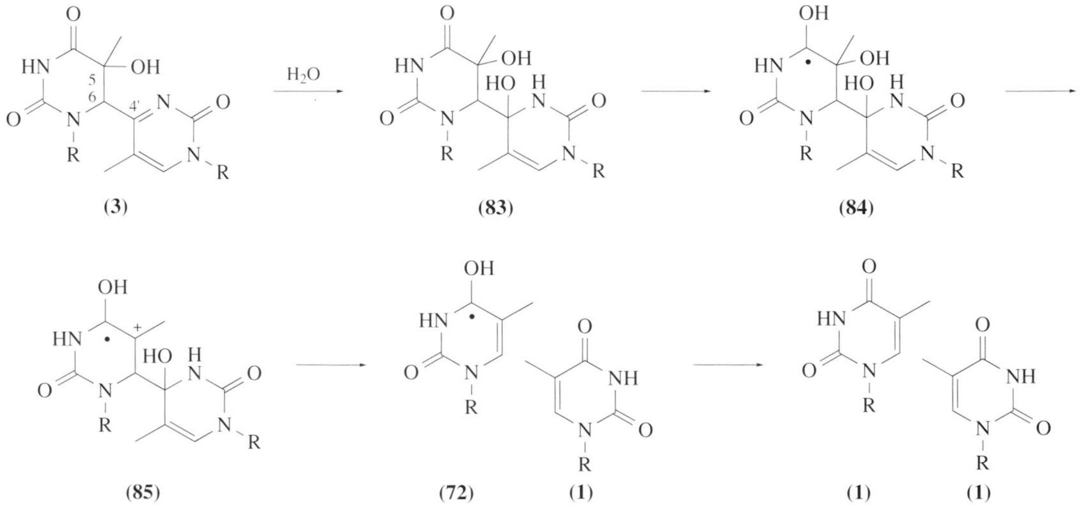

Scheme 13

The action spectrum for the enzyme and the activation of the enzyme by reduction both suggest that the enzymatic reaction occurs by reduction of the (6-4) photoproduct. The radical cation mechanisms are also unlikely because of competing oxidation of the enzyme as discussed in Section 5.15.2.5.5 for DNA photolyase. It should be possible to differentiate between the two remaining mechanistic proposals (A and D) by determining if oxygen from water is incorporated at C-4′ of one of the repaired thymines.

In contrast to the simple bisthymine model systems for the cyclobutane photodimer, there are currently no analogous simple model systems for the thymine (6-4) photoproduct. Such systems would greatly simplify investigations on the fragmentation chemistry of the (6-4) photoproduct radical cation and anion and facilitate the design of mechanistic probes for the enzymatic reaction.

5.15.4 THE DEWAR PYRIMIDINONE

The toxic effects of UV irradiation on *Streptomyces griseus*, *Streptomyces coelicolor*, *Staphylococcus epidermis*, and *E. coli* can be reduced by a subsequent exposure to long wavelength UV light.[126–128] This effect, called type III photoreactivation, is distinct from photoreactivation and is maximal at 314 nm, the λ_{max} of the (6-4) photoproduct. Comparison of the photolesion content of the DNA before and after the second irradiation demonstrated that the (6-4) photoproduct concentration was reduced.[129–132]

The reaction occurring during type III photoreactivation has been identified using dTpdT as a model system where it was discovered that irradiation of the (6-4) photoproduct (**86**) resulted in a photochemical electrocyclization reaction to give the Dewar pyrimidinone (**87**; Equation (6)).[133–135] This photolesion apparently is less toxic to cells that show type III photoreactivation than the (6-4) photoproduct.

$$hv\ (313\ \text{nm}) \qquad\qquad (6)$$

(86)   (87)

An enzyme catalyzed conversion of the Dewar pyrimidinone back to two thymines has not been identified. This photoproduct is repaired by the nucleotide excision repair system described in Section 5.15.6.

The Dewar pyrimidinone is the most base labile site in UV-irradiated DNA.[132] The mechanism of this reaction, which results in strand cleavage, has not yet been determined, nor have the reaction products been characterized.[133] A proposal is outlined in Scheme 14.

5.15.5 THE SPORE PHOTOPRODUCT

5.15.5.1 Formation of the Spore Photoproduct

The photochemistry of DNA, in the dehydrated-DNA-protein complex of the bacterial spore, is significantly different from the photochemistry of hydrated DNA in growing cells and UV irradiation of spores results in the formation of the spore photoproduct (**5**) as the major photolesion. It has been estimated that exposure of *Bacillus subtilis* spores to 1 h of summer sunlight in Dallas results in the formation of 1600 spore photoproducts per genome.[136]

The spore photoproduct can be isolated from UV-irradiated spores by acid catalyzed hydrolysis.[137,138] It can also be formed by irradiating plasmid DNA complexed with the SASP spore protein.[139] Irradiation of thymidine in ice or as a thin film gives the spore photoproduct as well as the six isomeric cyclobutane photodimers.[140–142] This demonstrates that formation of the spore photoproduct is an intrinsic photoreaction of constrained thymine and does not require some as

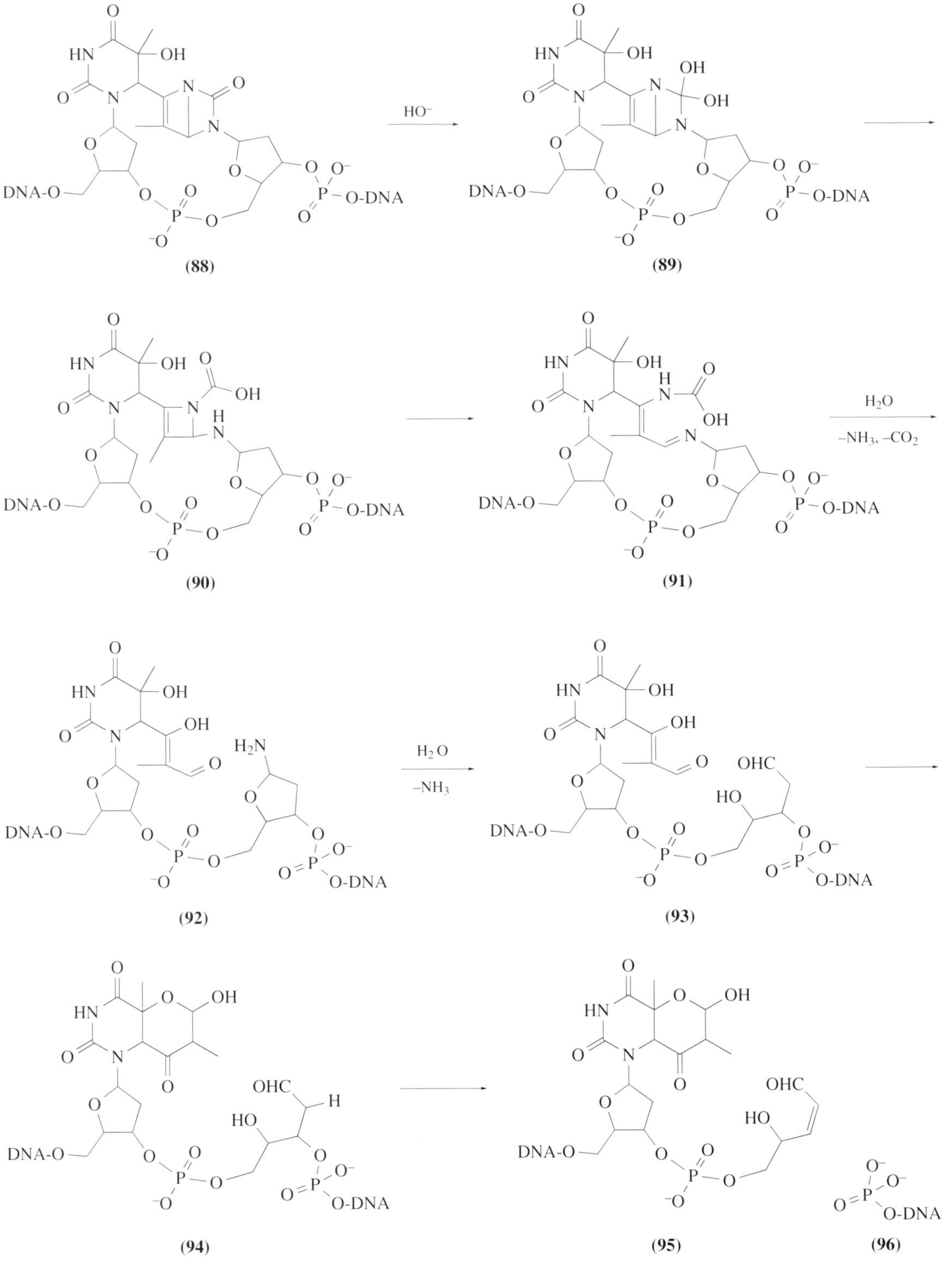

Scheme 14

yet unidentified biological cofactor. The spore photoproduct and the dinucleotide spore photo-product have both been synthesized.[143–145]

Mechanistic studies have not yet been carried out on the spore photoproduct forming reaction. A mechanistic proposal is outlined in Scheme 15. A Norrish type II hydrogen atom abstraction from the methyl group of thymine by the adjacent photoexcited thymine would give the radical pair

(98) and (99). Tautomerization of (98), followed by radical coupling, would complete the spore photoproduct formation. The protonated thymine radical anion (98) has been previously prepared by pulse radiolysis and by cleavage of the cyclobutane photodimer radical anion. In both cases, the C-6 protonated tautomer (100) is more stable than the oxygen protonated tautomer (98).[146-148]

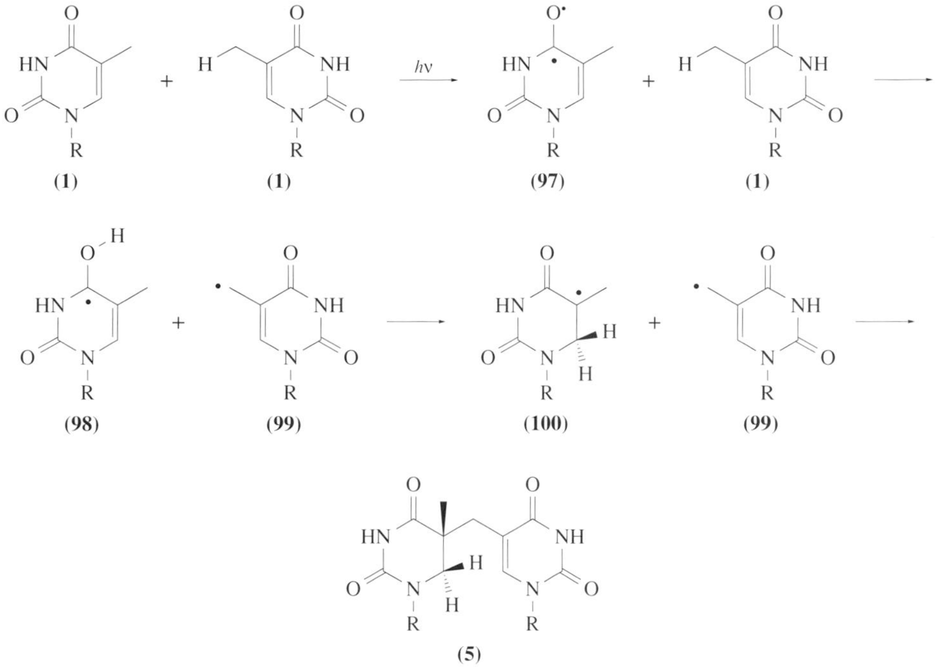

Scheme 15

Hydrogen atom abstraction by photoexcited pyrimidines is precedented. Irradiation of (101) at 254 nm results in the production of uracil (102) and ethylene. A plausible mechanism for this reaction is outlined in Scheme 16.[149-151] Mechanistically similar chemistry occurs when 5-*tert*-butyluracil is irradiated.[152]

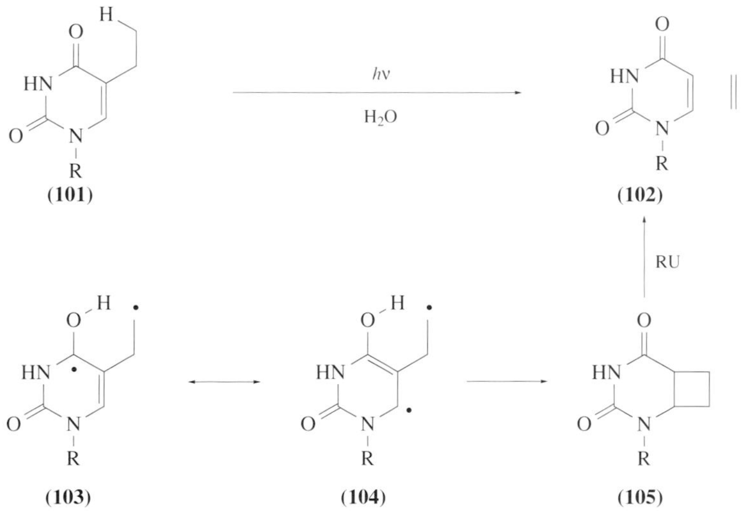

Scheme 16

Related hydrogen abstraction chemistry has been described for 4-thiothymine (Scheme 17). Each of the observed photoproducts (108), (110), and (112) results from an initial hydrogen atom abstraction by the sulfur of photoexcited thiothymine (106).[153] It is surprising that none of the thiospore photoproduct is formed in this reaction.

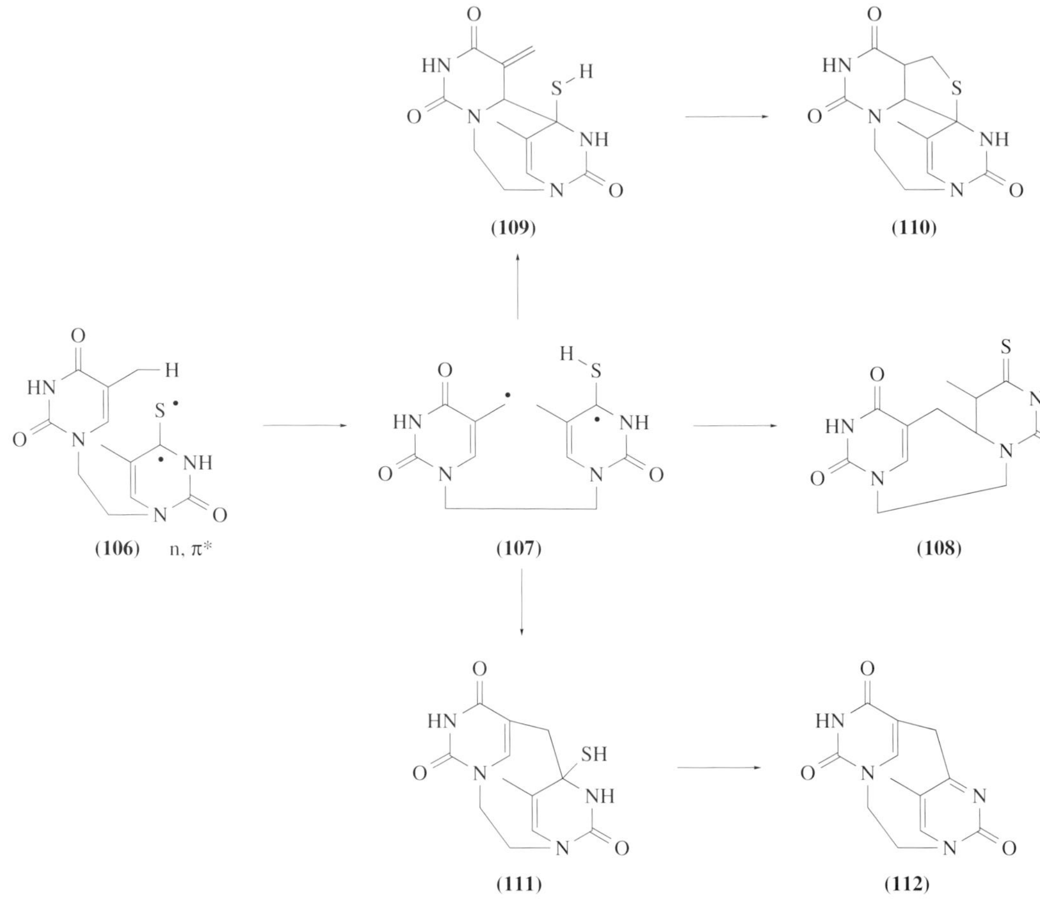

(109)   (110)

(106)   n, π*   (107)   (108)

(111)   (112)

Scheme 17

5.15.5.2   Repair of the spore photoproduct

In the germinating spore, the spore photoproduct is converted back to two thymines by the enzyme spore photoproduct lyase.[154,155] The gene for this enzyme has been cloned[156] and overexpressed at a high level.[157] The purified enzyme is monomeric, has a mass of 40 kDa and contains an Fe_2S_2 cluster.[157] Spore photoproduct lyase activity has not yet been demonstrated in a cell free system.

A mechanistic proposal for the cleavage reaction is outlined in Scheme 18. This mechanism involves the conjugate addition of a putative active site cysteine to the pyrimidine followed by electron transfer from the resulting enol to the iron sulfur cluster. Beta-scission of the bond linking the two pyrimidines followed by hydrogen atom transfer would give (**1**) and (**116**). Electron transfer from the reduced iron sulfur cluster followed by elimination of the active site cysteine would complete the reaction. Thiols are known to add to C-6 of uracil and to catalyze hydrogen–deuterium exchange at C-5.[158] However, the spore photoproduct is unaltered by prolonged treatment with a variety of thiols under basic conditions demonstrating that the bond linking the two pyrimidines is not cleaved by a simple retro-Michael reaction and suggesting the more complex mechanism outlined in Scheme 18.[159]

5.15.6   NUCLEOTIDE EXCISION REPAIR PATHWAY

The nucleotide excision repair mechanism, catalyzed by the UvrABC endonuclease is a general strategy for repairing modified bases in DNA (Figure 2).[160,161] The $(UvrA)_2(UvrB)_1$ complex binds to diverse types of base damage by a mechanism that is not yet clear. UvrA then dissociates from the complex and UvrC binds. Phosphodiester hydrolysis on the 3′ side and on the 5′ side of the lesion, catalyzed by UvrB and UvrC, respectively, generates an oligonucleotide containing the

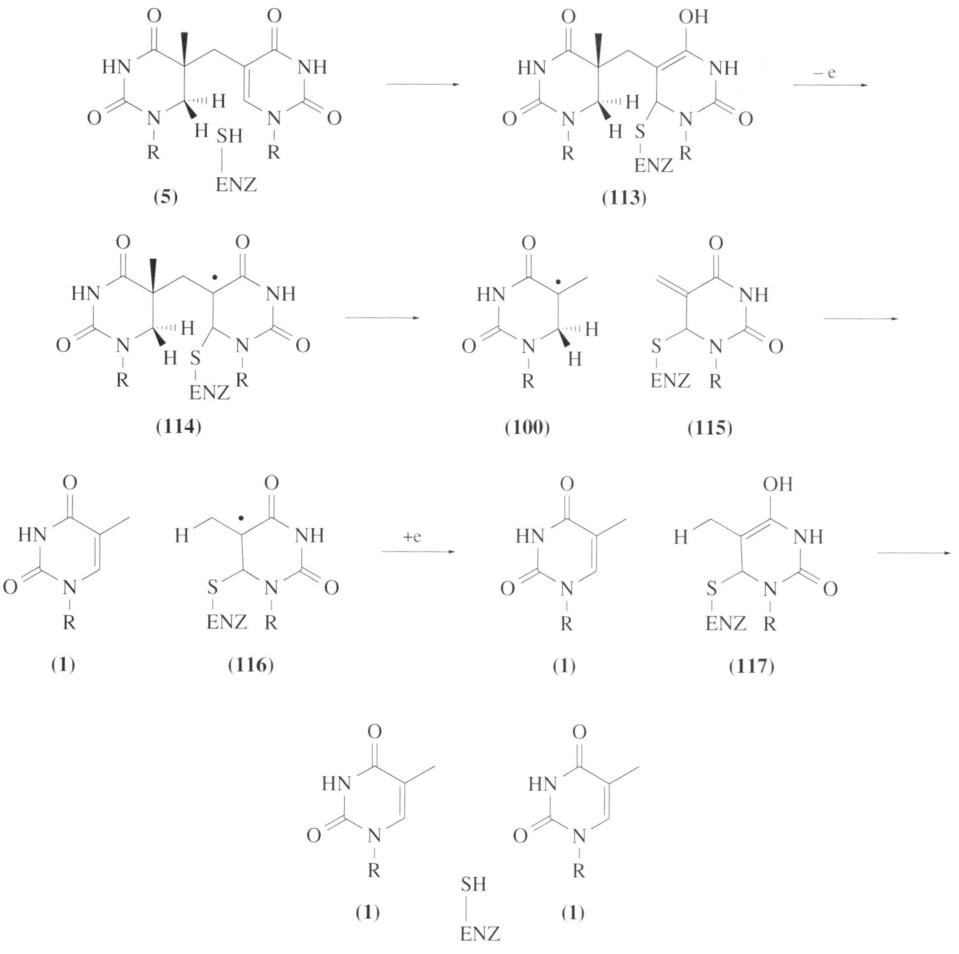

Scheme 18

damaged base. The release of this oligonucleotide is catalyzed by DNA helicase II, and the resulting gap is filled in by DNA polymerase I and the nick sealed by DNA ligase.

The UvrABC endonuclease hydrolyzes the eighth phosphodiester bond on the 5′ side of the cyclobutane photodimer and the (6-4) photoproduct and the fourth or fifth phosphodiester bond on the 3′ side of the photolesion.[162]

Using a 49-mer duplex containing the cyclobutane photodimer, the (6-4) photoproduct and the Dewar pyrimidinone at positions 21 and 22 from the 5′-end, the relative rates of excision of the different photolesions have been measured. Both the (6-4) photoproduct and the Dewar pyrimidinone are excised nine times faster than the cyclobutane photodimer.[163,164]

Although it is known that the spore photoproduct is also repaired by an excision repair mechanism, none of the details has yet been established.[165,166]

5.15.7 THE BASE EXCISION REPAIR PATHWAY

The base excision repair pathway for the repair of the cyclobutane photodimer is outlined in Scheme 19.[167,168] This is a more specialized repair pathway than photoreactivation or the nucleotide excision repair pathway and is found only in *Micrococcus luteus*, *Saccharomyces cerevisiae*, and in bacteriophage T4 infected *E. coli*. T4 endonuclease V is a multifunctional enzyme that catalyzes the cleavage of the N—glycosyl bond on the 5′ side of the photodimer to form an imine between the amino terminal threonine and the deoxyribose.[169] The enzyme then catalyzes a *syn* elimination[170] of the 3′ phosphate using glutamate 23 as the base. The structures of the free enzyme and of the enzyme complexed with its substrate have been determined.[171,172] Exonuclease III catalyzes the indicated

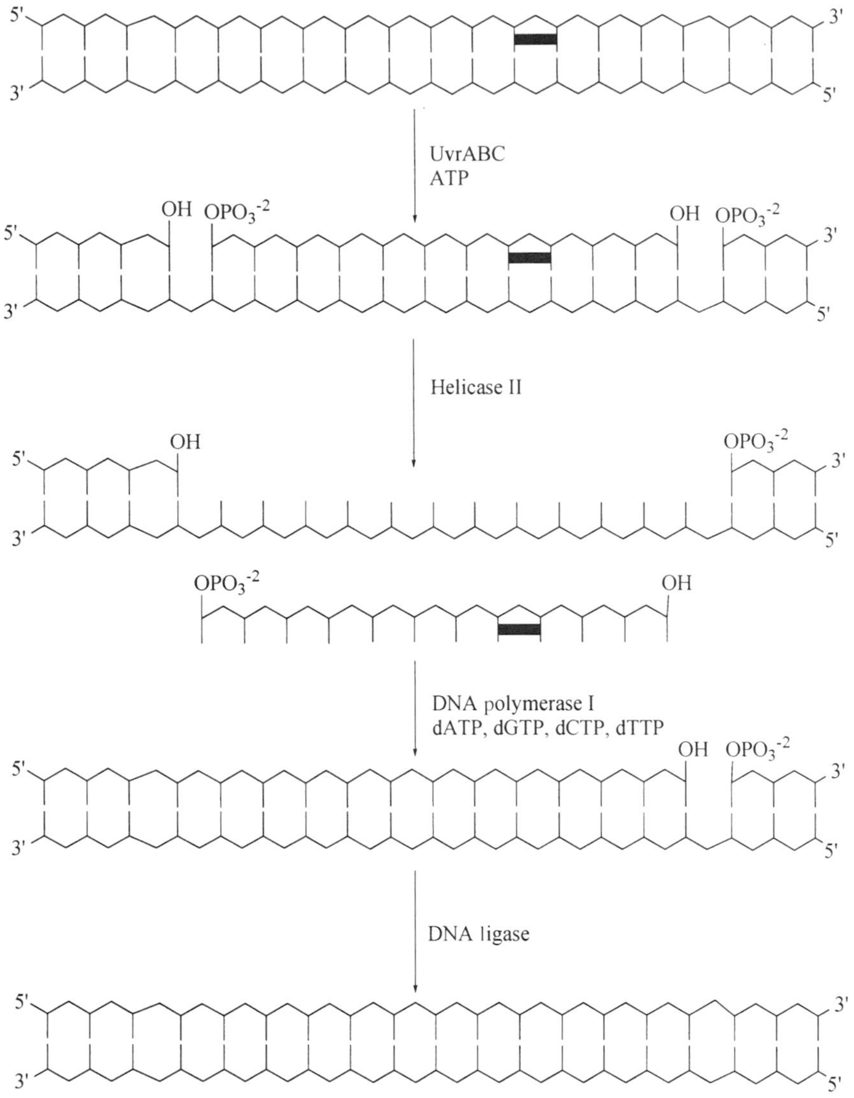

Figure 2 The nucleotide excision repair pathway.

phosphodiester hydrolysis (**121**)→(**122**)[173] and DNA deoxyribophosphodiesterase catalyzes the second phosphodiester hydrolysis to release the photodimer. The resulting two base gap is filled in by DNA polymerase and the nick is closed by DNA ligase.

5.15.8 MUTAGENESIS

If DNA photolesions are not repaired before DNA replication occurs, the template function of DNA is impaired and error prone synthesis past the damaged site by DNA polymerase can result in mutagenesis.[174,175]

The mutagenesis resulting from the presence of the thymine cyclobutane photodimer, the thymine (6-4) photoproduct and the thymine Dewar pyrimidinone has been determined in *E. coli*. Each of these photolesions, in the -AA[T-T]AA- sequence, has been site specifically incorporated into heteroduplex M13 vector DNA.[176] This DNA has been replicated in *E. coli* and progeny have been isolated and sequenced at the site of the photolesion. The strain of the *E. coli* used was defective in both DNA photolyase and UvrABC excision repair and was SOS induced to increase the rate of translesion synthesis.

For the thymine cyclobutane photodimer, translesion synthesis occurred without mutagenesis with the TT cyclobutane photodimer functioning as a template for the synthesis of AA. The thymine

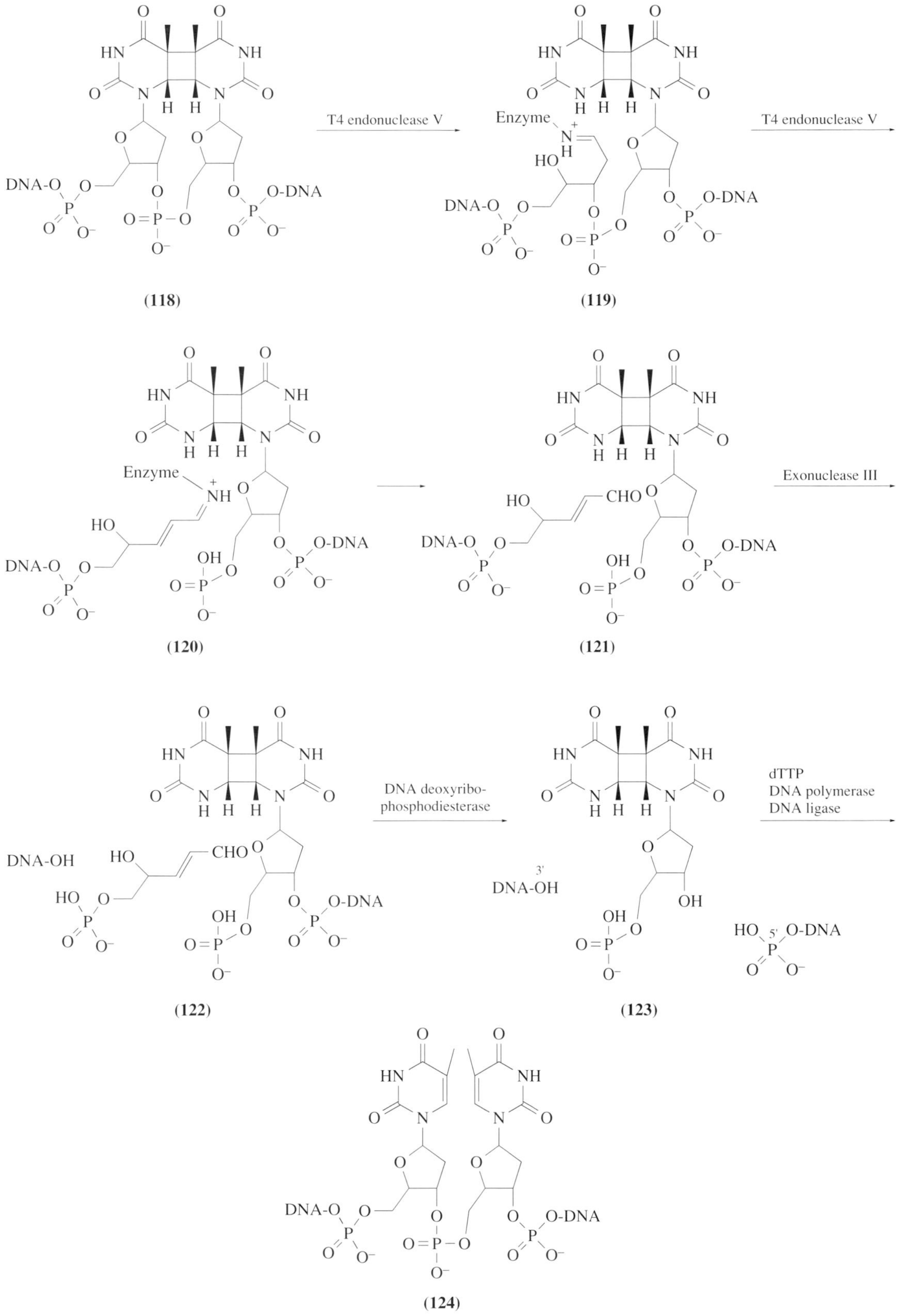

(118)

T4 endonuclease V →

(119)

T4 endonuclease V →

(120)

→

(121)

Exonuclease III →

(122)

DNA deoxyribo-phosphodiesterase →

(123)

dTTP
DNA polymerase
DNA ligase →

(124)

Scheme 19

(6-4) photoproduct was highly mutagenic and functioned as a template for the synthesis of AA (66.6%, no mutagenesis), GA (1.7%), AG (30%), and TG (1.7%). The thymine Dewar pyrimidinone was also highly mutagenic and functioned as a template for the synthesis of AA (72.7%, no mutagenesis), GA (3%), TA (4.5%), AG (4.5%), GG(1.5%), CG (3%), AC (6%), and AT (4.5%).

The distribution of mutants is different when the photolesion is in a -TA[TT]AT- or -AG[TT]GG sequence[177-179] and when the DNA is replicated in *Saccharomyces cerevisiae* rather than in *E. coli*.[180] The mutagenesis is therefore influenced by the sequence flanking the photolesion and by the details of the interactions of DNA polymerase with the DNA at the site of translesion synthesis. The mutagenesis code is not yet understood at the molecular level.

While these data suggest that the (6-4) photoproduct and the Dewar pyrimidinone are much more mutagenic than the cyclobutane photodimer, the biological consequences of these photolesions will be influenced by their relative rate of synthesis and repair. For example, the mutagenic potential of the (6-4) photoproduct relative to the cyclobutane photodimer, predicted by these studies, will be reduced because the (6-4) photoproduct is formed with a quantum yield that is about ten times less than the quantum yield for the formation of the cyclobutane photodimer and is repaired nine times faster than the cyclobutane photodimer by the UvrABC endonuclease.

The most frequent mutation in UV irradiated DNA results from TC sequences functioning as a template for the synthesis of AA (instead of AG). This is a mutation that can be explained at the molecular level. Hydrolysis of the TC (125) photodimer would give the TU (126) photodimer which can function as a template for the synthesis of AA during translesion synthesis (Equation (7)).[181] However, while the TC cyclobutane photodimer (125) is much more susceptible to hydrolysis than cytosine, this explanation requires that the rate of *in vivo* hydrolysis of (125) is faster than the rate measured *in vitro*.[182,183]

$$(7)$$

Nothing is yet known about the mutagenesis caused by the spore photoproduct.

5.15.9   CONCLUSIONS

The synthesis, biosynthesis, and chemical reactions of thymine in the ground state are very well understood. In contrast, our knowledge of the mechanisms of formation and repair of thymine photolesions is still quite incomplete.

Mechanistic studies on DNA photolyase are at an advanced stage. The remaining problems are the detection and characterization of intermediates by transient spectroscopy, the determination of the structure of the enzyme substrate complex and the possible use of time resolved crystallography to characterize intermediates. In contrast, the mechanism of the direct repair of the (6-4) photoproduct, and the mechanism of formation and direct repair of the spore photoproduct, are still at a very early stage of exploration. The details of how the UvrABC endonuclease recognizes a wide range of DNA modifications on DNA is also not yet understood.

The relevance of DNA photodamage to skin cancer in humans as well as the many unsolved mechanistic problems in this area, makes DNA photochemistry an interesting and important area of investigation.

ACKNOWLEDGEMENTS

The author would like to thank Cynthia Kinsland, Sean Taylor, Jun Xi and Ryan Mehl for critical reading of the manuscript. The Cornell component of the research described in this chapter was funded by a grant from the National Institutes of Health (GM 40498).

5.15.10 REFERENCES

1. K. H. Kraemer, *Proc. Natl. Acad. Sci. USA*, 1997, **94**, 11.
2. H. N. Ananthaswamy and W. E. Pierceall, *Photochem. Photobiol.*, 1990, **52**, 1119.
3. S. L. Parker, T. Tong, S. Bolden, and P. A. Wingo, *Can. Cancer J. Clin.*, 1996, **46**, 5.
4. J.-S. Taylor, *Pure Appl. Chem.*, 1995, **67**, 183.
5. H. Morrison (ed.), "Bioorganic Photochemistry," Wiley, New York, 1990, vol. 1.
6. S. Y. Wang (ed.), "Photochemistry and Photobiology of Nucleic Acids," 1976, vol. 1 and vol. 2.
7. I. G. Gut, P. D. Wood, and R. W. Redmond, *J. Am. Chem. Soc.*, 1996, **118**, 2366.
8. P. D. Wood and R. W. Redmond, *J. Am. Chem. Soc.*, 1996, **118**, 4256.
9. M. J. Ellison and J. D. Childs, *Photochem. Photobiol.*, 1981, **34**, 465.
10. M. H. Patrick, *Photochem. Photobiol.*, 1977, **25**, 357.
11. D. G. E. Lemaire and B. P. Ruzsicska, *Photochem. Photobiol.*, 1993, **57**, 755.
12. M. H. Patrick, *Photochem. Photobiol.*, 1977, **25**, 357.
13. G. J. Fisher and H. E. Johns, *Photochem. Photobiol.*, 1970, **11**, 429.
14. A. A. Voityuk, M.-E. Michel-Beyerle, and N. Rosch, *J. Am. Chem. Soc.*, 1996, **118**, 9750.
15. G. J. Fisher and H. E. Johns, *Photochem. Photobiol.*, 1970, **11**, 429.
16. S. Y. Wang, *Nature*, 1961, **190**, 690.
17. N. L. Leonard and R. L. Cundall, *J. Am. Chem. Soc.*, 1974, **96**, 5904.
18. D. T. Brown, J. Eisinger, and N. H. Leonard, *J. Am. Chem. Soc.*, 1968, **90**, 7302.
19. D. Burdi, S. Hoyt, and T. P. Begley, *Tetrahedron Lett.*, 1992, **33**, 2133.
20. D. J. Fenick, H. S. Carr, and D. E. Falvey, *J. Org. Chem.*, 1995, **60**, 624.
21. J.-S. Taylor, I. R. Brockie, and C. L. O'Day, *J. Am. Chem. Soc.*, 1987, **109**, 6735.
22. N. Camerman and A. Camerman, *J. Am. Chem. Soc.*, 1970, **92**, 2523.
23. F. E. Hruska, L. Voituriez, A. Grand, and J. Cadet, *Biopolymers*, 1986, **25**, 1399.
24. A. Kelner, *Proc. Natl. Acad. Sci. USA*, 1949, **35**, 73.
25. A. Sancar, F. W. Smith, and G. B. Sancar, *J. Biol. Chem.*, 1984, **259**, 6028.
26. G. B. Sancar, F. W. Smith, M. C. Lorence, C. S. Rupert, and A. Sancar, *J. Biol. Chem.*, 1984, **259**, 6033.
27. A. Sancar, *Biochemistry*, 1994, **33**, 2.
28. T. P. Begley, *Acc. Chem. Res.*, 1994, **27**, 394.
29. P. F. Heelis, R. F. Hartman, and S. D. Rose, *Chem. Soc. Rev.*, 1995, 289.
30. T. Carell, *Angew. Chem., Int. Ed. Engl.*, 1995, **34**, 2491.
31. I. Husain and A. Sancar, *Nucleic Acds Res.*, 1987, **15**, 1109.
32. S.-T. Kim and A. Sancar, *Photochem. Photobiol.*, 1993, **57**, 895.
33. G. Payne and A. Sancar, *Biochemistry*, 1990, **29**, 7715.
34. S.-T. Kim and A. Sancar, *Biochemistry*, 1991, **30**, 8623.
35. G. Payne, P. F. Heelis, B. R. Rohrs, and A. Sancar, *Biochemistry*, 1987, **26**, 7121.
36. M. S. Jorns, G. B. Sancar, and A. Sancar, *Biochemistry*, 1984, **23**, 2673.
37. J. L. Johnson, S. Hamm-Alvarez, G. Payne, G. B. Sancar, K. V. Rajagopalan, and A. Sancar, *Proc. Natl. Acad. Sci. USA*, 1988, **85**, 2046.
38. R. F. Hartman and S. D. Rose, *J. Am. Chem. Soc.*, 1992, **114**, 3559.
39. G. Payne and A. Sancar, *Biochemistry*, 1990, **29**, 7715.
40. S.-T. Kim, P. F. Heelis, T. Okamura, Y. Hirata, N. Mataga, and A. Sancar, *Biochemistry*, 1991, **30**, 11 262.
41. H.-W. Park, S.-T. Kim, A. Sancar, and J. Deisenhofer, *Science*, 1995, **268**, 1866.
42. S.-R. Yeh and D. E. Falvey, *J. Am. Chem. Soc.*, 1992, **114**, 7313.
43. C. Pac, J. Kubo, T. Majima, and H. Sakurai, *Photochem. Photobiol.*, 1982, **36**, 273.
44. M. P. Scannell, S.-R. Yeh, and D. E. Falvey, *Photochem. Photobiol.*, 1996, **64**, 764.
45. H. Diogo, A. R. Dias, A. Dhalla, M. E. Minas da Piedade, and T. P. Begley, *J. Org. Chem.*, 1991, **56**, 7340.
46. A. A. Lamola, *Mol. Photochem.*, 1972, **4**, 107.
47. S. E. Rokita and C. T. Walsh, *J. Am. Chem. Soc.*, 1984, **106**, 4589.
48. M. S. Jorns, *J. Am. Chem. Soc.*, 1987, **109**, 3133.
49. R. F. Hartman and S. D. Rose, *J. Org. Chem.*, 1992, **57**, 2302.
50. C. Pac, K. Miyake, Y. Masaki, S. Yanagida, T. Ohno, and A. Yoshimura, *J. Am. Chem. Soc.*, 1992, **114**, 10 756.
51. P. F. Heelis, R. F. Hartman, and S. D. Rose, *Photochem. Photobiol.*, 1993, **57**, 442.
52. A. A. Lamola, *Mol. Photochem.*, 1972, **4**, 107.
53. S. Sasson and D. Elad, *J. Org. Chem.*, 1972, **37**, 3164.
54. C. Pac, J. Kubo, T. Majima, and H. Sakurai, *Photochem. Photobiol.*, 1982, **36**, 273.
55. I. Rosenthal, M. M. Rao, and J. Salomon, *Biochim. Biophys. Acta*, 1975, **378**, 165.
56. P. J. W. Pouwels, R. F. Hartman, S. D. Rose, and R. Kaptein, *Photochem. Photobiol.*, 1995, **61**, 563.
57. D. J. Fenick, H. S. Carr, and D. E. Falvey, *J. Org. Chem.*, 1995, **60**, 624.
58. P. J. Wagner, M. J. Lindstrom, J. H. Sedon, and D. R. Ward, *J. Am. Chem. Soc.*, 1981, **103**, 3842.
59. D. Burdi and T. P. Begley, *J. Am. Chem. Soc.*, 1991, **113**, 7768.
60. D.-Y. Yang and T. P. Begley, *Tetrahedron Lett.*, 1993, **34**, 1709.
61. L. Melander and W. H. Saunders, "Reaction Rates of Isotopic Molecules," Wiley, New York, 1980, p. 95.
62. W. R. Dolbier, Jr., in "Isotopes in Organic Chemistry," eds. E. Buncel and C. C. Lee, Elsevier, Amsterdam, 1975, vol. 1, p. 27.
63. B. K. Carpenter, "Determination of Organic Reaction Mechanisms," Wiley, New York, 1984, p. 83.
64. P. F. Cook (ed.), "Enzyme Mechanism from Isotope Effects," CRC Press, Boca Raton, 1991.
65. R. A. McMordie and T. P. Begley, *J. Am. Chem. Soc.*, 1992, **114**, 1886.
66. P. J. W. Pouwels, R. F. Hartman, S. D. Rose, and R. Kaptein, *Photochem. Photobiol.*, 1995, **61**, 563.
67. A. A. Voityuk, M.-E. Michel-Beyerle, and N. Rosch, *J. Am. Chem. Soc.*, 1996, **118**, 9750.
68. J. R. Van Camp, T. Young, R. F. Hartman, and S. D. Rose, *Photochem. Photobiol.*, 1987, **45**, 365.
69. D. G. Hartzfeld and S. D. Rose, *J. Am. Chem. Soc.*, 1993, **115**, 850.

70. A. G. Cochran, R. Sugasawara, and P. G. Schultz, *J. Am. Chem. Soc.*, 1988, **110**, 7888.
71. M. S. Jorns, *J. Am. Chem. Soc.*, 1987, **109**, 3133.
72. R. F. Hartman and S. D. Rose, *J. Am. Chem. Soc.*, 1992, **114**, 3559.
73. R. Epple, E.-U. Wallenborn, and T. Carell, *J. Am. Chem. Soc.*, 1997, **119**, 7440.
74. S.-R. Yeh and D. E. Falvey, *J. Am. Chem. Soc.*, 1991, **113**, 8557.
75. D. J. Fenick, H. S. Carr, and D. E. Falvey, *J. Org. Chem.*, 1995, **60**, 624.
76. R. F. Hartman, J. R. Van Camp, and S. D. Rose, *J. Org. Chem.*, 1987, **52**, 2684.
77. P. F. Heelis, R. F. Hartman, and S. D. Rose, *Chem. Soc. Rev.*, 1995, 289.
78. P. J. W. Pouwels, R. F. Hartman, S. D. Rose, and R. Kaptein, *Photochem. Photobiol.*, 1995, **61**, 575.
79. I. D. Podmore, P. F. Heelis, M. C. R. Symons, and A. Pezeshk, *J. Chem. Soc., Chem. Commun.*, 1994, 1005.
80. D. J. Fenick and D. E. Falvey, *J. Org. Chem.*, 1994, **59**, 4791.
81. D. F. Burdi, Ph.D. Thesis, Cornell University, 1992.
82. J. C. Fishbein and W. P. Jencks, *J. Am. Chem. Soc.*, 1988, **110**, 5087.
83. R. A. McMordie, E. Altmann, and T. P. Begley, *J. Am. Chem. Soc.*, 1993, **115**, 10 370.
84. R. A. McMordie and T. P. Begley, *J. Am. Chem. Soc.*, 1992, **114**, 1886.
85. R. F. Hartman, J. R. Van Camp, and S. D. Rose, *J. Org. Chem.*, 1987, **52**, 2684.
86. B. K. Carpenter, "Determination of Organic Reaction Mechanisms," Wiley, New York, 1984, p. 97.
87. J. R. Jacobsen, A. G. Cochran, J. C. Stephans, S. D. King, and P. G. Schultz, *J. Am. Chem. Soc.*, 1995, **117**, 5453.
88. R. F. Hartman, J. R. Van Camp, and S. D. Rose, *J. Org. Chem.*, 1987, **552**, 2684.
89. A. A. Voityuk, M.-E. Michel-Beyerle, and N. Rosch, *J. Am. Chem. Soc.*, 1996, **118**, 9750.
90. M. P. Scannell, D. J. Fenick, S.-R. Yeh, and D. E. Falvey, *J. Am. Chem. Soc.*, 1997, **119**, 1971.
91. P. F. Heelis, R. F. Hartman, and S. D. Rose, *Chem. Soc. Rev.*, 1995, 289.
92. M. Witmer, E. Altmann, H. Young, T. P. Begley, and A. Sancar, *J. Am. Chem. Soc.*, 1989, **111**, 9264.
93. D. F. Burdi, Ph.D. Thesis, Cornell University, 1992.
94. R. F. Hartman, S. D. Rose, P. J. W. Pouwels, and R. Kaptein, *Photochem. Photobiol.*, 1992, **56**, 305.
95. P. J. W. Pouwels and R. Kaptein, *Appl. Magn. Reson.*, 1994, **7**, 107.
96. T. Okamura, A. Sancar, P. Heelis, T. P. Begley, Y. Hirata, and N. Mataga, *J. Am. Chem. Soc.*, 1991, **113**, 3143.
97. P. F. Heelis, R. F. Hartman, and S. D. Rose, *Photochem. Photobiol.*, 1993, **57**, 1053.
98. R. F. Hartman and S. D. Rose, *J. Am. Chem. Soc.*, 1992, **114**, 3559.
99. R. Eppel, E.-U. Wallenborn, and T. Carell, *J. Am. Chem. Soc.*, 1997, **119**, 7440.
100. T. Okamura, A. Sancar, P. Heelis, T. P. Begley, Y. Hirata, and N. Mataga, *J. Am. Chem. Soc.*, 1991, **113**, 3143.
101. S.-T. Kim, P. F. Heelis, T. Okamura, Y. Hirata, N. Mataga, and A. Sancar, *Biochemistry*, 1991, **30**, 11 262.
102. S.-T. Kim, M. Volk, G. Rousseau, P. F. Heelis, A. Sancar, and M.-E. Michel-Beyerle, *J. Am. Chem. Soc.*, 1994, **116**, 3115.
103. S.-T. Kim, A. Sancar, C. Essenmacher, and G. T. Babcock, *J. Am. Chem. Soc.*, 1992, **114**, 4442.
104. S.-R. Yeh and D. E. Falvey, *J. Am. Chem. Soc.*, 1991, **113**, 8557.
105. C. Pac, J. Kubo, T. Majima, and H. Sakurai, *Photochem. Photobiol.*, 1982, **36**, 273.
106. W. A. Franklin, K. M. Lo, and W. A. Haseltine, *J. Biol. Chem.*, 1982, **257**, 13 535.
107. D. L. Mitchell and R. S. Nairn, *Photochem. Photobiol.*, 1989, **49**, 805.
108. A. J. Varghese and S. Y. Wang, *Science*, 1968, **160**, 186.
109. I. L. Karle, S. Y. Wang, and A. J. Varghese, *Science*, 1969, **164**, 183.
110. R. E. Rycyna and J. L. Alderfer, *Nucleic Acids Res.*, 1985, **13**, 5949.
111. S. Iwai, M. Shimizu, H. Kamiya, and E. Ohtsuka, *J. Am. Chem. Soc.*, 1996, **118**, 7642.
112. R. O. Rahn and J. L. Hosszu, *Photochem. Photobiol.*, 1969, **10**, 131.
113. A. A. Lamola, *Photochem. Photobiol.*, 1969, **9**, 291.
114. M. L. Meistrich and A. A. Lamola, *J. Mol. Biol.*, 1972, **66**, 83.
115. T. Todo, H. Ryo, K. Yamamoto, H. Toh, T. Inui, H. Ayaki, T. Nomura, and M. Ikenaga, *Science*, 1996, **272**, 109.
116. T. Todo, S.-T. Kim, K. Hitomi, E. Otoshi, T. Inui, H. Morioka, H. Kobayashi, E. Ohtsuka, H. Toh, and M. Ikenaga, *Nucl. Acids Res.*, 1997, **25**, 764.
117. S.-T. Kim, K. Malhotra, J.-S. Taylor, and A. Sancar, *Photochem. Photobiol.*, 1996, **63**, 292.
118. S.-T. Kim, K. Malhotra, C. A. Smith, J.-S. Taylor, and A. Sancar, *J. Biol. Chem.*, 1994, **269**, 8535.
119. G. Prakash and D. E. Falvey, *J. Am. Chem. Soc.*, 1995, **117**, 11 375.
120. R. O. Rahn and J. L. Hosszu, *Photochem. Photobiol.*, 1969, **10**, 131.
121. P. F. Heelis and S. Liu, *J. Am. Chem. Soc.*, 1997, **119**, 2936.
122. J. Liu and J.-S. Taylor, *J. Am. Chem. Soc.*, 1996, **118**, 3287.
123. P. Clivio, J.-L. Fourrey, *J. Chem. Soc., Chem. Commun.*, 1996, 2203.
124. S.-T. Kim, K. Malhotra, C. A. Smith, J.-S. Taylor, and A. Sancar, *J. Biol. Chem.*, 1994, **269**, 8535.
125. J. Stubbe, *Adv. Enzymol. Relat. Areas Mol. Biol.*, 1990, **63**, 349.
126. J. Jagger, H. Takebe, and J. M. Snow, *Photochem. Photobiol.*, 1970, **12**, 185.
127. M. Ikenaga, M. H. Patrick, and J. Jagger, *Photochem. Photobiol.*, 1970, **11**, 487.
128. I. Husain, W. L. Carrier, J. D. Regan, and A. Sancar, *Photochem. Photobiol.*, 1988, **48**, 233.
129. M. H. Patrick, *Photochem. Photobiol.*, 1970, **11**, 477.
130. M. Ikenaga, M. H. Patrick, and J. Jagger, *Photochem. Photobiol.*, 1971, **14**, 175.
131. M. Ikenaga, M. H. Patrick, and J. Jagger, *Photochem. Photobiol.*, 1970, **11**, 487.
132. D. L. Mitchell, *Mutat. Res.*, 1988, **194**, 227.
133. J.-S. Taylor and M. P. Cohrs, *J. Am. Chem. Soc.*, 1987, **109**, 2834.
134. J.-S. Taylor, D. S. Garrett, and M. P. Cohrs, *Biochemistry*, 1988, **27**, 7206.
135. J.-S. Taylor, H.-F. Lu, and J. J. Kotyk, *Photochem. Photobiol.*, 1990, **51**, 161.
136. N. Munakata and C. S. Rupert, *Mol. Gen. Genet.*, 1974, **130**, 239.
137. A. J. Varghese, *Biochem. Biophys. Res. Commun.*, 1970, **38**, 484.
138. Y. Sun, K. Palasingam, and W. L. Nicholson, *Anal. Biochem.*, 1994, **221**, 61.
139. W. L. Nicholson, B. Setlow, and P. Setlow, *Proc. Natl. Acad. Sci. USA*, 1991, **88**, 8288.
140. A. J. Varghese, *Photochem. Photobiol.*, 1971, **13**, 357.

141. A. J. Varghese, *Biochemistry*, 1970, **9**, 4781.
142. J. Cadet and P. Vigny, in "Bioorganic Photochemistry," ed. H. Morrison, Wiley, New York, 1990, vol. 1, p. 96.
143. R. Nicewonger and T. P. Begley, *Tetrahedron Lett.*, 1997, **38**, 935.
144. D. E. Bergstrom and K. F. Rash, *J. Chem. Soc., Chem. Commun.*, 1978, 284.
145. S.-J. Kim, C. Lester, and T. P. Begley, *J. Org. Chem.*, 1995, **60**, 6256.
146. S. Steenken, J. P. Telo, H. M. Novais, and L. P. Candeias, *J. Am. Chem. Soc.*, 1992, **114**, 4701.
147. P. M. Cullis, P. Evans, and M. E. Malone, *J. Chem. Soc., Chem. Commun.*, 1996, 985.
148. I. D. Podmore, P. F. Heelis, M. C. R. Symons, and A. Pezeshk, *J. Chem. Soc., Chem. Commun.*, 1994, 1005.
149. I. Pietrzykowska and D. Shugar, *Acta Biochim. Polon.*, 1970, **17**, 361.
150. E. Sztumpf-Kulikowska and D. Shugar, *Acta Biochim. Polon.*, 1974, **21**, 73.
151. E. Krajewska and D. Shugar, *Acta Biochim. Polon.*, 1972, **19**, 207.
152. I. Basnak, D. McKinnell, N. Spencer, A. Balkan, P. R. Ashton, and R. T. Walker, *J. Chem. Soc., Perkin Trans. 1*, 1997, 121.
153. P. Clivio, D. Guillaume, M.-T. Adeline, and J.-L. Fourrey, *J. Am. Chem. Soc.*, 1997, **119**, 5255.
154. N. Munakata and C. S. Rupert, *Mol. Gen. Genet.*, 1974, **130**, 239.
155. T.-C. Van Wang and C. S. Rupert, *Photochem. Photobiol.*, 1977, **25**, 123.
156. P. Fajardo-Cavazos, C. Salazar, and W. L. Nicholson, *J. Bacteriol.*, 1993, **175**, 1735.
157. C. Kinsland, C. Costello, W. L. Nicholson, and T. P. Begley, unpublished results.
158. E. Vega, G. A. Rood, E. R. de Waard, and U. K. Pandit, *Tetrahedron*, 1991, **47**, 4361.
159. R. Nicewonger and T. P. Begley, unpublished results.
160. A. Sancar, *Annu. Rev. Biochem.*, 1996, **65**, 43.
161. E. C. Friedberg, G. C. Walker, and W. Siede, "DNA Repair and Mutagenesis," ASM Press, Washington, DC, 1995, chapter 5.
162. G. M. Myles, B. Van Houten, and A. Sancar, *Nucleic Acids Res.*, 1987, **15**, 1227.
163. C. A. Smith and J.-S. Taylor, *J. Biol. Chem.*, 1993, **268**, 11 143.
164. D. V. Svoboda, C. A. Smith, J.-S. Taylor, and A. Sancar, *J. Biol. Chem.*, 1993, **268**, 10 694.
165. N. Munakata and C. S. Rupert, *Mol. Gen. Genet.*, 1974, **130**, 239.
166. T.-C. Van Wang and C. S. Rupert, *Photochem. Photobiol.*, 1977, **25**, 123.
167. E. C. Friedberg, G. C. Walker, and W. Siede, "DNA Repair and Mutagenesis," ASM Press, Washington DC, 1995, chapter 4.
168. P. W. Doetsch and R. P. Cunningham, *Mutat. Res.*, 1990, **236**, 173.
169. M. L. Dodson, R. D. Schrock, and R. S. Lloyd, *Biochemistry*, 1993, **32**, 8284.
170. A. Mazumder, J. A. Gerlt, L. Rabow, M. J. Absalon, J. Stubbe, and P. H. Bolton, *J. Am. Chem. Soc.*, 1989, **111**, 8029.
171. K. Morikawa, O. Matsumoto, M. Tsujimoto, K. Katayanagi, M. Ariyoshi, T. Doi, M. Ikehara, T. Inaoka, and E. Ohtsuka, *Science*, 1992, **256**, 523.
172. D. G. Vassylyev, T. Kashiwagi, Y. Mikami, M. Ariyoshi, S. Iwai, E. Ohtsuka, and K. Morikawa, *Cell*, 1995, **83**, 773.
173. C. D. Mol, C.-F. Kuo, M. M. Thayer, R. P. Cunningham, and J-A. Tainer, *Nature*, 1995, **374**, 381.
174. J.-S. Taylor, *Acc. Chem. Res.*, 1994, **27**, 76.
175. F. Hutchinson, *Photochem. Photobiol.*, 1987, **45**, 897.
176. C. A. Smith, M. Wang, N. Jiang, L. Che, X. Zhao, and J.-S. Taylor, *Biochemistry*, 1996, **35**, 4146.
177. N. Jiang and J.-S. Taylor, *Biochemistry*, 1993, **32**, 472.
178. C. W. Lawrence, S. K. Banerjee, A. Borden, and J. E. LeClerc, *Mol. Gen. Genet.*, 1990, **222**, 166.
179. J. E. LeClerc, A. Borden, and C. W. Lawrence, *Proc. Natl. Acad. Sci. USA*, 1991, **88**, 9685.
180. P. E. M. Gibbs, A. Borden, and C. W. Lawrence, *Nucleic Acids Res.*, 1995, **23**, 1919.
181. N. Jiang and J.-S. Taylor, *Biochemistry*, 1993, **32**, 472.
182. D. G. E. Lemaire and B. P. Ruzsicska, *Biochemistry*, 1993, **32**, 2525.
183. M. J. Horsfall, A. Borden, and C. W. Lawrence, *J. Bacteriol.*, 1997, **179**, 2835.

5.16
Microbial Dehalogenases

SHELLEY D. COPLEY

University of Colorado, Boulder, CO, USA

5.16.1   INTRODUCTION

About 15 000 chlorinated organic compounds are used for a vast number of purposes in the manufacturing, agricultural, and pharmaceutical industries.[1] The large-scale use and disposal of these compounds has led to widespread environmental pollution. Some of the most notorious pollutants, including dioxins, DDT, chlorofluorocarbons, and PCBs (polychlorinated biphenyls) (see Scheme 1), belong to this category, and the environmental problems associated with these chemicals have caused great public concern about chlorinated organic compounds in general. Although some chlorinated organic compounds are quite innocuous, many have undesirable properties such as persistence in the environment, a tendency to bioaccumulate, toxicity, carcinogenicity, and possibly disruption of endocrine function in humans and wildlife.[1,2]

Interest in the potential for biodegradation of chlorinated organic compounds in various natural environments and in the use of biodegradation for remediation and wastewater treatment has fueled research into the pathways, genes, and enzymes involved in biodegradation of many chlorinated organic compounds. Certain strains of a wide variety of bacteria, including *Pseudomonas, Arthrobacter, Mycobacterium, Moraxella, Xanthobacter, Azotobacter, Agrobacterium, Sphingomonas,* and *Corynebacterium* have been found to be capable of mineralization of chlorinated organic compounds. (Mineralization refers to complete degradation to CO_2, H_2O, and HCl.) Others carry out cometabolic processes in which chlorinated compounds are adventitiously transformed by

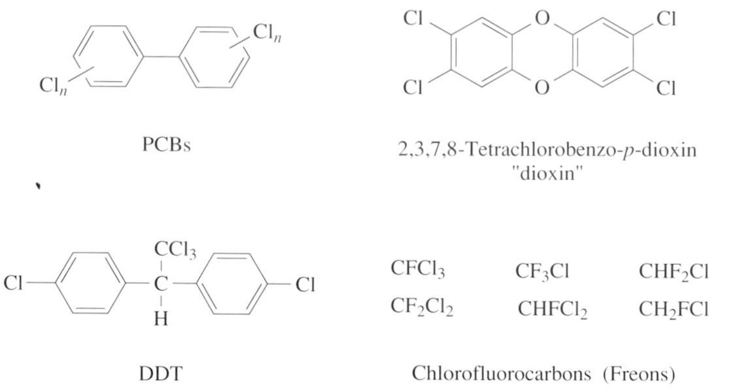

PCBs

2,3,7,8-Tetrachlorobenzo-*p*-dioxin
"dioxin"

DDT

$CFCl_3$ CF_3Cl CHF_2Cl

CF_2Cl_2 $CHFCl_2$ CH_2FCl

Chlorofluorocarbons (Freons)

Scheme 1

nonspecific enzymes without providing any metabolic benefit to the microorganism. It is important to emphasize that only certain strains of these bacteria have the metabolic capability to degrade chlorinated pollutants. For example, strains of *Pseudomonas* that degrade important pollutants such as PCBs,[3] 4-chlorobenzoate,[4] and lindane[5] have been identified, but most strains of *Pseudomonas* do not have these abilities. In almost all cases, strains that degrade a particular pollutant have been isolated from previously contaminated soil, water, sediment, or sewage sludge samples by selective enrichment for growth on the pollutant as a sole carbon source.

Microbial dehalogenases play a critical role in the biodegradation of chlorinated organic compounds. This chapter will describe the mechanisms of enzymes involved in the microbial biodegradation of both chlorinated aliphatic and aromatic compounds. The focus will be on enzymes for which considerable mechanistic and structural information is available, although some mention will be made of particularly intriguing enzymes that have not yet been studied intensively. As a prelude, some of the interesting issues regarding these enzymes will be introduced here to provide a context for the discussions of individual enzymes.

The evolutionary origin of dehalogenase enzymes is a subject of great interest and debate. The question of whether dehalogenase enzymes have evolved recently in response to the influx of chlorinated compounds into the environment, or whether they are ancient enzymes whose function was to detoxify naturally occurring chlorinated compounds, is difficult to answer. There are over 2000 known naturally occurring chlorinated and brominated compounds (and even a few fluorinated compounds) belonging to many structural classes, including alkanes, terpenes, amino acids, steroids, fatty acids, lipids, heterocycles such as pyrroles and indoles, phenols, benzenes, and dioxins.[6] Many of these compounds are toxic to bacteria, so it is likely that enzymes which detoxify these compounds via dehalogenation arose long ago. However, there is no doubt that the influx of large quantities of chlorinated organic compounds, many of which do not resemble natural products, has created novel evolutionary pressures. In addition, the process of isolating microorganisms by selective enrichment for growth on a certain compound as a sole carbon source exerts pressure of a kind that is certainly never present in the environment. Under these unusual conditions, evolutionary changes that allow microorganisms to more effectively dehalogenate chlorinated organic compounds seem inevitable.

Although the timing of the development of modern dehalogenases is often uncertain, in many cases the evolutionary progenitors of dehalogenases can be identified by sequence homology analysis. Such comparisons can also provide important clues about mechanisms and structures. X-ray crystallographic studies also provide important information. In some cases, crystal structures have indicated an evolutionary relationship to a class of proteins with which the dehalogenase has only a very low level of sequence identity. Thus, important evolutionary relationships might not have been recognized based only upon sequence homology analyses.

One striking feature of most dehalogenases is that they are rather sluggish. Values for k_{cat} are typically in the range of $1\ s^{-1}$, which is quite slow for key metabolic enzymes. (For comparison, the values of k_{cat} for triosephosphate isomerase, hexokinase, and crotonase (enoyl-CoA hydratase) are all about $1000\ s^{-1}$.[7]) Consequently, when microorganisms are grown on halogenated compounds as a sole carbon source, they must produce extraordinary quantities of these dehalogenases because all of the carbon for cell growth and energy production must be processed by these inefficient enzymes. The critical dehalogenase can account for as much as 30–40% of the soluble cellular protein.[8]

The poor catalytic abilities of dehalogenase enzymes may be due to a combination of three factors. First, the chemical reaction may be intrinsically difficult. Second, many of these enzymes have broad substrate specificity, and such enzymes are typically inefficient because the entropic advantages derived from binding a substrate in a perfectly fitting active site are necessarily sacrificed to allow turnover of structurally different substrates. Finally, these enzymes have undoubtedly not evolved to a peak of catalytic perfection with respect to the turnover of a specific substrate. Many of the substrates have only been introduced in large quantities into the environment in the last 50 years. Even in contaminated sites, there will be multiple carbon sources for growth, so optimization of the efficiency of a dehalogenase may not provide a selective advantage. Indeed, in contaminated soils, broad substrate specificity may be more favored by evolution than high turnover for a specific substrate.

5.16.2 ALIPHATIC DEHALOGENASES

Halogenated aliphatic compounds are used as solvents, refrigerants, and pesticides and as intermediates in the production of many pharmaceuticals, pesticides, fragrances, and polymers, including PVC. In addition, a variety of chlorinated aliphatic compounds are produced during the chlorination of drinking water supplies. A number of dehalogenase enzymes involved in microbial degradation of halogenated aliphatic compounds are described in this section.

5.16.2.1 Haloalkane Dehalogenases

Haloalkane dehalogenases have been purified from a number of microorganisms that can grow on halogenated alkanes as a sole carbon source. The enzyme from *Xanthobacter autotrophicus* GJ10 will be emphasized here, but others have been purified and characterized from microorganisms such as *Corynebacterium* sp. Strain m15-3,[9] *Rhodococcus* sp. Strain HA1[10] (originally identified as an *Arthrobacter* sp.[11]), *Rhodococcus erythropolis* Y2,[12] and *Ancylobacter aquaticus*.[8]

Haloalkane dehalogenases catalyze the substitution of chlorine, bromine, and iodine substituents with hydroxyl groups derived from water. They do not turn over fluorinated substrates. All of the enzymes reported are broad-specificity enzymes. For example, the *Rhodococcus* Strain HA-1 enzyme can dehalogenate at least 50 substrates, including terminally substituted, nonterminally substituted, branched-chain, and unsaturated haloalkanes.[10] The *X. autotrophicus* and *A. aquaticus* enzymes (which are identical[8]) have a more restricted substrate specificity, but can still handle chlorinated alkanes with a chain length of up to five and brominated alkanes with a chain length of up to ten carbon atoms.[8,13] The most unusual substrates are two chlorinated cyclohexadienes formed during the degradation of lindane (γ-hexachlorocyclohexane) by *Pseudomonas paucimobilis* UT26.[14] "Relative rates" of conversion of various substrates (either for a single concentration or for concentrations at the solubility limits) have been reported for several enzymes.[9,10,13,15] Terminally substituted haloalkanes and relatively short haloalkanes appear to be the best substrates, although the lack of data for the kinetic parameters k_{cat} and K_m for some enzymes and problems with insolubility of many of these hydrophobic compounds make these comparisons difficult. For cases in which kinetic parameters are available, the K_m values are reasonable, but the k_{cat} values are quite low. For example, for the *X. autotrophicus* enzyme, the K_m for 1,2-dibromoethane is 10 μM, but k_{cat} is only 3 s^{-1}.[13]

The *X. autotrophicus* haloalkane dehalogenase was the first dehalogenase for which a crystal structure was solved.[16] The protein is composed of two domains. The main domain is made up of eight beta sheets surrounded by alpha helices and is typical of the α/β hydrolase fold class of enzymes.[17] The smaller domain is a "cap" made up of five alpha helices with intervening loops. The active site is a hydrophobic cavity with a volume of 37 Å^3 located between the domains with no apparent connection to solvent to allow entry of substrate or exit of products. Most of the catalytic residues are provided by the main domain.

The mechanism of haloalkane dehalogenase was elucidated in an unusual and clever way. The two mechanisms shown in Figure 1 had been proposed based upon the residues observed in the active site cavity and the observation of Janssen *et al.*[15] that the reaction catalyzed by the haloalkane dehalogenase from a Gram-positive actinomycete proceeded with inversion of stereochemistry. Verschueren *et al.*[18] reasoned that protonation of His289 (the putative active site base) at low pH should interfere with catalysis and cause accumulation of different species at the active site for each of the mechanisms under consideration. If the mechanism involved general base-catalyzed attack of water upon the substrate (Figure 1(a)), then the enzyme–substrate complex should be stable at low pH. However, if the mechanism involved nucleophilic attack of Asp124 upon the substrate, followed by hydrolysis of the alkyl-enzyme intermediate (Figure 1(b)), then the alkyl enzyme intermediate should accumulate at low pH.

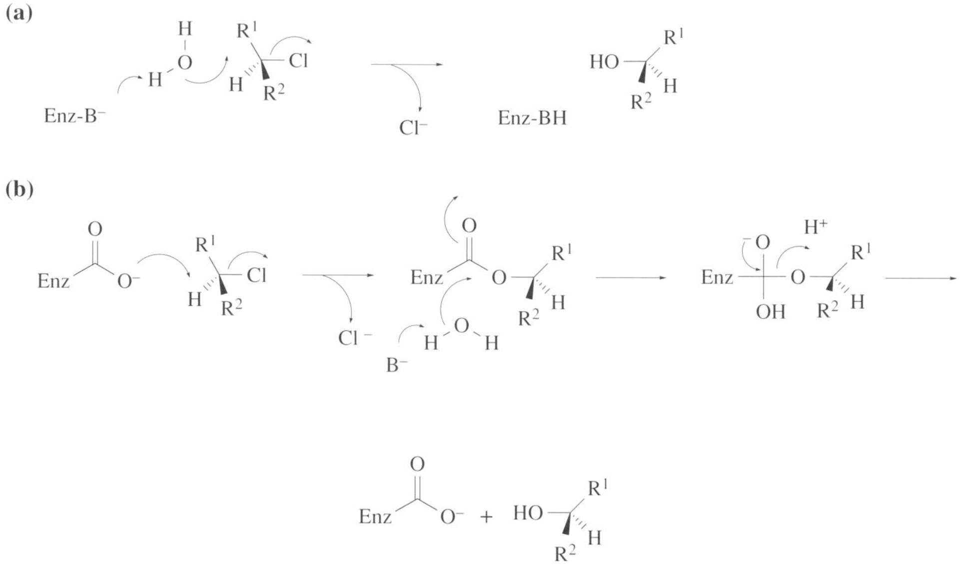

Figure 1 Possible mechanisms for haloalkane dehalogenase. (a) General-base catalyzed attack of water. (b) Displacement of chloride by an active-site carboxylate, followed by hydrolysis of the alkyl-enzyme intermediate.

Verschueren et al.[18] soaked crystals of the enzyme in mother liquor containing 10 mM 1,2-dichloroethane at different pH values, collected data for 48 h, and analyzed the data by difference Fourier analysis. Crystals soaked at pH 5 for 24 h at room temperature showed clear electron density consistent with the presence of both a covalent alkyl-enzyme intermediate and a chloride ion at the active site. Crystals soaked at pH 6.2 for 96 h at room temperature showed only chloride ion in the active site, demonstrating that the alkyl-enzyme intermediate could be hydrolyzed at the higher pH and that the product alcohol is lost from the active site. These data provide strong support for the alkyl-enzyme intermediate mechanism shown in Figure 1(b).

The alkyl-enzyme intermediate mechanism is also supported by the results of additional experiments. When the enzyme is incubated in $H_2^{18}O$ in the presence of 1,2-dichloroethane, incorporation of ^{18}O into a pentapeptide containing Asp124 is observed, while negligible incorporation occurs in the absence of substrate.[19] This result would be expected for the mechanism shown in Figure 1(b), since one of the oxygens of the catalytic aspartate is replaced by a solvent oxygen during each turnover. Site-directed mutagenesis experiments have confirmed that Asp124 is the active site nucleophile,[19] that His289 is the base which catalyzes the attack of water upon the alkyl-enzyme intermediate,[20] and that Trp175 and Trp125 form a chloride binding site.[21] The His289Gln mutant,[20] which would be expected to be defective in the hydrolysis of the alkyl-enzyme intermediate, provided additional evidence for the existence of this intermediate. When the His289Gln enzyme is incubated with substrate, a stoichiometric amount of halide is released and electrospray LC/MS shows that the enzyme contains the alkyl moiety of the substrate attached to the tryptic peptide containing Asp124.

Pre-steady state kinetic studies[13] have shown that the rates of the chemical steps, particularly hydrolysis of the alkyl-enzyme intermediate, are quite slow. However, the rate-limiting step is a slow enzyme isomerization that allows halide release.[13,22] This finding is particularly interesting in light of the crystal structure, which shows that the active site cavity is closed and that there is no solvent channel between the active site and the surface. Schanstra et al. have proposed that the enzyme isomerization involves a conformational change in the cap domain that allows water to enter the active site and solvate the halide ion.[23] This hypothesis is supported by studies of the Phe172Trp mutant enzyme. Phe172 is in the cap domain and interacts with the beta chlorine of 1,2-dichloroethane. Pre-steady-state kinetic analysis showed that the mutation causes changes in the rates of several steps. Most importantly, the rate of the slow conformational change is increased from 4 s^{-1} for 1,2-dibromoethane in the wild-type enzyme to 75 s^{-1} in the Phe172Trp enzyme. In fact, the conformational change in the mutant enzyme is no longer rate-limiting; instead, the hydrolysis of the alkyl-enzyme intermediate (which occurs at 9–$10 \, \mathrm{s}^{-1}$ in both wild-type and mutant) becomes rate-limiting. Crystallographic analysis of the mutant enzyme showed that the mutation results in a movement of the protein backbone that disrupts a hydrogen bond in the cap region. These data are consistent with the hypothesis that the rate-limiting step in the wild-type enzyme involves a movement in the cap region, and that this movement is facilitated by the loss of a hydrogen-bonding interaction that increases the flexibility in this region.

Not surprisingly, the cap region is also important in controlling the substrate specificity of the enzyme. Several mutant enzymes with improved ability to handle 1-chlorohexane were obtained by expressing the dehalogenase gene in *Pseudomonas* GJ31 (which can grow on long-chain alcohols) and selecting for growth on 1-chlorohexane.[24] All of the mutations were in the N-terminal part of the cap domain. Two mutants resulted from point mutations, while the others resulted from large deletions or tandem duplications. Thus, it appears to be fairly simple to obtain altered substrate specificity by altering the cap domain while holding constant the array of catalytic residues provided by the α/β hydrolase fold domain and Trp175 in the cap domain.

There are additional interesting aspects of the crystal structure that deserve comment. First, the active site of a broad specificity enzyme might be expected to be large and open to the solvent so that parts of large substrate molecules could protrude from the active site if necessary. However, the active site in haloalkane dehalogenase is a closed cavity which is just the right size for small substrates such as 1,2-dibromoethane, but is too small to accommodate large substrates such as 1-chlorohexane.[16,18] Perhaps deformations of the cap domain allow binding of large substrates to occur at some cost to the catalytic process. (The k_{cat}/K_m for 1-chlorohexane is 1000-fold lower than that for 1,2-dichloroethane.) A crystal structure with 1-chlorohexane bound at the active site would provide insight into this intriguing issue.

Another interesting feature is the chloride binding site which is provided by Trp125 and Trp175.[18] The chloride ion binds at the intersection of the planes of the two tryptophan residues, one of which (Trp125) is provided by the α/β hydrolase fold domain and the other (Trp175) by the cap domain. This is an unusual type of chloride binding site. Most known halide binding sites contain positively charged or polar residues.[25,26] However, this type of binding site may be ideally suited for an active site that is designed to bind highly hydrophobic substrates and optimally should not bind halides tightly lest dissociation of the halide becomes rate-limiting for catalysis.

5.16.2.2 Haloalkanoate Dehalogenases

Hydrolytic dehalogenases are found in many strains of soil microorganisms that are able to grow on chlorinated aliphatic acids. Many of these enzymes are encoded on plasmids,[27,28] and some are found on mobile genetic elements,[29,30] which may account for the widespread occurrence of these enzymes. The products of the dehalogenation reactions are readily metabolizable hydroxyacids (such as lactate, which is produced from chloropropionate), so the presence of these enzymes allows the microorganisms to take advantage of a novel carbon source with little additional effort.

Haloalkanoate dehalogenases vary with respect to the length of the carbon acid that is the preferred substrate. Those enzymes that only catalyze the dehalogenation of haloacetates are often designated by the specific term "haloacetate dehalogenase," while those that utilize longer carbon acids are classified together as haloacid dehalogenases. Most of the haloalkanoate dehalogenases that have been studied are α-haloacid dehalogenases.

Haloacid dehalogenases have been studied for three decades, and it has become clear that there are four classes of enzymes that vary according to the configuration of haloacid that serves as substrate and to the stereochemical course of the dehalogenation reaction itself.[31] Enzymes in classes I–III carry out the dehalogenation reaction with inversion of stereochemistry. Class I enzymes utilize only L-haloacids, class II enzymes utilize only D-haloacids, and class III enzymes, interestingly, can use both L- and D-haloacids. In contrast, class IV enzymes carry out the dehalogenation reaction with retention of stereochemistry. Like the class III enzymes, these enzymes also utilize both L- and D-haloacids. In many cases, a particular microorganism can express multiple haloacid dehalogenases. For example, *P. putida* PP3 expresses both class III and class IV enzymes when grown on DL-2-chloropropionate,[32] and a *Rhizobium* species expresses three dehalogenases (classes I, II, and III) when grown on 2,2-dichloropropionate (the herbicide Dalapon).[33]

The existence of these four classes of enzymes is interesting from an evolutionary perspective. At least in the case of the class I, II, and III enzymes that carry out the dehalogenation reaction with inversion of configuration, rather minor changes in the shape of the active site of an ancestral enzyme might allow a new enzyme to accommodate a different stereoisomer. However, class I and class II do not appear to be evolutionarily related. The class I and class II enzymes from *Pseudomonas putida* AJ1 have no homology,[34] and neither do the class I and class II enzymes from a *Rhizobium* sp.[35] The genes encoding the class III and IV enzymes have not yet been cloned, so sequence information is currently unavailable.

Mechanistic studies have been carried out only for the class I L-2-haloacid dehalogenase from *Pseudomonas* sp. YL. Sequence homology analysis and site-directed mutagenesis provided the first mechanistic clues. A multiple sequence alignment of a number of class I enzymes was used to identify conserved residues. Site-directed mutagenesis of all of the conserved charged and polar residues in the *Pseudomonas* sp. YL enzyme (36 out of 232) identified several residues that were required for catalytic activity.[36] Among these were two aspartate residues (Asp10 and Asp180) that were possible candidates for an active site nucleophile in a mechanism analogous to that catalyzed by haloalkane dehalogenase. (Asp10 was also shown to be essential for activity in *Pseudomonas* sp. Strain CBS3 haloacid dehalogenase I.)[37]

The possibility of the alkyl-enzyme intermediate mechanism was investigated using techniques similar to those described above in the case of haloalkane dehalogenase.[38] Under single-turnover conditions in $H_2^{18}O$, the hydroxyl group of the D-lactate produced from L-2-chloropropionate contains ^{16}O, suggesting that the oxygen in the hydroxyl group is derived from the enzyme rather than from the solvent during the first turnover. After multiple turnovers, electrospray LC/MS analysis showed that Asp10 had incorporated two atoms of ^{18}O.

Additional support for the existence of an alkyl-enzyme intermediate was provided by the observation that the enzyme was inactivated by treatment with hydroxylamine in the presence of substrate.[39] This technique is useful in diagnosing the existence of an alkyl-enzyme intermediate because hydroxylamine, a potent nucleophile, can attack the alkyl-intermediate in place of water and result in the formation of catalytically incompetent adducts at the site of the normal catalytic aspartate. Electrospray LC/MS analysis showed evidence for the presence of adducts formed by attack of hydroxylamine on the alkyl-enzyme intermediate (see Figure 2).

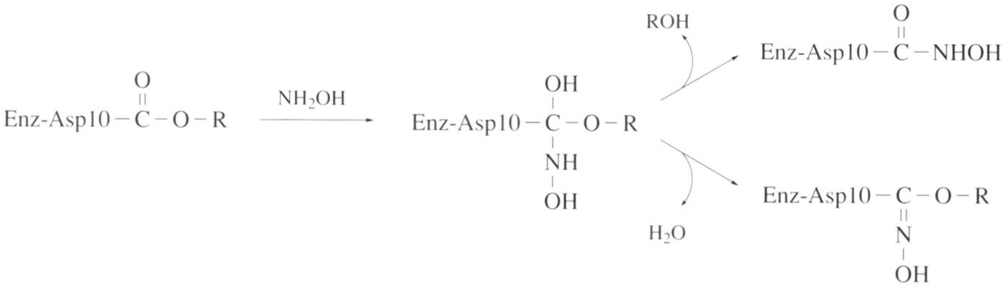

Figure 2   Adducts proposed to form at the active site of *Pseudomonas* sp. YL haloacid dehalogenase treated with hydroxylamine in the presence of substrate.[36]

The experimental evidence suggests that the mechanism of the class I haloacid dehalogenases is very similar to that of haloalkane dehalogenase. Interestingly, there is no overall sequence homology between the class I haloacid dehalogenases and haloalkane dehalogenase. The sequence identity between the *Pseudomonas* sp. Strain YL L-2-haloacid dehalogenase and the *Xanthobacter auto-trophicus* GJ10 haloalkane dehalogenase is only 12%.[40] Furthermore, the recently reported crystal structure of the haloacid dehalogenase[41] shows that the two proteins are not structurally related. The overall topology, as well as the position of the catalytic aspartate, are sufficiently different that the haloacid dehalogenase does not belong to the α/β hydrolase fold family. The *Pseudomonas* haloacid dehalogenase is a homodimer of 26 kDa subunits. Each monomer contains a main domain consisting of a six-stranded parallel beta sheet flanked by three alpha helices on one side and two on the other, and a subdomain consisting of four alpha helices. (For comparison, the main domain of haloalkane dehalogenase consists of an eight-stranded, mainly parallel beta sheet, flanked by three helices on each side, and a subdomain (the cap domain) consisting of five alpha helices.) In both proteins, the active site is located between the main domain and the subdomain and the catalytic aspartate is supplied by the main domain. However, while the active site of haloalkane dehalogenase is small, closed, and hydrophobic, the active site of haloacid dehalogenase is open and polar, in keeping with the ability of this enzyme to bind long substrates bearing a charged carboxylate group. Furthermore, there is no histidine in the vicinity of the active site that would correspond to His289 of haloalkane dehalogenase (which catalyzes the attack of water on the alkyl-enzyme intermediate).

The mechanisms of the class II and III enzymes have not yet been addressed. Since these enzymes also catalyze dehalogenation with inversion of stereochemistry, the most exciting question is whether they also use an active site aspartate as a nucleophile in a manner analogous to the class I L-2-haloacid dehalogenases and haloalkane dehalogenase. It would be interesting if Nature had arrived at the same catalytic strategy in all of these cases. An additional question which will have to await a crystal structure is how the active site of the class III and IV enzymes can accommodate both D- and L-substrates.

The class IV enzymes, which catalyze dehalogenation with retention of stereochemistry, will clearly have a different mechanism from the other haloacid dehalogenases. The stereochemical evidence suggests that the reaction will occur via a double-displacement mechanism (see Figure 3). The active site nucleophile has been suggested to be a cysteine residue based upon observations that the enzyme is inactivated by treatment with sulfhydryl reagents such as *N*-ethylmaleimide and *p*-chloromercuribenzoate.[32] However, this issue requires further investigation, since modification of a cysteine residue in the vicinity of the active site can inactivate the enzyme even if the cysteine is not involved in catalysis. An illustration of this was provided by Au and Walsh, who showed that a haloacetate dehalogenase that is inactivated by sulfhydryl reagents catalyzes a reaction that proceeds with inversion of stereochemistry. Since inversion of stereochemistry is consistent with a direct displacement mechanism, the cysteine that is presumably modified in the active site is unlikely to be involved in catalysis.[42]

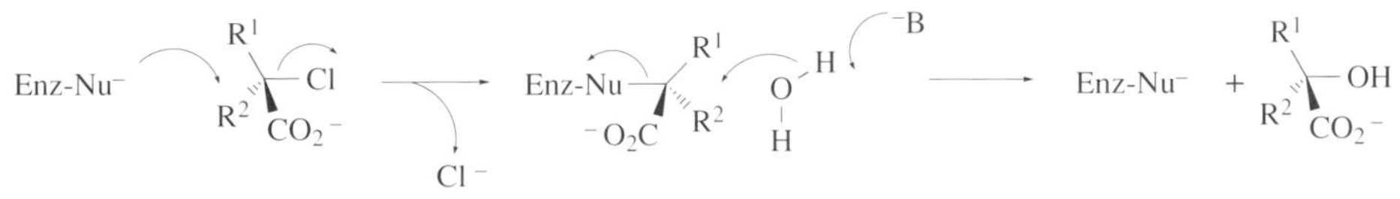

Figure 3 Possible mechanism for class IV haloacid dehalogenases that catalyze substitution reactions with retention of stereochemistry.

Most haloacid dehalogenases can dehalogenate chloro, bromo and iodocompounds, but cannot dehalogenate the corresponding fluoro-compounds. This finding is not surprising in two respects. First, fluorine is by far the worst leaving group among the halogens. Second, there are very few fluorine-containing natural products, so there has been little need for microorganisms to catalyze the cleavage of a C—F bond. However, *Pseudomonas* sp. strain A2 expresses a fluoroacetate dehalogenase that catalyzes the removal of fluoride.[42] The fascinating question of how this enzyme is able to cleave the strong C—F bond has not yet been answered.

5.16.2.3 Dichloromethane Dehalogenase

Dichloromethane (DCM) dehalogenase is found in a number of strains of Gram-negative facultative methylotrophic bacteria and a strain of nitrate-respiring *Hyphomicrobium* that are able to use this compound, as well as other dihalomethanes, as a carbon and energy source.[43] DCM dehalogenase converts DCM to formaldehyde, which can be oxidized to CO_2 to provide energy or assimilated into biomass.[44] The enzyme is specific for dihalomethanes. Halomethanes,[45] tri-halomethanes,[46] and larger chloroalkanes[47] do not serve as substrates, although some of these compounds are potent competitive inhibitors.

The hydrolysis of dichloromethane in solution is extremely slow (the $t_{1/2}$ for hydrolysis at pH 7 and 25 °C is 700 yr),[48] so the ability of microorganisms to degrade this compound is crucial for its removal from the environment. A comparison of k_{cat}/K_m (6×10^4 M^{-1} s^{-1}) with the rate constant for a comparable uncatalyzed reaction (6×10^{-6} M^{-1} s^{-1}) suggests that the enzyme achieves a rate enhancement of 10 orders of magnitude over the uncatalyzed reaction.[45]

DCM dehalogenase has been purified from several strains of bacteria. The enzyme is a homohexamer of 33 kDa subunits.[43,47] The genes encoding the enzyme have been cloned and sequenced. In most cases, the DCM dehalogenase gene is located on a large plasmid.[43] Two classes of enzymes have been identified that differ in catalytic efficiency. Type A enzymes have a k_{cat} of 0.9 s^{-1} and type B enzymes have a k_{cat} of 3.3 s^{-1}.[43] The type B enzymes allow the microorganism to grow faster on DCM,[49] suggesting that dehalogenation of DCM is the rate-limiting step for growth on DCM. The interesting question of whether the type B enzymes have evolved by recent optimization of the type

A enzymes was settled by sequence analysis. The two classes have only 56% sequence identity, suggesting that they diverged millions of years ago.[50]

Notably, the reaction catalyzed by DCM dehalogenase is dependent upon glutathione, although glutathione is not consumed.[51] This fact, and the observation that the enzyme shows significant sequence identity with enzymes in the theta class of the glutathione S-transferase superfamily,[50] provided important clues to the mechanism of the enzyme. The proposed mechanism for the enzyme is shown in Figure 4.[51] The enzyme catalyzes a simple S_N2 reaction in which the thiol moiety of glutathione serves as the nucleophile. The thioether product then undergoes nonenzymatic hydrolysis to form hydroxyglutathionylmethane, which decomposes to yield formaldehyde and regenerate the glutathione.

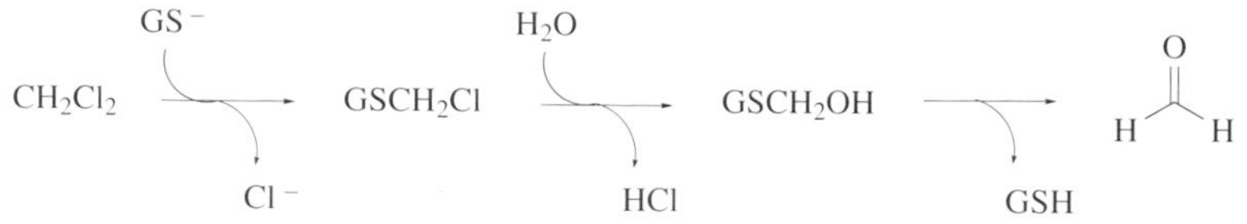

Figure 4   Mechanism of dichloromethane dehalogenase.

The validity of the mechanism shown in Figure 4 was established by elegant experiments by Wackett and co-workers, who analyzed the conversion of the substrate analogue CH_2ClF to formaldehyde.[52] When a tight-binding inhibitor ($ClCH_2CN$) is added during the reaction of CH_2ClF with DCM dehalogenase, disappearance of CH_2ClF ceases, but formaldehyde production continues for several minutes, suggesting that the formaldehyde results from nonenzymatic decomposition of an intermediate produced by the enzyme. ^{19}F-NMR experiments showed that a compound with a chemical shift consistent with that of authentic S-fluoromethylglutathione ($GSCH_2F$) accumulates and then disappears during turnover of CH_2ClF. (This intermediate accumulates because fluorine is a very poor leaving group. The comparable intermediate formed from CH_2Cl_2 would be expected to hydrolyze more quickly.) These data support the suggestion that the role of the enzyme is simply to catalyze the nucleophilic attack of glutathione upon CH_2Cl_2 to form $GSCH_2Cl$, which then decomposes nonenzymatically to form formaldehyde.

The reaction catalyzed by DCM dehalogenase is characteristic of the glutathione S-transferase enzymes, which catalyze the nucleophilic attack of glutathione upon an electrophilic substrate to form a glutathione conjugate. DCM dehalogenase further resembles the theta-class glutathione S-transferases in that a conserved serine is required for enzyme activity.[53] In the theta-class enzymes, a hydrogen bond between a serine hydroxyl group and the sulfur atom of glutathione is believed to stabilize the thiolate in the active site.[54] This interaction is important for catalysis because thiolates are orders of magnitude more nucleophilic than thiols.[55] The observation that mutant enzymes in which Ser12 has been converted to Ala or Thr are inactive is consistent with a comparable role for Ser12 in DCM dehalogenase.

An interesting aspect of the conversion of DCM to formaldehyde and Cl^- is that the thioether intermediate ($GSCH_2Cl$) is so much more susceptible to nonenzymic hydrolysis than the substrate (CH_2Cl_2). Secondary alkyl halides are not so sterically hindered as to preclude S_N2 reactions, so the slow rate of hydrolysis of CH_2Cl_2 can be attributed mainly to the poor nucleophilicity of water. If the hydrolysis reaction proceeded via a standard S_N2 reaction, the reaction of CH_2Cl_2 would be faster than that of $GSCH_2Cl$ because the glutathionyl substituent is less electron-withdrawing and more bulky than a chlorine substituent. Therefore, it appears that this reaction occurs via an S_N1 pathway. S_N1 reactions are known to be accelerated by heteroatom substituents such as O and S. For example, the rate of solvolysis of $MeOCH_2Cl$ is approximately 10^5 times faster than the solvolysis of $MeCl$.[56] Thus, the conversion of CH_2Cl_2 to a chloromethylthioether is quite clever because the hydrolysis of the latter is fast enough that an enzyme is not required for catalysis.

5.16.2.4   Haloalcohol Dehalogenases

Haloalcohol dehalogenases (also called halohydrin hydrogen-halide lyases) catalyze the reversible formation of epoxides from vicinal haloalcohols. The prototypical reaction shown in Equation (1) is the conversion of 1,3-dichloropropane to epichlorohydrin and chloride, although a range of other haloalcohols also serve as substrate.[57,58] These enzymes have been found in a number of microorganisms, including a strain of *Flavobacterium*[59] and *Pseudomonas* sp. Strain AD1[60] (both

Gram-negative) and the Gram-positive *Arthrobacter* sp. strain AD2[58] and *Corynebacterium* sp. Strain N-1074.[57] The *Corynebacterium* strain contains two distinct haloalcohol dehalogenases.

$$\text{Cl}\diagup\overset{\text{OH}}{\diagdown}\diagup\text{Cl} \; \rightleftharpoons \; \text{Cl}\diagup\diagdown\overset{\text{O}}{\diagdown} + \text{HCl} \qquad (1)$$

Only the enzymes from *Arthrobacter* and *Corynebacterium* strains have been purified. The *Arthrobacter* enzyme is a homodimer of approximately 29 kDa subunits.[58] One enzyme from *Corynebacterium* sp. Strain N-1074 (abbreviated H-lyase A) is a homotetramer of 27 kDa subunits, and the other, H-lyase-B, is a tetramer of 26 and 25 kDa subunits in various ratios.[61] (The 26 kDa and 25 kDa subunits of H-lyase B have identical sequences except that the smaller subunits lack the first few amino acids at the N terminus.[61] The smaller subunits appear to originate from an alternative translation initiation site in the mRNA encoding the enzyme.)

The dehalogenation reactions catalyzed by these enzymes do not require cofactors or O_2. The simplest mechanism for this reaction would be a base-catalyzed intramolecular attack of the hydroxyl group upon the adjacent carbon to displace chloride. However, this mechanism is difficult to reconcile with the observation that H-lyase B from *Corynebacterium* sp. Strain N-1074 converts 1,3-dichloro-2-propanol initially to *R*-epichlorohydrin, but that it rapidly racemizes the product.[62] H-lyase A also produces nearly racemic epichlorohydrin after a 60-min incubation,[57] but whether this reflects true racemization at the active site or initial production of one stereoisomer followed by rapid racemization of the epichlorohydrin cannot be distinguished from the published data. A mechanism that would account for these findings has not yet been proposed.

Both of the *Corynbacterium* sp. Strain N-1074 dehalogenases have been cloned and sequenced.[61] Neither enzyme has significant homology with haloalkane dehalogenases, haloacid dehalogenases, dichloromethane dehalogenase, or any other proteins except that a limited amount of homology is found in the C-terminal region with some proteins in the insect type short-chain alcohol dehydrogenase family.[61] Since this region is not involved in catalytic activity in the alcohol dehydrogenase family, and, indeed, is not even conserved in all members of the family, the significance of this finding is unknown.

The *Arthrobacter* haloalcohol dehalogenase appears to be relatively ineffective in catalyzing the dehalogenation reaction. The K_m for 1,3-dichloro-2-propanol is 8.5 mM, k_{cat} is 2 s^{-1} (assuming two equivalent active sites), and k_{cat}/K_m is only 250 M^{-1} s^{-1}.[58] The *Corynebacterium* H-lyase A has similarly unimpressive kinetic characteristics. The H-lyase B enzyme, however, is considerably better, with a k_{cat} for 1,3-dichloro-2-propanol of 126 s^{-1} (assuming two equivalent active sites), a K_m of 1 mM, and a k_{cat}/K_m of 1.3×10^5 M^{-1} s^{-1}.[62]

Because of the widely different catalytic abilities of the *Corynebacterium* H-lyase A and H-lyase B, these enzymes provide a fascinating system for further studies of structure and mechanism. H-lyase A and H-lyase B have little overall similarity, although the C-terminal 40 amino acids have 45% identity.[61] (The N-terminal 195 amino acids have only 15% identity.) It thus seems likely that there will be a conserved structural region in the C-terminal part of the protein. The possibility that the overall structures will be similar should not be discounted either. A number of hydrolytic enzymes, including haloalkane dehalogenase, that have insignificant sequence identity have a common α/β hydrolase fold structure.[17] Crystal structures of these two enzymes will be needed to firmly establish their evolutionary relationship and to help explain the greater efficiency of H-lyase B.

5.16.2.5 Other Interesting Aliphatic Dehalogenases

There are several additional dehalogenase enzymes that catalyze intriguing chemical reactions but for whom mechanistic and structural studies have not been done. Some of the most interesting are described here.

An unusual dehalogenase has been discovered in *Pseudomonas paucimobilis* strain UT6 (which may be reclassified as *Sphingomonas paucimobilis*), a Gram-negative soil bacterium isolated from soil to which γ-hexachlorocyclohexane (lindane) had been applied for 12 years.[63] The first enzyme in the degradation pathway catalyzes two successive elimination reactions to form first γ-pentachlorocyclohexene and then 1,3,4,6-tetrachloro-1,4-cyclohexadiene. This elimination reaction is particulary interesting because most known aliphatic dehalogenases catalyze substitution reactions. Furthermore, the gene for the enzyme (*linA*) has been cloned and sequenced and showed no sequence homology with any known proteins as of 1991.[63] The mechanism and structure of this enzyme are therefore likely to be unique.

Two enzymes that catalyze the dehalogenation of *cis*- and *trans*-chloroacrylic acid, respectively, have been isolated from a Coryneform bacterium designated FG41.[64] The product in both cases is malonate semialdehyde. These enzymes are interesting because they catalyze a rare hydrolytic dehalogenation of chlorinated alkene substrates. The *cis*-specific enzyme is a homodimer of 19 kDa subunits, while the *trans*-specific isomer is a multimeric enzyme composed of 7.4 and 8.4 kDa subunits. Comparison of the N-terminal sequences of the enzymes indicates that they are not related. Two possible mechanisms for this reaction are shown in Figure 5. It will be interesting to discover whether these two enzymes use similar catalytic strategies.

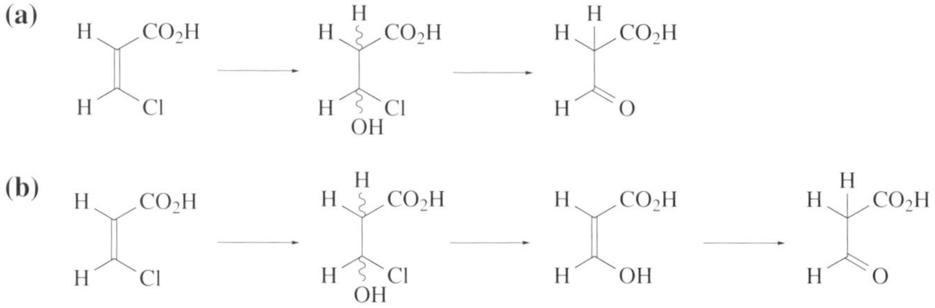

Figure 5 Possible mechanisms for dehalogenation of chloroacrylic acids. (a) Addition of water to the double bond, followed by nonenzymic decomposition to malonate semialdehyde. (b) Vinylic substitution followed by tautomerization.

A tetrachloroethane reductive dehalogenase that converts tetrachlorethene to trichloroethene and then to *cis*-1,2-dichloroethene has been purified and characterized from the strictly anaerobic Gram-negative bacterium *Dehalospirillum multivorans*.[65] This is one of only two dehalogenases that have been purified from anaerobes, and is particularly important because its substrates, tetrachlorethane and trichloroethene, are ubiquitous groundwater pollutants and are resistant to biodegradation by aerobic microorganisms. The function of the tetrachlorethane dehalogenase in *D. multivorans* differs from those of the other dehalogenases described so far in that it plays a central role in energy production. Tetrachlorethene and trichloroethene serve as the terminal electron receptors in a respiratory process in which electrons are provided by H_2. This process is apparently coupled to ATP synthesis by some type of chemiosmotic mechanism which has not yet been defined. The enzyme has relatively high specificity: tetrachlorethene, trichloroethene, and tetraiodoethene are the only identified substrates, although other halogenated ethenes, ethanes, and aromatic compounds have been tested. Furthermore, it has a k_{cat} of 150 s^{-1}, and is thus one of the most active of the dehalogenases yet described. The enzyme contains 1 mol of vitamin B_{12}, 9.8 mol of iron and 8 mol of acid-labile sulfur per mol of enzyme. The mechanism proposed by Neumann *et al.* is shown in Figure 6.[65] It begins with the transfer of one electron from a membrane-associated hydrogenase via a low-potential iron–sulfur cluster to the cob(II)alamin cofactor. After reduction to the cob(I)alamin state, the cofactor binds covalently to the haloalkene and displaces chloride. Addition of a proton and transfer of another electron through a second, high-potential iron–sulfur cluster causes release of the reduced product and regenerates the cob(II)alamin form of the cofactor. Only limited experimental evidence supports this mechanism so far. The spectral properties of the resting enzyme are consistent with the proposed cob(II)alamin form of the B cofactor. When the enzyme is reduced with titanium(III)citrate (which presumably reduces the cobalt to the +1 oxidation state), it is inactivated by propyl iodide, suggesting that the cobalt in the cob(I)alamin form of the cofactor can be alkylated and, therefore, may also be alkylated during the normal catalytic cycle. Studies of the redox potentials of the B_{12} and iron–sulfur cofactors, as well as efforts to clone and sequence the gene, are underway and will provide more insight into this unusual and important enzyme.

5.16.3 AROMATIC DEHALOGENASES

Chlorinated aromatic compounds are a particularly important class of environmental pollutants because of their resistance to biodegradation and consequent tendency to persist in the environment. Examples of these compounds include PCBs (polychlorinated biphenyls), dioxins, and numerous

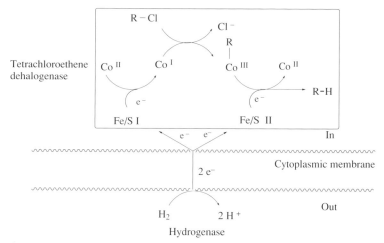

Figure 6 Mechanism proposed by Neumann *et al.* for tetrachlorethene dehalogenase. Reproduced by permission of the American Society for Biochemistry and Molecular Biology from *J. Biol. Chem.*, 1996, **271**, 16 518.[65]

pesticides, such as DDT, atrazine, pentachlorophenol, and 2,4-dichlorophenoxyacetic acid (2,4-D). Particularly problematic compounds such as PCBs and DDT have not been used in the USA since the early 1970s. However, significant quantities of these compounds still exist in many places, including the Great Lakes and the Hudson River (the site of a General Electric plant that legally dumped PCB-containing waste into the river for decades).

The removal of chlorine atoms from chlorinated aromatic molecules can take place by three fundamentally different processes. The first is a hydrolytic reaction in which the chlorine is replaced by a hydroxyl group derived from H_2O. The second is a reductive reaction in which the chlorine is replaced by a hydrogen. The third is an O_2-dependent reaction in which the chlorine is replaced by a hydroxyl group, but the oxygen is derived from O_2 rather than from H_2O. While the mechanisms of these three classes of aromatic dehalogenase enzymes will certainly be different, it is becoming clear that, within each of these classes, there is more than one way of catalyzing each transformation. Thus, there is considerable mechanistic diversity in nature with respect to accomplishing these difficult chemical transformations. This fascinating area is only beginning to be explored.

5.16.3.1 Hydrolytic Aromatic Dehalogenases

4-Chlorobenzoyl CoA dehalogenase is found in several strains of bacteria that degrade 4-chlorobenzoate,[66-69] a breakdown product of some PCBs. The enzyme responsible for removal of chloride from 4-chlorobenzoate was originally believed to be a 4-chlorobenzoate dehalogenase. Early attempts to study this enzyme were thwarted by the poor activity obtained in crude extracts. The solution to this problem is an elegant example of the use of sequence homology analysis to assign the function for an unknown gene.[70] The genes for 4-chlorobenzoate degradation in *Pseudomonas* sp. CBS3 were known to encode three essential polypeptides with molecular masses of 57, 30, and 16 kDa. The sequences of these polypeptides were compared with sequences available in the National Biomedical Research Foundation database. Six proteins were found to have a stretch of sequence identity with the 57 kDa polypeptide. All six of these proteins catalyzed the adenylation of a carboxylate in their respective substrates in a reaction that required ATP and Mg^{2+}, and, in one case (4-coumarate CoA ligase), this reaction was followed by attack of coenzyme A upon the acyl adenylate to form a CoA thioester. These results suggested that the 57 kDa protein might catalyze the thioesterification of 4-chlorobenzoate to form 4-chlorobenzoyl CoA, which would then be the actual substrate for the dehalogenase. This hypothesis was easily confirmed. Addition of ATP, Mg^{2+}, and CoA to cell extracts gave substantially improved activity. (The same conclusions were reached by Löffler *et al.*, who examined a large number of possible cofactors for the enzyme and discovered that activity in crude extracts was stimulated by the addition of ATP, Mg^{2+}, and CoA.[71]) Subsequently, all three polypeptides required for dehalogenation of 4-chlorobenzoate were isolated and characterized from *Pseudomonas* sp. CBS3.[72] The 57 kDa protein is 4-chlorobenzoate CoA ligase, and catalyzes the conversion of 4-chlorobenzoate to 4-chlorobenzoyl CoA using CoA, ATP, and Mg^{2+}. The 30 kDa protein is 4-chlorobenzoyl CoA dehalogenase, and the 16 kDa protein is a thioesterase that converts 4-hydroxybenzoyl CoA to 4-hydroxybenzoate. This pathway is also found in other microorganisms, including *Arthrobacter* sp. Strains 4CB1[73] and SU.[74]

Several possible mechanisms have been considered for the unusual reaction catalyzed by 4-chlorobenzoyl CoA dehalogenase. An aryne mechanism was eliminated by the absence of a deuterium kinetic isotope effect when 4-chlorobenzoyl CoA-d_4 was used as a substrate.[75] Mechanisms based upon the $S_{RN}1$ and $S_{ON}2$ reactions were deemed unlikely because of the absence of transition metal or organic cofactors on the enzyme.[75] The remaining possibilities are variants of the S_NAr reaction. In the mechanism shown in Fig. 7(a), a base at the active site would activate water for attack upon 4-chlorobenzoyl CoA. The mechanism shown in Figure 7(b) is similar to that of haloalkane dehalogenase and proceeds by attack of an active site carboxylate upon 4-chlorobenzoyl CoA to give an aryl-enzyme intermediate, followed by base-catalyzed hydrolysis of the aryl-enzyme intermediate to give 4-hydroxybenzoyl CoA and regenerate the free enzyme.

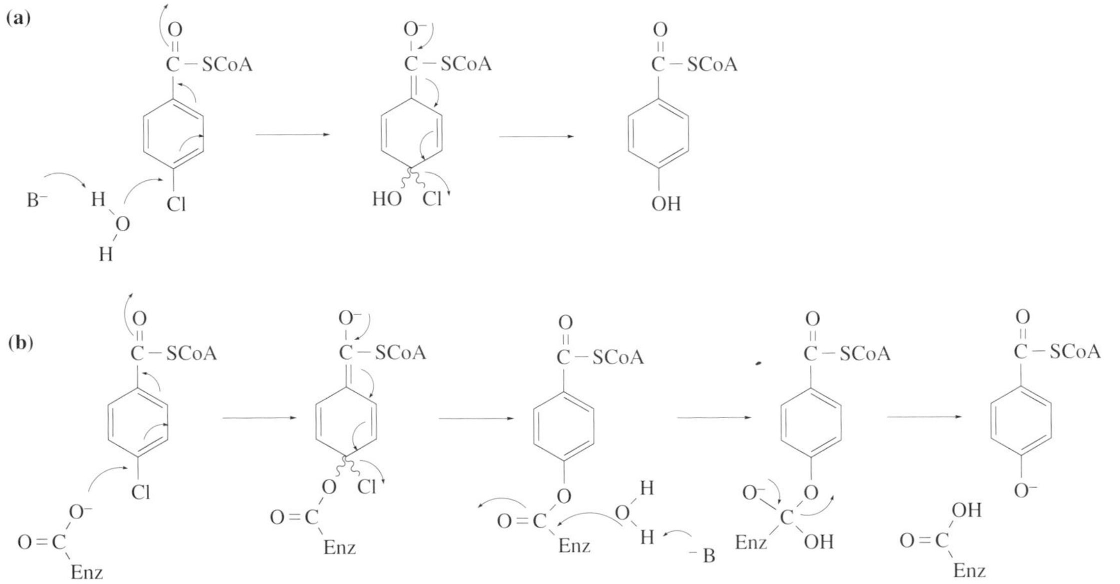

Figure 7 Some possible mechanisms for 4-chlorobenzoyl CoA dehalogenase. (a) General-base catalyzed attack of water to form a Meisenheimer complex, which decomposes to form 4-hydroxybenzoyl CoA and chloride. (b) Attack of an active-site carboxylate to form a Meisenheimer complex, followed by expulsion of chloride and hydrolysis of the aryl-enzyme intermediate.

Single-turnover experiments in $H_2{}^{18}O$ analogous to those used to differentiate between general-base catalyzed attack of water and formation of an alkyl-enzyme intermediate for haloalkane dehalogenase were difficult to interpret in the case of 4-chlorobenzoyl CoA dehalogenase.[75,76] With both the *Pseudomonas* and *Arthrobacter* enzymes, both ^{18}O and ^{16}O were found in the product isolated after single-turnover experiments (25% ^{18}O with the *Pseudomonas* enzyme,[76] and 50% with the *Arthrobacter* enzyme[75]). Although a number of explanations has been considered for this finding, the reason for the incorporation of the ^{18}O has not been determined. A possible explanation was recently suggested by Zheng and Ornstein, who proposed that the tetrahedral intermediate formed by attack of water upon the aryl-enzyme intermediate could undergo an isomerization at the active site that would result in the incorporation of some oxygen from the solvent into the product during the first turnover.[77]

Two other methods were successful in determining that the mechanism of the enzyme involves the formation of an aryl-enzyme intermediate as shown in Figure 7(b). For both the *Pseudomonas*[76] and *Arthrobacter* enzymes,[78] the appearance and disappearance of a radiolabeled covalent adduct can be detected during turnover of [carbonyl-^{14}C]4-chlorobenzoyl CoA by rapid flow-quench. This intermediate is presumed to be the aryl-enzyme intermediate shown in Figure 7(b). In addition, Crooks *et al.* showed that the *Arthrobacter* enzyme is inactivated by hydroxylamine in the presence of substrate.[75] Hydroxylamine would be expected to attack the aryl-enzyme intermediate in place of water, resulting in the formation of adducts such as those described above for haloacid dehalogenase (see above) at the position of the required active site carboxylate.

The sequence of the *Pseudomonas* 4-chlorobenzoyl CoA dehalogenase shows 16–32% identity with several 2-enoyl-CoA hydratases and two Δ^3-*cis*-Δ^2-*trans*-enoyl-CoA isomerases (both of which are involved in fatty acid β-oxidation) and dihydroxynapthoate synthase (which is involved in vitamin K_2 biosynthesis).[79] Because the mechanisms of all of these enzymes involve delocalization of electron density into the carbonyl moiety of a CoA thioester, some structural similarities might be expected. In fact, the recently solved crystal structures of the *Pseudomonas* sp. CBS3 4-chloro-benzoyl CoA dehalogenase–4-hydroxybenzoyl CoA complex[80] and the rat liver mitochondrial enoyl-CoA hydratase[81] show remarkable similarities. Both enzymes have an unusual trimeric structure in which the active site is made up of residues from the N-terminal domains of adjacent subunits, while the C-terminal domains are primarily involved in trimerization. (The 2-enoyl-CoA hydratase is a dimer of trimers, while 4-chlorobenzoyl CoA dehalogenase is a simple trimer.) The 4-chloro-benzoyl CoA dehalogenases from the Gram-positive *Arthrobacter* sp. Strain SU and Strain 4CB1 (which are identical[82]) have 50% identity to the *Pseudomonas* sp. CBS3 enzyme.[74] Therefore, the overall structural fold is likely to be similar, but there may be some interesting differences that warrant exploration of the crystal structure of the *Arthrobacter* enzyme.

The identity of the active site carboxylate in the *Pseudomonas* enzyme was suggested first by site-directed mutagenesis experiments in which Asp145 was shown to be required for catalytic activity[83] and subsequently confirmed by the crystal structure of the enzyme–4-hydroxybenzoyl CoA complex, which showed that Asp145 is in close proximity to the hydroxyl group of the bound product.[80] Similar studies have demonstrated that the base responsible for catalysis of the hydrolysis of the aryl-enzyme intermediate is His90.[80,83] Thus, these catalytic residues are the counterparts of the Asp124 and His289 in haloalkane dehalogenase. The third member of the catalytic triad in halo-alkane dehalogenase, Asp260, which interacts with the catalytic His289, does not seem to have a counterpart in 4-chlorobenzoyl CoA dehalogenase. Another notable difference between the active sites of the two enzymes is that there is no chloride binding site in 4-chlorobenzoyl CoA dehalogenase corresponding to that provided by Trp125 and Trp172 in haloalkane dehalogenase.

The rate-limiting step for dehalogenation of 4-chlorobenzoyl CoA is not yet known. For the mechanisms shown in Figure 7(b), it seems most likely that the rate-limiting step would be the attack of the nucleophile upon the substrate, since this is the rate-limiting step for nearly all S_NAr reactions.[84] However, the effect of the leaving group on the rate of the reaction demonstrates that this is not the case. For S_NAr reactions, the leaving group ability increases in the order Br < Cl < F.[84] This effect occurs because the attack of the nucleophile to form the Meisenheimer complex is rate limiting, and the Meisenheimer complex and transition state leading to it are best stabilized by fluorine. Thus, if the first step of the reaction catalyzed by 4-chlorobenzoyl CoA dehalogenase were rate limiting, the rate of the reaction should increase in the order 4-bromobenzoyl CoA < 4-chlorobenzoyl CoA < 4-fluorobenzoyl CoA. Contrary to this expectation, the k_{cat} for 4-fluoro-benzoyl CoA is dramatically lower (400-fold for the *Arthrobacter* enzyme[85] and over 2000-fold for the *Pseudomonas* enzyme[86]) than that for 4-chlorobenzoyl CoA, and the k_{cat} for 4-bromobenzoyl CoA is two-fold larger. Thus, attack of the active-site aspartate on 4-chlorobenzoyl CoA is not rate limiting.

Presteady-state kinetic studies carried out by Liu *et al.* show evidence for the accumulation of two discrete intermediates during substrate turnover.[86] The assignment of these intermediates is uncertain, but it appears that the rate-limiting step corresponds to the conversion of intermediate II to the product. Intermediate II has been proposed to be either the aryl-enzyme intermediate or the tetrahedral intermediate resulting from attack of water upon the aryl-enzyme intermediate. Thus, it appears that the rate-limiting step is one of the steps involved in hydrolysis of the aryl-enzyme intermediate. The marked effect of leaving group on the rate of the enzymic reaction might be explained in either of two ways. Either the nature of the halide at the active site dramatically influences the rate of hydrolysis of the aryl-enzyme intermediate, or the rate-limiting step for the chloro- and bromocompounds involves hydrolysis of the aryl-enzyme intermediate, but for the fluorocompound is the cleavage of the C—F bond in the Meisenheimer complex.

The reaction catalyzed by 4-chlorobenzoyl CoA dehalogenase is remarkable for its difficulty. The nonenzymatic rate for this reaction cannot be measured because of the lability of the thioester substituent and because carboxylates such as acetate act as general bases rather than nucleophiles in S_NAr reactions.[87] However, a second-order rate constant can be estimated based upon that for attack of methoxide upon *p*-nitrochlorobenzene (which is considerably more activated for nucleo-philic attack than 4-chlorobenzoyl CoA), which is 8.5×10^{-6} M^{-1} s^{-1} in methanol at 50 °C.[88] The rates of aliphatic substitution reactions with methoxide are about 100-fold faster than comparable reactions with carboxylates.[89] Thus, if the relative nucleophilicities of acetate and methoxide are similar in nucleophilic aromatic substitution reactions, the rate constant for the hypothetical attack of acetate on *p*-nitrochlorobenzene would be approximately $8.5 \times 10^{-8} \text{ M}^{-1} \text{s}^{-1}$. For the *Arthrobacter* 4-chlorobenzoyl CoA dehalogenase, k_{cat}/K_m is $3.8 \times 10^4 \text{ M}^{-1} \text{s}^{-1}$.[90] Disregarding solvent and temperature effects, the reaction catalyzed by 4-chlorobenzoyl CoA dehalogenase is at least 12 orders of magnitude faster than the hypothetical noncatalyzed reaction.

The rates of $S_N Ar$ reactions are strongly dependent on the electron-withdrawing ability of the substituents. Values of ρ (from the linear free energy relationship $\log k = \rho \sigma^-$) for $S_N Ar$ reactions can be as high as 8.5.[84] The remarkable rate acceleration achieved by 4-chlorobenzoyl CoA dehalogenase is due in part to its ability to polarize the thioester substituent and thereby enhance its electron-withdrawing ability. UV–visible, Raman, and NMR spectroscopic studies are consistent with a significant shift of electron density from the aromatic ring into the thioester when substrate analogues such as 4-methoxybenzoyl CoA are bound to the active site.[91] The crystal structure of the *Pseudomonas* enzyme-4-hydroxybenzoyl CoA complex shows that the backbone amide hydrogen atoms of Gly114 and Phe64 are likely to provide hydrogen bonds to the carbonyl group of the thioester substituent.[80] Furthermore, the positive end of a helix dipole is located near the thioester carbonyl. These interactions appear to provide the polarization that is observed in the spectroscopic studies.

Another important factor in catalysis would be expected to be optimization of the strength of the nucleophile. This could be achieved effectively by desolvating the carboxylate moiety of Asp145. In fact, the crystal structure of the enzyme–4-hydroxybenzoyl CoA complex shows that the active site pocket around Asp145 is very hydrophobic and contains no fixed water molecules.[80]

4-Chlorobenzoyl CoA dehalogenase uses an interesting combination of two catalytic strategies, each of which are seen in other enzymes. The polarization of the carbonyl to facilitate attack of a nucleophile upon a carbon in conjugation with a thioester is seen in enoyl-CoA hydratase[92] and Δ^5-3-ketosteroid isomerase.[93] The use of an active site carboxylate to form an alkyl-enzyme intermediate which is subsequently hydrolyzed by attack of water at the carbonyl carbon is seen in epoxide hydrolase,[94] haloalkane dehalogenase,[18] and class I haloacid dehalogenase.[38,39] There is no homology between these enzymes and 4-chlorobenzoyl CoA dehalogenase. It is interesting that the same catalytic strategy is used in these enzymes, given that there is no sequence homology and that the steric and electronic requirements of $S_N 2$ and $S_N Ar$ reactions are so different.

Xun *et al.* have reported the purification of a 2,6-dichlorohydroquinone dehalogenase from *Sphingomonas chlorophenolica* that requires ferrous ions for activity.[95] However, subsequent work in the author's laboratory has shown that the enzyme is not, in fact, a hydrolytic dehalogenase. The product formed from DCHQ has not yet been conclusively identified, but the enzyme's requirement for O_2, the properties of the product, and comparisons of the protein sequence with others in the database suggest that the enzyme is most likely a ring-cleavage dioxygenase.[96]

Hydrolytic dehalogenation of heteroaromatic compounds is also possible. de Souza *et al.*[97] have characterized an atrazine chlorohydrolase from *Pseudomonas* sp. strain ADP that is activated by Co^{II}, Mn^{II}, and Fe^{II},[46] and a hydrolytic enzyme that catalyzes both dechlorination and deamination of *S*-triazine substrates has been purified from *Rhodococcus corallinus* NRRL B-15444R.[98] These two proteins have 41% sequence identity with each other, but no significant identity to any other proteins in the database.[97] Mechanistic studies have not yet been reported for either enzyme.

5.16.3.2 Reductive Dehalogenases

Reductive dehalogenation is particularly important for the degradation of highly chlorinated aromatic compounds, since replacement of several chlorine atoms with hydroxyl groups would create a product that would be extremely prone to oxidation. Furthermore, the energetic requirements for reduction of aromatic compounds become more favorable as the number of electron-withdrawing chlorine substituents increases. However, the use of reductive dehalogenation is not restricted to highly chlorinated compounds. Recently, a reductive 3-chlorobenzoate dehalogenase has been purified.

The only reductive aromatic dehalogenase that has been studied in detail is tetrachlorohydroquinone dehalogenase from *Sphingomonas chlorophenolica*. This enzyme catalyzes the conversion of tetrachlorohydroquinone (TCHQ) to trichlorohydroquinone (TriCHQ) and then to 2,6-dichlorohydroquinone (DCHQ). The reducing equivalents are provided by two molecules of glutathione for each step.[99] The enzyme is a monomer of 27 kDa[100] and contains neither transition metal nor organic cofactors.[101]

The gene for TCHQ dehalogenase has been cloned and sequenced.[102] TCHQ dehalogenase is not closely related to any known proteins, but it does have 26–30% identity with some microbial enzymes that belong to the theta class of the glutathione *S*-transferase (GST) superfamily.[101] Recent studies have suggested some striking mechanistic similarities between TCHQ dehalogenase and the theta class GSTs. These similarities, as well as some interesting differences, will be described here.

One of the major clues to the mechanisms of TCHQ dehalogenase was the observation that the enzyme undergoes oxidative damage during purification and that the damaged enzyme produces the glutathione conjugates 2,3,5-trichloro-6-*S*-glutathionylhydroquinone (GS-TriCHQ) and an unidentified isomer of dichloro-*S*-glutathionylhydroquinone (GS-DCHQ) (see Scheme 2), rather than the expected reduced products.[101,103] The oxidative damage can be reversed by treatment with 25 mM dithiothreitol or β-mercaptoethanol, suggesting that a sulfur-containing amino acid is involved. Consequently, mutant enzymes were constructed in which each of the two cysteines (Cys13 and Cys156) had been replaced by Ser. Both mutant enzymes have values of k_{cat} equivalent to wild-type, and the C156S enzyme appeared to be identical to the wild-type. However, the C13S mutant enzyme, like the oxidatively damaged enzyme, produces only GS-TriCHQ and GS-DCHQ from TCHQ. Clearly, Cys13 is required for the reductive dehalogenation of TCHQ to TriCHQ.

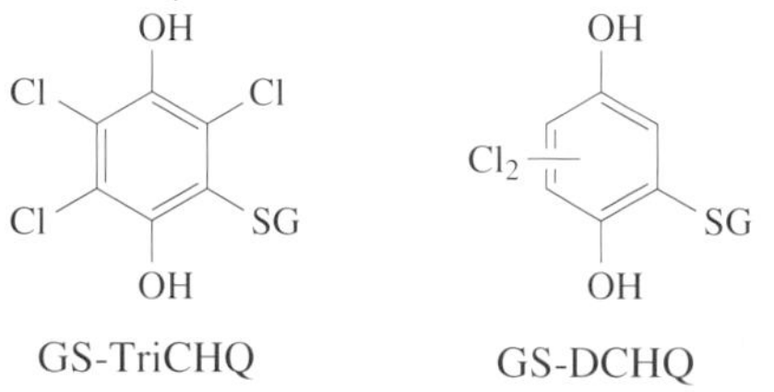

Scheme 2

The observation that the C13S mutant enzyme produces both GS-TriCHQ and GS-DCHQ provided another important clue. This enzyme, which apparently cannot catalyze the normal reductive reaction, is nevertheless capable of producing GS-DCHQ, which is reduced with respect to GS-TriCHQ. A mechanistic model that accounts for these findings is shown in Figure 8. The model postulates that the tautomer of GS-TriCHQ is an intermediate in the reductive dehalogenation reaction and that Cys13 is required to attack the sulfur of the glutathionyl substituent, releasing TriCHQ and forming a mixed disulfide between Cys13 and glutathione. In the final step, a second molecule of glutathione attacks the mixed disulfide and regenerates the free enzyme. When a functional thiol is not present at the active site (either in the C13S mutant enzyme or oxidatively damaged enzyme), this tautomer would be expected to decompose to give primarily GS-TriCHQ and GS-DCHQ (see Figure 9), the products formed by the mutant enzyme from TCHQ.

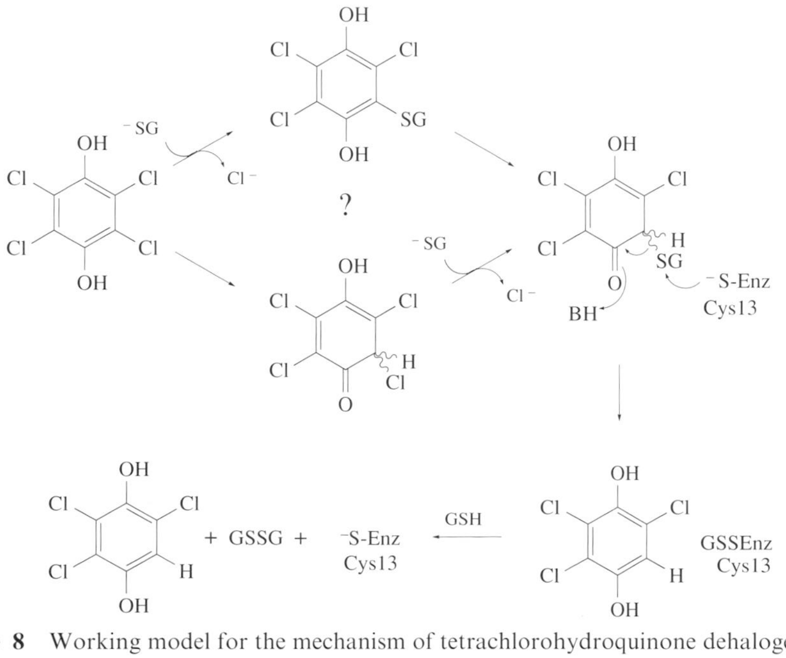

Figure 8 Working model for the mechanism of tetrachlorohydroquinone dehalogenase.

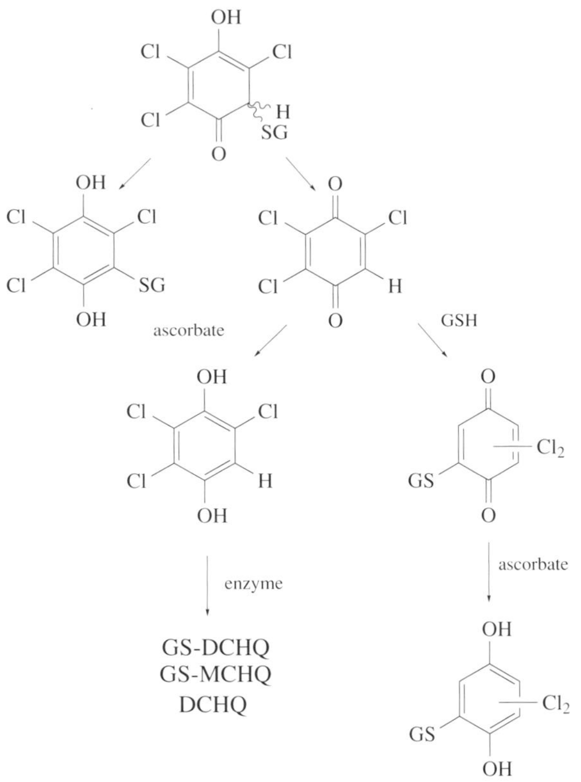

Figure 9 Expected pathways for decomposition of the tautomer of GS-TriCHQ in reaction mixtures containing glutathione and ascorbate (present to maintain hydroquinone substrates in their reduced forms).

The mechanism of the first part of the reaction, formation of the tautomer of GS-TriCHQ, is uncertain. Three possible mechanisms are currently being considered. The reaction may begin by formation of GS-TriCHQ, either by an S_NAr reaction or an $S_{RN}1$ reaction, followed by tautomerization of GS-TriCHQ. A third possibility is initial tautomerization of TCHQ, followed by an S_N2 displacement of chloride by glutathione.

Additional support for the proposed mechanism was provided by experiments that demonstrated the existence of a covalent adduct between glutathione and TCHQ dehalogenase during turnover of TCHQ. A rapid-flow quench instrument was used to quench reaction mixtures containing [glycine-2-^{3}H]glutathione, trichlorohydroquinone, and TCHQ dehalogenase and the protein was separated from small molecules by multiple cycles of dilution and concentration by ultrafiltration. A covalent adduct between glutathione and the enzyme accumulated to a maximum at about 200 ms and then declined. Analysis of tryptic digests of the enzyme quenched during turnover by electrospray LC/MS, confirmed that the glutathione was covalently attached to Cys13.[104]

The mechanism shown in Figure 8 resembles those of GSTs in that the enzyme catalyzes the nucleophilic attack of glutathione upon some as yet unidentified electrophilic form of the substrate. An additional parallel between TCHQ dehalogenase and the GSTs is that TCHQ dehalogenase appears to activate glutathione for nucleophilic attack in the same way as the theta-class GSTs. As discussed above in connection with DCM dehalogenase, a considerable part of the catalytic power of GSTs comes from their ability to ionize glutathione at the active site.[105] TCHQ dehalogenase also ionizes glutathione at the active site.[100] In the theta-class GSTs, the thiolate of glutathione is believed to be stabilized at the active site by a hydrogen bond with a nearby serine.[54] TCHQ dehalogenase has a serine (Ser11) at this position, and the S11A mutant enzyme lacks catalytic activity.[106] These results suggest that ionization of glutathione is critical for TCHQ dehalogenase activity and that it is likely achieved in a manner comparable to that seen in the theta-class GSTs.

An important difference between the reactions catalyzed by TCHQ dehalogenase and the GSTs is that two molecules of glutathione are required for reductive dehalogenation, but only one for glutathione conjugate formation. Therefore, the possibility of a second glutathione binding site in TCHQ dehalogenase is an intriguing issue. GSTs have only one glutathione binding site, and a perusal of GST structures does not suggest an obvious location for a second binding site. However, a second binding site is likely to be required, since thiol–disulfide exchange reactions are not rapid enough to support the observed rate of the enzyme reaction. (The rate constant for reduction of glutathione disulfide with dithiothreitol at pH 7.0 and 30 °C is about 0.2 M^{-1} s^{-1},[107] whereas $k_{cat}/K_{m,GSH}$ for TCHQ dehalogenase is 1600 M^{-1} s^{-1}.[100]) The location of the second glutathione binding site has not yet been identified, but it may involve a 20 amino acid insert that is not present in most GSTs, but is found in TCHQ dehalogenase and in β-etherase, another GST homologue that catalyzes the reductive cleavage of β-aryl ethers and should also require two equivalents of glutathione.[108]

Sequence homology analysis and mechanistic studies suggest that TCHQ dehalogenase resembles the theta-class GSTs in its ability to activate glutathione for attack upon an electrophilic substrate. However, TCHQ dehalogenase has additional abilities that GSTs do not have, including the ability to tautomerize a hydroquinone, to catalyze the attack of Cys13 upon the tautomer of GS-TriCHQ, to bind a second molecule of glutathione, and to catalyze a thiol–disulfide exchange reaction. The emerging picture of the mechanism of TCHQ dehalogenase brings up the interesting question of whether it has evolved from a protein with the general architecture and glutathione binding properties of a GST by the addition of some extra capabilities, or whether GSTs have evolved from ancient reductive enzymes by the loss of some capabilities.

There are undoubtedly other mechanisms for reductive dehalogenation. Romanov and Hausinger have recently reported the purification of a reductive 2,4-dichlorobenzoyl CoA dehalogenase from *Corynebacterium sepedonicum* KZ-4 and coryneform bacterium strain NTB-1 that utilizes NADPH as a reductant.[109] In addition, a reductive 3-chlorobenzoate dehalogenase has been purified from the anaerobic *Desulfomonile tiedjei* DCB-1.[110] This membrane-bound enzyme appears to be a heme protein and utilizes methyl viologen as a reductant *in vitro*. Although mechanistic studies have not yet been reported for either of these enzymes, the differing reductants imply significant differences between the mechanisms of these enzymes and TCHQ dehalogenase.

5.16.3.3 O₂-dependent Aromatic Dehalogenases

O_2-dependent aromatic dehalogenases fall into two classes: (i) dioxygenases that result in the loss of chloride and the formation of a catechol; and (ii) monoxygenases that replace the chlorine atom with a hydroxyl group. These enzymes are "accidental" dehalogenases in that their catalytic machinery operates to form an unstable product that decomposes spontaneously to release the halide. In spite of the fact that the dehalogenation reaction itself is fortuitous, these enzymes play a critical role in the biodegradation of many chlorinated aromatic compounds, including 4-chlorophenylacetate, 2-halobenzoates, and pentachlorophenol.

Enzymes that catalyze the introduction of two hydroxyl groups onto aromatic substrates are widespread in nature, as this is generally the first step in the degradation of aromatic compounds by aerobic microorganisms. Monooxygenase enzymes that catalyze the hydroxylation of aromatic compounds containing an activating substituent are also common. Thus, O_2-dependent dehalogenases are clearly examples of the recruitment of preexisting cellular enzymes with the appropriate catalytic abilities to perform a different function in the degradation of chlorinated aromatic compounds.

5.16.3.3.1 Dioxygenases

Dioxygenase dehalogenase enzymes are multicomponent systems comprised of an oxygenase component that catalyzes the installation of the hydroxyl groups on the aromatic substrate and a reductase component (consisting of either one or two proteins) that carries electrons from NADH through flavin and [2Fe–2S] cofactors to the oxygenase component. The enzymes that dehalogenate 2-halobenzoate and 4-chlorophenylacetate in *Pseudomonas cepacia* 2CBS[111] and *Pseudomonas* sp. CBS3,[112] respectively, are two-component enzymes. A three-component dioxygenase system that uses both 2-chlorobenzoate and 2,4-dichlorobenzoate has been purified from *Pseudomonas aeruginosa* 142.[113] In all of these cases, the deoxygenation reaction itself is believed to result in formation of a *cis*-dihydrodiol product which spontaneously undergoes either a *syn*-elimination of HCl or decarboxylation coupled with chloride release to give a catechol product (see Figure 10) which can then be cleaved and further metabolized. It seems reasonable to expect that the structures and mechanisms of these enzymes will be closely related to those of the family of arene dioxygenase enzymes, which includes enzymes such as phthalate dioxygenase, benzoate dioxygenase, and benzene 1,2-dioxygenase. Unfortunately, the mechanisms of these enzymes have not been characterized. The difficulty lies in the fact that the oxygenase components contain [2-Fe–2S] Rieske sites as well as nonheme iron.[114] The spectroscopic properties of the Rieske site obscure those of the nonheme iron, at which the hydroxylation reaction takes place. Thus, the identification of the active oxygen species at the active site and the sequence of events taking place during catalysis will be unusually challenging.

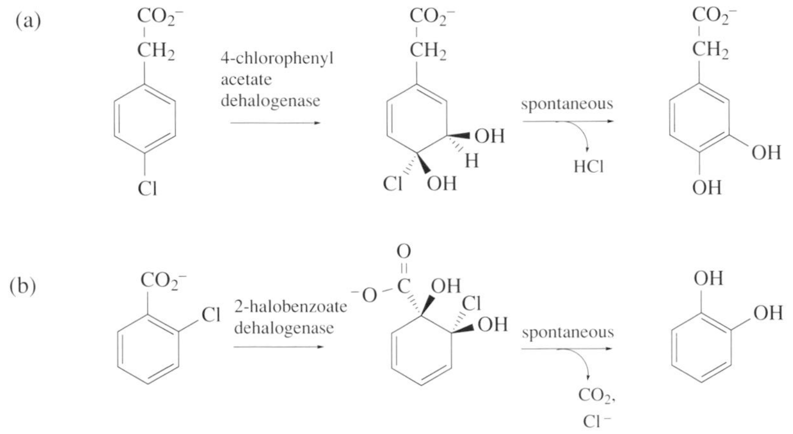

Figure 10   Reactions catalyzed by dioxygenase-type dehalogenases.

5.16.3.3.2   *Monooxygenases*

Monoxygenases that convert pentachlorophenol (PCP) to tetrachlorohydroquinone (TCHQ) have been identified. Biodegradation of PCP, a widely used wood preservative, is particularly challenging because it is highly toxic as well as xenobiotic. Nevertheless, a number of microorganisms that degrade PCP have been identified, including several strains of the Gram-negative bacterium *Sphingomonas chlorophenolica* (previously identified as strains of *Pseudomonas*[115] and *Flavobacterium*[116]) and the Gram-positive actinomycete *Mycobacterium* (previously identified as strains of *Rhodococcus*[117]).

PCP hydroxylase from *Sphingomonas chlorophenolica* Strain ATCC 39723 has been purified[118] and the gene cloned and sequenced.[119] It is a monomer with a molecular mass of 63 kDa and contains tightly bound FAD. The enzyme requires O_2 and utilizes NADPH as a reductant. PCP hydroxylase has quite broad specificity, and PCP does not appear to be the optimal substrate. This observation is consistent with the hypothesis that PCP hydroxylase has been recruited to participate in the biodegradation of PCP and that the enzyme is not optimally suited to this new function.

The *S. chlorophenolica* PCP hydroxylase is one of a large number of known FAD-dependent monooxygenases, including *p*-hydroxybenzoate hydroxylase, phenol hydroxylase, anthranilate hydroxylase, and melilotate hydroxylase.[120] The mechanisms of many of these enzymes have been studied in detail. In particular, there is a wealth of mechanistic information about *p*-hydroxybenzoate hydroxylase,[121–123] and a crystal structure of the *Pseudomonas fluorescens* enzyme is available.[124,125] PCP hydroxylase has 73% sequence identity with *p*-hydroxybenzoate hydroxylase in the N-terminal 75% of the protein,[126] which contains the active site residues, so it is reasonable to expect significant structural and mechanistic similarities between these two enzymes.

p-Hydroxybenzoate hydroxylase normally catalyzes the conversion of *p*-hydroxybenzoate to 3,4-dihydroxybenzoate, but it also catalyzes the hydroxylation of 3,5-difluoro-*p*-hydroxybenzoate and tetrafluoro-*p*-hydroxybenzoate with concomitant removal of fluoride and consumption of two (rather than one) molecules of NADPH.[122] This reaction provides an excellent model for the mechanism of PCP hydroxylase (see Figure 11). The formation of the C(4a)hydroperoxyflavin shown at the beginning of the reaction should occur by the typical reaction of reduced flavin with O_2 observed in the phenol hydroxylase family. Attack of the aromatic ring upon the C(4a)hydroperoxyflavin is facilitated by the hydroxyl group in the *para* position. The unstable intermediate resulting from the hydroxylation step eliminates HCl to form tetrachlorobenzoquinone, which can be reduced nonenzymatically by NADPH to give TCHQ. (Note that Figure 11 depicts the phenolate form of PCP, but it is not yet known whether PCP binds to the enzyme as the phenol or the phenolate. If it binds as the phenol, participation of a base to remove the hydroxyl proton during the hydroxylation step would likely occur.)

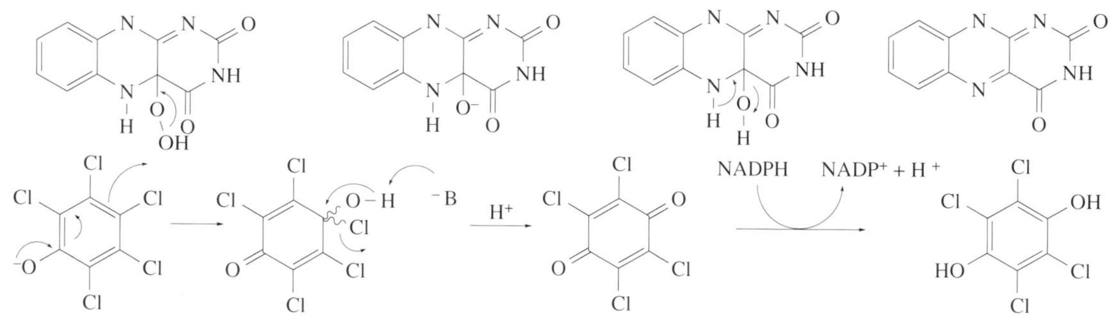

Figure 11 Possible mechanism of the hydroxylation reaction catalyzed by pentachlorophenol hydroxylase. The C(4a) hydroperoxyflavin shown would be formed by prior reaction of the reduced flavin with O_2.

Hydroxylation of PCP apparently occurs by a different mechanism in Gram-positive micro-organisms. The enzyme from *Mycobacterium fortuitum* CG-2 is membrane-associated and appears to be a P450-type enzyme based upon the increase in A_{448} of membrane fractions observed in the presence of CO (a typical spectroscopic signature of P450 enzymes) and the inhibition of the activity by SKF-525A, metyrapone, menadione, and parathion, all of which inhibit P450 enzymes.[127]

5.16.4 CONCLUSIONS

5.16.4.1 Diversity of Mechanistic Approaches

Only a few of the identified dehalogenases have been studied at a detailed mechanistic level. However, it is already clear that there is considerable diversity in Nature's approaches to the removal of chlorine from organic compounds. Substitution reactions, addition reactions, elimination reactions, oxidations, and reductions are all used in various contexts. A striking finding is that none of the well-characterized dehalogenases are related to each other except for DCM dehalogenase and TCHQ dehalogenase, which are related by virtue of their homology to GSTs. Thus, each of these enzymes appears to have arisen independently, no doubt from cellular enzymes that provided an appropriate array of catalytic groups for the needed reaction. As we learn more about the mechanisms and structures of these enzymes, a more complete picture of what types of protein scaffolds can be adapted to serve as dehalogenases and the types of adaptations that are specific to dehalogenases will emerge.

5.16.4.2 Recurring Themes

For the limited number of hydrolytic dehalogenases studied, it is striking that general base-catalyzed attack of water has not been observed. Instead, the most common approach seems to be the use of an active-site carboxylate to displace chloride and form an alkyl- or aryl-enzyme intermediate, followed by hydrolysis of the intermediate. This strategy is used by haloalkane dehalogenase, Class I haloacid dehalogenases, and 4-chlorobenzoyl CoA dehalogenase. The conservation of this approach is particularly notable because the mechanism and energetic requirements for the aliphatic substitutions catalyzed by haloalkane dehalogenase and haloacid dehalogenases are considerably different from those for the nucleophilic aromatic substitution reaction catalyzed by 4-chlorobenzoyl CoA dehalogenase. In addition, this catalytic strategy must have evolved independently for each of these dehalogenases, since the crystal structures of these three enzymes are significantly different. It is difficult to tell whether the multiple occurrences of this mechanistic strategy imply that attack of a carboxylate provides a more effective way to displace chloride than general-base catalyzed attack of water, or simply that a preexisting protein scaffold with an aspartate and a basic amino acid at the active site can be readily adapted to serve as a dehalogenase.

Another interesting theme is the use of a GST-like enzyme to catalyze the nucleophilic attack of glutathione upon an electrophilic halogenated substrate to form an unstable intermediate that then undergoes further transformations, either nonenzymatically in the case of DCM dehalogenase, or enzymatically in the case of TCHQ dehalogenase. It will be interesting to discover whether there is some particular characteristic of these substrates that makes glutathione an optimal nucleophile, or whether the use of glutathione has resulted simply from the evolutionary expediency of using the scaffold of a glutathione *S*-transferase to catalyze these two reactions.

5.16.4.3 Challenges for the Future

Mechanistic studies of microbial dehalogenases have only been carried out for a few enzymes, and many more fascinating enzymes await the attention of mechanistic enzymologists and protein crystallographers. The mechanisms of nonhydrolytic enzymes need to be addressed more thoroughly. A notable gap is that the mechanism of a metal-dependent dehalogenase has never been elucidated. The role of metal ions in metal-dependent dehalogenases is a particularly important area for future research.

Another area that deserves attention is dehalogenation in anaerobic microorganisms. Sediments in rivers and lakes, sewage sludge, and groundwater are anaerobic environments that are often contaminated with chlorinated pollutants. Chlorinated compounds undergo reductive dehalogenation reactions in anaerobic environments, and a few microorganisms that carry out these reactions have been isolated. In some cases, reductive dehalogenation appears to be coupled to energy metabolism and the chlorinated molecule apparently serves as a terminal electron receptor for an electron transport chain. However, only two enzymes that catalyze reductive dehalogenation have been purified from an anaerobe (tetrachloroethane dehalogenase from *Dehalospirillum multivorans* and 3-chlorobenzoate dehalogenase from *Desulfomonile tiedjei*), and the 3-chlorobenzoate dehalogenase has poor *in vitro* activity. The shortage of dehalogenases from anaerobic microorganisms can be attributed mainly to the difficulty of isolating pure cultures that degrade halogenated compounds from the complex microbial communities found in anaerobic ecosystems, but also to the technical difficulties of working with anaerobic species. However, the importance of reductive dehalogenation in anaerobic environments and the promise of new and different dehalogenase enzymes make this work a priority.

For the dehalogenase enzymes that have been studied intensively, the time is ripe for efforts to modify the naturally occurring enzymes to improve catalytic activity or to alter substrate specificity. Many of the dehalogenase enzymes described here are quite slow, and it should be possible in many cases to enhance the rate of the catalytic reaction. For those enzymes for which certain functional groups on the substrate are not required for catalysis, it should be possible to alter substrate specificity by altering the shape of the binding pocket. (The range of possible substrates will be somewhat restricted for enzymes such as TCHQ dehalogenase and haloalcohol dehalogenase which utilize hydroxyl groups in the substrate during the chemical transformation.) Studies with haloalkane dehalogenase have shown that enzymes with moderately better activity toward alternative substrates can be obtained both by selection and by directed mutagenesis based upon the known structure and mechanism of the enzyme. Both of these approaches should be exploited with other enzymes. Enzymes with enhanced activity or different substrate specificities may be useful in constructing new strains of bacteria for bioremediation or for use in enzyme-based waste treatment processes.

5.16.5 REFERENCES

1. B. Hileman, *Chem. Eng. News*, 1993, April 19, 11.
2. R. Stone, *Science*, 1994, **265**, 308.
3. D. A. Abramowicz, *Crit. Rev. Biotechnol.*, 1990, **10**, 241.
4. U. Klages and F. Lingens, *Zbl. Bakt. Hyg., I. Abt. Orig. C*, 1980, **1**, 215.
5. R. Imai, Y. Nagata, H. Senoo, H. Wada, M. Fukuda, M. Takagi, and K. Yano, *Agric. Biol. Chem.*, 1989, **53**, 2015.
6. G. W. Gribble, *Environ. Sci. Technol.*, 1994, **28**, 311A.
7. C. Walsh, "Enzymatic Reaction Mechanisms," Freeman, San Francisco, CA, 1979.
8. A. J. van den Wijngaard, K. W. H. J. van der Kamp, J. van der Ploeg, F. Pries, B. Kazemier, and D. B. Janssen, *Appl. Env. Microbiol.*, 1992, **58**, 976.
9. T. Yokota, T. Omori, and T. Kodama, *J. Bacteriol.*, 1987, **169**, 4049.

10. R. Scholtz, T. Leisinger, F. Suter, and A. M. Cook, *J. Bacteriol.*, 1987, **169**, 5016.
11. T. Leisinger and R. Bader, *Chimia*, 1993, **47**, 116.
12. P. J. Sallis, S. J. Armfield, A. T. Bull, and D. J. Hardman, *J. Gen. Microbiol.*, 1990, **136**, 115.
13. J. P. Schanstra, J. Kingma, and D. B. Janssen, *J. Biol. Chem.*, 1996, **271**, 14747.
14. Y. Nagata, T. Nariya, R. Ohtomo, M. Fukuda, K. Yano, and M. Takagi, *J. Bacteriol.*, 1993, **175**, 6403.
15. D. B. Janssen, J. Gerritse, J. Brackman, C. Kalk, D. Jager, and B. Witholt, *Eur. J. Biochem.*, 1988, **171**, 67.
16. S. M. Franken, H. J. Rozeboom, K. H. Kalk, and B. W. Dijkstra, *EMBO J.*, 1991, **10**, 1297.
17. D. L. Ollis, E. Cheah, M. Cygler, B. Dijkstra, F. Frolow, S. M. Franken, M. Harel, S. J. Remington, I. Silman, J. Schrag, J. L. Sussman, K. H. G. Verscheuren, and A. Goldman, *Protein Eng.*, 1992, **5**, 197.
18. K. H. G. Verschueren, F. Seljée, H. J. Rozeboom, K. H. Kalk, and B. W. Dijkstra, *Nature*, 1993, **363**, 693.
19. F. Pries, J. Kingma, M. Pentenga, G. van Pouderoyen, C. M. Jeronimus-Stratingh, A. P. Bruins, and D. B. Janssen, *Biochemistry*, 1994, **33**, 1242.
20. F. Pries, J. Kingma, G. H. Krooshof, C. M. Jeronimus-Stratingh, A. P. Bruins, and D. B. Janssen, *J. Biol. Chem.*, 1995, **270**, 10405.
21. C. Kennes, F. Pries, G. H. Krooshof, E. Bokma, J. Kingma, and D. B. Janssen, *Eur. J. Biochem.*, 1995, **228**, 403.
22. J. P. Schanstra and D. B. Janssen, *Biochemistry*, 1996, **35**, 5624.
23. J. P. Schanstra, I. S. Ridder, G. J. Heimeriks, R. Rink, G. J. Poelarends, K. H. Kalk, B. W. Dijkstra, and D. B. Janssen, *Biochemistry*, 1996, **35**, 13186.
24. F. Pries, A. J. van den Wijngaard, R. Bos, M. Pentenga, and D. B. Janssen, *J. Biol. Chem.*, 1994, **269**, 17490.
25. M. Qian, R. Haser, and F. Payan, *J. Mol. Biol.*, 1993, **231**, 785.
26. Z. Wang, A. B. Asenjo, and D. D. Oprian, *Biochemistry*, 1993, **32**, 2125.
27. H. Kawasaki, T. Toyama, T. Maeda, H. Nishino, and K. Tonomura, *Biosci. Biotech. Biochem.*, 1994, **58**, 160.
28. H. Kawasaki, K. Tsuda, I. Matsushita, and K. Tonomura, *J. Gen. Microbiol.*, 1992, **138**, 1317.
29. A. W. Thomas, J. H. Slater, and A. J. Weightman, *J. Bacteriol.*, 1992, **174**, 1932.
30. J. van der Ploeg, M. Willemsen, G. van Hall, and D. B. Janssen, *J. Bacteriol.*, 1995, **177**, 1348.
31. J.-Q. Liu, T. Kurihara, V. Nardi-Dei, T. Okamura, N. Esaki, and K. Soda, *Biodegradation*, 1995, **6**, 223.
32. A. J. Weightman, A. L. Weightman, and J. H. Slater, *J. Gen. Microbiol.*, 1982, **128**, 1755.
33. J. A. Leigh, A. J. Skinner, and R. A. Cooper, *FEMS Microbiol. Lett.*, 1988, **49**, 353.
34. P. T. Barth, L. Bolton, and J. C. Thomson, *J. Bacteriol.*, 1992, **174**, 2612.
35. S. S. Cairns, A. Cornish, and R. A. Cooper, *Eur. J. Biochem.*, 1996, **235**, 744.
36. T. Kurihara, J.-Q. Liu, V. Nardi-Dei, H. Koshikawa, N. Esaki, and K. Soda, *J. Biochem.*, 1995, **117**, 1317.
37. B. Schneider, R. Müller, R. Frank, and F. Lingens, *Biol. Chem. Hoppe-Seyler*, 1993, **374**, 489.
38. J.-Q. Liu, T. Kurihara, M. Miyagi, N. Esaki, and K. Soda, *J. Biol. Chem.*, 1995, **270**, 18309.
39. J.-Q. Liu, T. Kurihara, M. Miyagi, S. Tsunasawa, M. Nishihara, N. Esaki, and K. Soda, *J. Biol. Chem.*, 1997, **272**, 3363.
40. V. Nardi-Dei, T. Kurihara, T. Okamura, J.-Q. Liu, H. Koshikawa, H. Ozaki, Y. Terashima, N. Esaki, and K. Soda, *Appl. Env. Microbiol.*, 1994, **60**, 3375.
41. T. Hisano, Y. Hata, T. Fujii, J.-Q. Liu, T. Kurihara, N. Esaki, and K. Soda, *J. Biol. Chem.*, 1996, **271**, 20322.
42. K. G. Au and C. T. Walsh, *Bioorg. Chem.*, 1984, **12**, 197.
43. T. Leisinger, R. Bader, R. Hermann, M. Schmid-Appert, and S. Vuilleumier, *Biodegradation*, 1994, **5**, 237.
44. C. Anthony, "The Biochemistry of Methylotrophs," Academic Press, London, 1982.
45. L. P. Wackett, M. S. P. Logan, F. A. Blocki, and C. Bao-li, *Biodegradation*, 1992, **3**, 19.
46. L. P. Wackett, personal communication.
47. D. Kohler-Staub and T. Leisinger, *J. Bacteriol.*, 1985, **162**, 676–681.
48. W. Mabey and T. Mill, *J. Phys. Chem. Ref. Data*, 1978, **7**, 383.
49. R. Scholtz, L. P. Wackett, C. Egli, A. M. Cook, and T. Leisinger, *J. Bacteriol.*, 1988, **170**, 5698.
50. R. Bader and T. Leisinger, *J. Bacteriol.*, 1994, **176**, 3466.
51. G. Stucki, R. Gälli, H.-R. Ebersold, and T. Leisinger, *Arch. Microbiol.*, 1981, **130**, 366.
52. F. A. Blocki, M. S. P. Logan, C. Baoli, and L. P. Wackett, *J. Biol. Chem.*, 1994, **269**, 8826.
53. S. Vuilleumier and T. Leisinger, *Eur. J. Biochem.*, 1996, **239**, 410.
54. J. Rossjohn, P. G. Board, M. W. Parker, and M. C. J. Wilce, *Protein Eng.*, 1996, **9**, 327.
55. D. D. Roberts, S. D. Lewis, D. P. Ballou, S. T. Olson, and J. A. Shafer, *Biochemistry*, 1986, **25**, 5595–5601.
56. A. Streitweiser, Jr., "Solvolytic Displacement Reactions," McGraw-Hill, New York, 1962.
57. T. Nakamura, T. Nagasawa, F. Yu, I. Watanabe, and H. Yamada, *J. Bacteriol.*, 1992, **174**, 7613.
58. A. J. van den Wijngaard, P. T. W. Reuvekamp, and D. B. Janssen, *J. Bacteriol.*, 1991, **173**, 124.
59. E. W. Bartnicki and C. E. Castro, *Biochemistry*, 1969, **8**, 4677.
60. A. J. van den Wijngaard, D. B. Janssen, and B. Witholt, *J. Gen. Microbiol.*, 1989, **135**, 2199.
61. F. Yu, T. Nakamura, W. Mizunashi, and I. Watanabe, *Biosci. Biotech. Biochem.*, 1994, **58**, 1451.
62. T. Nakamura, T. Nagasawa, F. Yu, I. Watanabe, and H. Yamada, *Appl. Env. Microbiol.*, 1994, **60**, 1297.
63. R. Imai, Y. Nagata, M. Fukuda, M. Takagi, and K. Yano, *J. Bacteriol.*, 1991, **173**, 6811.
64. J. E. T. van Hylckama Vlieg and D. B. Janssen, *Biodegradation*, 1992, **2**, 139.
65. A. Neumann, G. Wohlfarth, and G. Diekert, *J. Biol. Chem.*, 1996, **271**, 16515.
66. P. Adriaens, H. P. E. Kohler, S. D. Kohler, and D. D. Focht, *Appl. Environ. Microbiol.*, 1989, **55**, 887.
67. H. Keil, V. Klages, and F. Lingens, *FEMS Microbiol. Lett.*, 1981, **10**, 213.
68. W. J. J. van den Tweel, N. ter Burg, J. B. Kok, J. A. M. de Bont, *Appl. Microbiol. Biotechnol.*, 1986, **25**, 289.
69. T. S. Marks, A. R. W. Smith, A. V. Quirk, *Appl. Environ. Microbiol.*, 1984, **48**, 1020.
70. J. D. Scholten, K.-H. Chang, P. C. Babbitt, H. Charest, M. Sylvestre, and D. Dunaway-Mariano, *Science*, 1991, **253**, 182.
71. F. Löffler, R. Mueller, and F. Lingens, *Biochem. Biophys. Res. Commun.*, 1991, **176**, 1106.
72. K.-H. Chang, P.-H. Liang, W. Beck, J. D. Scholten, and D. Dunaway-Mariano, *Biochemistry*, 1992, **31**, 5605.
73. S. D. Copley and G. P. Crooks, *Appl. Environ. Microbiol.*, 1992, **58**, 1385.
74. A. Schmitz, K.-H. Gartemann, J. Fiedler, E. Grund, and R. Eichenlaub, *Appl. Environ. Microbiol.*, 1992, **58**, 4068.
75. G. P. Crooks, L. Xu, R. M. Barkley, and S. D. Copley, *J. Am. Chem. Soc.*, 1995, **117**, 10791.

76. G. Yang, P.-H. Liang, and D. Dunaway-Mariano, *Biochemistry*, 1994, **33**, 8527.
77. Y.-J. Zheng and R. L. Ornstein, *Protein Eng.*, 1996, **9**, 721.
78. L. Xu and S. D. Copley, unpublished results.
79. D. Dunaway-Mariano and P. C. Babbitt, *Biodegradation*, 1994, **5**, 259.
80. M. M. Benning, K. L. Taylor, R.-Q. Liu, G. Yang, H. Xiang, G. Wesenberg, D. Dunaway-Mariano, and H. M. Holden, *Biochemistry*, 1996, **35**, 8103.
81. C. K. Engel, M. Mathieu, J. P. Zeelen, J. K. Hiltunen, and R. K. Wirenga, *EMBO J.*, 1996, **15**, 5135.
82. G. P. Crooks, Dissertation, University of Colorado, 1995.
83. G. Yang, R.-Q. Liu, K. L. Taylor, H. Xiang, J. Price, and D. Dunaway-Mariano, *Biochemistry*, 1996, **35**, 10 879.
84. J. Miller, "Aromatic Nucleophilic Substitution," Elsevier, New York, 1968.
85. G. P. Crooks and S. D. Copley, *J. Am. Chem. Soc.*, 1993, **115**, 6422.
86. R.-Q. Liu, P.-H. Liang, J. Scholten, and D. Dunaway-Mariano, *J. Am. Chem. Soc.*, 1995, **117**, 5003.
87. J. R. Gandler, I. U., Setiarahardjo, C. Tufon, and C. Chen, *J. Org. Chem.*, 1992, **57**, 4169.
88. J. Miller and W. Kai-Yan, *J. Chem. Soc.*, 1963, 3492.
89. R. G. Pearson, H. Sobel, and J. Songstad, *J. Am. Chem. Soc.*, 1968, **90**, 319.
90. G. P. Crooks and S. D. Copley, *Biochemistry*, 1994, **33**, 11 645.
91. K. L. Taylor, R.-Q. Liu, P.-H. Liang, J. Price, and D. Dunaway-Mariano, *Biochemistry*, 1995, **34**, 13 881.
92. R. L. D'Ordine, P. J. Tonge, P. R. Carey, and V. E. Anderson, *Biochemistry*, 1994, **33**, 12 635.
93. J. C. Austin, A. Kuliopulos, A. S. Mildvan, and T. G. Siro, *Protein Sci.*, 1992, **1**, 259.
94. G. M. Lacourciere and R. N. Armstrong, *J. Am. Chem. Soc.*, 1993, **115**, 10 466.
95. J.-Y. Lee and L. Xun, *J. Bacteriol.*, 1997, **179**, 1521.
96. L. Xu, K. Resing, P. C. Babbitt, S. L. Lawson, and S. D. Copley, unpublished results.
97. M. L. de Souza, M. J. Sadowsky, and L. P. Wackett, *J. Bacteriol.*, 1996, **178**, 4894.
98. W. W. Mulbry, *Appl. Environ. Microbiol.*, 1994, **60**, 613.
99. L. Xun, E. Topp, and C. S. Orser, *Biochem. Biophys. Res. Commun.*, 1992, **182**, 361.
100. D. L. McCarthy, S. Lawson, and S. D. Copley, manuscript in preparation.
101. D. L. McCarthy, S. Navarrete, W. S. Willett, P. C. Babbitt, and S. D. Copley, *Biochemistry*, 1996, **35**, 14 634.
102. C. S. Orser, J. Dutton, C. Lange, P. Jablonski, L. Xun, and M. Hargis, *J. Bacteriol.*, 1993, **175**, 2640.
103. W. S. Willett and S. D. Copley, *Chem. Biol.*, 1996, **3**, 851.
104. D. L. McCarthy, D. F. Louie, and S. D. Copley, *J. Am. Chem. Soc.*, 1997, **119**, 11 337.
105. R. N. Armstrong, *Chem. Res. Toxicol.*, 1991, **4**, 131.
106. S. Navarette, D. L. McCarthy, and S. D. Copley, manuscript in preparation.
107. R. P. Szajewski and G. M. Whitesides, *J. Am. Chem. Soc.*, 1980, **102**, 2011.
108. E. Masai, Y. Katayama, S. Kubota, S. Kawai, M. Yamasaki, and N. Morohoshi, *FEBS Lett.*, 1993, **323**, 135.
109. V. Romanov and R. P. Hausinger, *J. Bacteriol.*, 1996, **178**, 2656.
110. S. Ni, J. K. Fredrickson, and L. Xun, *J. Bacteriol.*, 1995, **177**, 5135.
111. S. Fetzner, R. Müller, and F. Lingens, *J. Bacteriol.*, 1992, **174**, 279.
112. A. Markus, U. Klages, S. Krauss, and F. Lingens, *J. Bacteriol.*, 1984, **160**, 618.
113. V. Romanov and R. P. Hausinger, *J. Bacteriol.*, 1994, **176**, 3368.
114. J. R. Mason and R. Cammack, *Ann. Rev. Microbiol.*, 1992, **46**, 277.
115. P. M. Radehaus and S. K. Schmidt, *Appl. Environ. Microbiol.*, 1992, **58**, 2879.
116. D. L. Saber and R. L. Crawford, *Appl. Environ. Microbiol.*, 1985, **50**, 1512.
117. M. M. Häggblom, L. J. Nohynek, N. J. Palleroni, K. Kronqvist, E.-L. Nurmiaho-Lassila, M. J. Salkinoja-Salonen, S. Klatte, and R. M. Kroppenstedt, *Int. J. Syst. Bacteriol.*, 1994, **44**, 485.
118. L. Xun and C. S. Orser, *J. Bacteriol.*, 1991, **173**, 4447.
119. C. Orser, C. C. Lange, L. Xun, T. C. Zahrt, and B. J. Schneider, *J. Bacteriol.*, 1993, **175**, 411.
120. W. J. H. van Berkel and F. Müller, in "Chemistry and Biochemistry of Flavoenzymes," ed. F. Müller, CRC Press, Boca Raton, FL, 1990, vol. 2, p.1.
121. B. Entsch, D. P. Ballou, and V. Massey, *J. Biol. Chem.*, 1976, **251**, 2550.
122. M. Husain, B. Entsch, D. P. Ballou, V. Massey, and P. J. Chapman, *J. Biol. Chem.*, 1980, **255**, 4189.
123. B. Entsch, B. A. Palfey, D. P. Ballou, and V. Massey, *J. Biol. Chem.*, 1991, **266**, 17 341.
124. H. A. Schreuder, P. A. J. Prick, R. K. Wierenga, G. Vriend, K. S. Wilson, W. G. J. Hol, and J. Drenth, *J. Mol. Biol.*, 1989, **208**, 679.
125. H. A. Schreuder, J. M. van der Laan, M. B. A. Swarte, K. H. Kalk, W. G. J. Hol, and J. Drenth, *Proteins: Struct. Funct. Genet.*, 1992, **14**, 178.
126. C. S. Orser and C. C. Lange, *Biodegradation*, 1994, **5**, 277.
127. J. S. Uotila, V. H. Kitunen, T. Saastamoinen, T. Coote, M. M. Häggblom, and M. S. Salkinoja-Salonen, *J. Bacteriol.*, 1992, **174**, 5669.

5.17
Catalysis by Antibodies

DONALD HILVERT

Eidgenössische Technische Hochschule, Zurich, Switzerland

5.17.1 INTRODUCTION

During the 1990s, antibodies capable of catalyzing a diverse set of chemical transformations have been prepared.[1,2] The strategy for generating these catalysts, as originally set out by Jencks,[3] is remarkably straightforward. Stable compounds that mimic the stereoelectronic properties of transition states are used as haptens to elicit an immune response. If the transition state analogue has been properly designed, some of the resulting antibodies also accelerate the corresponding chemical transformation.

This technology takes advantage of the programmable nature of the immune system and the extensive shape and chemical complementarity between the antigen and the induced antibody binding pocket.[4] Consequently, both the mechanism and selectivity of antibody catalysts can in principle be specified at the stage of hapten design. Moreover, because the immune response evolves over the course of weeks, the timescale for generation of such agents can be as short as several months.

In practice, the design of effective haptens has been informed by a variety of mechanistic strategies.[2] For example, conformationally locked molecules and bisubstrate analogues have been employed to induce antibodies that function as "entropy traps" capable of restricting the conformational degrees of freedom of flexible molecules and bringing two molecules together in the proper orientation for reaction. In contrast, charged haptens can create binding sites that stabilize anionic or cationic transition states by eliciting acidic or basic groups of complementary charge. Catalytically essential functional groups can also be induced when mechanism-based inhibitors are used as haptens.

Implementation of these strategies has yielded antibody catalysts for pericyclic processes, group transfer reactions, oxidations and reductions, aldol condensations, and miscellaneous cofactor-dependent transformations.[2] Considerable effort is now being expended on developing tailored antibody catalysts for transformations that cannot be achieved efficiently or selectively with more traditional chemical methods.[5] Typically, rate enhancements achieved in these systems are in the range 10^2–10^6 over background. Thus, while the best catalytic antibodies are certainly impressive, they are still inferior to enzymes in nature (where available!), often by many orders of magnitude. Significant improvements in catalytic efficiency are needed if practical applications in chemistry and medicine are to be realized.

As increasing numbers of catalytic antibodies are characterized using the tools of mechanistic enzymology and structural biology, it is appropriate to ask to what extent the instructions implicit in the structure of the hapten are actually realized in the functioning catalyst. Information gained from such analysis is needed to design more effective haptens and to enhance the catalytic efficiency of the first-generation catalysts. In this chapter, the mechanisms of several representative systems are highlighted to illustrate what has been learned from such studies and what challenges remain for the future.

5.17.2 PERICYCLIC REACTIONS

Utilization of binding energy to overcome unfavorable entropies of activation is perhaps the easiest task for antibodies to accomplish.[6] Restriction of rotational and translational degrees of freedom of reactants within a tailored pocket of appropriate shape should significantly increase the probability of many reactions. Concerted pericyclic reactions, which do not typically require general acid–base or nucleophilic catalysis, can be highly sensitive to such proximity effects. Indeed, several examples of antibody-catalyzed Claisen[7,8] and Cope[9] rearrangements, [2,3]-sigmatropic reactions,[10,11] and bimolecular Diels–Alder reactions[12–14] have been described.

5.17.2.1 Chorismate Mutase

The unimolecular rearrangement of chorismate (1) to prephenate (3) is a biologically important [3,3]-sigmatropic process—a Claisen rearrangement—in which formation of a carbon–carbon bond is accompanied by cleavage of a carbon–oxygen bond (Scheme 1). It is a key step in the biosynthesis of the aromatic amino acids tyrosine and phenylalanine in plants and lower organisms and is accelerated more than 10^6 fold by the enzyme chorismate mutase.[15]

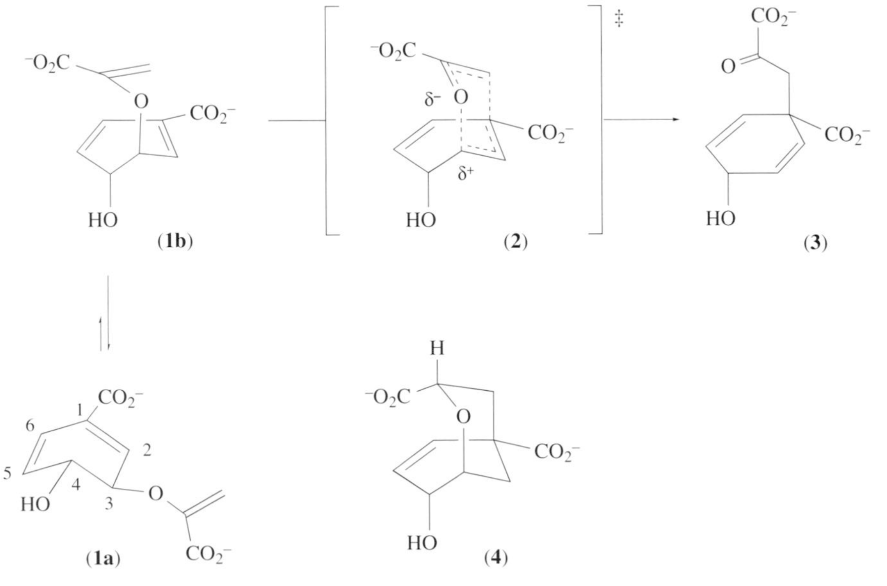

Scheme 1

The rearrangement of chorismate occurs spontaneously at an appreciable rate via a concerted but asynchronous pathway involving a chair-like transition state (**2**).[16–19] It is entropically unfavorable ($\Delta S^{\ddagger} = -9$ eu)[20] and requires the normally extended chorismate molecule (**1a**) to adopt the higher energy, pseudodiaxial conformation (**1b**).[18] Mimicry of the geometry of the conformationally restricted transition state has yielded good inhibitors of the natural enzyme.[21–23] The best of these, the oxabicyclic dicarboxylic acid (**4**),[24] is also an effective hapten for generating catalytic antibodies. The two antibodies that have been characterized in detail accelerate the rearrangement 10^2–10^4-fold over background[7,8] and are highly enantioselective, accepting only the natural ($-$)-isomer of chorismate as a substrate.[25,26]

Spectroscopic[27] and X-ray crystallographic[28] investigations of the chorismate mutase antibody 1F7 (which achieves a 200-fold rate enhancement) have provided direct insight into the origins of catalysis in these systems. Transferred nuclear Overhauser effects (TRNOEs) show that 1F7 constrains the flexible chorismate molecule in a diaxial conformation (**1b**) that approximates the chair-like geometry of the hapten.[27] Crystallographic data at 3.0 Å resolution[28] confirm that transition state analogue design is faithfully reflected in the catalyst's structural properties. The induced binding pocket exhibits excellent overall shape and charge complementarity to (**4**), with hapten recognition achieved through a combination of hydrophobic interactions, hydrogen bonds, and electrostatics (Figure 1). This complementarity dictates the preferential binding of the correct substrate enantiomer in a conformation appropriate for reaction.

The absence of acids, bases, and nucleophiles within the confines of the active site suggests that the 1F7-catalyzed rearrangement occurs via the same concerted pathway deduced for the uncatalyzed reaction.[16] The structural data are thus consistent with the antibody playing the role of "entropy trap",[20] preorganizing the flexible substrate into a reactive conformation and stabilizing the conformationally constrained transition state by a network of hydrogen-bonding, electrostatic, and van der Waals interactions. It is therefore surprising to note that the entropy of activation measured for this antibody-catalyzed reaction is actually *less* favorable than the spontaneous thermal rearrangement ($\Delta\Delta S^{\ddagger} = -10$ cal mol^{-1} K^{-1});[8] the observed rate acceleration is brought about entirely by lowering the enthalpy of activation. Interpretation of activation parameters is notoriously difficult, and while chorismate almost certainly retains some flexibility when bound at the active site of 1F7,[27] the unfavorable $\Delta S^{\ddagger}$ most likely reflects protein conformational changes and solvent reorganization that accompany the rearrangement.

Natural chorismate mutases are roughly 10^4-fold more efficient than 1F7.[29] Like the antibody, the monofunctional chorismate mutase from *Bacillus subtilis*[30] and the mutase domain of the bifunctional chorismate mutase–prephenate dehydratase from *Escherichia coli*[31] apparently restrict the conformational degrees of freedom of the substrate by fixing it in the pseudodiaxial conformation needed for reaction. They are likely to be substantially better entropy traps than 1F7, however, as they exploit a greater number of hydrogen-bonding and electrostatic interactions for ligand recognition (Figure 1). In addition, and probably more importantly, the natural enzymes appear much better able to stabilize charge separation in the transition state. In this regard, positioning of a cation close to the ether oxygen of the breaking C—O bond (Arg90 in the *Bacillus* enzyme and Lys39 in the *E. coli* enzyme) seems particularly important (Figure 1); mutagenesis experiments demonstrate that removal of this positive charge has a devastating effect on catalysis,[32–35] leading to the loss of nearly all activity. These results are thus consistent with isotope effect studies[17] and calculations[36] showing that the transition state for this reaction has considerable dissociative character. The side chain of ArgH95 occupies a similar position in 1F7 as the crucial cation in the natural mutases, but because it must neutralize the full negative charge on the enolpyruvate carboxylate (which binds in a relatively hydrophobic pocket),[28] it will be much less effective in accommodating additional negative charge on the ether oxygen at the transition state. In retrospect, it is not surprising that the antibody is suboptimal in its electrostatic properties. Hapten (**4**) effectively mimics the geometry but not the polarized character of transition state (**2**).

The mechanistic parallels observed for 1F7 and natural chorismate mutases, and the likely dominance of electrostatic effects in the latter, suggest that improved catalysts might be obtained by refinements of the approach used to obtain 1F7. Thus, antibodies more closely resembling the natural enzymes may be attainable through rational redesign of the transition state analogue or through protein engineering of the first-generation catalyst. Inclusion of additional charged groups in the hapten, for example, might elicit a more effective array of complementary charged active site residues. Alternatively, the crucial cationic group might be introduced into the binding site of 1F7 directly by site-specific mutagenesis. The availability of chorismate mutase deficient yeast[37] and *E. coli*[32] should greatly assist the screening of panels of mutants and, more generally, directing the evolution of molecular function through random mutagenesis and selection.[38]

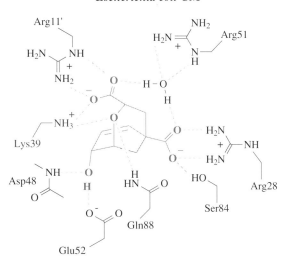

Catalytic antibody 1F7

$k_{cat}/k_{uncat} = 2 \times 10^2$

***Bacillus subtilis* CM**

$k_{cat}/k_{uncat} = 2 \times 10^6$

***Escherichia coli* CM**

$k_{cat}/k_{uncat} = 2 \times 10^6$

Figure 1 Active sites of the catalytic antibody 1F7 and two natural chorismate mutases (CM) illustrating critical contacts between the protein and a bound transition state analogue.

5.17.2.2 Oxy-Cope Rearrangement

Catalytic antibody technology is not limited to reactions for which natural enzymes are known, as shown by generation of antibodies that catalyze the abiological oxy-Cope rearrangement of the disubstituted 1,5-hexadiene (**5**) to (**7**) (Scheme 2).[9] This transformation is another [3,3]-sigmatropic process, and the cyclohexanol derivative (**8**) was used as a hapten to mimic the chair-like six-membered ring geometry of the corresponding pericyclic transition state. Antibody AZ-28, which binds (**8**) with a dissociation constant of 17 nM, accelerates the rearrangement by a factor of 5300-fold (k_{cat}/k_{uncat}). Although the aldehyde product inactivates the catalyst in a time-dependent manner, probably by reaction with surface lysine residues, it can be trapped conveniently with hydroxyl amine to give an oxime.

AZ-28 has been characterized in considerable detail.[39,40] Secondary kinetic isotope effects show that the chemical rearrangement is rate limiting and that there is significant bond formation between C-1 and C-6 at the transition state. The antibody exhibits a 15:1 kinetic preference for the *S* over

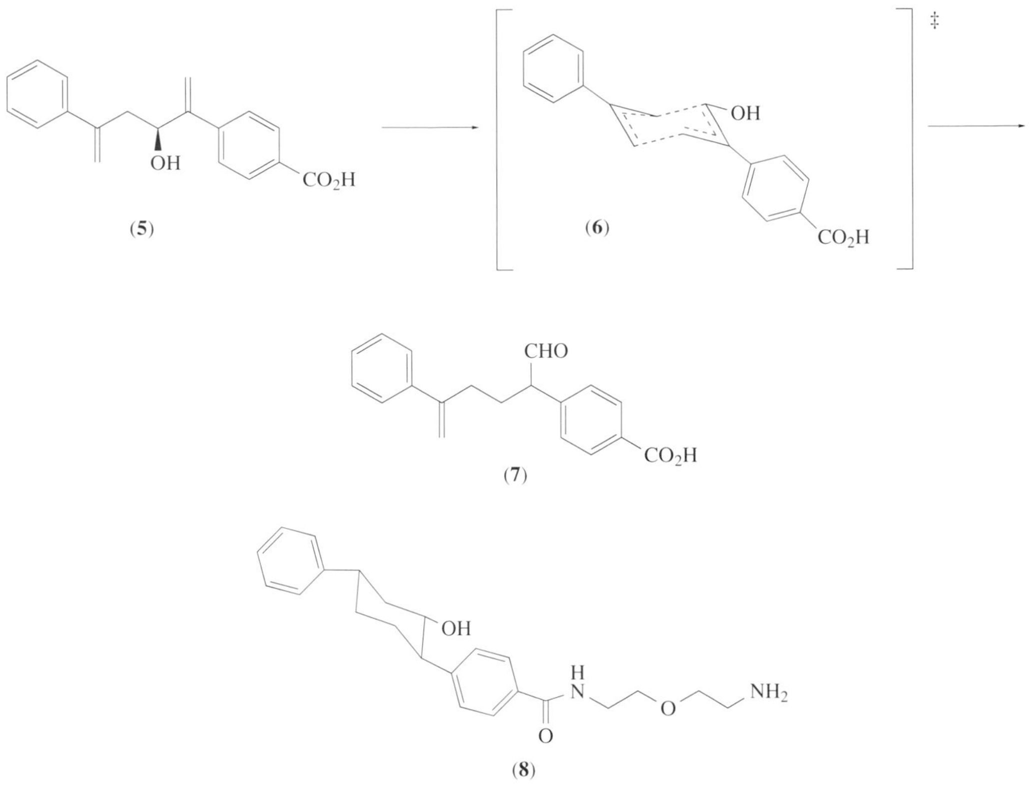

Scheme 2

the *R*-alcohol as substrate, and TRNOE experiments show that it, like the chorismate mutase 1F7, binds the normally extended substrate in a cyclic conformation with its alkene termini in close proximity. Despite such substrate preorganization, and again like 1F7, the $\Delta S^\ddagger$ for the AZ-28-catalyzed reaction is substantially less favorable than for the uncatalyzed rearrangement ($\Delta\Delta S^\ddagger = -20$ cal mol^{-1} K^{-1}).

The structure of AZ-28 complexed to (**8**) was determined to 2.6 Å resolution.[41] The hapten binds in a deep cylindrical cavity. The 5-phenyl substrate is tightly packed at the bottom of the pocket where it is surrounded by a large number of aromatic and hydrophobic aliphatic side chains. The cyclohexyl ring adopts the expected chair conformation with the hydroxyl group and the two aryl rings in equatorial orientations. It is fixed in place by hydrogen-bonding interactions between the hydroxyl group and the imidazole ring and backbone amide NH group of HisH96. This same imidazole also orients the 2-phenyl ring of the hapten, which is located near the entrance of the pocket, through π-stacking interactions.

Efficient catalysis of the oxy-Cope rearrangement of (**5**) depends principally on proper alignment of the $[4\pi + 2\sigma]$ orbitals of the hexadiene. In the antibody, extensive van der Waals interactions with the aryl rings and hydrogen bonding with the hydroxyl group evident in the complex with the hapten apparently fix the flexible hexadiene in a cyclic conformation appropriate for reaction, thereby contributing to the observed rate acceleration. Because the two aryl substituents at C-2 and C-5 are sp^2 rather than sp^3 hybridized as in the hapten, however, the ground state conformation of the substrate at the active site may not be optimal. Any reorganization within the pocket would necessarily contribute to the unfavorable entropy of activation observed for the k_{cat} step.

Alterations in the electron density on the hydroxyl substituent and the extent of π-orbital overlap with the 2,5-diphenyl substituents can modulate the reactivity of the substrate to a significant extent. Anionic substituent effects, for example, are known to accelerate the oxy-Cope rearrangements through hyperconjugation of electron density on oxygen.[42] Modulation of the electron density on the 3-hydroxyl group through hydrogen-bonding interactions with the side chains of HisH96 (and the nearby GluH50) may similarly influence the magnitude of the rate of the rearrangement within the antibody pocket. Mutagenesis experiments will be needed to clarify the importance of such effects in AZ-28.

In principle, rearrangement of substrate (**5**) could proceed via a concerted or stepwise (biradicaloid-like) pathway. In the case of a biradicaloid mechanism, conjugation of electron density at C-2 or C-5 with the aromatic substituents is expected to reduce the activation energy of the uncatalyzed reaction by as much as 10 kcal mol^{-1} relative to the unsubstituted diene.[43] The potential importance of effective overlap in the case of AZ-28 is indicated by studies of its germline precursor. Despite a 40-fold weaker affinity for transition state analogue (**8**), this protein achieves a 30-fold higher rate acceleration than AZ-28. These effects can largely be accounted for by the somatic mutation of active site residue SerL34 in the germline antibody to an asparagine in AZ-28. AsnL34 interacts directly with the cyclohexyl ring of the hapten and might therefore be expected to play a significant role in restricting rotations around the C-2—aryl and C-5—aryl bonds of the substrate. Because the orientation of the aryl groups is not fixed in the cyclohexane hapten, conjugation of the aryl groups with the alkenes in (**5**) presumably comes at the cost of proper alignment of the [4π+2σ] system for reaction. Small changes in structure of the antibody that improve orbital overlap could lead to significant increases in catalytic efficiency. Similarly, haptens in which the aryl substituents are coplanar with the cyclohexyl ring should increase the probability of identifying antibodies that achieve higher rates of reaction.

5.17.2.3   Cycloadditions

Bimolecular reactions are expected to be especially sensitive to entropic effects.[44] Diels–Alder reactions, for example, which involve concerted addition of a diene to a dienophile, have entropic barriers in the range −30 to −40 cal mol^{-1} K^{-1}. Although rare in biological systems, they are quite useful in synthetic organic chemistry, occurring through highly ordered, cyclic transition states with little charge separation. Antibodies raised against neutral bicyclic compounds that mimic the boatlike geometry of the transition state have achieved significant catalytic effects, including multiple turnovers, high effective molarities, and control over both reaction pathway and absolute stereochemistry.[12–14]

Product inhibition is a significant concern in the development of protein catalysts for synthetic bimolecular reactions. Successful approaches to minimize this problem in antibody catalysis have exploited chemical or conformational change to drive product release. The latter strategy is illustrated in the design of a hapten for the Diels–Alder reaction between the acyclic diene (**9**) and alkene (**10**) to give cyclohexene (**12**) (Scheme 3).[12] Substituted bicyclo[2.2.2]octene derivative (**13**) contains an ethano bridge that locks the cyclohexene ring into the requisite boat conformation, allowing induction of antibody pockets capable of preorganizing the diene and alkene for reaction. Product dissociation from the active site was expected to be facilitated by an energetically favorable conformational change of the cyclohexene from boat to twist boat.

Antibody 39-A11, which binds (**13**) tightly (K_d = 10 nM), accelerates the target reaction with multiple turnovers.[12] As expected, product binds about 1000 times less tightly than the transition state analogue (and only 75–100 times more tightly than either substrate). The efficiency of catalysis, however, is relatively low as judged by the effective molarity (0.35 M), the ratio of the pseudo-first-order rate constant for the reaction at the antibody active site (k_{cat}) to the second-order rate constant for the uncatalyzed process in free solution (k_{uncat}). The effective molarity corresponds to the concentration of one of the reactants that would be needed to convert the uncatalyzed bimolecular reaction into a pseudo-first-order process with a rate constant equivalent to that achieved in the antibody ternary complex. For calibration, Page and Jencks have estimated that the maximum kinetic advantage to be derived from converting a bimolecular to a pseudo-unimolecular reaction is of the order of 10^8 M for 1 M standard states.[44]

The structure of 39-A11 at 2.4 Å resolution shows that the bicyclo[2.2.2]octane hapten binds within a hydrophobic pocket and is oriented through hydrogen bonds to its carbamate moiety and one of the carbonyl groups of the succinimido moiety.[45] Extrapolation to the substrates suggests that the dienophile, which corresponds to the succinimido group of the hapten, will be oriented by π-stacking of the maleimide against TrpH50 and hydrogen bonding of one of its carbonyl groups to the side chain amide of AsnH35. Because the structure of the bicyclo[2.2.2]octane hapten mimics the transition state for both *endo* and *exo* addition of the maleimide, the binding mode of the diene is less certain. If the kinetically more favorable *endo* transition state is assumed, the diene can dock in a hydrophobic pocket in close proximity to the dienophile, with the position of the carbamate substituent fixed by a water-mediated hydrogen bond to TrpH50.

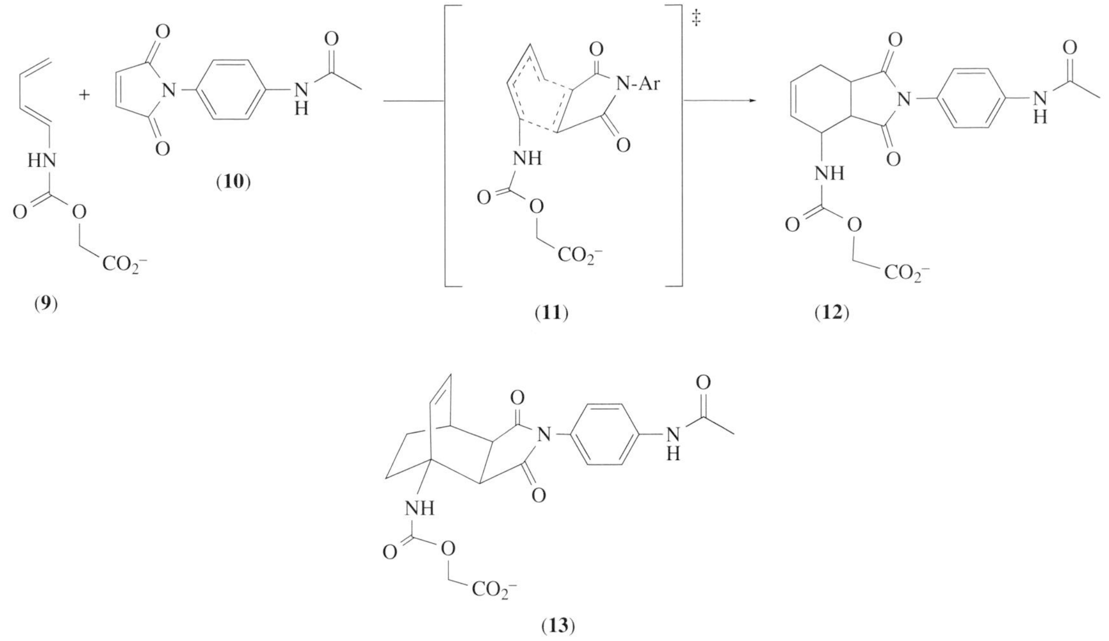

(9) **(10)** **(11)** **(12)**

(13)

Scheme 3

As for the antibody-promoted sigmatropic reactions discussed above, catalysis can be attributed to a combination of entropic and electronic effects. Antibody 39-A11 presumably binds the diene and dienophile in a reactive orientation and reduces translational and rotational degrees of freedom, albeit imperfectly. The residual degrees of freedom available to the diene may account, in part, for the relative inefficiency of this catalyst. Indeed, attempts to improve packing interactions at the active site by incorporation of large aromatic groups at positions L91 and L96 yielded increases in the k_{cat} parameter of 5–10-fold.[45] Suboptimal van der Waals and hydrogen-bonding interactions with the product (**12**) must also account for a dissociation constant intermediate between that of the substrates and the hapten. The hydrogen bond provided by AsnH35 to the dienophile may also contribute to catalytic efficiency. It will certainly activate the alkene as a dienophile, but whether such an interaction is translated into a rate acceleration is uncertain since it necessarily replaces analogous hydrogen bonds with water in bulk solvent. Its role in positioning the dienophile for reaction may therefore be more important than the electronic effect.

A second Diels–Alderase antibody 1E9 catalyzes the cycloaddition between tetrachlorothiophene dioxide (TCTD) and *N*-alkylmaleimides (**14**) (Scheme 4).[13] It was elicited with the stable hexachloronorbornene derivative (**18**) which mimics the high-energy intermediate (**15**). It achieves multiple turnovers, but product release is not driven by a conformational change. Instead, facile decomposition of the initially formed intermediate by chelotropic elimination of SO_2 gives a product which, after oxidation *in situ*, is structurally very different from the substrates and the transition states for cycloaddition or cycloreversion. Thus, *N*-ethylphthalimide (**17**) ($K_d = 0.5$ mM) binds only 40 times more tightly than the substrate *N*-ethylmaleimide (**14**, R = Et) but nearly 10^5-fold less tightly than hapten (**18**). Kinetic studies indicate that 1E9 is much more efficient than 39-A11, achieving effective molarities > 200 M. The limited solubility of TCTD precluded saturation of the antibody with both substrates, so the actual efficiency of 1E9 is likely to be at least an order of magnitude higher than this value.

Interestingly, sequencing studies show that 1E9 and 39-A11 are closely related to each other and to the antiprogesterone antibody DB3.[45,46] Preliminary structural data obtained at 1.9 Å resolution for 1E9 complexed to its hapten show excellent shape complementarity between the induced binding pocket and (**18**) (unpublished). By inference, TCTD and *N*-ethylmaleimide would completely fill the available cavity and the ensuing transition state would be tightly packed through extensive van der Waals interactions. Thus, 1E9 might be substantially better at orienting its substrates than 39-A11 due to its superior shape complementarity, achieved mainly through van der Waals rather than hydrogen-bonding or electrostatic interactions.

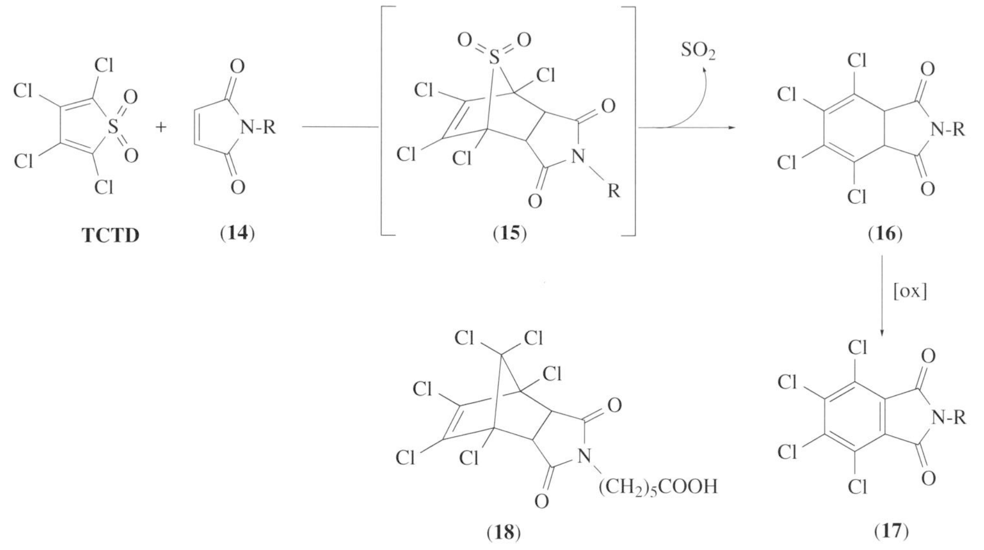

Scheme 4

A fundamentally different strategy was used to generate the Diels–Alderase antibody 13G5 which promotes the kinetically disfavored *exo* cycloaddition of *trans*-1,3-butadiene-1-carbamate (19) and *N,N*-dimethylacrylamide (20) to give a single diastereomer of the disubstituted cyclohexene (21) in high enantiomeric excess (Scheme 5).[47] In this case, a conformationally labile ferrocene derivative (22) was used as a mimic of a very early transition state. In the crystal structure of 13G5 determined at 1.95 Å resolution,[48] ferrocene binds in a single conformation at the active site. The structural data suggest that three antibody residues are primarily responsible for the observed catalysis and product control. Although all four *ortho* transition states for the Diels–Alder reaction can be accommodated within the spacious active site, docking experiments indicate that only the *exo* transition state leading to the (3R,4R) product stereoisomer can exploit the same hydrogen-bonding interactions seen in the ferrocene complex. This may explain why the *exo* transition state is preferred over the *endo* transition state even though it is predicted to be 1.9 kcal mol^{-1} higher in energy. In the proposed model, TyrL36 acts as a Lewis acid positioning and activating the dienophile for nucleophilic attack. AsnL91 and AspH50 form hydrogen bonds to the carbamate in the diene. As in 39-A11, the relatively small size of the substrates compared with the transition state analogue guarantees that packing interactions are not optimal, providing a possible rationalization for the modest effective molarity of this catalyst (7 M).[47] Repacking of the active site by site-directed mutagenesis to further restrict the conformational freedom available to both substrates could conceivably lead to substantial increases in chemical efficiency.

5.17.3 HYDROLYTIC REACTIONS

Many reactions of chemical and biological interest have charged transition states. Molecules that mimic the cationic or anionic nature of such species are often successful as haptens in generating antibody active sites which exploit complementary hydrogen-bonding or electrostatic interactions for catalysis.[2] Negatively charged phosphonate esters and phosphonamidates, for example, resemble the anionic tetrahedral transition state that arises during the hydrolysis of esters and amides. Such compounds have been used extensively as high-affinity inhibitors of hydrolytic enzymes.[49–52] They have also been productively employed as haptens for generating antibody catalysts for simple hydrolytic reactions.[2] More than 50 antiphosphonate and antiphosphonamidate antibodies have been described since 1986. They typically promote the hydrolysis of esters, carbonates, and (more rarely) amides, displaying classical Michaelis–Menten kinetics, rate accelerations up to 10^6-fold over background, and substrate specificity mirroring the structure of the original hapten. When the hapten contains chiral centers, high levels of stereoselectivity can also be attained.

In keeping with the large effort devoted to their production, esterolytic antibodies are also among the best characterized antibody catalysts.[53] X-ray structures of six different esterases have been

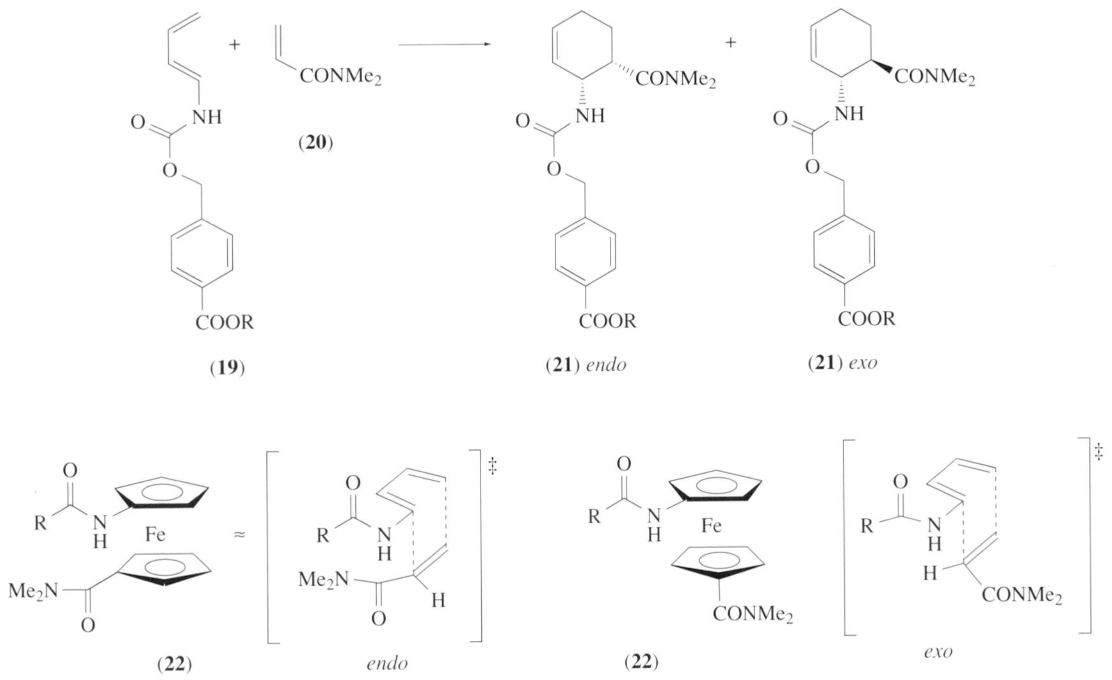

Scheme 5

reported,[54–57] and the structure of the amidase NPN43C9, which hydrolyzes an activated *p*-nitroanilide,[58] has been investigated by homology modeling.[59] Three of the esterases (CNJ206, 17E8, and 48G7) promote the cleavage of (nitro-) phenyl esters,[54,56,60] and three (D2.3, D2.4, and D2.5) accelerate the energetically more demanding cleavage of *p*-nitrobenzyl esters.[55] A comparative analysis of their properties provides valuable insights into the structural origins of their activity and selectivity.

5.17.3.1 A Typical Esterase

Antibody D2.3, which is one of the most efficient esterases produced to date,[61] is illustrative of this class of catalyst. It was generated with the nitrobenzyl phosphonate hapten (**27**) and selected from a panel of 970 hapten binders on the basis of its catalytic activity in a catELISA immunoassay.[62] In kinetic assays, D2.3 hydrolyzes nitrobenzyl ester (**23**) with multiple turnovers and a rate acceleration (k_{cat}/k_{uncat}) of 1.1×10^5 over the uncatalyzed reaction (Scheme 6); it was also found to hydrolyze the analogous *p*-nitrophenyl ester, albeit with somewhat lower efficiency ($k_{cat}/k_{uncat} = 7.2 \times 10^3$).[61] In both cases, the apparent rate enhancement correlates well with the ratio K_s/K_{TSA}, the relative affinity of the antibody for substrate vs. transition state analogue at pH 6.0. Interestingly, *p*-nitrobenzyl alcohol (**25**) inhibits catalysis ($K_p = 50$ μM) but *p*-nitrophenol does not ($K_p > 500$ μM), so that product inhibition eventually becomes a problem in the cleavage of the "specific" substrate that more closely resembles the hapten, whereas more than 1000 turnovers per active site are achieved with the "nonspecific" nitrophenyl ester.

The X-ray structures of the unliganded antibody and its complexes with the transition state analogue, a substrate analogue, and the alcohol product of the reaction yield an unusually complete description of the probable reaction pathway.[55,63] In the absence of ligand, the active site of D2.3 is characterized by a deep, well-defined, and relatively hydrophobic pocket. The cavity contains a large proportion of aromatic residues (four tyrosines, two tryptophans, and two phenylalanines), as well as an asparagine and a histidine. Hapten binding causes no significant conformational changes in the antibody. Main chain and side chain displacements are minimal, even in the often flexible H3 complementarity determining region (CDR), and despite the fact that (**27**) is 90% buried and solvent inaccessible in the complex. In what is emerging as a common theme used by esterolytic antibodies,[64] compound (**27**) binds with its aryl group buried deep in a conserved hydrophobic

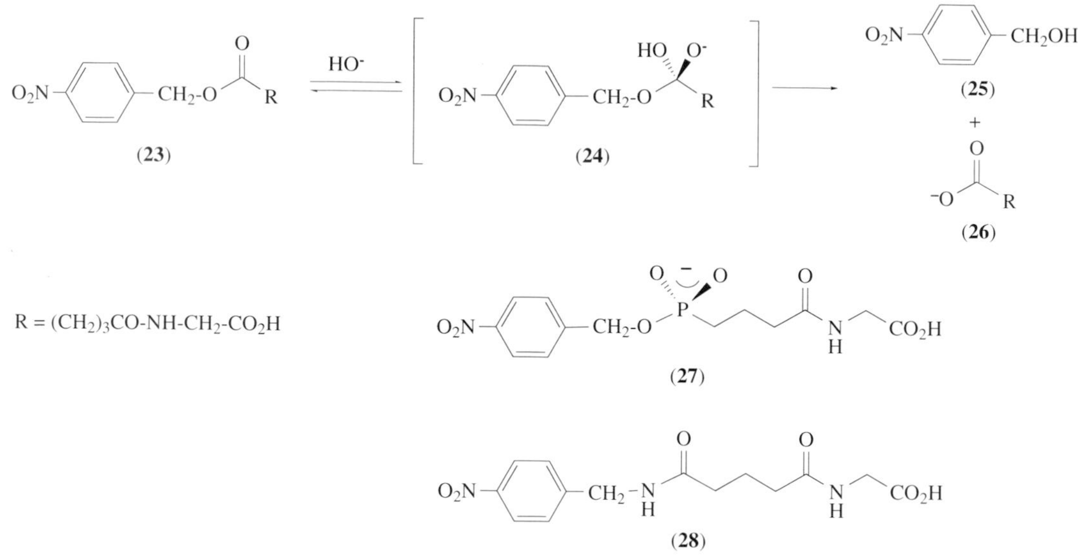

O$_2$N—⟨benzene⟩—CH$_2$-O-C(=O)-R
(23)

$\xrightarrow{\text{HO}^-}$

[O$_2$N—⟨benzene⟩—CH$_2$-O-C(OH)(O$^-$)-R]
(24)

$\longrightarrow$

O$_2$N—⟨benzene⟩—CH$_2$OH
(25)

+

O=C(O$^-$)-R
(26)

R = (CH$_2$)$_3$CO-NH-CH$_2$-CO$_2$H

O$_2$N—⟨benzene⟩—CH$_2$-O-P(O$^-$)(O)-CH$_2$CH$_2$CH$_2$-C(=O)-NH-CH$_2$-CO$_2$H
(27)

O$_2$N—⟨benzene⟩—CH$_2$-NH-C(=O)-CH$_2$CH$_2$CH$_2$-C(=O)-NH-CH$_2$-CO$_2$H
(28)

Scheme 6

pocket constructed from framework residues. Recognition of the negatively charged phosphonate moiety is achieved through multiple oriented hydrogen bonds with neutral antibody residues rather than by a complementary positive charge (Figure 2(a)). Its *pro-S* oxygen forms hydrogen bonds with the side chain amide of AsnL34 and the phenolic hydroxyl group of TyrH100d, and the *pro-R* phosphonyl oxygen is hydrogen bonded to TrpH95 and a water molecule. HisH35, which has been suggested to be an important catalytic residue in some esterolytic antibodies, is inappropriately oriented and too distant to contribute a stabilizing hydrogen bond to the phosphonate or to the intervening water molecule. Indeed, other residues that might be capable of promoting nucleophilic or general base catalysis are more than 5 Å away from the bound ligand, suggesting that such mechanisms are unimportant D2.3. Finally, the *N*-glutarylglycyl portion of the hapten adopts a well-defined conformation at the entrance of the cavity. It is fixed in place by hydrogen bonds with the carbonyl group of GlyL91 and the hydroxyl group of TyrL96, both located in CDR L3.

In keeping with the mechanistic information implicit in the structure of the transition state analogue (**27**), the crystallographic data suggest that D2.3 is a relatively simple catalyst that facilitates direct hydroxide attack on the scissile carbonyl of bound substrate. In this view, AsnL34 and TyrH100d constitute an "oxyanion hole"—analogous to that found in serine proteases—which stabilizes the tetrahedral and anionic intermediate and flanking transition states relative to the bound ground states through hydrogen-bonding interactions. The high degree of preorganization of the oxyanion binding residues, achieved through a network of van der Waals interactions and additional hydrogen bonds, may contribute to the relatively high efficiency of this catalyst.

These conclusions are strengthened when the structures of D2.3 complexed with an amide analogue of the substrate (i.e., **28**) and the alcohol product (**25**) are considered.[63] The orientation of amide (**28**) within the pocket is similar to that of the hapten and likely mimicks the geometry adopted by ester (**23**) in its Michaelis complex with the antibody. The amide NH makes no hydrogen bonding contacts with the antibody, but the scissile carbonyl group is within hydrogen-bonding distance of TyrH100d. This interaction shields one face of the substrate carbonyl and, at the same time, activates it for attack by a hydroxide ion. A channel of water molecules is evident in the structure adjacent to the alkyl chain of the bound substrate, indicating the probable route for attack of hydroxide on the *Re* face of the carbonyl. The location of the product alcohol superimposes on that of the corresponding portion of the hapten, except that its hydroxyl group makes a hydrogen bond with AsnL34. These interactions account for product inhibition by (**25**) ($K_i \approx 50$ μM).[61] The absence of significant product inhibition for *p*-nitrophenol, which lacks a methylene group and is therefore shorter than (**25**), can then be explained by its inability to take simultaneous advantage of the hydrophobic interactions with its aryl group *and* the hydrogen-bonding interaction with AsnL34.

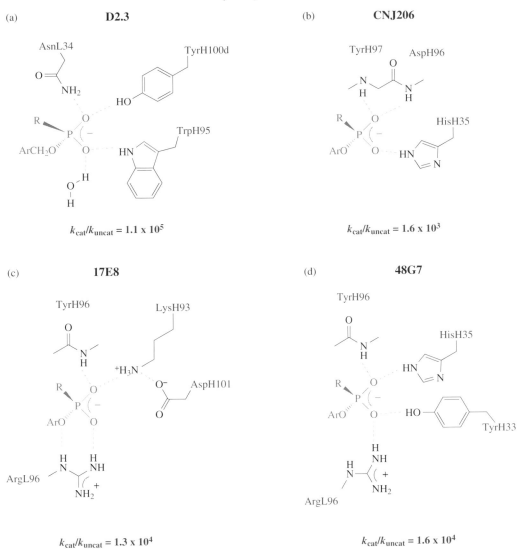

Figure 2 Schematic depiction of the principal interactions between four hydrolytic antibodies (a) D2.3, (b) CNJ206, (c) 17E8, (d) 48G7, and the phosphanate moiety of the bound hapten.

The observation that the negatively charged phosphonate group of (**27**) makes three hydrogen bonds with the antibody, whereas the corresponding neutral amide (**28**) (and by analogy the ester substrate (**23**)) makes only one such interaction, nicely accounts for the ratio of binding constants of transition state analogue and substrate (1.1×10^5) and also correlates with the magnitude of the rate acceleration achieved by D2.3 with its best substrate (1.3×10^5). For comparison, the single oxyanion-binding residue Asn155 in the bacterial protease subtilisin contributes 10^2–10^3-fold to the catalytic efficiency.[65,66] Although the *N*-glutarylglycinate portion of the substrate analogue adopts a slightly different conformation than its counterpart in the hapten, it makes analogous hydrogen bonds with GlyL91 and TyrL96, limiting its conformational freedom within the pocket. These interactions also appear to be important for catalysis, since a substrate like *p*-nitrobenzyl acetate which cannot make such interactions is cleaved approximately 20-times less efficiently than (**23**).[62]

5.17.3.2 Comparison with Other Esterases

All of the structurally characterized esterolytic antibodies appear to exploit the same basic hydrolytic mechanism as D2.3. Their relative efficiencies must consequently reflect their relative ability to stabilize the tetrahedral and anionic transition states relative to the bound ground state.

Antibodies D2.4 and D2.5 catalyze the same reaction as antibody D2.3, hydrolysis of (**23**), but with 4- and 50-fold lower efficiency (k_{cat}/k_{uncat}), respectively (Table 2).[61,62] Sequencing and X-ray studies show that all three catalysts are closely related,[55] with the most extensive differences localized in the H3 CDR loop which is used, in part, to construct the oxyanion binding site. Despite four mutations and an insertion that alters the overall conformation of this loop, antibody D2.4 uses an essentially identical set of hydrogen bonds to contact the phosphonate transition state analogue as D2.3.[55] Less effective preorganization of this longer loop could conceivably account for the modest decrease in activity relative to D2.3. The much less efficient D2.5 is more similar to D2.3 in its sequence, but the critical oxyanion-binding residue AsnL34 is mutated to a serine.[55] In the D2.5–hapten complex, SerL34 is too far from the phosphonate for direct hydrogen bonding. A bridging water molecule may somewhat ameliorate the deleterious consequences of mutation, but the loss of the oriented hydrogen bond could easily explain the 50-fold drop in activity.

The esterolytic antibodies CNJ206,[67] 48G7,[68] and 17E8[69] were raised against the analogous (*p*-nitro)phenyl phosphonate and hydrolyze (*p*-nitro) phenyl esters with varying efficiency (Tables 1 and 2). These antibodies, which were generated in different laboratories with different haptens, are far more divergent in sequence than the D2 antibodies, yet their active sites[54,56,60] show remarkable structural similarities among themselves and with the D2 antibodies, including the overall mode of hapten binding and a common hydrophobic pocket constructed from conserved framework residues that can accommodate the aryl leaving group. They differ in the specific interactions they employ to bind the negatively charged phosphonate (Figure 2).

Antibody CNJ206, for example, recognizes the phosphonate moiety via hydrogen bonds with HisH35 and the peptide NH groups of two consecutive residues in CDR H3 (Figure 2(b)).[56] This is the least efficient of the catalysts under consideration. It is also the least preorganized. Ligand binding is accompanied by domain movements and a significant conformational change in CDR H3.[56,71] Utilization of a significant fraction of the available binding energy for structural reorganization may explain the relative inefficiency of this catalyst. Studies on 48G7 and its germline precursor support this conclusion.[54,72]

Affinity-matured 48G7 is about an order of magnitude more efficient than CNJ206. It also has a 30 000-fold higher affinity for its hapten (**33**) than its germline counterpart and 20-fold greater catalytic efficiency as a consequence of nine somatic mutations.[54,72] The mutated residues lie outside the combining site and therefore do not directly contact the hapten. Structural studies of the affinity-matured protein with and without the transition state analogue show that hapten binding occurs via a lock-and-key mechanism with very little conformational change.[73] In contrast, hapten binding nucleates a significant reorganization of the combining site of the germline antibody.[72] Interactions with the phosphonate, which include hydrogen bonds from the backbone NH of TyrH96 (as in CNJ206) and from the side chains of TyrH33, HisH35, and ArgL96, are the same in the two complexes and cannot explain the differences in affinity or catalytic efficiency (Figure 2(d)). Instead, the somatic mutations must play an indirect role in preorganizing the pocket for catalysis, limiting side chain and backbone flexibility inherent in the germline antibody.

Despite the larger numbers of antibody residues devoted to phosphate recognition, the use of cationic interactions, and an inherently more reactive substrate, the chemical efficiency of 48G7 is lower than that of that of D2.3 by at least an order of magnitude. The contributions of HisH35, ArgL96, and TyrH33 were investigated by site-directed mutagenesis.[54] Changing these residues, respectively, to Glu, Gln, and Phe, reduces k_{cat} by 30-, 11-, and 3.2-fold. The effects for ArgL96 and TyrH33 are surprisingly small, and it was proposed that one residue may compensate for the other when one of the interactions with the oxyanion is lost. The high solvent accessibility and/or conformational mobility of these residues could also account for the results.

Similar themes are evident in the anti-(**36**) antibody 17E8 which catalyzes the enantioselective hydrolysis of unsubstituted phenyl esters of *N*-formyl-ʟ-norleucine and *N*-formyl-ʟ-methionine with rate enhancements of $\approx 10^4$ over the uncatalyzed reaction.[69] Although a two-step mechanism with transient acylation of an antibody serine residue by substrate was originally proposed for this catalyst on the basis of hydroxylamine partitioning data[69] and preliminary analysis of the crystal structure,[60] this hypothesis now seems unlikely in light of subsequent mutagenesis experiments.[74] Instead, this antibody, like the others described above, appears to stabilize the transition states resulting from direct attack on the scissile carbonyl relative to the ground state through oriented hydrogen bonds and electrostatic interactions. In this case, in addition to a hydrogen bond from the backbone NH of TyrH96, the side chains of two cationic residues, LysH93 and ArgL96, provide complementary electrostatic/hydrogen-bonding interactions with the *pro-S* and *pro-R* phosphonate oxygens, respectively (Figure 2(c)). The interactions between the guanidinium group of ArgL96 and the phosphonate suggest that this residue helps to stabilize the oxyanion intermediate as well as the incipient negative charge formed on the aryl leaving group. When this residue is replaced with a

Table 1 Ester hydrolysis by antiphosphonate antibodies.

Antibody	Hapten	Reaction
D2.3, D2.4, D2.5	(27) R = (CH₂)₃CO-NH-CH₂-COOH	(23) → (25) + (26)
CNJ206	(29) R = (CH₂)₄COOH	(30) → (31) + (32)
48G7	(33) R¹ = (CH₂)₄COOH	(34) R² = O(CH₂)₂COOH → (31) + (35)
17E8	(36) R¹ = NH-CO(CH₂)₃COOH, —CH₃	(37) R² = NH-CO(CH₂)₃COOH, X—Me, X = CH₂ or S → (38) + (39)

Table 2   Kinetic parameters for representative hydrolytic antibodies.

Antibody	Substrate	pH	k_{cat}/K_m $(M^{-1} min^{-1})$	k_{cat} (min^{-1})	k_{cat}/k_{uncat}	Ref.
D2.3	(27)	8.3	185	3.1	110 000	61
D2.4	(27)	8.3	64	1.0	36 000	61
D2.5	(27)	9.0	36	0.39	1900	61
CNJ206	(30)	8.0	5 000	0.4	1600	67
48G7	(34)	8.2	14 000	5.5	16 000	68
17E8	(37), X = CH$_2$	8.7	390 000	100	13 000	69
43C9	(40), X = O	9.3	28 000 000	1500	27 000	70
43C9	(40), X = NH	9.0	140	0.08	250 000	58

tyrosine, catalytic activity declines to an undetectable level.[74] Selection experiments with a humanized version of 17E8 displayed on phage showed that AspH101 also makes a significant contribution to catalysis.[74] This residue forms a salt bridge with LysH93, and it has been suggested that its role is to raise the pK_a of the buried ε-amino group so that the protonated form of the lysine can better stabilize the oxyanion intermediate formed during substrate hydrolysis.

Selection of 17E8 variants with higher affinity for the transition state analogue by phage display failed to yield more effective catalysts.[74] Indeed, most of the mutants were substantially less active than the parent antibody. By contrast, a mutation (TyrH100a to Asn) which causes a twofold decrease in affinity increased the apparent second-order rate constant k_{cat}/K_m by a factor of five. Like the somatic mutations that yielded a more efficient 48G7, TyrH100 is not in direct contact with the hapten. These results underscore how a complex network of interactions contributes to the efficiency of each catalytic antibody. They also highlight the difficulty of improving these agents by altering only residues that line the combining site.

5.17.3.3   Nucleophilic Catalysis

Although 17E8 and the other esterases discussed above appear to operate by the simple hydrolytic mechanism programmed by phosphonate transition state analogues, some antibodies raised against these compounds do occasionally exploit more complex mechanisms. Antibody 43C9 is a well-characterized example. It was generated using the phosphonamidate hapten (43) and accelerates the hydrolysis of the analogous *p*-nitroanilide (40) and related substituted phenyl esters by factors up to 2.5×10^5 at pH 9.0 (Scheme 7).[58,70] Considerable biochemical evidence supports a two-step cleavage sequence,[75] involving acylation of an active site histidine followed by hydrolysis of the resulting acyl imidazole intermediate by hydroxide (Figure 3). The acyl intermediate does not accumulate under most conditions, but was detected by electrospray mass spectrometry at pH 5.9.[76]

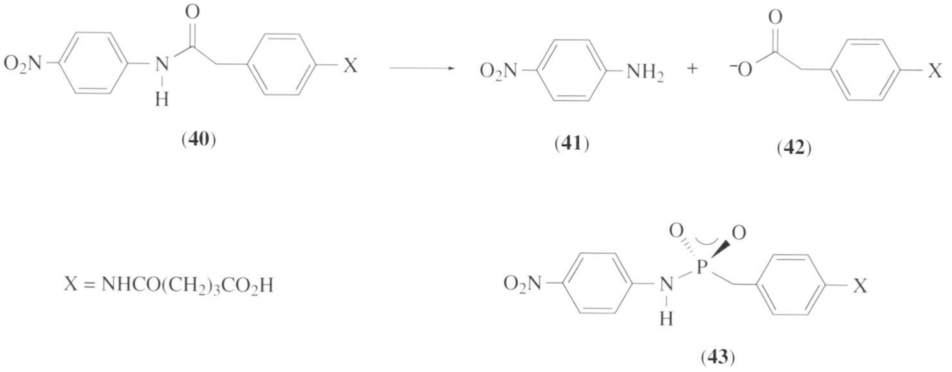

(40)   (41)   (42)

X = NHCO(CH$_2$)$_3$CO$_2$H

(43)

Scheme 7

Detailed structural data are not yet available for 43C9, but homology modeling indicates that this antibody and 17E8 share a large number of active site residues.[59] Mutagenesis experiments[59,77] have underscored the importance of three residues in the putative active site: HisH35, ArgL96, and HisL91. HisH35 and ArgL96 are also present in 17E8 (and in 48G7) and probably play structural

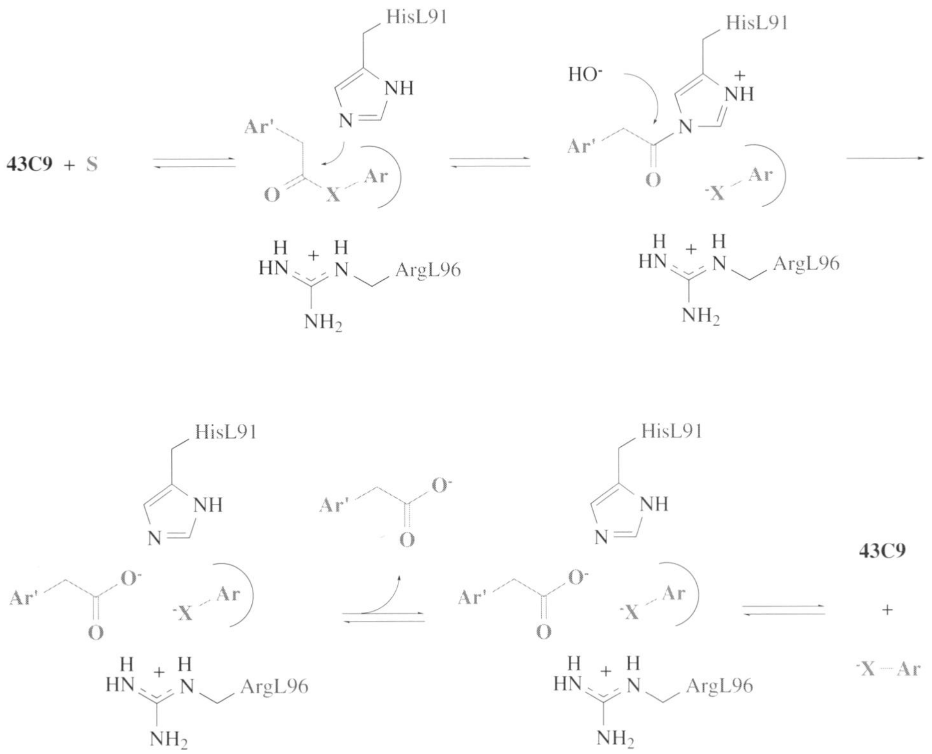

Figure 3 Proposed mechanism of the amidase antibody 43C9.

and catalytic roles, respectively. Substitutions at position H35 yield proteins with substantially weaker affinity for hapten and products, whereas replacement of ArgL96 with glutamine reduces esterase activity to background levels without affecting affinity for substrate. HisL91 is unique to 43C9 and is likely to be the nucleophilic histidine. Mutagenesis of this residue to glutamine decreases catalytic efficiency at least 50-fold with little effect on ligand binding. Failure to detect the intermediate by mass spectrometry when the HisL91Gln mutant was incubated with the ester analogue of (**40**) provides additional circumstantial evidence for its putative role in catalysis.[75]

Nucleophilic catalysis, which is also exploited by serine proteases, reduces a difficult chemical reaction to a series of more manageable steps and may explain why 43C9, but not 17E8 or the other esterolytic antibodies, is able to cleave an amide. The relative sophistication of 43C9 shows that unprogrammed mechanisms analogous to those used by natural enzymes can arise, albeit through serendipity. Contextual effects are clearly important, however, since most of the antibodies raised against (**43**) do not have amidase activity. Furthermore, attempts to augment the activity of the otherwise closely related 17E8 esterase by introducing the TyrL96His mutation into its active site proved unsuccessful.[73] Although 43C9 promotes the hydrolysis of an activated amide, it is still quite primitive compared to highly evolved protease like subtilisin. It catalyzes the hydrolysis of *p*-nitroanilide (**40**) (pH 9.0 and 37 °C) more than 30 000 times less efficiently than subtilisin catalyzes the cleavage of succinyl-Ala-Ala-Pro-Phe-*p*-nitroanilide (pH 8.0 and 25 °C),[58,65] and it cannot cleave substrates with leaving groups poorer than a *p*-nitroanilide.[70] The high efficiency of the natural enzyme reflects the synergistic interaction of multiple catalytic residues.[78,79] Acid–base chemistry, in particular, is used to facilitate proton transfers and to stabilize the amide leaving group. Analogous functionality is lacking in 43C9.

5.17.4 ALDOL CONDENSATIONS

The high efficiency of most natural enzymes—as exemplified by the protease subtilisin—can be attributed to the effective use of arrays of acids, bases, and nucleophiles to achieve catalysis. As seen in the previous section, charged haptens can be utilized to elicit sensibly configured networks of hydrogen-bonding residues and charged amino acids complementary to a transition state. Occasionally, as in the amidase 43C9, unplanned but useful catalytic groups can also be obtained.

Other experiments have shown that charge complementarity between antibody and antigen can be exploited intentionally to induce acids, bases, and nucleophiles. A variety of addition, elimination, and substitution reactions have been successfully catalyzed in this way.[80-82] Engineering of active sites with constellations of multiple catalytic residues, however, remains a significant challenge—particularly for multistep reactions.

An alternative approach takes advantage of mechanism-based enzyme inhibitors.[83,84] These are molecules that inactivate their targets by exploiting an enzyme's normal catalytic mechanism to potentiate their own decomposition to give a reactive species that can be trapped covalently by an active site residue. When such compounds are used as haptens, the process of antibody induction may involve an analogous chemical reaction within the binding site. Antibodies that become covalently modified can then be selected. A subset of these should be equipped with the catalytic groups necessary to promote transformations of similarly reactive substrates.

This strategy has yielded catalysts for several different kinds of reaction. Antibodies have been identified, for example, that react covalently with haptenic phosphonate diesters like (44).[85] These immunoglobulins, which apparently contain a reactive nucleophile, catalyze the hydrolysis of structurally analogous esters ($k_{cat}/k_{uncat} = 10^2$–10^4), presumably via a two-step mechanism involving a covalent acyl–antibody intermediate. The mechanism-based galactosidase inhibitor (45) was similarly used to trap phage-displayed antibodies with galactosidase activity covalently.[86] One of the catalysts identified in these experiments, Fab fragment 1B, accelerates the hydrolysis of *p*-nitrophenyl-β-galactose with a rate acceleration (k_{cat}/k_{uncat}) of 7×10^4. Aldolase antibodies generated against the 1,3-diketone (46)[87] are particularly interesting from a mechanistic standpoint.

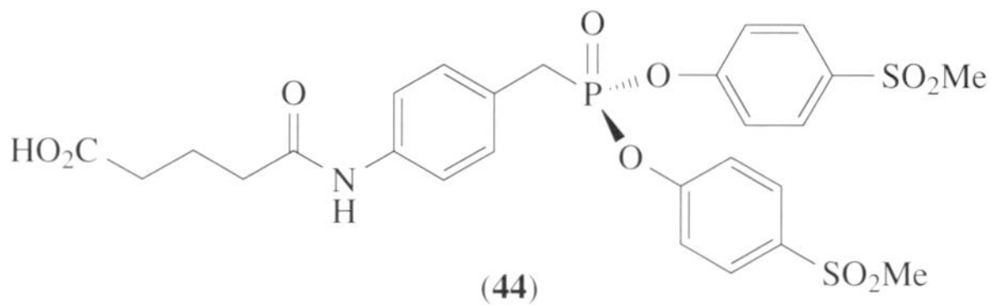

(44)

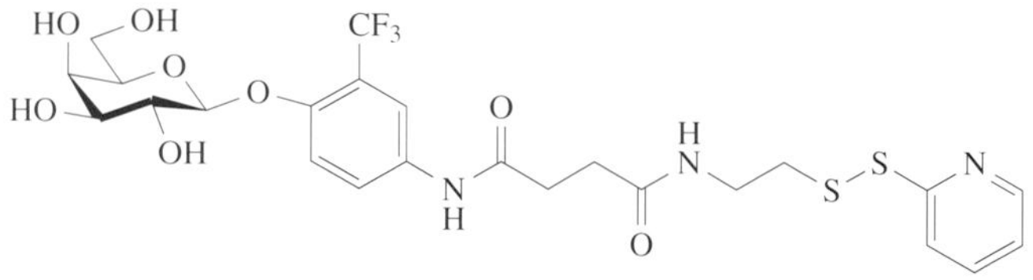

(45)

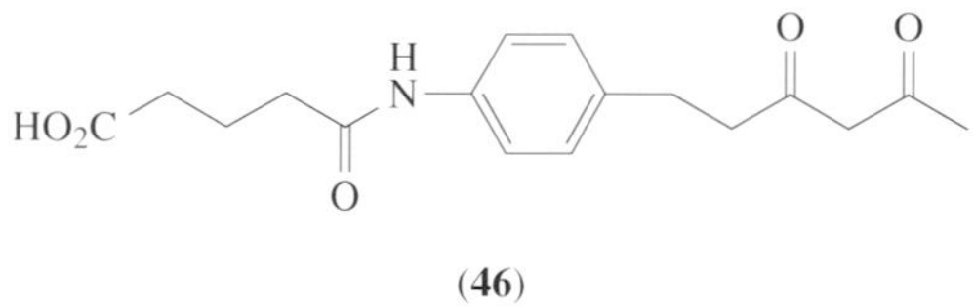

(46)

Aldol condensations are broadly useful carbon–carbon bond forming reactions in organic synthesis. One class of natural enzymes, the class I aldolases, exploits a catalytically essential and unusually reactive lysine residue for catalysis (Figure 4).[88,89] The ε-amino group of this residue forms a Schiff base with a ketone, thereby activating it as an aldol donor. The iminium ion acts as an electron sink and lowers the activation barrier for proton abstraction from the Cα atom to give a nucleophilic enamine that attacks an electrophilic aldehyde to form a new carbon–carbon bond. Subsequent hydrolysis of the Schiff base releases product and regenerates the active catalyst. Diketones like (46) inhibit class I aldolases by trapping the active site lysine as a Schiff base before rearranging to a more stable vinylogous amide (Figure 5).

Two of the 20 antibodies generated in response to (46) were likewise found to form vinylogous amides with the hapten.[87] This species, which is easily detected by its characteristic absorption band at 316 nm ($\varepsilon = 15\,000$ M^{-1} cm^{-1}), can be trapped by reduction with sodium cyanoborohydride. Titration experiments show that the reactive lysine residue has an unusually low pK_a (5.5 and 6.0 for antibodies 33F12 and 38C2, respectively). This lysine was also found to react with ketone substrates to form an enamine that reacts efficiently with a variety of aldehydes to form aldol products.[87] Catalytic activity is inhibited completely and irreversibly by reduction of the key imine intermediate.

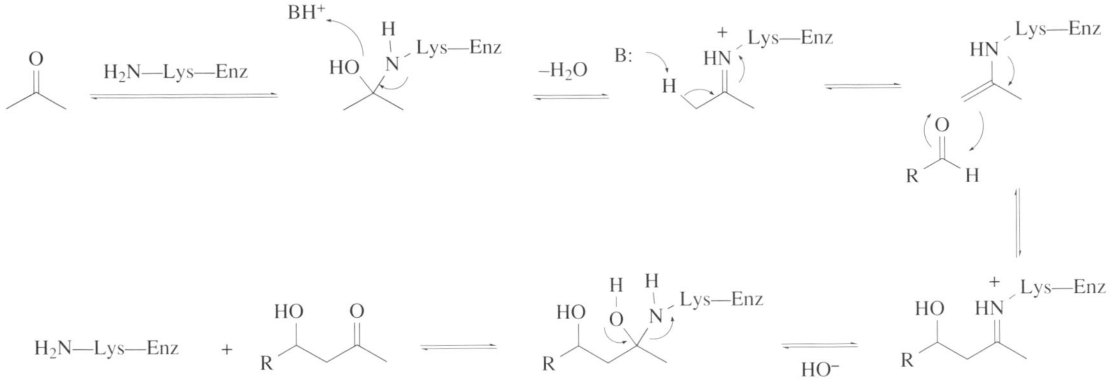

Figure 4 Reaction mechanism of class I aldolases and the catalytic antibodies 33F12 and 38C2.

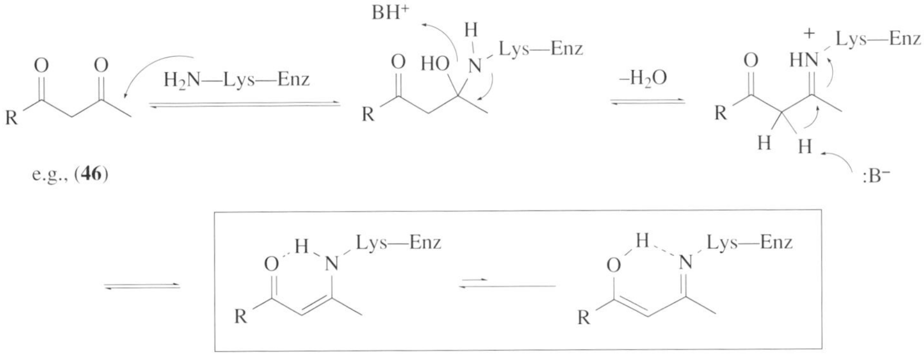

$\lambda_{max} = 316$ nm ($\varepsilon = 15000$ M^{-1} cm^{-1})

Figure 5 Inhibition of class I aldolases with diketones like (**46**) results in covalent modification to yield a vinylogous amide.

More than 100 different aldol additions or condensations have been promoted with these antibodies, illustrating their unusually broad scope.[90] Aldehyde–aldehyde, ketone–aldehyde, and ketone–ketone condensations are all subject to catalysis. The k_{cat} values for these transformations range from 10^{-3} min^{-1} to 5 min^{-1}, corresponding to rate accelerations (k_{cat}/k_{uncat}) of 10^5–10^7. Although the hapten is achiral, the induced antibodies were found to be remarkably stereoselective.[87,90] In the reaction between racemic (**47**) and acetone, for example, antibody 38C2 yielded a roughly 1:1 mixture of (4*S*,5*S*)-(**48**) (>95% *de*) and (4*S*,5*R*)-(**48**) (83% *de*) after 30% conversion (Scheme 8). The diastereofacial selectivity of the second catalyst, 33F12, was somewhat

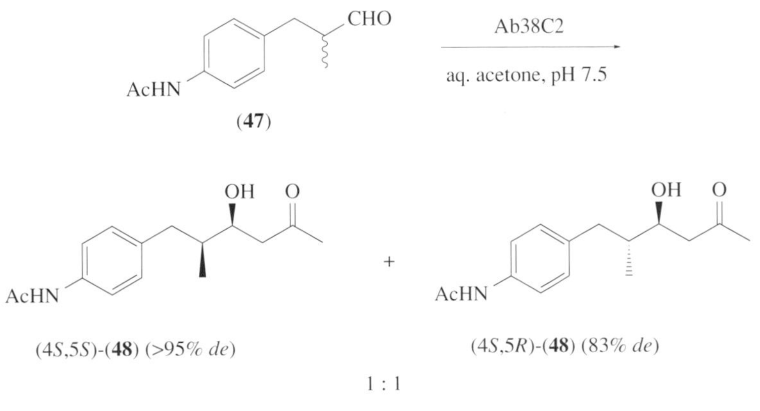

(4*S*,5*S*)-(**48**) (>95% *de*)   (4*S*,5*R*)-(**48**) (83% *de*)

1 : 1

Scheme 8

lower for the energetically more demanding formation of the anti-Cram–Felkin product (4S,5R)-(48) (65% de). In another example, antibody 38C2 was shown to catalyze the Robinson annulation of (49) to (50) in >95% ee (Equation (1)).[9i]

(49)

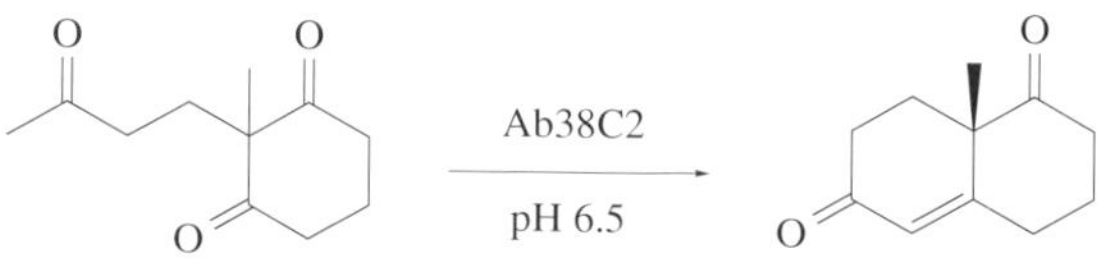

Ab38C2
pH 6.5

(S)-(+)-(50)
>95% ee

(1)

Consistent with the proposed enamine mechanism, both aldolase antibodies also catalyze the decarboxylation of structurally related β-keto acids (Figure 6).[92] Mechanistic studies, including inhibition of the decarboxylation reaction by 2,4-pentanedione, trapping experiments with cyanide, and ^{18}O incorporation experiments, strongly support a mechanism involving an imine intermediate. These catalysts thus mimic acetoacetate decarboxylase, which is also known to exploit Schiff base formation to achieve catalysis, although they are five to six orders of magnitude less active than the natural enzyme (k_{cat} = 1560 s^{-1}).[93]

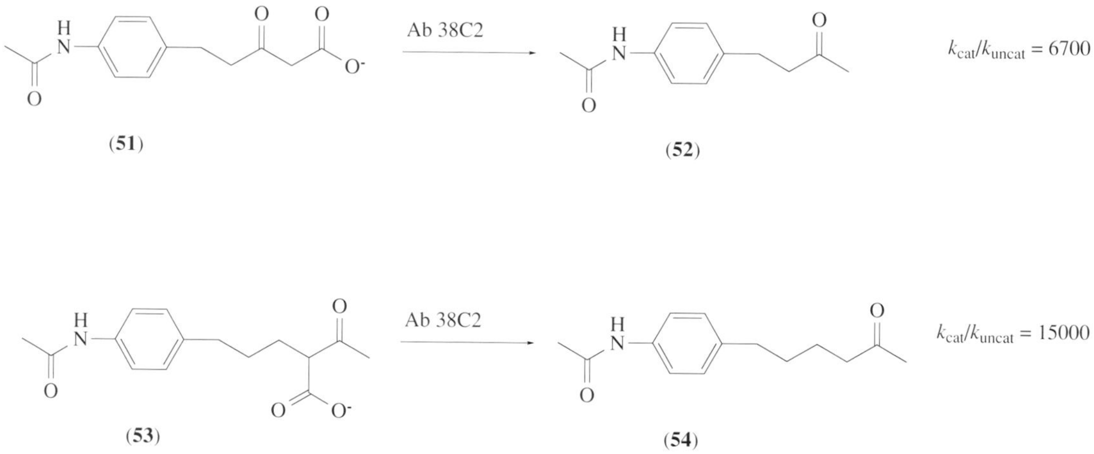

(51)   Ab 38C2   (52) k_{cat}/k_{uncat} = 6700

(53)   Ab 38C2   (54) k_{cat}/k_{uncat} = 15000

Figure 6   Decarboxylation reactions catalyzed by the aldolase antibody 38C2.

The crystal structure of 33F12 in the absence of ligand reveals only a single lysine in the binding site.[90] This residue, LysH93, is the same lysine found in the esterase antibody 17E8 discussed above. In 33F12, it is deeply buried in a hydrophobic pocket. AspH101, which forms a salt bridge with the positively charged ammonium group of this residue in 17E8, is absent in 33F12, and there are no other charged residues within a 7 Å radius. Nor does LysH93 appear to participate in hydrogen bonds with any main chain carbonyl oxygens. The absence of such interactions must be responsible for this group's unusual reactivity. The hydrophobic microenvironment surrounding the critical lysine is expected to disfavor protonation, perturbing its pK_a and making it a substantially better nucleophile. In class I aldolases, by contrast, the pK_a of the critical lysine is believed to be perturbed through its proximity to additional positively charged residues.

The entrance to the binding pocket is a narrow elongated cleft, while the pocket itself is more than 11 Å deep, expanding with depth.[90] Although information about how substrates bind is lacking, it appears that simple van der Waals interactions are exploited to accommodate substrates within the active site. The unliganded molecule provides no insight, however, into how the polar intermediates and transition states involved, for example, in carbinol amine formation and breakdown or in the subsequent carbon–carbon bond forming step, are stabilized. Sequestered water molecules or the hydroxyl groups of tyrosine or serine residues that form the walls of the pocket are plausible candidates. Nor is it clear why the 33F12 pocket is so promiscuous or whether conformational plasticity might contribute to the broad substrate specificity of this catalyst. Structural characterization of the antibody complexed to substrates and products will be important for further delineating its remarkable catalytic properties.

5.17.5 CATALYTIC EFFICIENCY AND MECHANISM

As the foregoing examples show, enzymes and antibodies share many common features. In both cases, hydrogen bonds, electrostatics, hydrophobic effects, and van der Waals interactions are employed to achieve highly specific molecular recognition. Catalysis results when the available binding energy is exploited to discriminate between ground states and transition states and to stabilize the latter preferentially.

Nevertheless, catalytic antibodies are still primitive compared to their natural counterparts. No antibody has yet been described that rivals the efficacy of an analogous enzyme when the two systems are compared under comparable conditions. Why is this? Does the relatively low efficiency of antibody catalysis reflect the short evolutionary history of these molecules, shortcomings in the strategies used to generate them, or intrinsic structural limitations on catalysis imposed by the immunoglobulin scaffold?

Rate accelerations for a large number of antibody-catalyzed reactions correlate with the ratio of equilibrium binding constants of the substrate and transition-state analogue.[94] Thus, for unimolecular reactions $k_{cat}/k_{uncat} \approx K_m/K_{TSA}$, and for bimolecular reactions $k_{cat}/k_{uncat} \approx K_{m1}K_{m2}/K_{TSA}$, where K_m is the equilibrium constant for antibody binding of the substrate and K_{TSA} is the equilibrium constant for the transition state analogue. These relationships are based on theoretical considerations relating enzymic catalysis and transition state theory,[95] and underscore the fundamental soundness of the transition state analogue approach to catalyst design. To the extent that the transition state analogue accurately mimics a reaction's true transition state, the chances of identifying highly active catalysts depend on optimizing affinity for the hapten.

In practice, of course, no stable molecule can ever successfully capture all the features of true transition states with their partial charges and fractional bond lengths. It is therefore unrealistic to expect that an antibody optimized to bind such an imperfect analogue will ever achieve maximal stabilization of the corresponding transition state. Moreover, shape and chemical complementarity between the binding pocket and the hapten can be achieved in multiple ways, only some of which will be suitable for catalysis. Thus, a significant fraction of the antibody binders generated in response to a given transition state analogue are often found to be catalytically inactive, despite their high affinity for the hapten.[62] The breakdown in the relationship between transition state analogue affinity and catalytic efficiency is pointedly illustrated by the failure to obtain more active versions of the esterase 17E8 by mutagenesis and affinity selection on phage.[73] None of the tighter binding variants was more active than the parent antibody, but ironically a weaker binding clone was fivefold faster.

Even if perfect transition state analogues were available, it is doubtful that the binding affinities needed for high reaction rates can be reliably achieved given the selection mechanisms available to the immune system. Enzymes bind transition states with apparent dissociation constants as low as 10^{-23} M.[96] Since substrate and product affinities are generally in the micromolar to millimolar range, enormous residual binding energy is available for selective transition state stabilization. By contrast, equilibrium dissociation constants for antibody–hapten complexes are typically in the range 10^{-6}–10^{-9} M.[4] While adequate for the purposes of the immune system, these smaller binding energies necessarily limit the extent to which transition states and ground states can be distinguished. Furthermore, because antibodies are not selected on the basis of catalytic function, binding generally involves elements common to both the ground state and transition state and hence is not directly useful for catalysis. The use of aromatic groups as dominant epitopes in many transition state analogues is illustrative of the problem: studies with polyclonal antibodies have shown that the catalytic activity of a series of esterases correlates inversely with the size and hydrophobicity of the aryl phosphate haptens used to raise them.[97] When a significant fraction of the available binding energy is directed to portions of the ligand distant from the actual site of reaction, product inhibition also becomes a practical concern.

Identification of higher affinity antibodies (and antibodies that utilize their available binding energy more effectively) is clearly necessary if we hope to create more active catalysts. To that end, more extensive screening of the immune response to a given hapten represents one of the simplest modifications of current methods that can be implemented. Only a minuscule fraction of the primary immunological repertoire, which is estimated to contain roughly 10^8 different antibodies,[98] is sampled in a typical catalytic antibody experiment. Given that the most active catalysts (which may exploit unusual germline sequences or serendipitous mutations) are also likely to be the rarest, more effective screening of large numbers of antibodies will increase the probability of finding potent catalysts. Combinatorial libraries of immunoglobulin fragments with more than 10^6 members can now be constructed[99] and produced in microorganisms or displayed on phage particles.[99–102] In principle, these libraries can be screened directly for catalytic activity using sensitive chemical,[103] biological,[37,68,104] or immunological[62] assays. In this way, statistically meaningful comparisons of different haptens and their propensity to induce active catalysts can be made. Information gained from these studies will be crucial for the development of more effective transition state analogues. The striking similarities evident in the antiphosphonate antibodies,[64] on the one hand, and the Diels–Alderase and progesterone-binding antibodies,[45,46] on the other, raise the additional concern that existing strategies for hapten presentation only yield antibodies of limited structural diversity. Recruiting larger numbers of antibodies should also facilitate exploration of different subsets of the primary immunological repertoire with respect to their catalytic potential.

The difficulty of finding highly active antibodies is compounded by the fact that most reactions of chemical interest require the carefully orchestrated participation of acids, bases, and nucleophiles. While induction of properly oriented catalytic groups within the antibody combining site is not impossible, it represents an improbable event that becomes less likely as the number of groups needed to act in concert increases. As discussed above, charge complementarity has been exploited with some success to increase the probability of eliciting catalytic residues, but arrays as sophisticated as the catalytic triad in serine proteases will be difficult to generate with a single hapten. "Heterologous immunization" has been introduced as a means of circumventing this limitation.[105–107] In this approach, two haptens, each containing a different electronic or structural feature, are used in succession to elicit an immune response. Preliminary results suggest that antibodies derived from a heterologous immunization experiment are more active than those generated in response to a single hapten,[105–107] but more work is needed to assess the generality of these findings and to ascertain whether the antibodies in question are in fact more richly functionalized than their counterparts obtained in the standard way. Perhaps the most significant advantage of this approach is its potential for introducing catalytic groups without resorting to complex hapten synthesis.

The third hapten-based strategy to functionalized antibody active sites—"reactive immunization"—may ultimately prove to be the most direct and powerful.[85,87,108] As illustrated by the versatile aldolase antibodies described in the previous section, the use of mechanism-based inhibitors is particularly attractive as a means of rationally engineering covalent catalysis into the antibody pocket.[87] Although the antibodies obtained in this way have not yet matched the potency of analogous enzymes,[85–87,92,108] reactive immunization certainly represents one of the most promising approaches for obtaining a functional catalytic apparatus in a single immunization step. In combination with other design strategies, it may well yield qualitatively better catalysts than those generated against stable, unreactive haptens.

While an intrinsic structural bias against catalysis could emerge from future studies of catalytic antibodies, it seems more likely at this stage that imperfect transition state analogue design and the reliance on binding rather than catalytic function as the basis for selection during maturation of the immune response are the primary reasons for low efficiency. If true, incremental improvement of first-generation antibody catalysts may be possible. The fact that small refinements in the same basic active site can lead to substantial increases in efficiency, as seen in the comparative studies of germline and mature antibodies[41,45,54] and more generally for the family of antiphosphonate esterolytic antibodies,[64] is a promising precedent.

In principle, antibody active sites can be repacked or equipped with catalytic functionality by site-directed mutagenesis.[59,77,109,110] The availability of increasing numbers of high-resolution structures makes this approach attractive, although our limited understanding of structure–function relationships within the immunoglobulin scaffold may still preclude engineering large enhancements in efficiency. Because residues that are not directly involved in ligand recognition can greatly influence catalytic efficiency,[54,111] identification of suitable target residues by inspection of crystal structures is nontrivial.

Nature has optimized the catalytic efficiency of all enzymes by the process of natural selection. An analogous approach can also be pursued with antibodies. Immunoglobulins have been produced in living cells as single-chain F_v or Fab fragments and shown to participate directly in cellular metabolism.[37,104] These experiments establish the feasibility of evolutionary studies involving recursive cycles of mutagenesis and selection based on catalysis rather than simple affinity for the transition state analogue, and it will be interesting to see to what extent individual starting antibodies can be improved. Selection schemes can be imagined for a wide range of metabolic processes as well as for transformations involving the creation or destruction of a vital nutrient or toxin. Consequently, if successful, directed evolution of antibody catalysts could be broadly useful.

5.17.6 CONCLUSION

Immunization with transition state analogues has proved to be a powerful and general method for creating catalysts with tailored activities and selectivities. The structural and mechanistic data reviewed here show, however, that this is only the first step to enzyme-like catalysts. Improved transition state analogues, refinements in immunization and screening protocols, and development of general strategies for augmenting the efficiency of first-generation catalytic antibodies remain obvious but difficult challenges for this field. Rising to these challenges and more successfully integrating programmable design with the selective forces of biology will enhance our understanding of enzymic catalysis and yield useful tools for a wide range of practical applications in chemistry and biology.

ACKNOWLEDGMENT

The author is indebted to the National Institutes of Health for support of the research in his laboratory.

5.17.7 REFERENCES

1. P. G. Schultz and R. A. Lerner, *Science*, 1995, **269**, 1835.
2. R. A. Lerner, S. J. Benkovic, and P. G. Schultz, *Science*, 1991, **252**, 659.
3. W. P. Jencks, "Catalysis in Chemistry and Enzymology," McGraw-Hill, New York, 1969, p. 288.
4. E. A. Kabat, "Structural Concepts in Immunology and Immunochemistry," Holt, Rinehart, and Winston, New York, 1976.
5. P. G. Schultz and R. A. Lerner, *Acc. Chem. Res.*, 1993, **26**, 391.
6. W. P. Jencks, *Adv. Enzymol.*, 1975, **43**, 219.
7. D. Y. Jackson, J. W. Jacobs, R. Sugasawara, S. H. Reich, P. A. Bartlett, and P. G. Schultz, *J. Am. Chem. Soc.*, 1988, **110**, 4841.
8. D. Hilvert, S. H. Carpenter, K. D. Nared, and M.-T. M. Auditor, *Proc. Natl. Acad. Sci. USA*, 1988, **85**, 4953.
9. A. C. Braisted and P. G. Schultz, *J. Am. Chem. Soc.*, 1994, **116**, 2211.
10. S. S. Yoon, Y. Oei, E. Sweet, and P. G. Schultz, *J. Am. Chem. Soc.*, 1996, **118**, 11 686.
11. Z. S. Zhou, N. Jiang, and D. Hilvert, *J. Am. Chem. Soc.*, 1197, **119**, 3623.
12. A. C. Braisted and P. G. Schultz, *J. Am. Chem. Soc.*, 1990, **112**, 7430.
13. D. Hilvert, K. W. Hill, K. D. Nared, and M.-T. M. Auditor, *J. Am. Chem. Soc.*, 1989, **111**, 9261.
14. V. E. Gouverneur, K. N. Houk, B. Pascual-Teresa, B. Beno, K. D. Janda, and R. A. Lerner, *Science*, 1993, **262**, 204.
15. P. R. Andrews, G. D. Smith, and I. G. Young, *Biochemistry*, 1973, **12**, 3492.
16. S. D. Copley and J. R. Knowles, *J. Am. Chem. Soc.*, 1985, **107**, 5306.
17. L. Addadi, E. K. Jaffe, and J. R. Knowles, *Biochemistry*, 1983, **22**, 4494.
18. S. D. Copley and J. R. Knowles, *J. Am. Chem. Soc.*, 1987, **109**, 5008.
19. S. G. Sogo, T. S. Widlanski, J. H. Hoare, C. E. Grimshaw, G. A. Berchtold, and J. R. Knowles, *J. Am. Chem. Soc.*, 1984, **106**, 2701.
20. J. Görisch, *Biochemistry*, 1978, **17**, 3700.
21. P. R. Andrews, E. N. Cain, E. Rizzardo, and G. D. Smith, *Biochemistry*, 1977, **16**, 4848.
22. P. A. Bartlett, Y. Nakagawa, C. R. Johnson, S. H. Reich, and A. Luis, *J. Org. Chem.*, 1988, **53**, 3195.
23. H. S.-I. Chao and G. A. Berchtold, *Biochemistry*, 1982, **21**, 2778.
24. P. A. Bartlett and C. R. Johnson, *J. Am. Chem. Soc.*, 1985, **107**, 7792.
25. D. Hilvert and K. D. Nared, *J. Am. Chem. Soc.*, 1988, **110**, 5593.
26. D. Y. Jackson, M. N. Liang, P. A. Bartlett, and P. G. Schultz, *Angew. Chem., Int. Ed. Engl.*, 1992, **31**, 182.
27. P. A. Campbell, T. M. Tarasow, W. Massefski, P. E. Wright, and D. Hilvert, *Proc. Natl. Acad. Sci. USA*, 1993, **90**, 8663.
28. M. R. Haynes, E. A. Stura, D. Hilvert, and I. A. Wilson, *Science*, 1994, **263**, 646.
29. B. Ganem, *Angew. Chem., Int. Ed. Engl.*, 1996, **35**, 936.
30. Y. M. Chook, H. Ke, and W. N. Lipscomb, *Proc. Natl. Acad. Sci. USA*, 1993, **90**, 8600.
31. A. Y. Lee, A. P. Karplus, B. Ganem, and J. Clardy, *J. Am. Chem. Soc.*, 1995, **117**, 3627.
32. P. Kast, M. Asif-Ullah, N. Jiang, and D. Hilvert, *Proc. Natl. Acad. Sci. USA*, 1996, **93**, 5043.
33. D. R. Liu, S. T. Cload, R. M. Pastor, and P. G. Schultz, *J. Am. Chem. Soc.*, 1996, **118**, 1789.
34. S. Zhang, P. Kongsaeree, J. Clardy, D. B. Wilson, and B. Ganem, *Bioorg. Med. Chem.*, 1996, **4**, 1015.
35. S. T. Cload, D. R. Liu, R. M. Pastor, and P. G. Schultz, *J. Am. Chem. Soc.*, 1996, **118**, 1787.
36. O. Wiest and K. N. Houk, *J. Org. Chem.*, 1994, **59**, 7582.
37. Y. Tang, J. B. Hicks, and D. Hilvert, *Proc. Natl. Acad. Sci. USA*, 1991, **88**, 8784.

38. G. MacBeath, P. Kast, and D. Hilvert, *Science*, 1998, **279**, 1958.
39. E. M. Driggers, H. S. Cho, C. W. Liu, C. W. Katzka, A. C. Braisted, H. D. Ulrich, D. E. Wemmer, and P. G. Schultz, *J. Am. Chem. Soc.*, 1998, **120**, 1945.
40. H. D. Ulrich, E. M. Driggers, and P. G. Schultz, *Acta Chim. Scand.*, 1996, **50**, 328.
41. H. D. Ulrich, E. Mundorff, B. D. Santarsiero, E. M. Driggers, R. C. Stevens, and P. G. Schultz, *Nature*, 1997, **389**, 271.
42. M. J. Steigerwald, W. A. Goddard, and D. A. Evans, *J. Am. Chem. Soc.*, 1979, **101**, 1994.
43. M. J. S. Dewar and L. E. A. Wade, *J. Am. Chem. Soc.*, 1977, **99**, 4417.
44. M. I. Page and W. P. Jencks, *Proc. Natl. Acad. Sci. USA*, 1971, **68**, 1678.
45. F. E. Romesberg, B. Spiller, P. G. Schultz, and R. C. Stevens, *Science*, 1998, **279**, 1929.
46. M. R. Haynes, M. Lenz, M. J. Taussig, I. A. Wilson, and D. Hilvert, *Isr. J. Chem.*, 1996, **36**, 151.
47. J. T. Yli-Kauhaluoma, J. A. Ashley, C.-H. Lo, L. Tucker, M. M. Wolfe, and K. D. Janda, *J. Am. Chem. Soc.*, 1995, **117**, 7041.
48. A. Heine, E. A. Stura, J. T. Yli-Kauhaluoma, C. Gao, Q. Deng, B. R. Beno, K. N. Houk, K. D. Janda, and I. A. Wilson, *Science*, 1998, **279**, 1934.
49. M. M. Mader and P. A. Bartlett, *Chem. Rev.*, 1997, **97**, 1281.
50. P. A. Bartlett and C. K. Marlow, *Biochemistry*, 1983, **22**, 4618.
51. M. A. Phillips, A. P. Kaplan, W. J. Rutter, and P. A. Bartlett, *Biochemistry*, 1992, **31**, 959.
52. A. P. Kaplan and P. A. Bartlett, *Biochemistry*, 1991, **30**, 8165.
53. J. D. Stewart, L. J. Liotta, and S. J. Benkovic, *Acc. Chem. Res.*, 1993, **26**, 396.
54. P. A. Patten, N. S. Gray, P. L. Yang, C. B. Marks, G. J. Wedemayer, J. J. Boniface, R. C. Stevens, and P. G. Schultz, *Science*, 1996, **271**, 1086.
55. J.-B. Charbonnier, B. Golinelli-Pimpaneu, B. Gigant, D. S. Tawfik, R. Chap, D. G. Schindler, S.-H. Kim, B. S. Green, Z. Eshhar, and M. Knossow, *Science*, 1997, **275**, 1140.
56. J.-B. Charbonnier, E. Carpenter, B. Gigant, B. Golinelli-Pimpaneu, D. Tawfik, Z. Eshhar, B. S. Green, and M. Knossow, *Proc. Natl. Acad. Sci. USA*, 1995, **92**, 11 721.
57. T. S. Scanlan, J. R. Prudent, and P. G. Schultz, *J. Am. Chem. Soc.*, 1991, **113**, 9397.
58. K. D. Janda, D. Schloeder, S. J. Benkovic, and A. Lerner, *Science*, 1988, **241**, 1188.
59. V. A. Roberts, J. Stewart, S. J. Benkovic, and E. D. Getzoff, *J. Mol. Biol.*, 1994, **235**, 1098.
60. G. W. Zhou, J. Guo, W. Huang, R. J. Fletterick, and T. S. Scanlan, *Science*, 1994, **265**, 1059.
61. D. S. Tawfik, A. B. Lindner, R. Chap, Z. Eshhar, and B. S. Green, *Eur. J. Biochem.*, 1997, **244**, 619.
62. D. S. Tawfik, B. S. Green, R. Chap, M. Sela, and Z. Eshhar, *Proc. Natl. Acad. Sci. USA*, 1993, **90**, 373.
63. B. Gigant, J.-B. Charbonnier, Z. Eshhar, B. S. Green, and M. Knossow, *Proc. Natl. Acad. Sci. USA*, 1997, **94**, 7857.
64. G. MacBeath and D. Hilvert, *Chem. Biol.*, 1996, **3**, 433.
65. P. Bryan, M. W. Pantoliano, S. G. Quill, H.-Y. Hsiao, and T. Poulos, *Proc. Natl. Acad. Sci. USA*, 1986, **83**, 3743.
66. J. A. Wells, B. C. Cunningham, T. P. Graycar, and D. A. Estell, *Philos. Trans. R. Soc. London, A*, 1986, **317**, 415.
67. R. Zemel, D. G. Schindler, D. S. Tawfik, Z. Eshhar, and B. S. Green, *Mol. Immunol.*, 1994, **31**, 127.
68. S. A. Lesley, P. A. Patten, and P. G. Schultz, *Proc. Natl. Acad. Sci. USA*, 1993, **90**, 1160.
69. J. Guo, W. Huang, and T. S. Scanlan, *J. Am. Chem. Soc.*, 1994, **116**, 6062.
70. R. A. Gibbs, P. A. Benkovic, K. D. Janda, R. A. Lerner, and S. J. Benkovic, *J. Am. Chem. Soc.*, 1992, **114**, 3528.
71. B. Golinelli-Pimpaneu, B. Gigant, T. Bizebard, J. Navaza, P. Saludjian, R. Zemel, D. S. Tawfik, Z. Eshhar, B. S. Green, and M. Knossow, *Structure*, 1994, **2**, 175.
72. G. J. Wedemayer, P. A. Patten, L. H. Wang, P. G. Schultz, and R. C. Stevens, *Science*, 1997, **276**, 1665.
73. G. J. Wedemayer, L. H. Wang, P. A. Patten, P. G. Schultz, and R. C. Stevens, *J. Mol. Biol.*, 1997, **268**, 390.
74. M. Baca, T. S. Scanlan, R. C. Stephenson, and J. A. Wells, *Proc. Natl. Acad. Sci. USA*, 1997, **94**, 10 063.
75. J. D. Stewart, J. F. Krebs, G. Siuzdak, A. J. Berdis, D. B. Smithrud, and S. J. Benkovic, *Proc. Natl. Acad. Sci. USA*, 1994, **91**, 7404.
76. J. F. Krebs, G. Siuzdak, and J. Dyson, *Biochemistry*, 1995, **34**, 720.
77. J. D. Stewart, V. A. Roberts, N. R. Thomas, E. D. Getzoff, and S. J. Benkovic, *Biochemistry*, 1994, **33**, 1994.
78. J. A. Wells and D. A. Estell, *Trends Biochem. Sci.*, 1988, **13**, 291.
79. P. Carter and J. A. Wells, *Nature*, 1988, **332**, 564.
80. S. N. Thorn, R. G. Daniels, M.-T. M. Auditor, and D. Hilvert, *Nature*, 1995, **373**, 228.
81. D. Y. Jackson and P. G. Schultz, *J. Am. Chem. Soc.*, 1991, **113**, 2319.
82. K. M. Shokat, C. J. Leumann, R. Sugasawara, and P. G. Schultz, *Nature*, 1989, **338**, 269.
83. R. H. Abeles and A. L. Maycock, *Acc. Chem. Res.*, 1976, **9**, 313.
84. C. Walsh, *Tetrahedron*, 1982, **38**, 871.
85. P. Wirsching, J. A. Ashley, C.-H. L. Lo, K. D. Janda, and R. A. Lerner, *Science*, 1995, **270**, 1775.
86. K. D. Janda, L.-C. Lo, C.-H. L. Lo, M.-M. Sim, R. Wang, C.-H. Wong, and R. A. Lerner, *Science*, 1997, **275**, 945.
87. J. Wagner, R. Lerner, and C. Barbas, *Science*, 1995, **270**, 1797.
88. A. J. Morris and D. R. Tolan, *Biochemistry*, 1994, **33**, 12 291.
89. C. Y. Lai, N. Nakai, and D. Chang, *Science*, 1974, **183**, 1204.
90. C. F. Barbas, III, A. Heine, G. Zhong, T. Hoffmann, S. Gramatikova, R. Björnstedt, B. List, J. Anderson, E. A. Stura, I. A. Wilson, and R. A. Lerner, *Science*, 1997, **278**, 2085.
91. G. Zhong, T. Hoffmann, R. A. Lerner, S. Danishefsky, and C. F. Barbas, III, *J. Am. Chem. Soc.*, 1997, **119**, 8131.
92. R. Björnestedt, G. Zhong, R. A. Lerner, and C. F. Barbas, *J. Am. Chem. Soc.*, 1996, **118**, 11 720.
93. L. A. Highbarger, J. A. Gerlt, and G. L. Kenyon, *Biochemistry*, 1996, **35**, 41.
94. J. D. Stewart and S. J. Benkovic, *Nature*, 1995, **375**, 388.
95. R. Wolfenden, *Acc. Chem. Res.*, 1972, **5**, 10.
96. A. Radzicka and R. Wolfenden, *Science*, 1995, **267**, 90.
97. M. B. Wallace and B. L. Iverson, *J. Am. Chem. Soc.*, 1996, **118**, 251.
98. F. W. Alt, T. K. Blackwell, and G. D. Yancopoulos, *Science*, 1987, **238**, 1079.
99. C. F. Barbas, A. S. Kang, R. A. Lerner, and S. J. Benkovic, *Proc. Natl. Acad. Sci. USA*, 1991, **88**, 7978.
100. D. J. Chiswell and J. McCafferty, *Trends Biotechnol.*, 1992, **10**, 80.

101. T. Clackson, H. R. Hoogenboom, A. D. Griffiths, and G. Winter, *Nature (London)*, 1991, **352**, 624.
102. J. D. Marks, H. R. Hoogenboom, A. D. Griffiths, and G. Winter, *J. Biol. Chem.*, 1992, **267**, 16 007.
103. J. W. Lane, X. Hong, and A. W. Schwabacher, *J. Am. Chem. Soc.*, 1993, **115**, 2078.
104. J. A. Smiley and S. J. Benkovic, *Proc. Natl. Acad. Sci. USA*, 1994, **91**, 8319.
105. O. Ersoy, R. Fleck, A. Sinskey, and S. Masamune, *J. Am. Chem. Soc.*, 1996, **118**, 13 077.
106. H. Suga, O. Ersoy, S. F. Williams, T. Tsumuraya, M. N. Margolies, A. J. Sinskey, and S. Masamune, *J. Am. Chem. Soc.*, 1994, **116**, 6025.
107. T. Tsumuraya, H. Suga, S. Meguro, A. Tsunakawa, and S. Masammune, *J. Am. Chem. Soc.*, 1995, **117**, 11 390.
108. K. Janda, C.-H. Lo, T. Li, C. F. Barbas, P. Wirsching, and R. A. Lerner, *Proc. Natl. Acad. Sci. USA*, 1994, **91**, 2532.
109. D. Y. Jackson, J. R. Prudent, E. P. Baldwin, and P. G. Schultz, *Proc. Natl. Acad. Sci. USA*, 1991, **88**, 58.
110. E. Baldwin and P. G. Schultz, *Science*, 1989, **24**, 1104.
111. I. M. Tomlinson, G. Walter, P. T. Jones, P. H. Dear, E. L. L. Sonnhammer, and G. Winter, *J. Mol. Biol.*, 1996, **256**, 813.

Author Index

This Author Index comprises an alphabetical listing of the names of the authors cited in the text and the references listed at the end of each chapter in this volume.

Each entry consists of the author's name, followed by a list of numbers, for example

Templeton, J. L., 366, 385[233] (350, 366), 387[370] (363)

For each name, the page numbers for the citation in the reference list are given, followed by the reference number in superscript and the page number(s) in parentheses of where that reference is cited in the text. Where a name is referred to in text only, the page number of the citation appears with no superscript number. References cited in both the text and in the tables are included.

Although much effort has gone into eliminating inaccuracies resulting from the use of different combinations of initials by the same author, the use by some journals of only one initial, and different spellings of the same name as a result of the transliteration processes, the accuracy of some entries may have been affected by these factors.

Subject Index

PHILIP AND LESLEY ASLETT
Marlborough, Wiltshire, UK

Every effort has been made to index as comprehensively as possible, and to standardize the terms used in the index in line with the IUPAC Recommendations. In view of the diverse nature of the terminology employed by the different authors, the reader is advised to search for related entries under the appropriate headings.

The index entries are presented in letter-by-letter alphabetical sequence. Compounds are normally indexed under the parent compound name, with the substituent component separated by a comma of inversion. An entry with a prefix/locant is filed after the same entry without any attachments, and in alphanumerical sequence. For example, 'diazepines', '1,4-diazepines', and '2,3-dihydro-1,4-diazepines' will be filed as:-

diazepines
1,4-diazepines
1,4-diazepines, 2,3-dihydro-

The Index is arranged in set-out style, with a maximum of three levels of heading. Location references refer to volume number (in bold) and page number (separated by a comma); major coverage of a subject is indicated by bold, elided page numbers; for example;

triterpene cyclases, **299–320**
 amino acids, 315

See cross-references direct the user to the preferred term; for example,

olefins *see* alkenes

See also cross-references provide the user with guideposts to terms of related interest, from the broader term to the narrower term, and appear at the end of the main heading to which they refer, for example,

thiones
 see also thioketones

Agkistrodon piscivorus piscivorus, venom, phospholipase
  A₂, 110
Agrobacterium spp., β-glucosidase, 289
AICA *see* imidazole-4-carboxamide, 5-amino- (AICA)
alanine, L-β-farnesyl amino- (faA)
  analogues, 333
  inhibitory activity, 333
β-alanine, biosynthesis, 87
alanine methyl ester, 2-methyl-3-bromo(*N*-
    benzaldimino)-, rearrangement, 210
Albery–Knowles analysis, triose-phosphate isomerase, 8
aldol additions, in antibody catalysis, 439
aldolase antibodies, mechanisms, 438
aldol condensation, in antibody catalysis, 437, 439
alkaline phosphatase
  binding, 158
  biochemistry, 153
  classification, 140
alkaline phosphomonoesterase *see* alkaline phosphatase
alkaloids
  biosynthesis, 47
  see also ergot alkaloids
alkylation
  electrophilic, **315–41**
    chain elongation, 318
alkyl-enzyme intermediates
  hydrolysis, 57
  in microsomal epoxide hydrolases, 55
  in soluble epoxide hydrolases, 55
  synthesis, 56
allosamidin(s), binding inhibition, 297
allylic fluorides, synthesis, 335
aluminum naphthoxide, applications, in Claisen
    rearrangements, 350
amides, cleavage, 437
p-aminobenzoate synthase, regulatory mechanisms, 344
aminomutases, adenosylcobalamin-dependent,
    coenzymes, 209
ammonia, production, in muscle, 80
ammonium ions, analogues, 337
AMP *see* adenosine monophosphate (AMP)
AMP:pyrophosphate phospho-D-ribosyltransferase *see*
    adenine phosphoribosyltransferase
AMP aminase *see* AMP deaminase
AMP aminohydrolase *see* AMP deaminase
AMP deaminase
  biological functions, 80
  catalytic site structure, 81
  chemical mechanisms, 81
  cooperative kinetics, 81
  isozymes, 81
  kinetic isotope effects, 81
  kinetic properties, 80
  localization, 81
  molecular electrostatic potential surfaces, 84
  regulatory mechanisms, allosteric, 80
  transition state inhibitors, 83
    slow-onset, 83
  transition state measures, 84
  transition state structure, 82
AMP deaminase deficiency
  and muscle weakness, 81
  prevalence, 80
AMP pyrophosphorylase *see* adenine
    phosphoribosyltransferase
α-amylase
  sequences, 298
  substrate distortion, 299
  superfamily, 298
β-amylase, occurrence, 301
γ-amylase *see* glucan 1,4-α-glucosidase
amyloglucosidase *see* glucan 1,4-α-glucosidase
anaerobic ribonucleotide reductase activating enzyme
    (ARR-AE), catalysis, 218

anaerobic ribonucleotide reductase (ARR)
  activation, 217
  glycyl radicals, 218
Ancylobacter aquaticus, haloalkane dehalogenases, 403
3,5-androstadien-17-one, 3-hydroxy-, isolation, 33
4-androstene-3,17-dione, from, 5-androstene-3,17-dione,
    33
5-androstene-3,17-dione, as starting material, for 4-
    androstene-3,17-dione, 33
anthranilate synthase(s), regulatory mechanisms, 344
antibiotics, fosfomycin, 53, 62, 66
antibodies
  active, 442
  antiphosphonamidate, 430
  antiphosphonate, 430
  binding affinities, 441
  catalytic, 354, 358
  as entropy traps, 423
  and enzymes compared, 441
  esterolytic, 430
  phosphonates, 442
  progesterone-binding, 442
antibody 1E9
  catalysis, 429
  sequencing, 429
antibody 1F7
  Fab' binding pocket, 364
  structure, 364
    studies, 425
antibody 13G5, functions, 430
antibody 17E8
  active site residues, 436
  catalysis, 434
  characterization, 434
  mechanisms, 436
  mutants, 436
antibody 33F12
  catalysis, 439
  crystal structure, 440
antibody 38C2, catalysis, 439
antibody 39-A11
  binding, 428
  sequencing, 429
  structure, 428
antibody 43C9
  active site residues, 436
  amide cleavage, 437
  mechanisms, 436
antibody 48G7
  characterization, 434
  efficiency
    catalytic, 434
    chemical, 434
antibody catalysis
  applications, 423
  characterization, 424
  development, 423
  efficiency, 423, 441
  future research, 441, 443
  mechanisms, **423–45**
    aldol condensations, 437
    Claisen rearrangements, 424
    cycloaddition reactions, 428
    Diels–Alder reactions, 428
    hydrolytic reactions, 430
    oxy-Cope rearrangement, 426
    pericyclic reactions, 424
  and product inhibition, 428
  rate enhancements, 423
antibody catalysts
  design, strategies, 425
  esterases, 431
  future research, 442

enzyme-catalyzed reactions
  intermediates in, 6
  mechanisms, 6
enzyme–inhibitor complexes, NMR spectra, 24
enzyme inhibitors, mechanism-based, 438
enzymes
  activating, 206
  active sites
    reactive intermediate stabilization, **5–29**
    structure, 12
  and antibodies compared, 441
  catalytic efficiency, 437
    evolution, 8
    and natural selection, 442
  electrophilic reactions, **315–41**
  hydrolytic, 430
  isoprenoid, 318
  microsomal, 62
    mechanisms, 65
    structure, 65
    in xenobiotic metabolism, 65
  in organic synthesis, 59
  proficient, 368
  reaction mechanisms, 205
enzyme therapy, severe combined immunodeficiency
      disease, 73
enzyme–transition state complexes, transition state
      binding energy, 76
epoxide hydrolases, 53
  classes, 54
  enantioselectivity, 58
  families, 53
    α/β-hydrolase-fold superfamily, 53
  leukotriene A₄ hydrolase, 54
  mechanisms, kinetic, 55
  possible stereospecific synthesis, of epoxides, 59
  regioselectivity, 58
  roles
    in metabolism, 58
    in organic synthesis, 59
  use in chiral vicinal diol production, 59
epoxides
  nucleophilic additions to, mechanisms, 52
  nucleophilic openings, **51–70**
  ring-opening
    electrophilic assistance, 63
    enzyme-catalyzed, 53
    and phenanthrene 9,10-oxide, 62
  S_N1 processes, 52
  S_N2 processes, 52
  as substrates, 62
3,4-epoxy-1-butene, conjugate addition, 52
(7E,9E,11Z,14Z)-(5S,6S)-5,6-epoxyicosa-7,9,11,14-
      tetraenoate:glutathione leukotriene-
      transferase (epoxide-ring-opening) *see* leukotriene
      C₄ synthase
equilenin, spectral characteristics, 39
equilenin, 17β-dihydro-, spectral characteristics, 40
ER *see* endoplasmic reticulum (ER)
ergot alkaloids, biosynthesis, 331
Escherichia coli
  anaerobic ribonucleotide reductases, 217
  base excision repair pathway, 393
  chorismate mutase, 345, 360, 425
  cytidine deaminase, 87
  cytosine deaminase, 93
  dCTP deaminase, 94
  deoxyribodipyrimidine photo-lyases, 375
  ebg genes, 289
  formate *C*-acetyltransferase, 217
  geranyl*trans*transferase, 323
    mutants, 323
  isopentenyl diphosphate isomerases, 336

lacZ gene, β-galactosidase, 289, 290
maltodextrin phosphorylase, 305
mutagenesis, 394
mutants, distribution, 396
pentalene synthase, 330
photoproducts, thymine-derived, 373
protein farnesyltransferase overproduction, 334
ribonucleoside-diphosphate reductase, 171
ribonucleotide reductases, 164
  inactivation, 168
  subunits, 185
  thiyl radicals, 177
  structure, 430
  substrates, 103
esters
  hydrolysis, **101–37**
    enzyme families, 103
    homogeneous vs. heterogeneous catalysis, 103
    mechanistic diversity, 102
    occurrence, 101
ESTHER database, access, 112
4,6-estradien-3,17-one, isolation, 34
4,7-estradien-3,17-one, from, 5,7-estradien-3,17-one, 33
5,7-estradien-3,17-one, as starting material, for 4,7-
      estradien-3,17-one, 33
17β-estradiol, spectral characteristics, 40
1,3,5(10)-estratrien-17β-ol, 3-amino-, spectral
      characteristics, 39
3,5,7-estratrien-17-one, 3-hydroxy-, isolation, 32, 33
ethane, 1,2-dibromo-, K_m, 403
ethane, 1,2-dichloro-, interactions, 404
[1-²H,³H]ethane
  stereochemical studies, 221
  as substrate, 219
ethanol, 2-chloro-, oxidation, 192
ethanol, 2,2,2-trifluoro-, mutase studies, 357
ethanolamine ammonia-lyase
  catalysis, 273
  occurrence, 273
  reactions, 264
  stopped-flow kinetic studies, 273
ethene, *cis*-1,2-dichloro-, biosynthesis, 410
ethene, tetrachloro-, as terminal electron receptor, 410
ethene, trichloro-, as terminal electron receptor, 410
β-etherase, catalysis, 417
ethyl diphosphate, 2-(dimethylamino)-, inhibitory
      activity, 335
ethylene glycol
  dehydration, 188
  oxidation, 188
ethyl radicals, 1,2-dihydroxy-
  biosynthesis, 189
  dehydration, 189
excision repair pathways, mechanisms, 372
exo-1,4-α-glucosidase *see* glucan 1,4-α-glucosidase
exoglycanase, occurrence, 289, 292
exo-β-1,4-glycanase, occurrence, 290
exonuclease III, catalysis, 393
exo-α-sialidase(s)
  Brønsted constants, 294
  inhibitors, 296
  mutagenesis studies, 296
  occurrence, 294
  roles, 294

faA *see* alanine, L-β-farnesyl amino- (faA)
faAVIA, inhibitory activity, 333
Fab fragments, catalysis, 438
FADH *see* flavin adenine dinucleotide (reduced) (FADH)
farnesylamine, inhibitory activity, 333
farnesyl diphosphate
  analogues, as substrates, 332

WITHDRAWAL